# On the Nature of Ecological Paradox

Michael Charles Tobias • Jane Gray Morrison

# On the Nature of Ecological Paradox

 Springer

Michael Charles Tobias
Dancing Star Foundation
Los Angeles, CA, USA

Jane Gray Morrison
Dancing Star Foundation
Los Angeles, CA, USA

ISBN 978-3-030-64525-0     ISBN 978-3-030-64526-7   (eBook)
https://doi.org/10.1007/978-3-030-64526-7

This Springer imprint is published by the registered company Springer Nature Switzerland AG
The registered company address is: Gewerbestrasse 11, 6330 Cham, Switzerland

*To the Future and Wellbeing of Life On Earth*

# Foreword

*On the Nature of Ecological Paradox* is unquestionably a masterpiece, in the tradition of two great treatises by Ludwig Wittgenstein and Baruch Spinoza. Nothing like it has ever been written before. It mirrors the dizzying careers of two people deeply in love. Michael Charles Tobias and Jane Gray Morrison have – collectively – conducted field research on every continent in an effort to understand and communicate the fall-out on Earth from human infliction, a mournful epoch geologists call the Anthropocene, which equates with the sixth (possibly seventh) global extinction event in the 4.2 billion years of documented life on this planet. While their field work – articulated in hundreds of published essays, over 75 books and 100 films, lectures, keynote addresses, and several professorships between them – has clinically analyzed one ecological crisis after another, often prophetically, Tobias and Morrison have also focused on solutions, or at least pathways toward amelioration. Their canvas of theoretical, empirical and applied research encompasses the natural sciences, deep demography and family planning, energy strategies, geopolitics, the history of ideas, conservation biology, systematics, taxonomy, ecological anthropology and animal liberation. But, in addition, Tobias and Morrison have for their entire careers been equally focused on ecological aesthetics, art and literary history; philosophical traditions; paleontology; psycholinguistics; comparative ethics, both in practice and in the ideal, as well as the vast array of psychoanalytic theories and tools to better grasp the human condition going back to its far distant hominid origins.

Some of their most notable past films and books include *Antarctica: The Last Continent* (1986); *Ahimsa – Non-Violence* (1987); *Black Tide*(1989); the ten-hour dramatic television series *Voice of the Planet*(1990); *World War III: Population and the Biosphere at the End of the Millennium* (1996); *A Day in the Life of India* (1997); *Nature's Keepers: On the Frontlines of the Fight to Save Wildlife in America* (1997); the sprawling, 1836-page modern-day Don Quixote, *The Adventures of Mr. Marigold* (2005); the massive *Sanctuary: Global Oases of Innocence* (2008); the PBS feature film trilogy *Mad Cowboy* (2005), *No Vacancy* (2006) and "Hotspots" (2008); *God's Country: The New Zealand Factor* (2011), and seven recent books with Springer Nature Publishers (New York/Switzerland), including *Anthrozoology: Embracing*

*Co-Existence in the Anthropocene* (2017), *The Theoretical Individual: Imagination, Ethics and the Future of Humanity* (2018), *The Hypothetical Species: Variables of Human Evolution* (2019), and, most recently, *Bhutan: Conservation and Environmental Protection in the Himalayas* (2021, with Dr. Ugyen Tshewang). This atlas of endeavors just scratches the surface of their body of work. A short précis of these myriad and deeply provocative intellectual odysseys can be gleaned from a book Michael wrote together with conservation biologist Dr. Paul R. Ehrlich of Stanford University, *Hope On Earth: A Conversation* (2014, University of Chicago Press).

It was on one of these innumerable projects, *No Vacancy* (the feature film and accompanying edited volume) that I worked with Michael and Jane and our film crews throughout the US and Mexico, the Netherlands, France, Italy, Nigeria, Ghana, Iran, India and Indonesia, to try to grasp the extent of the continuing human population as well as to foster voluntary, humane population stabilization. I have devoted my career, and as President of the non-profit Population Communication,[1] to helping couples have children only when wanted and space the births of their children for the health of the mother and child, and to informing governments on growth curves, and have commissioned reports on population stabilization from countries with total fertility rates over 3 with populations over 20 million. I have designed 56 operational research projects in 24 countries.

I have been contemplating ecological paradoxes – in one form or other – for most of my professional career.[2] Because, as Tobias and Morrison have microscopically delineated in this massive page-turner, everything we think, feel and touch seems to be embedded in a paradox – propositions, beliefs, assumptions, theories which, when thoroughly explored, may often prove to be self-contradictory, or something other than expected or perceived. In other words, our belief systems, born out of centuries, millennia of common practice, community standards, scientific paradigm, observation, endless statistics, fixed ideas and presumptions – even the cumulative phenotypic evidence of individual and population behavioral patterns – when enshrined within eco-system dynamics and feedback loops are constantly revealing new truths, nuances more complex than we ever imagined, and which are critical to the health and robustness of biodiversity. Such paradox is the core ambiguity shading our future as a species, a topic central to Tobias and Morrison's engrossing study. What a gift we have in this book that brings forth a story so beautifully written that is unlike anything written before.

*On the Nature of Ecological Paradox*, divided into three major sections totaling 100 essays – Part I, *Tractatus Ecologia Paradoxi*, Part II, *Ecological Memories and Fractions*, and Part III, *A Natural History of Existentialism* – each lavishly illustrated with over 325 images from every artistic and intellectual domain (1943 citations), traces every dominant divining rod in the history of ecology, but does so in a continually unexpected context. This book structures within many dimensions the paradox of the Green Revolution pioneered by Norman Borlaug. My burgeoning

---

[1] www.populationcommunication.org

[2] https://www.smith.edu/libraries/libs/ssc/prh/transcripts/gillespie-trans.pdf

career as a soil scientist was disillusioned as Borlaug saw populations exploding from 2.2 billion when I was born to over 7 billion today. In 1986, he wrote to me, "Many put great emphasis on ecological problems and to me one of the great unrecognized parts of the equation that bears on ecological systems is human population pressure and yet, one seldom hears the ecologists speak out about the population monster." Norman Borlaug's successful battle to give birth to the Green Revolution ultimately served as but one more desperate stop-gap measure that could not prevent inequitable distribution of staples, or the sheer expansion of populations forever undermining gains in high-yielding varieties, an equation for unstoppable chronic malnutrition in large population pockets across the globe. There were 2 billion people on the planet in 1930. Today there are 7.7 billion. Since the publication of the authors' book, *World War III*, 2.2 billion more people have been added to the planet. At the turn of the century there will be over 10 billion. Indeed, if current total fertility rates continue, our species might even exceed 11.5 billion.

And yet, as Tobias and Morrison point out in so many ways, while most species' well-being is measured, in part, by the degree of their reproductive success, in the case of *Homo sapiens*, we have clearly outsmarted ourselves. What are key drivers, stressors, psychological components and ethical inflection points that might yet enable our species to grow up, to make peace with the world, to learn from our all but 330,000 year+ history? We are newborns who have, as Nikos Kazantzakis warns in one of the chapters, discovered fire and are ready to burn up the world. Is this destiny socio-biological? Are there compassion fatigue/regeneration bifurcation points where we can, as individuals, and as a collective, transcend moral bankruptcy? Total self-destruction? Or is our destiny implacable, against the unflexing, albeit resilient combination of biomes, over 40% of which our species has already appropriated for its own fast increasing wants and desires?

The stranglehold of the self-interested political and economic structures is illuminated by the scientists, politicians, writers, artists and world leaders exquisitely detailed in Michael and Jane's prose. As is the senseless destruction of habitat that supports the myths of exponential growth, revealing the sheer cruelty to most other species and the billions of other humans caught in a path of evolution we ourselves have commandeered.

*On the Nature of Ecological Paradox* represents an intellectual and ethical breakthrough, by the very courage of its principles and sturdiness of its arguments. And by these two leading thinkers bringing such wide experience, data and wisdom to the most profound ecological and moral crises of our time. They have faced the abyss head-on, though not without flinching. There is a personal through-story in this great book that lends an emotional underpinning to all that they describe, explore and theorize. They passionately and persuasively bring to the open tables of science, mathematics and the many histories of consciousness a dramatic, sometimes terrifying focus upon the deeply frustrating reality that, despite sound reasoning and a plethora of information widely promulgated regarding the many human-imposed plights to the planet, it is more often the fact that what is good for the earth and all that dwell therein is not what humans want. By that predisposition, our future – and that of all others - is indeed imperiled. Few intellects in the world today have so comprehensively explored this baffling paradox. Tobias and Morrison's psychoanalytic approach is critical to an emerging perspective from the humanities and sciences that will not shy away from the absolute troubling truth of ourselves. And while there are countless conceptual frameworks for addressing these etiologies we face as a society, no one book has ever collated so much of the historic and contemporary fabric of these difficult deliberations. It is an ecological corollary, in a sense, to Will and Ariel Durant's *Story of Civilization* (though, obviously, in 1 volume, not 11.)

The ultimate paradox is that we are tiny, insignificant, probably irrelevant, dust motes in an obscure galaxy. And yet, we are so full of ourselves, so full of passion, longing and hope. All of these predicates and presumptions would assume, as a fundamental premise of human existence, that we would cherish every day and every life form. But we don't. Most people are so troubled and haunted by circumstances not of their making that it's just enough to get through the day, get some sleep at night and have a good meal, if possible. We've lost the connection to what we truly need to cherish with every breath. We either take it for granted or destroy it.

The great tragedy enshrined within the mesmerizing pages of *On the Nature of Ecological Paradox* is that most of us more or less have a general idea of what's going on, whether in the guise of sheer human chaos, millions of acres across the planet horribly on fire, the number of natural calamities, human violence, civil wars, extinctions, climate change, and the scope of medical fall-out of pandemics, all on the rise. Here, this syndrome of anthropocenic causes and consequences is conveyed through what can only be characterized as a culminating journey of a half century of deep reflection and field research. With its collective eye looking directly before it, as well as the near future, Tobias and Morrison unstintingly encourage us to think clearly and feel deeply, as to how we arrived at this moment, and what our remaining options may provide for. They offer us uncluttered avenues that reject an otherwise mathematically certain species demise – which by all accounts we currently, selfishly favor – and, instead, project cognitive and ethical scenarios that this and future generations can cultivate, and thereby usher in the time when a more sane, compassionate co-existence is within our grasp.

President, Population Communication                                   Bob Gillespie
Pasadena, CA, USA

# Prologue

For a generation Michael Tobias and Jane Morrison have been among the leading voices helping us understand the intricacies of the interconnectiveness of the web of life. Their perspective stands out because it goes beyond science and includes our religious beliefs and philosophical foundations. They earned their credibility by traveling to the very places they speak of, seeing droughts, famines and population dislocations on the one hand, and on the other experiencing the sublime through the beauty of the earth's remotest places and the people who live there.

I met Michael and Jane when I was serving as the secretary of the Smithsonian Institution because of their interest in the work of the Smithsonian Environmental Research Center. They were intrigued by the findings of the Smithsonian's Global Earth Observatory (GEO), a project that brings together multiple entities around the world observing changes in forests over time, and sharing the information using a common format. It was apparent that the GEO concept resonated with their interest in understanding how the earth's systems are responding to the overarching effects of humankind on the natural world.

During the course of their lives they have written, together and individually, over 70 books, and produced a similar number of film documentaries. This new book distills their insights as examined through the lenses of art, human sensibility, philosophy, religion, animal rights, statistics and conservation, among others. No stone is left unturned in the search for examples that can help us save ourselves, both physically and spiritually.

They probe the question as to why our species, with a brain large enough to appreciate the beauty and complexity of the natural world, seems unable to contain the damage done by its own choices and actions. Whether in the end we can find the will to save our fragile planet and the life on it, or if we will be the agent of our own destruction. Hence, the ecological paradox, the focus of this book.

As the authors note, in a 1993 essay, the scientist/sociologist E.O. Wilson was asked if humanity is suicidal. His answer, "no, we aren't suicidal, but we are death for much of the rest of life and, hence, in ultimate prospect, unwittingly dangerous to ourselves." Their new book points out in the interim between 1993 and today, while we have made some efforts to save species, protect wilderness, and develop new technologies to combat climate change, by and large we have lost ground. Sadly, we are reaping what we have sowed in the form of apocalyptic forest fires in the Western US, Siberia and Alaska, "300 hundred-year" rainstorms occurring annually, protracted droughts, massive garbage patches and dead zones in the oceans, huge glacial melting events in Greenland and Antarctica, increasing frequencies of hurricanes and typhoons, and a pandemic. Ironically, each of these events was predicted beforehand, including the pandemic, where Michael and Jane were among the first to note the linkage of disease spread from animals to humans to the "wet markets" of China.

Because it has proven difficult to mobilize the collective will of humanity and its governments to take the actions needed, the authors conclude that our hopes lie with the individual and free will. Individuals who not only deeply understand the need for transformation but that it must be done at speed. As the book concludes, "Otherwise, and all too clearly, our worst emergent nightmare, collectively, will be shown to have been the very paradox of our presence on earth."

Smithsonian Institution                                                  G. Wayne Clough
Washington, DC, USA
Georgia Institute of Technology
Atlanta, GA, USA

# Acknowledgments

The authors are most grateful to their editor Kenneth Teng, Publishing Editor in Life Sciences, as well as the entire production team involved at Springer Nature Publishers, including Mr. Vignesh Viswanathan, Project Coordinator. In addition, we want to extend our gratitude to Dr. G. Wayne Clough for his Prologue, and to Robert W. Gillespie for his Foreword. In addition, we thank all of our friends and colleagues throughout the world with whom we have collaborated on so many projects for many decades. Ms. Sudha Kannan, Project Manager, after Mr. Vignesh Viswanathan, Project Coordinator, (unless Ms. Kannan does not want her name mentioned).

# Contents

## Part III   A Natural History of Existentialism

# About the Authors

**Michael Charles Tobias**, long-time president of the Dancing Star Foundation (dancingstarfoundation.org), earned his Ph.D. in the history of consciousness at the University of California-Santa Cruz, specializing in global ecological ethics and biological systems as well as the interdisciplinary humanities. His wide-ranging work embraces the global ecological sciences, art, comparative literature, the history of ideas, philosophy, and natural history. He is the author (in addition to an editor) of over 60 books – fiction and non-fiction – and has written, directed, and/or produced over 50 films (mostly documentaries, docudramas, and TV series). His works have been read, translated, and broadcast throughout the world. He has conducted ecological field research in nearly 100 countries. Dr. Tobias has been on the faculties of such colleges and universities as Dartmouth, the University of California-Santa Barbara (as both distinguished visiting professor of environmental studies and the regents' lecturer), the University of New Mexico-Albuquerque (as the visiting Garrey Carruthers Endowed Chair of Honors), Canisius College (a former visiting professor of anthrozoology), and the Martha Daniel Newell visiting scholar at Georgia College and State University. An honorary member of the Club of Budapest, Dr. Tobias is also a member of the Russian International Global Research Academy as well as the Russian Public Academy of Sciences. Dr. Tobias' many works include the 10-hour dramatic television series, "Voice of the Planet," and such films as "Ahimsa: Non-Violence," "Hotspots," "No Vacancy," "Mad Cowboy," and "A Day in the Life of India," and such books as *World War III – Population and the Biosphere at the End of the Millennium*, *The Adventures of Mr. Marigold*, *Anthrozoology: Embracing Coexistence in the Anthropocene*, and *The Hypothetical Species: Variables of Human Evolution*.

**Jane Gray Morrison** is an ecologist whose work has taken her to over 30 countries. As a filmmaker, Ms. Morrison has produced numerous films for such networks as Discovery, PBS (where she co-directed "A Day in the Life of Ireland" for Irish Television and WNET/New York), "Hotspots" (www.hotspots-thefilm.com), and Turner Broadcasting, for which she served as senior producer for "Voice of the Planet," a 10-hour dramatic series based upon the history of life on Earth. As a

Goodwill Ambassador to Ecuador's Yasuní National Park, she produced the short film "Yasuní – A Meditation on Life," which premiered at the United Nations Rio+20 Summit in 2012. Her books include Sanctuary: Global Oases of Innocence (www.sanctuary-thebook.com); "Hotspots" (www.sanctuary-thebook.com), Donkey: The Mystique of Equus Asinus, God's Country: The New Zealand Factor, and No Vacancy (co-edited). She has co-written six previous books published by Springer. Since 1999, Jane Morrison has served as the executive vice president of Dancing Star Foundation, a nonprofit organization that focuses on animal rights, the interdisciplinary humanities, and social and ecological justice movements as they concern humankind's relationship to the natural world. In New Zealand, over a period of 15 years, she co-created an ecological refuge for rare and threatened native and endemic flora and fauna.

# Part I
# Tractatus Ecologia Paradoxi

# Chapter 1
# Introduction

## 1.1 Ecological Interpolations

We call this treatise, or *Tractatus*—in homage to both Baruch (Benedict de) Spinoza (1632–1677) and Ludwig Wittgenstein (1889–1951)—*On The Nature of Ecological Paradox*, knowing full well that its "causes and consequences" are the real clue to an Anthropocene that has, to date, been analyzed for its inflictions on the natural world, which are unprecedented in their *cruelty*.

SARS-CoV-2 (or Covid-19), climate change, extinctions, global pollution, human progress for some but not for most, vast inequities between species: These are only some of the most recent ecological pandemics that are, at heart, a prime lesson in paradox and pain. Ultimately, a healthy, well-balanced individual should, theoretically, make for a healthy planet. However, even in health, as in sickness, humanity has shown so many hostile predilections with respect to other life forms, that today we have no baseline for assuring reasonable correlations between human behavior and the options for successful furtherance and interactive vitality among other species.

Such escalating facets of the human experience question the very substance and verity of our notions of evolution and natural selection against a pressing context of ethics. Ethics that are phronetic (born of practical wisdom), dispositional as the vicissitudes of life dictate, silent, pragmatic, consequential, teleological, eudaemonist, self-effacing and rationalizing, absolutist, situationist, overtly volitional, and the like. These are just some of the philosophical traditions key to any thorough examination of the history of the natural sciences within a contemplative and unstinting context of analysis. And its many tentacles of speculation are fraught with paradox.

Paradoxical in that a hominid that for so long has prided itself on species success has done so only at an extreme cost to the rest of the biological planet. One most recent example involves the so-called "wet markets" throughout much of the world, where shoppers demand that their food be animals, and that those sentient beings be

© Springer Nature Switzerland AG 2021
M. C. Tobias, J. G. Morrison, *On the Nature of Ecological Paradox*,
https://doi.org/10.1007/978-3-030-64526-7_1

**Fig. 1.1** A Troubadour: "Nature instructing the Poet," from a M.S. in the King's Library in Paris. (Reproduced from Costello and Pickering,[3] Private Collection, Frontispiece, Photo © M.C.Tobias)

slaughtered right there before them (in the most violent, inhumane, and unsanitary of circumstances). Cruel because that *is* cruel—emblematic of our species' insensate grip upon the majority of life forms; and, ironically, the cruelty, at the stage of gene commingling into mutant virus bats, pangolins, mink, possibly other species; a resulting global pandemic that infects humans, thus totally backfiring, at least for our kind. But it is not the first time—nor will it be the last (the ultimate backfire is yet to occur)—that a paradox of the most heinous and vile dimensions has suddenly (seemingly) dominated the biosphere (Fig. 1.1).

## 1.2  Troubadours

It is by that amalgamation into a mere word, *cruelty*, Old French, "indifference to, or pleasure taken in, the distress or suffering of any sentient being",[1] stemming etymologically from the twelfth and thirteenth centuries, that we perforce from the

---

[1] https://www.etymonline.com/word/cruelty

very beginning grapple with its psychological implications, collapse of virtue, this singular indictment of the human species. A word, cruelty, that most prominently corresponded to take one salient, if seldom discussed, example to the Spanish Inquisitional flames that claimed so many lives of a culture of southwestern France, the Vaudois, which had given birth to the ascetic, pure, and sensuous musical traditions of the nomadic lyrical gypsies of Provençal. We single them out, that time and place in Western history, because it was a fashion of music subtly delivered in order to please, to quiet the heart, and it was nearly obliterated by forces who refused to be pacified by poets. Dante was greatly influenced by their fancies, the Utopian songs and lyricism that are their legacy, and the Inferno to which many were subjected.

This far-off and elusive aesthetic, two centuries of Arcadia-enraptured Occitan-singing musicians who had sought a safe haven in Spain, across the Pyrenees, only to be viciously turned upon (again) and many burned at the stake over their musical and poetic faith, represents one of the most emphatic dialectics in Western tradition; an iconic example of our species-wide psychoses of which much will be written; not unlike the many martyred saints throughout Christian tradition. In this case, the tragic victims of an ill-turning human nature in the Late Middle Ages were Troubadours.

Their melancholic soliloquies, sonnets, and cansos (love songs to one's beloved) would centuries later inform Monteverdi, Vivaldi, Bach, Mozart, and Handel, and even Goethe's first and most despairing of creations, *The Sorrows of Young Werther* (1774). We feel the fragile presence of the Troubadours in equal measure across the span of literature and natural history reflections, from Petrarch to Cervantes, from Shakespeare to Voltaire, Jean-Jacques Rousseau, and ultimately Buffon, Beethoven, and Darwin; so star-crossed were the heavens and hells to which their melodies and dreams fell prey.

Theirs is an unwritten history emblematic of The Birth of Innocence; the Fall of Man. Within its diverging paths are two distinct philosophical and moral poles: sagas writ painfully in so many leys d'amors, the pastorela, descort, and alba, this latter a shepherd's song of woeful countenance, as dawn broke ranks with darkness; and a natural history of Europe and, by sharp implication, all of human time, human contradiction, and ensuing turmoil.[2]

And at this very moment, a concatenation of global crises is not only bringing out the best in our species but also epitomizing how unprepared the collective really is to effectively deal with individual and community suffering, from Washington, D.C. to Siberia, from Sweden to the Antarctic. The breakdowns are not restricted to, though possibly made more intractable by, democracies. But in every nation, institutional vulnerabilities are now glaringly seen to track with our apparent inability to effectively change large numbers of minds and hearts; to engage in civil and transformational dialogue that can surmount the hurdles inherent to theory; let alone

---

[2] See *Protestant Endurance Under Popish Cruelty: A Narrative Of The Reformation In Spain*, by J. C. M'Coan, Esq., of the Middle Temple, Binns And Goodwin, London, 1859; See also, *The Troubadours -A History Of Provençal Life And Literature In The Middle Ages*, by Francis Hueffer, Chatto & Windus, London, 1878, particularly Chapter XIX, "Siege Of Autafort -Bertran's Death," and Chapter XXI, "The Vaudois And Albigeois," –"The Reformation Of The Thirteenth Century."

effect widespread parities or distributional justice across diverse constituencies. These gaps in the collective measures of all things—gaps between experience, education, cultural norms, and attitudes—are of greatest concern in a nuclear world, and at the very moment, the fast-waning biodiversity—the underlying, essential pillar of all—is least accounted for. That accountability factor shines the spotlight on human narcissism. As will be explored herewith, that one unconditional, human consciousness, and the one species that all but remains its principal beneficiary, provide vastly insufficient room to negotiate a truce with earth; to mentor the human personage in meaningful, ethical guises other than itself.

Sang Christine De Pise, "…Victim of thy cruelty!... May it not an emblem prove/ Of untold but tender love?"..[3]

And the twelfth-century Troubadour Jaufre Rudel, "I am happy to weep while singing,/it is the only way/to calm this sadness/which engulfs me./I sigh one hundred sighs a day…".[4]

It is this forlorn centrality, within the dream of peace, the centered rural life of all those who have been subsumed by the seasons, lulling late-afternoons, mottled lights of the winding orchards of Medieval France; reticent personages pondering their world, simply enjoying the fellowship of their flocks in the solace of shade trees, that so strikingly invokes its tortured aftermath. Any testament to those prior times necessarily invokes the psychological and anthropological reconciliation, from *inside* ourselves, that contradictory relationship of our kind to nature, however uncomfortable or bathetic its revelations.

The fact so much pain, murder, exile should have correlated with the very birth of Renaissance music and poetics does much to subsume human nature in the perplexity of all those exquisite visions, paradise scenes, celebrations and degrees of confidence, and sheer curiosity at our world that heralded the rise of the ecological sciences.

Except that now, our temptation to escape into the pensive elegiacs of lore, far removed from Black Plagues and Spanish Inquisitions (we would like to think), is an ill-afforded nostalgic luxury for dreamers. Nature has equipped us with only so much tolerance for each other, beyond all the ecstatic love affairs and intimations of some destiny, leading across our neighbor's fields to the pivotal horizon-teeming questions: What have we done? Who are we?

## 1.3   A Counter-Intuitive Discourse

*Ecological Paradox* hopes to shine an even light on some of humanity's foibles where self-interest easily undermines species and, by implication, the biosphere's survival, a guaranteed backfire on our kind. We do so throughout the book, with the

---

[3] *Specimens of the Early Poetry of France – From The Time Of The Troubadours And Trouveres To The Reign Of Henri Quatre*, by Louisa Stuart Costello, William Pickering, London, 1835, p. 103.

[4] *The Music Of The Troubadours*, *Provençal Series* amo ut intellegam, *Volume 1*, edited by Peter Whigham, Ross-Erikson, Santa Barbara, CA, 1979, p. 39.

best of intentions, or from a critical perspective on incremental change. Paradoxical behavior, contradictory assumptions and goals appear fundamental to human evolution. We harbor no illusions that in one generation, or in response to unprecedented catastrophe, our species is going to suddenly change. If it were to embrace some kind of revolutionary collective wake-up call, ready-or-not, a spiritual epiphany spanning a critical, now-or-never mass, history renders clear that it won't be good for all constituents. Given our habits of domination over most other vertebrates (a few exceptions earlier noted), and our near total vulnerability to our own bodily microbes and viruses, and the invertebrates that vast outnumber us (an estimated 400 trillion individual krill in Antarctica's waters during spring and summer, for example—500 million tons)[5] both in quantity and their generally rightful claims to biological tenure, it is a fretful, murky picture for *Homo sapiens*.

This book is divided into three parts meant to, in British philosopher Gilbert Ryle's famed expression, explore category mistakes[6]; to explore through a number of disciplinary provocations, the dualistic chasms of which Ryle was so enamored, in order to fathom and translate a crucial abyss, illogical obsessions without correction that have intellectually attacked the world. Literally attacked it. A situation in which our species has been blind, perverted, propounding a line of reasoning ever dangerously rooted to narcissistic, mechanistic blind-spots. Those intellectual and moral caesuras are profound, at the heart of this counter-intuitive discourse; and first promulgated most demonstrably in French philosopher René Descartes' (1596–1650) declaration of human mental and emotional superiority to all other life forms. There we were, amid one European war within another war, at the proposed zenith of God's creation, a central premise of so many philosophical and history of ideas family trees: "Je pense, donc je suis; Latin: cogito, ergo sum, I think, therefore I am," from *Discourse on the Method* (1637, Leiden, the Netherlands).

Ryle, and every student of animal rights and conservation, will inevitably collide with the arrogance inside the furnace of Descartes' unwavering belief in the human mind. Our everyday skepticism, he asserted, must help inform all natural science studies. But with so unvaried a faith in the capacity of the human intellect to surmount any paradox or contradiction as to all but obfuscate the world of nature and, in so doing, engender an implacable division or dualism between people and all other organisms (presumed no more than machine-like in their appearance of intricacy and behavior). That intellectual bifurcation point—and Ryle was most certainly determined to demolish it—was an essential and slow-to-become archaic flash point in the Renaissance obsession with grasping the role of humanity on earth; denuding every mystery, mapping every inch of the world in order to better control it.

Science has long abandoned mechanistic thinking per se. But its distinct opposite orientations, the so-called pathetic fallacy or anthropomorphic mythopoetic rituals and artifice galvanizing our species' self-righteousness, lift up the human spirit in a

---

[5] See David Attenborough, "Antarctica," "Seven Worlds One Planet," BBC World, 2019; see also, "Virtual World of Krill," Australian Antarctic Division, August 1, 2017 http://www.antarctica.gov.au/news/2017/virtual-view-krill,Accessed March 3, 2020.

[6] See Ryle's *The Concept of Mind*, University of Chicago Press, Chicago, Ill.,1949.

multitudinous impersonation that misses a fundamental point of departure for this work. Our self-consciousness is deeply distressed by the atavistic, unprecedented suspicion that we are missing, missing everything. That we are lost in ourselves, so sanctimonious when it comes to our actual plight, that our psyches have been usurped by something immutable, agonizing, inflictive, all but impossible to grasp in the species-wide embrace of the willful surrender of crucial memories of the time we were still connected. Perhaps thirty thousand years ago. One-hundred thousand years ago. Just when or how we slipped away cannot be deciphered. But it happened, as sure as continental drift. We lost the rest of the world outside of our little, fast-proliferating species of hominids. Lost our way toward the queasily approaching future. We jettisoned our life rafts. Aggressively chose to subdue and consolidate our power. And that was it.

So it is our intention, however inadequately realized in this book, to explore some varied inroads of this human ontology: aesthetics, ethics, anthropology, metaphysics, history, military debacles, struggling jurisprudence, conflict, that are ecologically kinetic in some form, prone to mismatches, destructive combinations, or strikingly optimistic correlations, delusional though they may be.

The first part of *On the Nature of Ecological Paradox*—"Tractatus Ecologia Paradoxi"—examines places, individuals, and certain moments in history that are illustrative of an ecological power over us, or through us. Instances we have selected from personal field research and academic concerns that suggest glimmers and interpreted injunctions of the bifurcation point: that moment when the sheer power of contradiction devolving from humanity's place in the natural world becomes an overwhelming source of concern, both delight and horror. Other areas of consideration summon philosophical ambiguity, our sheer fascination with those frontiers and rudiments, in mathematics, evolutionary science, biosemiotics, of what is best and most curious about humanity, as it struggles to coexist in and with a biosphere it has only recently begun an acquaintance.

The second part—"Ecological Memories and Fractions"—is a roving series of deep and peripatetic attenuations of topic: subjective and anecdotal meditations. These are deeply personal divagations but are steeped in aspects of the humanities and sciences, pithy teasers, absent any resolution, that mirror both the perplexity and a brief history of revelations humanity has elicited in itself on behalf of her lost families in nature. Each is a moment out of time that inflects at some level the abiding ecological paradoxes that add up to something poetic and possibly redeeming about humanity's plight, and her efforts to redress certain fundamentals that have gone wrong. In these pages we find optimism, we believe in hope, we want to believe the prime dualisms driving the world to ecological ruin can yet be sorted out. That the Anthropocene may be our present global crisis, but by faith, common sense, acutely ethical science, and a courageous collective, individuals still have the prowess, possibly the time, to stave off utter, global ruination. These brief glimpses into a huge variety of human experiences are left deliberately unfinished. They are jarring intangibles which in totality are intended to evince and evidence the quintessence of paradox: the perception of great beauty, the acknowledgment of great loss, of violence without surcease, and of the possibility of undying love.

Part three is, admittedly, a sobering conclusion, a meditation on what we perceive to be the only too real cartography of doomsdays our species is courting. We call it "A Natural History of Existentialism" although everything about it is terribly unnatural, other than the very inclination to have written it. That exercise has been painful, certainly striking of a certain misanthropy, a suspicion at best of our species' dangers to itself and to all others. From its ashes and ruins we hope—through the data and the arc of fast-looming icebergs before our species' Titanic—to invite consideration of desperate, last-minute alternatives that may still be within our grasp. We make some recommendations, but few predictions.

Ultimately, this is a work of painfully deferred hesitancies, written under the overarching influence of a global pandemic. After all, we are people, and throughout the book we are alleging that our species is the sole problem. Not an incoming asteroid or Black Plague out of our control. It is us. We can hardly hope to write from an objective perspective outside the self-inflicted disaster for which we are all culpable, perhaps some more than others, but guilty as a species, which verges on the very invalidation of natural history, calling upon metaphysics to somehow exhume fresh air and some novel, tenable, passionate twenty-first century ethical Renaissance. Nothing less can fix it. If it can be rectified at all. We must face up to our own psyches in the unmerciful collaboration that is human history in its war against the planet.

This is not a book of conclusions and omniscient narration, or even a study guide to the perplexed, which presupposes a level of understanding and the fact of having digested so much bad news as a dutiful precursor to a happy ending. The psychoanalytics that have evolved in Act One have not been exorcised from the human psyche. While art has continued to effectively curate our quixotic and teasing landscapes, celebrate the beauty we presume to recognize, and render commentary on all the shortcomings of collective human behavior, there is no true beginning or end, either in paleontology, physics, or so-called futurism, which makes for an exhausting Act Two, and as yet unimaginable Act Three. Dramatic effect is not an issue, to be sure; but any principled promotion of values that may gain traction in coming months and years requires a pronounced collaborative wherewithal to see it through. There is no telling. Humanity is quite capable of deceiving herself with numbers that prove whatever we want to prove. Aesthetics are equally deft in the flight from numbers, just as politicians and economists, philosophers and cosmologists, left to their own minions, can make any scenario or system of cognition work, or founder.

A book holds down any number of temptations that are marginalized for want of time, space, patience, or the actual formulae for working them out, a situation that might suggest an eschewal on principle of happy endings. That does not recognize our disposition in trying to make a certain case. With respect to its inevitable, existentialist-like pronouncements, there is the ever-present risk of putting forth an argument that unavoidably by paragraph two, page one, has readers daunted and dismayed. Nonetheless, somehow, without recourse to a didactic or pedagogical authority it is our hope to actually *discover hope* in the process of moving beyond the tyranny and orthodoxy of the ambiguous (Fig. 1.2). Timing is inevitably a flirt, and the present era, with its swiftly normalized mantras and jeremiads, will work against any novel introduction of data or interpretive gradations. It has all been said, but not much has changed, over thousands of years of exhausting historical precedent.

**Fig. 1.2**  Victoria crowned pigeon (*Goura victoria*), (New Guinea. Photo © M.C.Tobias)

That makes our current condition anything but simple. We can't just resort to obvious antiphons, dyads, or traditional structures dictating the nature of narrative and ethical dialectics to solve our problems. There is too much at stake to cotton to the comforts of logic and its systematic conclusions upon escalation. We wouldn't know where to begin, let alone conclude. And while science insists on itself and its forms, we believe that we are now in the midst of great scientific change. Paradigm shifts that go beyond what we can possibly intuit. Hence, we have taken random variables more seriously than we might have a century ago. There is good reason, we believe, to take many things seriously. But we do not second-guess what others are likely to do. As Jacques Maritain wrote of the "father of positivism" and of modern scientific method, so-called, Auguste Comte (1798–1857), behold a philosopher, declared Maritain, who rightly claimed that science must "remain chastely distant" from the taint of metaphysics, "characterized above all by the elimination of every ontological preoccupation…".[7] We differ, and stray, from that sensibility and approach.

Thought processes are fascinating. Inspiration may be learned, or not. There are no proofs for the most important things in life, except to convene in harmony where it seems to benefit the most numbers, and to refrain from human outburst, as decorum, the *quality of kindness* and common sense should suggest (Fig. 1.3).

---

[7] See Jacques Maritain, *Philosophy of Nature* , Philosophical Library, New York 1951, p. 51.

**Fig. 1.3** "Virgin and Child in a Landscape," ca. 1597. (By Aegidius Sadeler II, Netherlandish, after Dürer). (Private Collection). (Photo © M.C.Tobias)

# Chapter 2
# On the Nature of Paradox

## 2.1 Fundamental Current Contradictions

When mathematicians with equations for infinity tempt fate, when logicians go half-mad and statisticians start dangerously losing weight, or at the point etymologists and psychoanalysts converge in strikingly similar bouts of frustration, and science races to prove itself mistaken, or an adored pet pig to one by another be perceived as bacon, at that heaving moment, a seismic sigh, the earth, which cannot right herself, if we have cited her properties as having shifted, is left to lawless rules of contradiction and paradox. At least in our minds. And to the extent we will juggle the world's strange, entrancing, or off-putting meanings, a fundamental problem, a human problem, refuses to be lifted.

All too admittedly, any single stash of papers, a book, a portfolio of sketches, a diary of the subsumed and epigrammatic past, lies dormant among the top-heavy paperweights and ruins. In August of 2010, a Google algorithm declared that "129,864,880 books" had been "published in the modern world."[1] According to its methodology, divining such a number, a logical series of shortcuts to undermine redundancy, impeding duplication, were adopted, as well as specific steps to eliminate those relics deemed somehow irrelevant. The same Shakespeare in thousands of languages, or would that mean thousands of different Hamlets? Thumb-worn copies of the Bible in its more than 3300 translations, or Pinocchio's 300+ translations, Alice in Wonderland's 174, or the more than 140 translations of Don Quixote?[2] Six editions, during Darwin's lifetime, of his *On the Origin of Species*, and then the thousands, perhaps hundreds of thousands of marked up copies of Darwin's various editions, with often illuminating marginalia, as one would expect to find in any bird,

---

[1] "Google: There Are 129,864,880 Books in the Entire World," by Ben Parr, August 5, 2010, https://mashable.com/2010/08/05/number-of-books-in-the-world/Accessed February 20, 2020.

[2] See UNESCO's Index Translationum, http://www.unesco.org/xtrans/bsstatlist.aspx?lg=0, Accessed March 3, 2020.

© Springer Nature Switzerland AG 2021

M. C. Tobias, J. G. Morrison, *On the Nature of Ecological Paradox*, https://doi.org/10.1007/978-3-030-64526-7_2

flower, study guide, or textbook. CliffsNotes wherein every kid enamored of Huck Finn has likely left his/her own doodles and interpolations upon the pages.

The point goes to the proliferation and compulsion of human ideas; of frenetic, compulsive expression; its phenomenologies (first-person insights that are the subjective anecdotes that comprise any science and its inevitable thicket of interpretive experience) of translation. All those unending tensions between subjective and objective obsessions with endless layers of meaning, with levels of confidence, or "confidence intervals"[3]; of "negotiation" (Umberto Eco),[4] of "communication" (George Steiner),[5] of "individualist escapism" (Italo Calvino on James Purdy)[6]; the very rudiments of neurophysiology of the last remaining hominin mind. A brain/mind matrix of inscrutable layering, elusive and unknown to us, by turns enlightening and obfuscating, that somehow may be thought of as its own eco-dynamic flux fraught with the temptation, easily given in to, to act out contradictory behavior, or react to the world cynically.

The variables of such underlying contradictions (or ambiguities) have been discussed for thousands of years, involving such fundaments as life and death, the soul and matter, senses and reason, change and permanence, being, becoming, yes, no, here, there, not and why not, etc. These contraries, antonyms, oppositional forces, knowns and unknowns are no less intriguing and mysterious than those same combinatorial effusions which highlight poetic function, form, beginnings and ends, throughout the biochemical world we think of as earth, which *translates and transliterates* every inkling of DNA, every cell, molecule, and atom (Fig. 2.1).

So heavily entrenched is our species in copious domestic and international news, information, gigabytes, and fiscal transactions, that the weight of our presence hinders even the possible hint of our ever stepping outside *our* frameworks, the human context, the mind within. Or, as a website for Bloomberg Philanthropies states, "In God We Trust; Everyone Else Bring Data."[7]

When the twelfth richest man in the world according to Forbes,[8] Mayor Michael Bloomberg, joined his colleagues on the stage of his first democratic debate in Las Vegas (February 19, 2020), amid the scathing flurries of exhausted rhetoric, transparent mantras, desperate efforts at identity differentiation, one candidate, the

---

[3] See Wolfram Mathworld, http://mathworld.wolfram.com/ConfidenceInterval.html.

[4] *Mouse or Rat? Translation As Negotiation*, Weidenfeld & Nicolson, London, 2003.

[5] *After Babel Aspects of Language and Translation*, Oxford University Press, New York and London, 1975.

[6] *Hermit In Paris – Autobiographical Writings*, Translated from the Italian by Martin McLaughlin, Pantheon Books, New York, p.49.

[7] See "Bloomberg Philanthropies" – "Our Approach," https://www.bloomberg.org/about/our-approach/Accessed, February 20, 2020; this quote is allegedly taken from a statement by William Edwards Deming, a U.S. statistician – "In God we trust; all others must bring data." See online article by same title, by Adam Breckler, November 18. 2013 –"5 min read" - https://medium.com/@adambreckler/in-god-we-trust-all-others-bring-data-96784d01e9be, Accessed February 20, 2020.

[8] 2020 List, https://www.forbes.com/real-time-billionaires/#578cd3173d78, Accessed February 20, 2020.

**Fig. 2.1** *Lycogala epidendrum* at night (Groening's slime), Białowieża forest, Poland. (Photo © M.C. Tobias)

underdog Bloomberg, referenced (by our accounting) two most telling words otherwise lost in the angry miasmas of dramatic stagecraft: "India" and "methane." While passed over by his immediate peers, these two fleeting mere mentions by Bloomberg were not lost, not by a long shot. His foundation's multimillion-dollar embrace of global environmental concerns, particularly that of the oceans, and of zero-carbon civilization has resulted in Bloomberg co-founding and chairing the Task Force on Climate-Related Financial Disclosures (TCFD). Its charter declares, "Increasing transparency makes markets more efficient, and economies more stable and resilient."[9] But, when observing such social calls for change (good ones), we are nonetheless only too aware of the distractions. Humanity's desperate struggles to maintain a functioning human economy all but undermine the economies of nature, no matter how persistent the chorus echoing all things sustainable. Good intentions have not been enough to alter huge tides.

This is a fundamental contradiction. It underscores a powerful and paradoxical issue for individuals and nations: the origins and expenditures of wealth accumulation. Bloomberg took the debate stage the same week as the world's wealthiest man, Jeff Bezos of Amazon, pledged US$10 billion to help combat climate change. Such newsflashes harken back, in so many guises, to all the words of Adam Smith, Karl Marx, and Lenin; of Franklin D. Roosevelt's New Deal; of the Marshall Plan (European Recovery Program, worth $100 billion in 2018 dollars), and of the 1879

---

[9] https://www.fsb-tcfd.org, Accessed February 20, 2020.

treatise by Henry George (*Progress and Poverty: An Inquiry into the Cause of Industrial Depressions and of Increase of Want with Increase of Wealth: The Remedy*). They are good and useful pledges which strike not only at economic class warfare, of the rampant inequitable divides forever growing among people, communities, and nation states, and of the suffering and silent indifference by most. That silence is recorded in so many realizations, angrily repeated, by proponents of change; change that can exist somewhere between the extremes of increment and revolution.

But in that span of so many appeals for a constructive future, phrases, intentions, and deeds are easily weaponized in the political sphere, turning ultimately to the worst of walkouts on the planet, namely, human distraction with its own woes and inequities. Kin altruism defies interdependency, the heart of biological sustainability. A species that has invested solely in itself, on the pretext of survival, utilizing an entirely arbitrary if cunning mechanism, money, the Geary-Khamis dollar, or that international PPP—purchasing power parity[10]—meant to envision fair metrics amid differing values of the same product or system from nation to nation. Countless models of PPP have been extrapolated, including *The Economist*'s Big Mac Index, a specific metric intended to reveal what the overall costs of said hamburger are, including the entire regional supply chain endogenous sunk (fixed) cost; labor, land, distribution, and the expenditures associated with the sum total of environmentally depreciating footprints (e.g., packaging, fuel and transport, other extractive components, water abstraction, pollution), as well as its foreign exchange rates, from country to country.[11]

But Bloomberg more than merely mentioned India and methane. He singularly pointed to their respective importance in terms of the general wellbeing and likely mid-term scenarios for the health of the planet. His democratic competitors showed no interest.

Yet, to any ecologist, India—the largest, most polluted democracy in the world, with a dreadful animal rights record, despite its much-touted worship of cows; a local population that will eventually surpass that of China, at around 1.5 billion denizens—represents an environmental tipping point. As does methane, one of the most aggressive of all greenhouse gasses, at least 80 to 100 times that of carbon dioxide. These two categories of human trespass upon the natural world, India and methane, are emblematic of the Anthropocene, a topic central to ecological paradox, to the full psychoanalytic survey of those causes and consequences of human distraction and narcissism, which take from the planet, rather than helping her; and are core to the varied message construct of this book, just one more grappling in real time, between the dust and ash that, at least metaphorically, encinctures all of those 129,864,880+ other book titles.

---

[10] "International Dollar Geary-Khamis Define, Examples Explained," Business Case Web Site. 24 February 2016.

[11] See "What is the Big Mac Index?" by Justin Kuepper, The Balance, International Investing, November 26, 2019, https://www.thebalance.com/what-is-the-big-mac-index-1978992, Accessed February 20, 2020.

Every generation has sought to document its alarm bells, disasters and lamentations, deferrals, regrets, altercations at all levels, mysteries and rituals, aesthetic reveries, the senses and powers of the imagination, the pure pursuit of information, wealth, immortality, games, sex, and beliefs… to help smother the pains and, on occasion, by generosity, virtue, and sustained excellence to accomplish something better than ourselves. Solutions, in other words, to systemic or altogether new problems facing humanity. Our very solace taken in, and reverence for, Nature would always have seemed the right antidote for our sorrows, a key to our social and conceptual metamorphoses, and our durability as a species. All fine and well but for the fact that never, as far as we know, has humankind been confronted by its own total annihilation. Not even by the Toba Supervolcano of 74,000 years ago, thought until recently to have exerted a nearly fatal genetic/population bottleneck on *Homo sapiens*; a catastrophe so severe that there might not have been time for allele frequencies or infrequencies, as the case may be, to experience species-level rebalancing through genetic drift over a short time span of evolution, even a matter of generations. Contrary to the Toba suppositions, however, there is data which shows that African populations were largely unaffected by that mega-eruption in, what is today, Indonesia. If so, we need to look elsewhere for population bottlenecks. The twenty-first century is unambiguous in that respect.[12]

## 2.2   The Extinction Debates

But now, something has changed exponentially, as we come to view the very tragedy that is ourselves, and perhaps, or not, take heed of both warnings and manifestations. Whether we can actually do anything about it globally remains a mire of theoretical suppositions, data (or as yet unknown data) sets, all those small steps that translate into various positive conservation (by definition) countermeasures, modest progress in hundreds of millions of consumer choices; forward-leaning economic and civic arenas of awareness and action. At the 2020 Davos Economic Forum the key takeaway was "climate change," the trending words—"positive" and "impact"—as well as the *Global Risks Report 2020* in which was highlighted humanity's interconnect vulnerabilities.[13]

A much more reticent, overwhelmingly depressing and largely private but increasing through-story is that uncomfortable sense of something akin to a blurry Biblical Last Days-like intuition, the worst conceivable Rapture, foundering almost

---

[12] "Ancient Humans Weathered the Toba Supervolcano Just Fine," by Jason Daley, Smithsonianmag. com, March 13, 2018, https://www.smithsonianmag.com/smart-news/ancient-humans-weathered-toba-supervolcano-just-fine-180968479/Accessed February 20, 2020.

[13] See "What's everyone talking about at Davos 2020?" by Katie Clift, January 23, 2020, https://www.weforum.org/agenda/2020/01/what-are-people-talking-about-at-davos/Accessed February 21, 2020; see also 15th edition of the Forum's The Global Risks Report 2020, https://www.weforum.org/global-risks/reports, Accessed March 3, 2020.

embarrassed, and certainly ashamed, before the vague horizon lines of what could be our own species' extinction. Whether thinking about any of the relevant verses, from Micah 4:1–3 to 2 Peter 3:10,[14] or simply judging by the collective bad news— glaring headlines devolving into hackneyed everyday life—our frenetic worries track with an ever-increasing human population size, and those desperately plodding conservation gains and even greater losses that have become so many new norms. With them has arisen a philosophical category of deniers, cynics, peer skepticism, and outright protests erupting over everything but biodiversity loss. These costly if morally necessary civic distractions have been highlighted by such rounds of punch and counterpunch as in the case of one esteemed University of Arizona evolutionary biologist, Guy McPherson. He is noted for coining the phrase "Near Term Human Extinction," possibly coming as early as 2030, which has had the effect, in some, of inciting philosophical havoc and name-calling ("doomist cult hero," "fringe characters") with regard to such ultra-realists in the science community; accused of potentially leading society toward a road of apathy and inaction, an alleged infraction deemed by the op-ed world to be as grave as many in the US Republican Party calling climate change a hoax. But at the same time, how dare optimism, as a general category, be challenged? To question hope has been likened to a kind of sick heresy, anti-intellectualism, like mounting an uncomfortable inquiry into carbon offset schemes and their mere alleviation of guilt in sync with the escalation of corporate greenwashing.[15]

Just as the individual is hard-pressed to concede his/her own death, so we, the collective, prefer to imagine extinction-level events in museums where dinosaur skeletons are hailed; or Armageddon is tailored for audiences on enormous movie screens, in epic digital smithereens. Many will put themselves to sleep reading charming and somehow comforting British countryside murder mysteries. Is it the murder, murder's easy surrogacy, the countryside, or some weird combination in the blood? Indeed, "countryside"—said to occupy 98% of the planet, by human reckoning—has become the very focus of a softened, mellowing ecological interest, the February 20, 2020, premiere of the (short-lived) Guggenheim Museum in New York exhibition on countryside[16] which described it as, "the modern conception of leisure, large-scale planning by political forces, climate change, migration, human and nonhuman ecosystems, market-driven preservation, artificial and organic coexistence, and other forms of radical experimentation that are altering landscapes across the world" (Fig. 2.2). One can't help reading into that a distinct, if inevitable, anthropocentric accommodation, which of course is the point of it all. Of course, throughout all of that planning and experimentation, there is also murder, murder in

---

[14] See https://bible.knowing-jesus.com/topics/Last-Days

[15] See Michael E. Mann, "Doomsday scenarios are as harmful as climate change denial, The Washington Post, July 12, 2017. For distractions hampering global efforts to accelerate work to save endangered species, see, "Coronavirus disrupts global fight to save endangered species," by Christina Larson, June 6, 2020, Associated Press, https://apnews.com/e3cddd53e453a22158663f-0eeb116194, Accessed June 8, 2020.

[16] See https://www.guggenheim.org/exhibition/countryside

**Fig. 2.2**  Transitional Countryside, Central Texas. (Photo © M.C. Tobias)

the forests, the air, the water, the soils. Even the final remarkable days of Van Gogh's life; having just painted one of his greatest (arguably unfinished) works, "Tree Roots," at Auvers-sur-Oise, just north of Paris, which, he wrote to his brother Theo, the artist hoped would "express something of life's struggle"[17]; however, he would shoot himself that very night, and die two days later, July 29, 1890.

But when it comes to seeing a koala burn to death, and Syrian infants dying in refugee camps, there is no more of that comfort from novels and museum exhibitions. Rather, the very state of near emotional paralysis in which most animal rights/liberation and other nonviolence activists and ecological ethicists have been trapped for centuries.

---

[17] See "A Clue to van Gogh's Final Days Is Found in His Last Painting," by Nina Siegal, The New York Times, July 28, 2020, https://www.nytimes.com/2020/07/28/arts/design/vincent-van-gogh-tree-roots.html?smid=em-share

# Chapter 3
# Ecological Problems and Paradigmatic Solutions

## 3.1  Science Policy and Human Nature

Ever more taxing, emotionally, to analyze: The numbers, jurisprudence, pragmatic compromises, overall vision for how "to fix it" are not adding up, because, as a whole species we have no psychological or even faintly residual phenotypic record of having ever dealt with so unflexing an ecological ultimatum, never mind that it is solely our doing and that one would naturally assume we should therefore know some route toward rectifying it. We don't. The Holocaust was a microcosm of what, presently, is occurring in every geographical quadrant, often at levels we can neither visualize nor process. And if we could process (we usually can if we but attempt to do so), there are only so many solid and sustained examples of human restraint, historically, by which to gauge and extrapolate positive outcomes, where science policy and human nature are so frequently at loggerheads.

We can try and fiddle with the numbers, tilting reality and our very response to the hard sciences, toward any pre-existing bias. There are countless ways to "prove" that one plus one equals three.[1] But there is no evidence-based reason that can reverse the quite obvious contention that the sixth mass extinction is not likely to also sweep away *Homo sapiens* in its undiscriminating embodiment of a planet-wide biological crash. Either it is too late to meaningfully address what we have been doing for countless millennia, or, all has come down, in our case, to the failed notion of adaptive radiation being potentially self-guided. The big debate of moment-by-moment traits, the byproducts of some combination of nurture and nature, taking matters into their own hands. Informed re-evolution, re-genesis. Or, muddling through its philosophical hurdles, the Jack London inflection point

---

[1] See, for example, "1 + 1 = 3: Synergy Arithmetic in Economics," by Mark Burgin and Gunter Meissner, DOI: https://doi.org/10.4236/am.2017.82011, Scientific Research, AM> Vol.8 No.2, February 2017, https://www.scirp.org/journal/paperinformation.aspx?paperid=73964, Accessed February 21, 2020.

© Springer Nature Switzerland AG 2021

M. C. Tobias, J. G. Morrison, *On the Nature of Ecological Paradox*,
https://doi.org/10.1007/978-3-030-64526-7_3

**Fig. 3.1**  A Dog in the Eastern Himalayas. (Photo © M.C. Tobias)

between the canine hero, Buck, of his wildly popular *Call of the Wild*, and the author's proposed reversal in a new novel to be named *White Fang*, whereby "Instead of devolution of decivilization ... I'm going to give the evolution, the civilization of a dog," he said (Fig. 3.1).[2]

Attempting to clarify tongue-twisting breakthroughs of science, de-extinction technologies have thus far engendered the seven-minute wonder in the crossing of a rare ibex and a goat; or back-bred echoes of former selves; or, worse, fashioned freakish hybrids. Most DNA molecules have completely withered within a million or so years, so cloning of now extinct species with older lineages is not a foreseeable possibility.[3] Along with such constraints come the added hurdles of thinking through re-wilding scenarios, with their many perceived compromises in restoring partially healthy ecosystems. Our struggle to steer evolution is abetted by tens of thousands of finely nuanced research projects that have accelerated our capacities for best practices in the design of more high-confidence eco-restorative techniques. They incorporate literally hundreds of millions of data points from field researchers in

---

[2] See Labor, Earle; Reesman, Jeanne Campbell (1994). Jack London. Twayne's United States authors series. 230 (revised, illustrated ed.), Twayne Publishers, New York, p. 46.

[3] See "The sixth mass extinction, explained," The Week, Staff, February 17, 2019, https://theweek.com/articles/823904/sixth-mass-extinction-explained, Accessed February 21, 2020. See also, "Loss of land-based vertebrates is accelerating, according to Stanford biology Paul Ehrlich and others," by Lindsay FIlgas, Stanford Woods Institute for the Environment, June 1, 2020, https://news.stanford.edu/2020/06/01/loss-land-based-vertebrates-accelerating/Accessed June 5, 2020. "...scientists estimate that in the entire 20th century, at least 543 land vertebrate species went extinct. Ehrlich and his co-authors estimate that nearly the same number of species are likely to go extinct in the next two decades alone." Importantly, "they call for all species with populations under 5,000 to be listed as critically endangered..." A case in point being the Sumatran Rhino, of which no more than "80 individuals" are left.

every conceivable realm of the ecological and evolutionary sciences. Such collective energies, one would think, should hone, advantage, and ethically improve our re-adaptation and system resilience know-how. Such methodologies include mainland island and island biogeographical invasive species understanding advances; corresponding engineering breakthroughs; an avalanche of insights into biosemiotics, behavioral ecology and molecular biology; as well as important rapid advances in field acquisition data technologies, such as improved UAVs (conservation drones) and Global Navigation Satellite Systems (GNSS) that have been successfully employed, for example, in so-called polygon-to-point deployments in analyzing and monitoring protected areas and corridors.[4]

In recent years, the most comprehensive initiative to provide transparent overviews of the collective efforts at eco-restoration is found in the World Database on Protected Areas (WDPA),[5] a cooperative effort stemming from the United Nations Environment Programme and the International Union for Conservation of Nature, and launched online by Protected Planet.[6] Such research, stated goals, and mechanisms applied by all those, in every discipline, who care, are both a learning tool and generally recognized work in progress, entailing an enormous catch-up if our species is to engage in anything approaching successful ecological redemption. This is the case whether in reaching certain goals set by the Aichi Biodiversity Targets of the Convention on Biological Diversity,[7] the 2030 UN Sustainability Development Goals,[8] "core indicators" of the Intergovernmental Science-Policy Platform on Biodiversity and Ecosystem Services (IPBES),[9] the *United Nations List of Protected Areas*, updated from 2014 to 2018, *Protected Planet Report, 2018* (PPR)[10] and the 2018 analysis of the comparable effectiveness of management, according to numerous IUCN designated criteria, of those protected areas.[11]

---

[4] See "A Polygon and Point-Based Approach to Matching Geospatial Features," by Juan J. Ruiz-Lendínez, Manuel A. Ureña-Cámara and Francisco J. Ariza-López, December 5, 2017, International Journal of Geo-Information, https://pdfs.semanticscholar.org/67f9/2b05d2c02b1f6eab9813d2fbeb dbb8899478.pdf, Accessed February 21, 2020.

[5] https://www.protectedplanet.net/c/world-database-on-protected-areas

[6] https://www.protectedplanet.net.

[7] https://www.cbd.int/sp/targets/

[8] https://sustainabledevelopment.un.org/?menu=1300

[9] http://www.ipbes.net

[10] See https://www.protectedplanet.net/c/united-nations-list-of-protected-areas/united-nations-list-of-protected-areas-2018;     and     https://www.protectedplanet.net/c/protected-planet-reports/report-2018, Accessed March 3, 2020.

[11] See 2018 *United Nations List of Protected Areas – Supplement on protected area management effectiveness*, United Nations Environment Programme (Eric Solheim, Executive Director) and UNEP-WCMC (Director Neville Ash), UNEP-World Conservation Monitoring Centre, Cambridge, UK 2018, in Collaboration with the CBD (Executive Secretary, Cristiana Pasca Palmer), the IUCN (Director General, Inger Andersen), the WCPA (World Commission On Protected Areas, Chaired by Kathy MacKinnon) and UN environment, Editors: Marine Deguignet, Heather C. Bingham, Neil D. Burgess and Naomi Kingston, https://wdpa.s3.amazonaws.com/UN_List_2018/2018%20List%20of%20Protected%20Areas_EN.pdf

As the Executive Summary of the PPR points out, progress is being made, with a doubling in 5 years of marine protected areas to over 3% globally (though a far cry from the hoped-for CBD Aichi Target 11 of 10% coastal and marine protections by 2020) and, terrestrially, an increase to over 15%, close to the 17% goals set for 2020 (a target hard to envision given the global economic downturns of that year). But of significant interest, since 2008, the number of protected areas in the world overall has gone from just over 206,000 to "238,563 protected areas from 244 countries and territories, covering more than 46 million km$^2$" (as represented in the 2018 United Nations List of Protected Areas), including the Areas Beyond National Jurisdiction (ABNJ—i.e., the *High Seas* as designated in the 1982 UNCLOS treaty.)[12]

In those areas where the WDPA protected area monitoring effectiveness has been unclear or ambiguous, data gap analyses—as reflected in the monitoring criteria of the GD-PAME, a searchable database known as The Global Database on Protected Area Management Effectiveness—[13] were not included in the 238,563 total. As of late 2019, there were 2071 such ambiguous units.[14] Encouraging though the overall number of protected areas around the world may seem, in the first ever comparable study of underfunded biodiversity locales, it was made clear that "biodiversity declines have progressed rapidly, and further delays in improving finance are likely to lead to even greater global extinction risks…".[15] Conversely, what would it take to rapidly alter the course of inadequate funding for global biodiversity? The problem, at its source, is the paucity of actual fieldwork on species diversity, a gap that impedes the critical arguments needed to confidently solicit requisite levels of private, corporate, and government financing. According to one finding "only 33,536 of the 91,000 species on the IUCN Red List" have been "comprehensively assessed." The word "comprehensive" is a human word; and relative in every human sense.[16]

By the most conservative estimates, there are "8.7 million" species on earth.[17] But a later study argues that the number 8.7 million fully underestimates, to begin with, microbial species, which researchers have now placed at "upwards of 1 trillion species." And, to the point, have suggested that for a price tag of $500 million to $1

---

[12] See https://www.un.org/depts/los/convention_agreements/texts/unclos/unclos_e.pdf.

[13] See https://pame.protectedplanet.net

[14] ibid., https://wdpa.s3.amazonaws.com/UN_List_2018/2018%20List%20of%20Protected%20 Areas_EN.pdf, p.8.

[15] "Targeting global conservation funding to limit immediate biodiversity declines," by Anthony Waldron, Arne Mooers, Daniel C. Miller, Nate Nibbelink, et.al., Proceedings of the National Academy of Sciences 110(29), July 2013, DOI: https://doi.org/10.1073/pnas.1221370110, Pubmed, PY 2013/07/01, VL110,https://www.researchgate.net/publication/244482672_ Targeting_global_conservation_funding_to_limit_immediate_biodiversity_declines, Accessed February 21, 2020.

[16] See "Matches and Mismatches Between Global Conservation Efforts and Global Conservation Priorities," by David F. Willer, Kevin Smith, and David C. Aldridge, Frontiers in Ecology and Evolution, 07 August 2019, https://doi.org/10.3389/fevo.2019.00297, https://www.frontiersin.org/ articles/10.3389/fevo.2019.00297/full, Accessed February 21, 2020.

[17] See "How many species on Earth? About 8.7 million, new estimate says," Census of Marine Life, Science News, August 24, 2011, https://www.sciencedaily.com/releases/2011/08/110823180459. htm

billion per year (for 50 years) every species could be identified, the scaling and actual species numbers accurately calculated.[18] Even at that scale, and in view of several conservation cost/benefits assessments, it is clear that money alone will not necessarily save species, although in the case of one parrot taxon, approximately US$2.26 million spent over 25 years brought the Lear's Macaw from the Critically Endangered to Endangered status in Brazil.[19] That might seem like a negligible step, but it is the hardest, most critical one on the path toward species restoration. Numerous other taxa-specific cost/benefits studies have been done with respect to North American mountain lions, California condors, and New Zealand kakapos, as examples. The costs per species are always in the millions of dollars. And while various research estimates for saving all presently identified globally endangered species, at between $59 and $76 billion annually, sound more than possible (given the value of estimated free natural services exceeding $125 trillion, and the actual ability of mints to print emergency dollars), the bigger looming costs are outside cost/benefit algorithms (Fig. 3.2).

## 3.2   Trigger Effects and Moral Half-Resolutions

The inflictions that drive species to endangerment and extinction are simply too myriad to simplistically capture in linear economic projections.[20] In 2012, the US government spent approximately $1.7 billion to work toward saving its regional endangered species.[21] At the same time, however, that meager investment was calculated to generate some $1 trillion per year in benefits.[22] Because the field data is constantly being appended, the real cost/benefits over time has proven difficult to track. Moreover, at every cusp of spending decisions, there are predictable other, competing emergencies that drive dollars down toward the inevitable triage factor in policy-making.

---

[18] See "There Might Be 1 Trillion Species on Earth," by Stephanie Pappas, May 5, 2016, LiveScience, https://www.livescience.com/54660-1-trillion-species-on-earth.html, Accessed February 21, 2020.

[19] See "How much does it cost to save a species from extinction? Costs and rewards of conserving the Lear's macaw," by Antonio E. A. Barbosa and José L. Tella, July 10, 2019, https://doi.org/10.1098/rsos.190190, Royal Society Open Science, Royal Society Publishing, https://royalsocietypublishing.org/doi/10.1098/rsos.190190, Accessed February 21, 2020.

[20] See "Conservation Will Cost $76 Billion," by Dan Cossins, TheScientist, October 11, 2012, https://www.the-scientist.com/the-nutshell/conservation-will-cost-76-billion-40357, Accessed February 21, 2020.

[21] See "How Much Did the U.S. Spend on the Endangered Species Act in 2012?" by John R. Platt, November 1, 2012, Extinction Countdown, Scientific American, https://blogs.scientificamerican.com/extinction-countdown/how-much-did-the-us-spend-on-the-endangered-species-act-in-2012/ Accessed February 21, 2020.

[22] See "The Endangered Species Act Is Criticized for Its Costs. But It Generates More than $1 Trillion a Year," by Justin Worland, July 25, 2018, Time Magazine, https://time.com/5347260/endangered-species-act-reform/Accessed February 21, 2020.

**Fig. 3.2**  A California Condor *(Gymnogyps californianus)*. (Big Sur, Photo © M.C. Tobias)

*Average* costs for simply doing the review of each of the "more than 500 species petitions" pending before the US Fish & Wildlife Service are "$140,000" per species.[23] Other research suggests that globally to provide sufficient information on the status and necessary recovery financing of IUCN Red Listed species would cost approximately US$174 million, plus other costs of US$114 million annually to maintain baselines on existing data.[24] Others have argued that "protecting all the world's threatened species will cost around US$4 billion a year."[25] By percentage of

---

[23] See "How much would you pay to protect an endangered species," by Jonathan Wood, The Hill, October 8, 2018, https://thehill.com/opinion/energy-environment/410151-how-much-would-you-pay-to-protect-an-endangered-species, Accessed February 21, 2020.

[24] See PlosOne, "Assessing the Cost of Global Biodiversity and Conservation Knowledge," by Diego Juffe-Bignoli, Thomas M. Brooks, Stuart H. M. Butchart, et.al., August 16, 2016, https://doi.org/10.1371/journal.pone.0160640, Accessed February 21, 2020.

[25] See "Cost of Conserving Global Biodiversity Set at $76 Billion," by Daniel Cressey, October 12, 2012, Scientific American, https://www.scientificamerican.com/article/cost-conserving-global-biodiversity-set-76-billion/Accessed February 21, 2020.

US GDP, such numbers are almost an aside. US expenditures for healthcare in 2015 were $3.2 trillion (17.8% of GDP.)[26] But spending on "threatened" species is different than those expenditures targeting the "endangered," "vulnerable," "critically endangered," "at risk," and so on. Grammatically, they are all threatened, all at risk. But financial instruments are blunt objects insensitive to the subtleties of differentiating categories of risk that differ from country to country, taxon to taxon, population to population, policy expert to policy expert.

But accounting for 30% or higher medical care in America than elsewhere in the world, such (approximately) $10,000 per person costs annually, scaled up to 7.8 billion people, would far exceed the total global economy. Whereas estimated conservation costs to protect hundreds of billions of individual taxa—quadrillions of individuals if one extrapolates to invertebrates throughout all the protected areas of the world—suggest that single-matrix-targeted (multi-tiered) conservation biology looks impossibly *inexpensive*, by any reconciling of benefits and costs common-sense analysis, incorporating intangibles, particularly at the invertebrate and microbial levels.

The deep moral realities are conclusive: Human economics is at best a vague source of guidance for grasping how relatively clear the path for a conservation conscience should be; at least did we inhabit a world where budgets were balanced according to an ethical proposition, posed by utilitarian philosopher John Stuart Mill (1806–1873), which forcefully argued for "the greatest amount of good for the greatest number." Then integrate added value factors predicated upon the shortest time frames, most expansive and diverse habitats (e.g., shrublands and rocky landscapes, bryophytes and deserts ignored no longer, for starters) and a net zero-triage methodology, however incomprehensible at first. These components would inexorably amplify the topology of virtue, provisioning the broadest possible medical responses to the biodiversity crisis we have unleashed.

---

[26] See "CDC Health Expenditures," https://www.cdc.gov/nchs/fastats/health-expenditures.htm, Accessed February 21, 2020.

# Chapter 4
# Protected Area Dilemmas

## 4.1 The Paradox of Ecological Comparisons

Russian scientists, going back to the 1960s, were among the first, and most outspoken, to bluntly summarize the acceleration of extinctions, under-represented groups within protected area networks, and current and future challenges to halting the irreversible damage being occasioned in the largest country in the world. By situating that nation's vast biodiversity and range of biomes within a global human context of (seemingly) irreconcilable altruisms, waning sustainability options, and the essentialities of science and policy prioritizations, Pavlov and Shatunovsky made the frustrating expanse of human contradictions to overcome painfully clear.[1]

Such fickle biological fates at the mercy of countries and transboundary conservation urgencies and abnegating partners are exacerbated by realizations of counter-intuitive scale, as well as a great variance in measurement modalities and scales of ranked importance. No one is likely to prioritize research or protected areas in those countries with the most freckled or green-eyed individuals; but fruit fly diversity may be another thing entirely. Botswana ranks number 1 out of 152 countries on the Megafaunal Conservation Index.[2] Elsewhere in Africa, abnegated wildlife has long represented a life-affirming and/or systemic cultural failure. Poachers, principally, have undermined every political advantage sought by both one and multiparty constitutional rule; whether during the single-party power of conservationist leader

---

[1] See "Biodiversity Conservation in Russia," by D. S. Pavlov and M. I. Shatunovsky, Area Studies-Russia (Regional Sustainable Development Review) – Vol. 1, in Nicolay Pavlovich Laverov, (editor), *Encyclopedia of Life Support Systems*, EOLSS/UNESCO, Oxford, UK, 2009, pp.159-165, https://www.eolss.net/Sample-Chapters/C16/E1-56-08.pdf

[2] See "The countries that are best at conservation are the ones that depend on wildlife tourism," by Neha Thirani Bagri, Quartz, May 9, 2017, https://qz.com/979450/the-countries-that-are-best-at-conservation-are-the-ones-that-depend-on-wildlife-tourism/Accessed February 22, 2020.

© Springer Nature Switzerland AG 2021
M. C. Tobias, J. G. Morrison, *On the Nature of Ecological Paradox*,
https://doi.org/10.1007/978-3-030-64526-7_4

Kenneth Kaunda in Zambia (1972–1991),[3] or throughout the "wildlife wars" continuing across Central, East, and Southern Africa.[4] Post-civil war Mozambique may have seen a stunning, poacher-free revival at Gorongosa National Park; even as tiny Gombe National Park in Tanzania, one of the world's primate success stories—thanks, of course, to Jane Goodall and friends—has of late witnessed declining chimpanzee populations due to "human encroachment."[5] The bottom biological line for Africa, no matter how one presents the diverse tallies of conservation endeavors, is the continuing reality that over 40% of the human population in the continent's 54 nations is under the age of 15, and population growth remains over 2% per year, suggesting an increase by 2050 from the present 1.340 billion people to 2.4 billion.[6] That number figures starkly beside the current count of surviving mountain gorillas: slightly fewer than 800. But in context, 800 is a huge improvement over the 2003 census, which numbered 380 individuals left.[7]

On a very different note, not brighter, by any means, just quantitatively insightful, at least for terrestrial vertebrates, there are an estimated 24 billion mostly captive chickens at any one moment (not including hundreds of thousands of feral chickens, and a known 5 species of wild jungle fowl whose IUCN Least Concern status suggests large numbers); and trillions of bristlemouth fish distributed across 12 species in the oceans, particularly numerous in the genus Cyclothone.[8] But we'll discuss later on the true reality for those chickens.

Other large numbers of vertebrates—from brown rats (*Rattus norvegicus*) and house mice (*Mus domestics*) to red-billed quelea in sub-Saharan Africa (*Quelea quelea*, a small migratory weaver of the Ploceidae family)—together exceed the total human population by at least 10 billion individuals. But the numbers shrink rapidly with any species over 50 kilograms; just as intact faunal assemblages within

---

[3] See "Political Institutions and Conservation Outcomes: Wildlife Policy in Zambia," by Clark C. Gibson, Swiss Political Science Review 6(1): 87-121, 2000, https://onlinelibrary.wiley.com/doi/pdf/10.1002/j.1662-6370.2000.tb00287.x, Accessed February 22, 2020.

[4] See "On the frontline of Africa's wildlife wars," by John Vidal, The Guardian, May 7, 2016, https://www.theguardian.com/environment/2016/may/07/africa-frontline-of-wildlife-wars, Accessed February 22, 2020.

[5] See "For the famed chimps of Gombe, human encroachment takes a toll," by Anthony Langat, February 28, 2019, Mongabay, https://news.mongabay.com/2019/02/for-the-famed-chimps-of-gombe-human-encroachment-takes-a-toll/Accessed February 22, 2020.

[6] See "World Population Review," "Africa Population 2020," http://worldpopulationreview.com/continents/africa-population/Accessed February 22, 2020.

[7] See Mountain Gorilla Conservation Fund, http://www.saveagorilla.org/60-Questions.html,Accessed February 22, 2020.

[8] The chicken statistic cited in, "Noah Strycker's 'Thing with Feathers'," by Brendan Francis Newnam, Dinner Party, March 8, 2014, http://www.dinnerpartydownload.org/noah-strycker/: The Cyclothone number taken from, "This strange, glowing fish is more abundant than rats, chickens, and humans combined," by Cody Sullivan, June 30, 2015, Business Insider, https://www.businessinsider.com/bristlemouth-fish-is-the-most-abundant-vertebrate-on-earth-2015-6, which in turn borrows from, "An Ocean Mystery in the Trillions," by William J. Broad, The New York Times, June 29, 2015, https://www.nytimes.com/2015/06/30/science/bristlemouth-ocean-deep-sea-cyclothone.html, Accessed February 22, 2020.

**Fig. 4.1**  A Glacier in Southern Alaska. (Photo © M.C. Tobias)

habitats larger than 100 square kilometers are seen to be scarce when measured against smaller protected areas, which are the majority, 60% of them in Europe, and often extremely small. They account for "less than 10%" of all 46,414,431 square kilometers protected areas ("14.87% of the land [excluding Antarctica] and 7.27% of the sea.")[9] As tiresome as such numbers in array can be, they indisputably excite volatile emotional realities for the human species on a tightrope. In the United States (3.797 million square miles= 2.43 billion acres), there are 111 million acres in 44 states designated as "wilderness." In total, approximately one-third of all US public lands—some 235 million acres—are described as "wildlands," under some aegis of protection. Including Alaska, that represents less than 5% of America's land base, 2% in the lower 48 states (Fig. 4.1).[10]

Even many of those areas are not free of human interference, as hunting and fishing are frequently allowed. The US Wilderness Act of 1964 also allows for controlled fires under regimens designed to manipulate insect and disease vectors deemed harmful. Pre-existing water rights are left intact; and, obviously, "weeds," air and water pollution, all the impacts of climate change and other subtler anthro-

---

[9] See *2018 United Nations List of Protected Areas, Supplement on protected area management effectiveness*, pp. 40-43, https://wdpa.s3.amazonaws.com/UN_List_2018/2018%20List%20 of%20Protected%20Areas_EN.pdf

[10] See https://www.wilderness.org/articles/article/wilderness-designation-faqs, Accessed February 22, 2020.

pogenic influences, while frequently monitored, interact across every designation.[11] As the National Park Service in America decreased by 11% its staffing between 2011 and 2018, and the Congressional Natural Resources Committee in 2017 passed legislation imposing additional hurdles to designating future national monuments, as originally envisioned by the Antiquities Act, it became clear that Republicans in Washington were turning their backs on the environment. One nation, one trend oriented toward undermining 50 years of NEPA, the critical National Environmental Policy Act, as well as nearly every significant piece of legislation in EPA history.

Such measures represent a form of reprehensible irrationality at the biological level. Even within ranks, there are social consequences. For example, there was the grassroots formation of an "Alt National Park Service," devoted to continuing the values of the service for future generations, as embraced by NPS employees in collective ethical compliance with the substance of the original National Park Service Organic Act of August 25, 1916, signed into law by President Woodrow Wilson. Perversely, the NPS's first superintendent, Stephen Mather, was an industrial magnate who believed in as many automobiles and roads heading into the parks as possible, coinciding with the demise of resident indigenous peoples in the case of one of America's most famous parks, Yosemite, and its former Southern Miwok occupants. And while the 417 park units covering 84 million acres have been slowly expanded in the century following Wilson's mandate, most notably by President Jimmy Carter's doubling of the acreage in the late 1970s in one fell swoop, trends into the twenty-first century have not boded well.

## 4.2   Psychology and Policy

Policies of national, psychological sovereignty have resulted in regard, or disregard, to economic and political vagaries predicating an outright negligence of moral duties on one continent after another toward the environment. Our species' topographical hubris can be measured according to any number of distinct biases, some in accord with true nurturance, but most, deceptive and/or disinterested. The basis for environmental sophrosyne (excellence, soundness), as practiced, monitored, and policed by governance, is a fast-moving target subject to all the flaws built in to any collective tide of constituency discords. Realism and idealism, both, easily warp political aspirations; while real *people* can accomplish nearly anything. As ranked by the amount of terrestrial protection, the US is number 114, with 12.99% of its sovereign borders protected, as of 2018. Given the numbers, one sees how vague, diluted, and unclear the word "protection" really means. Rankings for New Zealand, Canada, Mauritania, Bhutan, and particularly Suriname are all soft data, riddled with complex, on-the-ground discrepancies in data assessment. To its significant

---

[11] See, for example, community monitoring in Colorado, https://wildernessworkshop.org/wilderness-monitoring/

credit, UN protected areas data management is continually soliciting updated information from those closest to their subject. While Russia comes in at number 126 on the ranking (9.73% of its massive land cover), Libya and Afghanistan at the bottom of 192 ranked nations, the top three are: New Caledonia (54.40%), Venezuela (54.14%), and Slovenia (53.62%). Bhutan is 4th.[12] Governments of those top three protection icons cannot simply rest on laurels (nor can Bhutan—with much more justification than nearly any other nation to do so).[13] The scars of New Caledonia's nickel exploitation coincide with a major biodiversity hotspot, while Venezuela's deforestation rate is now the third highest in South America, mirroring the country's instability and an economy in chaos. Slovenia remains the darling of European land conservation, but is grappling with two significant health problems stemming from air quality: "particulate matter and ozone".[14]

In Bhutan, the pillars of her national policy of gross national happiness[15] have entailed various pieces of legislation and biocultural norms such that, for 2 months each year, vegetarianism is all but the law, though a microanalysis reveals huge gaps, hoarding, lapses, even stockpiling of meat in time for the influx of the more than 200,000 Indian tourists who drive to the country each year, many wanting meat at mealtime. At the same time, more than 60% of Bhutan's forests remain intact, the decision taken many years ago by a 16-year-old king. Such overall contradictions, one hopes, will ultimately be resolved in favor of an increasingly non-violent national disposition, in keeping with a Buddhist nation.

Or consider another extraordinary act of globalization in favor of protecting the planet, namely, the 17.8 million acres of the Southern Suriname Conservation Corridor in the hands of 3000 indigenous Trio and Wayana Indians. An area of essentially 100% forest cover four times the size of New Jersey, second only in expanse to the Kayapo indigenous reserve in Brazil. Such conservation achievements translate into Suriname's unambiguous carbon credits under the UN REDD+ program, Reducing Emissions from Deforestation and Forest Degradation,[16] for the value of carbon sequestration on world markets. The European Natura 2000 Network Danube Delta program, widespread European Re-Wilding endeavors, and the Y2Y—Yellowstone to Yukon Conservation Initiative in the United States—have been following a similar

---

[12] See "Terrestrial Protected Areas," Index Mundi, https://www.indexmundi.com/facts/indicators/ER.LND.PTLD.ZS/rankings, Accessed February 22, 2020.

[13] See *Himalayan Transitions: An Assessment of Conservation Priorities in Bhutan*, Unpublished Manuscript in Review, by Dr. Ugyen Tshewang, Michael Charles Tobias, and Jane Gray Morrison, Dancing Star Foundation Research Project, 2021.

[14] See "Slovenia Country Briefing – The European Environment – state and outlook 2015, February 18, 2015, https://www.eea.europa.eu/soer-2015/countries/slovenia, Accessed February 22, 2020.

[15] See A Compendium of Gross National Happiness (GNH) Statistics, National Statistics Bureau of Bhutan; See also: The Last Shangri-la? A Conversation with Bhutan's Secretary of the National Environment Commission, Dr. Ugyen Tshewang, Michael C. Tobias, 2011, Forbes.

[16] See www.un-redd.org/aboutredd

**Fig. 4.2**  Dr. Russell Mittermeier Atop the Voltzberg in Suriname. (Photo © M.C. Tobias)

model.[17] As Dr. Russell Mittermeier, Chief Conservation Officer of Global Wildlife Conservation, has described, Suriname is now in a very strong position to sell the best drinking water in the world to countries throughout the Caribbean that are dependent on expensive and problematic desalinization. With a modest population of approximately 563,000 people, Suriname has blazed a remarkable trail. Vanuatu, Andorra, San Marino, and others are following.[18] Indeed, compared with the US at present, these countries are ecological islands of sanity (Fig. 4.2).

## 4.3  Human Nature and Red Foxes

We are not one human nature. Hence, the well-timed and dramatic work by Paul R. Ehrlich, *Human Natures: Genes, Cultures, and the Human Prospect.*[19] For all we know, humans may even have, after all, the SorCSi "tameness gene" (found in 1600

---

[17] See http://www.rewildingeurope.com/areas/

[18] See "New Conservation Corridor Latest Environmental Triumph for Suriname," by Dr. Russell A. Mittermeier, April 14, 2015, https://www.conservation.org/blog/new-conservation-corridor-latest-environmental-triumph-for-suriname/Accessed March 3, 2020.

[19] Island Press/Shearwater Books, Washington, D.C., 2000.

Russian red foxes) lurking somewhere amid our 25,000 to 50,000 genes.[20] It has been estimated that humanity weighs 300 million tons, while all remaining wildlife is less than 100 million tons, and the farm animals we slaughter over 700 million tons and growing.[21] All those sentient beings' pain constitutes that largest aggregate of suffering in the last 65 million years, what we have elsewhere termed the "pain points," referring to the geography of industrial slaughterhouses, on land and at sea.[22] Add to that the centuries-old crime against forests. Humanity in all her guises has cut down 50% of all known trees on the planet, or roughly 3.4 trillion sylvan individuals since the origin of all human civilizations.[23] If there is a gene tameness, it might account for what both Leonardo da Vinci and Rajendra Pachauri, former chairman of the Intergovernmental Panel on Climate Change, both embraced, namely, vegetarianism.[24] That tameness, or "reciprocity potential" (for compassion and virtue), can happen in an instant, without legal hair-splitting or cumbersome international treaty negotiations.

In ancient Jain doctrine, the foremost decree held to the proposition: "Parasparopagraho jivanam," all life forms are interdependent and must care for each other. Such pantheistic wisdom is hard-pressed against a syndrome now widely known as "treaty congestion," well over 500 international environmental treaties agreed to by nations since the formation of the UN—and a total of 3716 environmental agreements in general—but without requisite or realistic incentives for compliance. In Harvard Professor Lawrence Susskind's groundbreaking 2008 essay "Strengthening the Global Environmental Treaty System" he made it clear that "Today, there is no official body with responsibility for improving the global environmental treaty-making system," one of the most troubling of human paradoxes:

---

[20] See "red fox genome assembly identifies genomic regions associated with tame and aggressive behaviours," by Anna V. Kukekova, Jennifer L.Johnson, Goujie Zhang, et al., Nature Ecology & Evolution, August 6, 2018, ISSN 2397-334X (online), Accessed August 6, 2018; and "Open Questions: How many genes do we have?" by Steven L. Salzbert, BMC Biology 16, Article number: 94 (2018) BMC Biol 16, 94, August 20, 2018, https://doi.org/10.1186/x12915-018-0564-x, Accessed February 22, 2020.

[21] See "Industrial farming is one of the worst crimes in history," by Yuval Noval Harai, The Guardian, September 25, 2015, https://www.theguardian.com/books/2015/sep/25/industrial-farming-one-worst-crimes-history-ethical-question

[22] See M. C. Tobias and J. G. Morrison, God's Country: The New Zealand Factor, A Dancing Star Foundation Book, Zorba Press, Los Angeles CA and Ithaca, NY, 2010.

[23] See "Seeing the forest and the trees, all 3 trillion of them," by Kevin Dennehy, September 2, 2015, YaleNews, https://news.yale.edu/2015/09/02/seeing-forest-and-trees-all-3-trillion-them, Accessed February 22, 2020.

[24] See "Get Back In The Car: Vegetarian IPCC Chairman Rajendra Pachauri Says Less Meat Will Slow Global Warming," News Staff at Scientific Blogging, September 6, 2008, https://www.science20.com/news_releases/get_back_in_the_car_vegetarian_ipcc_chairman_rajendra_pachauri_says_less_meat_will_slow_global_warming_more; See also, "Go vegetarian to limit climate change: IPCC report," by Graham Lloyd, The Weekend Australian, August 8, 2019, https://www.theaustralian.com.au/nation/go-vegetarian-to-limit-climate-change-ipcc-report/news-story/411604a4b88159d85859e72fa5aee888, Accessed February 22, 2020.

That treaties intended to save life, for however many reasons, do not succeed, because of humans.[25] And yet, the countering argument: Change does occur. Consider the decision by the former Prime Minister of New Zealand John Key, to create a marine reserve in New Zealand's northern waters the size of France.[26] It can happen just like that. The five largest marine reserves in the world, covering millions of square kilometers, consecrated into law within the last seven years, in France, Britain, Australia, and the United States. Rapid protective mechanisms.

The dialectic at the heart of ecological paradox is unsparing. Since October 1945, when the UN was founded, the human population has soared over 300%, from approximately 2.5 billion to over 7.8 billion. During that same period the Green Revolution came of age, spawning great faith in technology to feed humanity through more genetic hybrid innovation, more sustainable land tenure and agricultural reforms that all promise greater parity, empowerment of women, environmental justice, and sustainable land use. Instead, other paradoxes—famine amid plenty, of a parched and violent world—have cascaded. Since the mid-1980s, our species has exceeded the appropriation of 40% of all the products of photosynthesis on the planet, or NPP, net primary production. No species has ever so trespassed, throwing into confusion the very nature of globalization, given our one species' surreal sense of superiority and accompanying dominion by force over every continent.[27] The "pain points," in combination with the NPP dominion, and the as yet inordinately modest patchwork of protected areas—and patchy protection itself—has combined both in human natures, and in the world we know, to exterminate more than 50% of life forms, including approximately 75% of the large ones, and—excluding the few prolific other vertebrates earlier mentioned—50% of the smaller ones.

## 4.4   Legal Standing for Nature

The late Juri Lotman, in his and Wilma Clark's famed 1984 essay "On the Semiosphere",[28] paved the way for recognizing a biosphere teeming with abundant signals *between species*—the basis of interspecies communications. By that insight all our notions of globalization and politics, of law and self-expression, have been

---

[25] See Donald K. Anton, "'Treaty Congestion' in International Environmental Law," International Law Reporter, January 24, 2012, http://ilreports.blogspot.com/2012/01/anton-treaty-congestion-in.html; and Lawrence Susskind, "Strengthening the Global Environmental Treaty System," Issues in Science and Technology, Volume XXV, Issue 1, Fall 2008, http://issues.org/25-1/susskind/; See "IEA Project Contents" for "Environmental agreements currently in the database," https://iea.uoregon.edu/iea-project-contents, Accessed May 13, 2020.

[26] See New Zealand to turn Kermadec into vast marine reserve, BBC News, September 29, 2015, http://www.bbc.com/news/world-asia-34387945

[27] See "Human Appropriation of the Productions of Photosynthesis,"by Peter Vitousek, Paul R. Ehrlich, Anne H. Ehrlich and Pamela Matson, 1986, http://dieoff.com/page83.htm; See also, also: http://earthobservatory.nasa.gov/GlobalMaps/view.php?d1=MOD17A2_M_PSN

[28] See Sign Systems Studies, 33 (1):205-226 (2005) https://philpapers.org/rec/LOTOTS-2

deeply challenged to account for a symbiotic biosphere of individuals with looming legal standing. It was Christopher D. Stone, in 1972, who published his groundbreaking work to this effect, *Should Trees Have Standing? Law, Morality, and the Environment.*[29]

For 75 years, the United Nations has modeled itself upon the broadest evolutionary principles of non-violence, non-friction, and of perpetual peace, as Immanuel Kant enshrined the phrase in the late 1790s, in the tradition of lawmaker Hugo Grotius, or Huig de Groot (1583–1645), the father of the first international peace treaty in the world. That was, of course, the 1648 Peace of Westphalia ending the Eighty as well as the Thirty Years' Wars and based upon his brilliant vision of *natural law.* Grotius was a neighbor of two of the greatest Dutch naturalists: the painter Vermeer, whose "View of Delft" has long been prized as the most perfect rendering of a resilient, urban landscape, and the microscopist Antonie van Leeuwenhoek, whose astonishing optic epiphanies paved the way for the application of outstanding earth science to the notion of a biosphere and noosphere, as centuries later envisioned by Vladimir Vernadsky and Eduard Suess. Westphalia engendered European self-determination and the first concept of balance of power. It also guaranteed Dutch independence. Grotius' insistence on nonviolence was a true precursor of John Ruskin's book *Unto This Last* (1860) and Gandhi's consequential reading of those four political economic essays.

In his seminal work *De jure belli ac pacis* (*On the Law of War and Peace*, 1625), Grotius envisioned what today we call the Geneva Convention, and the rule of law, among all nations. To grasp that challenge, as ambassador to the Soviet Union, George Kennan, in his somewhat legendary "A Modest Proposal," (July 16, 1981) stated, "Adequate words are lacking to express the full seriousness of our present situation."[30] Applied ecologically, this is no longer the case. We have words and we have the data. What we continue to lack, however, is a complete philosophical and global method of peace-making in the manner of a Hugo Grotius that is sufficiently comprehensive to impede the biological unraveling in this epoch, the Anthropocene.[31] Kennan described what he believed to be a basic flaw of US foreign policy, and by implication, foreign policies from nation to nation; namely, public opinion, "a force," as he enumerated it, "that is inevitably unstable, unserious, subjective, emotional and simplistic."[32] We first presented these ideas—linking conservation biology and global protected area data, to the history of legal and public responses to increasing demographic pressure—at the UN 70th Anniversary conference in

---

[29] See http://www.environmentandsociety.org/mml/should-trees-have-standing-law-morality-and-environment, Accessed July 29, 2020.

[30] See http://www.nybooks.com/articles/archives/1981/jul/16/a-modest-proposal/

[31] See "Anthropocene: the human age," by Richard Monastersky, Nature, March 11, 2015, http://www.nature.com/news/anthropocene-the-human-age-1.17085

[32] George Urban, September 1976, "From Containment to Self-Containment: A conversation with George Kennan", Encounter, p.17.

**Fig. 4.3** The "Vermont" and "Wawona," Mariposa Grove, Yosemite National Park. (Photograph from Southern Pacific Railway, ca. 1912, Private Collection)

Moscow in late 2015, and again for the 75th Anniversary.[33] Since 2015, we have witnessed sporadic but increasingly overburdened policy-making junctures in every sector, but most harrowingly in the ecological, life-sustaining realms.

Through the lens of a George Kennan, hope is significantly endangered. In so many ways the dialectic of hope is an active conversation, every day, on everyone's mind, as our species stands in the crosshairs. Of war, of ruin, of hostilities we know

---

[33] See "Ecological Challenges in a Global Context," November 4, 2015, https://mahb.stanford.edu/library-item/ecological-challenges-in-a-global-context/; See also, https://en.globalistika.ru/mediateka, and https://www.youtube.com/watch?v=jzKFe1hoNHc.

about and those we do not. Of 17 "UN Sustainable Development Goals"[34] for so many economically desperate human souls, people who are hungry, living on US$ 1.25/day, children dying before their first birthday, crises in quality education, energy, healthcare, gender equality, sanitation, opportunities for youth, and systemic racism, the rallying cry for which, "I can't breathe" certainly speaks to the planet; but more so, to human nature. Eventually, most people, hardened survivors in so many ways, will be living in cities, as these dilemmas escalate even further, and proximity between humans—actual physical proximities, as measured in yards and inches—intensifies.

What then? Those moments we all feel, suddenly and quite without warning, like vertigo. And we say, it's really happening, a desperate epiphany commensurate with the call for more wildness, and for its legal standing (Fig. 4.3).

---

[34] See "About the Sustainability Development Goals," https://www.un.org/sustainabledevelopment/sustainable-development-goals, Accessed July 24, 2020.

# Chapter 5
# The Paradox of Protection

## 5.1  Contradictions of Environmental History

The history of environmentalism from country to country is that commingled rallying cry, fraught with the desperate need to live in a healthful environment, and to believe that such health and the life of other species and their habitat will be safeguarded from our kind. But a paradox of protection has been enshrined in the geopolitics of any and all legal standing. For example, in 1909, 6 years after having initiated the consecration of the National Wildlife Refuge by the creation of the Pelican Island National Wildlife Refuge in Florida, Ex-President Teddy Roosevelt and his son, Kermit, went on their well-known expedition to East Africa and slaughtered 512 charismatic vertebrates. Their companions on the trip shot and trapped another 10,500+ animals, all of which ended up aimlessly stuffed and stacked so as to gather dust at the Smithsonian.

Roosevelt and his son with colleagues did something similar in 1913–1914 in the Amazon, using the pretext of the Roosevelt-Rondon scientific expedition to the "River of Doubt" to enable them to justify—with science—their addiction to hunting.

To be clear, Roosevelt's National Wildlife Refuge System was the work of a passionate genius with the ability to sway officialdom in Washington (Fig. 5.1). But for its 150 million acres, 568 refuges and 38 water management districts, the reason so fewer people visit these blessed geographies (approximately 40 million per year) versus the hundreds of millions who venture out to the 62 US National Parks is explicitly due to the fact hunting occurs in many of the existing refuges. The vast majority of citizens want to be able to meander in a refuge without fear of being shot, or of hearing gunshot, or seeing someone gut a catfish or even knowing that animals can be slaughtered.

That knowledge is critical: Most Americans will never be able to visit most of these special places, confined by democratic processes we cherish to those heavenly

© Springer Nature Switzerland AG 2021
M. C. Tobias, J. G. Morrison, *On the Nature of Ecological Paradox*,
https://doi.org/10.1007/978-3-030-64526-7_5

spheres of cartography, more akin to an ideal than dirt in one's palm. We want to
believe that these places are safe for all beings; the deer and the antelope; the bison
and the eagle; the wild turkey and the rainbow trout.

One thing for certain: Roosevelt's killing sprees abroad were not intended to
signal to Americans, or perpetuate at home, anything contrary to the sacrosanct
philosophy inherent to the National Wildlife Refuge System which was and is, "To
administer a national network of lands and waters for the conservation, management, and where appropriate, restoration of fish, wildlife, and plant resources and
their habitats within the United States for the benefit of the present and future generations of Americans" (National Wildlife Refuge System Improvement Act of
1997). Nowhere in that mission statement is there mention or condoning of populist
killing. Tragically, Roosevelt's son, Kermit, would take his own life with a gun, up
in Alaska, June 4, 1943.

But history, in a sense, has already judged, by assessment of contemporary mores, its popular sentiment regarding violence. It is fact that, according to numerous polls, less than 6% of Americans hunt.[1] The US Department of Interior has represented that, "In 2016, more than 103 million Americans – a staggering 40 percent of the U.S. population 16 years and older – participated in some form of fishing, hunting or other wildlife-associated recreation such as birdwatching or outdoor photography."[2] But what it is failing to capture in that "staggering" percentage is that the overwhelming majority of those Americans, and eco-tourists from abroad, are visiting parks and refuges for aesthetic recreation, not in order to kill or feed their family.

This discrepancy, and miscommunication of these inherent values, is paramount, and track with nearly every great ecological ideal that has somehow, by incremental changes, made it into law. But in the summer of 2020, the US Fish & Wildlife Service put out a notice to the American public to volunteer any comments regarding the Service's notion of opening up for hunting 96 National Wildlife Refuges, eight of which had never previously witnessed any killing.[3]

John Muir (1838–1914) once wrote, "Any fool can destroy trees. They cannot run away; and if they could, they would still be destroyed - chased and hunted down as long a dollar could be got out of their bark hides, branching horns, or magnificent bole backbones."[4] Applying this same logic at the epicenter of America's best and brightest conservation ethical activism, Roosevelt, however tormented he may have been personally, did not advocate for the slaughter of animals in our last remaining protected areas (for which he had been one of the first to even think in those terms, politically.)

Indeed, to fire upon migratory birds that had been utilizing those newly entrusted refuges, ponds, and lakes for centuries, millennia, without fear of sudden disturbance, would be biologically insidious; counter to the most basic conservation ethic, which Aldo Leopold laid down at the heart of his own great achievement, the Gila Wilderness in New Mexico, the world's first wilderness area, established June 3, 1924.[5]

---

[1] https://www.motherjones.com/politics/2013/02/hunting-demographics-charts-guns/Accessed April 10, 2020.

[2] U.S. Department of the Interior, U.S. Fish and Wildlife Service, and U.S. Department of Commerce, U.S. Census Bureau. 2016 National Survey of Fishing, Hunting, and Wildlife-Associated Recreation.

[3] As Pertaining to: Website: https://www.regulations.gov/document?D=FWS-HQ-NWRS-2020-0013-0001; Docket ID – FWS-HQ-NWRS-2020-0013.

[4] Quoted from "The American Forests," by John Muir, Atlantic Ideas Tour, August 1887, https://www.theatlantic.com/ideastour/nature/muir-excerpt.html, Accessed May 17, 2020.

[5] Not surprisingly, hunting is allowed in the Gila Wilderness. Numerous so called "game management units" have been given over to the wildly enthusiastic protagonists of archery, rifle and muzzleloader kills of "trophy" bull elk. Some outfitters even "guarantee" the size of the antlers. See https://www.fs.usda.gov/activity/gila/recreation/hunting, Accessed July 24, 2020. The Internet is rife with photos of proud, smiling people beside their trophies – elk crushed in death to the ground, that previously stood up to five feet at the shoulders, and weighing as much as 1100 pounds.

**Fig. 5.2**  Farallon Islands National Wildlife Refuge. (Photo © M.C. Tobias)

Between Muir (who certainly has had his own detractors, principally stemming from his own racist blind-spots),[6] Roosevelt, and Leopold, the politics and paradox of protection had been firmly enshrined in American environmental history (Fig. 5.2).[7] That there are singular contradictions in those policies, exacerbated by the negative feedback loops of an ever-increasing population and its consumerist pressures upon the fast-waning integrity and biodiversity of habitat, is consistent with that crisis individuals divine in their species, with little or no guidance on how to "fix it."

---

[6] "Sierra Club reflects on its racist roots and looks towards a new future," by Susanne Rust, Bettina Boxall, and Rosanna Xia, Los Angeles Times, July 23, 2020, https://www.latimes.com/california/story/2020-07-23/sierra-club-denounces-racism-of-john-muir, Accessed July 24, 2020. The story features at the top the famed photograph of Sierra Club co-founder John Muir "father of the national parks," and Theodore Roosevelt together standing atop Glacier Point in Yosemite.

[7] See, for example https://www.pbs.org/wgbh/americanexperience/features/earth-days-modern-environmental-movement/Accessed June 16, 2020.

# Chapter 6
# The Ecclesiastes Factor

## 6.1 The Satiation of Choices: Imperatives and Priorities

The present currency regarding the prospects of our species' extinction is tantamount to some vague clamor in the unconscious abetted by startling data sets; an attitudinal stylistic corpus feeding upon a premonitory event, or series of inordinately devastating biological diminutions, both inchoate and evolutionary, and presumed by a critical mass of mood-swing and supposition to be looming, sooner or later.

Our species focuses upon these trends with mental targets typically deriving from numbers. A six-fold increase—40 gigatons to 252 gigatons—of melted ice around the Antarctic continent, for example, in the last 40 years, one gigaton (in our minds) equaling a million metric tons, each ton equivalent to one thousand kilograms, and so on down to the scant divisibility of a sub-atomic particle. All targeting the vastly premature break-up of Antarctica, a result of vast, cumulative anthropogenic intrusiveness into the global cycles of climate and chemical turnover (Fig. 6.1). Such that a Pine Island Glacier lost an amount of ice nearly three times the size of San Francisco in a matter of days in mid-February, 2020.[1] Immense, at least, by earthly standards. We lodge the first question herewith: Does Pine Island Glacier meaningfully translate into the kinds of choices our evolution has prepared us for? If not, what are the missing genes, synapses, moral equivalencies exerted in a language we can understand, that evade us? Has natural selection, as framed by the evolutionary standards we have come to believe in, enjoined from implementation that process by which mentation might be truly capable of processing—not just internalizing—non-local events of global importance? Or events that on the surface seem to be affecting other species, but somehow magically, not ours?

---

[1] Eric Rignot, Jérémie Mouginot, Bernd Scheuchl, Michiel van den Broeke, Melchior J. van Wessem, and Mathieu Morlighem. *Four decades of Antarctic Ice Sheet mass balance from 1979–2017*. PNAS, January 14, 2019 DOI: https://doi.org/10.1073/pnas.1812883116; See https://www.youtube.com/watch?v=dYRcFPI2hhs. Accessed February 13, 2020.

© Springer Nature Switzerland AG 2021
M. C. Tobias, J. G. Morrison, *On the Nature of Ecological Paradox*,
https://doi.org/10.1007/978-3-030-64526-7_6

**Fig. 6.1** Paradise Bay, Western Antarctic Peninsula. (Photo © Robert Radin)

Similarly, in approximately the same time frame of four human decades+, *Homo sapiens* have altered, according to the WWF 2018 "Living Planet Report,"[2] some "75% of terrestrial ecosystems," "recorded an overall decline of 60% in [the population of] species," and perturbed "83%" of freshwater systems. The sum of these machinations comes at the price of our circus-like perturbations, the skewing, and outright destruction of an estimated "US$125 trillion" of annual nature's services; an equivalent purchasing power parity (PPP) that just happens to be on a par with the estimated average per capita dollar amount assigned to the current human population of nearly 7.8 billion individuals. Our species commenced with $0.00 PPP and did quite well, leaving scarcely a footprint. What happened? And equally important, in terms of the escalating Anthropocene, how does our collective organism—this allegedly sapient species—perceive, cognize, and internalize what it is that is overwhelming our home, this planet? Is our self-destructiveness, and by it our annihilation of so many of our living cohorts among some estimated 100 million other species, merely one of many narratives, the pejorative one? Or merely some illusory paradox that is fundamental to all of evolution, natural and sexual selection, of nature herself, and therefore not a quintessential problem we need worry about, a passive illusion?

These queries are now stubborn and crucial ones, which we shall take up in this work; an enigma that science in unable to address at the personal level (which is critically important) spanning that exceptionally modern gulf between innocence and experience. The appellations from poetry, art history, and the brief span, thus far, of human evolution, encompass an ineluctable surge toward the aforecited

---

[2] https://wwf.panda.org/knowledge_hub/all_publications/living_planet_report_2018/

Anthropocene, etymologically Mycenaean, Archaic, Classical, Hellenistic, or Koine, and Medieval Greek. By 1873, the Italian geologist and priest Antonio Stoppani acknowledged what he termed the Anthropozoic era,[3] and by 1999 the term "homogenocene" had come in to use,[4] a geological catastrophe unleashed by the sum total of reckless personalities all galvanized into but one species, and succeeding previous times, by comparison melodious and softly modest, the Holocene.[5] It is unclear whether Antonio Stoppani, George Rolleston, Vladimir Vernadsky, Pierre Teilhard de Chardin, even Albert Schweitzer, or his cousin Jean-Paul Sartre, could actually have had any empirical evidence at hand to have divined so broadly the "rupture" in biological integrity resulting from the all-encompassing human inflictions upon global ecosystems we now understand to be true.[6] Shakespeare seems to have anticipated, at least, the dialectic between a living and a dead world in his final singly written play, "The Tempest" (1610/1611), as the grand prestidigitator Prospero uses and counters the indigene, some kind of savage, all-clever Caliban. The island of shipwrecks ultimately sends its otherwise doomed wilderness players home to a refined and celebratory Naples. That a time would come when England's tears were considered by the Puritans running Parliament so great as to forbid the performance of a Shakespeare, a ban lasting 18 years, was just one more glitch among many of humanity's tempests.[7]

*But none would ever ascend to today's* Sixth Extinction Spasm, the first to ever be caused by but one species.

## 6.2 The Sum Total of Destructions

Picasso heroically represented the breach in his "Guernica" and famously spoke of the many destructions inherent to the very process of art. Oswald Spengler and later Arnold Toynbee easily anticipated the chronicles of ill-fitting *Homo sapiens* laid out

---

[3] Crutzen, P. J. (2002). "Geology of mankind". Nature. 415 (6867): 23. Bibcode:2002Natur.415...23C. doi:https://doi.org/10.1038/415023a. PMID 11780095

[4] Michael, Samways (1999). "Translocating fauna to foreign lands: here comes the Homogenocene"(PDF). Journal of Insect Conservation. doi:https://doi.org/10.1023/A:1017267807870

[5] Steffen, Will; Grinevald, Jacques; Crutzen, Paul; McNeill, John (2011). "The Anthropocene: conceptual and historical perspectives" (PDF). Phil. Trans. R. Soc. A. 369: 843; See also, Crutzen, Steffen (Winter 2017). "The Anthropocene: Are Humans Now Overwhelming the Great Forces of Nature". Ambio. Vol. 36, No. 8: 619.

[6] See "Was the Anthropocene anticipated?" Clive Hamilton, Jacques Grinevald, First Published January 28, 2015. https://doi.org/10.1177/2053019614567155, The Anthropogenic Review, Sage Journals, https://journals.sagepub.com/doi/abs/10.1177/2053019614567155?journalCode=anra#, Accessed December 3, 2018; See also, Modifications of the External Aspects of Organic Nature Produced by Man"s Interference. London, 1880.

[7] Firth, C.H.; Rait, R.S., eds. (1911). "September 1642: Order for Stage-plays to cease". Acts and Ordinances of the Interregnum, 1642–1660. London: His Majesty's Stationery Office. hdl:2027/inu.30000046036137. OL 6559925M.

by historians Clarence Glacken, Roderick Nash, and Jared Diamond. Nikos Kazantzakis performed, in every mounting addition to his prolific poetic and philosophical investigations, an autopsy of contradictions that both defined the human spirit, while recognizing humanity's mayhem upon itself. His characters of Odysseus and Zorba not only were instinct with dualisms that account for their greatness as literary individuals but also incite the imaginative havoc of the worlds with which they necessarily grappled all around, and in them (Fig. 6.2).

**Fig. 6.2** "St. Jerome and his Lion," from *Photographs Of Engravings, Etc., Of Albert Durer From The Pinnacothek At Munich*, 1879. (Private Collection). (Photo © M.C.Tobias)

And just as an Albrecht Dürer felt compelled to return to the subject of St. Jerome with his lion at least six times; a compulsion one can easily factor into the evolution of biophilia across every Renaissance, with its tempting transcendence beyond blunt bifurcation points in the history of anthrozoological thought—the heroic lion ultimately gracing the set of "The Wizard of Oz,"—the explosions of meaning, their fractions, margins, and colorful shrapnel in works by the likes of Willem de Kooning and Jackson Pollock are equally adept at advocating for a theory of paradox that keenly tells of the human psyche. All of the grand clichés and invocations come back to haunt us: That which we wish for; intuition before logic; greener pastures; patience is a virtue; time will heal all… have been fractured.

But again: Do we, can we, perceive rightly was is happening? The sum total of destruction?

Genetics, mathematics, neurophysiology, biochemistry, psychiatric studies, the history of ideas, anthropology, computational ecology, every discipline that speaks to evolution all chime in so as to better grasp the preconditions of this self-doubting. There is much companionship in this misery of philosophical print; nearly as many autobiographies as there are slaughterhouses.

Each increasingly dire premonition throughout human history yields to a specific story of its author and enshrouding cultural context, the cumulative harbingers of widely suffused revelations from aboriginal and Biblical hermeneutics. That nihilism which in an Ecclesiastes can collate life with death in imagery given to a similitude of opposites; of genealogies all verging upon the extinction of family names, annihilation of the least identity, the deeply embedded quashing of all vanities, so much later fortified in Percy Bysshe Shelley's "Ozymandias."

Only the vague taxonomies of structural nomenclature dating to before Linnaeus are able—however contested and changeable, like those of the Medieval Zen philosopher Dogen's famed Asiatic mountain and river sutras without end[8] or the late eighteenth-century Rev. William Gilpin's elusive reference to "some third species"—to focus our nomadic concentrations upon actual organisms of interest. But to be anointed in this manner, consorting with the prolix identities of a biosemiotic family tree (a biocultural phylogenetic array of both ancestral and current interspecies relations), each member of which will die out, in turn, is to court the most despairing of existentialisms.

That fatalistic disposition accompanies our entire known history, as a species. Every massacre, each action motivated by disturbance, hindrance, and ill-will furthers its verisimilitude. We cannot possibly be who we think we are. As Vladimir Lenin pointed out that "imperialism" was the highest form of capitalism, our biological studies, commencing long before the work of Aristotle, have only confirmed in reckless perpetuity our species' hubris and self-interest. Such biographical delusions which, in the twenty-first century, have been readily seized by the notion of our teleological purpose, our sovereignty, and—combining these two stark

---

[8] See the superb translation by Arnold Kotler and Kazuaki Tanahashi, https://www.upaya.org/uploads/pdfs/MountainsRiversSutra.pdf, Accessed December 5, 2018

untruths—our superiority to all other sensate beings, represents the most serious of all ecological heresies. We know it from the vantage of despoliation to be the central paradox of all. And while the nature of paradox consists of seemingly contrary intuitions colliding in some indisputable act of consciousness that fixates on the tangible, rights the odds, it is time and reflection that together constitute some other intangible. Because these appear from any vantage point to be contraries—substance and the insubstantial—the paradoxical unraveling of our presumptions of fixity has never acted as a call for unity, except, at rare moments. By the brutal show of force in the rigid, almost military-like garden designed by, for example, Niccolò Tribolo for the rapacious young Cosimo I de' Medici (1519–1574) at his Tuscan Villa di Castello; a benign but clearly manipulative arrogation, of shrubs shouldering row upon rectilinear row. Taken to other, human extremes, one senses about our tolerance levels for paradox the frustrated, angry, submissive euphoria of an entire nation calling for death camps that will exterminate the Other.

In parading this seriously pathological self-image—self-conscious biomes, economic, political, egotistical, that connote our attempt at subordinating fellow life forms (not just Constitutions or Magna Cartas)—we have encouraged one another, and by mathematical import, our communities, to presage a message of progress and, implicitly, enough time to mete out our enunciations, hazards, and eclipses. From this maelstrom of fictions, these psychic grandeurs of success, as against our idealized prelapsarian earth, no conclusion about ourselves that would in any way hint at the genocidal or the suicidal is likely to be gleaned. We have learned to gloss over darkness, relying on collectivities, village life, and the great urbanizing devolvement unto anonymity which cities afford us all. The megacity—ten, twenty, thirty million people crammed together—represents the ultimate vanishing act for individuals. Simply, such conurbations empower us to hide the traumatic truth of human beings within the group, as Elias Canetti trenchantly perceived in his devastating book *Crowds and Power* (1960).[9] All of those identities and the data associated with their uncountable memories may be likened to bits of data, whole lifetimes that amount to more than general macromolecules without purchase but nonetheless, in all their microscopy, mirroring the miracle of life, as was persuasively contemplated by Werner Holzmüller in his compelling work *Information in biological systems: the role of macromolecules.*[10]

Such truth might not be so alarming in the case of a Monet spending most of his later years at his Giverny gardens, indulging the light upon lily ponds in over 250 oil paintings, countless other sketches. But in biological disambiguation, personal aesthetic relish or philosophical claims cannot rationally be sequestered from the species-wide out-of-control syndrome that has broken with the overarching steady flow of ecological coherence. Our lack of cogency is our destiny, the strangest paradox of all.

---

[9] *Masse und Macht*, Chaasen Verlag, Hamburg, 1960; Victor Gollancz Ltd and Viking Press, 1962.

[10] Translated by Manfred Hecker, Cambridge University Press, Cambridge, U.K., 1984.

## 6.3   The Ecology of Ecclesiastes

The interpretive sphere of such ruminations arrives on the shores of paradox that is at once paralyzing, but potentially liberating. The choice is there, and it is ours to enact. In "Ecclesiastes" (unlike the subsequent "Song of Solomon") liberation connotes a myriad of contradictions. Chapter 1.4: … "the earth abideth for ever." Chapter 2.16: …"And how dieth the wise man as the fool." Chapter 3.19: "For that which befalleth the sons of men befalleth beasts:"…Chapter 9.12: "For man also knoweth not his time; as the fishes that are taken in an evil net, and as the birds that are caught in the snare; so are the sons of men snared in an evil time, when it falleth suddenly upon them." Chapter 11.3: "Whether a tree falls north or south, it stays where it falls." Chapter 11.4: "He that observeth the wind shall not sow; and he that regardeth the clouds shall not reap".[11]

Such pronouncements by the alleged "Son of David," a king of Jerusalem, Ketuvim = writings, one of 24 books of the Hebrew Old Testament or Tanakh, harbor a sprawling psychological allegory that is the same history of the "City of Peace," in which, according to Eric H. Cline, there have been at least "118 separate conflicts during the past four millennia" (massacres, divisions, uprisings of every persuasion, total chaos, utter destruction repeated and re-invented) (Fig. 6.3).[12] One must surmise that the author or authors of "Ecclesiastes" had endured much of that tumult. Perhaps this poignant burgeoning of timeless despair came in the wake of the brutal suppression of a revolt in Jerusalem by the Persian king Artaxerxes III (identified as the ruthless Ahasuerus of the *Book of Esther*). He had much of Jerusalem burnt to the ground, killing and/or exiling all of the resident Jews. This, in turn, was a telling precursor to the conquest by Alexander the Great. The young warrior's fury was followed by further outbursts of killing, one furnace of egos after another, and all this misery only leading to the ultimate fall of Jerusalem in 200 BCE, as the armies of Antiochus III the Great besieged the doomed city.[13]

We are utterly removed from these sagas, even as such blockbuster bloodshed continues in Israel, Palestine, and the entire Middle East. Civil strife from the streets of Paris to Charleston, from Rangoon to Portland.

And so to contemplate, for example, Rembrandt's relatively dismissive 1660 painting "Ahasuerus and Haman at the Feast of Esther" in Moscow's Pushkin Museum, a banquet, the king (Esther's husband) yielding little if any evidence of his genocidal madness, provides no clue to the cultural mayhem encincturing the times

---

[11] See *The Holy Bible According To The Authorized Version with The Marginal Readings And Parallel References Printed At Length And The Commentaries Of Henry And Scott Condensed By the Rev. John McFarlane*, London, Glasgow, William Collins, 1862, Ecclesiastes, Chapters I-XII, pp. 680–689.

[12] See Cline's book, *Jerusalem Besieged – From Ancient Canaan to Modern Israel*, Ann Arbor, University of Michigan Press, 2005, https://www.press.umich.edu/164087/jerusalem_besieged, Accessed December 5, 2018.

[13] See Taagepera, Rein (1979). "Size and Duration of Empires: Growth-Decline Curves, 600 B.C. to 600 A.D." Social Science History. 3 (3/4): 121. doi:https://doi.org/10.2307/1170959.

**Fig. 6.3** Nineteenth-century Jerusalem, Chromolithograph, 1862 *Collins's Pictorial Family Bible*, Glasgow

that gave rise to "Ecclesiastes," nor would do so in our own time as biodiversity is perverted and destroyed across the level playing fields of a seemingly ill-fated planet. Neither Rembrandt, nor today's scholars, can understand the totality of futilities that goaded the author(s) of the gloriously tired, troubled, schizophrenic dozen chapters comprising that which has been described as the single most important piece of writing in Western Civilization. The word itself is from the Greek word "ekklesiastes," its protagonist, Kohelet, referring to an individual who both convenes and also speaks to the assembled.

We have full license, literary, scientific, and moral, to presume meaning, all of its "under the suns" and other repetitive laments. Translations, however exponentially recessive, nonetheless convey a meaning altogether relevant to today; the colossal cautionary tale which is human egotism in the face of our mortality. Those few pieces of lines quoted above suggest the fragility of human observation, of science; terse, unambiguous decrees that allow for sufficient flexibility in the face of disaster, not unlike those final words in Wittgenstein's Tractatus (*The Tractatus Logico-Philosophicus*),[14] "Whereof one cannot speak, thereof one must be silent."[15]

---

[14] By Ludwig Wittgenstein, With an Introduction by Bertrand Russell, Harcourt, Brace & Company, Inc., New York and London: Kegan Paul, Trench, Trubner & Co., Ltd., 1922.

[15] p. 189.

In addition to Wittgenstein's powerful mathematical cantos spread generously throughout his *Tractatus*, one of the other most provocative books of the twentieth century, George Steiner's *After Babel—Aspects of Language and Translation*[16] consoled as much as it disturbed, particularly in its incessant Ecclesiastes-like reminder of our mortality, and our hopes of exploiting coordinated symbolisms to evade devastating truths. "It is our syntax, not the physiology of the body or the thermodynamics of the planetary system, which is full of tomorrows. Indeed, this may be the only area of 'free will', of assertion outside direct neurochemical causation or programming," he wrote.[17] Umberto Eco seems to concur with Steiner in his chapter on "The Plants Of Shakespeare" in his mystifying work *Mouse or Rat? Translation as Negotiation.*[18] Eco early on acknowledges the crucial reality of "homonymy" ("when a single term refers to two different things or concepts")[19] and bolsters its literal cause through "rules of contextual disambiguation"[20] which ultimately means nothing less than that "hypothesis about the probable structure of the world pictured by the original text…".[21] These instant and pressing enigmas are at the root of humanity's penchant, indeed, genetic and molecular obsessions with expression; and equally so, with the unavoidable conclusions that our many-storied peregrinations, myths, and dreams are so magnificently contradictory as to arouse the sensation that not just this or that, but that everything human must be paradoxical.

---

[16] Oxford University Press, New York and London, 1975.

[17] ibid., p.227.

[18] Weidenfeld & Nicolson, London, 2003.

[19] ibid., p.11.

[20] ibid., p.15.

[21] ibid., p. 19.

# Chapter 7
# Pathologies of Self-Image

## 7.1 Quagmires of Consciousness

One early hint at this high probability comes from Medieval bestiaries depicting Adam naming the animals from Genesis, a motif of inordinate influence upon Hieronymus Bosch in whose supreme masterpiece "The Garden of Earthly Delights" (1490–1510) many painted creatures "defy any classification, and result therefore in 'empty' aberrational memories. The three-headed bird with a feathered 'eye' on its peacock-like tail, just below the Creator's feet, for example...".[1] Not only did Bosch detail new existence, but posed a virtual dilemma for human consciousness in that our original taxonomies, at least in terms of the Biblical rendition in Judeo-Christian traditions, failed scientifically to take sufficient account of the true extent of wildness all around us. A century after Bosch, a single painting by Roelandt Savery, his 1624 "Bouquet of Flowers," said to be a still-life, in fact contains a living cornucopia: 63 flower and 44 animal species (Fig. 7.1).[2]

But even these expanded peripheries of biological say-so do not yet hint at an even more satisfying scenario that might account for a human species on the cusp of something grander than itself, the essence of ethics, but also of molecular biology, wherein crazy numbers—2 million to 5 trillion molecules in a single human cell, multiplied by the nearly 38 trillion cells on average per human body—begin to form an argument. We have earlier explored the scenario in our *Hypothetical Species* (2019) hinting at the possibility of twin species.[3] V. B. Sapunov contrasted and

---

[1] *The Land of Unlikeness – Hieronymus Bosch, The Garden of Earthly Delights*, by Reindert Faleknburg, W Books, Zwolle, The Netherlands, 2011, p. 122. See also, *A Dark Premonition – Journeys to Hieronymus Bosch*, By Cees Nooteboom, Translated from the Dutch by Laura Watkinson, Schirmer/Mosel, Munich, Germany, and Verona, 2016.

[2] As pointed out in Micheline Walker's essay, "Roelandt Savery: From Flowers to the Dodo," November 30, 2012, online Art blog, https://michelinewalker.com/2012/11/30/roelandt-savery-from-flowers-to-the-dodo/

[3] *The Hypothetical Species: Variables of Evolution*, by M. C. Tobias and J. G. Morrison, Springer, New York, 2019.

© Springer Nature Switzerland AG 2021

M. C. Tobias, J. G. Morrison, *On the Nature of Ecological Paradox*,
https://doi.org/10.1007/978-3-030-64526-7_7

**Fig. 7.1** "Adam gives the animals names in Paradise," 1604, Jan Saenredam, after Abraham Bloemaert. (Private Collection). (Photo © M.C.Tobias)

compared biological and sociological traits of early humans to gauge the likelihood (and potential) for two species of early human to coexist, a fact (and a mere 87 protein differential) less clear to paleontologists in 1994 when Sapunov wrote his essay.[4]

[4] See V.B.Sapunov's powerful essay, "Ecological twin species and some obscure questions of hominidae evolutions," Human Evolution, Sapunov, V.B. Hum. Evol. (1995) 10: 193. https://doi.org/10.1007/BF02438971, Received20 July 1993, Accepted21 March 1994, DOI, https://doi.org/10.1007/BF02438971 Academic Publishers, Print ISSN,0393-9375, Online ISSN1824-310X, https://link.springer.com/article/10.1007/BF02438971#citeas, Accessed January 18, 2019. See

The convergence of disciplines upon a period some 40,000 years ago in early human history shows all of the same signs of a genus (*Homo*) splitting into subgroups, both sociologically and physically, that are apparent today. That differentiation may be minute, something in the gut, or the jaw, or amount of hair; a dietary shift going into its third sequential generation; even subtle dialect drifts, abbreviations that become normal. Whatever the physical manifestations of a twin human species, there are moral components that do not fully share an existence until their convictions are challenged. The Anthropocene is a marathon test like no other.

Terrence W. Deacon, in his vast approach to problems of consciousness, wherein these tests, proclivities, and their thresholds are borne out, asserts three classic hurdles in the history of (human) philosophy that are germane to both the nature of language and the quintessential dilemmas confronted by any form of introspection. Deacon explores "the binding problem," "the grounding problem," and "the problem of 'agency' ".[5] Such epical hurdles involve the unlikely emergence of the apperceptive faculties and their compliance with subjective, unified experiences of consciousness, self-consciousness, and every aspect of evolution as it pushes and stretches our inner being linguistically, conceptually, and in every other arena involving efforts to understand what is happening in ourselves, in the worlds around us, and in their myriad interfaces.

Those transitional zones in which inner and outer natures collide, coalesce, or do things differently comprise the uncertain weather at the heart of each odyssey of sentience and, for all we know, other-than-sentience. At once, nothing is as it seems, and everything is precisely as it seems. Our choices of approbation or resistance are lodged somewhere between computational ecology, statistics, several pathways of normative ethics, probability theories within biology, physics, and mathematics. From Asian palm civets (Paradoxurus) and the Australian water mole (*Ornithorhynchus Paradoxus*—the —platypus, one of three "paradoxical" species painted by the artist Joseph Wolf), to large trilobites of the middle Cambrian-Paradoxidian, of the family Paroxididae, the fossils of the Stockingford Shales in the UK, being most famed, particularly the Lyell Collection. All are of a paradox (Fig. 7.2).

And like all paradox, something happens counter-intuitively, ironically, against the grain. In the case of the mammalian monotremes—like Joseph Wolf's platypus—lack of teeth as adults, and the development of electrolocation are just a few of their rare and counter-intuitive physiological qualities. But the root of paradox isn't supposed to emerge as the word itself suggests it does; the collapse of a species, for example, that had up until a certain breaking-point, has been fortunate in finding excess food sources. Or the 73,000-year-old abstraction painted in ochre, probably with a crayon-like tool, on to a tiny piece of rock and discovered in the

---

also, "Scientists are reconstructing the relationship between modern humans and Neanderthals," by Klaus Wilhelm, Max Planck Society, June 14, 2017, PhysOrg, https://phys.org/news/2017-06-scientists-reconstructing-relationship-modern-humans.html.

[5] *The Symbolic Species – The Co-evolution of Language and the Brain*, Terrence W. Deacon, W. W. Norton & Company, Inc., New York, 1997, pp. 438–439.

**Fig. 7.2** *Paradoxurus quadriscriptus = P. hermaphroditus* [Asian Palm Civet], 1856, by Artist Joseph Wolf, Proceedings of the Zoological Society of London

South African Blombos Cave. All the best dating techniques in the world—from standard radiocarbon to uranium series, potassium-argon, argon-40/argon-39, to paleomagnetism[6]—can bring us within a single millennium of the decisive artistic execution by someone of that noteworthy impression, a primeval example of human art, consciousness, or something else. But, in fact, while its discovery marked the oldest known target to date of such aesthetic forensics, we can come no closer to understanding what it is: Symbol, love letter, urgent telegram, warning, epiphany, doodle? The fact that we should even care is also in a class of paradoxes.[7]

These classes or sub-classes, as in any taxonomy or topology, are systems; paradoxical systems which comprise both organization, procedure, method, as well as events. Such categorically difficult forms of experience are couched in anomalies and perceptions. Mathematically, they are inscribed in equations designating time and space, divisibility or indivisibility. In logic, language, and philosophy in general, they inhabit, we must assume, every hypothesis, thesis, antithesis, and synthesis. They manifest as antinomies—Immanuel Kant declared four of them in

---

[6] See "Best Ways to Get a Date," by Michael W. Robbins, "Finding Time," in The American Scholar, Autumn 2018, pp. 42–43.

[7] See Letter | Published: 12 September 2018, Nature Journal, "An abstract drawing from the 73,000-year-old levels at Blombos Cave, South Africa,"Christopher S. Henshilwood, Francesco d'Errico, Karen L. van Niekerk, Laure Dayet, Alain Queffelec & Luca Pollarolo , https://www.nature.com/articles/s41586-018-0514-3, Accessed September 15, 2018.

his *Critique of Pure Reason* (1781)—and strike of bountiful contradictions, of intellectual, fructifying oppositions that defy dialectical logic. They hold to a persistent evolution toward some kind of transcendent reality or surreality, or awareness, that surpasses the latitude of normal rules and laws that may be either externally, or internally, reconciled. Or both. What, in Socratic dialogues, was known as the elenchus tactic of refutations[8] in disciplines of natural history may take the form of defiance of the rule of competitive exclusion, as with the unexplainable proliferation of planktonic species.[9]

## 7.2 Synderesis

But perhaps the most elusive and intriguing of all human classes of paradox involve the *conscience*. The term "synderesis" derives from writings by St. Jerome (347–419) as he elucidates the four beings found in Ezekiel's vision in the Septuagint at Ezekiel 1:10.[10] Synonymous with the "spark of conscience" said to have remained in Adam's heart post-paradise, it is the trail of tears leading to commiseration, empathy, and humility; and which led John Milton, in the final Book XII of *Paradise Lost*, to suggest "When this worlds dissolution shall be ripe, With glory and power to judge both quick & dead, To judge th' unfaithful dead, but to reward/His faithful, and receave [sic] them into bliss, whether in Heavn' or Earth, for then the Earth Shall all be Paradise, far happier place/Then this of Eden, and far happier daies…."

Douglas Langston incisively abetted recruitment of every conscious and subconscious discipline—as might be ascertained in a Milton, or Boethius' *De consolation philosophiae* (524 AD)—by affirming one of the early mathematical puzzles of Medieval theories of virtue and the human psyche. He did so by recognizing that the Franciscan friar William of Ockham (c. 1287–1347), on the subject of the conscience, agreed with philosopher-theologian Duns Scotus (c.1266–1308), who held that the "conscience can provide the entry into the seeming circularity of performing virtuous actions in order to develop intentions that seem to be required for

---

[8] See Gregory Vlastos, 'The Socratic Elenchus', Oxford Studies in Ancient Philosophy I, Oxford 1983, 27–58.

[9] See Scheffer, M., Rinaldi, S., Huisman, J. and Weissing, F. J., 2003, "Why plankton communities have no equilibrium: solutions to the paradox," Hydrobiologia 491, 9–18.

[10] See https://www.iep.utm.edu/synderes/, Accessed September 11, 2018. See St. Jerome, By Jean Steinmann, Translated By Ronald Matthews, Geoffrey Chapman, London 1959; See also, Traditio , Vol. 57, 2002 , Origen, Plato… Https://Www.Jstor.Org/Action/Showpublication?Journalcode=Tr aditio,Journal Article, "Origen, Plato, And Conscience ("Synderesis") In Jerome's Ezekiel Commentary," Douglas Kries, Traditio, Vol. 57 (2002), Pp. 67–83, Published By: Cambridge University Press, Https://Www.Jstor.Org/Stable/27832010?Seq=1#Page_Scan_Tab_Contents, Accessed September 11, 2018; See also Oscar James Brown, *Natural Rectitude and Divine Law in Aquinas: An Approach to an Integral Interpretation of the Thomistic Doctrine of Law*, Pontifical Institute of Medieval Studies, Toronto, 1981, pp. 175–77.

performing the virtuous actions in the first place."[11] Such passing insights, gathering energy between Aristotle and Aquinas, are exceedingly relevant to any discourse that proposes to investigate the long-held Socratic predicate that "a cause increases the probability of its effect…".[12] The rise of the great Greek and Latin understanding of "scintilla conscientiae," the sparks of conscience in human life and consciousness, touches upon our lives today in mores that explode the narrow categories of science and spirituality; of all forms of expression and perception, precisely because of their unavoidably paradoxical nature.

In one recent study of human behavior, investigators examined both aggressive and social dominance, and the degree to which information feedback stokes either penchant. Aggressive individuals tend to block out feedback. Whereas studies pertaining to aggression in other species have shown a particularly canny and demonstrative reliance, widespread, upon exogenous information-laden inputs. This suggests that humans may be inherently violent; other species not.[13] Such conclusions are assuredly ill-boding with respect to the continuing Anthropocene.

In the case of Simpson's Paradox, probabilities of differing classes or groups of data, when juxtaposed, reverse the probability earlier projected.[14] In other words, the larger the data class, the more dissimulated its projections and sums; much like mean averages diffusing stark contrasts. It suggests, at one critical level, that an ecologically overpopulated species such as our own will tend to lose distinctions, and that which is emphatic will be suffused. This proffers enormous genetic, emotional, and political implications beyond the scope of full enumeration at this point. Except to recognize that with a human population verging upon 8 billion, there is unlikely to emerge any countervailing factors that might serve as biological anodynes for the current reality of collective human adulterations.[15]

Lacking any consistent understanding of a sustainable human population—estimates have ranged from zero to 500 million and beyond—we are left with numeric correlations. The human global population was roughly 1 billion in 1804, when

---

[11] See "Medieval Theories of Conscience," First published Mon Nov 23, 1998; substantive revision Thu Jul 23, 2015, Stanford Encyclopedia of Philosophy, https://plato.stanford.edu/entries/conscience-medieval/, Accessed September 11, 2018.

[12] See "Causation, Probability, and the Continuity Bind," by Anthony F. Peressini, British Journal for the Philosophy of Science 69 (3):881–909, 2018: https://philpapers.org/asearch.pl?pub=158, Accessed September 11, 2018.

[13] See Volume 24, Issue 23, P2812–2816, December 1, 2014, "The Social Dominance Paradox," by Jennifer Louise Cook, Hanneke E.M. den Ouden, Cecilia M. Heyes and Roshan Cools, Open ArchivePublished:November 20, 2014DOI:https://doi.org/10.1016/j.cub.2014.10.014, Cell Biology, https://www.cell.com/current-biology/abstract/S0960-9822(14)01290-1, Accessed September 11, 2018.

[14] See Simpson, Edward H. (1951). "The Interpretation of Interaction in Contingency Tables". Journal of the Royal Statistical Society, Series B. 13: 238–241.

[15] See G. U. Yule (1903). "Notes on the Theory of Association of Attributes in Statistics". Biometrika. 2 (2): 121–134. doi:https://doi.org/10.1093/biomet/2.2.121; See Simpson, Edward H. (1951). "The Interpretation of Interaction in Contingency Tables". Journal of the Royal Statistical Society, Series B. 13: 238–241.

Lewis and Clark set out to map Jefferson's Louisiana Purchase, a financial transaction Jefferson enacted to re-shape a continent's carrying capacity as a flood of immigrants determined to make America home. It was the first and only noteworthy population policy exhibited by lawmakers in US history, with the exception of ultra-conservative Congressional rejections of China's population policy and that of the United Nations Population Fund's stance on abortion worldwide. By 1850, when Karl Marx was firmly elucidating a shooting gallery of "contradictions" in capitalism, humanity numbered approximately 1.2 billion.[16]

Presently, with 10, 11, 12, 13 billion humans projected by the end of the twenty-first century, as in no other social and scientific research arena of critical public importance are there such vague and often meaningless calculations. This is because the so-called ecological horizon fractions are assessed according to different metrics, and expressed as matters of policy recommendation where bias is intense. For example, what data sets would one utilize to determine that human carrying capacity had been exceeded as a function of the time when surrounding species extinctions began to accelerate? There are too many unanswerable gaps in this kind of imagined correlation to even venture an educated guess. To reiterate: At what point (precisely) did humans begin destroying their surroundings?

One could take any number of measures of human catalytics to formulate a sustainability quotient—such as the year in which assembly-line slaughter of animals began to shape public perception of animals as mere units of production. But the number of such correlations that form the basis for grasping the entirety of human ecological paradox is far beyond calculation. No amount of data is sufficient to reasonably establish the causes and consequences of such a paradox. Instead, the mind stumbles around it, while in its very physiological throes. If, as a baseline consequence, the Anthropocene is determined to be the ultimate result, we are still unable to verify the vast array of causes, breached boundaries, the multitudes of trespass, degrees of sullying, the nature and kind of biochemical catastrophe, the sheer multiplicities of extinctions and population diminutions across every taxon.

## 7.3   The Separation of Concept and Calculation

While Aristotle acknowledged the basic forerunners of all paradox, his four fundamental causes—material, formal, efficient, and final[17], the causal relations that ignite any theory, hypothesis, synthesis, or intuition, are too many to associate in a

---

[16] https://www.jstor.org/journal/philpublaffa, "The Fundamental Contradiction of Capitalist Production," Gary Young, Philosophy & Public Affairs, Vol. 5, No. 2 (Winter, 1976), pp. 196–234, Published by: Wiley, https://www.jstor.org/stable/pdf/2264873.pdf?seq=1#page_scan_tab_contents, Accessed January 23, 2019.

[17] See *Metaphysics*, Book 5, section 1013a, translated by Hugh Tredennick, Cambridge, MA, Harvard University Press; London, William Heinemann Ltd. 1933, 1989.hosted at perseus.tufts. edu., Accessed September 11, 2018.

rational framework; let alone extrapolate for our purpose of understanding the underlying forces guiding or moderating human conscience. Hence, as with Aristotle, emerges by way of alternative explanations, the realm of metaphysics. That unknown country between concept and calculation. Or, as alleged within the realm of *idealization*, a curious alternative to metaphysics, which Michael Strevens offers as an intriguing approach. He writes, in describing the notion of an idea or data pertaining to a "target phenomenon" as having been "idealized away" that "This streamlining hints at one important function of idealization: it serves the pragmatic end of keeping calculations simple, either by making the mathematics more tractable or by reducing the empirical demands of the model (for example, lowering the number of and precision of the initial conditions and parameters whose values must be determined in order to derive consequences from the model)".[18]

Our goal is to understand that which is allegedly evident, while the narrative is self-realizing, propelled by rules of grammar, habit, and cumulative experience that coalesce to engender a salient explanation of the human condition. But such a condition is but one condition, a circumstance, given to ratiocination and replication, that describes in some accessible manner that which is an event within the volume of propositional logic and calculus, prior to all variables. Such logic is the preexistent value without gradient or combination. A starting point that hopes to preempt bias, self-interest, or self-promotion. Anything less would simply connote a self-referential tautology.

The problems involved in computational ecology, such as in the case of "plankton patchiness" and the breakdown in "conceptual modeling",[19] only hamper our facility for grasping the dilemma confronting humankind. The numbers, in so defying easy reach or a layman's translation, compel that same aforementioned dissimulation, whereby horrors and bliss tend to travel together, leveling out somewhere mid-center, in the blandest climes of geography and contemplation. This is precisely where the litany of ecological disasters becomes the normal context for all remaining life forms.

Conceptualization itself defies such rationales and interpolation in as much as logic—etymologically deriving from mid-fourteenth-century philosophical distinctions between true and untrue reasoning—cannot be said to be tied to, or derived from, even un-delineated mental activity. For all we know, it is an emotional product; or psychologically divorced from the human brain, while centered in the mind. These are all concepts, and their byproducts do not contain links that we either see or can ever detect. We do not know where concepts come from, and whether their

---

[18] See Strevens' essay, "The Structure of Asymptotic Idealization," p. 3, http://www.strevens.org/research/expln/Asymptotics.pdf, Department of Philosophy, New York University, Accessed January 13, 2019.

[19] See "Computational ecology as an emerging science," by Sergei Petrovskii and Natalia Petrovskaya, Interface Focus, 2012 Apr 6; 2(2): 241–254. Published online 2012 Jan 5. doi: https://doi.org/10.1098/rsfs.2011.0083, PMCID: PMC3293204, PMID: 23565336, The Royal Society Publishing, https://www.ncbi.nlm.nih.gov/pmc/articles/PMC3293204/, Accessed January 13, 2019.

outcomes have anything to do with any reality. And so, with the ancient and conventionally ascribed forms of logic we are lost altogether: No constraints oriented toward truthful outcomes are sufficient to layer one's perception of the world with anything like a truth. And while there are undoubted consequences resulting from sitting upon a chair (we can be spoken of as apparently sitting on such a chair), any number of situations are likely to affect that picture, not limited to a direct hit of a lightning strike, or summer temperatures in southern California exceeding 150 degrees.

## 7.4  Between Rodin and Tragedy

Seated on the chair, poised pensively like Auguste Rodin's 1888 six-foot tall "The Thinker" in bronze, the nature of paradox does not begin and end in human thought. As Rodin himself wrote "He thinks not only with his brain, but with every muscle of his arms, back, and legs, with his clenched fist and gripping toes."[20] We see a striking similitude in Alberto Giacometti's various Solitaire sculptures, less literal, given to that ancestral absence of identity propounded in the famed third century BCE Etruscan bronze statue named "Ombra della sera" (Evening Shade) from the Museo Etrusco Guarnacci, in Volterra, Tuscany.

The nature of paradox is human nature fully illustrated. It is the heart of the fragment, waiting only to be addressed by additional information that transforms irony, a fallacy, an incomplete thought into something whole. This act of intellectual accretion simulates closure, but only in mind. Everything we contemplate grips us with an absolute yet ecologically untenable process. That is the comprehension that guides us; its every word issued from our lips. We have no other outlet. Our music may imitate the wild howl; all those modern petroglyphs embarking backward into the sanctum of elegant confinements and romantically warmed hearths. Humanity's professed love of life, and of individuals, may lay claim to all of those perceivable companions and candidates around us. But ultimately, we remain trapped by every memory and reflex that has accumulated; by hundreds of millions of years of neurological trial and error behind us (Fig. 7.3).

This dilemma, a closed circuit of introspection, or life-support system in which all variables are exogenous, causally independent, is exacerbated by its most pronounced and relative weakness: It cannot recognize its inability to enter a new realm of perception. To do so would require recourse to indigenous components. Dramatically, that would mean breaking the fourth wall of natural history and fronting the prospect that what happens outside ourselves, by the agency of our own behavior, also happens inside. That our bacteria have essentially willed our destiny in a collective, if unknown, frame of contexts that we continually step into; and that that stepping mirrors everything around us, Nature.

---

[20] *Rodin – A Biography*, by Elizabeth Ripley, J.B.Lippincott Company, Philadelphia, 1966, p.40.

**Fig. 7.3** Oaxacan Painted
Sacred Vase, Mixtec
Civilization, AD 1250–
1521, Constantine George
Rickards Collection, *Ruins
of Mexico*, Vol.1, London,
1910

It would be a simple matter to argue that every major scientific and technical breakthrough has, indeed, broken up that fourth wall. Aristotle noted the changes in visible constellations during the course of even a relatively short journey and rightly attributed this to a spherical earth. As early as 40,000 BCE in what is Moldovia, mammoth bones seem to have been utilized as tent poles.[21] These two insights—the earth is not flat; the protection offered by a tent against a brutal blizzard, though probably not a large carnivore—are emblematic of circumstances embroidering our recent cognitive emergence within the vast tapestries of primeval knowledge-based systems throughout every biome.

Our relatively neonatal surmises gain by the second but are utterly incommensurate with their physical manifestations. Therein lies the colossal mismatch that now tragically[22] defines what we call the ecological paradox.

---

[21] See *Eureka – An Illustrated History of Inventions From the Wheel to The Computer,* by Edward de Bono, Hold, Rinehart and Winston, New York, 1974, p. 132.

[22] And, as Angelos Terzakis wrote of such neonatal circumstances that is mankind, "For there exists an age of tragedy. It is youth." See *Homage To The Tragic Muse*, by Angelos Terzakis, Translated by Athan Anagnostopoulos, Foreword by Cedric H. Whitman, Houghton Mifflin Company, Boston, 1978, p. 186.

# Chapter 8
# Paradoxical Frontiers

## 8.1 Intra-Fallacy

A lasting paradox courts any number of intra-fallacies, as the computational facility of our own brains must resort to physics and arcane mathematical tangents for explanations; but physics and the mind are not conversationally compatible, the interconnects being removed from each object of respective attention, thus undermining the possibility, for example, of the creation of a garden, the writing of a letter, or the exercise of free will in the analysis of a fallacy. How do we therefore proceed? How do we do anything?

Neither data nor *explanations*, as a species of knowledge, have ever actually had much sway over substantive human behavior. Consider the fall-out from such a condition? Causality as fickle as all causes and consequences; information volatile, ephemeral, no more staying power than any other spontaneous eruptions of neurons. We know nothing, and explanations only accent our dependence on nothingness. Yet, at the same time, one could argue predestination, whether for Greek tragedian players, or all those in today's whorls of natural selection, unclear as to the genetic or narrative relationship between individuals, populations, and species. Between concept and behavior; consciousness and the conscience.

Spinoza's *Ethics* made rapturously clear the analytical breakdown of any one thing, given the supremacy of the love, and intuitive certainty, of God, in the philosopher's universe. For Spinoza, every course of reason invites a near infinity of options (read: explanations). It might be convenient to have laid a brackish pond or the Hundred Years' War into the hands of an alleged Creator. Our computing powers, even with all of the terabytes of additional boosts, are out of range, out of our depths, within no position to have even the most fundamental conversation with a humble shrub.[1] This is why Henry David Thoreau's prolific *Diaries* now read with

---

[1] *The Chief Works Of Benedict De Spinoza*, Vol., II, De Intellectus Emendatione – Ethica, Revised Edition, Bohn's Philosophical Library, Translated From The Latin With An Introduction by R. H.

© Springer Nature Switzerland AG 2021

M. C. Tobias, J. G. Morrison, *On the Nature of Ecological Paradox*,
https://doi.org/10.1007/978-3-030-64526-7_8

such lofty plausibility: His every observation, however detailed, was steeped in a foreknowledge of inconceivability. Thoreau provided us a pivotal framework for understanding paradox, where minutiae are swept away by principles and, simultaneously, vice versa. One could also see this dynamic in ancient Asiatic contexts, whether Jain, Hindu, or Buddhist schools of Yoga, for example.

The history of skepticism is thoroughly adulterated by the very cornerstones of fallacious thought and action. First-order logic, existential quantification, rational connectives, a syntax containing symbols, categorical propositions, major and minor premises, and the conclusive moral syllogisms that may result are all subject to abstraction and emptiness. Such fallacy-embroidered, calculus-inducing strategies are merely parallel depictions of arguments, affirmations, appeals and denials, correlations and equivocations, inferences predicated on arbitrary causations—cum hoc ergo propter hoc; post hoc ergo propter hoc—with, and/or after this, therefore this, and so on. In other words, the history of logic, in the face of skepticism, and underscored by ecological vagaries, is as dense with human confusion as is that span of known paradoxes. The gradations of truth and falsity values, as implicated by levels of non-confidence—the burden of skeptically undermined reasoning, lie within the nature of the references, projections, and conclusions. A fallacy incites little reason to be overwrought. Whereas a paradox can describe something seriously wrong, and simultaneously wanting for any conceivable solution. Such *want* intimates endless turning points. Which is not to simply turn dismissive, but to acknowledge that human perception of it is a variable independent of time. It can take the span of multiple generations before an apprehension of any ethical consensus, for example, is united.

The fallacy of bifurcation, or of the false dilemma, is a case in point. The world is neither all green nor all blue. There are myriad other propositions that neither conform to, nor should abide by, the assertion that an axiom is either this or that. The premise is wrong. The assertion is wrong. The axiom is also wrong. But to declare something is wrong also invokes fact-value distinctions that can be described as moralistic, or informationally fallacious. Those, in turn, are subject to a massive critique in the form of a shortcoming known as "proof by assertion." Or of "proving too much." In other words, there is nothing that human consciousness, in good faith, can communicate without falling subject to material consequences that are mental in nature and easily prone to every sophistry. The so-called "Nirvana Fallacy" is the ultimate reductionist problem for those concerned with fallacious thinking: If it is not perfect, then it is inaccurate, the equivalent, by degree, of an irrational number.

Every fallacy can be described by a regression analysis: Independent variables, two or more, acting upon a single dependent variable. The crux of such a tool is the second-order resulting line, through-story, interpretive outcome. Semantics, flimsy comparisons, rhetorical randomness, aimless objections, unclear circumstances, flawed suppositions, all shape interpolations that are no more definitive than random acts of collusion or re-assessment. Art conservation, attribution, provenance

M. Elwes, G. Bell And Sons, Ltd., London, 1912, pp.259–260.

are each subjects of debate. A Rembrandt is a Rembrandt is a Rembrandt. Yet, hagiographies and explanations about Rembrandt conflict.

As with heuristics and exegesis, a subjective variable will always prove to be the most tantalizing and efficacious. Explanations and conclusions drawn from personal conviction are untroubled, as a class, in their predictions and presumptions. The personal attestation is the one that most readily transcends blocking juggernauts of logic. And these are the heart and soul of a useful paradox.

With paradox, a vastly more substantiated ballast of logic, concern, ethic, or physical circumstance is intimated, than with a fallacy, whose contours and implications, at worst, are no more significant than a lie, or contradiction lacking shelf life. But the paradox level of confusion and skewed prospect involve any number of serious manifestations both in mind and physical being. Counter-intuition cascades across all mathematical and physical barriers; blurred horizons are transformed into pellucid, graspable precision; new perspectives abet old paradigms; until it proves impossible to witness the world without the assurance of antimony and the contradictory delusion of any and all resolution.

A paradox accounts for, and then moves beyond, the level of fallacy. Most distressing and appealing of all, paradox tutors liberation to move beyond past ages of constraint and the tyrannies of logic. There is no philosophy without paradox. No science in absence of contrasts. What was despair is to become possibility; and where whole ecosystems had been vanquished by our kind, the existential impulses and subjective ecological qualia that lend us pertinacity in the face of obliteration also turn toward the devising of probabilities for the biosphere we never even considered. Hence, the everlasting compulsions of exobiology.

On the cusp of a paradoxical frontier, one knows that fallacy turns to likelihood, much in the manner of that most simplistic of all basic equations, that $1 = 0.9999999....$ Such higher-power infinitesimals were simply ignored by Renaissance mathematicians, and the greatest atomic timekeepers of our age—those measuring radioactive decay and half-lives—for whom the concept of "almost zero" has been sufficient. Indeed, a pure concept of one, the equivalent of an argument in favor of a scientific limit, was not fully addressed until the emergence of the "sequential compactness theorem" first postulated in 1817 by Bernard Bolzano and fully articulated by Karl Weierstrass (1815–1897) with its language of consequences, substrings, bounded finitudes, "probabilist inference," limit points, closed intervals, and deducibility.[2] Not unlike the formal acceptance in logic of the number one, zero has its own remarkable concatenation of paradoxical frontiers, from art to physics. Writes Robert Kaplan in his engaging book *The Nothing That Is—A Natural History of Zero*, "One physicist says: 'The reason that there is Something rather than Nothing is that Nothing is unstable'".[3]

---

[2] See Bar-Hillel, Yehoshua (1950) "Bolzano's Definition of Analytic Propositions" Methodos, 32–55. [Republished in Theoria 16, 1950, pp. 91–117; reprinted in *Aspects of language: Essays and Lectures on Philosophy of Language, Linguistic Philosophy and Methodology of Linguistics*, Jerusalem, The Magnes Press 1970 pp. 3–28.

[3] Kaplan, Oxford University Press, New York, 2000 p. 180.

But instabilities also have a life of their own that comport with "late succes-sional" forests[4] and every conceivable form of ecological flux, inherent to all biodi-versity, even immortal cell lines. Like language and the meanings we attribute to it, environmental contradictions would appear to be fundamental to all integral sys-tems of life, although we cannot know this; we only glean it. We actually have no language to describe Nature (Fig. 8.1).[5]

Whether involving mathematics, legislative policy, or ecosystem dynamics, we can verify that any quantifiable idea harbors the notion of a commons, a public arena, a polis, or mall. There is a multiplicity that enables statistical through-stories—the "mean, median, mode and range" of basic numeric assemblages.[6] When we apply any concept involving multiples (e.g., the number of birdsongs or weather changes detected, choices made) to our own personal lives, to an entire watershed, wilderness area, national park, or to a whole nation with borders, we

**Fig. 8.1** Beach meadow filaments, Southeast Alaska. (Photo © M.C. Tobias)

---

[4] See "What is a Climax Forest?" by Michael Snyder, Northern Woodlands, September 1, 2006, https://northernwoodlands.org/articles/article/what_is_a_climax_forest, Accessed January 29, 2019.

[5] For an interesting discussion of this centuries-old debate in the contexts of different possibilities for meaning, see Antony Adolf's essay, "Multi-Lingualism and the (In)Stability of Meaning: A Neo-Heideggerian Perspective," https://www.brunel.ac.uk/creative-writing/research/entertext/documents/entertext032/Antony-Adolf-Multi-Lingualism-and-the-InStability-of-Meaning-A-Neo-Heideggerian-Perspective-An-Essay.pdf, Brunei University, London.

[6] See John Bibby, 1974. "Axiomatisations of the average and a further generalization of monotonic sequences". Glasgow Mathematical Journal. 15: 63–65. doi:https://doi.org/10.1017/s0017089500002135.

sense the profound varieties of changeability. We can take in microcosm that 1 day, call it a sample, and by nightfall have attempted to divine those diurnal spans of time that were more predictable than others. But by the following morning, everything within and without descends upon consciousness with a novelty that has no clear claim to the previous day.

In the midst of our lives, enmeshed in various sorts of numbers and ecologies, we can only read the status reports. There is no certainty. To believe otherwise is to become part of a fallacy which, by turns, guarantees a paradox. The Self.

## 8.2 Ecosystem-Scaled Paradox

Take the case of "the Netherlands Fallacy," a proposition regarding alleged sustainability and ecological independence in the sense of viable isolationism, which in reality is anything but the case. The delusion at the eleventh hour, eutropically speaking, hinges on the presumption that the Netherlands, or any country, is capable of sustaining itself without any external inputs. Hundreds of billions of dollars are equally a non-starter, totally unstable, a doomed stasis without a future, unless that money has a latitude of expenditures that contract with other parties. All that money cannot survive in solitude. Its language is the degree to which it can be expended, dispersed. Currency requires exchange. In that sense it is genetic and evolutionary; a mathematical plurality to the degree that instability, and the summations of chaos most accurately represent the natural state of affairs. Prior to the invention of money, there were other currencies. Indeed, not until January 15, 1520 in the Kingdom of Bohemia did the silver mining at Joachimsthal (Joachim's Valley) produce the first thaler, or dollar. But the history of the exchange of goods, pre-minting, involved a slew of other tradeable objects: "Amber, beads, cowries, drums, eggs, feathers, gongs, hoes, ivory, jade, kettles, leather, mats, nails, oxen, pigs, quartz, rice, salt, thimbles, umiacs, vodka, wampum, yarns, and zappozats (decorated axes)" (Fig. 8.2).[7]

---

[7] See Ehrlich, P. R.; Holdren, J. P. (1971). "Impact of Population Growth" (PDF). Science. 171 (3977): 1212–1217. doi:https://doi.org/10.1126/science.171.3977.1212. For history of money list, see, "Origins of Money and of Banking," by Roy Davies, July 23, 2019, and based upon A History of Money from ancient times to the present day, 3rd edition, , by Glyn Davies, University of Wales Press, Cardiff, UK, 2002, http://projects.exeter.ac.uk/RDavies/arian/origins.html, Accessed June 5, 2020; see also, "Key Challenges To Ecological Modernization Theory: Institutional Efficacy, Case Study Evidence, Units of Analysis, and the Pace of Eco-Efficiency, by Richard York and Eugene A. Rosa, Organization & Environment, Vol. 16, No. 3, September 2003, pp. 273–288, Sage Publications, Inc., https://www.jstor.org/stable/26162475, JSTOR, https://www.jstor.org/stable/26162475?seq=1, Accessed June 5, 2020. See also, "Scaling down the 'Netherlands Fallacy': a local-level quantitative study of the effect of affluence on the carbon footprint across the United States," by Matthew Thomas Clement, Andrew Pattison and Robby Habans, Environmental Science & Policy, Volume 78, December 2017, pp. 1–8, Elsevier, https://doi.org/10.1016/j.envsci.2017.09.001, ScienceDirect, https://www.sciencedirect.com/science/article/abs/pii/S1462901117304379, Accessed June 5, 2020.

**Fig. 8.2** Early twentieth-century merchant with articles of trade, East Africa. (Museu De Historia Natural de Maputo)

And that fallacy of the source of currency translates into a mismatch between value and ecological sourcing. The further geographically one must go to reach the source, theoretically, the most costly and precious the object. Yet, exogenous value as measured by price has often superseded value-at-place-of-origin, a totally irrational function. Substitution economics—the obscuring of true value through commodity pricing that exploits labor in those areas of the planet with lower purchasing power, or inflated home currencies—also clarifies the misnomers of borders, and the true cost ecologically of misreading carrying capacity, and thus failing to incorporate often severe depreciations of natural capital. There is a limit to the number of Netherlands the biosphere can accommodate, as a singular primate species continues its runaway exploitation unabated. Within the Netherlands, there can be only so many Netherlands.

Our languages of ecosystem viability elude us at the philosophical juncture where research and empirical data sets are supposed to cumulatively form the basis for fair and accurate judgment. The history of economics is always falling short of accounting for current economics. No moment-to-moment predictions can simply fall back upon the conclusions reached by historians of the Great Depression. That is because we nurture an ideal that is meant somehow inconceivably to stem the chasm between the individual and its species. That gulf is comprised of no particular substance or view. Rather, all substance, as manifested in the perfect view, a commentary suffused by near-instantaneous circumstances. One recognizes its contours and coloration, like some greater bird-of-paradise watching sunset over the jungles of Indonesia's Aru Islands, or that meadowed sward of a sloping hospice, the Selva di Val Gardena, in the Italian Dolomites, photographed by James Buchan on the cover of National Geographic's publication *Greatest Landscapes*.[8] But such ideal calculus corresponds with a Self that wields a particular language that has evolved out of ecosystems, all ecosystems, only to challenge itself, its selves for countless reasons, some of which we will explore. That Self forges new languages predicated upon the human history since the Holocene in which destruction not sympathies for Others, love of life as a principle, has dominated the New Nature that is humanity. In every fallacy, like that of the Netherlands', there is a perpetual and novel promise which whispers its delusional originality to whomever will listen.

This variability of linguistic development, in consciousness, has meant a distinct and pantheistic perversion that we know too well. It spins and skews and coordinates by way of a promulgation at the species level of deception. In the case of the Netherlands there is that deeply affecting dichotomy. At once we are compelled by the Dutch Renaissance and all the beauty and innovation it produced. But then, today, we need not second-guess a prevarication from Vermeer and all that Simon Schama ruefully dissected of Holland's Golden Age.[9] Instead, we find a blurred motif that the Dutch, for example, have supposedly been able to successfully maintain the 18th largest economy or so in the world without ecological repercussions. This mirrors a certain infantile Weltanschauung held by the early seventeenth-century Calvinist Church in Holland with respect to Jews. Writes, Schama, "Paradoxically, the church's predilection for describing its own flock as the reborn Hebrews did not dispose it to favor the real thing," notwithstanding a paranoid Church looking to the Jews, "children of the Covenant" to somehow help usher in the much anticipated "Last Days".[10] While seventeenth-century Dutch antisemitism was meted out, sadly pervasively, sometimes, by clear edict, in other instances by fanatical elitism, there still remained a level of tolerance that enabled Jewish communities to prosper. Today, it would be difficult to argue that there is not a similar cultural paradox perpetually played out in both the Dutch and the global perception of all things Holland. It is not about Calvinists and Jews but all of the Dutch and their environment. As *The Economist* has declared some years ago, "The Green image of the Dutch is at odds with the reality".[11]

Hence the name, "the Netherlands Fallacy." A national crisis of pollution in large measure brought about by the enormous number of standing bovines, pigs, and sheep, with all their manure tainting the soil, and ground water; not to mention its

---

[8] *Stunning Photographs That Inspire And Astonish*, Foreword by George Steinmetz, Text by Susan Tyler Hitchock, National Geographic, Washington D.C., 2016.

[9] See *The Embarrassment Of Riches – An Interpretation Of Dutch Culture In the Golden Age*, Simon Schama, Alfred Knopf, New York, 1987.

[10] ibid., Schama, p. 591.

[11] "Dirty Dikes," February 4, 2012, https://www.economist.com/europe/2012/02/04/dirty-dikes, Accessed January 21, 2019.

burden upon freshwater abstraction and the diversion of natural nutrients and proteins to a mono-governed bioscenario favoring meat consumption. This kind of green-spin, with rather deplorable underpinnings is precisely the favored internal hat-trick played by countries most notably like New Zealand and China; and, with but a handful of exceptions, replicated in principle and in reality, by every nation. The "American Fallacy" all but leads the way.

By 1990 the Dutch public was only too familiar with their "manure problem," as it was called.[12] Today, Dutch manure policy has put a sustainability-sounding spin on its altruistic participation in a global carcass-weight equivalent market of "325 million tonnes" of globally grown meat, from "poultry, pigs (and) cattle".[13] Despite the nation's massive surface and groundwater contamination, largely unregulated agri-ammonia atmospheric vapors and subsequent forest depletions, some of the highest levels of human impact on soils anywhere in the world, the language used to grapple with the lethal combination of regional consumption, imports, exports, and growth is unwavering in its anthropocentrism. With roughly 2,790,000 acres of arable land in 2015, as early as 2004, it had been estimated that nearly half a million acres of Dutch soils were contaminated with heavy metals, impairing soil function; and another 3 million+ acres were "phosphate saturated soils".[14] All that manure and its chemical fates must be grasped within the orbit of the outright slaughter of Dutch vertebrates. "In 2015, the Netherlands exported 1.6 billion euros worth of living animals, 6.8 billion euros of meat, 1.8 billion euros of fish, 6 billion euros of dairy and 0.2 billion euros worth of other animal products" (Fig. 8.3).[15]

Quite a contradiction from the country's 20 national parks, and such villages as Giethoorn, in the municipality of Brederwiede, a mere 119 kilometers from Amsterdam. While no cars are allowed in Giethoorn, as one of the most popular destinations of choice for the 17 million tourists per year to Holland (approximately one tourist per Dutch citizen), Giethoorn's population of fewer than 2700 residents sees a swarm of pedestrians and canoe addicts at overwhelming numbers, in the same realm of ratios between resident populations and visitors, as that of Mont Saint-Michel in France, and Bethlehem. Such visitation magnets translate into vast

---

[12] See "Legislation to Abate Pollution from Manure: The Dutch Approach," by Wim Brussaard Margaret Rosso Grossman, North Carolina Journal of International Law and Commercial Regulation, Winter 1990, Volume 15, Number 1, https://scholarship.law.unc.edu/cgi/viewcontent. cgi?referer=https://www.google.com/&httpsredir=1&article=1411&context=ncilj, Accessed January 27, 2019.

[13] According to the OECD-FAO Agricultural Outlook from 2011, as cited in The Dutch Ministries of Economic Affairs and Infrastructure and the Environment publication, Manure: A Valuable Resource, Wageningen, The Netherlands, 2014, http://edepot.wur.nl/294017.

[14] "Quick Scan Soils in The Netherlands - Overview of the soil status with reference to the forthcoming EU Soil Strategy," P.F.A.M. Römkens and O. Oenema (eds.), p.4, http://webdocs.alterra. wur.nl/pdffiles/alterraRapporten/rapport%20948.pdf, Accessed January 27, 2019.

[15] "Animal cruelty in Dutch animal agriculture and why this also concerns people outside the Netherlands," The Green Vegans, https://thegreenvegans.com/animal-cruelty-in-dutch-animal-agriculture-and-why-this-also-concerns-people-outside-the-netherlands/Accessed January 27, 2019.

**Fig. 8.3** Free-grazing bovines in the central Netherlands. (Photo © M.C. Tobias)

levels of consumption which, in turn, can be separated out, bio-computed, into all the same adulterants as if those tourists had lived there. In the case of Giethoorn, hundreds of thousands of those tourists come from China, at great carbon cost. While Mont Saint-Michel's 50 or so residents recently endured a decade of controversy as 52 French environmental directives were debated with respect to maintaining the religious epicenter's maritime character in the face of 2.5 million pilgrims each year.[16]

Such discordant voyeurism in the name of pleasure, aesthetics, human curiosity surely combats all those qualities embraced by landscape architect and environmental historian Randolph T. Hester in his groundbreaking work on "applied ecology and direct democracy".[17]

---

[16] See "Mont Saint-Michel's Lost Causeway Stirs Local Passions," re-printed from Le Monde in The Guardian, https://www.theguardian.com/world/2012/may/15/france-conservation, Accessed February 2, 2019.

[17] See Hester's *Design for Ecological Democracy*, The MIT Press, Cambridge, Massachusetts, 2006, p.5.

## 8.3 Microcosmic Narratives

Place even 100 people on a small island and the microcosmic dilemmas cascade, even in the best interests of rendering sustainable a human occupancy, one that subjects its present concerns to an overarching inclusion of future generations, as defined in 1987 by Gro Harlem Brundtland, Norway's Prime Minister, in the document "Our Common Future" (*The Brundtland Report*). For all the reasons that span an aggressive cognition, large-sized (50 kilograms+) vertebrate housing, fuzzy ethical bearing, fierce communication skills, unclear liberation as a result of bipedalism, and so on, it is not clear that one generation of such creatures can spare their next. Their intergenerational inflictions are a continuum, despite the rare anthropological glint of seeming benign habitation at the hamlet level of demography. While such a portrait nurtures the misanthropic profile without a likelihood of transition, we think, act, express, debate, and get on with it in spite of our deeply troubled reservations.

We will not outlive those concerns, nor solve them. But a future human subspecies might. Their small island, or mainland island, communities shall have endured slurries of nitrogen and phosphate spread across acreage devoted to much higher numbers (than humans) of sheep and so-called dairy and beef cattle and their dung heaps; strategies for mitigating and avoiding algal blooms and blight; uncontrolled run-offs; new types and cycles of disease outbreaks; an increasing disparity between males and females given a heating planet and declining sperm counts worldwide; energy needs in the hundreds of thousands of kilowatt hours annually, tens of thousands of liters of red diesel used each year in farm operations. Of course, these categories broaden—medically, in terms of treatment of dead bodies, and potential spread of pathogens; and culturally, with respect to how the dead are perceived; what myths cling to their bodies and depositions.

What is sustainable? How much biomass per square kilometer should be harvested? When does a farm become a marginal one? What synthetic chemicals should be permitted, if any, and by what standards of calculation in terms of trophic, greater ecosystem impacts? Epizootic sea louse transmission of diseases on salmon and trout can be expected to react in unpredictable varieties of behavior as climate change reaches core biotic levels (e.g., 4 degrees Celsius increase over preindustrialization levels).[18] De-lousing of fish in industries in Scotland have relied upon a mixture containing a legally defined and controlled "corrosive substance," namely, hydrogen peroxide.[19]

---

[18] See "Wild salmonids and sea louse infestations on the west coast of Scotland: sources of infection and implications for the management of marine salmon farms," JR Butler, Pest Manag Sci. 2002 Jun;58(6):595–608; discussion 622–9. PubMed PMID:, 12138626 DOI: https://doi.org/10.1002/ps.490, https://www.ncbi.nlm.nih.gov/pubmed/12138626, Accessed January 27, 2019.

[19] See "The Isle of Gigha: An Assessment of a Sustainable Rural Community," by Brendan Craig, 2005, http://www.globalislands.net/userfiles/scotland_9.pdf, Accessed January 27, 2019.

In the overall chain of contemporary human existence, the thousands of other ingredients cannot be ignored, from scrap metals, to everything imported into a community from the outside, utilitarian artifacts comprising alloys and plastics, pollutants, and other foreign bodies.

Bio-invasives on an island become overnight critical offenders of the indigenous sustainability, with rare exceptions. As human population levels vary, and the model of successful renewal is achieved, at what ethical price? When does rural sustainability necessitate a revisiting of basic moral preambles, most obviously in the case of dietary and conservation shifts away from ever harming other animals? There are plentiful examples of pseudomorphic transitional Faustian communities, as Oswald Spengler characterized such windows on human cultures.[20] In the case of western Scotland, a cursory glance at restaurants reveals a keen awareness of vegetarian and vegan options, in addition to a great tradition of whiskey distilleries. But the majority of all food produced and served for the cadres of eco-tourists upon whom the small island communities depend, continues a heavy reliance on killing, with cows, sheep, chickens, salmon, halibut, and trout being six of the most martyred species.[21]

## 8.4  Utopian Contradictions

When Sir Thomas More published his popular *Utopia* (1516) it is estimated that members of one in four households in France and probably as many throughout the Burgundian Empire, today's Benelux, could read its Latin text exhorting and satirizing "the Best State of a Republic and the New Island Utopia" (in its sub-title). The reading public had been spoiled with several million copies of hand-printed and hand-illustrated Horae, the Christian devotional genre known as the Book of Hours.[22] They could not have failed to grasp some of the basic contradictions instated by More. For example, the size and extant of the theoretical island's territory versus its having 54 cities, each of equal size and no closer than 24 miles apart, yet not so distant that one could not hike from one city to another in a day. The math doesn't add up, and this Utopia is, if anything, a densely teeming human nightmare

---

[20] See Spengler's 2-volume *Der Untergang des Abendlandes*, 1918–1922, Oskar Beck, Munich, 1922.

[21] See such studies as "Environmental Sustainability," University of the West of Scotland, https://www.uws.ac.uk/about-uws/uws-commitments/sustainability/Accessed January 27, 2019; and "Western Highlands – Natural Heritage Futures Update," Scottish Natural Heritage, https://www.nature.scot/sites/default/files/2017-06/A306342%20-%20Natural%20Heritage%20Futures%20-%20Western%20Highlands%20update.pdf, Accessed January 27, 2019; "Demographic Change in Remote Areas," the James Hutton Institute, https://www.hutton.ac.uk/research/projects/demographic-change-remote-areas; See also, 1.3.7 Ecosystem Services, New Zealand Department of Conservation, "*Stewart Island/Rakiura Conservation Management Strategy and Rakiura National Park Management Plan 2011-2021*"Accessed, January 27, 2019.

[22] See *Medieval Must-Haves – The Book of Hours*, by Sandra Hindman, Les Enluminures, Paris and New York, 2018, p. 11.

of congestions, like its literary alter ego, the region of Arcadia in southwestern Greece, which, from 370 BC had its capital in Megalopolis. The city's theater sat 20,000 people. Today, the town's population is under 6000 and it has essentially been usurped by modernity. The A7 freeway completed in 2002 goes from Kalamata to Athens, bypassing the ancient relic of Greek pastoralism.

More's Utopia had to be an island, or a mountaintop, or inaccessible jungle, polar outpost, or bastion of sand dunes. Because human geography in the early sixteenth century was correlated strictly with a global cartography more unknown than known. Most mainlands had been reached; even their "dark" interiors scoped out. Only the most wild and dangerous distances yet to be crossed might be appropriated in the imagination as some surrogate for the Garden(s) of Eden. In spite of his contradictory island/urban Utopia, More envisioned an implacable rule of law, a strange but mostly human state of governance, not that More himself had much reason to place his faith in any political system, given that he was martyred to one.

But in the United States it is a very short fuse indeed from the illusion of ecological freedom to a continuing history that encompasses all that oppression, genocide, and a blatant record of environmental schizophrenia. Seconds before any picture postcard might emerge, American sectors—energy, transportation, protected species, and habitats—implode, igniting a legacy of Superfund sites, CITES violations, roadkill, pollution-related morbidity statistics, and outright warfare between locals and wildlife. Her carbon footprint, per capita consumption of other animals, and contribution to the global commons of plastics and chemical pollutants are unprecedented.

Historic antidotes, like the formation of the National Park Service, can be viewed in light of discriminating altruisms that have never swayed from the aggressive exploitation of local, regional, and global natural capital, and all those deafening matters of human encroachment and skewed distribution of assets—principally echoed in the injustice to all those who were here long before the white supremacist conquerors—that have literally obfuscated the American wildlands, even if the purer aesthetic imagination is able to transcend such brutalities of modern life.

## 8.5   What Do We Know? What Do We Actually Perceive?

The ecological paradox that underlies all such destruction in the name of progress, profit, and protein irrationality, correlates with a fallacy prophetically described by the French economic thinker Frédéric Bastiat in an 1850 essay entitled "Ce qu'on voit et ce qu'on ne voit pas," referring to "that which we see and that which we don't see." Break a window and the cost will be passed along to the economy at large in the form of an expenditure, and somebody's profit. Profits and losses are assumed, under the many laws of averages, to ultimately reach a state of equilibrium, the larger the economy. Even when a nation or corporation runs a deficit, needing to borrow money, that state of affairs can be characterized as healthy, according to the "broken-window fallacy." Or as one economist has written, "Economic theory suggests that persistent trade deficits will be detrimental to a nation's economic outlook by negatively impacting employment, growth, and devaluing its currency. The United States, as the world's largest deficit nation, has consistently proven these theories wrong".[23]

Russian mathematician Andrey Kolmogorov (1903–1987) examined this class of contradictions to develop an explanation that was quickly adopted by ecologists looking at predator-prey relations (the Lotka-Volterra equations, named after American chemist Alfred Lotka and Italian physicist Vito Volterra) in order to better understand ecological flux, and steady-state systems that persist despite boom-and-bust population dynamics, whether among lemmings or white-tailed deer, to name but a few cases.[24]

All ecological numbers should give us pause, in the same way that young Ludwig Wittgenstein, in his initial versions of his *Logisch-Philosophische Abhandlung*, underscored—on its title page—his "Motto": That "anything a man knows… can be said in three words." And later, in his Philosophical Investigations, he adds, "Don't think, but look" (PI66). In *Abhandlung*, he even reminds us to toss away the ladder after we have ascended, using his ideas and propositions as steps, all of which, he adds, are "nonsensical." This is philosophy and language, not as thought and words, per se, but as action.[25] With all numbers and, by implication, all thoughts and words, the many methods of extracting data, and the very presence of human researchers, skew the outcome of scientific investigations with the same unassimilable syndromes that give so much verisimilitude to Werner Heisenberg's 1927 *Principle of Uncertainty*, or Indeterminacy, pertaining to the location of an observer, and the velocity of the observed object.

Stated differently, an Anthropic Principle injects bias at every level of human perception, from our notions of the cosmos to that infinitesimal inclusion of ourselves in the equations of galaxies, beginnings, and ends. A logical basis for determining this indeterminacy, these blurred peripheries between reality and unreality,

---

[23] See "The Pros and Cons of a Trade Deficit," by Adam Hayes, May 15, 2015, https://www.investopedia.com/articles/investing/051515/pros-cons-trade-deficit.asp, Accessed January 21, 2019.

[24] See *Lotka, A. J. (1920).* "Analytical Note on Certain Rhythmic Relations in Organic Systems". Proc. Natl. Acad. Sci. U.S.A.*6: 410–415.* doi:https://doi.org/10.1073/pnas.6.7.410. PMC 1084562. PMID 16576509. *See also,* Volterra, V. (1931). "Variations and fluctuations of the number of individuals in animal species living together". In Chapman, R. N. Animal Ecology. McGraw–Hill; and Abrams, P. A.; Ginzburg, L. R. (2000). "The nature of predation: prey dependent, ratio dependent or neither?". Trends in Ecology & Evolution. 15 (8): 337–341. doi:https://doi.org/10.1016/s0169-5347(00)01908-x. For a superb recent overview of the theories post-Lotka and Volterra, see "Dynamics from a predator-prey-quarry-resource-scavenger model," by Joanneke E. Jansen and Robert A. Van Gorder, Theoretical Ecology 11, 19–38(2018), September 14, 2017, Springer Link, https://link.springer.com/article/10.1007/s12080-017-0346-z, Accessed April 28, 2020.

[25] See *Protractatus, An early version of Tractatus Logico-Philosophicus* by Ludwig Wittgenstein, Edited by B. F. McGuinness, T. Nyberg and G. H. von Wright, with a translation by D. F. Pears and B. F. McGuinness, an historical introduction by G. H. von Wright and a facsimile of the author's manuscript, Cornell University Press, Ithaca, New York, 1971; *Philosophical Investigations*, 1953, G.E.M. Anscombe and R. Rhees (eds.), G.E.M. Anscombe (trans.), Oxford: Blackwell.

what Leonardo transformed into his sfumato, Rembrandt an equally graded chiaroscuro, or words and numbers and their transcendence in mind, could be construed out of seemingly inaccessible locations on the known surface of earth. Judith Schalansky's *Atlas of Remote Islands* offers a telling case in point. As Schalansky writes, "Geographical maps are abstract and concrete at the same time; for all the objectivity of their measurements, they cannot represent reality, merely one interpretation of it. The lines on a map prove themselves the artists of transformation... ".[26] This asymptotic condition of human mind grappling within the theatrum orbis terrarium, the theater of the world,[27] cannot easily resolve some of the great island riddles. For example, debate continues as to the meaning of the humanly uninhabited Semisopochnoi Island, part of the US Rat Islands, and other parts of the Aleutians closest to the 180th meridian by longitude. This ensemble of a seabird paradise is considered both the western-most and eastern-most point of North America. Or how did it come to pass that Robinson Crusoe Island, among the Juan Fernándes of Chile, should now be occupied by over 600 humans? On Russia's Lonely Island, the last word of the last human occupant's hastily written note in the resident logbook is "illegible,"[28] much like the impossible effort to grasp "biological affinities" among the "Early Proterozoic fossils" that continue to defy morphological associations, or, in some cases, even the most fundamental assessment by those studying primeval chemical evolution on earth.[29]

If there is human probity, moral value, it is not the byproduct of ecological numbers but of something more. We articulate its feelings, its beauty, and the passion to see so-called biological integrity retained at all costs. But our populations are easily swept up into a matrix of ever-shifting moral targets, an analytical chaos of costs and benefits, and self-referential Trees of Life that grant all manner of priority to our own species' wants, separate and alone, cut off from the vast majority of our biotic peers amongst the endlessly debated domains—Eukaryota, Bacteria, Archaea—and/or kingdoms of life: "Monera (bacteria), Protista, Fungi, Plantae, and Animalia".[30]

---

[26] See *Atlas Of Remote Islands – Fifty Islands I have Never Set Foot On and Never Will*, by Judith Schalansky, Translated from the German by Christine Lo, Penguin Books, New York, 2010, p. 10.

[27] ibid., p.23.

[28] ibid., 26.

[29] See *Earth's Earliest Biosphere – Its Origin And Evolution*, Edited by J. William Schopf, Princeton University Press, Princeton, New Jersey, 1983, Chapter 14, "Early Proterozoic Microfossils," by Hans J. Hofmann and J. William Schopf, pp.321–360.

[30] See "UCMP Phylogeny Wing: The Phylogeny of Life," by Ben Waggoner, http://www.ucmp.berkeley.edu/alllife/threedomains.html, Accessed January 21, 2019. See also the six, seven and eight kingdom of life theories, Cavalier-Smith, Thomas (2006)."Rooting the tree of life by transition analyses". Biology Direct. 1: 19. doi:https://doi.org/10.1186/1745-6150-1-19. PMC 1586193. PMID 16834776.

## 8.6 Amaurotic Compromises

So, in spite of our total living dependency upon that integrity of biological peerage—which we are still, daily, discovering and re-defining—our perpetually fulminating paradox is our *core blindness—amaurosis,* to Others. Blindness, in turn, to the reciprocity indebtedness which is far more interesting and obligate than all of the avalanches of incoming scientific data combined. A fine case in point: While archaeologist Harvey Weiss and team were able to publish astonishing findings regarding the specific episode of "climate change" that evidently wiped out the Akkadian Empire in or about the year 2200 BC, as evidenced from "the lifeless soil of Tell Leilan" in Syria (as described by Elizabeth Kolbert).[31] We have now seen COP24 come and go in Katowice, Poland—the 2018 United Nations 24th Climate Change Conference of the Parties to the United Nations Framework Convention on Climate Change.[32] And with its largely forgotten administrative, bureaucratic conclusions, we must ask: What status of collective good news, of calm and global human demeanor in the face of self-destruction, can be lauded? As Thomas Jefferson wrote, in one of his final letters before his death, to James Madison, "…the long succession of years of stunted crops, of reduced prices, the general prostration of the farming business… have kept agriculture in a state of abject depression…In such a state of things, property has lost its character of being a resource for debts".[33]

Biomes change in character and nutrition; in the proportion of blunt and blended chemicals; by nature of the fluctuations in humidity, pressure and that all flamboyant dew point that softens or hardens the life of invertebrates, and all those who stake their claim upon them. And while social insects, gregarious dinoflagellates within the tissue of corals, *shy* trees and the most industrious of other mammals, such as the beaver, are primed to offer their humble manipulations of physical structure to the world in the interests of co-symbiosis, the variables of human evolution invoke implacable strategies that succeed in terrorizing all others. This is a unique combination of human nature(s) and their bio-heritage which, even to ourselves, at this time in biological history, must strike as senseless.

---

[31] *Field Notes from a Catastrophe – Man, Nature, And Climate Change*, by Elizabeth Kolbert, Bloomsbury Publishing, New York, 2006, pp. 94–95.

[32] https://cop24.gov.pl/

[33] Written from Monticello, February 17, 1826, in *The Writings Of Thomas Jefferson*, Collected And Edited By Paul Leicester Ford, G. P. Putnam's Sons, New York, London, 1899, Volume X, p.377.

## 8.7  Anthropical Sub-texts

In that context, the multitude of well-recognized paradoxes are each of an anthropical nature. From Euclid's Elements (ca. 300 BCE) to Kant's Critique of Pure Reason (1781) human logic has dictated that from premises flow logical consequences. But that pure, transcendental deduction presupposes such premises, and consequences—also of a stark logic—irrespective of whatever empirical knowledge might alter our premonitions or hindsight. This purity (allegedly) rests upon our intuitive grasp of time and causality, and the supposed truth of such insights. It occurs in a specific space. All of the logical predicates that follow are subject to mentation and/or perception. It is a transcendent logic that is, in essence, the opposite of what, in Vedic, Brahmanical, and Buddhist traditions are known as Pratītyasamutpāda (Sanskrit), or dependent origination, a causal chain in Nature and in Mind whereby all phenomena (dharma) are variously linked.

But regardless of the debates and their outcomes, the logic is human logic. Even in the case of ghost lineages, whether an avian phylogeny is confirmed by fossil evidence, the impression of ancestry versus contemporaneity is a human impression. But what of a different order of impressions: In 2017, an estimated 821 million of the 7.8 billion *Homo sapiens* were under-nourished.[34] How do we accommodate our species' own vital signs within a philosophical template that is wide enough to encompass both our own blinders and some set of observations free of those same blinders?

Given the varieties of human perspicuity and an equally decisive battlement of narcissism, a crucial test of the durability of standard weights and measures fails when it concerns human demographics. Demographers themselves continue to modify projections unself-critically with respect to decade after decade of miscalculations.

Census data, however accurate, is mired in a homocentric vision that quickly smooths over a mere two-hundred-year timeframe in which the human population soared from 1 billion to 7 billion, with a continuing escalation.[35] Most classical paradoxes, among numerous such classes, are able to intellectually and logically tease their way out of ultimate discord and discrepancy. These often-surprising challenges to the eventual escape from mathematical traps, and countless other strangely disparate premises and conclusions are found throughout physics, economics, medicine, statistics and probability, mechanics and cosmology, relativity, chemistry, linguistics, and philosophy, among other zones of human conceptualiza-

---

[34] "Climate Change, Population Growth, and Hunger in the 21st Century – A MAHB Dialogue with Gernot Laganda," by Geoffrey Holland, https://mahb.stanford.edu/blog/climate-change-population-growth-hunger-21st-century-mahb-dialogue-gernot-laganda/Accessed January 25, 2019.
[35] See    https://www.learner.org/courses/envsci/unit/text.php?unit=5&secNum=4;    See    also Population Reference Bureau's http://www.worldpopdata.org/Accessed January 27, 2019.

tion. All such disciplines tend toward an equilibrium. But, alas, only in the human imagination. Invalid proofs work their way through passages of thought and rethinking, until they become valid. Mirages eventually disappear. Morality can be turned into the embrace of weapon-bearing arms, as faithful Christians learned during the massive bloodshed accompanying the Crusades. And even Beethoven's fourth movement of his ninth symphony "Ode to Joy" would have its veritable renaissance of performances under the Nazi regime.[36]

Every human enterprise is material for a paradox, from fiction, abstraction, movement, perception, and love/hate, to all those objects, sequences, processes, and conclusions abetting discernment, analysis, calculation, tabulation, inference, extrapolation, judgment, hope, and faith. Of the hundreds of many formal paradoxes, not one escapes the ceaseless trench of human involvement. This is not a problem, but a reality to accept and then deal with. Our species has never surrendered to any paradox. And it would be impossible to avoid them. At the same time, simply ignoring that which is terminally ill by our hands is no substitute for working through complex problems that seem unique to humankind.

Alternatively, we might—as Sir Thomas More hoped—live in a state of fairness beyond all notions of class and distinction. This may be interpreted within Utopian tradition, or in a manner akin to what ecologist Aldo Leopold famously described when he characterized his "land ethic": "A thing is right when it tends to preserve the integrity, stability and beauty of the biotic community. It is wrong when it tends otherwise" (Fig. 8.4).[37]

Hence, as Aristotle declared in his *Politics*, a government of the many, not the few, should be most persuasive to rational thought, the agora a teeming gamut of overtly volitional, rational discourse.

Without venturing, at this stage, into the paradoxical nature of all politics (etymologically, the citizen of a city), there is the independent observer status of mind versus brain that also stymies a firm conceptual grasp of what we're even talking

---

[36] See *Beethoven's Ninth – A Political History*, by Esteban Buch, Translated by Richard Mille, The University of Chicago Press, Chicago, Ill., 2003, Chapter Ten. That Beethoven was, after all, vulnerable; a human being, member of a population, is borne out by one particularly curious piece of memorabilia: a handwritten "shopping list" given to his servant and "comprising six items including a mousetrap and a metronome." See "Inlibris at the 2020 Ludwigsburg Antiquarian Book Fair and the California International Antiquarian Book Fair," https://www.google.com/search?q=Inlibr is+at+the+2020+Ludwigsburg+Antiquarian+Book+Fair+and+the+California+International+Anti quarian+Book+Fair,%E2%80%9D&client=safari&rls=en&source=lnms&sa=X&ved=0ahUKEw jD5Oe5lZrnAhUDPH0KHX8aAmYQ_AUIDCgA&biw=1439&bih=821&dpr=2, Accessed January 23, 2020.

[37] See "Understanding the Land Ethic," by Jennifer Kobylecky, May 29, 2015, The Aldo Leopold Foundation, https://www.aldoleopold.org/post/understanding-land-ethic/Accessed January 23, 2019; See also, *A Sand County Almanac Illustrated,* by Aldo Leopold, Photographs by Tom Algire, Tamarack Press, Madison, Wisconsin, A special edition for the benefit of the Leopold Fund, 1978; See also, Leopold Archives: https://uwdc.library.wisc.edu/collections/aldoleopold/.

**Fig. 8.4** Engraving circa 1660 by Jan De Visscher after Nicolaes Pietersz Berghem. (Private collection). (Photo © M.C. Tobias)

about when we discuss "right" and "wrong," rational and irrational, particularly with respect to the "biotic community" at large. Nothing about our language groups has yet translated thought into ecologically gentle understanding at the level of a global human population. Not unlike the dour and devastating "units of production" earlier referenced, and implicated in so much of humanity's destructive paradigms, our language-and-logic-based causes and actions have yet to demonstrate the humility and restraint which a few individuals among our species have managed. This paradox of taxa between the one and the many of the same species is a critical minefield which remains entirely unknown in its multiple contradictions and evolutionary pathways.[38]

Within that taxa paradox, numerous individuals leap out within the imagination; individuals whose body of work suggests a self-governing, non-urban biotic com-

---

[38] See *The Theoretical Individual: Imagination, Ethics and the Future of Humanity*, by M.C.Tobias and J.G.Morrison, Springer Nature, New York and Switzerland, 2018; See also, *The Hypothetical Species: Variables of Human Evolution*, by M.C.Tobias and J.G.Morrison, Springer Nature, New York and Switzerland, 2019.

munity. Japanese artist Kitagawa Utamaro (1753–1806) is renowned for his colored wood-block prints, his "floating world" (ukiyŏ) poetry and calligraphy, and his pronounced scientific observations of insects, shells, and birds. His fold-out masterpiece *A Chorus of Birds* ("Momo-chidori kyoka awase") housed today in the Metropolitan Museum of Art in New York, not only illuminates in alchemical colors and forms thirty birds with whom the artist was familiar, but the volume also endeavors to explore what Utamaro named "Fude no Ayamaru,"—"The Brush Goes Astray".[39] That lovely depiction of what we could take to mean a brush with illogic is quite useful. It is its own community within nature, but one that does not obey the false strictures and obsessions of one particular bipedal, largely carnivorous hominid. This individual, the contingent realities, circumstances, fate within an environment and biography are all matter of conditioning, de-conditioning, or non-conditioning. Evolution as an assimilated fact had been acceded to by a sufficient bulwark of scientific loyalties even before its rubrics could be corroborated by genetics, or the advent of psychology, quantum mechanics, or deep learning. So, the question of conditions by which a cause and consequence come about was only partly described—conditional clauses of unsubstantiated faith.

## 8.8 Us?

We still have nothing like a clear understanding of the impacts of the world around us on human nature. No one agrees on the language to describe human nature, let alone the principles that guide us. If there even is an *us*. Or whether twin species understandings would be better served by a multiplication of possible hybrid beings and cultures within the vague aegis of what is nominally thought of as *Homo sapiens*, but may, in fact, be nothing more than a passing vogue ascribed to by evolutionary biologists who all vie with one another, during their brief moments in time, to excel at a temporarily shared language of description. An embryological/genetic set of explanations that may be entirely false, given their partiality and severe competition unto itself. These countless unknowns have been discussed within a specific context, one which Paul Horwich has defined "…a paradox, roughly and superficially, as an assembly of apparently reasonable considerations that engender conflicting inclinations about what to believe, and hence a form of cognitive tension".[40]

---

[39] From Translator's Note, *Utamaro: A Chorus Of Birds*, with an Introduction by Juia Meech-Pekarik, and a Note on kyŏka and translations by James T. Kenney, The Metropolitan Museum of Art, A Studio Book – The Viking Press, New York, 1981.

[40] See See Paul Horwich, "The Nature of Paradox" in his book, *Truth-Meaning-Reality*, Oxford University Press, New York, 2010, DOI:https://doi.org/10.1093/acprof:oso/9780199268900.003.0011. Oxford Scholarship Online, Chapter 2, Published to Oxford

We don't feel such tension when stopping to consider whether we can go from one room to another in our home, point A to point B, as in the case of the most famous of Zeno's nine paradoxes. That's because it would require an extremely compulsive, neurotic individual to engage in what has become known as proof by contradiction, reductio ad absurdum, an absolute refutation of philosophical amnesty in the interests of moving a few feet left or right. This spatial and emotional entrapment, a philosophical obsessive-compulsive disorder, one at first assumes, but one that also saves trillions of individual lives every day with distinct apprehensions, suspicion, doubt, superstition, fuzzy logic, and a continuum of hindsights, and harkens to that moment of supreme nausea tested by Jean-Paul Sartre in his first major fiction, *La Nausée*, when his character of Roquentin encounters a chestnut tree's bark, only to be plunged into a psychomotor abyss.[41] A first, and disastrous, encounter with the natural world. Or, as E. M. Cioran has written, in speaking of the relativity of utopia and apocalypse, giving us the former, a confession of clichés which are "closer to our deepest instincts"; and noting that "Not everyone can reckon with a cosmic catastrophe…".[42] Cutting aphorisms, decisive plots in a cemetery—these are our certainties (Fig. 8.5).

Scholarship Online: February 2010, http://www.oxfordscholarship.com/view/10.1093/acprof:oso/9780199268900.001.0001/acprof-9780199268900-chapter-2, Accessed January 23, 2019. See also, "*How* we know biodiversity: institutions and knowledge-policy relationships," by Rajeswari S. Raina and Debanjana Dey, Sustainability Science, Sustain Sci 15, 975–984 (2020). https://doi.org/10.1007/s11625-019-00774-w, January 2, 2020, https://link.springer.com/article/10.1007/s11625-019-00774-w?sap-outbound-id=5476E303368693D1AE8022A18261041F4DA33070&utm_source=hybris-campaign&utm_medium=email&utm_campaign=000_NISH01_0000002209_SPSN_AWA_KN01_GL_WED2020&utm_content=EN_internal_7025_20200606&mkt-key=42010A0557EB1EDA9BA8DCC3BF5B4FBD

[41] Gallimard Publishers, Paris, 1938; for quotation, see "The Chestnut Tree: The Experience of Contingency," http://twren.sites.luc.edu/phil120/ch10/nausea.htm.

[42] Émile Cioran, *History And Utopia*, Translated from the French by Richard Howard, Seaver Books, New York, 1960, p. 83.

**Fig. 8.5** Mid-nineteenth-century photograph of monastic ossuary in Rome, photographer unknown. (Private Collection)

# Chapter 9
# The Obsolescence of Presuppositions

## 9.1 The Ineffable Landscape

Etymologically, paradox—from the Greek para = distinct from, doxa = opinion—comports with every layer and component of the landscape that we infinitely perceive, but also with the fact, as historian John Stilgoe nicely elucidates that "Dictionaries travel badly" and that "Minor constituents recall fragments";[1] "… what is natural, how do we know that…" asks Sean Ryan in a most uncannily effective essay, "Theorizing Outdoor Recreation and Ecology.[2]" With equally ineffable and incongruent affinities, cultural leanings over millennia inform our most sovereign intuitions. How "only in the desert, they declared, could a man find freedom," wrote the equally hard-driving explorer Wilfred Thesiger of the rarified but steadfast Bedu peoples he'd encountered[3] (Fig. 9.1).

Trying to comprehend another's seemingly impossible but unequivocal lifestyle has equally baffling comparisons with the attempt to measure the value of rare species when any imputed benefit to another, to its biome, to the world at large, defies normal market categories, as we would think of them in economic terms. What, generally speaking, are the *ethical* contributions to the planet of a hippopotamus? A cycad? Contingent needs bear scrutiny under widely varying categories: values deriving from "unpriced commodities" as well as fundamental behavior and the norms, "social benefits," and essential "existence value" inherent to both "direct measurement" as well as any kind of use, passive or otherwise.[4] Concluding upon

---

[1] See John R. Stilgoe, *What Is Landscape?* The MIT Press, 2015, pp. x1, and 125.

[2] pp 99–124l, The Nature of Paradoxes/the Natural Paradox, Sean Ryan, https://link.springer.com/chapter/10.1057/9781137385086_5, https://link.springer.com/chapter/10.1057/9781137385086_1 #citeas, Accessed February 1, 2019.

[3] See Wilfred Thesiger, *Arabian Sands*, Longmans, London, 1959, p. 308.

[4] See *Conservation of Rare or Little-Known Species – Biological, Social, and Economic Considerations*, Edited by Martin G. Raphael and Randy Molina, in Chapter 10, "Economic Considerations," by Richard L. Johnson, Cindy S. Swanson, and Aaron J. Douglas, Island Press,

© Springer Nature Switzerland AG 2021
M. C. Tobias, J. G. Morrison, *On the Nature of Ecological Paradox*,
https://doi.org/10.1007/978-3-030-64526-7_9

**Fig. 9.1** Central Park as Seen from Metropolitan Museum of Art. (Photo © M.C. Tobias)

the crucial nexus of US conservation corridor strategies, motives, and projected outcomes in places like Georgia and Florida and Gorongosa in Mozambique, E. O. Wilson argues for "the alternative world we are trying to save" by reckoning upon the essential reality of our current Anthropocene and acknowledging "two parallel worlds on the planet." One of them is human, and all but overwhelming by the nefarious chaos of its inflictions.

The other is that set of strategies and beliefs and subsequent behaviors that turn to culture and policy for edification and longevity; those "final sanctuaries" that will by necessity become "our transcendent heritage... preserving our own deep history" and from which, if we are indeed fortunate, we may "take constant pleasure from the surprise, mystery, awe, wholeness, relief, and redemption they offer.[5]" But, again, what of the cycad and hippo?

Such imperative considerations are in, themselves, the territory of a John Donne's most abundantly mortality-mixing mood of phrase, when he argued "Here dead men speak their last, and so do I..." which he greatly elaborated upon in his book dealing with laws of nature and a most candid and (at the time) controversial work on suicide.[6] Donne's Paradox connotes the nihilistic origins of all inevitabilities:

Washington D.C., 2007, pp. 288–289.

[5] See *A Window On Eternity – A Biologist's Walk Through Gorongosa National Park*, by Edward O. Wilson, Simon & Schuster, New York, 2014, pp. 140–141.

[6] John Donne, "The Paradox," (deemed 1635), p. 70 of the two-volume Grolier Edition, From *The Text Of the Edition of 1633*, Revised by James Russell Lowell, Notes by Charles Eliot Norton,

Light turns to darkness and back again. Love lasts until its origins have vanished. Is it so among other species? Even the most intimate human family connections turn sour and grave, as in the case of the Indian young man suing his parents for not conferring with him before choosing to give birth to him who claims that life is suffering.[7] In 1879 Henry George's book *Progress and Poverty*[8] was acquired by millions of readers caught up in the most paradoxical of a Progressive Era, as more and more people became enfranchised in direct proportion to the multitudes plunged into poverty. That paradox fueled the vital correspondence between Leo Tolstoy and Mahatma Gandhi during their own respective efforts to resurrect a tenable proposition of equality and sustainability within community life, from South Africa to Czarist Russia. Certain inexplicabilities taunted both the great Russian Christian communitarian sage of *War and Peace*, and India's complicated liberator from England. With hindsight we can decipher other characteristics marking the whole long saga of noble savage conceptualizations from English surgeon Walter Hamond to Rousseau, Chateaubriand, and even mid-nineteenth-century American artists and poets from George Catlin to Henry David Thoreau. Each of them imagined a world where the most "primitive" among us were the most enlightened (Figs. 9.2 and 9.3).

It was Hamond who, in 1640, published his emphatically romanticized treatise, *A Paradox, Prooving, That The Inhabitants Of The Isle Called Madagascar, Or St. Lawrence, (In Temporall Things) Are The Happiest People In The World*.[9] As for Catlin's own dream of a personal paradise, at the time of his death in Jersey City in 1872, his entire collection of remarkable Native Americana, containing "over 500 paintings, all in neat and appropriate frames, and my notes completely prepared, explanatory of everything in it" remained unsaleable. For decades the great explorer, artist, and ethnographer had failed to persuade with his beloved Indian Gallery. No historical society, nor the Smithsonian (which would wait and acquire it for free years later), showed the slightest interest. Catlin perished knowing that not one of his paintings ever sold, despite his enormous contribution to history's understanding of nineteenth-century Indian life. He, who had advocated for all of Native Americana to be enshrined as a national park, would see his great dream vanish in much the manner of his successor, the photographer Edward Curtis who had initially deemed it fitting to name his 20-volume masterpiece *The Vanishing Race*.[10] History has its

---

Volume 1, New York, 1895. See Donne's *Biathanatos, A Declaration of that Paradoxe, or Thesis, that Self-homicide is not so naturally sin*…London: Printed for Humphrey Moseley, 1648.

[7] "India man to sue parents for giving birth to him," BBC News US, February 7, 2019, https://www.bbc.com/news/world-asia-india-47154287, Accessed February 7, 2019.

[8] *Progress and Poverty: An Inquiry into the Cause of Industrial Depressions, and of Increase of Want with Increase of Wealth – The Remedy*, Wm. M. Hinton & Co., San Francisco.

[9] …*Whereunto Is Prefixed, A Briefe And True Description Of That Island, With Most Probable Arguments Of A Hopefull And Fit Plantation Of A Colony There*. Published for Nathaniel Butter, London.

[10] *See George Catlin And His Indian Gallery*, by Brian W. Dippie, Therese Thau Heyman, Christopher Mulvey and Joan Carpenter Troccoli, Edited by George Gurney and Therese Thau Heyman, Smithsonian American Art Museum and W.W.Norton & Company, Washington D.C., 2002, from Christopher Mulvey's essay, "George Catlin in Europe," pp. 63–64.

**Fig. 9.2** Indigenous Amazonians, *John Nieuhoff's Remarkable Voyages & Travels*, London, 1703. (Private Collection). (Photo © M.C.Tobias)

revenge upon even the most vindicated of aficionados. Seven years after he was to put up money for Curtis to embark on his ambitious documentation of Indian life, J. P. Morgan—the richest man of his times—would die in a hotel room in Rome at the age of 75. All his collections, homes, and myriad assets, the equivalent of some $25 billion in today's dollars, as divided up among the 42 corporations he either owned outright, or in part, reverted to others. What was true about him was instantly dispersed. In a similar way, the 1775 poems of Emily Dickinson left both a truth about herself, and "the most teasing anonymity," as Joyce Carol Oates once wrote, in describing Dickinson as "the most paradoxical of poets." And continuing, "By way of voluminous biographical material, not to mention the extraordinary intimacy of her poetry, it would seem that we know everything about her; yet the common experience of reading her work… is that we come away seeming to know nothing.[11]"

Indigenous peoples, great artists, corporate stewards, all are swept away, as Donne nobly elucidated. In each instance of historical paradox, a presupposition proves to be both true and false. What we assumed becomes unassumable; what lived, dies; what dies, is reborn. Human consciousness, at best, makes connections,

---

[11] See "Soul At the White Heat: The Roman of Emily Dickinson's Poetry," by Joyce Carol Oates, Critical Inquiry 15, Summer 1987, The University of Chicago, p. 806, https://www.jstor.org/stable/1343529?seq=1#page_scan_tab_contents, Accessed February 2, 2019.

# A
# PARADOXE:
## PROVING
The Inhabitants of the Island called *Madagaſcar*, or St *Lawrence* ( in
things temporall ) to bee the hap-
pieſt People in the
World.

Confeſſe ( worthy Sir) that
I have undertaken an Ar-
gument, which at the firſt
light, will feeme to moſt
Men, Idle and Imperti-
nent, although I might
anſwer for my excuſe; that
I was therefore idle, be-
cauſe I would not bee idle;
for it may be objected unto
mee : Will you take upon
you to preferre this poore, naked, and ſimple Ignorant
people before the rich Gallant, underſtanding men of
D 2      *Europe.*

**Fig. 9.3** Opening of Walter Hamond's 1640 Treatise. (Public Domain)

sometimes fuzzy, at times acute. In the world of British cultural and economic his-
torian/author David Boyle's book *The Tyranny of Numbers: Why Counting Can't
Make Us Happy*[12] our obsession with numbers and statistical vogues, can lead to
fascinating but unmistakably painful, madness. Indeed, the more numbers and equations, the more paradoxes. In an essay for the *London Observer* (January 14, 2001)
he recalls "the eighteenth-century prodigy Jedediah Buxton in his first trip to the
theatre to see a performance of Richard III. Asked whether he'd enjoyed it, all he
could say was that there were 5,202 steps during the dances and 12,445 words spoken by the actors. Nothing about what the words said, about the winter of our discontent made glorious summer; nothing about the evil hunchback king."

In physics (particularly with respect to the so-called Copenhagen Interpretation
within quantum mechanics), philosophy, and ecology, this level of analysis strikes
of the famed "Schrödinger's cat" paradox, in which the cat is simultaneously alive
and dead, depending on the microscopic or macroscopic vantage point; the location
of the observer, and the instantaneous relationship between entropy and probability.
It can, indeed, drive one to a degree of, shall we say, scientific restlessness, as it
must have done for those who tried to puzzle their way through what Bertrand
Russell first envisioned in the late spring of 1901 in his set-theoretical paradox, or
antinomy, one of the origins of naïve set theory.[13] Russell applied strict logic to sets,
and to sets of sets, therein imagining a set of sets adhering to a rule that excluded
any such set of a set from including itself. But if such a set of sets were such a set,
then it must be itself, which, by its own inner logic precluded itself from being a
member of its own set. Hence, a paradox in every way equal to missing the forest
for the trees, or "Richard III" for its meanings.

A very different paradoxical frontier from that of Cicero's famed *Paradoxa
Stoicorum* (46 BC) focusing in particular on six paradoxes of the Stoic philosophers
of his time, who gave great exercise to those passions concerning virtue and vice,
wisdom and folly; and of what higher qualities their combining contradictions
might ultimately enshrine. Given the moral confusion in every age, one might be
inclined to embrace cynical idealism, a first order of defense against one's own
eventual doubts in the face of inevitable mismatches between personal genetics, and
evolutionary principles, for example.[14]

---

[12] Harper Collins/Flamingo, London, 2001-2.

[13] See Godehard Link (2004), *One hundred years of Russell's paradox*, p. 350, ISBN 978-3-11-
017438-0; See also *From Frege to Godel, a Source Book in Mathematical Logic*, 1879–1931, edited
by Jean van Heijenoort, Harvard University Press, Cambridge, Massachusetts, 1967.

[14] *The booke of Marcus Tullius Cicero entituled Paradoxa Stoicorum Contayninge a precise discourse of diuers poinctes and conclusions of vertue and phylosophie according the traditions and
opinions of those philosophers, whiche were called Stoikes. Wherunto is also annexed a philo-
sophicall treatyse of the same authoure called Scipio hys dreame*. Anno. 1569. Cicero, Marcus
Tullius., Newton, Thomas, 1542?-1607., Cicero, Marcus Tullius. *Somnium Scipionis*. Imprinted at
London: In Fletestreate neare vnto Sainte Dunstones Church by T. Marshe, [1569]; See D Mehl
(2002). C Damon; JF Miller; KS Myers, eds. *The Stoic Paradoxes according to Cicero (in) Vertis
in Usum*. Walter de Gruyter. p. 39. ISBN 3598777108; See M.O. Webb (1985). Cicero's Paradoxica
Stiocorum: A New Translation with Philosophical Commentary (PDF). Texas Tech University.

## 9.2 Hypercontradictory Naturalism

If there is a species-wide mismatch with Nature—a supposition with no possible proof, other than a host of empirically accessible data sets all lending support for this belief—then what? Biological remorse within "the age of anxiety," as W. H. Auden titled a book of poems in 1947?[15] Remorse that has no physical infrastructure, no known psychological or emotional capacity to meaningfully grapple with information that tells us at least one-third of all Himalayan glaciers will have melted within a few decades?[16]

Logicians speak of "hyper-contradictions," levels of truth, and the very logic itself of paradoxes; a set of propositions, both proven true and proven false whose outlines involve semantics, mathematics, physics, and ecology. But ultimately, their Medieval Clouds of Unknowing—Insolubilia—remain.[17] Throughout the biological sciences, paradoxical frontiers meet every fact head on. Survival itself proves to be the most vulnerable of all statements of fact, and thus is central to paradoxical thinking. We posit points of alleged truth within a given duration that normally comports with the life of a thought and that of its thinker. Finitude cloaks every durability, transforming our most hallowed pillars of certainty into a realm of the penumbra, and the sequential fictions by which we manufacture our identities, pasts and futures, cultures and fixations.[18]

A long history that has seen the merging of formal logical studies with various sciences continues to prompt philosophical debate over the essence of truth, the nature of a proposition, and the endless twists upon the slightest inference, induction, extrapolation or conclusion, with or without the reality of infinity.[19]

---

[15] Random House, New York.

[16] See https://www.dw.com/en/stark-warning-on-melting-himalayan-glaciers/a-47353374, Accessed February 4, 2019

[17] See "Hyper-contradictions, generalized truth values and logics of truth and falsehood," by Yaroslav Shramko & Heinrich Wansing, Journal of Logic, Language and Information 15 (4):403–424 (2006); and Journal of Philosophical Logic, Vol. 8, No. 1, "The Logic of Paradox," by Graham Priest, Vol. 8, No. 1 (Jan., 1979), pp. 219–241, Published by Springer, https://www.jstor.org/stable/30227165; See also, *Paradox and Paraconsistency: Conflict Resolution in the Abstract Sciences*, by John Woods, Cambridge University Press, 2002.

[18] See, for example, "Nomadic-colonial life strategies enable paradoxical survival and growth despite habitat destruction," Zong Xuan Tan, Kang Hao Cheong, eLife 2017;6:e21673 DOI: https://doi.org/10.7554/eLife.21673, https://elifesciences.org/articles/21673v4, Accessed September 21, 2018; and "A Paradoxical Evolutionary Mechanism in Stochastically Switching Environments," Nature.com, Article | OPEN | Published: 14 October 2016, Kang Hao Cheong, Zong Xuan Tan, Neng-gang Xie & Michael C. Jones, Scientific Reports, Volume, 6, Article number: 34889 (2016), https://www.nature.com/articles/srep34889, Accessed September 21, 2018.

[19] For a superb bibliography of Primary Sources concerning these topics, from 1897 to 1945, as well as Recent Sources, see "Paradoxes and Contemporary Logic," the Stanford Encyclopedia of Philosophy, https://plato.stanford.edu/entries/paradoxes-contemporary-logic/, Accessed February 6, 2019. In Jainism, for example, there are at least eleven forms of infinity. See "Zero and Infinity in Mathematics: Ayoga Kevali and Siddha in Jain Philosophy," by Medhavi Jain, II ISJS-Transaction, Vol. 4, no. 1, Jan-Mar., New Delhi, India, 2020, pp. 18–19.

But, or course, dealing with the finite is far more difficult, stressful, uncertain, than grappling with any infinity.

# Chapter 10
# Ecological Contradiction, Antinomy, and Counter-Intuition

## 10.1 To Fly or Not to Fly?

Amid so many preemptive deep syntactic and grammatical rules of speech and pre-emptive cognition that have amounted to no less than entire self-contradictory Weltanschauungs, human consciousness, as possibly *wired* and varied from other species, supplies an endless and inchoate set of mentally charged prescriptions for its own exclusive use. Such utilities are, in aggregate, unself-conscious. This poses the problem.

One classic case tantamount to this *problem* in other species concerns the cosmo-politan family of birds known as Rallidae, first organized in terms of their nomen-clature (clustered) in 1815 by Constantine Samuel Rafinesque.[1] Found on every continent but Antarctica, many rails, as they are generally called, can fly, though their relatively short wings do not favor it, and their much preferred flightlessness over time has become a biogeographical classic case study. "The paradox of the distribution of the rail family around the world," a former Secretary of the Smithsonian Institution, the late American ornithologist Sydney Dillon Ripley, muses, "has been its ability to disperse widely and, on the other hand, its having the countervailing tendency to evolve rapidly towards flightlessness."[2] The conundrum is solved by any number of adaptive, genetic, and avian personality-related insights that may, in fact, pertain to human evolution, as well (Fig. 10.1).

As clumsy flyers, rails are more readily blown off course, rendering migration no simple given. Their long-distance vagrancies dot global maps. They can run fast, as we have seen in the case of takahe (*Porphyrio hochstetteri*) on islands like Tiritiri

---

[1] Bock, Walter J. (1994). History and Nomenclature of Avian Family-Group Names. Bulletin of the American Museum of Natural History. Number 222. New York: American Museum of Natural History. pp. 136, 252.

[2] See *Rails Of The World – A Monograph of the Family Rallidae* by S. Dillon Ripley, With Forty-One Paintings by J. Fenwick Lansdowne, Daivd R. Godine, Boston, 1977, p. 26.

© Springer Nature Switzerland AG 2021
M. C. Tobias, J. G. Morrison, *On the Nature of Ecological Paradox*,
https://doi.org/10.1007/978-3-030-64526-7_10

**Fig. 10.1** Takahe (*Porphyrio hochstetteri*), New Zealand. (Photo © M.C. Tobias)

Matangi, in the Hauraki Gulf of New Zealand's North Island, where they had imprinted on the lighthouse keeper's partner, keeping pace of her rapid stride, hoping for hand-outs. To fly or not to fly? This is not a question glibly dismissed by some 127 extant rail species. Somehow wonderfully they freely disperse, and have done so for nearly 50 million years, from the time of the lower Eocene epoch.[3]

While every biological pivot that seeks to understand the life of avifauna finds it "paradoxical" that widely dispersed birds should have made an evolutionary issue of flying or not (a much easier evolution to understand among most of New Zealand's ground-dwelling avifauna, like the Kiwi) we should assume that rails look at humans in much the same way. Hence, the real paradox should not merely emerge in our findings pertaining to those many species, from the Inaccessible Island rail, to the largest of the Ratites, the ostrich, all of whom will not fly (as far as has been observed). Rather, perhaps our very observations themselves are a clustering patchwork, ideas migrating with the specific birds that have been studied, less any and all generalizations? This mode of observation would re-ignite the enchantment of individual differences, re-tooling our fundamental organization of ideas as concerns everything in Nature.

And why should it be any different for our ideas? It is not. Princeton mathematician André Weil writes of Descartes' methodical self-assurance that he was capable of knowing perfect numbers and, consequently, "everything there is to know about the material and immaterial world." This professed omnipotence encompassed what was thought of in the 1630s as "aliquot parts," "perfect," "amicable," and

---

[3] See "A Synopsis of the Fossil Rallidae," buy Storrs L. Olson, Smithsonian Institution, 1977, https://repository.si.edu/bitstream/handle/.../VZ_77_Synopsis_fossil_Rallidae.pdf?...

"submultiple" numbers.[4] As biological facets of the infinite heaps of the life force increasingly dawn upon our species, two phases of passing importance become clear. The species drifts in one direction, while its individuals have every random reason to veer somewhere else. These two geographic migrations are as varied as the personal parts. Biographies coalesce or seek out alienation, exile, what the Desert Fathers prescribed as a prelude to ultimate union with *their* Holy Father, a topic the authors have dealt with in some depth in *The Theoretical Individual*.[5]

## 10.2   Paradoxical Sub-sets

Like all empirical evidence, we reclaim in every act of perception a different version of the sorites paradox, or heap-of-sand enigma. At what point have enough grains been removed so that it is no longer a verifiable "heap"?[6] A second paradox also confronts us. Sand may be initially laid claim to in the words of "a heap of sand." But we are presuming many things about each granule. They are not the same size or weight. Moreover, each has multitudes of organisms attached. A tiny heap contains, easily, some ten thousand Loricifera, one of the most recently discovered of all known phyla of life (out of approximately 40): Meiofauna (meiobenthos) are vast swathes of invertebrate animalia whose discoverers include Reinhardt Mobjerg Kristensen and Robert Higgins. Higgins calls them bibliocryptozoans. Relatively simple tools of the naturalist's profession can discern them: "a homemade sieve fashioned out of a nylon mesh net fastened over a funnel; a spray bottle for rinsing it out; a petri dish to place the samples in; and a stereomicroscope of at least 25 times magnification." Like a pair of binoculars, the human interest in Others explodes with so many newly perceived neighbors that, of course, our position standing on a Florida beach with a bucket of sand is not what it appears. Gulls and countless other seabirds eyeing us, we are strange members of a vast community. The gulls are teeming with their own life forms. And even those meiofauna are additionally inhabited by yet others, who are also each inhabited, and so on until we are lost in a black hole of biological numbers.[7] There are approximately 137 species of Loricifera (most have not yet been studied) and unknown numbers of species that inhabit them. We earlier recognized in the Russell's Paradox or Antinomy, a set of sets, sub-sets of sub-sets, until is theorized a set that cannot be a member of itself, although that Self is the very set of sets it cannot by definition be party to. While numerous equations have since been proposed that bypass the apparent contradiction in the numbers game, ecologists remain confounded by the essential truth

---

[4] See André Weil's book, *Number Theory – An approach through history From Hammurabi to Legendre*, Birkháuser, Boston, Basel, Stuttgart, 1984, pp. 52–53.

[5] Tobias/Morrison, Springer, 2018.

[6] Sorensen, Roy A. (2009). "sorites arguments". In Jaegwon Kim; Sosa, Ernest; Rosenkrantz, Gary S. *A Companion to Metaphysics*. John Wiley & Sons, p. 565.

[7] See "Life On A Grain Of Sand," by Virginia Morell, Discover, April 1, 1995.

factors—no matter how much latitude a particular truth be granted—defining sets of numbers that translate into biological organisms.

From our limited perspective—the sum total of consciousness, the human conscious, qualia, biophilia, physiolatry, reciprocities within an anthrozoological, interdependent biosphere, deep ethology—we cannot attest to any span of truth. The distance between whole and partially whole truths is further widened by the duration before, during, and after the object of perception, and the event upon which it is expatiated in our contemplations and computations.

"A paradox is a situation where observations are not in accordance with experiences or expectations," says the American Mathematical Society (AMS). And "the moral here is that via dramatic examples, mathematics has the ability to show that one's intuitive expectations cannot always be trusted!"[8] Perceived paradoxical behavior fluctuates across every target of our concentration. There is the legacy of the South Island alpine parrot the kea (*Nestor notabilis*), which once had a government bounty on its head for farmer allegations pertaining to the bird killing sheep, to its more recent protected and celebrated status as an avian rarity, and a model of good humor, pranks, and much pleasure for observant humans[9]. Or the Josiah Willard Gibbs Paradox concerning the alleged violation by otherwise indistinguishable particles in a system that manage to alter entropy thereby slighting the Second Law of Thermodynamics in the exchange of intensive versus extensive quantities. As with every paradox suggested for the behavior of birds, gasses, entropy, medical sociology,[10] and humans, for starters, the aforementioned AMS protocols suggesting that human perception and interpretation may be flawed (not to be "trusted") is a welcoming addition to the literature on paradox in that it transports certainty into subjectivity, a healthy signal that our own histories and the summation of our wisdom literature, and scientific verities are to be called into question.

## 10.3   The Argument for Equilibria

There are significant instances, ambassadors of variance, throughout natural history, philosophy, and the sciences in general, where the lack of firm elucidation must counter our proclivity as a species to leap at conclusions. When Alfred Russel Wallace compared the flora of the European continent, of India, and of the

---

[8] http://www.ams.org/publicoutreach/feature-column/fcarc-machines3, Accessed February 8, 2019.

[9] See *Kea, Bird of Paradox: The Evolution and Behavior of a New Zealand Parrot*, Judy Diamond + Alan B. Bond, University of California Press 1999.

[10] See "The Ecological Paradox: Social and Natural Consequences of the Geographies of Animal Health Promotion," by Gareth Enticott, Transactions of the Institute of British Geographers, New Series, Vol. 33, No. 4 (Oct., 2008), pp. 433–446, Published by: Wiley on behalf of The Royal Geographical Society (with the Institute of British Geographers, https://www.jstor.org/stable/30135326

Indo-Malay islands, for example, he perceived vast disparities in numbers of species (parrots versus pigeons), despite ample time for evolution to have altered and equalized biological composition within insular regions.[11] In the absence of a clear story to explain the apparent discrepant representations of different genera, Wallace expanded his geographical understanding of biodiversity in general to conclude with a global picture, closer to the presumed truth of species equilibria. He enlarged the context of understanding; rewrote the atlas of mind, achieving a new equilibrium or center from which to gaze out at the world, renewed.

If there is an ecological argument for equilibrium, within the too obvious tumult that occasions every observation of, and feeling about, nature, then it would theoretically require, and ethically should suggest, a treatment (human behavior meted out) of individuals in no different a manner than entire populations. And then on to complete ecosystems and ecosystem behavior. A most curious approach to what E. Gordon Ericksen describes under the heading of "The Principle of Insufficiency: The Assertion Model" is his sense of the mirroring effect: "Knowing that, like a mirror, a person reflects a neighborhood, a town, a nation, a place of work, a home… "[12] And for philosopher/political activist Jerome M. Segal, "The adequacy of a specific conception of the self cannot be assessed without some prior conception of the kind of theorizing in which we are engaged."[13] The mirroring is like that theoretical premise of personal involvement which motivated Wallace during his time observing habitat throughout Indonesia and surrounding regions. A concept that hinged upon a sense of perfect equipoise spanning individuals and populations (a mirroring), which is why he was so firm a believer in, and co-author of, the overall concept of evolution, along with Darwin to whom he dedicated his book *The Malay Archipelago*, the ultimate sense that evolution mirrored and propelled every one of us, personally. While its impetus echoed the old and severely obsolete cliché Man is the Measure of All Things, its incarnations in Wallace and Darwin looked very different, as if to remedy the Cartesian fallacy of feeling superior with a primal acknowledgement of the privacy, personality, and biographical details enlisted in the act of epiphany.

Far more recently, Jared Diamond examined mechanisms inherent to equilibrium among bird populations on islands that were once connected to New Guinea's mainland. His underlying methodology was core species-curve mathematics, the numeric description—long held as a kind of gospel of biogeography—which argued for specific "relaxation time" between species number agitations, great flux, looming extinctions, and/or re-establishment of equilibrium, of a balance, once pre-existent,

---

[11] See *The Malay Archipelago – The Land Of The Orang-Utan And The Bird Of Paradise, A Narrative Of Travel With Studies Of Man And Nature*, Alfred Russel Wallace, Macmillan And Co., Ltd., London, 1906, p. 303.

[12] *The Territorial Experience – Human Ecology As Symbolic Interaction*, by E. Gordon Ericksen, Foreword by Herbert Blumer, University of Texas Press, Austin, TX, 1980, p. 103.

[13] Jerome M. Segal, *Agency And Alienation -A Theory Of Human Presence*, Rowman & Littlefield Publishers, Inc., Savage, Maryland, 1991, p. 55.

between the numbers of species, and the size and behavioral relations of respective species populations.[14]

In the case of bird groups living on currently insular islands (what Wallace had originally referred to as "fragments"), Diamond extrapolated time frames: "For almost all of the islands he studied, the relaxation time was in the realm of ten thousand years."[15] Conversely, with respect to the known 59 avian species that have gone extinct in New Zealand as a result of the human presence (among, no doubt many other New Zealand avifauna disappearances science has not yet recognized) it is estimated that the human time frame necessary to recapture such evolutionary life force would require 50 million years.[16] But the point is, regardless of the degree of imbalance, the premise holds to a human *conception* of balance, as if nature has worked everything out, and our imperfect sciences are so much reportage, gainfully employed observations that can predictably rely upon a cumulative bias of centuries during which lenses of perception improve; calculations are honed; equations are improved; outcomes achieve a level of plus-or-minus confidence always moving closer to certitude. Certitude, in turn, depends on real individuals.

Hence, the very charming outlook driving conservation which can be ascertained from a US forester's memorandum, back in March, 1967, dispatched by William Hurst to the supervisor at Kaibab National Forest, pertaining to a squirrel: "The Forest service has a moral, as well as a legal, obligation to protect this animal… every activity undertaken within his range should be carefully evaluated, giving the needs of the squirrel prime consideration. Coupled with this, we should search out all the information we can on his habitat requirements. If this information is inadequate, which I suspect it is, we should give leadership to a research program to learn more about the squirrel…"[17] All of conservation biology comes down to this basic protocol, predicated on the historical realities that can be discerned throughout the history of human interactions with other individuals, biographical subjects worthy of our profound admiration; creatures who mirror us, and whom we mirror. From life within caves and cliffsides, to the huge density of parks and royal forests throughout thirteenth-century England that served as private grounds for hunting and forest product extraction,[18] to the very "languages of landscape," as Mark

---

[14] See Brose, U., A. Ostling, K. Harrison, and N.D. Martinez. 2004. Unified spatial scaling of species and their trophic interactions. Nature 428:167–171.

[15] See *Noah's Choice – The Future Of Endangered Species*, by Charles C. Mann and Mark L. Plummer, Alfred A. Knopf, New York, 1995, p. 71.

[16] See "Deep Macroevolutionary Impact of Humans on New Zealand's Unique Avifauna," by Louis Valente, Rampal S. Etienne, and Juan C. Garcia-R., Current Biology, Volume 29, ISSUE 15, P2563–2569.e4, DOI: https://doi.org/10.1016/j.cub.2019.06.058, https://www.cell.com/current-biology/fulltext/S0960-9822%2819%2930785-7?_returnURL=https://linkinghub.elsevier.com/retrieve/pii/S0960982219307857?showall=true, Accessed June 5, 2020. And, the Authors write, in general, "If threatened species go extinct, up to 10 Ma [million years are] needed to return to today's levels." From their "Highlights".

[17] Quoted from Sylvester Allred's *The Natural History of Tassel-Eared Squirrels*, University of New Mexico Press, Albuquerque, 2010, p. 123.

[18] See *The English Medieval Landscape*, Edited by Leonard Cantor, University of Pennsylvania

Roskill titled his important treatise that attempts to lay out a "code," apprehended early on by Pausanias and Vitruvius in their grasp of "varietates topiorum" (subjects from nature).

We are speaking of a cipher linking human imagination to the countryside.[19] This so-called *code* is crucial to gaining insight regarding the reading of texts, sub-texts, species, sub-species, populations, genera, phyla, the worlds of biology in general, and all leading to this similitude of individuals at the fantastic core of interdependency and reciprocity potential.

Kim Sajet, in an essay entitled "Providing Solace in the Age of Discontent," and focusing upon what has been called the "Dutch Utopia" among a group of artists from the US between 1880 and 1914, writes, "Having no clearly defined agenda, crossing societal boundaries, and employing various methods of protest, reformers of the period were not a coherent group."[20] It is a code, both coherent and incoherent at once, that stands out in such artists' precursor, Peter DeWint (1784–1849) who was much noted as having devoted himself as a painter "For the common observer of life and nature."[21] A correspondence in all that nervous expression that made for Henri Jourdain's gorgeous illustrations accompanying the spectacular edition of Gustave Flaubert's *Madame Bovary* (L'Imprimerie Nationale, Paris 1911) or that greatest of all nineteenth-century British landscape publications *Scotland Delineated in a Series of Views, with Historical, Antiquarian and Descriptive Letterpress*, by John Parker Lawson.[22] Such artistic champions of the personage as Nature provide some insight into our ancestors, bringing true closure to the paradox of their lives by our very gratitude for their having lived. This is that Nature in every conceivable sine qua non of what Carl Jung described—in his considerations of "Two Kinds Of Thinking"—as the Logos of centuries, by which scholastics invested their faith, "a dialectical gymnastics"[23] (Fig. 10.2).

For the poets of early Western Civilization, sixth-century Africa (in the voices of Appuleius and Saint Augustine), Italy (Maximianus and Ennodius), western-most Europe (Columba, Aldhelm writing in Hisperic Latin, and the great poets Venantius Fortunatus and Isidore of Seville), the Jungian transformations are those of what

---

Press, Philadelphia, PA, 1982.

[19] Mark Roskill, *The Languages Of Landscape*, The Pennsylvania State University Press, University Park, PA, 1997, p. 14.

[20] In *Dutch Utopia – American Artists in Holland, 1880–1914*, Edited and with an introduction by Annette Stott, Organizing Curator, Holly Koons McCullough, Essays by Nina Lübbren, Emke Raasen-Kruimel, Kim Sajet and Annette Stott, Organized by the Telfair Museum of Art, Savannah, Georgia in association with the Singer Laren Museum, the Netherlands, Telfair Books, Savannah, Georgia, 2009, p. 21.

[21] *Peter DeWint 1784-1849 – 'For the common observer of life and nature'*, Edited by John Lord, With essays by Peter Bower, Jim Cheshire, John Ellis, John Lord and Ian Waites, Lund Humphries, Aldershot and Burlington, 2007.

[22] E. Gambart & Co., Joseph Hogarth., London., 1854. 2 vols. Folio.

[23] See *Symbols Of Transformation – An Analysis Of The Prelude To A Case Of Schizophrenia*, by C. G. Jung, Translated by R. F. C. Hull, Bollingen Series XX, Second Edition, Princeton University Press, 1970, p. 20.

**Fig. 10.2** Dunderawe Castle, From John Parker Lawson's *Scotland Delineated*, 1854. (Private Collection). (Photo © M.C.Tobias)

Jack Lindsay has authoritatively characterized as a person "…in continually-changing symbiosis, grasping to the best of his ability the potential elements in himself…interpreting a difficult world…"[24] But if there is one quality that emerges unscathed throughout such historical chaos, awaking the need to scientifically describe, or poetically evoke that humanistic impulse which sees in all of Nature that mirror of oneself, the word is *hope*.[25] Hope that corresponds with George Steiner's "rudimentary grammar of the unfathomable."[26]

---

[24] See *Song Of A Falling World – Culture during the Break-up of the Roman Empire, A.D. 350–600*, by Jack Lindsay, Andrew Dakers Limited, London, 1948, p. 285.

[25] See *Brier-Patch Philosophy By 'Peter Rabbit'*, Written by William J. Long, Illustrated by Charles Copeland, Ginn & Company, Boston, MA., 1906, p.273.

[26] See *Real Presences,* by George Steiner, University of Chicago Press, Chicago, Ill., 1989, p. 201.

# Chapter 11
# Heavy and Light Contingencies of Consciousness

## 11.1 Jean-Paul Sartre and the Contingencies of Being

Sartre, in the Conclusion of *Being and Nothingness*, affirms that "Being is without reason, without cause, and without necessity; the very definition of being releases to us its original contingency."[1] Sartre's translator, Hazel Barnes, offers an example of what she believes Sartre means by "contingency," namely, "The contingency of freedom is the fact that freedom is not able not to exist."[2] Intriguingly, this declaration (more sentiment than empiricism) dovetails with the thinking of a contemporary of Sartre, photographer August Sander (1876–1964), who wrote in a lecture entitled "Photography as a Universal Language" that "Humanity has constantly been compelled to solve the problems arising from altered surroundings." His storyline has our species uniquely driven by its own irrepressible range of motives, geographical settings, and compulsions regarding all those Others who, for Sartre, operate at the level of the most startling of all contingencies, a word whose Medieval Latin etymology confirms a sense of both circumstance and contingere = *befall*.[3]

We note that with migrating flocks of waterfowl, as distinguished from solitary migrants, such as the Wandering Albatross, the "geomagnetic data" helping bird navigation is augmented by the intergenerational knowledge base embedded in the flock.[4] There are other flocks in addition to the extrinsic one most immediately associated with birds. That concerns the flock of nerves, the synaptic commons fueling

---

[1] See Jean-Paul Sartre, *Being and Nothingness – An Essay on Phenomenological Ontology*, Translated and with an introduction by Hazel E. Barnes, Philosophical Library, New York, 1956, p. 619.

[2] ibid., p. 630.

[3] See *August Sander – Seeing Observing and Thinking, Photographs*, With texts by August Sander and Gabriele Conrath-Scholl, Die Photographische Sammlung/SK Stiftung Kultur, Foundation Henri Cartier-Bresson, Schirmer/Mosel, Paris, 2009, p. 25.

[4] See *The Migration of Birds – Seasons on the Wing*, by Janice M. Hughes, Firefly Books, Buffalo, NY, p. 156

© Springer Nature Switzerland AG 2021
M. C. Tobias, J. G. Morrison, *On the Nature of Ecological Paradox*,
https://doi.org/10.1007/978-3-030-64526-7_11

activity at the interior level of motor primacy. And "Since there can be a practically unlimited number of possible states within this network, the possible behaviors of the organism can also be practically unlimited."[5] Echoes of the aforementioned mirroring effect is one way to intimate the biospheric code inherent to every ecosystem in relation to every other system.

In a similar manner, scanning electron microscopy targeting the feather of a goose that has been prepared with a combination of circumstantial alloys, "conductive metal coatings," a "vacuum chamber," a "low pressure atmosphere of argon gas," etc.,[6] "allows the otherwise chaotic electrons to be controlled"[7] while yielding the image as well as the reality of varying bacterial and quill mite loads in any given feather. One feather might contain thousands of such mites (of which some 320 species have thus far been identified) and millions of bacteria.[8] Follow the bacteria.

If we liken flocks and parasitic loads to populations in general, noting the Others, as outlined in the myriad contingencies and circumstances that lend to life a certain natural economy, from the lone individual to its species and heaps of sand that, by analogy, constitute discrete populations, we might also adduce from the Middle Ages the nature of the monk inhabiting a Cistercian grange, or farming complex. That individual, and his/her brethren, according to Saint Benedict "should be occupied at certain times in manual labor, and again in fixed hours in sacred reading..."[9] Both sets of activities exert a force that is felt in the universe, however modest the nudge of electrons and their excitement may be. By such similes and similitudes, we read into our own histories and recognize that thermodynamics plays as much a role in the natural depictions we foster, originate, reproduce, and express among ourselves, as the metaphysical questions that perpetually tease philosophers, farmers, poets, and, unquestionably, all other species inhabiting the ontological orbit that we know to be life.

---

[5] See *The Tree Of Knowledge – The Biological Roots of Human Understanding*, by Humberto R. Maturana and Francisco J. Varela, Foreword by J. Z. Young, New Science Library, Shambhala, Boston & London, 1987, p. 159.

[6] See *Journeys In Microspace – The Art Of The Scanning Electron Microscope*, by Dee Breger, Columbia University Press, New York, 1995, p. 16.

[7] ibid.

[8] See "These mites should make birds quiver right down to their quills," by John Barrat, Smithsonian Insider 30 April 2014, https://insider.si.edu/2014/04/bird-mites-make-quiver-quills/, Accessed February 14, 2019.

[9] *Architecture Of Silence – Cistercian Abbeys Of France*, Photographs by David Heald, Text by Terryl N. Kinder, Harry Abrams Publishers, New York, 2000, p. 20.

## 11.2    Prefigurements of the Odyssey

Within that orbit—or whatever symbol, metaphor, allegory, or language we choose
to target it in mind—there are both light and heavy contingencies. What does that
mean? Operands and binary connectives linking ideas, their transmutation into
words and sentences, and the conceptual demarcations that suggest, ultimately, the
nature of a proposition: such is the realm of adjuncts and extraneities—of pro-
grammed and unprogrammed links within whatever realm of thinking applies. Such
propositions contain all the seas and land between, every ecosystem and idea. They
can be called tautological (unavoidable) or contradictory, but they are based upon
dyads, dialectics, thesis, and antithesis. Combine them in whatever alchemical construct or illusion, and the abstraction meets a chair in the person who sits upon it.
The same person who might otherwise languish in bed listening somewhere in a
dark place to a mosquito that is home to a rich diversity of microbiota, as has been
demonstrated in studies of the abdomen of the Binh Phuoc *Anopheles* mosquito
complex in Vietnam.[10]

Hundreds of organisms, and millions of others within those organisms, occupy
the room, the bed, and exponentially into the trillions—encompassing cellular level
counts—within that person seated upon a chair, or in the dirt of a backyard garden,
such that our statements of fact instantly devolve into the blurred edges of fiction.
Twenty years ago, the authors of *Life Counts: Cataloguing Life on Earth* proposed
a numeric uncertainty of between 10 million and 200 million species believed to
exist on earth, yet to be discovered. Considering that each human's large intestine
hosts some 70 trillion bacteria ("Humans as Habitat" models), the numbers so
resemble those heaps of sand (the Sorites Paradox) that any one of us can commence a remarkable odyssey, the journey beyond paradox, within one's backyard[11]
(Fig. 11.1).

While we may quantify the differences between light and heavy contingencies of
consciousness, we are still tens-of-thousands of years removed from infraspecific
relations, such as our distant ancestors would have understood daily. But that both
are contingent realities—open to possibility—there is little doubt. We must necessarily begin within the furthest reaching parameters imaginable of a proposition of
and from life.

Of course, the numbers are changing radically. Since 1953, for example, the
Netherlands has seen its coastlines reduced—as a result of enormous human dam

---

[10] See Frontiers in Microbiology, Front Microbiol. 2016; 7: 2095, Published online 2016 Dec 23.
doi: https://doi.org/10.3389/fmicb.2016.02095, PMCID: PMC5181100, PMID: 28066401,
"Diversity of the Bacterial Microbiota of *Anopheles* Mosquitoes from Binh Phuoc Province,
Vietnam," by Chung T. Ngo, Sara Romano-Bertrand, Sylvie Manguin, and Estelle Jumas-Bilak,
NCBI Resources, PMC US National Library of Medicine, https://www.ncbi.nlm.nih.gov/pmc/
articles/PMC5181100/. Accessed February 14, 2019.

[11] See *Life Counts: Cataloguing Life on Earth*, by Michael Gleich, Dirk Maxeiner, Michael
Miersch, and Fabian Nicolay, In Collaboration with: UNEP, IUCN, WCMC, AVENTIS, Translated
by Steven Rendall, Atlantic Monthly Press, New York, 2000.

**Fig. 11.1** Lesser Flamingo (*Phoeniconaias minor*), Southern Africa. (Photo © M.C. Tobias)

projects—from 1500 kilometers to approximately 800 kilometers, a staggering consideration of the human mechanism.[12] Since just 1970, as referenced earlier, current estimates suggest that humanity has extirpated at least 60% of all wildlife populations, excluding far greater damage to insects and spiders, particularly flying biomass.[13] With such escalating human interventions throughout the biosphere, not to mention increasing animal consumption by humans—the mindful apotheosis of hourly holocausts and the ecologically widespread moral and biomass equivalency of cannibalism—our propositions, introspection, and future human consciousness and proclivities for virtuous action appear trapped in the worst of all documented malignancies. A far cry from the pure odyssey just referenced earlier. Even our notorious, usually well-intended inquiries too easily lead to paradoxical backfire.

Such has been the notorious case of the scientific ledger regarding primate taxonomy in the Amazon, wherein researcher access to regions previously off the map has been greatly amplified as forest destruction and roadbuilding has allowed for much greater intrusiveness and often malevolent, human invasion.[14] Our only certain

---

[12] See Zeldzame vogels van Nederland, Avifauna van Nederland 1, by Arnoud B van den Berg and Cecilia A W Bosman, GMB Uitgeverij, Haarlem/Stichting Uitgeverij van de KNNV, Utrecht 1999, p. 7.

[13] See "Humanity has wiped out 60% of animal populations since 1970, report finds," By Damian Carrington, The Guardian, October 29, 2018, https://www.theguardian.com/environment/2018/oct/30/humanity-wiped-out-animals-since-1970-major-report-finds, Accessed February 15, 2019.

[14] *Handbook Of The Mammals Of The World*, 3. Primates, Chief Editors, Russell A. Mittermeier, Anthony B. Rylands and Don E. Wilson, Published by Lynx Editions in association Conservation International and IUCN, Barcelona, Spain, 2013, p. 14 from Dr. Mittermeier's Introduction describing how "Paradoxically, the destruction of the forests and the construction of highways into remote regions have facilitated this search, providing access to parts of Amazonia and the Congo forests, for example, that were previously inaccessible."

**Fig. 11.2** Chameleon in Madagascar. (Photo © "Hotspots" Production, Dancing Star Foundation)

hope of altering our course demands a new plurality of human natures and human taxa in general wherein diverse personal redemptions might serve as ecological proxies, to whatever extent, in the ethical reengineering of human population attitudes motivated by a willingness to act rationally, moving forward, at the taxonomic level of the individual (Fig. 11.2).

# Chapter 12
# The Paradise Paradox

## 12.1  Statistical Islands and Mainlands

Bayesian inference (from the Reverend Thomas Bayes, 1701?–1761) takes unknown variables to calculate a range of future prospects—equations counting on a high chance of likelihood, thereby granting the ingredients of any algorithm a level of confidence. Its contingent propositions of prior distributions and near-predictions of the future are mathematically among the most interesting of all decision theory related statistical probability functions. It would allow one to have quietly determine well in advance a maximum a posteriori estimation (MAP),[1] a point estimate. Perhaps something as obscure as German firefighters rescuing a plump rat that trapped itself in a sewage drain and was crying out for help—one of the very few *Rattus* genus individuals favored, out of an estimated 7 billion rats across the planet.[2]

Or, the statistical flash points concerning those few short-wooled little black wild sheep first translocated to Litla Dimun (Dímunarseyðurin), one of 18 main islands in the Faroes. For centuries, these kindly occupants of the North Atlantic lived in their very own exquisite paradise. The 1358 ft summit of Mt. Slættirnir is perpetually wrapped in cloud, and the native grasses and wildflowers along its gentle slopes provide deeply buffered sinkholes of soft biomass in which to sleep and graze above the sheer cliffs. The total of 203 inaccessible acres always made for the perfect hideaway from the rest of the world. Until islanders from "abroad" (the villagers of Hvalba and Sandvík) would row across each Autumn to chase the Soay-like wild breed of sheep and carry them off to be exploited on other larger islands among the Faroes. Those sheep that could escape were often, inexplicably, shot. They had few places to which to run.

---

[1] https://www.probabilitycourse.com/chapter9/9_1_2_MAP_estimation.php, Accessed February 27, 2019.

[2] "Firefighters Rescue Fat Rat From Trapped Sewer," https://www.wfla.com/news/viral-news/firefighters-rescue-fat-rat-from-trapped-sewer/1814247342, Accessed March 1, 2019.

© Springer Nature Switzerland AG 2021
M. C. Tobias, J. G. Morrison, *On the Nature of Ecological Paradox*,
https://doi.org/10.1007/978-3-030-64526-7_12

There has never been a more splendid example of a closed statistical system yielding revelatory windows on every form of human and nonhuman behavior. Litla Dimun is the quintessential final mathematical verdict on the intersection of species. By the 1860s the population of Litla Dimun *Ovis aries* were extinct, replaced by local domesticated Faroe cousins. Why would people feel compelled to kill them? Sheepherders have their own excuses.

Other annual roundups of sheep each June on St. Kilda, Soay, Hirta, Boreray, and the Dun, in the Outer Hebrides, continue to occur. It, too, is frequently a deadly roundup known as Ruagadh, and while it says much about the hunters, what stands out most about the sheep is the fact many of them commit suicide by leaping off cliffs into the sea, rather than letting the men and their dogs get them. So much for the touted Utopic circumstances of these Hebridean isles, as first written up by one Martin Martin in his 1697 account, *A Late Voyage to St. Kilda*. And notwithstanding all of the vacuous panegyrics of poets for centuries that lavished romantic praise on the Scottish isles (Figs. 12.1, 12.2, 12.3).

Such poetics—betraying the statistical fields of life and death, as easily calculated on such islands—burgeoned prior to the final evacuation of St. Kilda. All 15 men and 21 women on August 27, 1930, at 7 a.m. were dispatched to Scotland's more populated regions. Most if not all of St. Kilda's dogs were killed by drowning before the final human departure. Why were they not brought along, with the

**Fig. 12.1**  Outer Hebrides sheep. (Photos © M.C. Tobias)

**Fig. 12.2** Lambs in Southern New Zealand. (Photo © M.C. Tobias)

**Fig. 12.3** Ghent Altarpiece Polyptych by Hubert van Eyck and Jan van Eyck, 1432, Saint Bavo Cathedral, Ghent, Belgium. (Photo: Public Domain)

bovines and sheep, all boarding the HMS Harebell for the journey by sea to Argyll, where many of the locals were promised work in the town of Artdornish part of the year?[3]

We are accustomed to speaking of the *expulsion* from paradise in human terms, most memorably in Sienese painter Giovanni di Paolo's (1403–1482) tempera and gold on canvas of the "Creation and Expulsion from Paradise" that once was the predella, or base of the altarpiece in the church of San Domenico of Siena and painted in 1445.[4] Or in the deeply evocative illustrations of Milton's *Paradise Lost*, particularly the final Book XII, by John Martin, and later Gustave Doré (Milton's original edition being published in the Autumn of 1667). Line 645 of Book XII is most telling: "Some natural tears they dropt, but wiped them soon."[5] But as for the sheep, we search back beyond the van Eyck brothers' Ghent Altarpiece, to the night of the exodus from Egypt by the Israelites, during which the Paschal Lamb[6] was sacrificed. No more wrenching image of this than in the painting, "Agnus Dei—The Lamb of God," by Francisco de Zubarán, c.1635–1640. "Behold the Lamb of God who takes away the sin of the world," John the Baptist exclaims.[7] We humans get over it. The Lamb never does.

## 12.2   The Ghent Paradox

The city of Flemish Ghent itself lays claim to other inherent contradictions. In the case of one of Ghent's other famous offspring, Leo Baekeland (1863–1944), often described as "The Father of the Plastics Industry,"[8] his instincts regarding resins—terpenes, one of the largest classes of organic compounds on the planet—led to his phenol + formaldehyde experimentations, resulting in the first commercially exploitable thermosetting network polymer, a synthetic plastic, *polyoxybenzylmethylenglycolanhydride*.[9] Its initial application was produced in 1908 by the Weston Electrical Instrument Corporation for the Boonton Rubber Company and was called Bakelite, the name Baekeland had given to the proliferation of products resulting

---

[3] See *Island on the Edge of the World – The Story of St. Kilda*, by Charles Maclean, Taplinger Publishing Co., New York, 1980, pp. 11, 108, and 142.

[4] See *Paradiso – The Illuminations to Dante's Divine Comedy by Giovanni di Paolo*, by John Pope-Hennessy, Thames and Hudson Ltd., London, 1993, p. 33.

[5] From the Cassell & Company, Limited edition, Edited, With Notes And A Life Of Milton by Robert Vaughan, London, Paris, New York & Melbourne, 1890, p. 329.

[6] Exodus 12.1-4.

[7] John 1:29 – See Preacher's Institute, "About the Paschal Lamb," by Fr. John A. Peck, May 7, 2013, https://preachersinstitute.com/2013/05/07/about-the-paschal-lamb/. Accessed March 1, 2019.

[8] *Landmarks of the Plastics Industry*. England: Imperial Chemical Industries Ltd., Plastics Division. 1962. pp. 13–25.

[9] https://www.sciencehistory.org/historical-profile/leo-hendrik-baekeland, Accessed January 24, 2020.

from his condensation reaction. It had all commenced at his laboratory in Yonkers, New York. He had come to the United States, already a distinguished Professor at the Ghent Municipal Technical School, in the late 1880s.

By November 9, 1933, Bakelite had been granted National Historic Chemical Landmark status. And it is no small testament to the ingenuity of Baekeland that by 2015, an estimated "381 million tonnes" of plastic debris in countless chemical incarnations and formulae (e.g., high- and low-density polyethylenes, polystyrenes, terephthalates, polyvinyl chlorides) had been produced by one of the world's dominant industries, with the resulting infestation of vast swathes of the planet.[10] These organic polymers are a far cry from the tree sap, or the shellac secreted by female lac insects (predominantly, the *Kerria lacca* scale species) upon trees in Asia with which Baekeland had early on experimented.

The huge tonnage should not be taken as destiny, any more than the van Eyck brothers' masterpiece, or the lives of lambs themselves can be perceived to be immune to injury. History rewards common sense at some level. In the Nilgiris of India, there are expansive "Plastic Free Zones."[11] China's National Development and Reform Commission, and the Ministry of Ecology and Environment, have set in place a ban on single-use plastic bags, beginning in 2022. The UK has banned numerous small plastic articles, including cotton-buds with plastic stems, and plastic stirrers and straws. Throughout the marine environs between St. Kilda and the coastlines of Scotland, plastic pollution was recently described by the BBC as a "scourge" as aerial photographs and GIS maps began to reveal levels of devastation[12] and initial surveys of plastic impacts upon seabirds has come to light with calls for action.[13] At the same time, more and more vegan Irish Lamb Stews, vegan Greek Roasted Lamb, and other cruelty-free recipes have been commercialized.

But even the van Eyck's "Adoration of the Mystic Lamb," in preparation for its 588th-year anniversary (rounded out to 600 years), which occurred in the Spring of 2020 at the Museum of Fine Arts Ghent (the MSK's "Van Eyck: An Optical Revolution"), required preparatory restoration, after the arguably greatest painting in the world's stormy history of having been repeatedly touched up (overpainted), looted, partially lost, and damaged. Such are the twists and turnabouts of otherwise perfect inspirations (like the paradise of Litla Dimun, or St. Kilda, or "paradise" itself).

---

[10] See "Plastic Pollution," by Hannah Ritchie and Max Roser, September 2018, Our World In Data, https://ourworldindata.org/plastic-pollution, Accessed January 24, 2020.

[11] "The Nilgiris battling to stay plastic free," Staff Reporter, The Hindu, December 24, 2018, https://www.thehindu.com/news/national/tamil-nadu/the-nilgiris-battling-to-stay-plastic-free/article25815378.ece, Accessed January 25, 2020.

[12] "Map reveals 'scourge' of Scotland's coastal litter problem," BBC News, 28 August 2018, n.a., https://www.bbc.com/news/uk-scotland-45323832, Accessed January 25, 2020.

[13] See "Seabirds and marine plastic debris in Scotland -A synthesis and recommendations for monitoring," by Nina J O'Hanlon, Neil A James, Elizabeth A Masden, Alexander L Bond, Circular Ocean, Environmental Research Institute, North Highland College, www.circularocean.eu, May 2017, http://www.circularocean.eu/wp-content/uploads/2017/10/Circular-Ocean_Scotland_SeabirdsPlastic_May2017.pdf. Though this was a somewhat preliminary study, researchers found that 20% of the 69 common Scottish seabird species had ingested various levels of plastic.

## 12.3   No Painting Is an Island

Not surprisingly, part of that intense 8-year+ Belgium Royal Institute for Cultural Heritage (KIK-IRPA) multi-million-dollar restoration project involved the crucial use of resins. Indeed, varnishes employed by art restorers typically include one or more of crucial natural fossil or tree resins, including copal, or mastic and dammar; or (Cochineal) insect excretions (to achieve the bright varieties of crimson). In the case of the van Eyck brothers, one of their purportedly major breakthroughs "was the development—after much experimentation—of a stable varnish that would dry at a consistent rate" utilizing "linseed and nut oils, and mixed with resins."[14] Such components, among many others, are the very basis of most natural varnishes. Hence, the conservators used "reversible watercolour and resin-based glazing paints" to fill in the minute damages with which paradoxical human history had burdened and partly obfuscated the van Eyck brothers' masterpiece.[15]

Seen, then, from a distance, the artistic and scientific cradle that has always been Ghent is not without its own endemic environmental fates, as one might describe the ever-unpredictable odysseys of a city's locals, most of whom disappear within quietly reserved genealogical charts and cemeteries, while the odd few make history that is invariably ecologically compromised. A very similar History Paradox attended upon Baekeland's Polish/German/Swiss contemporary, the chemist Fritz Haber (1868–1934), whose own careful research and discoveries would backfire in ways no one who was not caught out in the trenches of World War I would ever fully grasp. Nor has the Green Revolution, so abetted by the Haber-Bosch process of synthesizing ammonia, ever been reconciled with the human population explosion it helped foster.

---

[14] See WebMuseum, Paris, "Gothic Innovation in the North," https://www.ibiblio.org/wm/paint/auth/eyck/, © 19 Sep 2002, Nicolas Pioch.

[15] See "Ghent Altarpiece: latest phase of restoration unmasks the humanized face of the Lamb of God," by Hannah McGivern, 18th December 2019, The Art Newspaper, https://www.theartnewspaper.com/news/facelift-for-the-mystic-lamb, Accessed January 24, 2020. See also, "Unsettled by the Dead Animals in Your Paint? Welcome to the World of Vegan Art Supplies," by Dylan Kerr, August 8, 2017, Art World, https://news.artnet.com/art-world/vegan-art-supplies-1039508#:~:text=(Ox%20gall%2C%20the%20dried%20extract,of%20things%20to%20forget%20about. See also, "The Bad Vegetarian Artist: Animal Products In Watercolour Supplies," by Lee Angold, November 24, 2017, https://leeangold.com/2017/11/24/the-bad-vegetarian-artist-animal-products-in-watercolour-supplies/; and, "Kaia Natural Watercolor – Animal derived ingredients in paint and art supplies," September 14, 2019. Taken collectively, animal deaths involved in the overall painter's arsenal may include: Cochenille lice, animal bones, cuttlefish, chicken eggs, Murex snails, human remains (at least during the Renaissance), cow livers, ox gall, animal glycerin, animal fats, animal collagen, rabbit skin glue, mink, camel and sable hairs. See https://www.naturalwatercolor.com/en_GB/a-57397538/blog/animal-derived-ingredients-in-paint-and-art-supplies/#description, Accessed July 8, 2020.

# Chapter 13
# Codex Sinaiticus

## 13.1   The Holy Mountain

Our preconceptions of paradise slide hierarchically from some perfect, pinioned place—a St. Kitts or Litla Dimun in their aboriginal "pristine" state—toward its ideal (a Van Eyck painting), which is a subset of an idea that renders Utopia to scale, human interactions with the Divine. Or, if one prefers, the idea is the subset of the ideal. At the same time, few species details were ignored on canvases of the Renaissance: a Jackass penguin, African lovebird, macaws, gray herons and the Eurasian bittern,[1] golden-lion tamarins, every tree, and flower species imaginable: a veritable pictorial inventory of Noah's Ark. Jan Breughel the Elder painted nearly 800 windows on these easily blurred semantics, paradise scenes scattered across so many floral arrangements, feasts of the Gods, summer harvests, and communities in the midst of their gaieties along a river bank.[2] He and his family—a veritable dynasty of paradise-related portraitures—were betrothed to a singular motif that engaged viewers in contemporaneous political, religious, and all entablatures human. His paradise collectives began in earnest with his 1594 painting, "The Creation with Adam," at the Doria Pamphilj Gallery in Rome.[3]

Brughel's work is always contextualized, and much of that scenery is Biblical, a fascinating link directly to the Sinai Peninsula and the oldest thriving monastic library and gallery of icons in the world.

As succinctly summarized by the Codex Sinaiticus Project, there are more than 400 leaves in the surviving Codex, including the hand-copied half of the Old Testament and Apocrypha, together known as the Septuagint, or the first Greek translation directly from the Hebrew originals, in addition to all of the New

---

[1] See *Jan Brueghel The Elder – The Entry of the Animals into Noah's Ark*, by Arianne Faber Kolb, The J. Paul Getty Museum, Getty Publications, Los Angeles, 2005, p. 27.

[2] See the 812 images in the http://janbrueghel.net/

[3] op.cit., See *Jan Brueghel The Elder – The Entry of the Animals into Noah's Ark*, by Arianne Faber Kolb, The J. Paul Getty Museum, Getty Publications, Los Angeles, 2005, p. 47.

© Springer Nature Switzerland AG 2021
M. C. Tobias, J. G. Morrison, *On the Nature of Ecological Paradox*,
https://doi.org/10.1007/978-3-030-64526-7_13

Testament. The missing half of the Old Testament from the Codex has never been found; the portions surviving include "2 Esdras, Tobit, Judith 1 and 4 Maccabees, Wisdom and Sirach."[4]

Configurative ideals, nostalgic quadrants, ideas of exile, reunion, and the sheer scope of human drama that is portrayed by the epic tale of Moses, the Israelites, and the Christ Passion all figure in our myriad Biblical ideals of paradise. And as David Hume once expressed, if there is an ideal, it must exist, somewhere, in the universe. This interpretation, despite the rigorous formalities imposed by numbers and equations, in human truth suggests a presumed human capacity in touch with the Cosmos, even if it is only our cosmos, the little one confined to our brains. It may be framed in ancient languages, pictograms, idioms, and cultural mores long buried in our subconscious. But, mathematically speaking, those same frames of narrative remain undimmed. A murky translation over time does not alter or diminish the purity or exactitude that was embodied the day the words were conceived and written thousands of years ago.

Therein exists the ideal, no less demonstrative and instructive than our ideas of carbon neutrality, minimum standards, the value of a living organism, or value in general. A setting, purpose, and compulsion that comport with that very compelling urge to find a sacred resting place that is both sublime and personal. The romance of an ideal precludes nearby oil refineries, train tracks, sawmills and city views, and the acoustical cacophonies of crowds: a site like Mount Sinai or the Flemish Brigadoons of Jan Breughel and family—isolated to the extent that our conceptual drift engages in an ideal that is as geographical as it is spiritual.[5]

In the case of the Plain of Er-Raha (in the Sinai),[6] the neo-Platonist philosopher Plotinus (ca. AD 205–70) had been asked by Emperor Gallienus to venture out to just such a remote region (in this instance, somewhere in the Roman Campania whose creative center was Naples) and to found something akin to Plato's Utopia. It was to be called Platonopolis, an Arcadia—no mere thought experiment—that should comport with Plato's highest "form" in a real place: an ideal juxtaposition Plotinus named *spiritual nature*.[7] Within a century of Plotinus' death, several major changes in northeastern Egyptian culture had rendered Er-Raha and the entire southern Sinai, ripe for an ascetic Renaissance. It would be championed by such anchorites as St. Anthony the Great beneath a mountain at the Gulf of Akaba, not far from Mount Sinai. Others involved in this resurgence included St. Basil and St. Ambrose, both of whom celebrated that spiritual nature in demonstrative and lyrical calls of the wild involving the immersion and subsequent transcendence of Self; St. Augustine, whose own *Confessions* (ca. AD 399) relayed visions of intoxication

---

[4] http://www.codexsinaiticus.org/en/codex/content.aspx, Accessed March 11, 2019.

[5] See "Jan Brueghel Complete Catalogue," https://www.janbrueghel.net/page/about-us, Accessed July 8, 2020.

[6] For example, see Frances Frith's famous, albeit desolate photograph in the Library of Congress, depicting Saint Catherine's Monastery in the upper heart of the rocky valley, ca. 1862; https://www.loc.gov/item/00652982. Accessed March 1, 2019.

[7] See "A Night Out In The Sinai," Chapter One, from *A Vision of Nature – Traces of the Original World*, by Michael Tobias, Kent State University Press, Kent, Ohio, 1995, pp. 12–14.

**Fig. 13.1** Saint Catherine's Monastery Beneath Mount Sinai, After Lithograph by Louis Haghe, Original Drawing by David Roberts, Cassell, Petter, Galpin & Co., London, ca.1889. (Private Collection). (Photo © M.C.Tobias)

with the Creation; Symeon Stylites (AD 390–459), called by followers "the most holy martyr in the air"—the Syrian hermit who lived atop his pillar for 37 years. And there were hundreds of others who flocked to this holy of holies, the Sinai, throughout the fourth and fifth centuries, a period known as the Era of Retreat[8] (Fig. 13.1).

Mount Sinai was the allure, its in situ extremes ideal for tortured souls, like St. Marcarius of Alexandria, who willingly lived in mosquito marshes of Scete (the Egyptian Nitrian Desert)[9] for half-a-year; for St. Mary of Egypt who wandered the

---

[8] See *The Paradise of the Fathers*, Volume 1, by Wallis Budget, Chatto & Windus, London, 1907.

[9] See http://desertfathers.blogspot.com/2010/11/area-known-as-scetis-scetes.html.

area naked for 47 years, between the towering headwalls of Wady Hamr and the most desolate of stretches near Wady Amârah. A paradise of revelatory masochisms for the monk, Agathonicus, who kept warm in snowy winters living and sleeping in the thickness of a herd of desert gazelles amid the tamarisk (Tarfah) grove at Wadi Feiran or Wady Es Sheikh. And for his colleague who is rumored to have lived years upon a ledge somewhere near the turquoise mines of Maghârah, a ledge so insecure and upon so sheer a cliff as to prevent him from ever sleeping.

In their various humilities and abnegations, they were seeking union with God, willing happily to die for that Supreme Being who bid them come to the wilderness of jackals, the Schokari sand racer snake, of locusts and Red-Dwarf honey bees, of Northern Wheatears and White storks; to that simmering topography of granite out-crops including the clustered three peaks comprising Jebel Zebir, Jebel Abu, and Jebel (Mount) Katarína, the site of the Monastery of Saint Catherine's, 8625', and the (historically debated) Mount Sinai, 7497'—a reiteration of that calling which so intoxicated Moses, Christ, Mohammed, St. John the Baptist, and countless others.[10]

## 13.2   The World of Saint John Climacus

But for those fragments known as the Dead Sea Scrolls discovered in the Caves of Qumran during the period 1946–1956, the amulets of Ketef Hinnom and the En Gedi Scroll, the Codex Sinaiticus remains the oldest known collection of materials bound together from the primeval Hebrew Bible. Written down sometime between AD 326 and 360,[11] it was eventually secured within the fortress which Justinian had built to protect the monks in 527 AD beneath Mount Sinai, where earlier had stood a church commissioned by the Empress Helena. And while the monastery's doctrinal orientation spearheaded Eastern Orthodox Christianity, the Roman Catholic world maintained extraordinarily romantic and formal ties to it. Indeed, Pope Gregory the Great (ca. AD 540–604) wrote a letter to the Abbot of St. Catherine's, the famed ascetic, St. John Climacus, dated September 1, 600, in which he likened all of humanity to castaways, save for those monks nestled safely at St. Catherine's Monastery, who lived in harmony upon the shore, in Paradise.[12] The Pope lamented the fact that his duties in Rome would not permit him to join Climacus and his fra-ternity there in Sinai.[13]

---

[10] See the Project Noah, https://www.projectnoah.org/missions/11481015, Accessed March 9, 2019; See also, *The Natural History of the Bible – An Environmental Exploration Of The Hebrew Scriptures*, by Daniel Hillel, Columbia University Press, New York, 2006, particularly Chapter Six, "The Desert Domain – Wanderings in Sinai and the Negev," pp. 118–139.

[11] See http://www.codexsinaiticus.org/en/codex/history.aspx, Accessed March 13, 2019.

[12] op.cit., Tobias, *A Vision of Nature*, p. 22.

[13] See *The letters of Gregory the Great*, translated, with introduction and notes, by John R.C. Martyn, Pontifical Institute of Mediaeval Studies, Toronto, 2004, Hathi Trust Digital Library, https://cata-log.hathitrust.org/Record/004917310, Accessed March 8, 2019.

Climacus (AD 579–649) himself exerted one of the most profound influences upon the climate of asceticism in the Sinai through his powerful little book, *Κλῖμαξ, Scala Paradisi* or Ladder to Paradise, composed in his early 20s or 30s (he'd first arrived at the Holy Monastery in 595 AD, an enthusiastic teenager).[14] Climacus' work illuminates the thirty steps of his ladder to/reunion with God. In it, he summarizes all of the virtues, despondencies, ambitions, and fearlessness that the ascension of the ladder—and by implication and historic evidence, the mountains surrounding the Monastery of Mt. St. Catherine's where Climacus wrote his book—must necessarily entail. "May this ladder teach you the spiritual combination of the virtues... And now there remain faith, hope, love, these three; but the greatest of all is love."[15] The famed icon and hundreds of variations on it show the monks of Mount Sinai climbing the steep ladder following Climacus, with God and St. Peter awaiting them, while angels in heaven, monks on the ground rally "round for encouragement," as a band of devils try plucking the monks from the ladder.

Unlike the impenetrable silences that draped the entire Sinai (in terms of modernity's discovery of its location) even as late as the 1880s when the London publisher of Virtue and Company commissioned Colonel Wilson's spectacular four-volume edition on Palestine,[16] today's Sinai, while bearing all the predictable resemblances to its Mosaic past, also purports to a checkered contemporary reality. Environmental conditions have changed. Just a mile or so West of where I (mt) had lived in a cave above the monastery, in the southern cliffs of Wady Sho'eib, or Jethro's Valley, today Egypt's South Sinai Governate hosts a metropolitan cluster within the El Tur Mountains, St. Catherine, population nearly 6000. There is a highway, a patchwork of dirt roads, hotels, B&B's, and every incentive for massive international tour groups to St. Catherine's. On just a single day, Sunday, November 26, 2017, "1295" tourists showed up at the monastery.[17]

When I resided there, up among the surrounding rock faces, it was truly the "Valley of Refuge." I saw one visitor in approximately 6 months. Today, not infrequently, the solemn rock walls of Râs Sufsafeh, of Jebel ed Deir, Wad yes Sheikh, and Wady Seba-iyeh are echo chambers during the not infrequent bombings by the Egyptian military to vanquish various rebel groups. It is a shamble of geopolitical obfuscation, harassing the fortress and her adjacent, millennia-old sites, various hermitages, the once flourishing convent of St. Episteme, heavenly gardens just

---

[14] See Jacob's *Ladder Divine Ascent*, P.Pincius, Venice, 1518; See also, *L'E'Chelle Sainte, ou Les Degrez Pour Monter Au Ciel, Composez par S. Jean Climaque, Abbe Du Monastere Du Mont Sinai*, Traduits Du Grec En Francois par Mr. Arnauld D'Andilly, With George and Louis Josse, Paris, 1688. And *Sermoni Di S. Giovanni Climaco – Abbate nel Monte Sinai*, Apresso Francesco, Milano 1585.

[15] See *The Ladder Of Divine Ascent*, by Saint John Climacus, Revised Edition, Holy Transfiguration Monastery, Boston, Mass., 1991, p. 229.

[16] See *Picturesque Palestine -Sinai and Egypt*, Edited by Colonel Wilson, Assisted By The Most Eminent Palestine Explorers, Four Volumes, London, Virtue And Company, 1880-1884. St. Catherine's section, Volume 4, pp. 98–120, by Rev. C. Pickering Clarke.

[17] See http://www.egypttoday.com/Article/9/34273/Saint-Catherine-receives-1-295-tourists-in-a-single-day, Accessed March 13, 2019.

**Fig. 13.2** A Monk's Retreat Abase Mount Sinai. (Photo © Eric Alfred Hoffman)

prior to the mouth of Wady Lejâ, Elijah's Chapel, the "Spring of Moses," Wady Zawâtin ("The Valley of Olive-trees"), and the very shadows of Jebel Músa overlooking that mythic landscape wherein the Israelites are said to have camped for four decades (Fig. 13.2).

In the same decade, Virtue and Company commenced publication of their steelengraved four-volume edition, picking up on the international glories attendant upon the sensational discovery of the Codex Sinaiticus at the monastery, Cassell, Petter & Galpin published David Robert's two-volume color masterpiece, *The Holy Land* (London, 1880) in which its famed author spoke of "freaks of nature"[18] "easily seized by fancy or modified by art; and the Mahometan as much entitled to the exercise of his imagination as the Monk." But this sentiment regarding a landscape of ladders for monks like Climacus and his followers was as spiritual and mythopoetic as it was historical. After all, wrote Roberts, "The whole career of the Israelites, from the passage of the Red Sea [in their Exodus] to the entrance into Palestine, was a display of miracle."[19] Indeed, Roberts added, "the traveler can still imagine the 'cloud, the lightning, and the trumpet'."[20] Those trappings of the Laws handed down

[18] Roberts, ibid., Volume 11, p. 25.

[19] ibid., Roberts, Volume II, p. 26.

[20] ibid., Roberts, p. 28.

to Moses on the granite walls above where the Monastery would be built were designed, said Roberts, "to establish the morality of mankind. It was the first instance, from the days of Noah, in which peculiar sins were marked by Divine condemnation. The general impulse of natural justice..."[21] had previously prevailed, but—here in this most alluring configuration of spiritual buttresses and topographically intimidating walls of granite piercing a merciless sky—something altogether new was occurring: a vegetable cornucopia of sustenance for the monks—cypresses and Nebek trees providing shade, and a monastery garden teeming with sweetly scented herbs, "oranges, lemons, almonds, mulberries, apricots, peaches, pears, apples, and olives, and all of the finest quality."[22] And the famous Spring itself, the water the Israelites drank to survive, flowing from somewhere in Mount Sinai, emptying into the courtyard of the Monastery adjoining the small basilica with its spherical ceiling of tile spelling out the "Transfiguration of Christ." This configured sanctuary, wrote Roberts, fit into the perfect equation:

"To the observer of Nature, the peninsula of Sinai is one of the most singular anomalies on the globe. It is an immense mass of mountains, without any of the discoverable purposes for which mountains seem to have been formed. It marks no boundary between nations; its summits collect no water to fertilize the surrounding region.... Yet are we not entitled to regard the problem as solved by Scripture, and by Scripture alone."[23]

Amongst contemporary photographers, Neil Folberg has dramatically evoked the engaging probity of the Holy Monastery's unique positioning,[24] in tune with the great and exaggerated El Greco painting of St. Catherine's from 1570, now housed at the Historical Museum of Crete in Heraklion, near to where El Greco (originally known as Theotokópoulos) was born.

But the greatest imagery of all comes from an intimate acquaintance with the more than 3300 manuscripts (many written on papyrus, or the skins of goats and antelope and other quadrupeds) and the many hundreds of icons going back to the sixth century, the most famous of them being the St. John Climacus icon. Writes John Rupert Martin, the world's authority on this ladder tradition, "...John Climacus, living at Sinai, became as it were a second Moses by ascending that mountain to receive instruction from God."[25] The most famed icon treating of Climacus' treatise dates to the twelfth century, more than 500 years after Climacus' demise. It remains on site at the Monastery. Writes Martin, "The resurgence of monastic iconography at this time can nowhere be observed more clearly than in the illustrated Climax

---

[21] ibid., Roberts, p. 30.

[22] ibid., Roberts p. 32.

[23] ibid., Roberts, p. 34.

[24] See *In a Desert Land – Photographs of Israel, Egypt, and Jordan*, by Neil Folberg, Abbeville Press, New York, 1987, particularly pp. 81–120.

[25] See *The Illustration Of The Heavenly Ladder of John Climacus*, by John Rupert Martin, Princeton University Press, Princeton, New Jersey, 1954, pp. 7–8.

manuscripts."[26] Martin examined 33 icons or fragments of icons—parchment folios—each with Penitential canons or rules laid down by the Church to accompany the images, which are to be found in libraries and museums from the Monastery of Mount St. Catherine's to the Vatican; from Mount Athos to the Bibliotheque National de Paris, and in every major rare books room from UCLA to Harvard's Widener. In fact, from the 1491 Venetian edition of *Sancto Ioanne Climacho, Altramente Schala Paradisi*[27] until the late 1800s there were dozens of editions.[28]

In every instance, the congruence of the painter-monk/mountaineering seer, the surrounding isolated gorges, and that metaphysical fortress, all added up to a majestically sequestered and contiguous spirituality. The iconographic legacies were the perfect illustrations for an essentially pivotal manuscript at their heart, namely, the *Codex Sinaiticus*. This perfect union of composites heralded humanity's utterly compelling and dire fixations with the hereafter.


## 13.3   The Incarnation of Fragments

The iconic world of Sinai had so universal an appeal that just 33 years after El Greco painted his illustrious Iraklion triptych of the Holy Mountain and its Monastery, a fellow Cretan, Georgios Klontzas, rendered his marvelous "Transfiguration with Scenes of Monastic Life" (1603, Candia, Crete?).[29] Neither El Greco nor Klontzas ever made the pilgrimage to Sinai but the legends spanning the vast timeframe between the composition of the Old Testament, and the book finally arriving in fragments and hundreds of pages (tens of thousands of lines) at the Monastery of St. Catherine's had long become a kind of certitude in Judeo-Christian history. Not only was the "Ladder to Paradise" and the many mythical qualities attributed to Saint John Climacus instinct with the very essence of a Biblically inflorescent Sinai,

---

[26] ibid., Martin, p. 3. The great Danish genius, Søren Kierkegaard devoted an entire manuscript to the name of Johannes Climacus, writing in a section entitled, "III: The Absolute Paradox (A Metaphysical Caprice)," "But one must not think ill of the paradox, for the paradox is the passion of thought, and the thinker without the paradox is like the lover without passion: a mediocre fellow. But the ultimate potentiation of every passion is always to will its own downfall, and so it is also the ultimate passion of the understanding to will the collision, although in one way or another the collision must become its downfall. This, then, is the ultimate paradox of thought: to want to discover something that thought itself cannot think." Second Period: Indirect Communication (1843–46)

Philosophical Fragments, p. 37, http://sorenkierkegaard.org/philosophical-fragments.html, Accessed August 11, 2019. For Kierkegaard, this passion to know is betrayed by the quest for knowledge, discovering its aspired to precision through faith and experience, not cognition.

[27] Publisher: Bernardinus Benabus and Matheo da Parma.

[28] See World Cat, https://www.worldcat.org/title/sancto-ioanne-climacho-altramente-schala-paradisi-fol-s-1-recto-incomecia-el-sermoe-sacto-iohane-climacho-al-pastore-translated-from-the-latin/oclc/561777726&referer=brief_results, Accessed March 3, 2020.

[29] See *Holy Image – Hallowed Ground, Icons From Sinai*, Edited by Robert S. Nelson and Kristen M. Collins, The J. Paul Getty Museum, Los Angeles, 2006, p. 232.

but so was Saint Peter, as conveyed in the remarkable sixth-century icon from the Monastery itself. There is the very stylish half-life-sized Saint, right out of Acts 5:15: possessed of "...the power to cure by the mere passage of his shadow."[30]

Other enduring qualities of the monastery's remarkable collection of paintings can be viewed and compared in the many photographs taken during four research trips to Sinai by Kurtz Weitzmann and George Forsyth from 1958 to 1965, working with Fred Anderegg of the University of Michigan. Many were reproduced in the Weitzmann section on Sinai in the book *A Treasury of Icons—Sixth to Seventeenth Centuries*.[31] The very cult of icon "veneration" blossomed at St. Catherine's Monastery, first in its utilitarian aspect—the icon as "religious object"—and then "to a new attitude of greater spirituality stressing the transcendental relationship between the image and the holy personage depicted."[32] When I first went to live in a cave above the monastery for many months one winter in the early 1970s, there were but a handful of scholarly works serving as stimulus/guides: Weitzmann's unprecedented devotion to and widely disseminated knowledge of the icons in Sinai, John Rupert Martin's exquisite study of the specific icons and commentaries portraying "the heavenly ladder," Colonel Wilson's four-volume *Picturesque Palestine*, the 1688 edition of Arnauld D'Andilly's translation of the "Ladder to Paradise" and a Life of Saint John Climacus, from the writings of various Greek historians, and a smattering of publications excerpting observations by many of the early travelers to Sinai, as well as the Archimandrite Porphyrius Uspensky's drawings during his trip to the Monastery in 1857 (Fig. 13.3).

In the early Renaissance, two pilgrims en route to Sinai met up in Jerusalem. Bernhard von Breidenbach (1440–1497) was a leading political figure in one of the Holy Roman Empire's most powerful states, that of the Electorate of Mainz. Felix Fabri (1441–1502) was an influential Swiss Dominican theologian. Both published remarkably detailed accounts of their journeys to the Holy Monastery.[33]

Such pilgrim's tales fueled the imaginations of an El Greco, a Breughel....

Of course, there was nothing quite like the biography of Constantin von Tischendorf himself. I had seen some of the 346+ folios of both the Old and New Testaments at the British Library. But behind glass. However, nothing could quite prepare for holding and studying the 1922 *Codex Sinaiticvs Petropolitanvs*

---

[30] ibid., Nelson and Collins, pp. 122–123.

[31] Kurt Weitzmann, Manolis Chatzidakis, Krsto Miatev and Svetozar Radojðic. Harry N. Abrams, Inc., New York, 1966, Plates 1–36, and p. IX.

[32] Breidenbach's *Peregrinatio in terram sanctam*, 1486; and Fabri's Fratris Felicis Fabri *Evagatorium in Terrae sanctae, Arabiae et Aegypti peregrinationem*, first published in 1484 – https://archive.org/stream/libraryofpalesti07paleuoft/libraryofpalesti07paleuoft_djvu.txt, and eventually consolidated and published in three volumes.

[33] Preserved In The Public Library Of Petrograd, In The Library Of The Society Of Ancient Literature In Petrograd, And In The Library Of The University Of Leipzig, Now Reproduced In Facsimile From Photographs By Helen And Kirsopp Lake, With A Description And Introduction To The History Of The Codex By Kirsopp Lake, Oxford, At the Clarendon Press, Oxford UK, 1922.

**Fig. 13.3** Fragments of the *Codex Sinaiticus Petropolitanvs et Friderico-Avgvstanvs Lipsiensis*, in the Facsimile Edition by Helen and Kirsopp, Oxford, Clarendon Press, 1922. (Photo © M.C. Tobias)

*Friderico-Avgvstanvs Lipsiensis, The Old Testament.*[34] The ancient library at Caesarea, still flourishing in the third, possibly early fourth century, is the likely location where scribes edited this oldest known Greek translation from the Hebrew before it was moved to St. Catherine's Monastery. Prior to that, it may have come from Alexandria. Replete with fabulous details—the vellum itself, the inks, notations, palimpsests, styles, corrections, and as yet unaccountable versions, word-tweaks, literal-versus-generic translations—the Codex and its enigmatic history is unquestionably one of the most perplexing and influential books ever written.

Tischendorf attributed the entire New Testament portion of the Codex, held by the Vatican, as having been written down by the scribe denominated as "D." What is generally not fully appreciated, brought to light by Tischendorf and noted at the end of Kirsopp's own introduction to the Facsimile,[35] is the likelihood "that D actually wrote the whole of Tobit and Judith, and probably the Pentateuch in the Old Testament, so that he was clearly a member of the scriptorium." This places the Codex to within a century or two centuries, probably no more, from the living Christ. And that, in turn, conveys some sense of the sacralities swirling around that garden, those courtyards, the library, the basilica, the monks' quarters, and the refectory of St. Catherine's. Of the evening shadows through the olive grove, and the sweet morning songs of the Sinai rosefinch (the national bird of Jordan).

Such sentiments are quintessentially logical, because of everything we imagine paradise to be: all those qualia and noumena remote from the here and now, but magnified in our thirst to escape imperfect realities. Of course, that same paradise is at odds with natural history. If evolution, as so many theologians after Darwin would argue, were pointed toward a purposeful, human odyssey, then why obsess so in the interest of renouncing our biological presence in favor of words targeting the unknown? Words amalgamated in the complex tapestry of edits and elusive source material that make up the *Codex Sinaiticus*? (Fig. 13.3).

Of particular interest to us is the Old Testament and its fragmented incarnation in the Codex. As many as 929 chapters were predominantly written during the post-exilic years of the fifth and fourth centuries BCE. Predictions (of a coming Messiah), ambiguities, and paradise lost. A God as real as any vertebrate, with mood swings, tremendous anger, captivating love, an ability to alter his/her/its thinking, to forgive, to promulgate premonitory warnings (Moses and the Commandments) (Fig. 13.4).

Browsing week after week through the monastery's library and viewing firsthand many of its icons were dazzling enough. I spent days soloing the headwalls of Sinai, and the countless other nearby granite peaks and escarpments, in both snow-covered and blisteringly hot weather. I held up in a cave, a deep crevice which could easily

---

[34] ibid., p. xxiii.

[35] See "The Cryptic African Wolf: *Canis aureus lupaster* Is Not a Golden Jackal and Is Not Endemic to Egypt," Eli Knispel Rueness,   Maria Gulbrandsen Asmyhr, Claudio Sillero-Zubiri,   David W. Macdonald, Afework Bekele,   Anagaw Atickem,   and Nils Chr. Stenseth, Thomas M. Gilbert, Editor, PLoS One. 2011; 6(1): e16385.Published online 2011 Jan 26, doi: https://doi.org/10.1371/journal.pone.0016385, PMCID: PMC3027653, PMID: 21298107, Accessed March 10, 2019.

**Fig. 13.4** Rare anonymous
seventeenth-century
Retablo Depicting Moses
at Mount Sinai. (Private
Collection). (Photo ©
M.C. Tobias)

have been inhabited during centuries past by any number of famed ascetics-turned-saints. The cool lambent nights were thick with the gorgeous communal howls of Egyptian jackals, taxa placed genetically within the "grey wolf species complex" as of 2011.[36] The hot mornings emerged slowly to the lovely cooing of the Sinai doves. All in retrospect so many years later couches memory in that enigmatic bundle which was the Codex Sinaiticus itself: 188 lines in four columns, on average, per leaf, until, some halfway through, turning to two much thicker columns. A mesmerizing tale told with uncials of Alexandrian text-type. Each of the quaternions comprising four conjugate leaves is signed in red ink by a scribe, of which there were probably four or five, as well as three groups of so-called "correctors." Nothing quite prepares the reader for the intensity of the manuscript, the density of its importunate style, and the obscurity of its provenance.

---

[36] See standard comparisons, see https://www.biblestudytools.com/micah/1-1-compare.html, Accessed March 13, 2019.

Nor can we assimilate the countless translations and editorial fixes of words and meanings.[37]

## 13.4  Structures of Biblical Consciousness

Dust, ash, earth; sea, water, ocean… Herein, the phenomenology of translations over edits, atop edits. Consider what one commentator/translator, Albert Pietersma, writes of various translations from the Septuagint of Psalms: "Understandably, it is especially idiomatic and figurative language that tends to suffer severely at the hands of a heavily word-based, interlinear mode of translation."[38] That "interlinear" comes from James Strong's 1890 *Exhaustive Concordance of the Bible* in which the syntactic and grammatical collations were keyed in to both the Greek, from the LXX (its normal abbreviation), or original Septuagint, and the original Hebrew, with some Aramaic. The 39 books of the Tanakh—the acronymic reference to the Old Testament's Five Books of Moses (the Torah), the Prophets, or Nevi'im, and the writings, or Ketuvim, which have collectively been translated into over 670 languages—represent legendary difficulties in terms of actually getting it right. This is a curious dilemma for what is the most widely read book in the world. And it must be remembered that the original Urtext (original) of the Old Testament is missing. Not one page of it has ever been found. Moreover, because the paleo-Hebrew-alphabet (known as Ashuri = Assyrian) is an abjad script (principally consonants), the terminologies do not fit conventional categories of translation, that is, literal, poetic, or generous. Interestingly, the earth is mentioned 1476 times in the Bible; and heaven, 570 times.

At least, for all of its paradoxical truancies, the search for paradise in Sinai mirrors human evolution: gaps and discrepancies in translation, tortured souls (a Christian portrayal of a horned, demonic Moses, inclement weathers, a Babel of vernaculars, fragments and more fragments; the beauty of its location, contested ideals, and a cumulative record of vanishing truths that remain with us. The Codex is a summary of every ecological and ascetic paradox stemming from prehistory up to and beyond year zero.

---

[37] See page 2 of 80, Psalms, To The Reader, http://ccat.sas.upenn.edu/nets/edition/24-ps-nets.pdf, 24-Ps-NETS-4.qxd, Accessed March 13, 2019; See http://ccat.sas.upenn.edu/nets/edition/. Accessed March 13, 2019.

[38] Cincinnati: Jennings & Graham; See the "Interlinear Bible," https://www.biblestudytools.com/interlinear-bible/. Accessed March 13, 2019.

# Chapter 14
# Russell's Paradox as Ecological Proxy

## 14.1 Surrogate Ecologies

Thomas Bayes,[1] son of a London Presbyterian minister, wrote three religious and mathematical treatises in his life that would change our concepts of *what is possible* forever more. The first, *"Divine Benevolence, or an Attempt to Prove That the Principal End of the Divine Providence and Government is the Happiness of His Creatures* (1731); a second defending Isaac Newton's concept of "fluxions" (instant or gradual change; variables of transition and derivatives of constants); and a third that would be published 2 years after he died, *"An Essay Towards Solving a Problem in the Doctrine of Chances.* Today, his ghostly presence haunts the mathematics and statistics of probability theory. Point estimates, within what Bayesian specialists refer to as a maximum a posteriori estimation, are coefficients within a purgatory between a predicted event and the power of belief in that event happening. Bayes' calculations are not interpretive. Rather, they outline the reality of the anthropic principle: cosmological calculations must somehow account for the perceptions and number crunching, not to mention the very ecological presence, of *Homo sapiens.* Our subjectivity is just one of the world's infinite objectivities. While it is still tenable to calculate a planet without humans, a post-human biodiversity implicates a fascinating matrix of ethical targets, each with probability distributions entirely relevant to choices made in the here and now. But every conception is a human conception. Every space a biased dimension (Fig. 14.1).

Even if we speculate upon the doomsday of doomsdays, the branch of mathematics dealing with faith is a *human* faith, a compilation of biases and a less than sanguine harbinger of what those fixed outcomes bode.

---

[1] Statistical Science 2004, Vol. 19, No. 1, 3–43 DOI 10.1214/088342304000000189 © Institute of Mathematical Statistics, 2004 "The Reverend Thomas Bayes, FRS: A Biography to Celebrate the Tercentenary of His Birth," by D. R. Bellhouse, http://biostat.jhsph.edu/courses/bio621/misc/bayesbiog.pdf, Accessed March 16, 2019.

© Springer Nature Switzerland AG 2021
M. C. Tobias, J. G. Morrison, *On the Nature of Ecological Paradox*,
https://doi.org/10.1007/978-3-030-64526-7_14

**Fig. 14.1** A glimpse at the Anthropocene: Downtown Old Delhi, India. (Photo © M.C. Tobias)

An example of that bias in the midst of human calculations comes from European histories of forest cover in what today would be France. Writes Carole L. Crumley, who speaks of a literal "attack by agriculturalists….A somewhat ephemeral assault on the woodlands of Europe was transformed into a war on the primary vegetative cover of the continent."[2] Her "evidence" for these changes in biodiversity encompasses climate change, iron mining, sheep grazing, changing population densities, even Neolithic "social relations," the analysis of pollen, and assessment of gravesites. Such determinations allow for some insight into ancient economic status and thus the political hierarchies driven by cultural differentiations; of privilege versus poverty; of forest management practices, if any; the evidence for "venerated species," of trade, "Roman exploitation," shipbuilding, and the chemistry of soils indicative of different regional occupations. Taken together, one can ascertain forms of pasturing and husbandry, as well as fundamental patterns of "agriculture, horticulture, and viticulture," and dramatic weather events, like flooding, and erosion.

The build-up over millennia of these anthropogenic presences accounts at once, in today's perspective, for a massive sylvan diminution. We who are alive in the present generation have inherited this sum of destructions. Whatever we say about it—"The commitment made by the first French Environmental Conference in 2012 'to make France a model biodiversity restoration country,'" for example—recognizes

---

[2] See *"Historical Ecology," Carole L. Crumley, in Regional Dynamics – Burgundian Landscapes in Historical Perspective*, Edited by Carole L. Crumley and William H. Marquardt, Academic Press, Inc., San Diego and London, 1987, p. 241.

that we are a member of that set, or class of distinctions that does not appreciably change from the Iron Age to the present.[3] The contrasts between the Burgundian Renaissance and its unabashed impact upon primary forest cover, and today's candid assessment by French authorities regarding the global sixth extinction spasm, and the modest tools the French can mobilize in defense of remaining ecological integrity both within and outside its political boundaries, are all part of the same subset.

What does that mean in terms of predicting coming years? Enter Bertrand Russell (1872–1970) and a moment in time, 1902, when he corresponded (now famously) with German philosopher Gottlob Frege (1848–1925).[4] Their friendship would converge upon the basis of Russell's curious predilection for a certain phenomenon he detected in logic, namely, that concerning "the set of all sets that are not members of themselves. Such a set appears to be a member of itself if and only if it is not a member of itself"[5] (Fig. 14.2).

## 14.2   The Confusion Borne of Trichotomy

Russell's approach to resolving this paradox (also implicating Frege's "trichotomy"—all numbers are positive, negative, or zero) has been clarified thus: "…we are confusing a description of sets of numbers with a description of sets of sets of numbers. So Russell introduced a hierarchy of objects: numbers, sets of numbers, sets of sets of numbers, etc. This system served as a vehicle for the first formalizations of the foundations of mathematics… Russell's paradox becomes: let $y = \{x: x$ is not in $x\}$, is $y$ in $y$?"[6]

That is one form of antinomy crucial to understanding the ecological paradox unleashed in the guise of humanity's own introverted use of biases to enact calculations that, in turn, build into theories by which we ascribe to the world

---

[3] https://www.diplomatie.gouv.fr/en/french-foreign-policy/sustainable-development-environment/french-policy-on-biodiversity/Accessed March 16, 2019

[4] See http://brianrabern.net/onewebmedia/FregeRussellCorr.pdf, March 16, 2019; See also "The Russell Paradox," in Gottlob Frege, The Basic Laws of Arithmetic, Berkeley: University of California Press, 1964, 127–143; abridged and repr. in A.D. Irvine, Bertrand Russell: Critical Assessments, vol. 2, New York and London: Routledge, 1999, 1–3; and "Paradoxes, Self-Reference and Truth in the 20th Century," in Dov M. Gabbay and John Woods (eds) (2009) Handbook of the History of Logic: Volume 5 – Logic From Russell to Church, Amsterdam: Elsevier/North Holland, 875–1013. See also, "A Guide to the Jean Van Heijenoort Papers, 1946-1988," Briscoe Center for American History, The University of Texas at Austin, https://legacy.lib.utexas.edu/taro/utcah/00245/cah-00245.html, and most importantly, his edited translations in, From Frege to Gödel: A Source Book in Mathematical Logic, 1879-1931, Harvard University Press, Cambridge, Mass., 1967

[5] See https://plato.stanford.edu/entries/russell-paradox/Copyright 2016 by Andrew David Irvine and Harry Deutsch.

[6] "What Is Russell's Paradox?" by John T. Baldwin and Olivier Lessmann, Scientific American, https://www.scientificamerican.com/article/what-is-russells-paradox/. Accessed, July 25, 2020.

**Fig. 14.2** Bertrand Russell. (Photo: Public Domain)

certain principles, all couched in ambiguous sets and subsets, ideas and ideals, each contingent upon the reality of human perception. Russell says he came upon this peculiar function of mathematical circularity while working on his *Principles of Mathematics*. Others, most notably logician, mathematician Ernst Friedrich Ferdinand Zermelo (1871–1953), would also formulate a similar mind experiment.

Moreover, as early as December 1873, the brilliant Russian-born mathematician Georg Cantor had written of the crisis of logic necessitating some heuristic revisioning of those differences between classes and sets, categories and types, semantics and true meaning, summary and schematic. He writes from Halle in southern Germany on December 9 of that year, "…the totality (x) cannot be correlated one-to-one with the totality (u); and I infer that there exist essential differences among totalities and value-sets that I was until recently unable to fathom…"[7]

Russell's great philosophical tease, which has (surprisingly) been lauded by both mathematicians and logicians, is quite a simple deduction, in fact; metaphysics has shown its extraordinary evolutionary implications by way of far less fanfare. Pre-Socratics Heraclitus and Parmenides both had grasped the essentials of the distinctions[8] and Aristotle[9] had fundamentally apprehended a through-story leading from

---

[7] See "The Nature of Infinity—and Beyond, An introduction to Georg Cantor and his transfinite paradise," by Jørgen Veisdal, https://medium.com/cantors-paradise/the-nature-of-infinity-and-beyond-a05c146df02c, December 17, 2018; Ann., 65 (1908), pp. 261–281.

[8] See "Stepping Into the Same River Every Week: Parmenides, Heraclitus, Chaos Theory, and the Nature of Change in Group Psychotherapy," Michael P. Frank, *Group,* Vol. 36, No. 2, Philosophy and Group Psychotherapy (SUMMER 2012), pp. 121–134.

[9] See "How Bertrand Russell discovered his paradox,"Grattan-Guinness, Historia Mathematica, Volume 5, Issue 2, May 1978, Pages 127–137, https://doi.org/10.1016/0315-0860(78)90046-0, Science Direct, Elsevier, https://www.sciencedirect.com/science/article/pii/0315086078900460, Accessed March 16, 2019.

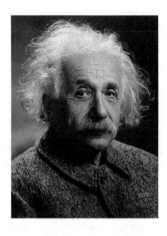

**Fig. 14.3** Albert Einstein in 1947. (Photo: Public Domain)

teleology and the dialectics of human purpose, to the basics of a biological meta-physics that today is enthusiastically debated like never before. It encompasses the nature of evolutionary progress—so-called, animal behavior—and plasticity within the realms of cellular discourse.[10] Most crucially, the breakdown into logical types and accompanying mathematical notations initiate suggestion that *Homo sapiens*, certainly in recent generations, has as a species become so irrationally destructive as to render a most fitting paradoxical category in which they are the sole occupants. The problem with this logical type is that it does not obey the population that solely inhabits its mathematical space. That dimension is the entire earth and the type goes to a single word, the Anthropocene (Fig. 14.3).

## 14.3 EPR Paradox

In any sobering ecological context, transposing one paradox into another almost seems to be part of the evolution of human consciousness. No other set of earthly circumstances has ever so blatantly elicited a confirmation of this hypothesis, than that presented in a short paper published in 1935 by Albert Einstein, Boris Podolsky, and Nathan Rosen, known as the Einstein-Podolsky-Rosen (EPR) paradox. The cur-rent extinction event makes manifest the logic of combinatorial antinomies, posited by a lifeform that is destroying itself.[11] This condition underscores an irrefutable

---

[10] See Hladký, V. & Havlíček, J. (2013). Was Tinbergen an Aristotelian? Comparison of Tinbergen's Four Whys and Aristotle's Four Causes. Human Ethology Bulletin, 28(4), 3–11.

[11] See "Can Quantum Mechanical Description of Physical Reality Be Considered Complete?" by Albert Einstein, Boris Podolsky and Nathan Rosen, Physical Review. 47 (10): 777–780, Bibcode: 1935PhRv...47..777E; doi:10.1103/PhysRev.47.777. See also John S. Bell, "On the Einstein Podolsky Rosen Paradeox," Physics, 1 (3): 195–200, doi:10.1103/PhysicsPhysiqueFizika.1.195, November 4, 1964. See also, *Bohm, D. (1952). "A Suggested Interpretation of the Quantum Theory in Terms of "Hidden" Variables. I". Physical Review. 85 (2): 166.* Bibcode:1952PhRv...85 ..166B. doi:https://doi.org/10.1103/PhysRev.85.166

ecological truth of entanglement. It functions both as metaphor—symptomatic of entanglement at the atomic (spinning) particles and wave function levels, beyond the reach of irreversibilities in classical physics—and as the mirroring of insights deriving from modern quantum mechanics, and the so-called Many Worlds Interpretation that transcends wave collapse in this or any other universal dimension.

The specific conclusion reached by Einstein, Podolsky, and Rosen has proved classic: "While we have thus shown that the wave function does not provide a complete description of the physical reality, we left open the question of whether or not such a description exists. We believe, however, that such a theory is possible."[12]

Ecological, heterogeneous, exogenous causes and obvious consequences of an extinction event over space and time show this probability factor, or function, to be a near 100% certainty, to the extent that extinctions are unambiguous after a conventional century in human time (though not other specie's time). The paradox hovers over two particles, an intervening wave separating local and nonlocal physical space, and the impact of measurements of any of these entities in isolation from the others. While the Theory of Special Relativity[13] may forbid the spread of information faster than the speed of light between these components, in the ecological record of life on earth, such speed and its all-encompassing theories involving entanglement are quite compellingly further adduced. The Anthropocene entities are affected in a finite realm of definable ecosystems; the time, space, and nature of the information conveyed present no obstacles. It is real, no matter the nonlocal variables. The only uncertainty principles are those concerning biological resiliency at all scales.

---

[12] ibid. Einstein, Podolsky and Rosen, 1935.

[13] For an excellent description of the EPR paradox as a thought experiment (gedankenexperiment) and its relation to Einstein's Theory of Relativity, see"EPR Paradox in Physics," by Andrew Zimmerman Jones, ThoughtCo., July 3, 2019, https://www.thoughtco.com/epr-paradox-in-physics-2699186, Accessed July 27, 2020.

# Chapter 15
# The Evolution of Innocence and Strategy

## 15.1 Ethics and the Biological

The minimal water vapor, the nitrogen, the argon—such gasses our minds tend to ignore. But the 20.95% normal oxygen level in today's atmosphere is our conditioned lifeline. Our biological opposites, in this realm, were earth's early obligate anaerobes, toxic by our living standards. Fermentation and respiration were restricted under aboriginal conditions to the aerobes. The culprits of this early global extinction event, the ones that produced the oxygen, represent the most classic example of storytelling in the annals of biochemistry; where photosynthetic cyanobacteria, the "culprits," are today recognized for engendering all of life's preconditions, first in the oceans, and 250 million years later, on land.[1] This Great Oxidation Event ("GOE") should remind us how sensitivities and tolerance levels have meaning on this planet; but less so in terms of mathematical values that are, after all, entirely the byproduct of human imagination.[2] What may be life-sustaining for one, kills the other. If ever there were two fundamental stanchions, dyads, bedpartners utterly at odds, it is within this framework that our narrative seeks out mechanisms like evolution to ease the courting procedures, and invite slow transitions to lessen the trauma. Generally speaking, evolution under unforced conditions renders less draconian these early biological conversations; until evolution itself dwindles into insignificance.

---

[1] See Mojtaba Fakhraee, Sean A. Crowe, Sergei Katsev. Sedimentary sulfur isotopes and Neoarchean ocean oxygenation. Science Advances, 2018; 4 (1): e1701835 DOI: https://doi. org/10.1126/sciadv.1701835, Accessed March 18, 2019.

[2] See Sosa Torres, Martha E.; Saucedo-Vázquez, Juan P.; Kroneck, Peter M.H. (2015). "Chapter 1, Section 2 "The rise of dioxygen in the atmosphere"". In Peter M.H. Kroneck and Martha E. Sosa Torres. *Sustaining Life on Planet Earth: Metalloenzymes Mastering Dioxygen and Other Chewy Gases.* Metal Ions in Life Sciences. 15. Springer. pp. 1–12. doi:https://doi.org/10.1007/978-3-319-12415-5_1

© Springer Nature Switzerland AG 2021                                                      135
M. C. Tobias, J. G. Morrison, *On the Nature of Ecological Paradox*,
https://doi.org/10.1007/978-3-030-64526-7_15

One constellation of examples has been beautifully laid out in that great summary, *Insects and Flowers*, by Friedrich G. Barth.[3] Aside from reminding us that the Biblical manna from heaven is actually the excretion ("chemically modified form") of various "aphids and scale insects,"[4] Barth's reflections upon "directed evolution"[5] and "selection pressure"[6] manifest striking industrial-like metaphors in the guise of "guide marks," "pollen models," and other strategies employed by flower species as diverse as "Foxglove," "Alpine butterwort," "Common toadflax," and Broomrape"[7] to better ensure the phenomenal hunter/gatherer successes of bumblebees, as one case. Barth suggests a colony of "60,000 workers" and brings to bear data on the "0.0021 g per animal per flight" of sugar; subtracting the "0.027 g brought back," the net gains, and the extrapolation to an astonishing "18 liters of honey" production per day by an entire colony.[8] Quite an impressive manufacturing success story that begs the question of where it is all leading to the Jewish Promised Land of Canaan, "flowing with milk and honey"; and a variety of commentaries on sacred wild bees, and deification of nature in Babylonian and Egyptian texts.[9]

Whether cyanobacteria, bees, or beavers, all such industry courts an obviously devastating dichotomy when taken to human levels. Its causes and consequences have been graphically assessed, particularly by such historians as Clarence J. Glacken. When discussing Renaissance theologians like John Ray on the topic of natural destruction, Glacken makes clear that Ray (and so many of his contemporaries) "rejected the belief in the exhaustion and dissolution of the world on philosophic, religious, and scientific grounds."[10] There was no reason, it would seem, to believe that the robust action of pollinators and the animated presence of livestock under human dominion were not eternal facts of life. Yet, in our time, "A total shutdown of the thermohaline circulation is considered extremely unlikely in the coming century," wrote Elizabeth Kolbert in 2006. Adding, "But, if the Greenland ice sheet were to start to disintegrate, the possibility of such a shutdown could not be ruled out"[11] (Fig. 15.1).

---

[3] *Insects and Flowers – The Biology of a Partnership*, Translated by M. A. Biederman-Thorson, Princeton University Press, Princeton, NJ, 1985.

[4] ibid., p. 83.

[5] ibid., p. 263.

[6] ibid., p. 265.

[7] ibid., p. 125.

[8] ibid., pp. 245–246.

[9] See Exodus 3.8, *The Holy Bible Containing The Old And New Testaments*, Authorized King James Version, The Christian Science Publishing Society, Boston, Mass., (*See "Why Milk and Honey?" by Jonathan Cohen, Stony Brook University, n.d., p.83; also see https://www.uhmc.sunysb.edu/surgery/m&h.html, Accessed March 18, 2019.

[10] *Traces on the Rhodian Shore – Nature And Culture In Western Thought From Ancient Times To The End Of The Eighteenth Century*, University of California Press, Berkeley and Los Angeles, 1967, p. 415.

[11] *Field Notes from a Catastrophe – Man, Nature, And Climate Change*, by Elizabeth Kolbert, Bloomsbury Publishing, New York, p. 57.

**Fig. 15.1** Greenland melting. (Photo © M.C. Tobias)

How quickly our global problematique is changing! Or, as the poet Christopher Fry once described, how in spite of all the time that has elapsed, the most amazing aspect of a human life is that "I am aware of being alive, *now*...."[12] As William Faulkner wrote in his majestic and ironic work, *Sanctuary*, "The insects had fallen to a low monotonous pitch, everywhere, nowhere, spent, as though the sound were the chemical agony of a world left stark and dying above the tide-edge of the fluid in which it lived and breathed. The moon stood overhead, but without light; the earth lay beneath, without darkness."[13] The groundwork for such sentiments had been laid out clearly by John Ruskin in his unforgettable *Seven Lamps of Architecture*, and we quote at length: "... I do not know anything more oppressive, when the mind is once awakened to its characteristics, than the aspect of a dead architecture... to see the shell of the living creature in its adult form, when its colors are faded, and its inhabitant perished – this is a sight more humiliating, more melancholy, than the vanishing of all knowledge, and in the return to confessed and helpless infancy."[14]

---

[12] *Death Is A Kind Of Love*, by Christopher Fry, Drawings By Charles E. Wadsworth, The Tidal Press, Cranberry Isles, Maine, 1979, p. 31 – though unnumbered.

[13] *Sanctuary*, William Faulkner, With A New Introduction By The Author, The Modern Library, New York, 1931 and 1932, p. 267.

[14] John Ruskin, *Seven Lamps Of Architecture*, Chapter Five, "The Lamp of Life," Thomas Y. Crowell & Company, Publishers, New York, 1880, p. 199.

Even the most fundamental alleged Law of Nature, gravity, knows no eternity; is compromised to the extent that it is exploited by every movement of every object, organic or inorganic. This leaves our histories and futures not only checkered, but also lacking all guidance other than those incremental convictions we might muster during our ephemeral stays in this human form. And to what end is more a matter of ethics than quantum mechanics or calculus.

## 15.2   On a Tightrope Between Innocence and Experience

Ethics in human context does not easily admit to innocence, and is naturally frustrated by the physics of incrementality. Rather, it literally gravitates toward strategy. Evolution is instructive in this matter. Our dispositions are ecological to the core, no matter what we say or do. They function according to the logic of honeycombs, the language of the dance, a proboscis and energy balance, the discrimination of odors, and the so-called "constancy" in flowering plants. Our memory, polarization filters, ability to recognize patterns, crowd paths, temperature regulation, and sexual attractants are no different than the evolution of a stamen: it is us by so many labyrinthine twists.[15]

As Arthur E. Murphy has reflected, in discussion of John Dewey's *Experience and Nature*,[16] Dewey frequently described how "the pervasive traits of human experience are traits of nature itself and can be used in metaphysics as a guide to its character." Indeed, Dewey wrote, "Man fears because he exists in a fearful, an awful world. The *world* is precarious and perilous."[17]

We speak constantly of that gap between innocence and experience; between naivety and strategy. If our bee-like industry is our downfall, which is a correct deduction by all indications (we do not so much pollinate as de-pollinate at industrial scale), then the chasm separating us from all others is wholly inexplicable. Unnatural selection, as Lois Wingerson has written of it,[18] defines our profound dilemma. All the beauty by which we describe our predicaments and hopes fail to arrest the ineluctable turning points of our brief span.

At the same time, the totality of our cognition horizons inculcates a reasoning faculty that, despite all malevolence to the contrary, situates the notion of *solutions* at the level of applied ethics; stops at nothing to embrace millions of like-minded

---

[15] ibid, Selected from Barth's Index, as an example of the comprehensive intimations of what, in humans is co-evolved as a result of such primal relations as those between insects and plants.

[16] Dover Publications, New York, 1958.

[17] "Dewey's Epistemology And Metaphysics," in *The Philosophy of John Dewey*, The Library of Living Philosophers, Edited by Paul Arthur Schilpp and Lewis Edwin Hahn, Southern Illinois University, Carbondale, 3rd Edition, 1989, p. 219.

[18] *Unnatural Selection – The Promise And The Power Of Human Gene Research*, Bantam Books, New York, 1998.

individuals in causes, protests, and emphatic changes in attitude and behavior. All of these coefficients represent a most peculiar circumstance that coincides with real-time evolutionary potential.[19] However, we shall investigate the ambiguous under-pinnings of *potential* later on.

---

[19] For a discussion of real-time evolution, see the Author's work, *The Hypothetical Species: Variables of Human Evolution*, Springer New York, 2019.

# Chapter 16
# Tatters and Poignancies

## 16.1   Epiphanies in Rome

Supposedly, here at the lavish basilica of Santa Maria on Rome's Via Lata, Saints John, Paul, Peter, and Luke each hid themselves for years at a time under the crypt. Of course, legends like this are founded under every piece of excavation and re-affirmation within the Eternal City. Inside Santa Maria, there is an exhibition on the crucifixion. Some things remain the same (Fig. 16.1).

Back out into the glare of day, the street is noisy with cars and pedestrians. We wander across, into the Galleria Doria Pamphilj, a palace that Nathaniel Hawthorne called "the most splendid in Rome" in his *Italian Notebooks* of 1858. The Pamphilj lacks the grand serenity, or views of Rome's Villa Medici Palace. Instead, it is centered along the mob-scene that is Via del Corso, one of Rome's busiest streets, leading directly to the Piazza Venezia abase the Capitoline Hill, with the most imposing building in all of the city, that blinding white Monument ("wedding cake") to the first king of a unified Italy, Vittorio Emanuele II.

Inside the block-long series of the Galleria's enormous real estate are the residencies of the living heirs. One walks to the modest little ticket booth (this is not the Louvre) past their cars adjoining the elegant courtyard garden. Everything is elegant, of course, including all of the original deteriorating silk on many of the furnishings. The guardian angels of all this magnificent foment are not your normal armed guards. A stunning informality watches over this lair; and "lovers" are now and then invited in at a discount, as are all youths. But it is not free, as in the case of the Prado at present. Such freedom can be crisscrossed with aesthetic twists, beyond the scope of Diego Velázquez' "Las Meninas." Ferdinand VII's personal toilet greets museum-goers to the Prado; the real toilet, not some replica, like Marcel Duchamp's "Fountain," of which 14 were made and signed, and still incite ironic commentary.[1]

---

[1] Signed and dated, "R. Mutt, 1917" by Duchamp in 1950. See "A 7-Hour, 6-Mile, Round-the-Museum Tour of the Prado," by Andrew Ferren, The New York Times, March 18, 2019; See also, "Marcel Duchamp and the Fountain Scandal," Philadelphia Museum of Art, March 27, 2017, https://press.philamuseum.org/marcel-duchamp-and-the-fountain-scandal. Accessed March 19, 2019.

© Springer Nature Switzerland AG 2021                                                       141
M. C. Tobias, J. G. Morrison, *On the Nature of Ecological Paradox*,
https://doi.org/10.1007/978-3-030-64526-7_16

**Fig. 16.1**  Inside the Doria Pamphilj. (Photo © M.C. Tobias)

But here at the Galleria, the art of the Renaissance admits to different concerns, as well as another great Velázquez. In a private alcove is the painter's entrancing portrait of Giovanni Battista Pamphilj (1574–1665), owner occupant of the palace, who would succeed Pope Urban VIII with the official title Pope Innocent X in 1644 and 7 years into his tenure, just South at the Vatican, would turn his home into the Galleria Doria it remains today.[2] Ten primary rooms, four wings,[3] and an inventory

[2] Amongst the legion of tourist guides that lavish details upon such matters, see, for example, http://walksinsideitaly.com/walk/2018/07/21/palazzo-doria-pamphilj-a-private-mansion-in-the-heart-of-rome. Accessed March 18, 2019.

[3] See http://www.doriapamphilj.it/roma/en/la-galleria-doria-pamphilj. Accessed March 18, 2019.

of 848 works of art.[4] And therein lies a Renaissance version of St. Catherine's Monastery in the heart of Western Civilization, across the street not only from the ancient basilica, but also, this being Rom, from men and women's clothing stores, a few minute's stroll from the Trevi Fountain, an Irish Pub (the Scholars Lounge), cappuccino bars and pizzerias, the Jewish Monti District, and the Viale di Fori Imperiali, which leads one past the Forum to the entrance of the Colosseum.

Curatorial accomplices are densely packed upon the walls, one above the other: "The Triumph of Art."[5] Genes for certain types of landscapes, which populate in abundance; genotypes expressing all those causes and consequences of choice, both by artists and patrons. Subject matter that is artistically and scientifically as paradoxical as can ever be described. Human nature floods the palace in a wanton self-reference, gorgeously choreographed, to what ends we cannot say. Individuals and aggregates. Dynamics of the biosphere that confuse and perpetrate. Like all of the greatest museums, the range of historic and ecological evocations is unending, by turns euphoric and tragic.[6]

Distinct paradise scenes by Jan Brueghel the Elder and Jacopo Da Ponte (Bassano); perfect unions of humanity and her precise vision of an organized nature, half wild, half human, as in a multiplicity of scattered sheets of painted copper and a joyful cascade of canvases and panels each lavishly framed. More than two dozen landscape scenes by Dughet Gaspard ("Paesaggio con cascate," etc.), scores of others by Dürer, Filippo Napoletano, Weenix Jan Baptist, Pasquale Chiesa, Herman van Swanevelt, Philips Wouverman, Giovannini Giovan Battista, Guiseppe Rosa, Mola Pier Francesco, and Jan Gossaert ll Mabuse. A small late eighteenth-century watercolor by Seganti Antonio, the view of the lake from the Villa Pamphilj—today, Rome's largest public park—four kilometers from the Galleria. This density of landscaping provides proof positives for every conviction and desire. Since each frame purports to a regulatory hurdle, the imagination is free to wander inside, of course, but that is the extent of it. These museums of grandeur are sending us perfectly mixed reminders. Enough heartbreak to overwhelm any frame, and all the pretty pigments, and brushes fine aspiring to discover calm and succor beauty.

Amid the myriad faces staring back at us, Sebastiano del Piombo's portrait of Andrea Doria, painted between 1520 and 1526, cannot but herald imagery of this infernal dichotomy we cling to from the first seconds of birth, this biosphere where neither lament nor odes of joy can scarcely part a chilling mist. The great and long-lived Admiral of the Genoan Republic, living into his mid-90s, and the ocean liner named after him that went down in 1956, killing 46 on board. So many equivalencies

---

[4] See *Collezione Doria Pamphilj – Catalogo generale dei dipinti*, by Andrea G. De Marchi, SilvanaEditoriale, Editorial Director Dario Cimorelli, Art Director Giacomo Merli, Milano 2016, pp. 450–455.

[5] See main Galleria website, http://www.doriapamphilj.it/roma/en/la-galleria-doria-pamphilj. Accessed March 20, 2019.

[6] See all paintings listed as on display: http://www.doriapamphilj.it/roma/en/tutte-le-opere-esposte-doria-pamphilj. Accessed March 20, 2019.

and lack of them.[7] Heads of John the Baptist floating before us, serene, without a sound. Still lives, with pernici (partridges) all dead, as their genre evinces (Natura morta); hunting scenes, crucifixions, prayer, violence meted out in every guise of the human story; besides the classic elements, also meted out, by Annibale Carracci (1560–1609) in his "Paesaggio con la fuga in Egitto" ("Landscape With the Flight into Egypt," FC236), one of the Galleria's most famous works, though the only flight is in the sky, a flock above, not terrestrially; and the landscape is anything but Egyptian, whose New Testament iconography is borne out by two vaguely evoked camels on a far ridge.

Meanwhile, the burro is lazily burdened only by a piece of clothing on its back, a halter and slight incline above the water. All is of a languorous pace. Commissioned for the chapel in the palace in 1603 by its owner, Cardinal Pietro Aldobrandini (1571–1621), this painting, one of six lunettes framed in their lunar hideaways, elicited in art historian Kenneth Clark the idea that Carracci had formulated the perfect, if academic, landscape;[7] never mind that it was intended to depict that period in the life of Mary when her very existence, and that of Joseph and of Christ, depended on quickly escaping from the wrath of Herod in Palestine.[8]

## 16.2  Museological Nostalgias

These are the conflicts and contradictions that have attended upon the impulse to share the tatters and poignancies of private lives with the public. Pope Julius II had, well over a century before Pope Innocent X, opened a sculpture gallery at the Vatican. And before that, Romans enjoyed Pope Sixtus IV's own sculptures. From Schloss Ambras in Innsbruck to the city of Basel, the example of meaningful collections, tested on the diffuse wants of public interest and wide eyes examining the past, had been a fixation. The obsessed were given to expeditions and meticulous collecting, in the style of Bologna's great naturalist, Ulisse Aldrovandi (1522–1605)—the Aldrovandi family was to the Emilia-Romana, what the Doria Pamphilj Landi was, and remains, for Genoa, Rome, and the Vatican.[9] Ulisse himself is thought of by many as the European father of natural history, 150 years prior to Linnaeus.

When the possession of explorer/collector John Tradescant the Younger of Kent's abundant cabinets of curiosity fell into the hands of the eccentric polymath, Elias Ashmole (1617–1692), Britain's tradition of open university museum holdings was also assured, in the formation of the Ashmolean Museum at the University of

---

[7] Goldstein, Richard, *Desperate Hours: The Epic Rescue of the Andrea Doria*. John Wiley & Sons, New York, 2003.

[8] *See Landscape into Art*, by Kenneth Clark, John Murray, London, 1949, Chapter Four "Ideal Landscape," pp. 54–73.

[9] See, for example, https://opusdei.org/en-us/article/life-of-mary-x-flight-into-egypt. Accessed March 20, 2019.

Oxford, which received and established its bequest in 1677. It was that tradition that later inspired the Danish botanist Nathaniel Wallich (1786–1854) to help establish the oldest museum in Asia in 1814, the Oriental Museum of the Asiatic Society, later known as the Indian Museum of Calcutta.[10] He also became a long-lasting superintendent of that city's first botanic gardens. Other public floral spaces, in Philadelphia, and a medicinal garden in New York, had already opened their doors.[11]

The brief history of museum culture gives us the zeal of not quite regret but concern. Some archaeologists hold that the oldest-known museum remains in the Dhi Qar Governorate of modern-day Iraq, dating to 530 BCE, to the last of Babylonia, the Ennigaldi-Nanna Museum, where objects, such as clay tablets, are described in three languages.[12]

Every museum-like revelation, of connections there in front of us that invite contemplation, should prepare us in a manner not unlike that deeply telling conjecture conveyed by Elisabeth Tova Bailey in her compelling speculations on the longevity of snails. She wonders whether their legendary pertinacity would allow them to survive an Ice Age, and asks malacologist Tim Pearce who is not entirely pessimistic on the subject. But when Bailey envisions the snails trying to outpace the advancing rivers of ice, she clearly *feels* the fast-arriving reach of the cold air, the weather turning into storm semipermanently, triggering who knows which primeval snail survival modes, as they bury themselves and prepare for the long hibernation. "I thought of a tiny snail with a glacier bearing down on it," she writes. But over the course of thousands—tens and hundreds of thousands—of years, their passionate life forces would be unlikely to match.[13]

And it will be of no small curiosity to visit Rome's eternal Galleria in a thousand years. Under water? Under ice? A desert? Redecorated in some form. Because the veins and muses of museums are the nuances of tumult that is ecological culture. No artist or history of historians can brave the biosphere without instructive consequences[14] (Fig. 16.2).

---

[10] See Paula Findlen, *Possessing Nature: Museums, Collecting, and Scientific Culture in Early Modern Italy*, University of California Press, Berkeley, CA, 1994.

[11] See the Kew Gardens site on Wallich at, https://web.archive.org/web/20120112235432/http://www.kew.org/collections/wallich/index.htm, Accessed March 20, 2019.

[12] See *American Eden – David Hosack, Botany, and Medicine in the Garden of the Early Republic*, by Victoria Johnson, Liveright Publishing Corporation, a Division of W. W. Norton & Company, New York, 2018.

[13] See Museums.EU, http://museums.eu/highlight/details/105317/the-worlds-oldest-museums, Accessed March 20, 2019.

[14] See Bailey's *The Sound of a Wild Snail Eating*, Algonquin Books Of Chapel Hill, North Carolina, 2010, p. 108.

**Fig. 16.2** Outdoor courtyard at Doria Pamphilj. (Photo © J.G. Morrison)

# Chapter 17
# The Echoes of Malhazine

## 17.1 Time-Out in Africa

It was the discovery a decade ago in a limestone cave near Mozambique's Lake Niassa that researchers found the earliest evidence yet for critical human inclusions of cereal grains in their diets, dating back some 100,000 years.[1] Much as Morocco, Kenya, Ethiopia, and South Africa have dominated most important discoveries pertaining to the earliest *Homo sapiens*, Mozambique has very much entered that geography of paleontological epiphany. Research just since 2014 has revealed some "200 new Stone Age surface sites"[2] (Fig. 17.1).

Forgetting other crucial turning points in the brief human timeline, to date—such as the earliest jewelry in the form of beads from approximately 110,000 years ago; or the footprints of three humans descending a volcano in today's Italy circa 325,000 BCE[3]—Mozambique entails an intertwining of all that is paradoxical in human nature with a rash of ecological and political fall-out, consequences of recent tragedy stemming from so much late twentieth century political tumult and not a little madness.

The end of the War of Independence in Mozambique against the Portuguese (1964–1975) was followed by a brutal Civil War (1976–1992) between two political foes, Frelimo (Frente de Libertação de Moçambique) and Renamo (Resistência Nacional Moçambicana) that left most areas of abundant, indigenous wildlife utterly

---

[1] See Julio Mercader et al. Mozambican grass seed consumption during the Middle Stone Age. Science, December 18, 2009.

[2] See, "Middle Stone Age Technologies in Mozambique: A Preliminary Study of the Niassa and Massingir Regions," in Journal of African Archaeology, Brill, Authors: Nuno Bicho, João Cascalheira, Jonathan Haws and Célia Gonçalves, Online Publication Date: Aug 2018, Volume 16: Issue 1, DOI: https://doi.org/10.1163/21915784-20180006, https://brill.com/abstract/journals/jaa/16/1/article-p60_3.xml, Accessed March 24, 2019.

[3] See "Timeline: Human Evolution," by John Pickrell, New Scientist, September 4, 2006, https://www.newscientist.com/article/dn9989-timeline-human-evolution/. Accessed March 24th, 2019.

© Springer Nature Switzerland AG 2021
M. C. Tobias, J. G. Morrison, *On the Nature of Ecological Paradox*,
https://doi.org/10.1007/978-3-030-64526-7_17

**Fig. 17.1**   School outing at the Museu De Historia Natural de Maputo. (Photo © M.C. Tobias)

devastated. Little wonder that the word gorongosa, in local languages refers to a "place of danger."[4] As David Quammen writes, "Back in the early 1990s, after fifteen years of civil war, with two armies treating Gorongosa as a battlefield—and killing its wildlife for meat to feed soldiers and for ivory to buy arms—the place was in wreckage."[5]

Then the Gregory Carr Foundation stepped in, devoted to ensuring that Gorongosa National Park would become a jewel of globally significant conservation biology for Mozambique. Indeed, in 2018, not a single snare was found in the park. It had become not only a lifeline for native and much endemic wildlife, but a parallel oasis for surrounding human locals. Yet, biogeographical size matters for the sustained populations of diverse species into the future. Gorongosa is modest in proportions, at 1456 mi$^2$; just slightly larger than Yosemite National Park at 1169 mi$^2$. By comparison, 45 other national parks across the African continent are larger, beginning with Namibia's 32,818 mi$^2$ Mudumu National Park, Tanzania's 21,235 mi$^2$ Selous Game Reserve, and Botswana's 20,077 mi$^2$ Central Kalahari Game Reserve. A small national park translates at once into the problem of habitat constraints for large

[4]See *A Window On Eternity – A Biologist's Walk Through Gorongosa National Park*, by E. O. Wilson, Photographs by Piotr Naskrecki, Simon & Schuster, New York, N.Y., 2014, p. 6.

[5]"Devastated by war, this African park's wildlife is now thriving," by David Quammen, National Geographic, December 13, 2018.

vertebrates. With Gorongosa and the nation's six other national parks, Mozambique remains far below—in physical extant and connecting corridor options—many other smaller countries in Africa.

Moreover, at 309,496 m$^2$, Mozambique is a third the size of Africa's largest nation, Algeria. The country's fortunes are largely dependent—not on a Kenya's ecotourism revenues, or not yet—upon small-farming production, an economy essentially beholden to agricultural subsistence; coal and mineral sands, and robust predictions for coming exports of liquefied natural gas.[6]

Overall size and demographics loom large. There are countervailing forces against conservation biology in Mozambique, a country that is economically 16th out of 54 African countries, and 25th in terms of national GDP within the world's second largest continent. In as much as Africa is rightly hailed as the last best hope, allegedly for large megafauna such as lions and elephants, hippos, and giraffes, Mozambique indicators are critical bellwethers of the Continent-wide outlook for wildlife. The population of Mozambique will likely become 40% urban by 2040,[7] hopefully, with a continuing stable government.

But most notably, as of 2017, Mozambique was considered to be the second poorest nation on earth, with a per capita income of US$429, just after the poorest, the Democratic Republic of the Congo (DRC), at US$439.00.[8] Ironically, it was in the far-east of the DRC that Africa's first national park was created, Virunga in 1925.[9] Three hundred and ten miles across the Mozambique Channel lies Madagascar, 227,800 mi$^2$ with 20 national parks, and a per capita income of US$1477.00 making it the 10th poorest country in the world.

The idea—advanced by most conservationists who look at any nation where human poverty surrounds protected areas—was to engender preconditions within and outside Mozambique's Gorongosa that made it not only a refuge for the megafauna of East Africa, but also a human rights refuge. This formula, as Jane Goodall early on described at Tanzania's Gombe Stream National Park (one of the smallest in Africa, at 13.5 m$^2$), turns out to be the only pan-rational concept that can work. In light of the estiated one million people+ who had been killed during the decades' long conflicts and no Crimes Against Humanity charges ever levied because of a nation-wide and unconditional amnesty law enacted throughout the Civil War, it would behoove conservationists to bring some semblance of ecological sanity to the troubled lands.

"Nation will not lift up sword against nation, And never again will they learn war" Isaiah, 2:4, repeated in Micah 4:3. Such was the case with Gorongosa National Park in northwestern Mozambique.

---

[6] See "Mozambique's Economic Outlook – Governance challenges holding back economic potential," by Deloitte Touche Tohmatsu Limited, December 2016, https://www2.deloitte.com/content/dam/Deloitte/za/Documents/africa/ZA_Mozambique%20country_report_25012017.pdf, Accessed March 24, 2019.

[7] ibid., Deloitte.

[8] See   https://www.focus-economics.com/blog/the-poorest-countries-in-the-world,   Accessed March 24, 2019.

[9] See https://virunga.org/about.

Proactive strategies called for to deal with ever-present droughts and floods is the same eco-dynamic compact that is Mozambique's ecological struggle.[10] Preparatory methodologies for mitigating extremes of both weather and climate are largely absent, or out of focus, in the country. Its political hierarchy harbors the same misperceptions about the severity of climate change and what to do about it as most other nations. Mozambique's bottom-line poverty does not realistically make for those core societal values (ecological rationality) that would be necessary to resolve a populous completely out of touch with even mid-term survival (e.g., the next two generations). Hence, a lack of the very grasp, let alone coordinated planning, of biodiversity loss; notwithstanding the euphoria expressed by those who come to parks like Gorongosa and can take legitimate pride in megafaunal numeric gains.

In those nations recovering from horrific internecine conflict, like Mozambique—which veered toward communism after the exile of most of its remaining Portuguese citizens, only to throw off that imposed socialist yoke in favor of a democratic process that has obtained for decades—other unanswered problems persist. Particularly, the perceived notion of fairness in the distribution of wealth in human communities. What is fair and equitable?[11]

## 17.2   Reversing the Tides

Now consider the academic orientation that in large measure is likely to influence regional politics in a country like Mozambique, where everything relating to the natural world is trapped on a tightrope: resource extraction paradigms at one end of the spectrum, no-kill, no-harm at the other. In one recent graduate program study in environmental planning, this range of activist ethics was broken down into protracted units of ecological manipulation, formal and relentlessly tied to the abstracted language of human development seen in study after study, nation by nation: "…the proposed functional geoecological zoning regionally that determined suitability of use and preservation in total protection zone; plant extraction area; agricultural production area, family agro-ecological production area and environmental recovery zone…"[12] Such well-intended tactics, the stock and trade of environmental managers, tend to be far removed from the actual daily travails of people

---

[10] See "Floods And Droughts In Mozambique – The Paradoxical Need Of Strategies For Mitigation And Coping With Uncertainty," by A. I. Mondlane, DOI: 10.2495/RISK040321, WIT Transactions on Ecology and the Environment, Volume 77, 2004, WIT Press, https://www.witpress.com/elibrary/wit-transactions-on-ecology-and-the-environment/77/14312, Accessed March 24, 2019.

[11] See "The Role of Equality in Negotiation and Sustainable Peace," by Cecilia Albin and Daniel Druckman, Chapter 7, pp. 131–152, in Psychological Components of Sustainable Peace, edited by Peter T. Coleman and Morton Deutsch, Springer New York, 2012.

[12] See "Geoecological Analysis directed to the Environmental Management and Planning of Sofala province – Mozambique," by Júlio Acácio António Pacheco, Federal University of Ceara, Graduate Program in Development and Environment, 2014, Open Access, BDTD, http://bdtd.ibict.br/vufind/Record/UFC_7e3e54b88a9febad4f365fd22ba36aea, Accessed March 24, 2019.

and other species. When disasters strike, like cyclone Idai which made landfall at Mozambique's port city of Beira on March 14, 2019, it might as well have been an asteroid, in terms of Beira's population of half-million people thoroughly unprepared; and stranded in the post-cyclone's further inflictions of increased cholera and malaria cases, and a grinding to the halt of much infrastructure, despite international aid pouring in. Some 1.7 million people in Mozambique, and nearly another million people in neighboring Malawi and Zimbabwe, were hit by the hurricane-force winds and flooding.[13]

Across the continent, in Mali meanwhile, while UN officials were formally visiting Bamako, Mali's capital, to discuss the rising tide of violence, at least 134 Peuhl herders—elderly, children, women—were shot, burnt, and/or hacked to death with machetes by a Dogon hunting militia. Months before, "Human Rights Watch had warned that "militia killings of civilians in central and northern Mali [were] spiraling out of control."[14] The systemic human aggression had continued to arise over a slew of ecological issues—access to drinking water and grazing land, among them.

Such numerous shocks to the many systems escalate across all of Africa, in a far less than stable dynamic. Environmental flash points that might traditionally have persuaded individuals to simply tune out the broader risks and statistics have, instead, by their magnitude and multiplicity, encroached in a continuum of human/ cultural and ecological whiplash. No one escapes the fall-out. Nor are there any reliable revisionist histories to make troubles go away. And, obviously—and in spite of at least 7 million years of habitation by our kind/our hominid ancestors—the biosphere cannot continue to absorb all of the breaches.

Mozambique has more than 30 million residents. Her Total Fertility Rate ("TFR" as of 2016) was at "5.24 births per woman" or more than double the global average.[15] According to sobering projections, by 2065, Mozambique could be staring down the gunsights of 163 million inhabitants; or, were the nation to immediately institute significant Family Planning reforms, a population possibly slightly less than today.[16] Kenya, which is roughly a third smaller than Mozambique, hosts a stunning 24 national parks, 16 national reserves, and another six marine parks/ reserves. She is the eighth wealthiest nation in Africa. Kenya lived through Mozambique's numeric profile—total population size and TFR—in 1998 and 2008 respectively and today numbers 50 million.

[13] See "Cyclone Idai: How prepared was southern Africa?" BBC News, by Jack Goodman and Christopher Giles, March 24, 2019, https://www.bbc.com/news/world-africa-47639686, Accessed March 24, 2019.

[14] See "Death toll from central Mali massacre up to 134, says UN," https://www.pbs.org/newshour/ world/death-toll-from-central-mali-massacre-up-to-134-says-un, by Baba Ahmed and Krista Mahr, Accessed March 24, 2019.

[15] See World Bank data, https://data.worldbank.org/indicator/SP.DYN.TFRT.IN, Accessed March 24, 2019.

[16] See http://populationcommunication.com/wp-content/uploads/2017/05/Mozambique-Legal-Length-Redux.pdf, Accessed March 28, 2019.

Juggling the deep story here comes down to a combination of demographic/cultural (tribal and religious) pressures upon women, and any government's struggles to balance, if at all, questions of national debt, per capita income, access to a sustainable quality of life (incorporating natural capital depreciations and all those economic externalities that typically involved non-renewable resource extraction) distributed across the entire population of the country, as well as equality between regional and state sovereignties. Such conjectures were posed by John Locke in his *Second Treatise on Government* (1689) which—upon the author's 20 years of flipping positions—arrived at a conception of natural rights and the power of human conscience.[17]

Such rights, conceived earlier by Aristotle, but widely disbanded throughout the Renaissance, were not burdened by the conceptual and physical burdens of a human population explosion that continues to rally the most inclement and pessimistic factors to which human communities are heir, in our times.

Take a measure of all such communities in the twenty-first century and we find the aforecited calming words of Isaiah and Micah incompatible with the implacable context of the Anthropocene. Indeed, Africa's biodiversity looms large in the crosshairs of that desperate effort currently being finalized by some "500 experts in 50 countries" under the banner of the Intergovernmental Science-Policy Platform On Biodiversity and Ecosystem Services (IPBES). At present, the publication papers number over 8000 pages, intimating the true extent of the Sixth Extinction Spasm[18] (Fig. 17.2).

## 17.3   The Search for a Sanctuary Amid Conurbations

All of these numbers swirl in and out of public policy, applied ethics, convictions that might gain sufficient traction to urge our species forward in a positive manner. But down on the ground, trying to cope with these convolutions, the realities are grey. They certainly weighed heavily upon my own mind as I (mt) bushwacked through the approximately 25 square miles of wildland in a suburb of Mozambique's capital, Maputo, with its nearly three million residents. I was followed by machine-gun and machete toting military trackers, clearing dense thickets, avoiding enormous wasp nests, and trying to decipher decaying fuses or rusted metal bombs that one best not step on. Those wilds were the walled-in remains of what had been a weapons depot for the Mozambique Defense Armed Forces, and where thousands of abandoned munitions still lay scattered.

---

[17] See *Two Treatises of Government (or Two Treatises of Government: In the Former, The False Principles, and Foundation of Sir Robert Filmer, and His Followers, Are Detected and Overthrown. The Latter Is an Essay Concerning The True Original, Extent, and End of Civil Government*, Published in London, 1689 by the Whig radical, friend and eventual co-trustee of Locke's estate, Awnsham Churchill.

[18] http://www.unesco.org/new/en/natural-sciences/special-themes/biodiversity/biodiversity-science-and-policy/ipbes/#:~:text=The%20Intergovernmental%20Science%2DPolicy%20 Platform,of%20biodiversity%20and%20ecosystem%20services, Accessed June 6, 2020; See also, See "The Rapid Decline Of The Natural World Is A Crisis Even Bigger Than Climate Change," by John Vidal, Huffpost, March 16, 2019.

**Fig. 17.2**   Scouting Malhazine. (Photo © M.C. Tobias)

On the afternoon of March 22, 2007, numerous landmines and other armaments behind these secretive walls spontaneously detonated, killing at least 83 and injuring hundreds of people in dwellings outside the walls; missiles penetrated neighborhoods that are among the poorest and most congested in the entire greater metropolitan region of Maputo. The ammunition was obsolete and had not been used since the time of the Civil War. Authorities claimed that extreme heat had caused the ammunition to detonate. And this is where I now moved quickly with ranking officers of the military, scouting Malhazine for the purpose of helping envision and bring into being the government's new dream: an ecological sanctuary where once the military had stored so much of its weaponry. That was the goal of Malhazine. To become the most environmentally sensitive major city park in all of Africa. An ecological peace-park.

At the invitation of the government, I went to confer with the Environmental Ministry, the military and civilian engineers, architects, and ecologists involved in this great vision. We scouted the park by vehicle and by foot. Everywhere we walked, large rats would later come to help the military disarm the bombs, Russian missiles, disintegrating ordnance, old ammunition dump sites, and landmines which we carefully skirted.[19]

---

[19] See "Mozambique plans to turn old ammo depot into nature reserve," by Christopher Torchia, October 12, 2015, Business Insider, https://www.businessinsider.com/ap-mozambique-plans-to-turn-old-ammo-depot-into-nature-reserve-2015-10, Accessed March 24, 2019.

I had been to the marine biological station at Inhaca Island, across the bay some 30 kilometers, a few hours by boat from Maputo, the Indian Ocean on its eastern side. The 299 bird species there, including much discussed Afrotropical "extra continental" land birds, made it one of the most densely packed avifauna paradises on the planet.[20] But in addition, Inhaca hosted remarkable assemblages of plant life in half-dozen main vegetation types, from mangrove to evergreen forest. I'd seen the unusual nests of leaves of the ant *Oecophyllum smaragdina*, as described by William Macnae and Margaret Kalk 60 years before when they had compiled the first comprehensive overview of biodiversity on Inhaca. These eaves were held together by silk gland secretions from larvae, stimulated by worker ants. And palms under which the diurnal Old World Fruit bats hung, taking off hours before sunset to help themselves to the various flies and abundant wasps and stink bugs. Geckos and tree frogs also cohabited in great numbers there, along with nesting sea birds.

All of which paled in comparison with the intertidal species—tropical, subtropical, and warm-temperate, as carefully documented back in 1958.[21]

But it was Malhazine that was key to the country's aspiration to create its own version of New York's Central Park, and thereby provide not only the cooling grace of shadowing arbors for the surrounding shanty town, but solid environmental education opportunities for the residents of Maputo. It would be the largest and most ambitious city park on the Continent. Uniquely, Malhazine would also provide an oasis/staging ground for transcontinental, migratory birds and flying insects that benefitted from the combined power of having Malhazine so close to an island of endemics.

By 2016, however, political scandals had driven Mozambique into financial turmoil and an actual currency crash. The International Monetary Fund was pulling out.

To this day, Malhazine remains a dream, like that in the Old Testament, of "swords into plowshares, and their spears into pruninghooks…" (Isaiah, 2.4); an ideal held hostage by human distractions, and a population size guaranteed to add cumulative counterweight to the survival needs, and mishaps by the end of each day. The quotidian vagaries of an economically marginalized history, where such woes are certainly as salient in the case of Mozambique's past, as are the definitively *positive* radar signals coming from the sanctuary and enormous sacred mountain of Gorongosa; and the birds of a friendly and beautiful little island in the southern Indian Ocean known as Inhaca. How the human psyche manages to endure fickle pros and cons that add up in the tired mind, or broken heart, on a long-suffering Continent, is not something easily given to rhetorical, geopolitical, or economic indicators. If experience be our guide, all humans need basically the same things.

---

[20] See *The Birds of Inhaca Island, Mozambique*, by W.F. de Boer and C.M.Bento, BLAS Guide #22, BirdLife South Africa and Avian Demography Unit, 1999, p. 51; See also Clancey's *The Rare Birds Of Southern Africa*, Written and Illustrated by Dr. P.A.Clancy, Winchester Press, Johannesburg, 1985; and *Birds of East Africa*, by V. G. L. van Someren, A.C.Allyn, Allyn Museum of Entomology, Sarasota, Fl., 1973.

[21] See William Macnae and Margaret Kalk, eds., *A Natural History of Inhaca Island, Mocambique*, Witwatersrand University Press, Johannesburg, South Africa.

**Fig. 17.3** Outside the Walls of Malhazine. (Photo © M.C. Tobias)

Except that one person lives in a palace with a thousand rooms; another in a tent with eight children. The middle-grounds where thoughts and hopes may come together are just possibly heightened and attuned for our senses, in protected places; sacred spaces; amidst the fanfare of wild creatures who co-evolved with all of our ancestors. Mozambicans know this to be true. So do New Yorkers. We have all inherited this collection of genes that are atavistically guided from somewhere before the paradoxes we have imposed on the legacy. Which is why a proposed sanctuary like Malhazine—mirroring other biological sanctuaries all the way across Africa, 13 national parks in Gabon, and nearby Obô Natural Park of São Tomé—is so crucial a key to moderating twenty-first century human contradictions (Fig. 17.3).

# Chapter 18
# A Cave at Taranga

## 18.1 A Question of Nirvana (Fig. 18.1)

At well over 1.353 billion people in 2020, India's human density transforms the sensation that otherwise separates urban from rural, rendering countryside suspect and disappointing. By the time of the country's first census in 1871, she numbered over 235 million. But India's ancient demographics—those coeval with the rise of pan-Asiatic theories of karma, the invention of Nirvana, and the widespread adoption of ascetic disciplines, throughout the sixth and fifth centuries BCE—reveal an estimated population of as many as 25 million people[1]. All those irrigation channels can still be viewed through the lens of archaeology, as can the tens-of-thousands of domesticated animal trails and human pathways from the towns to the forests. It has always been a crowded, human environment. For the founders of great religious traditions, 25 million individuals meant 25 million separate human natures, ever-changing philosophies, aspirations, fears, and peopled dilemmas.

As much as the solitary yogi ventured into a wilderness to meditate, the teeming marketplaces were always within earshot. This human geography helped inform the quintessence of Jainism, alleged by Jain scholars to be the oldest ethical tradition within India, and the one most indebted to the strong, inalienable bonds linking laypersons in the towns to Jain mendicants who received their food from those merchant classes. Moreover, Jains tended to remain in the villages, and then cities, averse to agricultural practices, as a rule, because agronomy was, at its core, not merely a practice of nurturance, but of consumption. Without limits, consumption turns to killing, and killing is antithetical to all things Jain, a traditional obsessed, and rightly so, with sensory qualifications: all organisms, bespeaks Jainism, have between one and five senses. Jains may only consume one-sensed beings, which they do so grudgingly, with a distinct penitent remorse. When a Jain monk is no

---

[1] Colin McEvedy; Richard Jones (1978). *Atlas of World Population History*. New York: Facts on File. pp. 182–185.

© Springer Nature Switzerland AG 2021
M. C. Tobias, J. G. Morrison, *On the Nature of Ecological Paradox*,
https://doi.org/10.1007/978-3-030-64526-7_18

**Fig. 18.1** A child in the Vegetarian City of Pushkar, Rajasthan. (Photo, © M.C.Tobias)

longer willing to be a burden on society, he or she will take no food from the house-holders, deliberately fasting unto death and presumed rebirth. This is how it has been for millions of years, say the Jains, echoing a refrain that melds the human condition with the ascertainable statutes defining the lives and destinies of every organism, every soul.

Sometime between the first and fifth centuries of the Common Era, the head of a Jain monastic order, Umasvati (or Umasvami) composed a compilation of sutras that remains something of a Bible for both Digambara and Svetambara sects of Jainism—the *Tattvarthadhigama Sutra*[2]. Actually, *That Which Is* concerns the mul-titude of lists, numbered and rigorous in their attempt to encompass, calculate, and order every facet of life and death. The goal, ultimately, of this prodigious effort was and remains liberation, or moksa (also known as Nirvana). The concept was much

---

[2] See *That Which Is – Tattvartha Sutra – A Classic Jain Manual For Understanding The True Nature of Reality*, Translated by Nathmal Tatia, With the combined commentaries of Umasvati/ Umasvami, Pujyapada and Siddhasenagani, With a foreword by L.M.Singhvi and an introduction to the Jaina fath by Padmanabsh S. Jaini, The Institute of Jainology, HarperCollins Publishers, San Francisco, CA, 1994. In addition, see the three commentaries on the Tattvartha Sutra "used as the Basis of the Commentary" in the translation by Nathmal Tatia, namely: *Svopajna Bhasya,* by Umasvati and recognized as authentic by the Svetambara sect, ed. Pandit Khubchandji Siddhantashstri, Shrimad Rajchandra Jain Shastramala, Bombay, 1932; *Svopajna Bhasya Tika*, by Siddhasenagani, Svetambara sect, 2 vols., Seth Devchand Lalbhai Jain Pustakoddhara Fund, series nos. 67 & 76; and *Sarvarthasiddhi,* by Pujyapada Devanandi, of the Digambara sect, 4th edition,Bharatiya Jnanapitha, Delhi, 1989.

refined in the earlier precipitous sagas attendant upon the lives of Mahavira (599–527 BCE) and his predecessor, Parsvanatha (872–772 BCE), the 24th and 23rd Jinas of the Jain faith. In the famed text, *Uttaradhyayana*, by Niryukti Bhadrabahu (433–357, the last head of a Jain monastic institution or sangha, prior to the division between Digambara and Svetambara monks) this place of deliverance, the Utopic finale for all such liberation ideals, was characterized as follows: a "place" of "freedom from pain, or perfection, which is in view of all; it is the safe, happy, and quiet place which the great sages reach. That is the eternal place, in view of all, but difficult of approach. Those sages who reach it are free from sorrows, they have put an end to the stream of existence"[3]. While there is great bestowal of generosity on this vision of the end, which is a true beginning, say the Jains, it is also clear that the emphatic openness to all—"in view of all'—predestines this topographical, emotional, and psychological quadrant—somewhere in the Universe—an urban place, a crowded nexus wherein every soul may partake.

For Mahavira, attainment of this *ethical vortex* was strictly predicated upon a supreme knowledge; a precise understanding of "all sentient beings" [4]. Our end, in other words, is predicated quite clearly on our connections with others. In Jainism, the word "all" (सब sab) is emphatic.

For both men/saints—Mahavira and Parsvanatha—whose lives are acutely documented in the Jain annals, faith itself derives from Nature[5]. But nature herself is insufficient, according to the copious Jain (or for that matter, Buddhist) strictures to simply convey the mendicant or layperson to this divine locus of spiritual contentment. Right Faith, Right Knowledge, and Right Conduct (the origins of Virtue) are implicit and relentless in their 73 manners of induction ("Exertion of Righteousness")[6]. These categories include not only the "Longing for liberation" but also a "Desire for the Law," "Moral and intellectual purity of the soul," "Begging forgiveness," "Study," "Control," "Austerities," "Questioning the teacher," "Turning away the world," "Renouncing company," and a slew of ascetical universals such as "Perfect Renunciation," "Freedom from passion," "Patience," "Simplicity," "Humility," and every conceivable nuance of "Watchfulness" and "Stability"[7].

---

[3] See *Uttaradhyayana by Bhadrabahu*, English translation by the great German Indologist, Hermann Jacobi, in the *Sacred Books of the East*, Volume XLV, Oxford University Press, Oxford, UK, 1895, Reprinted in 1980, *Jaina Sutras*, Part II, by Motilal Banarsidass Publishers PVT. LTD., Delhi Lecture XXIII, p. 128.

[4] See *Lord Mahavira And His Times*, by Kailash Chand Jain, Motilal Banarsidass Publishers PVT. LTD., Delhi, Revised edition, 1991, p. 95.

[5] ibid., Kailash Chand Jain, p. 94.

[6] ibid., p. 111.

[7] ibid., pp. 112–119.

## 18.2   The Cave

We sat before a cave at the Jain complex of 19 Svetambara and Digambara temples at Taranga in the northeastern Gujarati District of Mehsana, in India, whose construction had commenced in 1121. It was at that time that the King Kumarapala (1143–1174) of the Gujarati Chauylukya Dynasty converted to Jainism under the legendary Jain guru, Acharya Hemchandra. On this particular day, in the early 1980s, a word that a tiger had been seen near, if not at the same cave where two Digambara monks had spent the night, came as no surprise. Mahavira himself was said to have converted a lion to a vegetarian diet; a story that most certainly informed the many subsequent tales surrounding Saint Jerome and his companion lion. As of the most recent census of 2015, there are believed to be 523 Asiatic (Gir) lions left in India, and an estimated 3000 tigers, mostly within the 40 national parks throughout the country, and a few outside the parks. In addition to the Canine distemper virus, which is known to have killed many Indian lions in the past, both species of large feline are fundamentally victims of inevitable collisions with humans, in habitat that is almost entirely fragmented and insufficient to sustain robust lion populations.

Many years ago, Indian conservationists had hoped to translocate the Gir lions into regions removed from approximately 20,000 known indigenous cattle-droving pastoralist nomads in the area. Traditionally, a conservationist rule of thumb suggested that at least 4000 square kilometers of untrammeled, pristine habitat was necessary to be equated with *wilderness*, into which endangered species might be moved. But across India, there is no such wilderness (or in Nepal, for that matter). The largest regional semi-wild areas of the Indian subcontinent do not exceed 500 square kilometers, at best.

Which places concepts like reincarnation, and the pan-Asiatic ethical infatuation with zoo-narratives into a chilling topology, one fraught with competitive *Homo sapiens* who are distributed across India in over 600,000 villages. The Jain pantheon of sentient beings, in which a Jain monk speaks to the animals, in the manner of St. Francis, has been enshrined in the Jain canons for millennia, known as the Samavasarana, or Refuge to All. It comprises a divine pavilion of devas, heavenly beings, and has been painted in countless guises; menageries of species phylogenetically choreographed in a zoological fancy of freedom, discourse, interdependent joy, and participation. Often the architectonic is cosmological, embracing a circular vision of blissful reunion[8]. In the Jain yatras (mandala-like works of art), the disciple meditates upon the loka, or Jain universe, recognizing its fundaments of pudgala (matter) and the restless realm of jivas and nigodas, souls and atoms. It is a world

---

[8] See, for example, *The Peaceful Liberators – Jain Art from India*, by Pratapaditya Pal, with contributions by Shridhar Andhare, John E. Cort, Sadashiv Gorakshakar, Phyllis Granoff, John Guy, Gerald James Larson and Stephen Markel, Los Angeles County Museum Of Art, Los Angeles, 1994, p. 76, Cat. 105, "Hall of the Universal Sermon, (Samavasarana, detail) Rajasthan, Jaipur, c. 1800.

centered by a most curious Roseapple tree, and the cosmic mother's womb, or Jambudvi-pa. There are hells and heavens. The esthetic fundaments of this dream-like peaceable kingdom has its roots in the thousands of palm-leaf and paper or cloth-illustrated manuscripts that embroider Jain history.

Gorgeous sutras, particularly the *Kalpasutra* of Bhadrabahu, and a famed manu-script detailing the temporal coordinates for any would be pilgrim, the *Dravyasamgraha*. Most confounding of all to the uninitiated is that rejection of time, as spelled out in the Acharya Jinasena's *Mahapurana*, in which there really is no time, no beginning, no ending. Six mountains, six states of happiness, a general configuration not unlike some ancient version of Dante's own itinerary 500 years after Jinasena (ninth century).

In this extravagant cosmology, beauty is all of a paradox: it belies the inherent orientation toward the spare and ascetic renunciation of that very passion in which such art is clearly grounded. Hence, the Jain concept of passionless passion[9]. Whether that paradoxical emotion could effectively grasp the fact that the tiger's historic range within India was likely over 40,000 individuals, versus today's depau-perate, desperate, and dispersed numbers is a question of ethical psychoanalysis. We don't know, but Jain faith, generally speaking, tends toward a brutally realistic assessment of nature. Hence, the extraordinary lengths to which it goes to try and protect everything.

Presently, the two munis have emerged from the cave and walked into a cluster of householders at the adjacent temples. These were among the fewer than 80 remaining Digambara monks in all of the India at that time (mid-1980s). They were naked and had all but two possessions: a gourd for carrying boiled water, and a pic-chi, a hand broom made of fallen peacock feathers to gently remove any insects from their path. Their nudity, of course, is considered an ancient and sacred compo-nent of pan-Indian spirituality, unique to male Digambara ascetics. The household-ers, male and female Jains, placed strictly vegetarian food into the palms of the monks and they took (or one of them did) his one meal of the day. The other was fasting. These ascetics never stayed more than three nights in the same place. Always walking, spreading the message of nonviolence (ahimsa) from village to village. That was also the life of their esteemed predecessor, Mahavira, born a prince, who—like his younger contemporary and follower, Buddha—would renounce his wealth, everything, to adopt a comprehensive lifestyle of extreme asceticism oriented toward that goal of liberation during his 72 years of life.

---

[9] For a superb overview of such paintings, see, *Treasures of Jaina Bhandaras*, General Editors, Dalsukh Malavnia, and Nagin J. Shah; Edited by Umakant P. Shah, L.D.Series 69, L.D.Institute Of Indology, Ahmedabad, 1978. See also the three-volume, *Jaina Art And Architecture*, Published on the Occasion of the 2500th Nirvana Anniversary of Tirthankara Mahavira, Edited by A. Ghosh, Bharatiya Jnanpith, New Delhi, 1974; and see, *Kalpasutra – Eighth Chapter of the Dasasrutaskandha of Bhadrabahu*, with Hindi and English Versions and Coloured Reproductions of Original 16th Century Miniatures, Editor & Hindi Translator, Mahopadhyaya Vinaya Sagar, English Translation by Dr. Mukund Lath, Note on Paintings by Dr. (Smt.) Chandramani Singh, Published by D. R. Mehta, Prakrit Bharati, Jaipur, Rajasthan, India, 1977.

It must be pointed out that Buddha did not go anywhere near to the peripheries of Mahavira's asceticism, adopting, instead, a middle path (Mahayana) of compromises which for most of Buddha's followers from country to country would mean meat-eating. Whereas Jain, true Jains, are strict vegetarians. They consume only what they categorize in the one-sensed variety, a diet rigorously circumscribed in order to inflict—within the broadest realm of ethical trade-offs—the least pain[10]. This obvious convergence of nonviolence into practice is a principle that "has permitted the entire structure of the Jaina ethics and the way of life"[11].

Later that day, speaking with the ascetics, we were told by the older gentleman, probably in his 70s, "Khamemi sabbajive, sabbe jiva khamantu me, metti me sabbabhuyesu, veram majjha na kenavi," meaning, "I forgive all beings, may all beings forgive me. I have friendship toward all, malice toward none"[12].

This outright penance in the form of an ecological forgiveness (mettim bhavehi, universal love) is no plaintive cry in the face of the inconsolable realities of ecosystems ultimately stacked against a world without suffering and death (though they are). Rather, we interpreted the emboldened remark as one steeped in crucial mores that might be variously described: serene, unafraid, resolved, quietly at peace with endless turmoil. Assured in the face of zero assurance. The paradox of the mind caught out in the whorls of existence[13].

## 18.3   Whorls of Existence

Those *whorls* are broken down by Jain monks into a very few basics, all translating into ways to mitigate karmic dust which covers and obfuscates the soul. The muni intends throughout thoughts and actions to impede that accretion of such karma, of wrongful intentions and actions. This state of being is called kshayika-samyak-darshana, or, "true insight through the destruction of karma." The rules that apply toward such antitheses are the result of a vast body of literature relating to ahimsa, nonviolence, and the endeavors that help bolster it: Satya (truth), asteya (not stealing), brahmacharya (sexual abstinence, read: population control), and aparigraha (nonpossession). Eight mulagunas, or basic restraints, help acquaint the newcomer to Jainism with a path toward spiritual fulfillment that enables the individual to negotiate the labyrinth of pitfalls. The Digambara's very nudity (acelakka) is a

---

[10] See the PBS documentary, "Ahimsa-NonViolence," by Michael Tobias, 1987, as well as the book, *Life Force – The World of Jainism*, by Michael Tobias, Asian Humanities Press, Freemont, CA, 1981.

[11] See *A Sourcebook In Jaina Philosophy*, by Devendra Muni Shastri, Translated into English by Dr. T. G. Kalghatgi, Edited by Dr. T.S. Devodoss, Sri Tarak Guru Jain Granthalaya, Udaipur, India, 1983, p. 538.

[12] Translation by Dr. Padmanabh S. Jaini, in *Life Force*, ibid., p. 92.

[13] See *Asceticism In Ancient India*, by Haripada Chakraborti, Punthi Pustak Publishing, Calcutta, 1973, pp. 423, 425.

profound reminder, at once, to the monk, and to all he encounters, that something extraordinary has happened, is happening.

By implication that nudity—no tapas, or austerity, but a true freedom-in-renunciation—translates into one of Mahavira's most revered injunctions. He is said to have declared, "He who injures these (earth-bodies) does not comprehend and renounce the sinful acts; he who does not inure these, comprehends and renounces the sinful acts. Knowing them, a wise man should not act sinfully toward earth, nor cause others to act so, nor allow others to act so. He who knows these causes of sin relating to earth, is called a reward-knowing sage. Thus I say"[14]. The clear fact that Mahavira intoned the words "nor allow others to act so" provides a remarkable insight (if it is true) into the activist stance at the heart of Jain non-violence. Such proponents certainly are there. Most recently, in September 2018, the 51-year-old Digambara muni, Tarun Sagar fasted to death after an illness of over a decade. Over 200 monks fast to death each year, a Jain tradition known as Sallekhana[15]. We'd met him in Mumbai during a pilgrimage and he spoke about interceding in the build-up of nuclear weapons, and particularly he displayed great agitation over the Reagan Star Wars program. He was involved in animal rights, issues of ecological legislation, and a host of other pronounced and burning topical issues.

Like his ascetical peers, his ardor, to paraphrase Jean-Paul Sartre, was no *useless passion*. His path was meant to be viable, and accessible to everyone the world over. He was engaged in this world. His dharma-tirtha, or holy path, was the result of daily salutations (namaskara-mantras), compassion, empathy, and charity (jiva-daya); of care in walking (irya-samiti), forgiveness (kshama), universal friendliness to all species (maitri), affirmation (astikya—as opposed to a negative, world-weary predilection one might assume); the sharing with guests of everything (atithi-samvibhaga), critical self-examination at all times (alocana), a vast realm of behavioral restraints (gunavratas), aversion leading to renunciation of everything other than renunciation itself (vairagya), and constant meditation on these matters (Dhyana). These many assertions of his daily quest were at the basis of that aspired to liberation in this life, from this life, to this life—and hence the mathematical complexity of such a philosophy, with its ultimate probity resting in the realization

---

[14] *The Akaranga Sutra*, Book I, Lecture 1, Lesson 3, in Jaina Sutras, Translated from Prakrit by Hermann Jacobi, Motilal Banarsidass, New Delhi, 1980, pp. 4–5. First published by Oxford University Press, 1884, with Translations by Hermann Jacobi. Moreover, when Mahavira spoke of earth-bodies, he was explicitly referencing every life-form imaginable, from "seeds" and "sprouts" to "lichens," "water-bodies and fire-bodies and wind-bodies…" (1.8.1.11–12, Jacobi, 1884).

[15] See Times of India, September 1, 2018, https://timesofindia.indiatimes.com/india/jain-digambara-monk-tarun-sagar-dies-at-51/articleshow/65631337.cms, Accessed March 30, 2019; See also, "Jainism - Its relevance to psychiatric practice; with special reference to the practice of *Sallekhana*," *by* Ottilingam Somasundaram, A. G. Tejus Murthy,1 and D. Vijaya Raghavan, Indian J Psychiatry. 2016 Oct-Dec; 58(4): 471–474. https://doi.org/10.4103/0019-5545.196702, PMCID: PMC5270277, PMID: 28197009, PMC, NCBI, US National Library of Medicine, https://www.ncbi.nlm.nih.gov/pmc/articles/PMC5270277/. Accessed March 26, 2019.

that all souls are interdependent, which in Jain tradition bespeaks a most powerful refrain: parasparopagraho jivanam[16].

Jain principles eschew, in an infectious manner, the paradoxical presuppositions that doom most optimism in a world of such biological upsets. If life is *only* suffering (Dukkha in Buddha's Sanskrit) a key concept in Buddha's Four Noble Truths, as outlined in his bespoke earliest teachings merged over subsequent centuries by followers—*The Dhammacakkappavattana Sutta*—[17] then, naturally, the end of suffering is the liberation described as Nirvana. As with the Jain Digambara monks, Buddhist tradition treats liberation as Nirvana from that suffering (in the Third Noble Truth). But it is a Nirvana that is deemed to be vague, unlike the precise Nirvana at death. A multiplicity of interpretations that have resulted from country to country in differing styles of Buddhism add no small confusion to the ramifications, for example, of Maitreya Buddhism, in which an enlightened Buddhist disciple (bodhisattva) returns to earth in the future to enlighten others and lead them toward liberation.

Such eschatologies and teleologies have little relevancy at the moment of encounter in any hospital emergency room or out in the world where wounded beings are everywhere struggling without surcease. Meditation and activist compulsions swirl across consciousness at instantaneous velocities. Environmentalism is constantly challenged with an altogether novel crisis. Priorities and imperatives daunt and chill one's emotional and neural infrastructures.

The desire to become a Digambara ascetic immediately topples into the realization that the human population explosion, with its horrid tallies of consumer destruction, remains out of control[18].

---

[16] See *Environmental Meditation*, by Michael Tobias, Crossing Press, Freedom, CA, 1993, pp. 47–65; See also, Hermann Jacobi's essay, "On Mahavira and His Predecessors," in Indian Antiquary, IX, pp. 158–163.

[17] See Ajahn Sucitto. *Turning the Wheel of Truth: Commentary on the Buddha's First Teaching*, Shambhala Publishers, Vermont, 2010

[18] See "World population stabilization unlikely this century," Patrick Gerland, Adrian E. Raftery, Hana Ševčíková, Nan Li, Danan Gu, Thomas Spoorenberg, Leontine Alkema, Bailey K. Fosdick, Jennifer Chunn, Nevena Lalic, Guiomar Bay, Thomas Buettner, Gerhard K. Heilig, John Wilmoth http://www.demographic-challenge.com/files/downloads/452fbf0a4300800ec6cc4af4315c11ca/ science-1257469-full.pdf, Accessed March 28, 2019; See also, "What if fertility decline is not permanent? The need for an evolutionarily informed approach to understanding low fertility," Oskar Burger and John P. DeLong, The Royal Society Publishing, Philos Trans R Soc Lond B Biol Sci, 19, v.371(1692); 2016 Apr 19,PMC4822437, 20150157. https://doi.org/10.1098/ rstb.2015.0157, PMCID: PMC4822437, PMID: 27022084, https://www.ncbi.nlm.nih.gov/pmc/ articles/PMC4822437/. Accessed March 28, 2019.

## 18.4   Revelations That Cascade

That menacing epiphany cascades; an implosion of anxieties that crisscross the gamut, from panic to melancholy; from the individual to the species and back. The sheer weight of our species' numbers is dire enough for any individual contemplation. But certainly, the present human generation—if not countless before—have already been conditioned by so many disasters and fears as to have infused a bias without end into any equation that purports to the manifestation of virtue. While plentiful exemplars of goodness abound, the ecological crises associated with the Anthropocene have fronted a new logic that has every reason to doubt humanity's constancy and capacity for rational thought, for reason[19].

Imbibing sensory, emotional, and mental broken glass, such that a kind of private computation arises; one which can be articulated to oneself, and subsequently—in whatever collation of nefarious nuances—to those members of one's species (the public forum) provides access to the idea of reason. But that figment of reasonableness only compounds the debate as to whether there is a primeval structure guaranteeing rational thought. Whether human logic functions according to a synonymous reason. To what degree our ability to think helps us to survive. This latter association, whilst seemingly self-evident, is—from any ecological perspective subject to duration and cumulative turmoil—a most doubtful proposition. Virtually every thinker of note has weighed in on what it is thought all about. No one knows. Immanuel Kant's famed struggle with the concept of *pure reason* could write, "Our reason (Vernunft) has this peculiar fate that, with reference to one class of its knowledge, it is always troubled with questions which cannot be ignored, because they spring from the very nature of reason, and which cannot be answered, because they transcend the powers of human reason…reason becomes involved in darkness and contradictions, from which, no doubt, it may conclude that errors must be lurking somewhere, but without being able to discover them, because the principles which it follows transcend all the limits of experience and therefore withdraw themselves from all experimental tests. It is the battlefield of these endless controversies which is called Metaphysic"[20] (Fig. 18.2).

---

[19] See "Ecological anxiety disorder: diagnosing the politics of the Anthropocene" Paul Robbins, Sarah A. Moore, First Published January 7, 2013, https://doi.org/10.1177/1474474012469887

Volume: 20 issue: 1, page(s): 3–19, Article first published online: January 7, 2013; Issue published: January 1, 2013, Sage Journals, Cultural Geographies, https://journals.sagepub.com/doi/abs/10.1177/1474474012469887?journalCode=cgjb, Accessed March 31, 2019.

[20] Immanuel Kant's *Critique Of Pure Reason*, translated by F. Max Müller, Macmillan Company, London 1896, pp. xvii–xviii.

**Fig. 18.2** Mahavira detail from Kalpasutra Manuscript, ca. 1470s. (Private Collection). (Photo ©
M.C.Tobias)

## 18.5   Fearless Before Infinity

Paradox turns to inherent contradiction, which in turn suggests a pattern of proved
behavior in humans. The empirical evidence is clearly ill-boding for the biosphere.
Ambassadors of that contradiction, each of us, recognize and feel the self-inflicted
wounds, the frenzies from centuries of reasoning faculties that have backfired.
Weather and climate anomalies; every environmentally induced illness; epidemiological factoids indicting our rage for chemical synthesis, production, and massive
distribution; for every toxic extreme, extractive regime, and syndrome of slaughter,
such is the fate of thought in the human commons. Hence, the startling statistics

with respect to comorbidities, co-misery, co-depression, and singular suicide amongst our kind[21].

"I learned not to fear infinity," wrote the poet Theodore Roethke. "The far field, the windy cliffs of forever, the dying of time in the white light of tomorrow..."[22]. Perhaps, fearless, before such infinitudes—esthetic jubilation in the face of the end. Or, as Buddha's final whisper is often characterized, "This I tell you, Bhikkhus. Decay is inherent in all conditioned things. Work out your own salvation, with diligence[23]. But human reason transmogrified, allied with that inferno within oneself, was given voice in the guise of that famed prayer book published in 1609, by Thomas Dekker, citing Ephesians 4:26–32, "Wrath is a short madness; madness is the murderer of reason, so that anger transforms us into brute beasts. Give us courage, therefore, O Lord, to fight against this strong enemy, and not only to fight, but to overcome him – for it is harder to triumph over our raging affections than to subdue a city"[24].

The rage for progress (capitalism) has long blurred the lines of survival and of essential human needs, this much can easily be adduced. The sum-total of economic, gender, and social justice disparities equates with a blind and grasping effort on the part of every individual to achieve that Jain and Buddhist liberation. In the face of two Digambara monks emerging one morning from a cave, the coefficients of salvation seem almost at hand, certainly emblematically. They rouse enthusiasm for it, in spite of ourselves. But the global biological problématique cannot be understated, let alone ignored, except at our fast-emergent peril. Hence, a vast literature now exists on the correlations between climate change and mental health;

---

[21] See, for example, World Health Organization tabulations of depression globally – said to be the number one disabler of individuals: https://www.who.int/news-room/fact-sheets/detail/depression, Accessed March 31, 2019; And the WHO statistics on the approximately 800,000 suicides per year worldwide: https://www.who.int/mental_health/prevention/suicide/suicideprevent/en/. Accessed March 31, 2019.

[22] From "Meditative Sequences," quoted in *Contemplation and The Creative Process*, by Ann T. Foster, The Edwin Mellen Press, Lewiston, NY, 1985, p. 102.

[23] "Buddha's last words," in *The Wisdom of Buddhism*, edited by Christmas Humphreys, Random House, New York, 1961, p.94.

[24] *Four Birds of Noahs Arke*, republished b Appleton & Co., New York, 1925; See also, Sarah Hinlicky Wilson's review of a 2017 edition, "Review of 'Four Birds of Noah's Ark,' by Thomas Dekker," Lutheran Forum, October 9, 2017, https://www.lutheranforum.com/blog/2017/9/26/review-of-four-birds-of-noahs-ark-by-thomas-dekker, Accessed March 31, 2019.

between every form of ecological disaster (as yet) and the rates of depression and suicide[25]. An early node for such research followed Hurricane Katrina[26].

---

[25] "Chronic Environmental Change: Emerging 'Psychoterratic' Syndromes," by Albrecht G. In: Weissbecker I., ed., *Climate Change and Human Well-Being*. International and Cultural Psychology. Springer, New York, NY, https://doi.org/10.1007/978-1-4419-9742-5_3, First Online 22 June 2011, Springer, New York, NY; See also, "Flooding and mental health: A systematic mapping review," Fernandez, A., Black, J., Jones, M., Wilson, L., Salvador-Carulla, L., Astell-Burt, T., & Black, D. (2015), PLOS ONE, 10(4), e0119929; See also, Anderson, C. A. (2012). "Climate change and violence. In D. Christie (Ed.), *The encyclopedia of peace psychology*. Hoboken, NJ: Wiley-Blackwell. https://doi.org/10.1002/9780470672532.wbepp032; See also, Hobfoll, S. E. (1989). "Conservation of resources: A new attempt at conceptualizing stress," American Psychologist, 44(3), 513–524; See also, Lee, H. C., Lin, H. C., Tsai, S. Y., Li, C. Y., Chen, C. C., & Huang, C. C. (2006). Suicide rates and the association with climate: A population-based study," by Lee, H. C., Lin, H. C., Tsai, S. Y., Li, C. Y., Chen, C. C., & Huang, C.C., Journal of Affective Disorders, https://www.researchgate.net/publication/7265085_Suicide_rates_and_the_association_with_climate_A_population-based_study, Accessed March 28, 2019; See also, Hanigan, I. C., Butlera, C. D., Kokicc, C. N., & Hutchinson, M. F. (2012). Suicide and drought in New South Wales, Australia, 1970–2007. PNAS, 109(35), 13950–13955; "Global warming is breeding social conflict: The subtle impact of climate change threat on authoritarian tendencies," by Fritsche, I., Cohrs, J., Kessler, T., & Bauer, J.. Journal of Environmental Psychology 32(1): March 2012, https://doi.org/10.1016/.jenvp.2011.10.002; Centers for Disease Control and Prevention. National Center for Environmental Health. (2014). "Impact of climate change on mental health," https://www.cdc.gov/climateandhealth/effects/. Accessed March 28. 2019; See also, "How Climate-Related Natural Disasters Affect Mental Health," American Psychiatric Association, by Joshua C. Morganstein, M.D. Robin Cooper, M.D, July 2017 https://www.psychiatry.org/patients-families/climate-change-and-mental-health-connections/affects-on-mental-health, Accessed Marche 28. 2019; "There is evidence that increases in mean temperature are associated with increased use of emergency mental health services." "Mental Health and Our Changing Climate: Impacts, Implications and Guidance," March 2017, American Psychological Association, Climate For Life and EcoAmerica, Editors and Contributors, Ashlee Cunsolo, PhD, Director, Labrador Institute of Memorial University Victoria Derr, PhD, Assistant Professor, Environmental Studies, California State University, Monterey Bay Thomas Doherty, PsyD, Licensed Clinical Psychologist Paige Fery, Research Coordinator, ecoAmerica Elizabeth Haase, MD, Chair, Climate Psychiatry Committee, Group for the Advancement of Psychiatry Associate Professor, University of Nevada, Reno, School of Medicine John Kotcher, PhD, Post-doctoral Research Fellow, Center for Climate Change Communication, George Mason University Linda Silka, PhD, Psychologist, Senior Fellow, Senator George J. Mitchell Center for Sustainability Solutions Lise Van Susteren, MD, Psychiatrist, Private Practice Jennifer Tabola, Senior Director, Climate for Health, EcoAmerica, Washington DC and San Francisco, 2017, p. 25. https://www.apa.org/news/press/releases/2017/03/mental-health-climate.pdf, Accessed March 26.

[26] See "Trends in mental illness and suicidality after Hurricane Katrina," Ronald C. Kessler, Sandro Galea, Michael J. Gruber, Nancy A. Sampson, Robert J. Ursano, and Simon Wessely, Mol Psychiatry. Author manuscript; available in PMC 2008 Oct 1, PMCID: PMC2556982, NIHMSID: NIHMS69926, PMID: 18180768, PMC2556982, NCBI, US National Library of Medicine, Published in final edited form as:Mol Psychiatry. 2008 Apr; 13(4): 374–384. Published online 2008 Jan 8. https://doi.org/10.1038/sj.mp.4002119, https://www.ncbi.nlm.nih.gov/pmc/articles/PMC2556982/, Accessed March 26, 2019; See also, "Assessment of suicidal intention: The Scale for Suicide Ideation," Beck, Aaron T. Kovacs, Maria Weissman, Arlene, Beck, A. T., Kovacs, M., & Weissman, A. (1979). Assessment of suicidal intention: The Scale for Suicide Ideation. Journal of Consulting and Clinical Psychology, 47(2), 343-352.https://doi.org/10.1037/0022-006X.47.2.343, https://psycnet.apa.org/record/1979-27627-001, APA PsycNET,American Psychological Association.

## 18.6   Co-depressions

And the fact that in Sweden for nearly a decade a new illness has been recognized as climate angst ("Klimatångest")[27]. While occupational Sentinel Health Events have been intensively studied since the early 1980s[28], environmental epidemiology can be traced back to the writings of Hippocrates' essay "On Airs, Waters and Places"(460–370 BCE), which no doubt influenced Aristotle's thinking about biology[29]. From the earliest Eastern ascetical dietary strictures, to twentieth-century legislation like the 1938 US Federal Food, Drug and Cosmetic Act, and 1947 Federal Insecticide, Fungicide, and Rodenticide Act, there has been growing recognition of the Jain ethos of interdependencies. From the time of Leonardo (who would have been familiar with translations by neo-Platonists in Florence of the vegetarian writings of Porphyry, Theophrastus, Plutarch, and the Peripatos) to the earliest English anti-vivisectionists, dozens of major figures had weighed in on the importance to human health of vegetarianism, reiterating ancient Jain, Brahmanical, and Essene credos.

The nature of environmental stress on human and other life-forms has been of obvious and paramount importance to evolutionary studies. In 2005, Rudolf Bijlsma and Volker Loeschcke concluded their assessment of future ecological stress "resistance and adaptation" research at is involves evolutionary pressures. Their list of priority areas of research centered around "...selection lines...genetic correlations and trade-offs...the role of genotype-by-environment interactions... candidate gene approach...[and] revealed variation in candidate genes in natural populations..."[30].

In 2018, it had become clear to Bijlsma and Loeschcke that "inbreeding depression" had become one of the major points of critical focus for "population persistence" and for the future of biodiversity in general[31]. But no ecological baselines

---

[27] See Lagerblad, Anna (December 6, 2010). "Climate anxiety new phenomenon in psychiatry". Svenska Dagbladet. Schibsted.

[28] See for example, "Sentinel Health Events (occupational): a basis for physician recognition and public health surveillance."D D Rutstein, R J Mullan, T M Frazier, W E Halperin, J M Melius, and J P Sestito, Am J Public Health. 1983 September; 73(9): 1054–1062. PMCID: PMC1651048, PMID: 6881402, https://www.ncbi.nlm.nih.gov/pmc/articles/PMC1651048/, Accessed March 28, 2019; See also, The Diseases of Occupations, Sixth Edition, Donald Hunter, C.B.E., D.Sc., M.D., F.R.C.P., Hodder and Stoughton, London. ISBN 0-340-22084-8, 1978

[29] See Chapter 2 of Clarence J. Glacken's *Traces on the Rhodian Shore – Nature And Culture In Western Thought From Ancient Times To The End Of The Eighteenth Century*, University of California Press, Berkeley and Los Angeles, 1967.

[30] https://onlinelibrary.wiley.com/doi/full/10.1111/j.1420-9101.2005.00962.x, Accessed March 26, 2019. See, for example, "Environmental stress, adaptation and evolution: an overview," R. Bijlsma and V. Loeschcke, First published: 13 July 2005, Journal of Evolutionary Biology, https://doi.org/10.1111/j.1420-9101.2005.00962.x,https://onlinelibrary.wiley.com/doi/full/10.1111/j.1420-9101.2005.00962.x.

[31] See Evol Appl. 2012 Feb;5(2):117–29. https://doi.org/10.1111/j.1752-4571.2011.00214.x. Epub 2011 Nov 7., "Genetic erosion impedes adaptive responses to stressful environments," Bijlsma R· Loeschcke V, PMD, 25568035, PMCID:, PMC3353343, https://doi.org/10.1111/j.1752-4571.2011.00214.x, https://www.ncbi.nlm.nih.gov/pubmed/25568035, Accessed April 1, 2019.

have ever approached a human-induced mass extinction by which to extrapolate long-term adaptability for the one remaining *Homo*[32], although the likelihood of most other large vertebrates surviving many more generations under anthropocenic conditions, is highly questionable. We witness countless forms of breaking polarities, splits, electron shifts, and unstable atoms with every conceivable composition of positive and negative charges throughout the disciplines of cosmology, astronomy, astrophysics, chemistry, and research from quantum physics and electromagnetic research[33].

We see countless instances of fundamental bifurcation points across the spectrum of mathematics (the Hopf Bifurcation, wherein an entire system becomes unstable and then switches, for example, named after German mathematician Heinz Hopf, 1894–1971), and have done so at least since the time of Hippasos (fifth century BCE), the Greek philosopher who appears to have first detected the existence of irrational numbers. Legend has it that he drowned at sea as a result of the displeasure the Gods took in his discovery, which followed his contemplating the square root of 2, or 1.41421356237309504880168872420969

80785696718753769480731766797789... The numbers move toward no closure, a cusp of infinite descent, as ever smaller natural numbers (integers, whole numbers) are crammed into the incompletion, forever. Pi,$\pi$, as well as the famed Euler Number (2.71828182845904523536028747135517... —pertaining to natural logarithms) are other such irrational propositions. They cannot be written as simple fractions, a hallmark of the mathematically irrational[34], and mirror image of ecological paradox in practice, not mere theory.

---

[32] For in-depth discussion, see the Author's book, *The Hypothetical Species: Variables of Human Evolution,* Springer, New York, 2019.

[33] See, for example, A&A 380, L5–L8 (2001) https://doi.org/10.1051/0004-6361:20011505 c ESO 2001 Astronomy & Astrophysics Zeeman-split opposite-polarity OH lines in sunspot spectra: Resolution of a puzzle S. V. Berdyugina1 and S. K. Solanki, October 2001; http://www2.mps.mpg.de/dokumente/publikationen/solanki/j129.pdf, Accessed March 26, 2019; See also, http://hyperphysics.phy-astr.gsu.edu/hbase/quantum/zeeman.html, Accessed $\sqrt{}$March 26,2019; See also, Song C, Phenix H, Abedi V, Scott M, Ingalls BP, et al. 2010, "Estimating the Stochastic Bifurcation Structure of Cellular Networks," PLoS Comput Biol 6(3): https://doi.org/10.1371/journal.pcbi.1000699, https://journals.plos.org/ploscompbiol/article?id=10.1371/journal.pcbi.1000699, Accessed April 1, 2019; See also, Gregor F. Fussmann, Stephen P. Ellner, Kyle W. Shertzer, and Nelson G. Hairston Jr. Crossing the Hopf Bifurcation in a Live Predator–Prey System. Science. 17 November 2000: 290 (5495), 1358–1360. https://doi.org/10.1126/science.290.5495.1358.

[34] See Sir Thomas Little Heath's work, *A History of Greek Mathematics*, Clarendon Press, Oxford, UK, 1921.

## 18.7 Ecological Interrelationships

Other, so-called discrete mathematics leads directly to graph theory and all those structures, directed or undirected, that involve ecological interrelationships, equally peculiar by way of the complete mental bias that a human-manufactured image composes, suggesting a visual confirmation that may or may not have anything to do with reality, the most prominent version of a working anthropic principle. It encompasses not only our mathematical assumptions about earth and the Universe, but also helps to assuage and satisfy our esthetic questions and nostalgia[35].

Systems, whether atomic or biological, are subject to bifurcations and general theories regarding the point of transformation have shown how even the smallest input can manifest massive, qualitative change.

This is precisely why the medical graphs plotting the intersections of human epidemiology and ecology are such restless protagonists in distribution patterns aligned with peril and illness. Most prominently, such distributions are mental, a factor that weighs most heavily in human evolution. The US Preventative Services Task Force (USPSTF) describes eight primary tests utilized in the diagnosis of human depression but the characterizations scarcely account for what, in the human organism, is an unambiguous morbidity in response to a massive set of ill-boding and self-imposed environmental circumstances[36]. Such depression ratings, often utilizing what is called the "Double-Blind, Placebo-Controlled Clinical Trial," are humble efforts at best to cope with ecological panic attacks; Severe and Systemic Environmental Stress Saturation Syndromes, and all of the negative and punishment feedback loops that account for the widespread mental breakdown in the fact of unprecedented extinctions by our own kind.

Most educators have come together in an effort to inspire youth and give them every reason to endeavor toward hoped for outcomes, ecological success stories, a so-called sustainable future. This can mean, for example, getting "beyond beef"

---

[35] See N.V.R. Mahadev and Uri N. Peled, Threshold Graphs and Related Topics, North Holland, Volume 56, 1995; See also, the paper written by Leonhard Euler on the "Seven Bridges of Königsberg," and published in 1736, See Biggs, N.; Lloyd, E.; Wilson, R. *Graph Theory, 1736–1936*, Oxford University Press, 1986.

[36] See The Hamilton Depression Rating Scale, Back Depression inventory, Patient Health Questionnaire, Major Depression Inventory, Center for Epidemiologic Studies Depression Scale, Zung Self-Rate Depression Scale, Geriatric Depression Scale, and the Cornell Sale for Depression in Dementia. In addition are other related conditions including Postpartum Depression, Pediatric Depression, Cognitive Behavioral Therapy for Depression, and Respiratory Depression in the Postoperative period. In addition, there are innumerable approaches to ascertaining suicide in at risk populations. not to mention every analysis pertaining to suicide at risk populations, in addition to a slew of other psychiatric studies attempting to correlate such seemingly abstruse causes-and-effects between bacteria in the gut and depression, gum disease, cannabis use, and/or pomegranate juice in relation to Alzheimer's or other type memory declines.MEDSCAPE, https://emedicine.medscape.com/article/1859039-overview, Accessed March 26, 2019; See also, Screening Tests for Depression, Updated: Dec 23, 2018, Author: David Bienenfeld, MD; Chief Editor: David Bienenfeld, MD, et.al.

ethics into the corporate, profit-making worlds; transcending despair in oneself and ditching not only sedatives, but school; taking to the streets, instead, to protest the twenty-first century adult world's inaction on climate change (the Sunrise Movements), cruelty, and oppression in general[37].

## 18.8   Biological Restoration

But in a place far away and long ago, in the company of two Digambara ascetics, a far more plausible reconstructive scenario of biological restoration cries out to the future without drama or intrigue, data sets or jeremiads. It is simply that total galvanizing of nonviolence, ahimsa, in the moment, not in some probability factor of the future. Its most manifest realization is that of a person who is naked, walking, during an entire lifespan, calmly describing at every juncture, in each village, the essential reality of kindness to all beings. Their enumeration, said to have numbered 875,000 injunctions, by Mahavira's reckoning, was established in the *Jiva Vicara Prakaranam*, a manuscript written in the eleventh century by the Svetambara writer, Santisuri, who attributed life even to rocks; not just any rocks, but a labyrinth of crystals, lava, gold, and so on:[38] The ekendriya jiva, or Sthavar Jivas, those with one sense and four pranas, or powers, including touch, inhalation, and exhalation and a body itself. Their six classes, according to the Jains, included rocks, diamonds, salt, rain, and dew and fog droplets, as well as creatures living in hurricanes, tornados, trade winds, lightning, fire of all kinds; creatures in trees (vanaspatikaya) and finally those vegetables under the soil, like garlic and potatoes. *Try not to harm them*, said Mahavira's disciples. And especially refrain from ever afflicting the *trasa jivas*, those beings with more than one sense. Such interdictions commenced with one's behavior toward *dvindriyas*, the worms.

---

[37] See for example, "Grown-Ups Get a Scolding on Climate," Inspired by a Swedish teenager, students around the world on Friday will protest political inaction, By The Editorial Board; The editorial board represents the opinions of the board, its editor and the publisher. It is separate from the newsroom and the Op-Ed section.

March 12, 2019, https://www.nytimes.com/2019/03/12/opinion/climate-change-children-greta-thunberg.html, Accessed April 1, 2019; See also, "There is hope amidst environmental anxiety, says a scholar of ecotheology,"University of Helsinki, Public Release, Dec. 14, 2018. https://www.eurekalert.org/pub_releases/2018-12/uoh-tih121418.php, Accessed March, 26, 2019; See also, "Burger King is rolling out a meatless Whopper. Can McDonald's be far behind?" By Tim Carman, April 1 2019, The Washington Post, https://www.washingtonpost.com/news/voraciously/wp/2019/04/01/burger-king-is-rolling-out-a-meatless-whopper-can-mcdonalds-be-far-behind/?noredirect=on&utm_term=.a2d3341ac836, Accessed April 1, 2019. See also "The meat industry will be completely gone in 15 years, Impossible Foods CEO says," by Steve Goldstein, MarketWatch, June 24, 2020, https://finance.yahoo.com/m/22bde9f5-dab7-34d2-9616-7e204d78796a/the-meat-industry-will-be.html?.tsrc=applewf

[38] Santisuri, Jiva Vicara Prakaranam, along with Pathaka Ratnakara's Commentary, Edited by Muni Ratna-Prabha Vijaya, translated by Jayant P. Thaker, Madras 1950.

Unlike Buddha, who had set his heart on a generic otherworld, consumed by the conception of maya or illusion associated with earth, Mahavira was firmly planted on solid ground. He elevated that ultimate goal of a sentient being's soul to the state of "samatva," a sense of kinship with all fellow life forms, and recognized that "the kernel of what makes a human being human, is the same kernel that makes every other organism itself"[39]. Of the 363 known schools of Jain philosophy that thrived at the time of Mahavira, the most prominent of them had focused upon Kryiavada, that doctrine involving jiva, or the soul, and what pious intentions and behavior were required to stave off the avalanche of ajiva, or karmic impurities (corollaries of hinsa, violence) in the body, and—by inherent connectivity—in the soul[40].

By the time Mahavira died, aged 72, in meditation beneath a sala tree in Pava-puri, today part of Bihar State, surrounded by five palm-laden hills and a large lotus pond (his footprints still said to be found nearby), the weary, but wise poet/states-man on behalf of all life forms had achieved what later Greek philosophers would think of as a hylozoistic naturalism. It was the essence of a soul's reciprocity with all other souls, a kinship key to all ecology. While Buddhism saw nirvana as meaning both extinction, total affirmation, nothingness, and emptiness, Mahavira was more than satisfied to recognize in nirvana, the very earth from which its concepts, and Mahavira himself, originated (Fig. 18.3).

---

[39] See op.cit., *Life Force – The World Of Jainism*, Tobias, p. 67.
[40] ibid., p. 66.

**Fig. 18.3** Late 18th century Siddhachakra Yantra, Mahavira seated in the Center, A Jain Cosmological Meditation Painting. (Private Collection). (Photo © M.C.Tobias)

# Chapter 19
# A Village in Prince Christian Sound

## 19.1 Lunatic Bifurcations

A cruise ship consumes roughly 1,000 gallons of fuel per 1,000 tons of weight in the water. Two hundred and fifty tons per day, per average sea-faring vessel. One ship in particular, upon which we find ourselves charging toward Greenland weighs over 86,000 tons. No other seemingly living breathing monster other than a massive storm weighs 176 million+ pounds. Public notices for cruise ship pollution violations include problems with fecal coliform, chlorine, air quality, oily waste, sulfur content in fuel, overboard garbage, and other wastewater breaches[1]. In addition, there is the significant matter of ballast water management into, and of such ships, both in terms of water quality but also invasive species management[2].

Traveling up the 60-kmr long Prince Christian Sound on such a behemoth requires a profoundly dismal sense of irony and self-loathing, not unlike that tidal wave of a critique that slammed into an Autumn, 2016 art exhibition at Yale University entitled, "Yosemite: Exploring the Incomparable Valley"[3]. "A Photo Show Romanticizes Yosemite at the Expense of Native Americans," writes Ari Akkermans, "a freelance writer and art critic based in Beirut..."[4]. "It's through exhibitions like this that you can see the profound disconnect between institutions

---

[1] See "Pollution and Environmental Violations and Fines, 1992 – 2018," Source: Cruise Junkie dot Com, http://www.cruisejunkie.com/envirofines.html, Accessed April 5, 2019.

[2] See "Cruise Lines Readying for Ballast Water Treatment," by Wendy Laursen, 7/28/2016, The Maritime Executive, https://maritime-executive.com/editorials/cruise-lines-readying-for-ballast-water-treatment; See also, "Ballast Water Management, August 9, 2017, Smithsonian Environmental Research Center, https://nbic.si.edu/glossary/ballast-water-management/, Accessed April 5, 2019.

[3] https://artgallery.yale.edu/exhibitions/exhibition/yosemite-exploring-incomparable-valley, Accessed April 5, 2019.

[4] January 2, 2017, Hyperallergic, https://hyperallergic.com/348476/a-photo-show-romanticizes-yosemite-at-the-expense-of-native-americans/ Accessed April 5, 2019; See https://hyperallergic.com/author/ari-akkermans/. Accessed April 5, 2019.

© Springer Nature Switzerland AG 2021                                    175
M. C. Tobias, J. G. Morrison, *On the Nature of Ecological Paradox*,
https://doi.org/10.1007/978-3-030-64526-7_19

and the history they are entrusted with," Akkermans says[5]. When Akkermans dissects the Native American history of the Southern Miwoks who were directly caught up in the massive onslaughts upon Yosemite Valley and her progeny, in the face of a parallel narrative of art history, his thrust is inarguable. He declares, "violent conflicts" means extermination, "rapacious development" means extermination, and "forced change" means, well, also extermination. The wall texts refer almost casually to the annihilation of Native Americans, unwilling to let this distract viewers from the exhibit's noble message: "Over the course of the later nineteenth century, however, naturalist John Muir led a transformation of the nation's perception of Yosemite by drawing attention to its uniqueness"[6].

Muir's dilemma—aside from a biocultural and thus North American, historic, ritual, and esthetic blindness—was not so different from those of us who travel, not just by foot. There are ways, of course, to kayak or quietly sail to Greenland, though with little predictability in terms of getting into a sufficiently ice-free Prince Christian Sound. The weather variables escalate in a kayak, underscoring a seemingly basic ecological principle: People should not be here. Yes, history proves that Vikings readily (not easily) sailed to Greenland, then trekked in. But whether a 1,000 years ago or today, that represents an invasion. Whereas a cruise ship to Greenland typically is barred by ice from getting into most Sounds, and spends all of 24 hours several miles from shore careening southward past Cape Farwell. Given sufficient opportunity, most people today would be Vikings. Ecologically, nearly everyone (Digambara excluded) behaves like the most arrogant of legendary Vikings.

Comparisons come with curious data sets of baggage. But by comparison with that 176-million-pound sea creature, the impact differences between a monster ship and a few people by sailboat or kayak; even a boatload of Vikings require no special skills to appreciate.

Nonetheless, there is a greater orbit of paradox seething around all sides of such fertile grounds of debate. As ecologically vexing as it is, a ship carrying 2100 people to a village of fewer than 150 souls, Aappilattoq, nestled in an unprecedented consortium of nature at her esthetic best, is a nearly perfect equation for chaos theory and the unwanted. There, in the North Atlantic's equivalent Concordia, a cul-de-sac of towering granite walls of verglas at the rear of the sound. It forces upon the mindful a complicated psychological and emotional dilemma on all fronts: what are we doing here? We should not have the luxury to allow for such dilemmas, but across the planet our species engages in it with little dialogue and the promiscuous photoshopping of the advertising industries. Still, in the case of Greenland it is a troubling topic for conversation because it involves, ultimately, ethical values lodged in numerous false storylines of hunting cultures versus esthetic-and-pleasure grazing cultures. Neither are entirely or honestly identifiable by their means of survival (Fig. 19.1).

---

[5] op. cit., January 2, 2017.

[6] ibid, Akkermans.

**Fig 19.1**   Aappilattoq, Greenland. (Photo © J.G.Morrison)

Most of the inhabitants of Aappilattoq are in their early 30s and their village—again, to be clear, one of the most spectacular on earth—has a lovely array of quaintly orchestrated, freshly painted homes, businesses, a church, school, and cemetery. Almost picture perfect, except for the ship of camera-mad tourists gawking at the splendor all around from up on one of 11 passenger decks, with their quieted 86,000 horse-power signature, over 900 feet of length and—in addition to the 2100 passengers, more than 900 crew members. A floating urban environment approximately 20 times larger than Aappilattoq, and capable of moving on average at over 27 miles per hour through the oceans. Those on shore are labeled as obtaining most of their sustenance from hunting and fishing. The animals they kill are also mostly locals.

This geographical situation gives rise to controversy in our age, when in a previous century it did not. The very concept of wilderness, as historian Roderick F. Nash describes, is etymologically tied to the word *will*, and thus to "Last Will," to one's "own free will," and "to self-willed land"[7]. And to any number of etymologically perverse twists and subjugations in mind. Greenland, more than any other physiographical region, purports to the ultimate paradoxical place in real time. Saqqaq

---

[7] See *Leave No Trace: The Vanishing North American Wilderness*, Aerial Photography By Jim Wark, Forward and Essays By Rockerick F. Nash, Universe Publishing, New York, 2011, p. 15.

cultural trappings in Greenland date back to at least 4400 years ago, and one individual in particular known as "Inuk"—something of a physiological ambassador for the Saqqaq world, whose remains were found in the village port of Qeqertarsuaq, half-way up the West Coast. Inuk was a hunter. But during early Norse settlements, commencing in 985 CE, with Erik the Red's second exile, there accumulated great Nordic wealth, the importation of bovines, sheep, and goats from Iceland, and an eventual population zenith of nearly 2000 probably very satisfied human souls. There were churches and a widespread farming community with evidence for some "500 farms" on the southeastern coast, and approximately "100" to the southwest. All the comforts of home, for a couple of centuries[8]. Then came a great "silence" that seems to have enshrouded Norse culture in Greenland, corresponding to dramatic climate change.

## 19.2   A Teleology of Killing

The Inuit people who also occupied southern Greenland continued their hunting traditions which sustained them, without a dependency upon the vagaries of Greenland agriculture, which appears to have ultimately doomed the Nordic settlers. Although today, with a human population of approximately 57,000 there has evolved the expected kinds of changes that Denmark's own influences would have exacted. A curious mix, most certainly. The largest island in the world, with over 650,000 square miles still under ice, but melting fast for no fault of the indigenous residents, whose self-rule since 1979 and later separation from the EU has guaranteed a demographic bubble under the Sun.

In 1991, the Premier of Nuuk, Lars Emil Johansen wrote in the Foreword to the book *Nature Conservation in Greenland – Research, nature, and wildlife management*[9], "Only by bringing down wild animals and using them for food and heat, and by making clothing and boats from skins, can we survive in our part of the world. And that is why the people of Greenland have always been hunters. Technological development has not changed this." He adds that it is "no longer a matter of life or death" but rather a question of "identity"; an "attitude to nature and its resources" that differs from those points of view shared within industrialized nations. Indeed, such human changes have definitely altered Greenlanders' Weltanschauung. Today, whilst an increasing percentage of those hundreds-of-billions of hamburgers devoured throughout the world are being replaced with varieties of plant-based

---

[8] Groeneveld, Emma. "Viking Age Greenland." *Ancient History Encyclopedia*. Ancient History Encyclopedia, https://www.ancient.eu/article/1208/viking-age-greenland/. Accessed 06 Apr 2019.

[9] Kalaallit Nunaanniu nunanik piinillu allanngutsaaliuineq, Atuakkiorfik, Greenland, p. 7.

meat (what some have called the new era of "Peak beef" as in peak oil[10]), even attitudinal changes are coming to Greenland, as witnessed by a vegan-friendly café in Nuuk, Greenland's capital founded in 1728. But as of the early 1990s, approximately 100,000 seals (70% of them ringed seals) were killed each year in Greenland; as well as every possible rorqual and finback whale within the International Whaling Commission quotas, under the guise of "nutritional value"[11]. Approximately 400 musk oxen were slaughtered annually as of the early 1990s, as well[12]. And to this day, welcoming websites offer trophy hunting in Greenland[13].

Corroborating the attitudinal context for hunting amongst Greenlanders are Moravian immigrant observations throughout Cape Farewell dating to the early 1800s. Detailed notes pertaining to local famines, particularly within the Eastern Settlements; the fact that one hunter's industry might typically support five family members if the weather was not entirely overwhelming; and also unholy competition for marine mammals from seasonal Norwegian hunters who would shoot "hundreds of thousands of hooded seals and harp seals" using only "the blubber and skins" and leaving behind the meat. For local Inuit, "The result of this was more and longer periods with reduced catches, and as a whole the population must have seen the whole basis of their existence disappear"[14].

An estimated 1% of Greenland's ice-free perimeter is arable and with fast-escalating climate-change effects on the land, projections for increasing near-term production of potatoes, broccoli, carrots, strawberries, and even apples look promising. Potato cultivation in Greenland, however, is also fraught with serious fungal contaminants, in part as a result of dramatic temperature changes (increases)[15]. Documented animal husbandry has always accompanied non-indigenous lifestyles and diet in Greenland since the first Norse waves of immigration. In the twenty-first century, that has meant the slaughter of tens-of-thousands of sheep, goat, and

---

[10] See Tim Lewis's essay, "Have we hit 'peak beef'?", Environment, The Guardian, March 16, 2019, https://www.theguardian.com/environment/2019/mar/16/peak-beef-ethical-food-climate-change, Accessed April 7, 2019.

[11] op. cit., *Nature Conservation in Greenland*, by essay by Hans Jakob Helms, p.16.

[12] ibid., "The Musk Ox in Angujaartorfiup Nunaa," essay by Carsten Riis Olesen, p. 116.

[13] See, for example, https://visitgreenland.com/about-greenland/hunting-culture/, Accessed April 7, 2019. See also, the first major survey on Greenlander attitudes regarding climate change, hunting traditions and the future, "Life on thin ice – Mental health at the heart of the climate crisis," by Dan McDougall, Mon 12 Aug 2019, The Guardian, https://www.theguardian.com/society/ng-interactive/2019/aug/12/life-on-thin-ice-mental-health-at-the-heart-of-the-climate-crisis, Accessed January 12, 2020.

[14] See *Cultural Encounters at Cape Farewell – The East Greenlandic Immigrants and the German Moravian Mission in the 19th Century*, by Einar Lund Jensen, Kristine Raahuge and Hans Christian Gullov, Museum Tusculanum Press, University of Copenhagen, 2011, p. 123.

[15] See *Gerald Traufetter.* "Arctic Harvest – Global Warming a Boon for Greenland's Farmers". Der Spiegel; See also "Diseases in potatoes and grass fields," by Eigil de Neergaard et al 2009 IOP Conf. Ser.: Earth Environ. Sci. 6 372013, Eigil_de_Neergaard_2009_IOP_Conf._Ser.__Earth_Environ._Sci._6_372013(2).

bovines. They are not petting zoos. And as far as other biomass-related industries, the only known forest in all of Greenland, the Qinngua ("paradise") Valley just northwest of Prince Christian Sound (60.2739 N 44.5307 W) comprises 600 hectares adjoining the Kangikitsoq Fjord; a 12.4-mile-long valley along the banks of the Tasersuaq Lake wherein some 300 plant species thrive in the milder climate, including mountain ash, gray-leaf willows, and downy birches, some nearly 30 feet high; a protected area since 1930[16] (Fig. 19.2).

Less protected are the apex predators themselves—from human hunters to Orcas, in both of whom toxic concentrations of PCBs and other contaminants have long been studied[17].

A huge area of the ice sheet potentially affected with large arrays of transmigrated pollutants includes the largest national park in the world, that of Northeast Greenland[18]. With fewer than 50 full-time human inhabitants, the park, created in 1974, 102 years after the world's first such park, Yellowstone, was expanded to its current 375,000 square miles in 1988. It is "77 times" the size of Yellowstone[19]. Far to the west of the park is the US military base at Thule, where a ballistic missile early warning system was installed. In 1968, a B-52 carrying four nuclear warheads crashed near the base. It remains a question of controversy whether the emergency clean-up crews were ever able to find a portion of one of the weapons, and its uranium and plutonium contents.

---

[16] https://www.wondermondo.com/qinngua-valley/. Accessed April 7, 2019.

[17] See, for example, "Contaminants in Marine Mammals in Greenland – with linkages to trophic levels, effects, diseases and distribution," Doctor's dissertation by Rune Dietz, 2008, https://www2.dmu.dk/Pub/Doctor_RDI.PDF; See also, "The Inuit's Struggle with Dioxins and Other Organic Pollutants," by Bruce E. Johansen, American Indian Quarterly, Vol. 26, No. 3 (Summer, 2002), pp. 479–490, Published by: University of Nebraska Press, https://www.jstor.org/stable/4128495, JSTOR, https://www.jstor.org/stable/i384369. Accessed April 7, 2019; See also, "Greenland Ice Sheet Likely Contains High Levels Of Anthropogenic Pollutants," by James Ayre, August 6, 2017, Clean Technica, https://cleantechnica.com/2017/08/06/greenland-ice-sheet-likely-contains-high-levels-anthropogenic-pollutants-research-suggests/. Accessed April 7, 2019; "Pollution in Greenland," Numbeo, https://www.numbeo.com/pollution/country_result.jsp?country=Greenland; "Lead pollution recorded in Greenland ice indicates European emissions tracked plagues, wars, and imperial expansion during antiquity," Joseph R. McConnell, Andrew I. Wilson, Andreas Stohl, Monica M. Arienzo, Nathan J. Chellman, Sabine Eckhardt, Elisabeth M. Thompson, A. Mark Pollard, and Jørgen Peder Steffensen; PNAS May 29, 2018 115 (22) 5726–5731; published ahead of print May 14, 2018 https://doi.org/10.1073/pnas.1721818115, Proceedings of the National Academy of Sciences of the United States of America, PNAS, https://www.pnas.org/content/115/22/5726; "Poisons From Afar Threaten Arctic Mothers, Traditions," by DeNeen L. Brown, April 11, 2004, The Washington Post, https://www.washingtonpost.com/archive/politics/2004/04/11/poisons-from-afar-threaten-arctic-mothers-traditions/89af6e09-c411-4b6f-bc97-4839963e9208/?utm_term=.245c11aba081, Accessed April 7, 2019.

[18] See https://visitgreenland.com/the-national-park/. Accessed April 7, 2019.

[19] See https://www.cntraveler.com/stories/2014-05-12/largest-national-park-in-world, Accessed April 7, 2019.

**Fig. 19.2**   In Prince Christian Sound, Greenland. (Photo © J.G.Morrison)

## 19.3   The Severe Silences of Archaeology

All of these human incremental overlays—from traditional hunting cultures, military investiture commencing with the Vikings, global pollution, a serious tourist destination, and climate change—pose insoluble questions within the very context of what it is the human collective aspires to achieve. The creators and managers of Northeast Greenland National Park clearly try to send a message of protection; of the lowest possible impact by our species; a clear call for the appreciation of nature. To a large degree, the message succeeds. But strangely and sadly, nonviolence is not a uniform code, even within the most iconic patch of twenty-first century ecological imagination. "Special permission is given to mineral exploration and exploitation. Limited hunting is permitted for ptarmigan, Arctic foxes, and seals by permanent staff members of the station. Professional Greenlandic hunters may hunt polar bears"[20]. These vestiges of an implacable mindset stifle projections into the future; erase quietude; reverse consistency; underpin the historically persistent lack of objective standards in the human perceptive or cognitive agency. We lack both the purity of our idealisms and the constancy of those ethics we pretend to perform in

[20] See Greenlandtoday, "Trapping in the National Park," by Admin, August 15, 2017, http://greenlandtoday.com/trapping-in-the-national-park/?lang=en, Accessed April 7, 2019.

our daily lives. Individual variables, the geography of their expressions, and motifs are not merely a class of ideas and practices subject to judgments. Nor can they be explained simply by the unanimity of community ordinance and custom. The anthropology of a place like Greenland is as fickle and useless as a bullet to the polar bear's head; or the ephemeral pleasures gleaned by a cruise ship of fools (Fig. 19.3).

In a discussion of observations recorded during the Colonial period of southern Greenland, in the mid-to-late nineteenth century, with specific details regarding "ritual relations between the hunter and the hunted"[21], it is suggested that Thule

**Fig. 19.3** Above Northeast Greenland National Park. (Photo © M.C.Tobias)

---

[21] op.cit., *Cultural Encounters at Cape Farewell*…pp. 164–167.

culture shamans and their hunting minions felt deep respect for and sensitivity toward the animals they killed. So indoctrinated in the way of life of these post-Dorset (Paleo-Eskimo) Inuit peoples was this conceptualization, that it is said "it is thus generally assumed that seals will allow themselves to be killed by good hunters who observe their taboos, in fact, that they even call it 'going home' when they go to a breathing hole to let themselves be stabbed"[22].

Because we are all in this together, our use of "the camera and the tweezer"[23] is a realm of actions at odds with nature. So clearly so as to render the joys of imagined solitude, or even friendship, in one of the world's most magnificent places no more than a moment of stuffing (*bouche*), or the consumption of layer cake. A fiction that can support no conclusions about our species because such testimonies are like whiplash from one person, one community, one culture to another. And yet we perpetuate its infinite loops and twists, its lies and provocations in hopes of someday believing that it all makes sense.

But with no possible objectivity in science, in the arts, and certainly not in the underlying rationale for *hope*, our imaginations run no more riot than our daily selves and their mechanical expressions. A cruise ship in Greenland. A man on the moon. Tourists to the summit of Everest. Looters on Santa Monica's 3rd Street Promenade or of the famed Macy's of the Macy Parade. The cover-ups of Seveso, Chernobyl, of Bhopal. The lists of human hubris and self-deception can only escalate with our desperate demographics and boredom, entitlements and social chasms. Sociology quickly turns to anthropology which, in turn, and soon enough, settles into that chilling silence masking the whole of archaeology.

---

[22] ibid., p. 164, Citation from Knud Rasmussen, *Iglulik and Caribou Eskimo Tests. Report of the Fifth Thule Expedition 1921-ered1924*, vol. 7, 3.

[23] See the extraordinary book, *Objectivity*, by Lorraine Daston and Peter Galison, Zone Books, New York, 2007, p. 397

# Chapter 20
# The Grampians

## 20.1 Beauteous, Befallen (Fig. 20.1)

George Fennell Robson (1788–1833) came from the northeast of England, in Durham, and by the age of 20 had fallen swoon—as would every generation after him—to the highlands of Scotland. While he loved painting birds, his principal obsession was early morning and late afternoon lighting on the Grampians—their Lochs, summits, and distances, the imperturbable shadow play and incarnate longings before the human stance and its plausible exile. He championed an early nineteenth century calling that took from the legendary language of Dionysius Longinus's treatise *On The Sublime*[1] which was as much the language itself—dangling in "the space between heaven and earth"—as a study in effective metaphors and imagery[2]. Other painters of Robson's generation—Patrick Nasmyth, Sir Robert Ker Porter, Ramsay Richard Reinagle, John Fleming, and his engraver Joseph Swan[3] —each capitalized on Longinus' sentiments, as assimilated by Edmund Burke in his *Philosophical Enquiry into the Origin of Our Ideas of the Sublime and Beautiful* (London, 1757) and his *Vindication of Natural Society* published a year earlier.

Burke, a Dublin-born statesman, harbored a massive, historical punch whose traction derived from his keen abilities to merge both the language and content of scenery and the feelings they may elicit, with political realism of his time. His points-of-view on civilizations from India to France were very much in tune with those of Jean-Jacques Rousseau's *An Inquiry into the Nature of the Social Contract;*

---

[1] Translated from the Greek, With Notes and Observations, And Some Account of the Life, Writings, and Character of the Author, By William Smith, London, 1770, 4th edition.

[2] ibid., p.31.

[3] See *The Lakes of Scotland. A Series of Views From paintings Taken Expressly For The Work by John Fleming Engraved by Joseph Swan. With Historical and Descriptive Illustrations by John Leighton*, Published by Swan, Engraver, Glasgow, 1834.

© Springer Nature Switzerland AG 2021
M. C. Tobias, J. G. Morrison, *On the Nature of Ecological Paradox*,
https://doi.org/10.1007/978-3-030-64526-7_20

**Fig. 20.1** Drawing by George Fennell Robson, *Scenery of the Grampian Mountains*, 1819. (Private Collection). (Photo © M.C.Tobias)

*Principles of Political Right*[4]; and an early precursor to the writings of Oswald Spengler, particularly *The Decline of the West* (*Der Untergang des Abendlandes*, 1918)[5].

Burke (1729–1797) (unlike a Robson) could not ignore the paradoxical truth that bloodshed seemed to be the "cement" that held nations together[6]. But, at first glance, such political contexts seemed utterly incommensurate with the saturation of Romanticism in the Grampians; from the images of Robson to those of William Beattie and Thomas Allom[7]; J. Macwhirter's illustrations for *Caledonia*; (Described

---

[4] Translated From the French, London 1791, in which he famously summed up the transition from a state of nature to that of a political state with equally membered citizens, "The man who had till then regarded non but himself, perceives that he must act on other principles, and learns to consult his reason before he listens to his propensities." p.49.

[5] See Spengler, Oswald. *The Decline of the West*. Ed. Arthur Helps, and Helmut Werner. Trans. Charles F. Atkinson. Preface Hughes, H. Stuart, Oxford University Press, New York, 1991.

[6] See "Edmund Burke and the Scottish Enlightenment," by George McElroy, in Man and Nature, Volume 11, 1992, URL: https://id.erudit.org/iderudit/1012681ar; https://doi.org/10.7202/1012681ar, Canadian Society for Eighteenth-Century Studies, pp. 171–185, https://www.erudit.org/en/journals/man/1992-v11-man0306/1012681ar.pdf, Accessed April 9, 2019.

[7] See *Scotland*, by William Beattie, engravings by Thomas Allon, Geo Virtue, London, 1836. See also, *Romantic Gardens – Nature, Art and Landscape Design*, by Elizabeth Barlow Rogers,

by Scott Burns And Ramsay, Engraved by R. Paterson, William P. Nimmo, London and Edinburgh, 1878) the first comprehensive collection of *The Poets and Poetry Of Scotland*[8] with an astonishing 170 poets anthologized; to one of the earliest *Tourist's Guide to the Trosachs*[9]; and the twentieth century rise of biological investigations across all of Scotland, with such notable works as Alexander Edward Holden's *Plant Life in the Scottish Highlands—Ecology and Adaptation to Their Insect Visitors*[10]—the first study of one of the last and largest native forests in all of Britain, namely the 140 square miles of Scots pine at Rothiemurchus, and smaller groups at the Moor of Rannoch and near Loch Tulla[11]. As far as ornithological expeditions into the Highlands, they began in earnest by 1871 and were recorded in the annals of the Scottish Naturalist, and eventually, the book series known as *Scottish Birds*[12].

George Fennell Robson was the artist at the origins of this naturalist's Scottish history. A sampling of his observations suffices to incite some token of his genius: Image No. VIII. Ben Venue, he cites a Dr. Graham calling it the "climax of sublimity"; and in No. IX, also of Ben Venue, states that "The mind of the painter, stored with such materials, must unite them in one grand harmonious composition, which like the sculpture of the ancients, shall combine in a perfect whole the scattered beauties of imperfect individuals." Of Loch Lomond No. III, there on the edge peering out beyond old yews and oaks, remains scarcely visible of a disintegrated church above the quieted vale of "wheat and oats," Ben Lomond far beyond the darkening scene, 3240 feet above the islands of Inch Murrain and Tavannach, where stands the very man who owns much of the scene in question, His Grace the Duke of Montrose with his dog. This is no image of reversion or solitude-inspired rebellion, as in the case of the iconic Caspar David Friedrich "Wanderer Above the Sea of Fog," painted precisely 1 year prior to Robson's work in the Grampians. Rather, a celebration— true to Longinus—of a melancholic moment that may, after all, be key to appreciating the harsh beauty of the Highlands. The village of Luss is not visible but Robson makes clear that this most celebrated of all the Lochs is "one of the most interesting and delightful scenes in Britain; not that such a number of scattered objects is

---

Elizabeth S. Eustis, and John Bidwell, The Morgan Library & Museum, New York, in association with David R. Godine, Publisher, Boston and the Foundation For Landscape Studies, New York, 2010.

[8] By James Grant Wilson, Blackie & Son, London, 1877.

[9] Thomas Nelson And Sons, London, Edinburgh And New York, 1899.

[10] Photographs by Robert M. Adam, Oliver And Boyd, Edinburgh, 1952.

[11] ibid., pp. 83–95; see also, Smout, T. C., MacDonald, R., and Watson, F., *A History of the Native Woodlands of Scotland 1500–1920*, Edinburgh University Press, 2nd edition, Edinburgh, 2007.

[12] The Journal Of The Scottish Ornithologists' Club, Vol. 1, No. 1, Autumn 1958; See also, the Avibase for all reported bird species in Scotland, at: https://avibase.bsc-eoc.org/checklist.jsp?region=UKsc&list=howardmoore&region=UKsc&list=howardmoore; See also, https://www.gov.scot/policies/biodiversity/. Accessed April 9, 2019; and https://www.nature.scot/scotlands-biodiversity, Accessed April 9, 2019; and *Scotland's Nature In Trust – The National Trust for Scotland and its Wildland and Crofting Management*, by J. Laughton Johnston, Poyser Nature History, In association with the National Trust for Scotland, Illustrations by John Busby, Academic Press, London, 2000.

consistent with that unity which picturesque composition generally requires, but because the varied profusion of beauty, displayed by the whole, cannot fail to raise in the mind astonishment and delight."

Loch Katerine, Image No. XI, conversely, according to Robson, "meets with no competitor in the variety of rude magnificence displayed by its eastern shores, which baffles description, and leaves far behind it the boldest flights of the poet and the painter." Elsewhere among the infinite varieties of the Grampians, Robson comments on "the inhabitants of these remote tracts [who] are seldom visited by strangers, and live as if shut out from civilized intercourse. Their habitations are rude and comfortless huts; and their best food is milk, oatmeal, and potatoes, of which their store is often scanty; yet are these poor people given to hospitality, and if they see a stranger about to ascend the steep sides of the valley, will run and invite him to partake of their homely fare, before he ventures on so toilsome a journey." This commentary accompanies the image of Ben Lawers, No. XIX.

The killing of red deer is made much of, as multitudes of men forced them into enclosures which they called "elrig," a word connoting "strife"; and thus fenced in by stakes and thick brush-wood, the deer were slaughtered. Why? For fun, of course. These thousands of acres, and tens-of-thousands of deer were part of the "Duke's pleasure-grounds" (referring to "His Grace" the Duke of Atholl who owned "the most extensive trace of the Grampians frequented by them [the deer]" as seen in Image No. XXVII of Ben-Y-Gloe[13].

Between 1850 and 1854, John Parker Lawson, an Episcopal clergyman/historian of the natural history of the Bible, assembled his oddly titled, *Scotland Delineated*. The 1854 edition has been rightly characterized with unrivaled superlatives as "One of the major achievements of British topographical lithography…"[14]. A smaller,

---

[13] From *Scenery Of The Grampian Mountains; Illustrated by Forty-One Plates, Representing The Principal Hills From Such Points As Display Their Picturesque Features; Diversified By Lakes And Rivers: With An Explanatory Page Affixed To Each Plate, Giving An Account Of Those Objects Of Natural Curiosity And Historical Interest , With Which The District Abounds, by George Fennell Robson, The Engravings Executed By Henry Morton And Coloured From Original Drawings Made On The Spot By The Author*, Longman, Hurst, Rees, Orme, And Brown, London, 1819.

[14] As written by bookseller Sims Reed Ltd., London: *Scotland Delineated in a Series of Views, with Historical, Antiquarian and Descriptive Letterpress*, Turner, Roberts, Nash et. al., *Lawson, John Parker*. Published by E. Gambart & Co. / Joseph Hogarth., London, 1854. Sims Reed Ltd. has a copy of the work for sale at over US$80,000., and the bookseller's descriptive copy continues, "The hand-coloured lithographs are, in the main, after drawings of Scottish scenery commissioned especially for the work from the foremost landscapists of the day by Joseph Hogarth, the publisher. The outstanding plates combined with the outstanding scenery they depict give the work a dramatic grandeur that is unparalleled in British topographical illustration. Among the artists represented (in the order they appear on the title) are J. M. W. Turner (2 plates), Sir W. Allan (2 plates), Clarkson Stanfield (4 plates), George Cattermole (11 plates), W. L. Leitch (16 plates), Thomas Creswick (3 plates), David Roberts (14 plates), J. D. Harding (4 plates), Joseph Nash (2 plates), Horation MucCulloch (5 plates), D. O. Hill (1 plate) and W. A. Nesfield (2 plates). The lithographs were drawn by Harding, Carrick, Gauci, Needham and others." See https://www.abebooks.com/ servlet/BookDetailsPL?bi=7059000024&searchurl=bi%3D0%26ds%3D30%26bx%3Doff%26so rtby%3D1%26tn%3Dscotland%2Bdelineated%26an%3Dlawson%26recentlyadded%3Dall &cm_sp=snippet-_-srp1-_-title1.

subsequent edition of *Scotland Delineated*, on page 273, has Lawson describing what to his mind's eye is the most beautiful waterfall in all of Scotland, namely the Tummel. He quotes one Dr. MacCulloch, a historian who'd written on the Battle of the Grampians (Mons Graupius, AD 83), an early prelude to a recent revival of interest in Britain's most famous Roman governor: Gnaeus Julius Agricola, father-in-law to the ancient military historian, Tacitus.

Agricola had famously pressed his armies from southern England clear into Scotland between 77 and 84 AD. Today, some of the research about this period in Scottish history is known as the Roman Gask Project, referring to the Gask Ridge in Perthshire[15]. Wrote MacCulloch, "...it is beautiful in itself, and almost without the aid of its accompaniments...Whether low or full, whether the river glides transparent over the rocks to burst in foam below, or whether it descends like a torrent of snow from the very edge, this Fall is always varied, and always graceful....the general landscape is at the same time rich and romantic, nothing being left to desire to render this one of the most brilliant scenes which our country produces"[16].

A century prior to Lawson's masterpiece, and a half-century before Robson appears in the Grampians, the Reverend William Gilpin had written his own *Observations* on the "High-Lands of Scotland"[17]. His three watercolors of Loch Lomond are somber, exalted and give no evidence of a looming penchant on the part of the Scottish people to exploit those "modes of beauty" [of which, writes Gilpin] "we had great profusion"[18]. Although Gilpin's lyricism does presage a familiar tension when he promotes an eighteenth century common-place aesthetic that declares that "high places, and extended views have ever been propitious to the excursions of the imagination. As we surveyed the scene before us, which was an amusing, but unpeopled surface, it was natural to consider it under the idea of population"[19]. During Gilpin's time in the Grampians, another Reverend, Thomas Malthus was but a 10-year old. But the adult Malthus's subsequent influence on the social and economic chasms pushing Anglo-Saxon and Celtic cultures apart, and leading by every continuing vexation and racist outrage to the unprecedented period in Scottish history known as the "Highland Clearances" had not yet occurred. It was to be one of the most heinous country-wide policies of "forced eviction" ever visited upon a large population, on ethnic and invented economic grounds, but it was soon to do so. The political crises in Scotland were looming toward it.

---

[15] See http://www.theromangaskproject.org.uk/Pages/Introduction/Agricola-hecame.html.

[16] See *Scotland Delineated. A Series Of Views Of The Principal Cities And Towns, Particularly Of Edinburgh And Its Environs; Of The Cathedrals, Abbeys, and other Monastic Remains; The Castles And Baronial Mansions, The Mountains And Rivers, Sea Coast, And Other Grand And Picturesque Scenery*, by John Parker Lawson, Day And Son, Lithographers To The Queen, London, 1860.

[17] *Observations, Relative Chiefly To Picturesque Beauty, Made in the Year 1776, On Several Parts of Great Britain; Particularly The High-Lands of Scotland*, Printed for R. Blamire, London, Second Edition, 1792.

[18] ibid., p. 3.

[19] ibid., p. 33.

Indeed, as every child in Scotland learns almost by rote, just 30 years prior to Gilpin's tour, the infamous 40-minute Battle of Culloden sullied the moorlands in county Inverness with the blood of thousands. "Some 1000 of the Young Pretender's army of 5000 weak and starving Highlanders were killed by the 9000 Redcoats, who lost only 50 men"[20].

Then came the subsequent hunting down and murder of an additional thousand+ highlander clansmen who had supported Charles Edwards' Jacobite Rebellion, an attempt to take back the British throne and reestablish the Stuart Dynasty. That dream, not shared by every Scot, and however mixed in terms of clan martial relations and greatly varying qualities of life, was gone. So too, much of the practical underpinning of highland cultural norm in terms of chief/clan combinations of socialized/feudal land tenure. From April 16, 1746, onward the memory of the "Forty-Five Rebellion" would remain etched in a cultural population that was soon to be entirely fragmented, as the northern half of Scotland dissipated demographically, whilst the southern half of the country grew dramatically urban. Crofter subsistence, with particular dependencies on kelp, bovines, and potatoes, lost out to massive factory-worker institutions in the central lowlands, as well as an eventual potato blight, with a similarly catastrophic impact as in Ireland.

When Gilpin visited the Grampians, Scotland's total population was approximately 1.3 million, the majority in the South. By the period of Lawson's rhapsodic visit, Scotland had almost tripled in number. Today's numbers of people between the north and the south of Scotland mirror her rapid history of technology onslaught: 235,000 highlanders in 2017, and 5.2 million Scots to the South.

## 20.2   Every Conceivable Infliction

Ecological history records a parallel and equally salient evolution during the nineteenth century of artistic jubilation, namely, the ill-fitting infrastructure and by-products of industry that played its considerable role in shaping the political and economic realities of those who might otherwise wish to reside purely in the aesthetics of Scotland, on one hand; and those who simply wished to *survive* every conceivable infliction, as they surely mounted from one disaster to the next. For example, when, in the early 1860s, artists and tourists were gulping up the Grampian views, just to the South, in Glasgow, "100,000 people lived in one-roomed houses... [and] from time to time there were outbreaks of cholera and typhus, and thousands died"[21].

---

[20] See "Battle of Culloden," Encyclopedia Brittanica, Updated March 30, 2019, By the Editors, https://www.britannica.com/event/Battle-of-Culloden, Accessed April 8, 2019.

[21] See *Scotland – A Short History*, by P. Hume Brown, New Edition By Henry W. Meikle, Oliver And Boyd LTD., Edinburgh, 1955, p. 330.

By 1908, "thirty-nine million tons" of coal were produced in Scotland, corresponding to equally ungainly numbers in sectors ranging from steel to ship tonnage on the Clyde[22].

Today, Scotland has 32 councils ("unitary authorities"), one of which, Perth and Kinross hosts a hydroelectric plant, begun in 1946. The Tummel hydroelectric power scheme is located between Lochs Ericht, Rannoch, and Tummel. No such industry attended these roaring cascades at the height of the Grampian artistic and philosophical revolutions which veritably worshipped those falls.

To grasp the extent of transformation throughout the Highlands, one telling source is the *Royal Grampian Country: A Report Prepared and Published for the Scottish Tourist Board by the Department Of Geography*, University Of Aberdeen, Introduction by Kenneth Walton, Professor of Geography, February 1969. At that time, summer season visitations numbered between "300,000 and 350,000" tourists, as discussion ensued as to the maximum carrying capacity of such tourism (as it has in equally finite areas of great interest in the world, from Yellowstone, Yosemite, and the Grand Canyon to Chamonix and Milford Sound)[23]. In mapping the coexistence of diverse visitor activities by altitude and by month, a most instructive graphic emerges[24] in which, for example, there is skiing above, at around 3000 feet, red deer calving in the center, just above hill sheep lambing, grouse nesting and shooting, salmon, brown trout and sea trout angling, and pheasant nesting. There is also hind shooting, partridge, and wild duck shooting, and, amid the whole picture, from lower hills and valleys to high summits—tourism, which, by the end of World War II, outpaced small family poultry farmers. Tourists, at approximately one pound per head of profit for the farmer, slept in the newly dominating B&Bs, where prior hens laid eggs.

While areas like the "high Cairngorms, Caenlochan and Dinnet oakwood" were administratively given over to conservation reserves[25] traditional human settlements had already—in the late 1960s—witnessed tremendous decline. As depicted in *Royal Grampian Country*, "The inhabitants of the high glens, or of the crofters in the valleys compare their physical and social environment most unfavourably with that of the town"[26]. Such trends coincided and were deemed to do so with increasing fall-out, in terms of indigenous agriculture and the continuing rural-to-urban country-wide migration.

And so came the roads, toilets, tourist accommodations, buses, various amusements, ski tows, "après ski entertainment"[27], and the predictable cascade of other amenities, infrastructure, and diversion. Probably no more paradoxical sadness (and not a little terror) attends this brief history of the Scottish sublime than that of the

---

[22] ibid, p. 329.

[23] ibid., p.118.

[24] ibid., p.111.

[25] See *The Protection Afforded to Wildlife in Scotland*, by R. A. Haldane, 1966.

[26] p.58.

[27] ibid., p.106.

road A82, rated the most dangerous in Scotland, and the one whose course from Inverness to Glasgow parallels several Lochs, particularly Lomond, by way of a narrow, two-lane road. It is a deathtrap, allowing for few turnouts and very few spots to ever get off the road and appreciate the scenery. Communities themselves are all but trapped, it would seem, by this ill-conceived shooting gallery of a road[28]. Indeed, the tourist who rents a car and drives the A82 is likely, at length to wonder whether he/she had seen anything at all, both hands gripped in sweat, have clenched a steering wheel for unrelenting hour-after-hour. In late Spring and Summer, outdoor rest-stops in those Highlands provide relief for less than ten seconds: the time it takes for uncountable swarming midges (*Culicoides impunctatus*) to detect a large, warm-blooded target. One might have thought (hoped?) the persistently aggravated freeze of a long Winter would do these miniature demons in, for good. But, no: the wonders of biology. Their nearly 5000 fellow species, members of the Ceratopogonidae family, have been around for approximately 100 million years, inspiring what taxonomists have termed a kind of "phylogenetic chaos." If nothing else, plein air painting in May or June throughout the Highlands (and much of Western Wales, for that matter) is nothing less than pure masochism.

## 20.3   The Ecology of Fiscal Riot

Solutions? Declares the report, "…a re-institution of the fiscal conditions which helped to maintain a reasonable level of population in the late 18th century,"[29] excepting the circumstances of 1782–1783, in which famine afflicted most Highland Parishes, a period which to this day is perceived as having been a "turning point" in Scotland's history[30]. The first census in Scotland in 1755 revealed a population of "1,265,380," 30% of whom were already living within Scotland's four principal cities[31]. With more than four times that population today, the question of energy is obviously on everyone's mind. There are two nuclear power plants in Scotland, the one in Ayrshire having recently exhibited hundreds of cracks, underscoring the perils of aging nuclear power in a region known for seismic activity[32]. But public

---

[28] See https://www.obantimes.co.uk/2019/06/03/highlands-roads-among-top-10-most-dangerous-in-scotland-for-bikers/; see also, https://www.dangerousroads.org/europe/scotland/4002-a82.html. For a superb overview of all those forces in the arts, exploration and science that helped shape the period between the latter half of the 18th, and first part of the 19th centuries, see Barbara Maria Stafford's *Voyage into Substance: Art, Science, Nature, and the Illustrated Travel Account, 1760-1840*, MIT Press, Cambridge Mass, 1984,

[29] ibid., pp.162, 55.

[30] ibid., p.53.

[31] See Tyson, R. E., "Population Patterns: 1. to 1770", in M. Lynch, ed., *The Oxford Companion to Scottish History*, New York, 2001.

[32] See "Hunterston B: Pictures show cracks in Ayrshire nuclear reactor," by Kevin Keane, BBC Scotland's environment correspondent, BBC News, March 8, 2019.

opinion in Scotland shows a majority in favor of large-scale hydropower[33], the Grampian hydro-electric scheme, the aforecited Tummel Power Station LB51715 being the one large power entity within the Highlands[34]. Scotland, ultimately, is unlikely to ever reverse ecological gears to the extent that Loch Lomond would be turned into a Hoover Dam. But, Luddite-like, alternative energy reversions have recently occurred. Between April 21st, 2017, and the following April, all of the UK went for between one and 3 days with zero reliance on coal for its electricity, the first time since the Industrial Revolution began[35].

Perhaps, as the world convulses, the Grampians, as viewed by Gilpin, Robson, and so many others, will remain tranquil, intact, free of revolution, famine, and other, as yet uncharacterized catastrophes. Perhaps not (Fig. 20.2).

**Fig. 20.2** "On the Shores of Loch Lomond," Photograph by George Chance, ca.1900–1908. (Private Collection). (Photo © M.C. Tobias)

---

[33] "Scots support renewable energy," Cordelia Nelson, March 20, 2013, https://yougov.co.uk/topics/lifestyle/articles-reports/2013/03/20/scots-support-renewable-energy, YouGov.,

[34] See http://portal.historicenvironment.scot/designation/LB51715, Accessed April 9, 2019.

[35] See "First coal-free day in Britain since Industrial Revolution," BBC News, April 22, 2017, http://energynewsbd.com/details.php?id=948, Accessed April 9, 2019.

# Chapter 21
# The Yasuní Effect

## 21.1 The Great Bounty (Fig. 21.1)

Moving slowly by canoe through Ecuador's Amazonian provinces of Napo and Pastaza, and on into the heart of the country's Yasuní National Park, a UNESCO Biosphere Reserve since 1989, we are at once far more aware of and embarrassed by our presence than ever. Ungainliness amid a splendor at once exhausting and oppressive, fraught with tension where splendor gives way to so much that is out of bounds, unimaginable. Whatever proof-of-concept might have served as some silly human apotheosis across the span of time is now, instantly, rendered senseless.

Indeed, we stand out, we humans, in ways no less apparent than were we casting shadows on some of the world's largest sand dunes in the Rub' al Khali. If we ever belonged here the biopic prefigurements would have at once rejected almost everything—our attire, disposition, every version of heat, damp, and stinging nettle tolerance; re-molding the dumb clay of spoiled modernism, re-plastering the disputable brain, re-attuning every nerve ending to comport with a new atlas of insect sex, frenzies of pheromones, some hanging in the air for over 7 months, the shadow worlds of green anacondas, velvet worms, crab spiders, bullet ants, feathered cicada, narrow-mouthed and eyelash frogs, assassin bugs and stinky birds (Hoatzins). A furious phalanx of mandible-flaunting army ants (*Eciton burchellii*), desperate to have their way with every mottled patchwork of biomass, stampeding across the odd naked footprint of a tapir, or jaguar; trigger-ecstatic traces and triumphs of life and death cascading to right, to left, above and in the tiers and depths that defy delineation, with no obvious beginnings or ends.

Such sprawling sinkholes and battered but bedazzled, never-betrayed continuity, of transitional life forms clinging and infiltrating, underlie every library of metaphors. Our Self and social mirrors, both, are obliterated entirely by forces that form a through-story within all of the remaining neotropics, enveloping with the most precocious of chemical elixirs, philosophical echo chambers collectively akin to No Exit, once entered. The mind, countryside, and city minds, at least, are doomed in

© Springer Nature Switzerland AG 2021                                         195
M. C. Tobias, J. G. Morrison, *On the Nature of Ecological Paradox*,
https://doi.org/10.1007/978-3-030-64526-7_21

**Fig. 21.1** In the Heart of Yasuní National Park. (Photo © M.C.Tobias)

so merciless a world of slimes, saps, thorns, micros and macros, blisters and bites. Fifty-five million-year-old threats to bipedal movement; menaces masked delicately, swiftly, colorfully, or unseen. But at maximal cost. The burdens accumulate in a mental maelstrom, comminations without respite. Mostly, it is the *idea* of the Amazon that prompts both terror and her rewards.

Such rainforest, if anything, precludes all obvious awareness and intrusion, instantaneously drowning out consciousness that would ask, pathetically, for the least, confused exceptionalism. Rainforests are, by primordial definition, the purest forms of self-maintenance in absence of governance. This dense forest cover eschews political etiquette. Standouts like ourselves are insulting. Any physical manifestation that does not reign in ego, self-absorbing within a biological ethic instituted tens of millions of years ago, is either a random mutation with virtually no success in store, or simply the momentary bulk of debris that has accumulated from flood waters, a fire, or some other natural disturbance. Acoustics, light, shade, movement through an unharnessing by the second of all that refuses to be touched, performs the perfect cosmological set of non-equations, indeterminacies and subterfuge, the very obverse of an astrogeophysical black hole.

Lyanda Lynn Haupt writes of Charles Darwin's diaries during his years on the *Beagle*, and specifically in concert with some of his early solitary walks into the Brazilian tropics, "He touched, lightly, everything that didn't look like it could bite, he smiled just a little, and finally he reached for his pocket notebook, wanting to capture something of this strange, soft elation 'Twiners entwining twiners—tresses like hair'—'beautiful Lepidoptera…' until he was barely within the realm of words, '-Silence-.' And finally exclaimed, quietly to himself… '-Hosannah-.'"[1] Of course, he was wearing boots, and had a cutlass and other English essentials.

In early 2007, the President of Ecuador, Rafael Correa, pledged to protect the biologically profuse portions of northeastern Ecuador's Amazon[2]. It was known as the Ishpingo-Tambococha-Tiputini (ITT) Initiative, and for Correa, and environmentalists throughout the world, the ITT seemed like the dream materialized. Moreover, Correa had obtained his Ph.D. in economics from the University of Illinois and seemed deeply committed to the ideal of neutralizing the temptation to drill for oil[3].

One group of researchers declared that "keeping the northwestern Amazon—home to the Basin's highest biodiversity and the region least vulnerable to climatic drying—largely intact as a biological refuge is a global conservation priority of the first order. If the world's most diverse forests cannot be protected in Yasuní, it seems unlikely that they can be protected anywhere else"[4].

Alas, by August of 2013, the vision had been abandoned[5]. Like other economic theories, human reality had yet again proved too massively unified an international force to withstand even a single instance of altruism at the national level, of just one country's president stepping beyond the capitalistic paradigm of brute force and blunt domestic policy. This was the late biologist Garrett Hardin's "Tragedy of the Commons" scenario writ large[6]. It also corresponded with the Nobel Prize-winning

[1] See *Pilgrim on the Great Bird Continent – The Importance of Everything and Other Lessons from Darwin's Lost Notebooks*, by Lyanda Lynn Haupt, Little, Brown And Company, Boston, 2006, p. 62.

[2] https://web.archive.org/web/20120415204628/http://í-itt.gob.ec/preguntas-y-respuestas/los-beneficios/See also, http://www.sosYasuni.org/en/index.php?option=com_content&view=article&id=177&catid=1&Itemid=34, Accessed April 11, 2019.

[3] "Ecuador's Yasuni-ITT Initiative: Avoiding emissions by keeping petroleum underground," Carlos Larrea, Lavinia Warnars, https://doi.org/10.1016/j.esd.2009.08.003, Energy for Sustainable Development, Volume 13, Issue 3, September 2009, Pages 219–223, ScienceDirect, Elsevier, https://www.sciencedirect.com/science/article/pii/S0973082609000581; See also, Ecuador's Yasuní-ITT Initiative: The old and new values of petroleum Laura Rival ODID, Available online 9 October 2010, Ecological Economics 70 (2010) 358–365, http://www.loisellelab.org/wp-content/uploads/2015/08/Rival-et-al.-2010.pdf, Accessed April 11, 2019.

[4] See PLoS One. 2010; 5(1): e8767,Published online 2010 Jan 19. https://doi.org/10.1371/journal.pone.0008767, PMCID: PMC2808245, PMID: 20098736, *Global Conservation Significance of Ecuador's Yasuní National Park,* Margot S. Bass, Matt Finer, Clinton N. Jenkins, Holger Kreft, Diego F. Cisneros-Heredia,  Shawn F. McCracken, Nigel C. A. Pitman, Peter H. English, [1]Kelly Swing, Gorky Villa, Anthony Di Fiore, Christian C. Voigt, and Thomas H. Kunz Andy Hector, Editor, https://www.ncbi.nlm.nih.gov/pmc/articles/PMC2808245/. Accessed April 11, 2019.

[5] See "Ecuador approves Yasuni park oil drilling in Amazon rainforest," BBC News, 21 August 2013.

[6] See Hardin, G (1968). "The Tragedy of the Commons". Science. 162 (3859): 1243–1248. Bibcode:1968Sci...162.1243H.

economist William Nordhaus' lecture, "Climate Change: The Ultimate Challenge for Economics," as acutely described in David Leonhardt's essay, "The Problem With Putting a Price on the End of the World – Economists have workable policy ideas for addressing climate change. But what if they're politically impossible?"[7]

When we flew over one of the oil complexes in Yasuní with Dr. Ivonne Baki, former Ecuadorian Ambassador to the United States, and Ecuador's then Plenipotentiary Representative and Secretary of State for the Yasuní-ITT Initiative, the 1.7 billion barrels of crude oil beneath the jungle were worth twice the price of collapsed value in 2012, nearly four times their all but abandoned price thresholds at the gas pumps in 2020, though the 407 million tons of carbon consumption that oil represents, if refined and consumed, had not, and has not, changed. At the time, the ITT proposal's objective, according to Dr. Baki, could "be qualified as both holistic and revolutionary because, in addition to addressing the root of global warming and biodiversity loss, it also aspires to fight poverty and inequality within Ecuador; to stop deforestation and promote reforestation, to protect the National Parks and invest in research and sustainable development. Given the fact that the Yasuní-ITT Initiative is a government project, it offers an opportunity for oil-producing developing countries, such as Ecuador, to end their dependence on an extractive economy and seek dignified development opportunities through the sustainable use of its natural resources"[8].

Spending many days and nights in that tropic made it abundantly clear how ill-fitted we American urbanites were to the habitat. But how remarkably graceful and at home were the small local populations of Quechua-speaking Añangu, the Huaorai, Gagaeri, and Taromenane. Two known "uncontacted tribes" in the region had already apparently gone extinct, the Oñamenane and Huiñatare[9]. Perhaps the remaining macaws in the trees above their final campsites recall some of their words, remember their songs, a scenario which has indeed played out in other such instances. Their complete disappearance into the forest, Republics and Confederacies up in smoke, join at least two dozen other vanished North American indigenous societies, but no precise number is certain. At least 100 or so uncontacted groups remain worldwide, according to an abundance of data gathered by organizations like Survival International in London.

And while some communities, like the Añangu, had thoroughly engaged with the outside world, building community-owned lodges like that at the Napo Wildlife Center, others desperately tried to maintain their own worlds, and in some cases that desperation has led to necessary political engagement, particularly in defense of

---

[7] *The New York Times, April 9, 2019,* https://www.nytimes.com/interactive/2019/04/09/magazine/climate-change-politics-economics.html, Accessed April 13, 2019.

[8] See *Why Life Matters: Fifty Ecosystems of the Heart and Mind,* by Michael Charles Tobias and Jane Gray Morrison, Chapter 8, "Ecuador's Imperiled Paradise – One of the World's Most Important, If Least Known Battles: A Conversation with Dr. Ivonne Baki," p. 48.

[9] See Nuwer, Rachel (2014-08-04). "Future – Anthropology: The sad truth about uncontacted tribes". BBC http://www.bbc.com/future/story/20140804-sad-truth-of-uncontacted-tribes, Accessed April 13, 2019.

their homelands from oil industry and government incursions[10]. I (mt) witnessed this firsthand at the Rio+20 Summit in June 2012, where I had the opportunity to deliver a speech in defense of these peoples and their habitat, seated next to several world leaders, including Correa who there and then laid out his ambitious Yasuní-ITT economic agenda. In the audience of many hundreds there were tribal representatives from throughout the Amazon. A huge array of conference halls on the outskirts of Rio, with tens of hundreds of paramilitary protecting the 50,000 delegates and other participants. Honest passions, but empty words. Speeches headed nowhere, while the Amazon continued to burn.

## 21.2    The Paradox of Opposites in the Same Rainforest

Reynard Loki's April 10, 2019 examination of those oil tenures that overlap the nations of "Achuar, Kichwa, Waorani, Shiwiar, Andoa and Sápara" discusses lawsuits filed by Waorani in 2018 against the Ecuadorian ministries that they allege have enabled a government-held auction of "16 new oil concessions covering nearly seven million acres of roadless, primary Amazonian forest" far to the south of Yasuní[11]. "Roadless" represents a key ecological factor in the degree of ecological edge effect damage[12]. From above, the transition from pristine to destroyed is emphatic. The transit zones are dusted over with the indifference of mob violence, in so many hues and forensic blow-by-blow artifacts of human/mechanical transgression. A pit in one's heart explodes with manifest destinies of melancholy; of utter hopelessness when you see it on the ground and from a helicopter. All those endeavoring to save the last remaining rainforest on earth feel the same incredulity,

---

[10]"Indigenous Organizers Halted Plans For Oil Drilling in the Amazon," by Kimberly Brown, March 13, 2019, http://inthesetimes.com/article/21730/indigenous-organizers-halt-oil-drilling-amazon-ecuador, Accessed April 13, 2019; See also, "Indigenous peoples go to court to save the Amazon from oil company greed – Historic lawsuit launched by the Waorani people of Ecuador to save their homes – and our planet – from destruction," by Reynard Loki, Nation of Change, Earth/FoodLife, Independent Media Institute, April 10, 2019, https://www.nationofchange.org/2019/04/10/indigenous-peoples-go-to-court-to-save-the-amazon-from-oil-company-greed/. Accessed April 13, 2019.

[11]See "Indigenous Women Activists Fight to Save Ecuador's Land," by Julia Travers, March 26, 2019, in LABROOTS, https://www.labroots.com/trending/earth-and-the-environment/8370/indigenous-women-activists-fight-save-ecuador-s-land, Accessed April 13, 2019)

[12]"Road-Edge Effects on Herpetofauna in a Lowland Amazonian Rainforest," Ross J. Maynard, Nathalie C. Aall, Daniel Saenz, et. al., First Published March 1, 2016, https://doi.org/10.1177/1940082916009001114, Sage Journals, Tropical Conservation Science, https://crossmark.crossref.org/dialog?doi=10.1177%2F1940082916009001114&domain=journals.sagepub.com&uri_scheme=https%3A&cm_version=v2.0, Accessed April 13, 2019; See also, "Ecuador has begun drilling for oil in the world's richest rainforest, The country is poised to cash in on one of its most valuable assets. But at what cost?" By Jason G. Goldman Jan 14, 2017, VOX, https://www.vox.com/energy-and-environment/2017/1/14/14265958/ecuador-drilling-oil-rainforest, Accessed April 13, 2019.

**Fig. 21.2** A squirrel monkey (*Saimiri sciureus*), Yasuní National Park. (Photo © M.C.Tobias)

a morbid loathing teetering around the madness in others, those who would ignore the all-enduring labors of a planet throughout the tens of millions of years of her biological experiments to produce a Capped Heron (*Pilherodius pileatus*) or the genus morpho, with its 176 species and sub-species of unforgettable iconic, metallic blue butterflies—the iridescence occurring in the nanostructures (lamellae layering, tetrahedral scales) of the dorsal side of the wings. To see, or imagine, their destruction is impossible (Fig. 21.2).

As in the work of the mystery poet of Latin riddles, Symphosius, who lived sometime between the third and sixth century CE, known for his lengthy poem, "Aenigmata,"[13] words and feelings are trampled. "From many mothers I, though one, derive," wrote Symphosius. "When born, I see no parents left alive; they're trampled flat and wounded everywhere; their death creates the power I gain and share..."[14] The profoundly complicated Rumanian/Parisian philosopher E.M. Cioran wrote of our ability to "imaginatively follow in reverse the course of the individual as he comes into life and thereby retrace the various species..."[15] "Paradox also confuses because it asks us to live with simultaneous opposites," writes Charles Handy[16]. To

[13] See Erin Sebo, 'Was Symphosius an African? A Contextualizing Note on Two Textual Clues in the Aenigmata Symphosii', Notes & Queries, 56.3, (2009), 324–26, https://www.academia.edu/8106549.

[14] From *Song Of A Falling World -Culture during the Break-up of the Roman Empire (A.D. 350-600)*, by Jack Lindsay, Andrew Dakers Limited, London, 1948, p. 219.

[15] From *The Fall into Time*, Translated From The French By Richard Howard, Introduction By Charles Newman, Quadrangle Books, Chicago, 1970, p. 113.

[16] *The Age of Paradox*, Harvard Business School Press, Boston, Massachusetts, 1994, p. 47.

this end, the individual, and her/his species, is incapable. Incapable of everything that would matter in an ideal world, and that poses an inseparable crisis. But, as Calin O. Schrag long ago considered, "Why is the problem of fact vs. value a problem?... The true character of this crisis only becomes visible when one moves from the level of methodological analysis to archaeological inquiry"[17]. By then, of course, it is too late to realize effective, virtuous change that conforms to any aboriginal land ethic.

Down on the ground, in the middle of the dirt, the concrete, the metallic infrastructure of petroleum extraction—a grotesque edifice that fumes and gorges like a swollen canker—leads by pipe, truck, and river lorries to towns like Coca[18] barreling through from oblivion to oblivion, graveyard to sand dune. A solitaire stands in the way and is either mowed down or shrinks into some equal oblivion, amid exponentially endowed quadrillion, quintrillion, sextillion numerics that are the cells, leaf stomata, bacteria within the insects and spiders, the droplets of moisture, and vast entanglements of soil. These continua have self-honed into a stability that defies all encroachment, until the eleventh hour. And in Yasuní, an especially critical premonition of such ambiguity abases the human family tree, given that it is considered to be among the most biologically diverse micro-quadrants on earth, with as many as 100,000 invertebrate species per hectare, and other numeric superlatives across nearly every suite of plants and animals. Indeed, scientific consensus now has it that Yasuní, in fact, is as close as we are ever likely to get to something approaching the planet's epicenter of the collective life force.

## 21.3   Evolutionary Enigmas

Biologists speak of the C-value paradox that recognizes far greater amounts of DNA in what most had taken to be, for example, a simple unicellular *Polychaos dubium* (freshwater amoeboid) genome. Whereas nearly 99% of our human DNA is noncoding, meaning that our proteins, amino acids, and genes are largely redundant, or outright "junk." Not so throughout the rainforests of Yasuní. Here, 98% of the DNA is indeed meaningful, perhaps purposeful. These are just words. It has evolved to be necessary for physical thought, not abstract thought. The G-value paradox adds further complexity to what, ultimately, is our own ignorance of an experiential horizon that differentiates between what we take to be authentic and integral, and what is actually all around us biospherically, an integrity comprising proteins and a geography of folding that is heralded and authenticated by time, far more time than the 300,000+ years of *Homo sapiens'* ontology[19].

---

[17] See *Radical Reflection and the Origin of the Human Sciences*, Purdue University Press, West Lafayette, Indiana, 1980, p. 85.

[18] See "11,000 barrels of oil spill into the Coca River in the Amazon," by Jeremy Hance on 12 June 2013, https://news.mongabay.com/2013/06/11000-barrels-of-oil-spill-into-the-coca-river-in-the-amazon/. Accessed April 13, 2019.

[19] See Bioessays. 2007 Mar;29(3):288-99. "The relationship between non-protein-coding DNA and eukaryotic complexity," by Taft RJ, and Pheasant M, Mattick JS., PubMed, The U.S. National Library of Medicine, https://www.ncbi.nlm.nih.gov/pubmed/17295292, Accessed April 13, 2019.

The overall genetics conferring increasing biological integrity as a direct function of time as it relates to the intensive biodiversity that has accumulated with perfect success at Yasuní, and the tragically hopeless case of lawsuits filed by all but powerless indigenous people still living there. Their power is undermined precisely because of their inevitable efforts to lodge their voices in the arena of non-indigenous white noise, where far greater numbers of consumers vie for that same moment of power. All of this is metaphorical, and without absolute predictive qualification. Hence, "About half of the known proteins are amenable to comparative modeling; that is, an evolutionarily related protein of known structure can be used as a template for modeling the unknown structure. [But] For the remaining proteins, no satisfactory solution has been found"[20].

Protein structure prediction is as open to evolutionary traits and categories as any perceived biological or ethnographic tapestry. "Evolution creates mosaics of traits like this," says anthropologist Matt Tocheri at Lakehead University, reacting to news of a previously unknown "human species" from the Philippines 50,000 years ago: *Homo luzonensis*, from the Island of Luzon, and a specific cave known as Callao, where bones were found[21]. The late American molecular biologist Cyrus Levinthal took the numbers game associated with protein folding even further in the mathematical realm, suggesting that the number and surface areas of such genetic turns and twists were all but infinite in their varietal forms and prospects. That there was—despite a freedom by incredible orders of magnitude ($10^{143}$)—a curious paradox at work in a place like the Amazon: a primeval pattern of protein folds adhering to a standard conformation. With so many possibilities, he argued, it should, theoretically, require whole light years to randomly arrive at a standard, any standard, in the formation of bonds within the DNA molecule. Yet, proteins can fold in a nanosecond, without the laborious recourse to all of their options. This has been cited as the Levinthal Paradox[22]. Clearly, the resolution to such a mathematically bizarre empiricism says as much about our perception of numbers and the language we wield to describe and represent them as it does about the reality of the Amazon, or any ecosystem (Fig. 21.3).

These number discrepancies also mirror the case of the Moravec Paradox (named by Hans Moravec of the Robotics Institute of Carnegie Mellon University). Consider that the computational (artificial intelligence) prowess needed to logically repre-

---

[20] "From Summary, "Big-data approaches to protein structure prediction," Johannes Söding, *Science* 20 Jan 2017: Vol. 355, Issue 6322, pp. 248–249, https://doi.org/10.1126/science.aal4512 https://science.sciencemag.org/content/355/6322/248, Accessed April 10, 2019

[21] See "Evidence of New Human Species Found in Philippines," by Robert Lee Hotz, The Wall Street Journal, April 20, 2009; See also, https://www.nationalgeographic.com/science/2019/04/new-species-ancient-human-discovered-luzon-philippines-homo-luzonensis/. Accessed April 10, 2019.

[22] See Zwanzig R, Szabo A, Bagchi B (1992-01-01). "Levinthal's paradox". Proc Natl Acad Sci USA. 89 (1): 20–22. https://doi.org/10.1073/pnas.89.1.20. PMC 48166. PMID 1729690. See also, Levinthal, Cyrus (1969). "How to Fold Graciously". Mossbauer Spectroscopy in Biological Systems: Proceedings of a meeting held at Allerton House, Monticello, Illinois: 22–24. Archived from the original on 2010-10-07. Levinthal, Cyrus (1968). "Are there pathways for protein folding?" (PDF). Journal de Chimie Physique et de Physico-Chimie Biologique. 65: 44–45. Archived from the original, on 2009-09-02.

**Fig. 21.3** A Capped Heron (*Pilherodius pileatus*), Yasuní National Park. (Photo © M.C.Tobias)

sent, say, a macaw's, social insect's, or tapir's brilliant reasoning and poetic capacities requires far less energy than that required for programming basic sensorimotor functionality. How is that possible? Mechanically, it seems that everything is upside down. But that's because hundreds of millions of years have gone into the exquisite perfecting of the physical realities of a jaguar. By now such motion comes easily. But not if it has to be simulated from the beginning. Whereas *abstraction*, by evolutionary standards a newborn, is freakishly untoward, clumsy, without precedent within the rubrics of natural selection. These manuscript pages are, in the most real sense, unnatural, and therefore a kind of utterance, all at once, easily edited, taken in a single grasp. They do not require eons to prepare. Flight feathers or night vision are very different. They rely upon the comparability of evolutionary time to hope for something akin to simulation.

Our proxies fail us, not for lack of imagination but because poetic license is shackled the moment it hits the level of a species.

Alice's wonderland is a Yasuní effect: We stand amazed and unable to surmise the extent of our surroundings, like proverbially gawking tourists aiming cameras at the wilderness from our lazy canoes[23]. What do we expect?[24] What can, what should, we even hope for?

[23] See "Some paradoxes in biology," By Prospector PJ, Beyond pharmacy blog30 APR 2014

The Pharmaceutical Journal, A Royal Pharmaceutical Society publication, https://www.pharmaceutical-journal.com/opinion/blogs/some-paradoxes-in-biology/11137939.blog?firstPass=false, Accessed April 13, 2019.

[24] See "Yasuni: A Meditation on Life," A Yasuni-ITT-Dancing Star Foundation-SATRE Production, 2012, https://www.youtube.com/watch?v=AJP_jlw3brw.

# Chapter 22
# Sakteng

## 22.1 The Pressure from Within

At the time we first ventured to Bhutan back in the mid-1970s, there appeared on the surface to be striking discrepancies within the realms of economic perception outside Bhutan. The family of mathematical equations linking poverty to misery was simply inaccurate. The World Bank in 1974 considered it one of the poorest countries in the world. Yet, persistent Quality of Life Indicators suggested just the opposite, and this gap exists to this day. Most recently, for example, International Monetary Fund (IMF) researchers write, "Consistent with the Easterlin Paradox, available evidence indicates that Bhutan's rapid increase in national income is only weakly associated with increases in measured levels of well-being" based upon a determined economic GDP increase of some 700% during four decades of development, with only alleged modest increases in ascertainable satisfaction by citizens[1] (Fig. 22.1).

In 2014 the Easterlin Paradox was characterized as a "paradox lost," in as much as the "claim was tested using the time trend data available in the World Database

---

[1] p.1, "Gross National Happiness and Macroeconomic Indicators in the Kingdom of Bhutan," Prepared by Sriram Balasubramanian and Paul Cashin, 1 January 2019, WP/19/15 Gross National Happiness and Macroeconomic Indicators in the Kingdom of Bhutan, © 2019 International Monetary Fund WP/19/15 IMF Working Paper Asia and Pacific Department, Accessed April 17, 2019; See also, Easterlin, Richard A. (1974). "Does Economic Growth Improve the Human Lot?" in Nations and Households in Economic Growth. Editors: P.A. David and W.B. Melvin. Palo Alto, Stanford University Press: pp. 89–125. See also, Centre for Bhutan Studies and GNH Research (2016). A Compass Towards a Just and Harmonious Society: 2015 GNH Survey Report. Government of Bhutan: Thimphu. Clark, Andrew E. and Claudia Senik (2011). Will GDP Growth Increase Happiness in Developing Countries? Discussion Paper No. 5595. Bonn: IZA.

© Springer Nature Switzerland AG 2021
M. C. Tobias, J. G. Morrison, *On the Nature of Ecological Paradox*,
https://doi.org/10.1007/978-3-030-64526-7_22

**Fig. 22.1** Brokpa Children, Bhutan. (Photo © M.C.Tobias)

of Happiness, which involve 1531 data points in 67 nations that yield 199 time-series ranging from 10 to more than 40 years"[2].

Bhutan's Vajrayana Buddhist ethics can be divined, in part, within much of the Tibetan Tantric traditions spelled out in the 108 volumes of the Kanjur and 3850 texts and treatises of the Tengyur. This vast canon underlies the overwhelming cultural and spiritual dispositions of the government and her people, Vajrayana being closer to Mahayana than Theravada, or Hinayana Buddhism. As a state religion encompassing nearly 75% of the population, Buddhism circumscribes both the adoption of the country's Constitution in 2008 (100 years in the making), and its defining characteristics of Gross National Happiness (GNH), namely, those "key areas of GNH [which] fall within the domains of psychological wellbeing, health, time use, education, culture, good governance, ecological resilience, community vitality and living standards"[3].

---

[2] "The Easterlin illusion: economic growth does go with greater happiness," by Ruut Veenhoven; Floris Vergunst, International Journal of Happiness and Development (IJHD), Vol. 1, No. 4, 2014, InderScience Publishers, https://www.inderscience.com/info/inarticle.php?artid=66115, Accessed April 17, 2019; See also, Easterlin, Richard (2017). "Paradox Lost?". Review of Behavioral Economics. 4 (4): 311–339. doi:10.1561/105.00000068; See also, DeNeve, J., D. Ward, G. Keulenaer, B. van Landeghem, G. Kavetsos, and M. Norton (2018). "The Asymmetric Experience of Positive and Negative Economic Growth: Global Evidence Using Subjective Well-Being Data," Review of Economic Statistics: 100 (2), 362-375. For further discussion of the Easterlin Paradox, and Bhutan specifically, see Chapter 53.

[3] op.cit, Balasubramanian and Paul Cashin, p.10.

The nation's Constitution is a modern political miracle of fairness, and spiritual jurisprudence. The Chairperson of the drafting committee was Lyonpo Sonam Tobgye, who lives in Thimphu and was, until recently, Bhutan's Chief Justice on her Supreme Court. He is unquestionably one of the world's current great legal minds, an expert on Buddhist law. Like many in his country, he wonders about the challenges to a monarchy—a traditional Kingdom with a King everyone loves (the fifth King), one who is not focused on his "re-election" in a matter of years, but holds the perennial future of the country in his heart. As against the tumult and transiency that a democracy, however vital, represents. He recognizes that Bhutan's first democratic election, in 2008, was an important milestone for the nation. But he also knows that democracy, in and of itself, is not a guarantee of the best in collective human behavior, or the fair distribution of the most integral and ethical quality of life to everyone. At least that is our interpretation of his writing after many extended conversations. The United States, between 2016 and 2020, he cites as the perfect example of widely emergent flaws in the normal rubrics of a democracy. And who could disagree? India's 2019 religious test for immigrants from neighboring, largely Islamic, nations ignites an equally perilous challenge for the world's largest social experiment in alleged distributional justice.

In analyzing Article 5, Section 1 of the Constitution, Lyonpo Tobgye writes that "Bhutan has been successful in preserving its fragile eco-system for the benefit of the country as well as the world at large. [But] The people were understandably worried because economic development and population growth has had a direct impact on the environment...." And he singles out the critical income derived from hydropower for the Bhutanese, now threatened with climate change, whose greenhouse gas signatures are generated outside Bhutan, the greatest of all paradoxes for a nation spiritually devoted to the Buddhist fundament of an interdependent world. Writes Lyonpo Tobgye, "In Bhutanese tradition, it is said: 'Water, oceans, mountains, cliffs and wonderful trees are the abode of the local deities. Do not let it decline. Preserve them as ornaments'"[4] (Fig. 22.2).

This issue of deities in Bhutan is not mere folklore. So-called Green House Deities, of which the scholar Karma Ura identifies 392 (deities as well as spirits) throughout Bhutan[5], are believed to be able to "manifest in scientific, political and cultural terms"[6]. Their ontological primacy fuels protection in every sense; translates into attitudes and the bioheritage recognition, from town to town, region to region, of "citadels" and "sacred groves." Some "are located at the headwaters of springs and rivers" while others take sanctuary in mountains "perhaps holding in their folds unknown vital crops and seeds in the wild. Only passing herders have

---

[4] *The Constitution of Bhutan – Principles and Philosophies*, by Lyonpo Sonam Tobgye, Chairman of the Drafting Committee, ISBN 978-99936-658-7-8, p. 125; See "ༀ།འབྲུག་གི་རྩ་ཁྲིམས་ཆེན་མོ།," [The Constitution of the Kingdom of Bhutan] (PDF) (in Dzongkha). Government of Bhutan. 2008-07-18; See also, *The Judiciary of The Kingdom of Bhutan*, Royal Court of Justice, Thimphu, Bhutan, www.judiciary.gov.bt, n.d.

[5] See *Green House Deities and Environment*, by Karma Ura, Thimphu, Bhutan, 2004, p. 44.

[6] ibid., p.1.

**Fig. 22.2**  Former Bhutanese Chief Justice, Lyonpo Sonam Tobgye. (Photo © M.C.Tobias)

viewed them from close quarters. The chances of human contact with wild life in these places are certainly slim. This means, there is also less chance of humans spreading diseases and spreading new pathogens and chemical pollutants to these places"[7].

The linguistic and psychological relationship between Bhutanese and their ecological deities is intense. Interdependency with biodiversity in native Dzongkha spells out any number of inroads to the connections: takor natang = environment; ranjin natang = nature; ranjin natang lu gami = one who loves nature; nyingje = compassion; zoepa = nonviolence; gyalyong linger = sanctuary; ridag semchen = wildlife; larjung = renewable; sung chop = protect and preserve; Gyalyong Gakyid Pelzom = Gross National Happiness; jha = bird; semchen = animal; meegoe = yeti[8].

In 2007 we trekked 130 kilometers into one of Bhutan's most recently created protected areas, the Sakteng Wildlife Sanctuary in the far east of the country, bordering the state of Arunachal Pradesh in northeastern-most India. Sakteng is one of ten areas designated as a National Park, Wildlife Sanctuary, or Strict Nature Reserve, in addition to eight connecting biological corridors. Their totality equals 19,084 square kilometers, out of 38,393 square kilometers for the whole country, or roughly 50%[9]. In addition, more than 72% of Bhutan retains its primary forest canopy, a

---

[7] ibid., pp.3–4.

[8] Translations by Tandin Wangdi, see *Sanctuary: Global Oases of Innocence*, by M.C.Tobias and J.G.Morrison, A Dancing Star Foundation Book, Tulsa and San Francisco, 2008, p. 314.

[9] See *Bionomics in the Dragon Kingdom – Ecology, Economics and Ethics in Bhutan*, by Ugyen Tshewang, Jane Gray Morrison and Michael Charles Tobias, Springer Fascinating Life Sciences, New York, 2018, pp. 30-52.

continuing vision as proclaimed by the 4th King, His Majesty Jigme Singye Wangchuck, back in the 1970s when he was still a teenager. This, in turn, was in accord with the Bhutanese Zhabdrung Rinpoche's legal code of 1729, a firm precursor to the GNH policies of today, in which was articulated the "responsibility of any government to create happiness for the people"[10].

Among the indigenous 2000+ Brokpa of Sakteng and Merak, the two principal villages of the Sakteng Wildlife Sanctuary, GNH and feelings for, and a sense of profound duty toward, protection of nature are irrepressible reciprocities. When we first surveyed Sakteng (the second such thorough survey since the Wildlife Institute of India had sent two scientists there in 2004, one year after the Sanctuary's creation), the taxonomic numbers included 18 mammal species, 119 avians, and 203 plant species, all within 21 known forest communities[11]. By 2018 research had vastly increased this remarkable biotic repository: 622 vascular plant species, 30 mammals, and 227 birds[12].

While Sakteng Wildlife Sanctuary had, in great measure, been created to protect the Yeti (meegoe)—everyone we interviewed had seen it him/herself, or knew of a friend or family member who had also encountered it directly—there was, as yet, no direct scientific evidence of this apparently tall, bipedal vertebrate, comparable in size and strength to a Great Ape. However, the mass of local anecdotal assurances that it indeed existed, among so many other cryptic species, deities, and uncanny spiritual beings who saw to it that the entire region was sacred, promulgated its own inimitable peculiarities and sense of possibility. Why else, in this remote mountainous region, would there exist so unambiguous a convergence of opinion on the matter?

We re-emerged from the hallowed rhododendron epicenter of the eastern Himalayas (35 species in Sakteng, several endemic, out of a total of 46 across all of Bhutan) completely enchanted by the yaks, the yakherding Brokpa, and their universe. There were no tourists allowed in Sakteng at that time. No road. No hotel. One government bungalow for visiting guests.

In subsequent years, the Brokpa population has remained stable (a very low fertility rate) but the buffer zones surrounding Sakteng have burgeoned with a current human population exceeding 11,000. Moreover, a road has been built and will soon be fully covered in tarmac. "People reasoned that eco-tourism benefits only a few individuals…" [whereas] "A road benefits the entire community"[12]. By February 2019, the road had been completed and the outcries of happiness were manifest: "Sakteng Gup Sangay Dorji said that the road would expedite developmental projects in the gewog. 'We are extremely happy. For us it is dream come true. Many cried tears of joy seeing the first vehicle arrive in our village'"[13] (Fig. 22.3).

---

[10] Citing K.Ura, et.al., "An extensive analysis of GNH index," The Centre for Bhutan Studies, 2012, in *Bionomics*, ibid., p. 10.

[11] op.cit., *Sanctuary*, p. 290.

[12] ibid, *Bionomics*, p.39.

[13] See "Road to reach Sakteng's gewog centre," January 28, 2017, by Tempa Wangdi, Kuensel, http://www.kuenselonline.com/road-to-reach-saktengs-gewog-centre/. Accessed April 17, 2019.

**Fig. 22.3**  Leisurely Yak in Sakteng. (Photo © M.C.Tobias)

## 22.2    The Philosophy and Computational Ecology
of a Single Road

Issues of mobility, population pressure, development, and technology innovation
need not equate with the potential depredation of utterly sanctified places like
Sakteng, or the build-up of microplastics in the ocean, although to date that has been
the predictable correlation. Our hearts sank when we heard about the ambitious road
being constructed into Sakteng—just as we were alarmed, and then distraught,
decades ago when the TransAmazonian Highway began construction, or, as earlier
referenced, when oil drilling began in Yasuní. This has been the story of civilization,
whether by roadbuilding, the erection of barbwire fencing, the conquest of civilizations by endogenous cultures, Old Worlds, New Worlds, or dissemination of pesticides like dichlorodiphenyltrichloroethane (DDT) that made for Rachel Carson's
*Silent Spring*.

Conversely, we see technology, as applied to manufacturing alternatives to meat
and fish for human consumers, as a long-overdue and—we would like to hope—
predictable evolution of the human spirit. The *Economist* magazine labeled 2019 the
"Year of the Vegan," stating, "Fully a quarter of 25- to 34-year-old Americans say
they are vegans or vegetarians"[14].

---

[14] See "First Vehicles Reach Sakteng," KuenselOnline, February 4, 2019, by Younte Tshedup,
http://www.kuenselonline.com/first-vehicles-reach-sakteng/. Accessed April 17, 2019.

At the same time, a 16-year-old, Greta Thunberg, stood before a meeting of the Davos Economic Forum and told everyone they must "panic…the house is on fire." The audience stood and applauded[15]. To cheer when informed that the house is on fire is something new, if not a little insane. And in New Zealand, the bird with the least privacy and most intensively manhandled existence on the planet, the Kakapo (*Strigops habroptila*)[16], in one breeding season went from a population under 150 to over 200 due to the climate change-related strength of the indigenous rimu tree fruiting mast, the Kakapo's favorite food[17]. The numbers would soon sink to below 150, however. Elsewhere, Japanese farmers have collectively brought back over 160 White Oriental Storks (*Ciconia boyciana*) to their country (out of a global population of approximately 2500 individuals). Fluctuations send a clear signal that human cultures suffer their own internal eco-dynamics, in sync with those they inflict.

These challenges to our ethical sense of stability are not without precedent, regionally. Germany in World War II made that clear. But at the global scale, the local resident, the citizen, has suddenly become the tourist, the nomad, the wayfarer, the invading marauder. Mathematically, attempts to remind locals of the planetary nature of human cumulative destruction fail to leave much of an impression, like oblique asymptotes with no resolution at an ideal point of convergence. In other words, two story lines that are forever incapable of merging into one. This asymptotic paradox is as tactile in the number of random and nonrandom points on a curve intersecting a line as it is psychological: there are no rational proxies for grasping self-destruction at the level of the entire biosphere, no matter how many actual points, sequences, and events of destruction we recognize. And like the Sorites Paradox, we have all but lost the capacity for extrapolating from the one to the many[18].

[15] See "The Year of the Vegan," by John Parker, The Economist, "The World in 2019," https://worldin2019.economist.com/theyearofthevegan, Accessed April 17, 2019; See also, "The Coming Obsolescence of Animal Meat -Companies are racing to develop real chicken, fish, and beef that don't require killing animals. Here's what's standing in their way," by Alga Khazan, The Atlantic, April 16, 2019, https://www.theatlantic.com/health/archive/2019/04/just-finless-foods-lab-grown-meat/587227/, Accessed April 17, 2019.

[16] See "I want you to panic,' Swedish teen raises climate alarm at Davos," by Nina Larson, The Local, Sweden News in English, January 26, 2019, https://www.thelocal.se/20190126/i-want-you-to-panic-swedish-teen-raises-climate-alarm-at-davos, Accessed April 17, 2019.

[17] https://www.iucnredlist.org/species/22685245/129751169, Accessed April 17, 2019; See also, the New Zealand section of "Hotspots," www.hotspots-thefilm.com.

[18] See "Rare kakapo parrots have best breeding season on record," BBC News, April 17, 2019, https://www.bbc.com/news/world-asia-pacific-47960764, April 17, 2019.

# Chapter 23
# A River Somewhere in Georgia

## 23.1 Negative Convergence

By 1981, interdisciplinary approaches to grasping the nature of human evolution tended toward a short-list of behavioral-modifying candidates. James V. Neel, the Lee R. Dice University Professor of Human Genetics at the University of Michigan Medical School, asserted the following conditionals in an essay for a Smithsonian symposium: "Genetic structure," "Differential fertility," "Inbreeding," "Genetic consequences of village fissions," "Disease pressure," "Dietary patterns," "Loss of human diversity," "Changes in the nature of natural selection," "Our New Epidemiological Vulnerability," "More Inertia in the Gene Pool(s)," and an "Accumulation of Deleterious Genes Due to Relaxation of Inbreeding." This was the human saga as viewed through the lens of amino acids, polypeptides, nucleotides, and deoxyribonucleic acid (DNA).[1] While it satisfies the temptations of one narrative, it leaves out entirely that odyssey entailing "the inspiration of sentiment and irregularity"[2] as well as "an atom in the excited state."[3] And it certainly does not explicitly evoke often unseen horror, for example, "crabs [who] die from an inability to molt as they grow older," or "dispossessed" muskrats who—dispatched for one reason or other into exile—"hobbled across open spaces or through rushes and weeds, bleeding, trying to find food and shelter… dying from hunger or cold, their tails, feet and eyes freezing…."[4]

---

[1] See Chapter 3, "Some Base Lines for Human Evolution and the Genetic Implications of Recent Cultural Developments," by James V. Neel, pp.67–93, in *How Humans Adapt – A Biocultural Odyssey*, Donald J. Ortner, Editor, Foreword by S. Dillon Ripley, Epilog by Wilton S. Dillon, Smithsonian Institution Press, Washington D.C., 1983.

[2] See *A World With A View – An Inquiry into the Nature of Scenic Value*s, by Christopher Tunnard, Yale University Press, New Haven and London, 1978, p. 87.

[3] See "Theory Of Gravitation," by Hermann Bondi, in *Cosmology and Astrophysics*, Edited by Yervant Terzian and Elizabeth M. Bilson, Essays In Honor Of Thomas Gold, Cornell University Press, Ithaca and London, 1982, p. 73.

[4] See *Dismantling Discontent: Buddha's Way Through Darwin's World*, by Charles Fisher, www. DismantlingDiscontent.com, Elite Books, Santa Rosa, CA, 2007, p. 111

© Springer Nature Switzerland AG 2021
M. C. Tobias, J. G. Morrison, *On the Nature of Ecological Paradox*,
https://doi.org/10.1007/978-3-030-64526-7_23

So connected are all such referential points on the compass of biological experience that, as A.K. Coomaraswamy explored by way of "the native connection" of cross-cultural mathematical relationships, "the beginning of all series" [is] "the same as their end."[5] Jains, conversely, suggest that the finite mathematical points at any origination or conclusion are fictive. The genetics of human associations do not come close to intimating that open forum of interspecies biosemiosis in which humanity evolved. Foxes speak earnestly with snails; people with birds; lions with lambs. This conceptualization of our lives within a greatly expanded timeframe and order of molecular magnitudes eclipses steadfast paradigms, like Western Civilization, or singularity points within genetics, where we can experiment with our future.

In fact, at a moment's notice, our introspection explodes upon a fuzzy logic of frontiers that heed no formalism, no rules. Orientation flees conventional attitudes, embracing the most conceivably eclectic of all thought experiments, a meditation as anonymous as a signature the size of a river in Georgia, for example. With their eternally changeable whims, nocturnal enigmas, and nonlinear nuances, such rivers are without authorship or guidance. Their reality is that of some history that does not comport with river genetics, but the ebb and flow of utterly unpredictable human events at larger scales. The connections have no science by which to proffer predictions or coordinates. They may flood or dry up and disappear, like the Sarasvati of ancient Harappan culture in India's Thar Desert (Fig. 23.1).

We stand beside banks vastly overflown this winter, one of countless documented January floods, where the Oconee heads wide and bristling toward its convergence with the Ocmulgee, the western tributary of the Altamaha. It flows past the South Appalachian Mississippian mounds, dating to their eleventh-century constructions by their indigenous occupants, and heading—in total 357 miles—into the Atlantic near Darien. It is the largest freshwater system east of the Mississippi. Her traditional peoples still speak Chackasaw, Choctaw, Creek-Seminole, Hitchiti-Mikasuki, Apalachee, Alibamu, and Coushatta, all languages of the Western and Eastern Muskogean peoples. Overhead, at least 425 bird species are known to have made their way at one time or other.[6] And to all sides at least 84 dominant reptiles, including 46 snake species like the magnificent and at-risk Eastern Indigo, one of which lies dismembered, chopped in half, perhaps by a sadistic biker, there in front of us. On a sandy ridge 1000 feet away, above the high-flood mark, are burrows of the alligator snapping turtle (*Macrochelys temminckii*), some exceeding 200 pounds, making it the largest freshwater turtle in the world. We have seen aquatic salamanders, the Eastern Hellbender, one of 80 or so known amphibians in the state; Oconee burrowing crayfish, from among 20 crayfish species; dozens of the 125 glorious dragonfly species; and half of the 112 mammalian species known to exist in the state

---

[5] See "*Kha* and Other Words Denoting 'Zero,' in Connection with the Indian Metaphysics of Space," in *Coomaraswamy*, 2: Selected Papers, Metaphysics, Edited by Roger Lipsey, Bollingen Series LXXXIX, Princeton University Press, Princeton, NJ., 1977, p. 220

[6] See "Georgia Ornithological Society, GOS 2018, https://www.gos.org/2018-checklist, Accessed April 19, 2019.

**Fig. 23.1**   Oconee River through Milledgeville, Georgia. (Photo © M.C. Tobias)

(although the panther is a reach, with an underground circulation of night photos claiming to have caught a glimpse of this rare Floridian, and nearly as rare Texan/ Floridian hybrid).

As for all the fish, mussels, and snails—Frecklebelly Madtoms, the Ochlockonee Moccasinshell, the Tangerine Darter—we are lost in the maze of taxonomies; the Etowah Darter (*Etheostoma etowahae*) is believed to be the most endangered species in the state, along with the North Atlantic Right Whale.[7]

---

[7] See "Here Are The Most Endangered Animals In Georgia - There are 48 threatened or endangered animals in Georgia, including loggerhead turtles," by Deb Belt, Patch National Staff, Community Corner, May 21, 2018, https://patch.com/georgia/atlanta/here-are-most-endangered-animals-georgia, Accessed April 20, 2019.

As we venture down river, inching our way toward the Atlantic, we see river otters, an enormous pileated woodpecker searching patiently for carpenter ants in the rotted out holes, mid-canopy, of a Hickory glen, Mississippi Kites, a Red fox and Eastern Fox Squirrels, the odd wild turkey, and raccoons. Georgia, and the Oconee River specifically, presents to the adverting heart a lovely and bewildering biological labyrinth.[8] A beautifully planted garden millions of years ago, whose geology and human history have much altered its contours, and pruned it in ways both rude and reckless. Nonetheless, a garden remains.

It is, of course, the one American state named after a King, George II, when it was founded—the last of 13 colonies—by James Oglethorpe in 1733. A confusion of ecotones, overlapping systems, biotic revelations, as in the case of Broxton Rocks, near Douglas, 90 minutes south of where we presently sojourn. At Broxton one chances upon a remarkable microbiome of cascades plunging off mini-cliffs teeming with ground orchids, and a miasma of wiregrass and Japanese climbing ferns, the now rare longleaf pines beyond.[9] Broxton and the Oconee are but two moments in tens of millions of years' worth of bioreverie amid Georgia's myriad physiographic minutiae—highlands, lookout plateaus, peneplains, the Piedmont and Coastal plains, sounds, tidal creeks, salt marshes, 15 barrier islands[10]—from the Blue Ridge and 11 peaks over 4000 feet, to the so-called Austroriparian (humid-coniferous) Biotic Provinces.[11] The State, ripe with contradictions, is a repository of every paleological and illogical biological assertion one cares to posit. It might as well be the entire world (Fig. 23.2).

We have seen plentiful bones from extinct species that meandered throughout these near tropical groves and marshlands, like the giant *Bison latifrons*, the largest bovid ever documented, horns from tip to tip exceeding 200 centimeters, over four times that of a modern bison.[12] *Latifrons* vanished during the Last Glacial Maximum of the Pleistocene in Georgia, approximately 20,000 years ago. But for well over 200,000 years this 2700-pound herbivore enjoyed the thickets of fog and over 3600 plant species to sample, a monarch in its domain, co-habiting its luscious swamps with short-faced and cave bears, ground sloths and saber-toothed cats. Macrofossils of the Columbian mammoth and Giant Land Tortoise have also been unearthed in

---

[8] See "The Flora of the Oconee National Wildlife Refuge," Georgia Botanical Society, compiled by Marie B. Mellinger, © 1997, John Pickering, https://www.discoverlife.org/nh/cl/ONF/plants.html, Accessed April 20, 2019.

[9] https://www.nature.org/en-us/get-involved/how-to-help/places-we-protect/broxton-rocks/, Accessed April 19, 2019.

[10] See "Island Habitats," https://ugami.uga.edu/sapelo-island/island-habitats/, Accessed April 20, 2019.

[11] See "Physiographic & Biogeographic Regions Of Georgia," by Robert A. Norris, in *Georgia Birds*, by Thomas D. Burleigh, with Reproductions of Original Paintings by George Miksch Sutton, University of Oklahoma Press, Norman, OK, 1958 pp. 25–76.

[12] See "Bison lattifrons" in "Prehistoric Fauna" site by Roman Uchytel, https://prehistoric-fauna.com/Bison-latifrons, Accessed April 20, 2019.

**Fig. 23.2** Bones of the extinct *Bison latifrons*, Georgia College Paleontology Laboratory. (Photo © M.C. Tobias)

excavations at the Clark Quarry near Brunswick, Georgia.[13] We stroke some of their bones, laid out on tables at a laboratory at Georgia College in Milledgeville, where, it seems, a true combustion of enthusiasm for the natural world continues to reverberate. One of the college's famed alumni, the late author Flannery O'Connor, displayed from childhood an enduring fascination with all creatures, particularly peacocks, which she raised at her family's nearby farm in Milledgeville known as Andalusia. Today, the farm's several hundred acres are all but engulfed by a modern world attuned to noisy highways and dense commercial outlets. But once 1000 meters inside the confines of O'Connor's world, that nature, even from a few decades ago, all rings true.[14]

The fame O'Connor garnered as a child for her love of birds mirrors the fact that Georgia remains an epicenter for biodiversity studies. Not just by the likes of Al Mead, the Late Pleistocene expert in the region, or famed paleobiologist Melanie DeVore, or the geneticist Eugene M. McCarthy, a University of Georgia specialist on hybrids in birds and mammals (he has suggested that nearly half of all avifauna,

---

[13] See "The Pleistocene Vertebrate Fauna Of Clark Quarry, Brunswick, Georgia," Paper No. 1–6, by Robert A. Bahn and Alfred J. Mead, The Geological Society of America, Southeastern Section – 54th Annual Meeting, March 17–18, 2005, https://gsa.confex.com/gsa/2005SE/finalprogram/abstract_82857.htm, Accessed April 20, 2019; see also https://www.thedahloneganugget.com/news-subscribers/fossils-georgia-coast-next-archeology-meeting, Accessed April 20, 2019.

[14] See "Of Birds and Dreams and Flannery O'Connor," April 8, 2016, by Michael Tobias and Jane Gray Morrison, http://andalusiafarm.blogspot.com/2016/04/of-birds-and-dreams-and-flannery-oconnor.html, Accessed April 19, 2019.

and probably mammals, have hybridized naturally at one time or another). But then there is Wayne Clough, 12th Secretary of the Smithsonian whose virtual backyard as a child was Broxton Rocks.[15] And the late Eugene Odum, whose 1953 textbook *Fundamentals of Ecology*, co-authored with his brother Howard, and his later Institute of Ecology at the University of Georgia, all but engendered the very study of ecosystems in America as a new subset of conservation biology.[16]

Add to that legacy the earliest ornithologists in America, Mark Catesby in 1723,[17] the eloquent William Bartram, son of John, America's first official botanist up in Philadelphia,[18] and John Abbot, whose water colors executed between 1790 and 1809, though unpublished in his time, comprise the first comprehensive ornithological study of any one state in the entire United States.[19] Alexander Wilson came to Georgia in 1809, meandering throughout the Altamaha River basin, studying wintering birds in Georgia, and confirming his own notes with Abbot. Later, Thomas Nuttall—whose plant collecting in the early 1800s added a remarkable storehouse of data to science—also showed up in Savannah, as did John James Audubon. Because the state is so biologically abundant, and the largest state east of the Mississippi, it is little wonder that it aroused such enormous enthusiasm for students of the natural world, including John Muir, who on September 21, 1867, wandered into Georgia from North Carolina and proceeded to walk throughout the region, rhapsodizing about almost everything he laid eyes upon, as was his wont.[20]

---

[15] See Clough's book, *Things New and Strange: A Southerner's Journey through the Smithsonian Collections*, University of Georgia Press, Athens, GA, 2019.

[16] See Betty Jean Craige's *Eugene Odum: Ecosystem Ecologist and Environmentalist*, University of Georgia Press, Athens, Ga., 2001; see also "An Ode to Odum," Eugene Odum: Ecosystem Ecologist and Environmentalist, Reprint ed., Betty Jean Craige, University of Georgia Press, Athens, BioScience, Volume 53, Issue 12, December 2003, pp. 1229–1230, https://doi.org/10.1641/0006-3568(2003)053[1229:AOTO]2.0.CO;2, published, Oxford Academic, December 1, 2003, https://academic.oup.com/bioscience/article/53/12/1229/303121, Accessed April 20, 2019.

[17] See his two-volume masterpiece, *The Natural History of Carolina, Florida and the Bahama Islands*, London 1731–1743.

[18] See *Travels through North and South Carolina, Georgia, East and West Florida, the Cherokee Country, the Extensive Territories of the Muscogulges or Creek Confederacy, and the Country of the Choctaws. Containing an Account of the Soil and Natural Productions of Those Regions; Together with Observations on the Manners of the Indians. Embellished with Copper-Plates*, London 1791; see also "Georgia Historical Society," https://georgiahistory.com/education-out-reach/online-exhibits/featured-historical-figures/william-bartram/bartram-in-georgia/, Accessed April 20, 2019.

[19] op.cit., Burleigh, pp. 6–7.

[20] See "*Muir's walk across Georgia and on to the sea,*" https://muirsouthtrek150.weebly.com/georgia.html; see also https://www.exploregeorgia.org/blog/muir-his-appalachian-experience-and-the-modern-connection, Accessed April 19, 2019.

## 23.2   In Opposition to Reverie

Numerous writers and ecologists have sought to re-trace Muir's ecstatic trails as he combed voluminously throughout the southeast (as he did in California, Alaska, and elsewhere).[21] His literary reveries yield a picture that was enshrined five years after his time in Georgia in the massive two-volume *Picturesque America: Or, The Land We Live In*, wherein Savannah is portrayed to resemble the same tonal tropical paradise as the famed "Heart of the Andes" painting by Frederic Church in 1859 and entered into the exalted interiors of the Metropolitan Museum in New York.[22] But that paradise had long been tainted, in the case of Georgia. By the time Muir had reached the borders of the State, like most of the South, the geography and a large percentage of its inhabitants, human and otherwise, had been grotesquely oppressed, or eclipsed entirely.

That was the headline for a region whose customs of a century or more had tragically endorsed economic dependence on slave labor in the cotton fields. There were 3.2 million slaves in America by 1850, 1.9 million of those individuals picking cotton.[23] In Georgia, the Trustees Garden in Savannah had received seed specimens from Chelsea England almost immediately upon colonization. So-called *Georgia cotton* had been received back in England as early as 1741. By 1743, despite objections from the colony's founder, "the use of slave labor [in Georgia] was legalized." The vast industries abetting slavery had a rapacious impact on Georgia's ecosystems, which had co-evolved in an unadulterated spasm of aggressions and indifference. It was, in essence, that very cruelty meted out to much of that natural history abundance which Muir, and so many others like him, had come expressly to experience, whether they quite grasped the grotesque ironies or not.[24]

By the middle of the nineteenth century, Eli Whitney's 1793 invention of the cotton gin had enabled Georgia to become the veritable fountainhead of the Industrial Revolution, monetary profits from which propounded the lure of European immigration to unprecedented levels—"6.2 million" in the decade of the 1850s. By the time the Civil War would devastate commercial cotton, Georgia boasted over 1000 plantations, over 1000 acres each, populated in large measure by slave laborers doing all the back-breaking work.[25] Emancipation would decrease the size of family farms, but also provide for a liberated agriculture at odds with native biomes.

---

[21] See https://vault.sierraclub.org/john_muir_exhibit/geography/retracing.aspx, Accessed April 20, 2019.

[22] Edited by William Cullen Bryant, in D. Appleton And Company, New York, 1872, p. 117.

[23] See E. J. Donnell, *Chronological and Statistical History of Cotton,* James Sutton & Co., New York, 1872.

[24] See *Cotton Production and the Boll Weevil in Georgia: History, Cost of Control, and Benefits of Eradication*, by P.B.Haney, W.J.Lewis, and W. R. Lambert, The Georgia Agricultural Experiment Stations, College of Agricultural and Environmental Sciences, The University of Georgia, Research Bulletin Number 428, reviewed March 2009, p. 2, https://athenaeum.libs.uga.edu/bitstream/handle/10724/12179/RB428.pdf?sequence=1, Accessed April 20, 2019.

[25] ibid., pp. 3–4.

By 1903, Georgia had another problem, which has been characterized as "the most serious menace that the cotton planters of the South [had] ever been compelled to face," as potential losses from a single insect—*Anthonomus grandis grandis*—mounted in the hundreds of millions of dollars.[26]

The massive, indeed internationally coordinated, efforts to stop the rampaging boll weevil failed (it migrated over vast territory quickly and could build up seven generations in a single human year);[27] and in sync with the Great Depression, the Dust Bowl, and huge applications of arsenic in conjunction with molasses to the soil, entomologists furiously engaged in an all-out battle with a lovely looking little personage who persistently both imagined and evoked resistance to whatever chemicals were laid down. "By 1953," write Haney, Lewis, and Lambert, "there were 3259 brands of economic poisons and application devices from 439 companies registered under the 1950 Georgia Economic Poisons Act."[28] As farmers desperately moved, those that could, from one turf to the next, over seven inches of soil were, on average, lost everywhere that farming families trod. Such was the inferno of poor farming practices and ill-informed zeal bent upon perpetuating the doomed cotton legacy.

Throughout this devastating period, Georgian lumber mills wreaked as much or more havoc, as commercial sawmills (by the time of the Civil War) were already consuming "more than 1.6 million board feet per day" of longleaf pine (*Pinus palustris*), as well as vast swaths of the old-growth cypress (three species of the *Taxodium* genus in the Cupressaceae cypress family of southeastern American conifers).[29] Before commercial logging had begun, there were an estimated 90 million acres covered in longleaf pine. Today, less than 3% of these stands remain[30] (Fig. 23.3).

Additional stressors on John Muir's southeastern paradise were to cumulatively manifest. Georgia had a population of roughly 1.8 million people when he tramped around. A century later, by 1970, there were over 4.5 million inhabitants (and 10.5 million+ today), despite many having given up on a rural livelihood in the South, embarking—those who could—upon the Great Migrations during the first half of the twentieth century. This exodus resulted in over 6.5 million people fleeing southern areas of America for northern, urban sanctuaries. For African-Americans, prior to 1964, and Public Law 88–352 (78 Stat. 241)—the Civil Rights Act—a majority managed to evacuate their perennial nightmare of the South for northern states only if they had a relative (e.g., an "Auntie") who could shelter them, or were equipped

---

[26] ibid., p. 10.

[27] See Animal Diversity Web, "Anthonomus grandis," by Ben Thompson, https://animaldiversity. org/accounts/Anthonomus_grandis/, Accessed April 20, 2019.

[28] ibid., p.18.

[29] See New Georgia Encyclopedia, "Environmental History of Georgia: Overview," by Leslie Edwards, Atlanta, GA, May 25, 2004, https://www.georgiaencyclopedia.org/articles/geography-environment/environmental-history-georgia-overview, Accessed April 20, 2019

[30] "Photos Document the Last Remaining Old-Growth Pine Forests of the American South," by Jennifer Nalewicki, smithsonian.com, January 19, 2018, Smithsonian, https://www.smithsonian-mag.com/travel/photos-capture-longleaf-pine-forests-american-south-180967891/,      Accessed April 20, 2019.

**Fig. 23.3** Longleaf pine forest at Broxton Rocks Preserve, Coffee County, Georgia. (Photo ©
M.C. Tobias)

with an irrepressible courage in the face of one of the worst and most sustained documented oppressions in human history—America. That, in itself, must rank as among
the worst forms of ecological paradox ever rained down upon a human community.

But Georgia's own story continues: By 1960, there were an estimated "400 toxic
chemicals"—largely unregulated—being added to America's ecosystems, and,
within the state of Georgia, "70 percent of the municipal sewage" was untreated at
the point of entry into her 14 river basins and nearly 110 rivers and creeks.[31] The
poorest of the poor—again African-Americans—tended to be the downstream victims of such pollution and hence some of the first attempts at environmental justice
in the United States grew up in Georgia.

As of 1967 that situation began to change, as organizations like the Georgia
Conservancy mobilized public awareness of what, to many, reads as one of the most
devastating biocultural regional histories anywhere in North America.[32] In 2002, the
Georgia Water Coalition emerged, eventually adding the voices of some "200
organizations committed to ensuring that water is managed fairly for all Georgians
and protected for future generations."[33]

---

[31] ibid.

[32] See https://www.georgiaconservancy.org/policies/water

[33] See https://www.gawater.org/, Accessed April 20, 2019.

## 23.3   A Near-Perfect Mirror

Walking along the Oconee River, the major basin of which drains "5,330 square miles,"[34] provides a salient, if disturbing, lookout over this chemistry of ecological and human fiascos and, in the same instant, hopefulness.[35] A perfect mirror upon catastrophic propensities and perpetual turns-of-conscience that together may galvanize twenty-first-century contemplations of what it is we are—the one species, perhaps, that actually *doubts* its own kind, yet is too stubborn or dreamy to turn back. Where would we go? Flint, Michigan?

Such reflections are additionally addled by the fact of January flooding, whose aftermath translates to a swarm of high-intensity pollutants and debris fields throughout the waterlogged oak-pine forest systems to both sides of the restive river.[36] The garbage load to this day is distributed everywhere one tries to walk: pools of distinct sewage aftermath, toxic fecal coliform counts, storm wastes of every conceivable chemistry. Side pools show evidence of algal build-up from thousands of fertilizing-polluting points, nitrates, ammonia, their result being the abstraction of oxygen and killing of fish. Right there, a decaying largemouth bass, in the very zone of three protected riparian species: the Altamaha Arcmussel and Oconee Burrowing Crayfish, in addition to the Robust Redhorse.[37] In its laborious turns the life of the Oconee journeys due south of Milledgeville, beyond one of Georgia's most notable Public Liberal Arts Colleges, and also past the Bartram Forest Wildlife Management Area, which abuts the most historically notorious psychiatric hospital in the state.

Whatever the odd juxtaposition of history and contemporary life, nothing is as striking as the *mental* and ecological problems of that garbage in the river—in the microcosm of a perceptual instant, with eutrophic pools forming one after the other as photosynthesis declines as a measure of excessive algal growth.[38] The banks of the Oconee on this day represent a global syndrome, bringing back personal

---

[34] See "Georgia River Network, https://garivers.org/oconee-river/, Accessed April 20, 2019.

[35] See L. Shearer, L. *"Athens streams show 15-year decline; leaky sewers, runoff likely culprits," 2017,* onlineathens.com. http://onlineathens.com/local-news/2017-05-02/athens-streams-show-15-year-decline-leaky-sewers-runoff-likely-culprits, Accessed April 21, 2019.

[36] See "Georgia River Network," Biological Resources, https://garivers.org/oconee-river/, Accessed April 21, 2019. See also *"Policy Idea 2: Watershed Pollution in Athens Clarke County—Oconee River Fertilizer and Sewage Pollution," by Rachael Akinola, September 7, 2017,* https://medium.com/@RachaelFAkinola/policy-idea-2-watershed-pollution-in-athens-clarke-county-oconee-river-fertilizer-and-sewage-a84dfbd78fad, Accessed April 21, 2019.

[37] See "Georgia River Network," Biological Resources, https://garivers.org/oconee-river/, Accessed April 21, 2019.

[38] See "Eutrophication: Causes, Consequences, and Controls in Aquatic Ecosystems," by Michael Chislock, M. F., Enrique Doster, E., Rachel A. Zitomer, R. A.lan E. Wilson, 2013, Nature Education Knowledge 4(4):10, https://www.nature.com/scitable/knowledge/library/eutrophication-causes-consequences-and-controls-in-aquatic-102364466, Accessed April 21, 2019.

memories of Italy's Sarno, India's Ganga, and the Buriganga in Bangladesh, of Brazil's Doce and China's Yellow River. And they harken to that ugly outflow of the Mississippi, the pollution visible from space.[39]

But they also portend what Brazil's notorious Cubatao River,[40] not to mention the Thames,[41] might yet signal: significant ecological restoration. Somehow, on a cold late afternoon, these thoughts serve up a fitting metaphor for yet another fundamental, if odd, biological paradox, the evolution of which occurred in water, before transmigrating to land, good for some species, fatal to others, namely, oxygen herself.

## 23.4  The Oxygen Paradox

The oxygen paradox pertains to its "conversion into oxygen radicals, and other reactive oxygen species, that cause damage to cells, organs, and organisms."[42] A substance eukaryotes cannot live without, and, in many cases, cannot live with. Whether liquid, gas, or solid, there are scores of explanations within the evolutionary biochemical canons for the realities of oxygen and dissolved oxygen, as well as all those untidy reactions to $O_2$ which create potentially lethal "oxygen-based free radicals." Despite the apparent fundamental dialectic, life herself, with all the seemingly contradictory behavioral norms, provides for what we primates take to be a predictable lifespan, in an aggregate of multiples, and a confusion of senseless inflictions. Is this merely one more example, among an infinitude, of the *ecological paradox* we simply must accept? It was Galileo Galilei who said that "By denying scientific principles, one may maintain any paradox."[43]

---

[39] See http://1mississippi.org/mississippi-river-makes-most-endangered-list/, Accessed April 21, 2019.

[40] See "Cubatao Journal; Signs of Life in Brazil's Industrial Valley of Death, by James Brooke, June 15, 1991, New York Times, https://www.nytimes.com/1991/06/15/world/cubatao-journal-signs-of-life-in-brazil-s-industrial-valley-of-death.html, Accessed April 21, 2019.

[41] See *Restoration of the Tidal Thames*, by Leslie B. Wood, Adam Hilger Publishers, Briston, UK, 1982; and the Thames River Trust, http://www.thamesrivertrust.org.uk/thames-restoration-fund/, Accessed April 21, 2019.

[42] See SpringerLink, GeroScience, "The Oxygen Paradox, the French Paradox, and age-related diseases," quoted from the Abstract, by Joanna M. S. Davies, Josiane Cillard, Bertrand Friguet, Enrique Cadenas, Jean Cadet, Rachael Cayce, Andrew Fishmann, David Liao, Anne-Laure Bulteau, Frédéric Derbré, Amélie Rébillard, Steven Burstein, Etienne Hirsch, Robert A. Kloner, Michael Jakowec et al., December 2017, Volume 39, Issue 5–6, pp. 499–550, first online: December 21, 2017, https://link.springer.com/article/10.1007/s11357-017-0002-y, Accessed April 21, 2019.

[43] Cited in "The Oxygen Paradox," by Monica Butnariu, Journal of Pharmacogenomics and Pharmacoproteomics, 2012, 3:1 DOI: https://doi.org/10.4172/2153-0645.1000e104, https://www.ncbi.nlm.nih.gov/pmc/articles/PMC4838776/, https://www.omicsonline.org/the-oxygen-paradox-2153-0645.1000e104.php?aid=3762, https://www.longdom.org/open-access/the-oxygen-paradox-2153-0645.1000e104.pdf, Accessed April 21, 2019.

Others have been more generous in their appreciation for the conflict of logical thought (or thoughtlessness) and subsequent behavior. In biological Georgia, logical types are easily transmogrified, and then restored to some level of human submission. From our point of view, all is compromise. There are no tenable ideals.

Edward Teller, for example, suggested that "Two paradoxes are better than one; they may even suggest a solution."[44] While Søren Kirkegaard wrote, "This, then, is the ultimate paradox of thought: to want to discover something that thought itself cannot think." Or Friedrich Schlegel, who stated, "Irony is a form of paradox. Paradox is what is good and great at the same time."[45] Physicist Richard Feynman stated that "The 'paradox' is only a conflict between reality and your feeling of what reality 'ought to be.'" But whose reality? The environmental and artistic history of Georgia, as briefly touched upon, is so ripe with personalities and catalysts, for better or worse, that the greatest insight into paradox may well be the biological timelines which, from the perspective of cosmology, form unions over centuries in essentially simultaneous time. This is clearly one salient reason, perhaps, that Einstein was so intent upon ethics, religion, and the cosmological as his declared motives for pursuing science. Jorge Luis Borges wrote, "That one individual should awaken in another memories that belong to still a third is an obvious paradox."[46] From the perspective of two people holding hands, walking along a polluted river somewhere in Georgia, perhaps the greatest paradox of all was one attributed to Bertrand Russell: "The greatest challenge to any thinker is stating the problem in a way that will allow a solution."[47] That is the definition of a philosophy ripe for picking.

---

[44] See *Conversations on the Dark Secrets of Physics,* by Edward Teller, Wendy Teller and Wilson Talley, Plenum Publishers, New York, 1991, Ch. 9, p. 135 footnote.

[45] See Aphorism 48, as translated in *Dialogue on Poetry and Literary Aphorisms* Friedrich von Schlegel; Ernst Behler; Roman Struc, Pennsylvania State University Press, University Park, PA, 1968, p. 151.

[46] Jorge Luis Borges from *Evaristo Carriego,* Dutton Publishers, New York, 1983, Ch. 2.

[47] Attributed to Bertrand Russell in Stuart Crainer's *The Ultimate Book of Business Quotations* Capstone Publishers, Mankato, Minnesota, 1998, p. 258. For these and countless other quotations about "paradox," see https://citaty.net/citaty/677362-galileo-galilei-by-denying-scientific-princi-ples-one-may-maintain/ as well as, http://www.wiseoldsayings.com/paradox-quotes/page-2/, Accessed April 21, 2019.

# Chapter 24
# Jan van Goyen's Exquisite Obsession

## 24.1 Golden Memories

Jan van Goyen (Leiden 1596-The Hague, 1656) is considered the most prodigious of all Dutch Renaissance painters, with an infectious nature passed on to his three daughters, as well as his contemporaries. His work was in high demand, though ultimately he would die nearly insolvent, due to risky real estate investments and madcap speculation in tulip bulbs that backfired.[1] Restless, a born speculator, he was a man in love with the Dutch coastline to a degree that gives one to believe he was painting not the Netherlands but Canada, Chile, Russia, New Zealand, Norway, or Indonesia—the largest coastlines in the world. Arthur K. Wheelock, Jr. cites van Goyen as having initiated the Golden Age of landscape painting in the Netherlands.[2] Whereas all but 34 paintings are currently fully attributed to Vermeer—and an additional three others still debated[3]—van Goyen's output numbers in the thousands of works, with many paintings and sketches doubtlessly lost. The definitive three-volume catalogue raisonné by Hans-Ulrich Beck shows 1211 oil paintings, in addition to numerous varieties of many of those otherwise singular works, and 861 sketches, again, with varieties bringing the total sketches to well over 1000[4] (Fig. 24.1).

---

[1] See *Great Masters of Dutch and Flemish Painting*, by W. Bode, translated by Margaret L. Clarke, Duckworth And Co., London, 1909, pp. 141–151.

[2] See the short bio, "Goyen, Jan van," National Gallery of Art Online Editions, https://www.nga.gov/collection/artist-info.1354.pdf, Accessed April 22, 2019.

[3] See *Vermeer and the Delft School*, by Walter Liedtke with Michiel C. Plomp and Axel Rüger, The Metropolitan Museum of Art, New York and Yale University Press, New Haven and London, 2001; see also Shin-Ichi Fukuoka's *Vermeer: Realm of Light*, in which the Japanese biologist-art historian brilliantly opines that Vermeer might well have executed—in addition to his 34–37 oil paintings, many additional sketches of animal and plant parts, e.g., "the leg of a louse" or "cross section of a fruit," in some of his neighbor, Antonie van Leeuwenhoek's letters, "A Certain Hypothesis," pp. 240–251.

[4] See the three-volume *Jan Van Goyen 1596–1656, Ein Oeuvreverzeichnis* by Hans-Ulrich Beck, with a preface by Wolfgang Stechow, Davaco, Doornspijk, The Netherlands, 1987.

© Springer Nature Switzerland AG 2021
M. C. Tobias, J. G. Morrison, *On the Nature of Ecological Paradox*,
https://doi.org/10.1007/978-3-030-64526-7_24

**Fig. 24.1** "Coastal Scene," Detail, Attributed to Jan van Goyen, ca. mid-1630—1640s. (Private Collection). (Photo © M.C. Tobias)

So intensely evocative are his works that sensitive viewers are led inside to participate within the Dutch seventeenth century in a way that rarely happens. Historians may comment upon fabrics, eyes, settings, a mood cemented into some alternative reality that is permanently fixed in consciousness. Perhaps nostalgia aids in that accumulation of verities that help along the way. In the case of van Goyen, he quickly achieves something that has other explanations. For example, it has been written that "On his palette he would grind out a colour collection of neutral grays, ambers, ochre and earthen greens that looked like they were pulled from the very soil he painted."[5] So wet is Dutch earth that researchers have found "more than 90%

---

[5] Jan Van Goyen House > Paintings, https://www.fundatievoorhoeve.com/painting.html, Accessed April 25, 2019; see also https://buitenmuseum.com/en/jan-van-goyens-house/, Accessed April 24, 2019.

of the soils have groundwater within 140 cm of the soil surface during the winter."[6] In this land which comprises a veritable mosaic of dikes—"50,000 miles" worth—of enormous areas of land reclaimed from the North Sea, and local "water boards" dating to the Middle Ages to deal with constant flooding and resulting hydrological projects throughout the country (the current Delta Works and Maeslantkering, among many expensive others), there is a certain obsessive insecurity that can be attributed to a uniquely Dutch geographical bifurcation. It is an aqueous flash point, a cultural inflection affecting every generation that sees the interior as the familiar dark and cozy retreat, whereas the outdoors mingles pleasure with the historical fact of Holland having been once a landscape of predominant "polderland," a place of swamps that held in the cold, the damp, the dangerous.[7] An ecological paradox across the Netherlands prevails at the nexus of every hamlet, every doorway leading to the outside.

In the case of van Goyen, his son-in-law, Jan Steen, painted van Goyen's indoor scenes, while van Goyen expressly relied upon the exterior as an expression of his claim to popularity. This separation gathers affirmative intensity in both Dutch manuscript illuminations and gardens of the seventeenth century. Gardens remain free of inclement weather, but the surrounding seas tend to emphasize coming storms, the ferocity of waves-in-uprising, of ships seized by hailstorm and hue-and-cry, and, in turn, the psyche constantly in tune with the ever-looming inevitability of flood waters and sleet, of calamity forever emergent.[8] It is the interior domain—whether in the Netherlands or the Gobi Desert—where humanity's comforts and stability reside. Other species have their nests and burrows. Even wild boar in Africa know how to keep themselves relatively well-hidden beneath ground, usually out of reach of large felids. Others have their millions of years of well-honed camouflage. Only humans must focus the vast majority of their time indoors, where the safe haven is the most predominant motif and pastime, the matrix for cooking, reading, sleeping, eating, procreating (Fig. 24.2).

Curiously, van Goyen gave a certain sense of a safe zone to those choppy waters and pummeling skies. He managed to paint those exteriors most familiar to his neighbors, with a density of weather, crowding together calm predictability, social enjoyment, and a distinct structure to life, that re-engineered in mind that

---

[6] See "Soil maps of The Netherlands," by Alfred E. Hartemink and Marthijn P. W. Sonneveld, Geoderma 204–205, 2013, 1–9, SciVerse ScienceDirect, Elsevier, http://citeseerx.ist.psu.edu/viewdoc/download?doi=10.1.1.296.4604&rep=rep1&type=pdf, Accessed April 25, 2019.

[7] See "Lessons for U.S. From a Flood-Prone Land," by Andrew Higgins, November 14, 2012, The New York Times, https://www.nytimes.com/2012/11/15/world/europe/netherlands-sets-model-of-flood-prevention.html, Accessed April 25, 2019.

[8] See *The Dutch Garden in the Seventeenth Century*, Dumbarton Oaks Colloquium on the History of Landscape Architecture, XII, Edited by John Dixon Hunt, Dumbarton Oaks, Trustees For Harvard University, Washington D.C., 1990; and *The Golden Age of Dutch Manuscript Painting*, Introduction By James H. Marrow, Catalogue Essays By Henri L. M. Defoer, Anne S. Korteweg and Wilhamina C. M. Wüstefeld, George Brailler, Inc., New York 1990.

**Fig. 24.2** Interior of Jan van Goyen's House. (Photo © M.C. Tobias)

environment to reflect activities which seemed internal and safe, residues of the Golden Age. All those vistas, part water, part land, that engage residents who are to Sturm und Drang what sherpas are to avalanche and whole winters snowed in. What is undeniably fascinating about van Goyen's genius is his ability, in scene after scene, to invite the fascination which can only arise in the purview of differentiations. Every vision, no matter how similar to another, is entirely fresh. His focus is so intense, his love of a specific genre and the details therein so comprehensive, as to have lent van Goyen a posterity chiefly marked by his descendants laying claim to possibly the most personal of all landscapists in history.

## 24.2    The Atlas of a Romantic

Van Goyen had moved from the city of his birth, Leiden, to The Hague in 1632, where he resided in the town center along the Dunne Bierkade adjoining the Groenewegie. At that time, the city's population hovered around 15,000 and 70% of the northern Netherlands was urban. The conurbations of The Hague, Amsterdam, Rotterdam, and the like accented the vulnerabilities to water which an aesthetic worldview could exploit, minimize, or integrate dramatically into the practical everyday life of the country's residents. That is what van Goyen accomplished.[9] Citing Beck, the late Walter Liedtke suggests that van Goyen had an important influence on at least 40 contemporary and subsequent followers.[10]

Consider just a few of van Goyen's repeated impressions (titles of paintings): "The Pelkus Gate near Utrecht," "River Landscape," "Country House near the Water," "Castle by a River," "River View with a Village Church," "A Windmill by a River," "View of Dordrecht downstream from the Grote Kerk," "View of The Hague from the Northwest," "The Thunderstorm," "Polder Landscape," and "Village at the River." Three hundred and twenty years ago, a van Goyen's obsession with the surrounding waterways could be likened to a clairvoyant intimation of climate change. It is in the Dutch psyche, certainly. But to have fashioned from its eerie prospects an entire aesthetic of enduring intimacy—of pleasure within each coming storm— marks a radical shift in human apprehension, from fear and flight to loving, even customary, embrace; "an attempt to reassert the validity of the organic, as against the mechanical, of the individual personality as against the statistical aggregate, of the instinctual and traditional and historical as against the purely rational and measurable," wrote Lewis Mumford, speaking of the advent of Romanticism.[11]

## 24.3 A Philosophy of Vulnerability

In an entirely different mode of perceiving, the 29-year-old Albert Camus wrote in his meditation during World War II, *The Myth of Sisyphus*, "His scorn of the gods, his hatred of death, and his passion for life won him that unspeakable penalty in which the whole being is exerted toward accomplishing nothing. This is the price that must be paid for the passions of this earth."[12] The Dutch have always known of that *price*. One late Saturday night, on January 31, 1953, and into the next morning, the North Sea flooded, the so-called watersnoodramp event, killing

---

[9] See "Population developments in the Northern Netherlands (1500–1800) and the validity of the 'urban graveyard' effect," A.M. van der Woude, Annales de Démographie Historique Année 1982, 1982, pp. 55–75, https://www.persee.fr/doc/adh_0066-2062_1982_num_1982_1_1528, Accessed April 25, 2019; see also "A maritime worldview in Jan van Goyen's 'View of The Hague from the south east,'" by Noortje Knol, www.vliz.be/imisdocs/publications/252980.pdf; see also "Van Goyen's Puddles," by Lytle Shaw, *AA Files,* No. 65 (2012), pp. 76–86, published by: Architectural Association School of Architecture, https://www.jstor.org/stable/41762329, https://www.jstor.org/stable/41762329?seq=1#page_scan_tab_contents, Accessed April 25, 2019.

[10] See Walter Liedtke's two-volume *Dutch Paintings in the Metropolitan Museum of Art*, The Metropolitan Museum of Art, New York, Yale University Press, New Haven and London, New York, 2007, p. 224.

[11] See Mumford's *In the Name of Sanity*, Harcourt, Brace And Company, New York, 1954, p. 119.

[12] See Albert Camus, *The Myth of Sisyphus and Other Essays*, translated from the French by Justin O'Brien, Alfred A. Knopf, New York, 1955, p. 120.

a known 1836 Dutch residents.[13] But many centuries before that, in mid-November 1421, Dordrecht became "an urban island," as dozens of villages were washed away and 10,000 or more people killed ("early accounts tell of one hundred thousand deaths"). Ten thousand or 100,000, it was "a maritime apocalypse" that van Goyen and everyone around him would have heard about since their childhood. Indeed, writes Simon Schama, "hydraulic needs" engendered a "moral geography" that was embedded in the geopolitics, local tax schemes, and the very message of redemption in Church. And thus, one could "describe Dutch society as having a diluvian personality...this semiaquatic existence...."[14]

Little wonder that the first major work, graphically illustrated in the annals of Western hydrology, was published in Amsterdam in 1678 by Johannes Herbinius, with its famed frontispiece depicting a naked goddess riding a sea monster into a waterfall, as well as illustrations of a Scandinavian Charybdis, and a map following the four wild rivers into paradise. This was the transcendental imagination of the Netherlands which would produce a van Goyen, a Vermeer, a Ter Borg, each of whom painted transformational portraits, of landscapes, interiors, faces, and, in the case of Rembrandt, of oneself succumbing to the powers and visual artifacts of old age, of nature.[15]

In the realm of Dutch ornithology, observations of rare birds have occurred most frequently along the coastlines of Holland. Those provinces with the longest shorelines—the ones that have held out against continued lashings by storms and accompanying saltwater intrusions—have witnessed the most numerous avifauna sightings, a trend that has only accelerated in direct correspondence with climate change, and in a country of which more than half lumbers beneath sea level. Moreover, the Netherlands presently possesses only some 50% of the coastlines it had prior to the "large-scale dam construction programme" following the 1953 disaster.[16]

A catastrophe that recalls the observations by that astute Italian novelist, traveler, and poet Edmondo De Amicas (1846–1908), whose two-volume work *Holland* (1894) juxtaposes the profound stillness and silence of canals and streets and buildings and trees throughout the Hague, with his observations outside the city, at Scheveningen. It was there, along the North Sea, about which Amicas described the

---

[13] "Today in Dutch history: the 1953 North Sea flooding disaster," by Thomas Lundberg, I Am Expat, January 31, 2016, https://www.iamexpat.nl/expat-info/dutch-expat-news/today-dutch-history-1953-north-sea-flooding-disaster, Accessed April 25, 2019.

[14] Simon Schama's *The Embarrassment of Riches—An Interpretation of Dutch Culture in the Golden Age*, Alfred A. Knopf, New York, 1987, pp. 34–44.

[15] See *Dissertationes de admirandis mundi cataractis supra & subterraneis, earumque principio, elementorum circulatione, ubi eadem occasione aestus maris reflui, vera ac genuina causa asseritur, nec non terrestri ac primigenie paradiso locus situsque verus Palaestina restituitur, in tabula chorgraphoca ostenditur, & contra utopios, indianos, mesoppotamios, aliosque afferitux*, published by Jansson—Waesberghe Amsterdam.

[16] See *Zeldzame vogels van Nederland met vermelding van alle soorten—Rare birds of the Netherlands, with complete list of all species*, by Arnoud B van den Berg and Cecilia A W Bosman, GMB Uitgeverij, Utrecht, and Pica press, East Sussex, 1999, pp. 7–8.

absence of any shells, or life whatsoever in the sands, and "the superstitious terrors of the ancients, who believed it [the North Sea] to be driven by eternal winds and peopled by gigantic monsters." And he continued, "The sky is almost always gray, overcast with great clouds which throw dense changeable shadows on the waters; in places these are as black as night, and bring to other parts the sky is lighted up by patches and wavy streaks of bright light, which seem like motionless lightning or an illumination from mysterious stars." And he concludes a passage that could not describe any better van Goyen's own dozens of paintings and sketches of Scheveningen: "The ceaseless waves gnaw the shore in wild fury with a prolonged roar which seems like a cry of defiance or the wailing of an infinite crowd"[17] (Fig. 24.3).

To perceive amid oceanic hostility the comforts of home, a nation portrayed as hardened, resolved, and loyal to the tumult, orderly stowaways on the churning Continent, lends remarkable insight to the paradoxical Dutch. Living forever on the cusp of the Great Flood, as if forever flirting with all those qualities which give to the Biblical Noah such esteem in the mind of every sentient being who understands the value of the Ark. While many of his contemporaries prodigiously turned out commissions of sitters in their Sunday best, unthreatening vases adrift in flowering profusions, still lifes of dead ducks and hanging venison, entire villages buttoned up in woolies, communing happily on ice skates, or paradise scenes from the antediluvian, van Goyen had managed to make the menacing world of imposing towers over troubled waters, darkening skies, and the unstoppable threat of watery chasms inexplicably amenable, their brilliant light focusing upon the inner strength of character the Dutch were granted by their very landscapes.

---

[17] See *Holland*, by Edmondo De Amicis, 13th edition, translated by Helen Zimmern, Porter & Coates, Volume 1, 1895, Vol. 1, pp. 215–216.

**Fig. 24.3** "Fishermen's Children, Scheveningen," from Edmondo De Amicas' two-volume, *Holland*, Porter & Coates, Philadelphia, 1894. (Private Collection). (Photo © M.C.Tobias)

# Chapter 25
# Paradox of the Lamb

## 25.1 Iconographic Communions

The All Saints Picture (Allerheiligenbild), "the ultimate beatitude of all believing souls," as Erwin Panofsky describes it in reference to the Ghent Altarpiece ("Adoration of the Mystic Lamb" c. 1421–1431), is an expressly human devotional.[1] Within this deeply saturated schematic of human behavior is enshrined the Lamb, whose quintessence, as painted in its terror and sorrow, strikes as surely one of the most persistent universally insane paradoxes known to human kind (Fig. 25.1).

The animal's iconographic significance, at least in Western Judeo-Christian traditions, belies the zeal with which our one primate species among hundreds rallies to its slaughter. It's as if the majority of our receptor genes involved with killing are those very internal semions that have mutated. Their affiliation with the irrational and thoroughly contradictory nature of reverence has spilled out into a thoroughfare of sacrifice and cruelty. Its apotheosis is a genotype that spells cruelty, courting nullity whose fuel is that paradox that lives in abandonment of love. Its engine is heartlessness. Only a small number of those genes devoted to restraint and sparing life—whichever they may be—appear to be functional from the instant we are all cast on to the shores of earth in a weathered and disoriented condition. As pessimistic a view as this may be, looking to one another and finding, apparently, the confirmation bias that is comfort in misery, to empower the full ramification of this genotypic expression, so full of contrarian aptitudes and permanent vitriol, the lamb (and a far greater number of sacrificed adult sheep) is dying proof.

How many of the millions of tourists annually include in their visit to Ghent a meal comprising as many spare ribs as one can devour, as advocated in one regional

[1] See *Early Netherlandish Painting—Its Origins and Character*, by Erwin Panofsky, The Charles Eliot Norton Lectures, Harvard University Press, Cambridge, Massachusetts, 1953, Volume 1, p. 212.

**Fig. 25.1** Ghent Altarpiece Polyptych by Hubert van Eyck and Jan van Eyck, 1432. Panel of the Lamb, Saint Bavo Cathedral, Ghent, Belgium (Photo, Public Domain)

tourist ad?[2] As of 2016, according to "The Belgian Meat Round Table," "*The Belgian is and remains carnivore*': Home consumption of meat, poultry, and game reached 29 kilograms per capita in 2016. Furthermore, 60% of the Belgian meat consumers declared that they eat meat at least 4 times a week…."[3] As of 2018, there were significant decreases in Belgian meat consumption according to at least one major source, suggesting that the average household in Belgium consumed 17.2 kilograms of meat in a year,[4] while the European Union overall remains the largest importer in the world, consuming over 490 million pounds in 2018 of lamb, hogget, and mutton, primarily from Australia and New Zealand.[5] While Flemish lawmakers in 2018 made new reforms to the Belgian Animal Welfare laws, the first such reforms in 32 years,[6] reports suggest that 60% of all animal abuse cases are never followed up by Flemish courts.[7]

---

[2] https://visit.gent.be/en/see-do/st-bavos-cathedral-majestic-tower

[3] See Euromeat, Belgium. "Belgian Meat Round Table: Meat consumption decreased by 18% between 2008 and 2016."

[4] See https://english.fleischwirtschaft.de/economy/news/Data-Belgian-consumption-of-fresh-meat-still-on-the-retreat-39239

[5] See "International Lamb Profile," Agricultural Marketing Research Center, https://www.agmrc.org/commodities-products/livestock/lamb/international-lamb-profile, Accessed May 3, 2019.

[6] See "Flanders goes tougher on animal abuse," The Brussels Times, by Christopher Vincent, July 5, 2018.

[7] See "60% of reported cases of animal cruelty not followed up," Flanders News, n.a., VRT NWS, February 12, 2018.

## 25.2   Notwithstanding "Donderdag Veggiedag"

In Ghent proper, since 2009, the city has adopted Thursdays as their Veggie Day ("Donderdag Veggiedag"), in which the aldermen urge all residents to go vegetarian.[8] Indeed, Thursdays in Ghent witness "over 90 restaurants, 20 hotels and 50% of the city's population observing vegetarianism for the day." Moreover, "Ghent hosts the most vegetarian restaurants per capita of any city in the world,"[9] with the exception of Pushkar, in Rajasthan—a vegetarian city.

Such reforms are not lost in the maelstrom of quanta, but are certainly unavailing within the overall European Union context, thus far. But then the van Eycks' "Mystic Lamb" is at the heart of both false consciousness and suspect statistics. The winged polyptych of 20 panels (one of which was stolen and never recovered, in keeping with the artwork's fame for having witnessed "13 crimes and seven thefts")[10] has also endured re-interpretations in oil, including massive overpainting in the sixteenth century by two commissioned artists, Lancelot Blondeel and Jan van Scorel. Koenraad Jonckheere, a professor of Baroque art at Ghent University, has clarified how their "careful overpainting campaign subtly adapted the shapes to the taste of the time and to some extent neutralised the Van Eycks' intense and humanized identification of the Lamb into an expressionless animal...."[11] The restoration was begun in 2012 and members of the team have reported that the sixteenth-century reworking had turned the sacrificial animal into "an impassive and rather neutral figure" whereas the van Eycks' original lamb had been painted with "an intense gaze... large, frontal eyes, drawing onlookers into the ultimate sacrifice scene"[12] (Fig. 25.2).

---

[8] See Ghent International, "Thursday Veggie Day," https://stad.gent/ghent-international/city-policy/food-strategy-ghent/thursday-veggie-day, Accessed May 3, 2019.

[9] See "How the meat-loving city of Ghent became the veggie capital of Europe," by Tom Burson, MIC, October 31, 2017, https://www.mic.com/articles/185650/how-the-meat-loving-city-of-ghent-became-the-veggie-capital-of-europe, Accessed May 3, 2019; the Lilly Library at the University of Indiana-Bloomington had an exhibition on vegetarian books from July 21-September 10, 2016, a collection largely donated by antiquarian William Dailey, and including the first seventeenth-century Renaissance edition of Porphyry's *On Abstinence from Animal Food Περὶ ἀποχῆς ἐμψύχων, De Abstinentia ab Esu Animalium;* see also "The Ethics of Diet – A Catena," by Howard Williams, 1883, https://ivu.org/history/williams/porphyry.html. See also *The Heretic's Feast—A History of Vegetarianism,* by Colin Spencer, Fourth Estate, London 1993, pp. 104–107; Thomas Taylor's translation of the *Select Works of Porphyry,* Thomas Rodd, London, 1823; and *An Essay on Abstinence from Animal Food as a Moral Duty,* by Joseph Ritson, Richard Phillips, London, 1802.

[10] See "Jan Van Eyck and the Ghent Altarpiece," by Rune Pettersson, Draft, p. 13, https://www.researchgate.net/profile/Rune_Pettersson/publication/325253172_Jan_van_Eyck/links/5b07ceaca6fdcc8c252ce456/Jan-van-Eyck.pdf?origin=publication_detail, Accessed May 11, 2019; see also "Is This The World's Most Coveted Painting?" NPR Staff, December 23, 2019, https://www.npr.org/2010/12/25/132283848/is-this-the-worlds-most-coveted-painting, Accessed May 11, 2019.

[11] See "Van Eycks' original lamb uncovered on Ghent Altarpiece," by Lisa Bradshaw, Flanders Today, June 19, 2018, http://www.flanderstoday.eu/van-eycks-original-lamb-uncovered-ghent-altarpiece, Accessed May 11, 2019.

[12] ibid.

**Fig. 25.2**  A Lamb in Central England. (Photo © M.C. Tobias)

It was Ghent's native son, Charles V, who had financed the overpainting of at least 40% of the panels. This was a man who had inherited the Holy Roman Emperor at the age of 16, fought multiple wars for four decades, and then, exhausted, ultimately abdicated his throne in 1556, retiring to a monastery in Yuste in the Spanish autonomous community of Extremadura. He died 2 years later, aged 58, clutching a cross. Throughout his entire life he had been a faithful Catholic, to the extent that he endeavored, in demanding certain overpainting, that the lamb not be too easily construed as a graven image by the Protestants with whom he was forever in ideological combat. He had good reason to fear his enemies. At one point during the height of the Reformation, Protestant interlopers raided the Cathedral of St. Bavo and destroyed part of the frame holding the altarpiece's panels (Fig. 25.3).

The Lamb dominates the final book of the New Testament, Revelation, particularly the final chapters 14–22, wherein atop Mount Sion the Lamb, with 144,000 disciples, sings untranslatable praise for the creator, "a new song before the throne…"[13] John the Baptist 1:29, "Ecce agnus Dei, Behold the Lamb of God"; and from Isaiah 53:7, "he is like a lamb that is led to the slaughter" (Fig. 25.4). Much has been divined and construed from this seeming contradiction.[14] But there is no question that originally, "Each day Mass was to be celebrated before it in honor of God, the Virgin Mary and all the saints… Christ's offering of His own blood is evoked by the Lamb and the instruments of the Passion held by the angels…."[15]

[13] *The Holy Bible* King James Version, Revelation 14:1, The Christian Science Publishing Society, Boston, Massachusetts, n.d., p. 1537.

[14] See, for example, http://www.w1vtp.com/the1.htm, Accessed April 28, 2019.

[15] See *Hubert and Jan Van Eyck*, by Elisabeth Dhanens, Alpine Fine Arts Collection, LTD, New York, 1980, pp. 91, 97; see also *Jan Van Eyck—The Play of Realism,* by Craig Harbison,

**Fig. 25.3** Inside St. Bavo
Cathedral. (Photo ©
M.C. Tobias)

The original lamb, as painted in the central panel of this, arguably the most emotional, provocative, and influential painting in the history of Western Civilization, is a lamb of great sorrow. His destiny has not been altered, despite the superficial efforts to destroy, remove, or steal his look away. In the end, the magnificent incarnation of his spirit by the van Eyck brothers endures. Indeed, while the lamb carried primary symbolism to the heart of the Middle Ages worshippers, it is a very different substantiation carried out by the van Eycks. As Sir Martin Conway wrote of it, "It does not look like a symbol, it looks like a sheep; and instead of at once suggesting a mystic thought, it shocks the eye by its sturdy realism...."[16]

And while a small and devoted contingent of vegans attempt to rewrite the history of Ghent surrounding the Altarpiece, a world of Christians and Jews, of Muslims and people of every other conceivable faith, or lack of faith, continue to slaughter sheep and, as per the use of the young singular in Exodus 12:3–8, the

Reaktion Books, Ltd., London, 1991; and *Van Eyck*, by Ludwig Baldass, Phaidon Press, London, 1952.

[16] See *Early Flemish Artists and their Predecessors on the Lower Rhine*, by William Martin Conway, Seeley & Co., London 1887, p. 138.

**Fig. 25.4** Slaughterhouse for Sheep and Goat. (Photo © M.C. Tobias)

lamb.[17] Indeed, such paradoxical ethics amount to consumption patterns worldwide which saw 15.77 metric kilotons of slaughtered sheep in 2020, or over 32 million pounds (divided by 135 pounds—the average weight of a sheep at slaughter) equaling 237,000 murdered sheep, and a standing "head" figure of well over one billion.[18]

---

[17] See "The Reason You Won't Often Find Often Fine Lamb At A Jewish Passover Meal," One For Israel, https://www.oneforisrael.org/bible-based-teaching-from-israel/holidays/reason-wont-find-lamb-jewish-passover-meal/, Accessed May 11, 2019.

[18] See "Projected sheep meat consumption worldwide from 2014 to 2026," statista, https://www.statista.com/statistics/739976/sheep-meat-consumption-worldwide/, Accessed June 9, 2020.

# Chapter 26
# Botanical Equations for Paradox

## 26.1  The Language of Flowers

After the Impressionists had intimated every conceivable feeling, tinge, and resplendent form inherent to flowers, French Nobel Laureate Maurice Maeterlinck (1862–1949) published his masterpiece *Hours of Gladness*,[1] sumptuously illustrated by E. J. Detmold. Nothing like it had ever been composed; no paean to flowers or to the civilizations of botany was ever so replete with life's ceaselessly declarative quintessence. Maeterlinck writes in his 79-page chapter "The Intelligence of Flowers" "of the great systems of floral fertilization: the play of stamens and pistil, the seduction of perfumes, the appeal of harmonious and dazzling colours, the concoction of nectar...." He goes on to sing of the great champion within every plant, and how "the energy of its fixed idea, mounting from the darkness of the roots to become organized and full-blown in the flower, is an incomparable spectacle. It exerts itself wholly with one sole aim: to escape above from the fatality below, to evade, to transgress the heavy and somber law, to set itself free, to shatter the narrow sphere, to invent or invoke wings...."[2] A flying fish. A butterfly whose metamorphosis takes it from the moribund up toward the stars (Fig. 26.1).

A brief phylogenetic history of flowers reminds us of that undefinable biological ordination between the Ordovician and Silurian Periods, over 440 million years ago. That is when plants first appeared on land and flowered by way of an evolutionary strategy known as "dispersed spore."[3]

---

[1] Translated by A. Teixeira De Mattos, George Allen & Co., London, 1912.

[2] Maeterlinck, pp. 22–23.

[3] See "The Earliest Land Plants," by Patricia G. Gensel, in Annual Review of Ecology, Evolution, and Systematics, Vol. 39:459–477, December 1, 2008, https://www.annualreviews.org/doi/abs/10.1146/annurev.ecolsys.39.110707.173526, Accessed May 12, 2019.

© Springer Nature Switzerland AG 2021                                                                 239
M. C. Tobias, J. G. Morrison, *On the Nature of Ecological Paradox*,
https://doi.org/10.1007/978-3-030-64526-7_26

**Fig. 26.1**  Wildflower
Bouquet, Frontispiece, by
Mrs. C. M. Badger, Private
Collection. Illustrations
from *Wild Flowers, Drawn
And Colored From Nature*
by Mrs. C. M. Badger,
Charles Scribner,
New York, 1859. (Photo ©
M.C. Tobias)

Today, there are over 390,900 plant species on earth, of which 369,400 are flow-ering.[4] The numbers will have changed by this time tomorrow, with approximately 4 new plant species being formally discovered each day by researchers. Even those interpolations will necessarily be skewed, impoverished, as the actual totality of species on earth continues to be debated and impressively escalated. All the system-atics correspond to previous Linnaean binomial nomenclature (with a Swedish phy-logenic logic, linguistically Northern Scandinavian, but deliberately distinct from the Danish Bible, with its own Swedish Gustav Bava Bible of 1541; two genders, vestiges of Old Norse, a morphology like English, and a few phonetic anomalies somewhere literarily between military decorum and the modernism of August Strindberg—that is the structure of Swedish taxonomy). Linnaeus commenced his endeavor to name all of the animals by naming plants. His earliest instincts regard-ing what would become *evolution* was the evolution of reproductive organs across

[4] See "Kew report makes new tally for number of world's plants," by Rebecca Morelle, BBC News, May 10, 2016, https://www.bbc.com/news/science-environment-36230858, Accessed May 12, 2019.

the botanical spectrum. The outcome, post-Linnaeus, is a pan-floristic langue open to radical alteration (not least transgender, asexual, and the like).

Botany poses the ultimate series of scenarios in terms of projecting habitat vulnerability. A member of the Cuphea genus, rose, Asclepias, hold the secrets to inviolable scenarios of collapse. The more taxa are discerned, the larger the volume of at-risk species, and liable ecosystems. Scientists at the University of Arizona "recently estimated that there are roughly 2 billion living species on the planet, over a thousand times more than the current number of described species."[5] We may surmise that photosynthetic eukaryotes, Plantae, represent their most emphatic and mind-altering ambassadors on earth.

So when we speak of the destruction of flower habitat, we actually are not certain of its extent or acceleration, except to acknowledge and correlate the extrapolations from our understanding of devastating changes to other biomes like coral reefs and the neotropics, Geographic Information Systems-inferred total estimates of biomass, and even mathematical probability theories lending substance to species-specific extinction models,[6] the psychology of which must—whether we are up to admitting it—cotton to the currencies of communication among humans—from Swedish to Latin, Latin to English, English to thousands of other languages. Because the lingua franca of a plant is, typically, its floret and efflorescence, human personality must be taken into account when assessing botanical equations intended to be peer-reviewed. Every review has its psycholinguistic bias; every bias is human; every human is different. Botany, ultimately, is playing a game of floristic roulette in terms of evolutionary quotients (Fig. 26.2).

## 26.2 Ground Zero, Colorado

My (mt) early California and Colorado home turfs are noteworthy in regard to assessing botanical peril. California remains one of the 36 terrestrial hotspots, in a Mediterranean Floristic Province, and—as of 2016—222 endangered, 83 threatened species under the (ever deviating) aegis of the Federal Endangered Species Act, and out of a total of an estimated 6500 California plant species. By 2025, the state's human population is expected to increase from 39.937 million to 45 million in 2050. But other organizations who tally the continuing rise suggest 45 million by 2030. The numbers are entirely politicized because of immigration positions, and hence we exclude footnotes for those numbers because their variations are wholly

---

[5] See "A new estimate of biodiversity on Earth," The Quarterly Review of Biology, September 2017, University of Chicago, PHYS ORG, https://phys.org/news/2017-08-biodiversity-earth.html, Accessed May 12, 2019.

[6] See "Predicting When Plants Face Extinction Threat, Mathematical model will help conserve biodiversity, says UC Riverside's Carla Essenberg," UCR Today, by Iqbal Pittalwala, June 15, 2012, http://www.bl.uk/manuscripts/FullDisplay.aspx?ref=Add_MS_43725, Accessed May 13, 2019.

**Fig. 26.2** Medicato
Arborea, from P. Virgilii
Maronis, *Bucolica*, Robert
Jennings London, 1710.
(Private Collection). (Photo
© M.C. Tobias)

dependent upon subjective criteria and the conservatism, or not, of policy perceptions.

Multiply such impressions by internal mechanisms of impact (where blame is easily cast on race, industry, point of origin, even gender) and we find the current pre-conditions for listing or not listing species as Threatened or Endangered ripe with human folly. Folly × folly does not equal good science, and even worse public policy. Neither domain can quite account for the gaps in human nature, which, ultimately, come down to two fundamental realities: human hunger and the availability of plants; and the senses, which appreciate, reject (as in the case of allergenic plants), or exhibit tone deafness. But the basic scientific debates according to threat levels to plants (where science also confuses bioinvasives, e.g., weeds, with reality) tracks along linguistic lines, and Linnaeus and his predecessors across Europe have much to do with the gaps, fueled by almost perverse obsessions with the sexual nature of plant proliferation, that a Maeterlink assayed (and quite brilliantly) to improve upon.

As for Colorado, the state is ranked number 8 in the nation for at-risk plant species, with 16 species-level taxa federally listed as Threatened or Endangered. And of 117 plant species found in the state that happen to be "globally imperiled and vulnerable," 68 of these are actually endemic to Colorado.[7] Moreover, some 500 Colorado plant species are listed as being of "Special Concern." When you look through the dozens of "Imperiled" species ("G2") face to face on the Colorado Rare Plant Guide List,[8] their beauty, out of all those hundreds of thousands of plant species worldwide, goes instantly to your heart. Their endangerment is transposed to ourselves. Their fate is our fate. The gorgeous Fisher Towers milkvetch, or Golden blazing star, among them. But no plant species was more glorious than Colorado's state flower, the Rocky Mountain Columbine (*Aquilegia coerulea*), unanimously hailed in 1899.

By 1925, there was so much worry over the future of this delicate creature that Colorado's General Assembly "made it the duty of all citizens to protect this rare species from needless destruction or waste. To further protect this fragile flower, the law prohibits digging or uprooting the flower on public lands and limits the gathering of buds, blossoms and stems to 25 in one day."[9] Frankly, 25 per day per consumer seems strikingly strange, given that the population of the state in 1925 was just over one million, 5.7 million today, and the plant's status is currently listed as a G3, "Vulnerable." Of course it is vulnerable.

Moreover, Colorado, like every Western state, has seen its share of exploitive resource extraction, waste, and mismanagement. The very notion of a *ghost town* summons the cumulative record of reckless opportunism, and the fallout leaves its signature. In Montrose County, the Dolores River twists at peculiar angles through the much-attenuated (highly relevant) community of Paradox. There, and in surrounding regions, uranium tailings, brine accumulation, water with excessively acidic pH, and other human-induced impacts have left the area more depauperate than it might have been, an equation of inordinately resounding relevance given the town's low density of human occupation.[10] Like the native plants, at least two documented native fish species within the Dolores River expanse are also listed in the "sensitive" category, namely, the "roundtail chub and flannelmouth sucker."[11]

---

[7] See Colorado Parks and Wildlife, "Rare Plant Conservation," https://cpw.state.co.us/aboutus/Pages/CNAP-Rare-Plants.aspx, Accessed May 13, 2019.

[8] https://cnhp.colostate.edu/rareplants/list.asp?list=G2

[9] See https://www.50states.com/flower/colorado.htm, Accessed May 13, 2019.

[10] See Colorado River Basin Salinity Control Project, Paradox Valley Unit Final Environmental Statement, U.S. Department of the Interior, and Bureau of Reclamation, Upper Colorado Regional Office, Salt Lake City, 1978; see also "The Uranium Widows," by Peter Hessler, September 6, 2010, *The New Yorker*. https://www.newyorker.com/magazine/2010/09/13/the-uranium-widows, Accessed May 17, 2019; see also *Reclamation – Managing Water in the West, Supplement to January 2013 Scoping Report*, Paradox Valley Unit EIS, December 2016m https://www.usbr.gov/uc/wcao/progact/paradox/Supplement2013FinalSupplementalScopingReport_12.2016.pdf, Accessed May 17, 2019.

[11] See *Evaluation of Existing and Proposed Areas Of Critical Environmental Concern for the Uncompahgre Planning Area*, February 2013, p. 77. Prepared by US Department of the Interior, Bureau of Land Management, Uncompahgre Field Office, Montrose Colorado, https://eplanning.blm.gov/epl-front-office/projects/lup/62103/78795/90462/ACEC_Report_Final_01152013.pdf, Accessed May 17, 2019.

Such relatively unknown species, and remote areas of even a settled place within the Rocky Mountains, call out the contradictory nature of presence, and absence as concerns human errors of judgment; the bias of fishermen, plant hunters, hunters in general. Organisms beneath the conventional radar of public perception vie with all those "charismatic species" for protection, at a time when no amount of charisma can actually guarantee a truly safeguarding response. This has been likened to "the paradoxical extinction."[12]

## 26.3   Plant Pertinacity Curves

Notwithstanding such apparent analytical gaps, as translated into public policy, the scientific community, abetted by satellite data and hands-on field research at tens of thousands of campsites around the world, has a fairly adept grip on past botanical fallout. For example, it is estimated that roughly 3 trillion individual trees have been destroyed, accounting for approximately 50% of all known forests to have ever seized the day on this earth. While approximately 3 trillion trees also remain (plus or minus roughly 4% of the various estimates), "around 15 billion trees are cut down each year," a rate of destruction that first ensued roughly 12,000 years ago with the "onset of agriculture" and the rise of domestications.[13] For every tree cut down, immediately adjacent or co-dependent flowering plants are impacted. In a region like Central Asia, noted for numerous gene pools of wild relatives and landraces (local cultivars)—from grapes and apples to almonds, pears, and pomegranates— the current ecological stressors, from the sheer numbers of consumers to poorly managed agriculture, do not bode well for countless other botanical species, given the evolutionarily interdependent mosaics essential for robust ecosystems.[14] So fundamental to overall ecosystem health are flowering, seed-producing plants (angiosperms) that the definition of the 36 terrestrial biological hotspots is as follows: a place where "at least 1,5000 species of vascular plants found nowhere else on Earth...[have] lost at least 70 percent of its primary native vegetation."[15]

---

[12] See "The paradoxical extinction of the most charismatic animals," by Frank Courchamp, Ivan Jaric, Céline Albert, Yves Meinard, William J. Ripple, Guillaume Chapron, PLOS Biology Fifteenth Anniversary, April 12, 2018, https://doi.org/10.1371/journal.pbio.2003997, https://journals.plos.org/plosbiology/article?id=10.1371/journal.pbio.2003997, Accessed May 17, 2019.

[13] See "Global count reaches 3 trillion trees," Rachel Ehrenberg, September 2, 2015, Nature, https://www.nature.com/news/global-count-reaches-3-trillion-trees-1.18287, Accessed May 13, 2019.

[14] See *Conservation of fruit tree diversity in Central Asia: Policy options and challenges* edited by Isabel Lapeña, Muhabbat Turdieva, Isabel López Noriega and Wagdi George Ayad, Biodiversity International, World Bank Group, Rome, Italy 2014, https://www.thegef.org/sites/default/files/publications/Conservation_of_fruit_tree_diversity_in_Central_Asia_1734.pdf, Accessed May 13, 2019.

[15] See "Biodiversity Hotspots Defined," Critical Ecosystem Partnership Fund, Conservation International, https://www.cepf.net/our-work/biodiversity-hotspots/hotspots-defined, Accessed

In this Anthropocene epoch, pertinacity curves for countless plant species remain a great unknown.[16] With at least five previous global extinction events and countless local and regional glaciations, volcanic eruptions and other natural transmutations of the earth's surface, it remains for now impossible to divine whether there was a time when there were more plant species than today. An estimated half of all plants have naturally hybridized. A multitude of others have been bred so aggressively by humans for agriculture and other reasons that disassembling the genetic aggregates, breeds, varieties, races and landraces, countless clades, and other taxonomically understood, phylogenetic types (as Swedishly implied) is a notoriously tricky enterprise.

## 26.4   Botanical Frontiers

But one could start with Ani. In 1250 BC the famed Theban scribe's death hastened the writing of a manuscript emphasizing the importance of plants as offerings during human transition from life to the otherworld.[17] An analysis of Egyptian influence on human–plant relations, manifesting from the Papyrus of Ani, and its hieroglyphic "Lord of the palm-tree," "green twig" hosting every bird, its "Tree of Life," "plant of truth"(wheat), "sekemu"(vegetables), the sycamore fig-tree, and "the plant which cometh forth from Nu" (the God of the Celestial Ocean), leaves us a profound legacy that firmly roots science in the arts and spirituality.

Six hundred years later, the "hanging gardens of Babylon" were consecrated (one of the touted Seven Wonders) during the reign of Nebuchadnezzar II (605–562 BCE). No actual documents remain to provide any botanical reality or ecological context for these gardens but their legend says it all, triggered by their first mention centuries later by the Greek-Babylonian historian Berossus of Kos (ca. 258–253 BCE).[18] Plato lamented the rampant deforestation throughout Greece and the entire Mediterranean littoral, while Cicero would later suggest that human

---

May 17, 2019; see also *Hotspots: Earth's Biologically Richest and Most Endangered Terrestrial Ecoregions*, by Russell A. Mittermeier, Norman Myers, Cristina Goettsch Mittermeier, Project Coordinator, Patrice Robles Gil, CEMEX, Mexico City, Mexico 1999; see also "Hotspots," PBS, 2008, www.hotspots-thefilm.org

[16] See "Nature's Dangerous Decline 'Unprecedented'; Species Extinction Rates 'Accelerating,'" IPBES Science and Policy for People and Nature," May 7, 2019, https://www.ipbes.net/news/Media-Release-Global-Assessment, Accessed May 13, 2019; see also *Life Counts: Cataloguing Life on Earth*, by Michael Gleich, Dirk Maxeiner, Michael Miersch, and Fabian Nicolay, Atlantic Monthly Press, New York, 2000.

[17] See The Papyrus of Ani, known as *The Book Of The Dead, The Hieroglyphic Transcript of the Papyrus of Ani*, translated and with an Introduction by E. A. Wallis Budge, University Books, New York, 1960, p.613; see also *Flora: Gardens And Plants in Art and Literature*, by Edward Lucie-Smith, Ivy Press Ltd., East Sussex, 2000. See specifically, pp. 611, 155–156, 88, and 285.

[18] See "The Hanging Gardens of Babylon", http://www.plinia.net/wonders/gardens/hg4berossus.html, Accessed May 12, 2019.

manipulation of arable lands was so extensive as to have engendered a *human world within the world*. Greco-Roman philosophers debated the wisdom of this; Aristotle assessed well over 500 species, and many of the earliest saints, including Basel, Ambrose, and Augustine, celebrated plant life as the clearest evidence for the goodness of God and His Creation.[19] The associations in every religion were always implicit.

All of these intellectual debates, insights, and poetic lyricism seizing upon the obvious importance of plants to human life and death were fixated on those hanging gardens, granting them the stature of the divine, as the psyche similarly invested the gardens of the fifteenth-century Aztec Huastepec and Texcotzingo (covered today by the northern reaches of Mexico City) with an imperial conundrum of potency. The God Tlaloc was "He Who Makes the Plants Spring Up." Such juxtapositions of hallowed space with vast paradise mythologies are most populated by flowers. Every artist, like that most tangible *gene* of a gardener in each human being, at some point engages in her or his own meditation on the flowers of paradise: what they mean to us personally and perennially. When Sandro Botticelli painted "Primavera" (Spring, 1482–1485), his forest landscape teemed with 190 discernible flower petals. And when Sir Thomas More published his popular *Utopia* (1516) European readers had already assimilated the concept from childhood. The *horae*, or Christian devotional Book(s) of Hours, in which every page was typically designed with vertical and horizontal flower gardens of every real and imaginary variety, were to be found throughout secular life, which segued from the religious oracles of every church in every quadrant of Europe.

One in particular comes to mind: *The Hours of Mary of Burgundy*, produced in Flanders in the 1470s, written by the famed Burgundian scribe Nicolas Spierinc and illustrated by the greatest of the Flemish miniaturists/illuminators. Flowers and birds and the garden-as-Otherworld were the most notable backdrop for the religious storylines.[20] Renaissance households were fully conversant with gardens of Eden, and many of the flower species occupying them, the same flowers that grew in pots within their homes, in community gardens, and in those arable commons upon which so much of Medieval and Renaissance village life depended. One witnesses all the same obsessive and iconic motifs in the Dutch love of fifteenth- and sixteenth-century manuscript illuminations, alive with most of the same wildlife[21] (Fig. 26.3).

---

[19] See Clarence J. Glacken, *Traces on the Rhodian Shore – Nature And Culture In Western Thought From Ancient Times To The End Of The Eighteenth Century*, University of California Press, Berkeley and Los Angeles, 1967, Chapter 3, "Creating a Second Nature," pp.116–149.

[20] See *Codex Vindobonensis 1857*, Vienna, Österreichische Nationalbibliothek, Commentary by Eric Inglis, Harvey Miller Publishers, London 1995, p. 1.

[21] *See The Golden Age of Dutch Manuscript Painting*, Introduction by James H. Marrow, Catalogue Essays by Henri L. M. Defoer, Anne S. Korteweg, and Wilhemina C. M. Wüstefeld, George Braziller Inc., New York 1990.

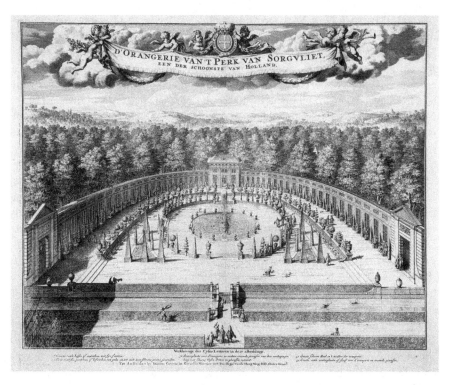

**Fig. 26.3** "Orangery of the Catshuis, The Hague, The Netherlands," by J. J. van den Aveelen, 1690, http://www.geschiedenisvanzuidholland.nl/verhalen/archiefstuk/447/Oranjerie-Sorghvliet,-Den-Haag-(1690)

In the case of the Dutch Republic, those well-versed denizens were equally assured of new botanical frontiers to all sides, geographically.[22] And often, as was so much the case in the Netherlands, the enthusiasm fueling human relationships with flowers and their spiritual as well as economic values might morph into a speculative frenzy over the market worth of certain tulip bulbs, valued at far more than their weight in gold. And when crowded city life began to weigh on one's soul, gardens modeled after some version of those original hanging gardens of Eden were readily at hand for a measure of time-out. In the Hague, on the way to the dunes of Scheveningen, the famed politician Jacob Cats (1577–1660) created the gardens of Sorghvliet, which stand to this day and roughly translate as "Leave your worries at the gate."[23]

---

[22] For information regarding the popularity of the horae, see *Medieval Must-Haves: The Book of Hours*, by Sandra Hindman, Les Enluminures, Paris and New York, 2018, p. 11.

[23] See "Sorghvliet – Leave your worries at the gate," by Jacqueline Alder, https://ikgidsudoorden-haag.nl/en/sorghvliet-leave-worries-gate/, Accessed May 13, 2019; see also *The Dutch Garden in the Seventeenth Century*, edited by John Dixon Hunt, Dumbarton Oaks Colloquia on the History of Landscape Architecture, XII, Washington DC, Trustees for Harvard University, 1990.

Central Park, in New York, hosts a similar resplendence of respite for some 30 million sojourners annually. But other recent biological soul retreats in Holland have witnessed tremendous blow-back. In the case of the marshlands east of Amsterdam, an attempt at re-wilding resulted in a grim vertebrate system collapse, as thousands of large herbivores starved to death, accompanied by the loss of countless trees and unknown numbers of flowering plant species.[24] As in the case of the Colorado state flower, the global human population explosion, with its accompanying habitat manipulation, has changed everything for conservation, since the time of the Flower Breughel. With so altered and adjudicated an environmental context, the backstory for artists has also changed.

The consistently astonishing Claude Monet, embarking on his famed "Water Lilies" period, is said to have declared that "It is maybe to flowers that I owe becoming a painter."[25] One might rightly assume that irises and sunflowers had an equally transformative impact upon van Gogh. At the height of European Romanticism, flowers had recalled a "superintending providence of the Almighty," of "Heaven's immortal spring," wrote Rebecca Hey in her famed work *The Moral of Flowers* (London 1833).[26] Charles Darwin's grandfather, Erasmus Darwin (1731–1802), a true disciple of Carol Linnaeus, had infused his own scientific passions with a sonance unique to his uncanny love of plants. In 1791 he published *The Botanic Garden* containing two lengthy and ecstatic poems, "The Economy of Vegetation" and "The Loves of the Plants," in which he endeavored, in his words, to "enlist Imagination under the banner of Science…to induce the ingenious to cultivate the knowledge of Botany…."[27] Darwin had been influenced by Richard Bradley, who in 1721 wrote an extensive "Account of the State of Gardening" throughout Great Britain.[28] Bradley, in turn, must have been familiar with the proliferation of earlier botanical codices and herbaria, such as those published by Otto Brunfels in 1532, Leonhart Fuchs in 1542, Hieronymus Bock in 1546, and Prince Eugen of Savoy in his ten-volume *Florilegium* in the 1670s, to name but a few of the remarkable *florae* of the Renaissance.[29]

Today, artists and scientists have easily inherited all of these influences, an abundance of ethnobotanical literature from every region,[30] biological epiphanies, and

---

[24] See "Dutch rewilding experiment sparks backlash as thousands of animals starve," by Patrick Barkham, The Guardian, April 27, 2018, https://www.theguardian.com/environment/2018/apr/27/dutch-rewilding-experiment-backfires-as-thousands-of-animals-starve, Accessed May 13, 2019.

[25] See https://fondation-monet.com/en/claude-monet/quotations/; see also *Monet Or The Triumph Of Impression*, Daniel Wildenstein, Taschen/Wildenstein Institute, Köln Germany 1996.

[26] pp.vii–viii.

[27] First Edition, J. Johnson, London, 1791, p.v.

[28] See *A Philosophical Account Of The Works Of Nature*, by Richard Bradley, London, 1721.

[29] For one of the most lavish and comprehensive works on the topic, see *A Garden Eden. Masterpieces of Botanical Illustration*, by H. Walter Lack, Österreichische Nationalbibliothek, Taschen, Cologne, Germany 2008. See also "Floral Art History," by David Wagner, https://www.davidjwagnerllc.com/janeJonesInsert_Apr2020.pdf

[30] See, for example, "Amazon tribe creates 500-page traditional medicine encyclopedia," by Jeremy Hance, June 24, 2015, Mongabay, https://news.mongabay.com/2015/06/amazon-tribe-creates-

hard-hitting anthropocenic revelations. We see manifested across the botanical art world so-called ephemeral realism, meditations on the life and death of real flowers within a public sculptured form, much like Tibetan sand paintings—here today, blown to the four corners tomorrow. And so-called reverse agriculture as revealed by contemporary flower artist Azuma Makoto.[31] Other stunning glimpses of angiosperms in full glory have utilized the latest techniques of scanning electron microscopy, as created by Oliver Meckes and Nicole Ottawa.[32] But we also recognize the historic retention of the Medieval mind as it continues to flower in cloistered, four-square courtyard gardens, like those in Utrecht, or the abbey of St. Gallen in Switzerland, whose Benedictine designs of the Middle Ages were refreshed at the Metropolitan Museum of New York's Cloisters.[33]

## 26.5 Flowers Among the Ancients

The contention that Neanderthals may have buried their dead wearing tiaras of dried flowers (an echo of the Egyptian burial rites) as suggested by early evidence from the cave of Shanidar 4 in modern-day Iraq has been largely proved false.[34] But the fact that human evolution, and our love of nature (physiolatry), has been comprehensively tutored by flowers remains a telling and indisputable fact of our lives.[35] One need simply remember the life of twentieth-century explorer/botanist Francis Kingdon-Ward (1885–1958), famed for his knowledge of Himalayan rhododendrons and blue poppies, and for trudging throughout Asia's outback in an ecological

---

500-page-traditional-medicine-encyclopedia/, Accessed May 14, 2019, as well as the most comprehensive book published on the Toda's of South India, a vegetarian collective that worships flowers, *The Toda Landscape – Explorations in Cultural Ecology,* by Tarun Chhabra, with a foreword by Anthony R. Walker, Harvard University Of South Asian Studies, Cambridge, Massachusetts and London, England, 2015.

[31] See "Japanese Artist Plants Colorful Flower Landscapes to Explore Nature's Cycle of Life and Death," by Emma Taggart on September 29, 2017, My Modern Met, ://mymodernmet.com/japanese-flower-art-azuma-makoto/. As well as Makoto's astonishing three-volume *Encyclopedia of Flowers* (2012), *Florilegio i fiori, il tempo, la vita* (2017), and *Flora Magnifica: Art of Flowers in Four Seasons* (2019) all with photographer Shunsuke Shiinoki.

[32] See "The hidden beauty of flowers: Microscopic images reveal the alien landscapes to be found on petals, pollen grains and leaves," by Victoria Woollaston, The Daily Mail, May 8, 2014.

[33] See "In medieval monastery gardens, an uplifting model for something we could all use: Refuge," by Adrian Higgins, June 10, 2020, https://www.washingtonpost.com/lifestyle/home/in-medieval-monastery-gardens-an-uplifting-model-for-something-we-could-all-use-refuge/2020/06/09/ad4d6522-a5c3-11ea-b473-04905b1af82b_story.html, Accessed June 11, 2020.

[34] See D.J. Sommer, The Shanidar IV 'Flower Burial': a Re-evaluation of Neanderthal Burial Ritual, Cambridge Archaeological Journal, vol. 9(1), pp. 127–129, 1999.

[35] See *The Cabaret of Plants: Forty Thousand Years of Plant Life and the Human Imagination,* by Richard Mabey, W. W. Norton & Company, 2016.

avant-garde of solitude, searching for his ephemeral botanical idylls while the rest of the world was distracted by wars, a Great Depression, DDT, and other follies.

"And let me never, never stray from thee!" wrote James Thomson in his *The Seasons*, at the culmination of the Fall, with the disquieting onset of Winter, as foliage fell fast and seemingly forever upon the face of the earth.[36] It was the same cry of ultimate destruction described in memorable conversation between the leaves of a great oak tree in *Bambi*, by Felix Salten, as those out at the very tip of a branch, in the midst of a howling, late Autumn wind, bid farewell to one another.[37] That life and death form the most lasting of all unions remains our most telling and inexplicable of paradoxes: We simply cannot plumb its meaning though we feel everything (Fig. 26.4).

**Fig. 26.4** *Papilio multicaudatus* feeding in the Southern Rockies on *Lilium lancifolium*. (Photo © M.C. Tobias)

---

[36] James Thomson, *The Seasons, To Which Is Prefixed the Life of the Author*, by P. Murdoch, Wilkie and Robinson, London, 1811, p. 184.

[37] See *Bambi: A Life in the Woods*, by Felix Salten, translated by Whittaker Chambers, Illustrated by Kurt Weise, Grosset & Dunlap, New York, 1931, Chapter VIII, pp. 86–89.

# Part II
# Ecological Memories and Fractions

# Chapter 27
# The Metaphysics of Photography

## 27.1 The Photographic Ecosystem

Syrian asphalt—bitumen (a natural petroleum), lavender oil, and a thin sheet of pewter. Who would have thought that in the mid-1820s in France this unlikely concatenation in a little artist's studio might together concoct an after-image, leaving posterity with riddles and the dilemma of self-examination against the context we would conspire to break? Two centuries later, the explosion of imagery in our lives has occasioned the human investiture of every corner of the planet, pixel manipulations and geographical information systems that, by their intrusiveness and uncanny durability, have normalized every life form, personal identity, catastrophe, and cartographic nuance. The only things left are metaphysical aftermaths which compete with photography in the mind to either preserve or desecrate: The photographic ecosystem (Fig. 27.1).

"What is *there*? What is it *like*?"[1] When you're *there* is it really *you*? Taking compulsive pictures, storing images in the memory bank that is now a global commons, all vulnerable?

This was core to the characteristics challenged by Leibniz's foremost "Principle" – his "Law of the Indiscernibility of Identicals."[2] These look-alikes within the thought process—the digital minions representing what was free, prior to the investitures by the human imagination—encompass the minutiae of Philosophy and Physics, subatomic Weak and Strong Forces/Principles, and counterintuitive, counterfactual, counter-philosophical absolutes. We have no way of understanding whether other species store memories to this degree, or for anything like the human purposes,

---

[1] What is it (i.e., whatever it is that there is) like? Hall, Ned (2012). "David Lewis's Metaphysics". In Edward N. Zalta (ed.) *The Stanford Encyclopedia of Philosophy*, Fall 2012 ed., Center for the Study of Language and Information, Stanford University. Fall, 2012.

[2] *Discourse on Metaphysics*, 1686; See Alyssa Ney, 2014, Metaphysics: an Introduction, Routledge, Oxford, UK

© Springer Nature Switzerland AG 2021
M. C. Tobias, J. G. Morrison, *On the Nature of Ecological Paradox*,
https://doi.org/10.1007/978-3-030-64526-7_27

**Fig. 27.1** "The Valley, from Eagle Peak, Yosemite," Underwood & Underwood 1902, The Wonders of the Yosemite Valley, and of California, Boston and New York, 1872. (Private Collection)

which especially in real-time we fail to fathom. Aftermaths devolve, resolving into an identity that cannot be identified, yet remains intact in the image. Rendered easily conquered, fenced, grazed, extracted, codified, then suburbanized. It happened by dint of sheer bad luck, unstoppable hiker enthusiasm following a *National Geographic* magazine story on the North Cascades in August 1949 (Vol. 96, No. 2), and it has happened thousands of other times, everywhere.

Man created in God's image: a paradox plaguing technical process, spiritual reckoning, the many fallacies of memory, prediction, and exploitation. All the antique photographs, nameless, placeless, studio arrangements, or random capture of motion-in-motion. The history of photography is one more dramatic prelude, another "starting point," as Aristotle described the metaphysical impulse, in

projecting all those changes by which "nature begins" and with which human consciousness struggles to articulate itself.[3]

The photograph hastens our love affairs with philosophical fiction, the Self as
participant in something greater than the Self—an ecosystem, of sorts. Yet, every
produced image brings us closer to an oblivion of the unproduced, the inchoate,
filled with lingering doubts, or withheld sensations doubting its scientific verity.
This question of ecological certainty, patriotism, or the egotistical aggrandizement
connoted by the manufacture of images involves conundrums of psychology, of
speculative realism, panoramic endemism, picturesque escapism, those postcards
and billboards, the rational and irrational, potential and actual, underlying principles of identity, integrity, and place. Ultimately, the landscapes of mind, those myriad, wistful losses of physical identity into the greater miasma of an aesthetic swamp
that is the photograph, leave us ecologically stymied. There could be no stranger
paradox than that of a landscape photograph. Taking into account the optical technology, what is it, really?

## 27.2   Picturesque Prolegomenae

All of these changing artistic and objective tides are, to quote Kant's definitive
*Prolegomena To Any Future Metaphysic* (1783), a gratifying antithesis of the critique of reality we would personally prefer to reject: "Thus the order and regularity
in phenomena, which we call nature, we ourselves introduce, and should never find
it there, if we, or the nature of our mind, had not originally placed it there."[4] Just
think about Eugène Atget's spectacular subtlety in his photograph of "La Marne à
la Varenne," (1925–1927, Museum of Modern Art, New York).[5] This gold-toned
print on paper, of the Abbott-Levy Collection, resounds with the black and white
paradise of a few slack canoes. The draping trees, the commune in the Maine-et-
Loire department, with its 100 bird and 17 dragonfly species, ring-necked snakes,
and 370 plants species.[6] All converge in the imagination of that estimated 100
meters of reflective waterway—a rarified scintilla of aqueous contemplation out of
514 kilometers of this eastern tributary of the Seine southeast of Paris—witnessed
upon no particular day or hour by a photographer about whom "few facts are
known."[7]

Just before the astonishing English composer Henry Purcell died in 1695 he
wrote a song, "From rosy bowers" that has been described as "horrifically

---

[3] See *Aristotle's Metaphysics*, translated with Commentaries and Glossary by Hippocrates
G. Apostle, The Peripatetic Press, Grinnell, Iowa, 1979, p. 73.

[4] *Kant's Critical Philosophy*, by John P. Mahaffy and John H. Bernard, Macmillan And Co.,
London, 1889, p.177.

[5] See *A World History Of Photography*, by Naomi Rosenblum, Abbeville Press, New York, p.278.

[6] See "Fauna and Flora in Meaux," https://www.boating-paris-marindeaudouce.com/fauna-flora-
meaux/, Accessed March 12, 2020.

[7] op. cit., Rosenblum, p. 278.

autobiographical." As the tuberculosis overwhelmed the 36-year-old genius, his song declared, "…'tis all in vain. Death and despair must end the fatal pain.'"[8]

Like a song, a photograph gives us the ephemeral yet lasting impact of an image in our minds. Photographer August Sander (1876–1964) acknowledged that "humans often change the results of biological process" and partly attributed the power of landscape photography to the recognition of the "human spirit of a time… captured with the help of the camera."[9]

Few photographs are as given to the comprehensive dislocation of aesthetic vulnerability to the rude, glib, metallic mechanics than that which we would, absent a camera in our hands, hail as sublime. This was the dislocation through technology that John Ruskin found so distasteful, as described in his writings, particularly in *The Elements of Drawing & the Elements of Perspective*.[10] Ruskin (1819–1900) knew well the incremental problems. He was not just the social critic and art historian, but a painter and amateur photographer, greatly conflicted by the irrepressible contagion of beauty in landscapes. In time he grew concerned that the tremendous verisimilitude of an Albert Bierstadt or Frederic Church landscape might be eclipsed by the mirror-imagery of photography, and contested the photographic soul. It could not replace the emotions of a great painter's patient interactions with a glacier or a tree. And this was abundantly evident in the Yosemite Valley and at Niagara Falls.

These two matrices of gravity have been viewed, and swarmed, of course, by hundreds of millions of enthusiasts during the past centuries, economic rationale, goes one lumbering argument, for their very protection. The tragedy of that theory of democratization for the people was the exile and assimilation of *the* people, the Miwok themselves.[11] Their culture's virtual extirpation from Yosemite's valley floor in 1923, as the first automobiles poured into the park, was memorialized by the height of government arrogance in the form of the Chandelier Tree "Drive-Thru". This was a 276-foot-tall *Sequoia sempervirens* at Leggett, California, with a hole carved out at the base, 6 ft by 6.9 ft. A savage puncture large enough for one of the park superintendent's ludicrously lauded automobiles to pass through. The overall crisis accompanying such desecration accompanies Columbus' deep-seated legacy, pitting the personable and private against the mob; aesthetic genocide, one after the other across the map of North America, and—at the species level—the entire planet.

---

[8] *Henry Purcell*, by Robert King, Thames And Hudson, 1994, p. 226.

[9] "Photography as a Universal Language," Lecture 5, April 12, 1931, Accompanying #14, "The Siebengebirge seen from the Westerwald, 1930s," *August Sander – Seeing, Observing and Thinking -Photographs*, with text by August Sander and Gabriele Conrath-Scholl, Die Photographische Sammlung/SK Stiftung Kultur, Foundation Henri Cartier-Bresson, Schirmer/Mosel, Paris 2009.

[10] *In The Ethics of the Dust, Fiction fair and foul, The Elements of Drawing*, by John Ruskin, Illustrated Cabinet Edition, Merrill and Baker, New York, 1877.

[11] *Legends of the Yosemite Miwok*, compiled by Frank LaPena and Craig D. Bates, illustrated by Harry Fonseca, published by the Yosemite Natural History Association, Yosemite National Park, California 1981.

Pointillism versus pointlessness; privacy and totalitarianism; hallowed ground and rubbish heaps, all confounding the *paradox of progress*.

## 27.3 Yosemite and Niagara

Yet, therein persists an undeniable spirit embodied in certain photographs, for example, of Yosemite and Niagara, two iconic landscapes photographed more than any other subjects in the world, save perhaps for a kiss beneath the Eiffel Tower, "a love song celebrating life/one flame remains illuminating loneliness," as Shimpei Kusano (1903–1988) so perfectly fixed in his "the white frog".[12] We have marveled at the photographs by W. E. Dassonville, in George Sterling's perfect *Yosemite: An Ode*;[13] the Flemish Renaissance-like compositions of John P. Soule's photographic engravings for Samuel Kneeland's *The Wonders of the Yosemite Valley, and of California*;[14] the "more than 200 illustrations including eight color plates from paintings by Chris Jorgensen in John H. Williams' *Yosemite and its High Sierra*;[15] the pictures by Thomas Moran to illustrate Thomas D. Murphy's *Three Wonderlands of the American West*;[16] the photographs bridging the Gerhard Richter divide between photorealism and the suasion of a blurred surreality underscoring our profound ecological uncertainty; the receding, uncredited photos illuminating *The Yosemite*, by John Muir;[17] and Galen Rowell's venerable version in color of Muir's *The Yosemite*.[18]

The evolution of these and so many other volumes of photographic tribute situate our scintillating idée fixe in a penchant provisioning for anonymity that is so at ironic odds with the masses of tourism following behind. That moment at sunset when the El Capitan falls spray the North American Wall, and Bridal Veil thunders toward landfall, adjoining the Leaning Wall, its miraculous prismatic mists on occasion cooling down the parking lot near the base.

These unassimilable juxtapositions of human conquest and our penchant for solitude and the capture of beauty were best enshrined in two volumes of photographs— *Yosemite Valley* and *Niagara Falls,* the text on the back of the 24 stereoscopic double

---

[12] In *frogs &. others.* [sic] poems by Kusano Shimpei, translated from the Japanese by Cid Corman and Kamaike Susumu, illustrations by Olhmo Hidetaka and Kuano Shimpei, a Mushinsha Limited book, Grossman Publishers, Charles E. Tuttle Co., Inc., 1969, p. 43.

[13] With a cover in color after the painting by H. J. Breuer, A. M. Robertson, San Francisco, 1916.

[14] Alexander Moore Lee & Shepard, Boston; Lee, Shepard & Dillingham, New York, 1872.

[15] John H. Williams, Tacoma And San Francisco, 1914.

[16] L.C. Page & Company, Boston, 1913.

[17] The Century Co., 1912.

[18] *The Yosemite* – the original John Muir text illustrated with photographs by Galen Rowell, each photograph accompanied by an excerpt from the works of John Muir and an annotation by Galen Rowell, introduction by the photographer, a Yolla Bolly Press book published by Sierra Club Books, San Francisco, 1989.

cardboard images in English and Russian. They commend, at one point, the fact that "The cliffs and cascades are to this day thick with old Indian legends, and traditions learned from the savage people who used to roam these mountain sides…." What are we to make of the few pictured strangers—riding mules, lounging beside violent cascades, or—gazes half-askance—floating aimlessly into an oblivion we envy? Unknown souls, all deceased we can be certain, who once ventured there and asked of the moment, and posterity in its wake, to simply leave them be, discretely captured in a mode new to American outdoor sensibility—the "sweetheart stereoviewer" first introduced to the public in 1851.[19] The Lawrence & Houseworth stereographs and those by Carleton Watkins (1829–1916)[20] also encompassed three-dimensional portraiture of Native Americans, a technique first employed by Sir Charles Wheatstone in 1838.[21] By the time Edward Sheriff Curtis, commencing in 1906 with J. P. Morgan funding, began a 20-year odyssey to photograph some 80 "tribes" west of the Mississippi, the 20-volume masterpiece initially entitled "The Vanishing Race," then changed to *The North American Indian*, the technological frenzy to capture images was to transform ecology.[22]

## 27.4  Psychic Divides

The lessons learnt were built into a monumental psychic divide. There was the aggressive will to power; to achieve acute detail in natural history, but also mytho-poetic recollections of that indefinable line between art and science. This latter motivation, that being-within-being wanting badly to confront the other, had captured the artistic prayers levied by the world's public in country after country. Such vanishing points, species, peoples, and feelings were paramount in an era where technology had captured a bias flawed by its mass marketing, manipulations, advertising, and no thought of what all these hundreds of millions of photographs might be doing to the fragile relations between humanity and nature, keeping in mind the mixed adage that familiarity may breed love affairs, or accelerate an indifference to multiplicities all proffering the same thing, as destruction proceeded apace.

The photographic suasion had catapulted the picture of a landscape behind a Burgundian chateau in 1826 by Joseph Nicéphore Niépce, and his heliogravure of a man leading a horse a year earlier, thought to be the oldest known photograph;[23] "of

---

[19] https://www.nps.gov/yose/blogs/curious-matters-where-two-become-one.htm, Accessed March 12, 2020.

[20] See https://www.donaldheald.com/pages/books/27829/carleton-e-watkins-photographer/collec tion-of-56-stereoscopic-photographic-views-from-watkins-pacific-coast-series-principally

[21] See "The California Indian in Three-Dimensional Photography," by Peter E. Palmquist, *Journal of California and Great Basin Anthropology*, Vol. 1, No. 1, Summer 1979, pp. 89–116, published by Malki Museum, Inc., https://hstor.org/stable/27824948

[22] See *Edward S. Curtis: One Hundred Masterworks*, curated by Christopher Cardozo, 2015.

[23] World's Oldest Photo Sold to Library," March 21, 2002 BBC News n.a. http://news.bbc.co.uk/2/ hi/europe/1885093.stm

the acutely detailed image of an orchid leaf in 1839, by Sarah Anne Bright (1793–1866);[24] and on to a genre of noumenal solemnity that throughout the nineteenth and early twentieth centuries evoked the most solemn, nostalgic, and noble of memories. The runaway lures of this new technology were splashed across the new world of illustrated books, with titles such as *The Gallery of Nature, Home Ballads, Song of the Sower, Story of the Fountain, The Riddle Of The Tsangpo Gorges, The Worst Journey in the World, My Climbs In The Alps And Caucasus, The Ascent of Mount St. Elias, Among The Waterfalls Of The World, Abbeys and Castles of Great Britain and Ireland, Ruins And Old Trees Associated With Remarkable Events In English History, and Winter In Arabia.* They all form a distinct genre that combines our lost hopes, vanished geographies, and thorough trouncing of the laws of physics and biology: Something else is at work in the powers of nostalgia, a subset of life that is enshrined in dreams of an opium eater, of sandcastles, and rabbit-holes. Or, of that inescapable sensation that life is/was always greener and better among horizons and heirlooms. The first principle of modern ecology. What does it say about the contemporary psyche?

## 27.5 Mortality and Memory

Today, in the works of such luminaries as Michael Kenna working in *Japan*,[25] or his *France*,[26] Tom Chambers' hallucinatory-like rapid-eye movement sensations of otherworldly Mexico captured in a work entitled *Dreaming In Reverse – Soñando Hacia Atrás*,[27] or, again, Kenna's floating imagery of *Huangshan*;[28] the works of Rocky Schenk, particularly his partially blurred illustrations to the Authors' novella, *Jan & Catharina*,[29] Sebastião Salgado's *An Uncertain Grace*,[30] and one of the most intoxicating photographic albums of the last 50 years, Don Hong-Oai's *Photographic Memories – Images From China And Vietnam*,[31] with dream dissolving photographs entitled "Lovers, Vietnam," (1966), "Floating In Spring, Hunan," (1998) and "Morning On The River Li, Guilin," (1986). These are fixities, central pillars of a

---

[24] See "How 1170-year-old 'Leaf' provoked a hunt for the world's first photographer," by Nick Clark, The Independent, July 5, 2015. https://www.independent.co.uk/arts-entertainment/photography/how-a-170-year-old-leaf-provoked-a-hunt-for-the-world-s-first-photographer-10367441.html, Accessed March 13, 2020.

[25] 1987–2002, Nazraeli Press, Tuson, Arizona 2001.

[26] Nazraeli Press, 2009, www.nazraeli.com

[27] Photo-Eye Editions, Santa Fe, NM, 2011.

[28] Nazraeli Press, Portland Oregon, 2010.

[29] Santa Monica, Smart Art Press, 1998.

[30] *An Uncertain Grace: Essays by Eduardo Galeano and Fred Ritchin*, by Sebastião Salgado, Aperture Foundation, New York, 1990.

[31] Custom & Limited Editions, New York and San Francisco, 2000.

transcendent wish that wishes away hours, combing the multitiered disparities and discontent of an age—all ages—of devout anxiety, desperation resulting in some great, albeit illusory, dream of exile.

Christopher Cardozo has told us that when Native Americans visit exhibitions of Edward Curtis, which Cardozo has been brilliantly championing for much of his life, they inevitably discern deceased members of their families. The continuity, Christopher says, choking up with excitement, is like a family tree, lost ancestors, future destiny, come to light, in the epiphanies of photographic moments. This was what both Aristotle and Kant were more than hinting at. How our imagination might be fully enlivened by the encounter with an image; the same field of excited electrons that catapults electrical impulses into our receptor visual cells within the occipital lobes, laying bare the ecological implications of who and what we are; of that which we can do to align our brief lifespans with the visual mapping in which our brains are forever engaged—those who can see, or imagine such scenes, reconnoiter other sights ecstatic, or gracefully bow out, leaving so many places alone (Fig. 27.2).

All those elusive images of Yosemite and beyond accrue in the corneas that contain the largest and densest number of nerve endings in our entire body. The brain, its basic anatomical features including the eye, makes our entrancing connections irrevocable, our stewardship a point of both wishful thinking and moral compass.

**Fig. 27.2**  "A Visitor," 1921, by Edward S. Curtis. (Private Collection)

This ethic is the inescapable behavioral poetic that, in true kinship with the evolutionary purpose of such anatomy, must perforce engage in the noblest of actions: succumbing to the imagery in such fashion and rigor as to protect and uphold the autonomy, anonymity, and sacrosanct grounds from whence such images have been created. Every National or Royal Geographic cover convenes upon this photographic credo. The dangers, of course, are that one's celebrity becomes its prolific undoing.

A problem naturally arises under the aegis of so august a manifesto of logic. In the early 1900s radiologists were troubled. Pathologists were rendering inaccurate diagnoses based upon any number of misreadings of X-rays. As Lorraine Daston and Peter Galison explore, mechanical images in medicine at once demanded a discerning scientific reception; objectivity became dependent on subjectivity. The core of machine intelligence, for example, photography, was no autonomous truth, but the "interpretive judgment" of observation. *"Trained judgment* came increasingly to be seen as a necessary supplement to any image the machine might produce."[32] And if this were the case for scientific instrumentation, what about landscape aesthetics? Write Daston and Galison, "Judgment by the author-artists joined the psychology of pattern recognition in the audience."[33] Hence the de facto second opinion, when an MRI suggests either a "mass" or some "narrowing" in the bones.

Beyond the infinity and pathology of photographic longings, devolving into the normal or abnormal, remains the steadfast iconic, the selfie, a preamble to whatever else philosophy or science might have to say about it. Or, the lone whisper, the last gasp, even a monumental arousal, simply disintegrating to a digitally catalogued nothing, dust beyond that, in which we recall the words of the late political theorist Jack Schaar, who once told us, "I love the mountains. That's why I never go there" (Fig. 27.3)

---

[32] See Lorraine Daston and Peter Galison, *Objectivity*, Zone Books, New York, 2007, p. 314.
[33] ibid., p. 315.

**Fig. 27.3** Milford Sound and Mitre Peak, Fiordland National Park, New Zealand. (Photo ©
M.C. Tobias)

# Chapter 28
# The Consolations of a Château

## 28.1  A Distinctly Indoor Paradise

In Latin, castellum, castle; in Old French, chastel; the chatelaine, its mistress, living rapturously in her fine country estate. Country, one of the most abstruse and elongated etymologies in the English language. It fixes in our minds all that structure overlooking, or opposite to, landscape. A place of undefined forest stretches, dragons (in Arthurian romance), of places all under the authority of the Crown and, hence, the opposition by those who poached on those lands. Poaching in general, forgetting class warfare, simply the archaisms of culling for one's own foodstuffs. Chalk-and-fen country, baronies, and tribal chiefdoms. Almost at once a linguistic repository of various patriotisms and battlefronts. In Diderot, it is either the dungeon or the château. Ballad poetry and romance, or the drudgery of human existence at its lowest ebb. Conversely, a structuralism aggregating our most fervent needs to escape others, at least until we become deathly ill (Fig. 28.1).

In France, he that can afford to "bids farewell to the landscapes copied by painters, that he may enjoy the real landscape of the good Creator, the eternal landscape which returns each year with the spring-roses, always younger and more smiling."[1] That Homeric landscape, as John Ruskin applied it in his *Modern Painters*, "which even an immortal might be gladdened to behold." With foaming white waters, "long-tongued sea-crows," and "meadows full of violets and parsley." Sweet cypress, groves of alder, grapes in bloom, and copious rocks.[2] And in the very central thickness of those perennials and hazy vistas "a record of history," the Rev. G. N. Wright described the Château of Blois, a monument that "can excite none but

---

[1] *France Illustrated, Comprising a Summer and Winter in Paris, Drawings by M. Eugene Lami, Descriptions by M. Jules Janin, Supplemental Vol. IV*, Peter Jackson, Late Fisher, Son, & Co., London and H. Mandeville Paris, 1840, pp. 178–179.

[2] "Of Classical Landscape," *Modern Painters*, Volume III, *Of Many Things*, John Ruskin, Merrill and Baker, New York, Illustrated Cabinet Edition, 1890, p. 233.

© Springer Nature Switzerland AG 2021
M. C. Tobias, J. G. Morrison, *On the Nature of Ecological Paradox*,
https://doi.org/10.1007/978-3-030-64526-7_28

**Fig. 28.1** Engraving of Dutch Floating Castle, detail, by David Vinckboons, engraver Nicolaes de Bruyn, 1601. (Private Collection). (Photo © M.C. Tobias)

melancholy reflections."[3] Or, in the case of the fine Castle of Chinon, Henry the Second of England's final, humiliating domicile, where he remained "prey to grief and disappointment," a resting place for the "broken-hearted," and "where his life of misery was terminated by a fever, in the year 1189."[4]

In 1790, mobs of the French Revolution intent upon eradicating all memories of the Ancien Régime sacked the great Benedictine Abbaye de Cluny in Saône-et-Loire, whose 800 years of construction had commenced as far back as the fourth century, subsequently burning down the largest library and Church in the world, beside St. Peter's Basilica in Rome. Tens of thousands of pieces of the lichen-rich "brick and ashlar" masonry, the stone hued from southern Burgundian quarries,[5]

[3] op.cit., *France illustrated, exhibiting its landscape scenery, antiquities, military and ecclesiastical architecture, &c.;: Drawings by Thomas Allom, esq., descriptions by the Rev. G. N. Wright*, M.A., Vol. III, Fisher, Son, &Co. The Caxton Press, London, H. Mandeville, Paris, 1840, p. 36.

[4] *France illustrated, exhibiting its landscape scenery, antiquities, military and ecclesiastical architecture, &c.;: Drawings by Thomas Allom, esq., descriptions by the Rev. G. N. Wright*, M.A., Vol. III, Fisher, Son, & Co. The Caxton Press, London, H. Mandeville, Paris, 1840, p. 45.

[5] See "Cluny Abbey, "by Christine Bolli, n.d., Khan Academy, https://www.khanacademy.org/humanities/medieval-world/romanesque1/a/cluny-abbey, accessed March 14, 2020.

were dismembered and distributed throughout France during the construction of the town of Cluny, just northwest of Mâcon, with a population nearing 5000 today.

From the ruined Teutonic Castle of Schaaken in today's Nekrasovo Kaliningrad oblast in Russia, to the sadly roof-bereft Robert Adam mansion in Ayrshire, Scotland, known as Dalquharran Castle (owners evidently removed the lead in the roof to thereby obviate the requirement of payment of local rates),[6] to all of those other romantically neglected sites across the world, countless architectural dreams of the ultimate recluse have succumbed to teeming jungle hybrids, half-human history and esthetic memory, the other halves, their own worlds of weeds away.

Somewhere in between the runic symbology of an abstracted form of enormous beauty and that which remains standing and inhabitable is that famed view of the Monastery of La Trappe[7]; or the Cisterican Abbey of St. Mary's, Melrose, in Roxburghshire along the Scottish Borders area, built in 1146, leading to an inexorable demise. Two photographs from the much crumbled Melrose were used in 1872 to rightly illustrate Cantos II and III—from Sir Walter Scott's "The Lay of the Last Minstrel"—"If thou would'st view fair Melrose aright, Go visit it by the pale moonlight; ...And, home returning, smoothly swear, Was never scene [sic] so sad and fair!"[8] Robert the Bruce's embalmed heart was exhumed there in 1921. In 1805 Phillip de Loutherbourg sketched and colored the genre for living ruins in his "Tintern Abbey," built in 1131 on the celebrated western banks of the River Wye in Monmouthshire, and the "Ruin at Basingstoke" in Hampshire[9] (Fig. 28.2).

By the late eighteenth century, both had become the celebrated haunts of a brisk tourist industry bent upon wildness and weekend poetics. William Gilpin, who described its ruins as "beautiful" but "ill-shaped," "an elegant Gothic pile," had famously rendered a photo-realistic watercolor of the Abbey-Church in 1770 for his *Observations on the River Wye and Several Parts Of South Wales, &c. Relative Chiefly to Picturesque Beauty; Made in the Summer of the Year 1770.*[10] And in William Wordsworth's "Lines Composed a Few Miles above Tintern Abbey, On Revisiting the Banks of the Wye during a Tour. July 13, 1798," the poet wrote "Of joyless daylight; when the fretful stir/Unprofitable, and the fever of the world, Have hung upon the beatings of my heart – How oft, in spirit, have I turned to thee."

This combining of the chateâu genre with the reality of humanity's history of harvests, of existentially satisfying esthetics—a new departure for all who were melancholic—and of looking beyond, also recognized a great demarcation of a

---

[6] See "The Architecture of Robert Adam (1728–1792), The Castle Style", sites.scran.ac.uk, published by Cadking Design Ltd., Edinburgh, Scotland, http://www.rls.org.uk

[7] See *A Visit to the Monastery of La Trappe, in 1817: With Notes Taken During a Tour Through Le Perche, Normandy, Bretagne, Poitou, Anjou, Le Bocage, Touraine, Orleanois, and the Environs of Paris* by W. D. Fellowes, Esq., William Stockdale, London, 2nd ed., 1818, p. 9.

[8] Provost & Co., London, 1872, pp. 23 and 44; quotations from pp. 24–25.

[9] See *Scènes Romantiques Et Pittoresques, D'Apres Les Dessins Faits Exprès Pour Cette Enterprise Par P.J. de Loutherbourg*, Robert Bowyer, Galerie Historique, London 1805.

[10] William Gilpin, Third Edition, printed for Blamire, London, 1792, p. 47

**Fig. 28.2** "Tintern Abbey in Monmouthshire," drawn by P.J. de Loutherbourg, in his *The Romantic and Picturesque Scenery of England and Wales*, Robert Bowyer, London, 1805. (Private Collection). (Photo © M.C. Tobias)

truth. As writes Ian Niall, "It might not have occurred to the farm labourer that when the horse went he would be the next to be made redundant…"[11] (Fig. 28.3).

## 28.2   The Romance of a Ruin

But that obsolescence has not happened, not yet, not entirely. Ruins have remade countryside, which, in turn, have granted wilderness a habitable domain for a constructive if unwieldy human wilderness of remnants and phantasms. From Pontchartrain, in Yvelines (the municipality of Jouars-Pontchartrain), built in the sixteenth century and described by one enthusiast as "a long, low silhouette of a building so integrated with its environment as to seem a perfect abstraction of the lake and the hillside forest on either side,"[12] to at least 6000 other French châteaux

---

[11] *English Country Traditions, Wood Engravings by Christopher Wormell*, written by Ian Niall, The Victoria and Albert Museum, The Hand Press, Kent, 1988, p. 56.

[12] *The Chateaux of France*, by Editors of Réalités-Hachette and Daniel Wheeler, The Vendome Press, New York, Paris, Lausanne, 1979, p. 120.

**Fig. 28.3** Château and Poitevins, in Poitou. (Photo © M.C. Tobias)

and castles, there is an undeniable urging to become something other than merely human. The vast Chambord, or Vaux-le-Vicomte (whose contractor-finance minister, Nicolas Fouquet, Marquis de Belle-Île, would be scandalized and sentenced to life imprisonment), institutionalizes what elsewhere across the landscape proliferate more intimate, personable domains, like the Bagatelle in Picardie, La Dame Blanche, noted for its rare soils and fecund grapes producing one of the great dry white wines of the Bordeaux region.

From the Renaissance Montal in Aquitaine to the spectacular Anjony in the mountains of Auvergne, or Château Rouffillac in the heart of the Perigord Noir, a half-hour drive instructively from Lascaux, and once owned by the greatest French zoologist of the twentieth century, Pierre-Paul Grassé and his family, these enormous edifices are no ziggurats or pyramids, but the spacious homes of people in love with countryside. Whether in neoclassical villas outside Florence or Vicenza (the latter city home to the sixteenth-century master architect Andrea Palladio), or aboard rambling ensembles from Montecito (e.g., the 37-acre historic estate of Madame Ganna Walska, known as Lotus Land) to Morocco, to northern Thailand, the integrative facilities of our species have always sought to architecturally console its spirit by palpably giving vent to combinatorial powers. What we think of as *our* Nature, roaming her peripheral boundaries in hopes of appropriating them for our own bedrooms, is a primal peregrination that has never dimmed (Fig. 28.4).

In its final gasp, a potted Bella Blue Prunella or August Grove Geranium, quietly minding her business, showing off such miniature resplendence upon a lonely balcony over bustling Milan or strident Mumbai, still mirrors our desperate harkening.

**Fig. 28.4** "Lanthony Abbey," in the "retired vale of Ewias." Photo by F. Bedford, from *Ruined Abbeys and Castles of Great Britain*, by William And Mary Howitt, A.W. Bennett, London 1862. (Private Collection)

Da Vinci, relocating in 1516 to Amboise at the bequest of King Francis 1, lived next door to the King's château on a 3-acre miniature paradise known as Château de Cloux, where he spent his final 3 years. During that time, the King invited da Vinci to the nearby village of Romorantin, where he was asked to create, essentially, a utopian village and utopian palace.[13] It was the kind of challenge once offered him near Florence—to remake Nature, in a sense—but here, in central France, again he failed—had not the time in his body—to accomplish so overwhelming a task, an artistic and scientific riddle. The romance and indefinable melancholy of a fortressed inner nature—a few classics on the bookshelf, and paintings on the walls—

---

[13] See *Leonardo Da Vinci*, by Walter Isaacson, Simon & Schuster, New York, 2017, pp. 498–516.

lends a nearly stifling poetic to the reality of ephemeral lives, and the inevitable dissipation of a dream that is human life, all of life. But da Vinci most certainly would have considered its fated ensemble until his dying day, May 2, 1519.

# Chapter 29
# Book of the Dead

## 29.1  Zoomorphic Perplexities

The *Papyrus of Ani*, or *Book of the Dead*, was composed c. 1250 BCE, the period of the nineteenth dynasty of the Egyptian New Kingdom, and rediscovered in 1888 by Sir E. A. Wallis Budge (1857–1934). Budge was a Keeper of the Egyptian and Assyrian Antiquities at the British Museum, a man who could read handwritten hieroglyphs. But he experienced a strange problem, upon endeavoring to translate the manuscript of Ani.[1] Known to some of his friends as "Father of the Skulls,"[2] Budge's fluency in multiple ancient human languages in Sudan, Egypt, Syria, Ethiopia, and Mesopotamia, among others, brought him into a confluence, alien in his time, of zoomorphic ensembles unlike any other linguistic and philosophical puzzle in human history. In chapter CLXXV, a dying man asks the god (of) Anu, "How long have I to live?" to which the god replies, "Thou shalt exist for millions of millions of years…," and all signified by birds conducting themselves in various situations. In chapter LXXXIV, a deceased individual thinks himself Shu, who embodies the belief in the eternal life of the soul, enshrined in motifs of other species[3] (Fig. 29.1).

The myriad hymns, Heart-souls, signs of life, funeral passages, solar mounts, doorways, dead bodies, raised hands, gods and goddesses, sun-disks and horns, shrines and mummies are all dominated by a celestial ensemble of—predominantly—

---

[1] See Budge, E. A. Wallis (1920). By Nile and Tigris: A Narrative of Journeys in Egypt and Mesopotamia on Behalf of the British Museum Between the Years 1886 and 1913. Main 4th floor: London, J. Murray.

[2] https://research.britishmuseum.org/research/search_the_collection_database/term_details. aspx?bioId=93650

[3] *The Book of the Dead: The Hieroglyphic Transcript of the Papyrus of Ani, the Translation into English and an Introduction by E. A. Wallis Budge, Late Keeper of the Egyptian and Assyrian Antiquities in The British Museum*, University Books, New Hyde Park, New York, University Books, Inc., 1960, pp. 67–68.

© Springer Nature Switzerland AG 2021                                           271
M. C. Tobias, J. G. Morrison, *On the Nature of Ecological Paradox*,
https://doi.org/10.1007/978-3-030-64526-7_29

**Fig. 29.1** Sheet from the Papyrus of Ani, British Museum #Q6373

birds: falcons, vultures, owls, eagles, pelicans, human-headed birds, the odd contradiction of so many feathered stories populating the somber walls of tombs within pyramids, within a desert; a resoundingly non-aerial environ oriented to the great Pharaonic dream of taking flight.

In his book of poetry *Flight Among The Tombs*, Anthony Hecht cites Christopher Smart, who writes, "Let Mattithiah bless with the Bat, who inhabitieth the desolations of pride, and flieth amongst the tombs."[4]

As with so many ornithological archives in which the passerines and non-passerine birds are easily identified, it is the context that renders commentary troubling. The once common southern Mediterranean ostrich was adorned in various psalters and bestiaries as a camel, because of its cloven hooves.[5] In the "Animal Scroll of the Choju Jimbutsu Giga" ("Frolicking Animals and People" attributed to Toba Sojo, early twelfth century in Japan) there is little reason to suppose concerted differentiation between species. They are all of a singular community.[6]

The Egyptian fictions were non-fictions, just as so many tales of wonder and of the infinite—from El Dorado and the Hanging Gardens of Babylon, to Atlantis, the realms beyond the Boreas, Pinocchio's Land of Toys, and Shambhala—all occupy a conversant realm in the imagination that might as well be deemed non-fiction,

---

[4] *Flight Among The Tombs: Poems* by Anthony Hecht, wood engravings by Leonard Baskin, Alfred A. Knopf, New York, page prior to Table of Contents, 1996.

[5] See the "Bestiary assida (ostrich)," in *Birds in Medieval Manuscripts* by William Brunsdon Yapp, Schocken Books, New York, 1982, p. 55.

[6] See *Painting in the Yamato Style*, by Saburo Ienaga, translated by John M. Shields, Weatherhill/ Heibonsha, New York, Tokyo, 1973, p. 44.

because, for the believer, they are. Frequently, the owl and the griffon commingle to tease from our rational powers a susceptibility to the fantastic.[7] Even during the US Pacific Railroad Survey from the Mississippi River to the Pacific, 1853–1856, the explorers were fixated on practical matters—where railroad ties might be laid down in the interests of a manifest destiny—but their *real* obsession was the study and detailing of birds; hundreds and hundreds of pages of acute observation, for no other apparent purpose than the provisioning of the 33rd Congress, House of Representatives, gorgeous color renditions of avians from the 38th, 39th, and 41st Parallels. Why, one supposes (other than having been instructed to do so)?[8]

The profound irony of that, and of many other such government reports, is the very symbol of the train traveling through the garden, as historian Leo Marx wrote of it in his *The Machine in the Garden: Technology and the Pastoral Ideal in America*.[9] Marx borrowed the trope from Thoreau's *Walden*, but it lives, for example, in each of the original 40,000+ Superfund sites around the United States, targeted by CERCLA (1980), the Comprehensive Environmental Response, Compensation, and Liability Act enacted by the 96th US Congress and administered by a now ravaged Environmental Protection Agency, struggling to clean up at least 1600 of the most grievously polluted sites from among the 40,000+. Those are the NPL—National Priorities List, the ones that James Florio, a Democrat from New Jersey, recognized as absolutely lethal to life forms, birds, and people in particular, when he introduced his "Hazardous Waste Containment Act of 1980" into the House on April 2 of that year. Five months later the bill was passed, with subsequent passage 2 months later in the Senate.

## 29.2  Modern Waste Lands

And yet, today, 40 years later and counting, a polluted, industrialized America—the world's emblem of wealth and achievement—has endured far more setbacks than gains, as photographer and environmental activist David T. Hanson, in his book *Waste Land*, has illustrated, with aerial images of 67 of the most polluted Superfund sites in the United States. Hanson went to 45 states between 1985 and 1986 to document his targets, selected from "400,000 toxic waste sites and 500,000 abandoned mines" across America. That so much disgrace can be quantified in such multitudes alone condemns the spirit of the Pharaohs; of the Medieval and Renaissance love affairs with all things natural. But, as Hanson points out, all of those "mines, smelters, and wood-processing plants to landfills and illicit dumps, large petrochemical

---

[7] See *The Book of Legendary Lands*, by Umberto Eco, translated by Alastair McEwen, Rizzoli ex libris, New York, 2013.

[8] *Report of Explorations for a Railway Route, Near the Thirty-Fifth Parallel of Latitude: From the Mississippi River to the Pacific Ocean*, by Lieutenant A. W. Whipple, United States Pacific Railroad Survey, Washington, D.C., 1853–1854.

[9] Oxford University Press, New York, 1964.

complexes, aerospace water-contamination sites, nuclear weapons plants, and nerve gas disposal areas" reverberate with ever-escalating despair.[10]

There are no overall estimates on the number of birds killed in US Superfund sites each year, but a random sampling of troubling headlines makes clear how tragic the continuing matrix of death is: "Birds fall from sky amid massive chemical cleanup,"[11] "At least 3000 geese killed by toxic water from former Montana copper mine,"[12] "EPA Responds to Dead Birds Around Michigan Superfund Site,"[13] "Government Officials May Have Mishandled DDT Superfund Site,"[14] "After 3 decades, birds still fall dead from sky in St. Louis,"[15] "Thousands of bird deaths draw focus on brimming toxic pit,"[16] and on and ever on.

In T. S. Eliot's lean epic "The Waste Land," V. "What the Thunder said," lines 357 and 359, "Where the hermit-thrush sings in the pine trees...But there is no water," the poet makes a footnote expressly to delineate this species as *Turdus aonalaschkae pallasii*, which Eliot apparently saw and heard in Quebec Province, and saw fit to recognize during a catastrophic coming drought that would surely kill its kind.[17] The species is also called the Dwarf Hermit Thrush. At least 1450 bird species around the world are currently deemed to be threatened and "one in eight of the world's birds faces extinction."[18] In his Foreword to Blake L. Twigden's 1991 masterpiece, *The Fifty Rarest Birds of the World*, writing from Soestdijk Palace, then Prince of the Netherlands, Bernhard of Lippe-Biesterfeld, stated simply: "I urge you to play a part in the new conservation age. The loss of the delightful creatures in this book would be a tragedy for mankind...."[19] While 182 bird species are thought to have gone extinct since 1500,[20] several of those painted by Twigden are

[10] See "This bird's eye view of America's most polluted sites will break your heart," by Marcus Baram, Fast Company, September 24, 2018, https://www.fastcompany.com/90240249/this-birds-eye-view-of-americas-most-polluted-sites-will-break-your-heart; see David T. Hanson's *Waste Land*, preface by Wendell Berry, afterword by Jimena Canales, essay by David T. Hanson, Taverner Press, 2018, https://www.tavernerpress.com/wasteland.html

[11] https://www.usatoday.com/story/news/nation/2014/08/05/birds-fall-from-sky-amid-massive-chemical-cleanup/13609579/

[12] https://www.theguardian.com/us-news/2017/jan/23/geese-die-montana-toxic-mine-epa

[13] https://connect.edrnet.com/s/article/EPA-Responds-to-Dead-Birds-Around-Michigan-Superfund-Site-1489082045049

[14] https://www.scientificamerican.com/article/government-officials-may-have-mishandled-ddt-superfund-site/

[15] https://www.lansingstatejournal.com/story/news/2014/08/03/after-3-decades-birds-still-fall-dead-from-sky-in-st-louis/13525865/

[16] https://missoulian.com/news/state-and-regional/thousands-of-bird-deaths-draw-focus-on-brimming-toxic-pit/article_18521284-41ad-5142-9e28-b13c2fc0ad66.html

[17] See *The Complete Poems and Plays of T. S. Eliot*, Faber and Faber, London 1969, p. 79.

[18] BirdLife International Data Zone, "Key messages and case studies," http://datazone.birdlife.org/sowb/state/theme4

[19] *The Fifty Rarest Birds Of The World*, by Blake L. Twigden, Osborne Editions International, Auckland, New Zealand, 1991.

[20] BirdLife International Data Zone, http://datazone.birdlife.org/sowb/casestudy/we-have-lost-over-150-bird-species-since-1500

either thought to have vanished or are down to as few as ten individuals, as in the case of the Mauritius (Echo) Parakeet, the world's rarest surviving parrot in the wild.[21]

## 29.3 In Memoriam of the Birds

In 1930 Colonel R. Meinertzhagen, working under the authority of the Egyptian Government, published his two-volume *Nicoll's Birds of Egypt*,[22] the first survey of its kind since poet Percy Bysshe Shelley's distant relative, George Shelley, published his own *A Handbook to the Birds of Egypt* in 1872. Devoting a considerable section to "The Birds Of Ancient Egypt" (written by R. E. Moreau) and representing images of birds carved into a tomb at Saqqara, a mural of the Griffon Vulture from Thebes, a pelican drawing from the Tomb of Mera, of a mallard from the Tomb of Tehuti Hetep, a swan from the Tomb of Ptahhotep, a night heron from the Tomb of Baqt, and so on, the 90 identified species from ancient Egypt are all detailed as beneficiaries, somehow, of a culture that celebrated bird life, and certainly figured most prominently in the translations of Wallis Budge, the Sacred Ibis, the Shoebill, Cormorants, Egrets, Red-footed Falcon, Geese, and the Chanting Goshawk among them.

But not unlike the famed image/inscription of the "Hero Pursuing Two Ostriches" ca. twelfth/eleventh century B.C. and the "Winged Lion Griffin Attacked by Hero," both from the Middle Assyrian period of Mesopotamia, among the Morgan Library in New York's most provocative seals, so too in ancient Egypt the alleged love of birds was a cul-de-sac, a politically promulgated delusion representing an entire social scheme by which the upper class were buried with quite literally millions of mummified birds. Thus, for example, the nightjar in Arabic is known as Teyr-el-mat, "corpse fowl."[23]

This hideous contradiction was perpetuated under the confused sovereignty of the rulers of Tenochtitlan (current Mexico City) throughout the sixteenth century, where one of the most oppressive of all zoos was created for the perverse enjoyment of the Aztec rulers, the last of whom was Motechuzoma II Xocoyotzin (also known as Montezuma, "Angry Like a Lord"). Many of the hundreds of animals were fed some 500 turkeys slaughtered each day, as well as "portions of the carcasses of

---

[21] Jackson, H.; Jones, C. G.; Agapow, P. M.; Tatayah, V.; Groombridge, J. J. (2015). "Micro-evolutionary diversification among Indian Ocean parrots: temporal and spatial changes in phylogenetic diversity as a consequence of extinction and invasion". Ibid. 157 (3): 496–510. doi: https://doi.org/10.1111/ibi.12275

[22] Hugh Rees LTD, London, 1930.

[23] ibid., Vol. 1, p. 62.

human sacrifices," as literally depicted in the Codex Tudela.[24] When Cortés' troops destroyed the Aztec capital on August 13, 1521, the famed four-story House of Birds was set ablaze. Most of the precious occupants burnt to death. A few managed to escape, flying back into the jungles.[25]

The spirituality and deification of avifauna in ancient Egypt was a colossal blind spot perpetuated for millennia. All those archaeologically lionized tombs are both sacred and profane, the killing fields, and superfund sites of birds. It is amazing that any avifauna species survived the massacres in the name of civilization.[26] One thing is certain: Notwithstanding Egyptian hypocrisies over 3000 years ago, the twenty-first century is proving to be the most hostile century in the history of human–avian interactions. We have learned nothing from the *Book of the Dead*, an implicit and intoxicated tribute to our most sacrosanct pact with and subsequent betrayal of Nature: taking everything from, but learning nothing of, history (Fig. 29.2).

---

[24] See "Montezuma's Zoo: A Legendary Treasure of the Aztec Empire," by Natalia Klimczak, Ancient Origins, March 18, 2020, https://www.ancient-origins.net/ancient-places-americas/mont-ezuma-zoo-legendary-treasure-aztec-empire-005090, Accessed July 13, 2020. See also "Aztec capital falls to Cortés," This Day In History, History.com Editors, February 9, 2010, https://www.history.com/this-day-in-history/aztec-capital-falls-to-cortes, Accessed July 13, 2020.

[25] See *History of the Conquest of Mexico: With a Preliminary View of the Ancient Mexican Civilization, and the Life of the Conqueror, Hernando Cortés,* Volume II by William Hickling Prescott, edited by John Foster Kirk, George Routledge and Sons, London, 1884, p. 278.

[26] "Ancient Egyptians May Have Corralled Millions of Wild Birds to Sacrifice and Turn Into Mummies," Nicoletta Lanese, Live Science, November 14, 2019, https://www.livescience.com/ancient-egypt-millions-of-wild-bird-mummies.html, Accessed March 15, 2020. The data taken from "Mitogenomic diversity in Sacred Ibis Mummies sheds light on early Egyptian practices," by Sally Wasef, Sankar Subramanian, Richard O'Rorke, Leon Huynen, Samia El-Marghani, Caitlin Curtis, Alex Popinga, Barbara Holland, Salima Ikram, Craig Millar, Eske Willerslev and David Lambert, PLOS One, November 13, 2019, https://doi.org/10.1371/journal.pone.0223964, https://journals.plos.org/plosone/article?id=10.1371/journal.pone.0223964, Accessed March 15, 2020.

**Fig. 29.2** Egyptian Vulture Migrating through Yemen. (Photo © M.C. Tobias)

# Chapter 30
# Ecological Double-Binds

## 30.1   Survival Against Our Best Interests

In the tradition of our late colleague Gregory Bateson, whose work on schizophrenia in the 1950s first elicited the phrase "double-bind,"[1] (a time-sensitive, high-pressure, lose–lose paradigm), French philosopher René Girard described "a contradictory double imperative, or rather a whole network of contradictory imperatives."[2] If the imperative is to survive, an underpinning of every caveat implicit in the annals promulgated by Darwin and his peers (and continued famously in the debates between Richard Dawkins and Stephen Jay Gould), then clearly there are flaws in the biological and fossil records. Records are dead records. Living testaments make for a very different set of schematics and scientific questions.

We can't ask questions of the dead, although we count on our own living instincts and observations as meaningful surrogates. But every comparison between present and past, life and death, fails, in the end, because consciousness, however self-fulfilling, is incapable of filling that gap. No theory—whether some version of Aristotelian or Newtonian kinematics, punctuated equilibrium, gradualism, or any of a dozen other keynotes in the analysis of eco-dynamics, from the position of driving forces toward complexity, orthogenesis (directed evolution), to neutral, molecular hypotheses, or postmodern prokaryotic synthesis—can account for the double-binds of being human (Fig. 30.1).

Lewis Carroll's character of the Gnat in *Through the Looking-Glass, and What Alice Found There* described a certain "Bread-and-butter-fly" whose fragile morphology relies on a food-source, "Weak tea with cream in it," which was not easily had, and, moreover, when it was obtainable tended to melt the fly's wings. In either

---

[1] See Bateson, G., Jackson, D. D., Haley, J. & Weakland, J., 1956, "Toward a theory of schizophrenia." Behavioral Science, Vol. 1, 251–264.

[2] René Girard, *Deceit, Desire, and the Novel: Self and Other in Literary Structure*, translated by Y. Freccero, The Johns Hopkins University Press, Baltimore, 1965, pp. 156–157.

© Springer Nature Switzerland AG 2021                                                         279
M. C. Tobias, J. G. Morrison, *On the Nature of Ecological Paradox*,
https://doi.org/10.1007/978-3-030-64526-7_30

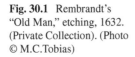

**Fig. 30.1** Rembrandt's
"Old Man," etching, 1632.
(Private Collection). (Photo
© M.C.Tobias)

case, evolution would ensure the fly's starvation. Alice is astonished but reminded by the Gnat that "It always happens."[3] Or, as John Gay (1685–1732) wrote in his *Fables*, "…Tis vanity that swells thy mind. What, heav'n and earth for thee design'd! For thee! Made only for our need; That more important Fleas might feed."[4] A sentiment most prolifically illustrated in Ivan Krylov's own *Fables* (1815). Krylov (1769–1844) was most likely the pictorial and philosophical link between Aesop, Jean de La Fontaine, and Lewis Carroll (1832–1898). And not surprisingly, there is significant evidence linking Charles Darwin's theories to the philosophical disposition of Lewis Carroll in his fiction.[5]

So contradictory is the evidence in evolution for *anything* actually working out as it should that French philosopher François de La Rochefoucauld (1630–1680) wrote, "There are some who never would have loved if they never had heard it

---

[3] See Lewis Carroll, *Through The Looking-Glass and What Alice Found There: With Fifty Illustrations by John Tenniel*, Macmillan And Co., London, 1898, pp. 58–59.

[4] *Fables*, John Gay, J. and R. Tonson and J. Watts, London, 1753, Volume 1, p. 189.

[5] See "What are You? The influence of Charles Darwin's evolutionary theory in Lewis Carroll's Alice's Adventures in Wonderland and Through the Looking-Glass," by Michelle W. H. Smit, Leiden University, September 30, 2019, https://openaccess.leidenuniv.nl/handle/1887/79963

spoken of."[6] At the primacy of human evolution is the same leveling paradox between life and death. In *On the Sublime*, Longinus, describing the writings and sentiments of Homer, declares that "…when man is overwhelm'd in misfortunes, death affords a comfortable port, and rescues him from misery."[7] Several decades ago, in a fascinating effort to characterize evolution as showing no empirical justification whatsoever for "violent human behavior," Daniel R. Brooks and E. O. Wiley wrote, "Finally, we have discovered that some things we thought required evolutionary explanations do not even exist as products of the evolutionary process."[8]

The linear component of any evolutionary theory will impress upon posterities marked run-ins with the gnats and fleas and inordinately omnipresent moments of death that hindsight enshrines, thus complicating our ability to establish biohistorical guidelines for the Anthropocene, the sum total of every paradoxical event in our species' past and present. A good example of that comes from a cursory analysis of pre-colonial indigenous peoples on nearly every continent. Amerindians were once entirely immune to Old World visitors and suffered zero incidence of "smallpox, measles, diphtheria, trachoma, whooping cough, chicken pox, bubonic plague, malaria, typhoid fever, cholera, yellow fever, dengue fever, scarlet fever, amebic dysentery, influenza, and a number of helminthic infestations."[9] But subsequent history—the unchecked distribution of human populations everywhere—has made every ecological double-bind yet more paradoxical as human social interactions both define and threaten to destroy our species, breaking down old immunities, a common progressive reality commensurate with aging,[10] while rendering the earth a mixed chorus of biotic, strong, weak, and constitutive incommensurabilities. Of organisms both akratic (lacking self-control) and unfitted.

---

[6] La Rochefoucauld, "Maxims," maxims 230, 136, from *Reflections or, Sentences and Moral Maxims, by Francois Duc De La Rochefoucauld, Prince de Marcillac, Translated from the Editions of 1678 and 1827 with Introduction, Notes, and Some Account of the Author and His Times*, by J. W. Willis Bund, M.A. LL.B and J. Hain Friswell, Simpson Low, Son, and Marston, London 1871, http://www.gutenberg.org/files/9105/9105-h/9105-h.htm, Accessed March 17, 2020.

[7] *Dionysius Longinus On The Sublime: Translated from the Greek, With Notes and Observations, And Some Account of the Life, Writings, and Character of the Author*, by William Smith, E. Johnson, London, 1770, p. 38.

[8] D. R. Brooks and E. O. Wiley, *Evolution as Entropy: Toward a Unified Theory of Biology*, University of Chicago Press, Chicago, 1986, pp. 305, 307.

[9] *Ecological Imperialism: The Biological Expansion of Europe, 900–1900*, by Alfred W. Crosby, Cambridge University Press, New York, 1986, pp. 197–198.

[10] See Noga Ron-Harel, Giulia Notarangelo, Jonathan M. Ghergurovich, Joao A. Paulo, Peter T. Sage, Daniel Santos, F. Kyle Satterstrom, Steven P. Gygi, Joshua D. Rabinowitz, Arlene H. Sharpe, Marcia C. Haigis. "Defective respiration and one-carbon metabolism contribute to impaired naïve T cell activation in aged mice." Proceedings of the National Academy of Sciences, 2018; 201804149, DOI: https://doi.org/10.1073/pnas.1804149115, Accessed June 11, 2020.

## 30.2  Ethical Prioritizing

In considering every conservation effort to prioritize, arduous ethical choices on a battlefield predicated upon scientific methods arrive at a homogeneous philosophy that prizes efficiency above all else, regardless of such outcomes as those painful criteria driving triage and self-destruction. If a feral cat—the beloved Cheshire cat—is deemed (in many countries of the world) a predator, and the means of sterilizing and containing the cat are ignored (2–6.3 million feral cats move across the whole of Australia, and are believed to "kill hundreds of millions of native birds"[11] each year), then who precisely takes the moral high ground against the efforts to poison all of those cats? In the United States it is estimated that there are 86 million cats, but, according to ornithologist Pete Marra, in charge of the Smithsonian Migratory Bird Center, the majority of them stay indoors and cannot be thought culpable for an estimated 33 species extinctions worldwide thus far documented to have involved free-roaming cats. That said, researchers believe that the several million outdoor cats in America kill approximately "2.4 billion birds and 12.3 billion small mammals" annually.[12] No continent is free of the sum total of eco-dynamics, including what it is convenient if malicious to label noxious *weeds* and *bioinvasives*, but the double-bind ethically undermining every philosophical best practice is the reality that "any group of organisms can become bio-invasive...."[13] The ratio of indigenous to bioinvasive species per country (relying on human boundaries) is unknown, but is likely to be high. For example, in 2007 a New Zealand government study indicated that out of the existing 90,000 native species, 30% were invasives.[14] Moreover, some have argued that this prodigious transmission can, in fact, represent

[11] See "Feral Cats in Australia Sentenced to Death by Sausage," by Windy Weisberger, Live Science, April 30, 2019, https://www.livescience.com/65356-australia-feral-cats-poison-sausages.html, Accessed March 17, 2020.

[12] See also "The Moral Cost of Cats -A bird-loving scientist calls for an end to outdoor cats 'once and for all,'" Smithsonian Magazine, by Rachel E. Gross, September 19, 2016; republished August 14, 2020. https://getpocket.com/explore/item/the-moral-cost-of-cats?utm_source=pocket-newtab, Accessed August 14, 2020. At least "1.4 million" stray cats that end up in U.S. shelters are euthanized every year.

[13] ibid.; see also Prognosis Disaster: The Environment, Climate Change, Human Influences, Vectors, Disease and the Possible End of Humanity? by David Arieti, Jacob Nieva, and Randolph Swiller, Authorhouse Publishers, 2011, p. 62.

[14] That percentage included 25,000 non-native plant species, 2000 invertebrate species, and 54 non-native mammals, including humans. See "Environment New Zealand 2007," Ref: ME847.

**Fig. 30.2** *Ulex europaeus*, flowering evergreen shrub (Gorse)—perceived by many to be New Zealand's most prolific botanical bioinvasive, beneath old Rimu Forest, Southern New Zealand. (Photo © M.C. Tobias)

a boon for biodiversity (Fig. 30.2).[15] We saw a human world obsessed with a "flattening the curve" paradigm in combatting Coronavirus[16].

But to ultimately transcend the double-bind which, inherently, recognizes the philosophy of invasives—whether a virus, a feral cat, a flea, or a human being moving from continent to continent—we are condemned to the paradoxical embrace of some version(s) of biological chaos in the name of community, renascence, and re-evolution.

To transcend such a double-bind (and there are many others) would still result in a single-bind, with no demonstrated neural mechanism in humans for substituting biological altruism for our self-destructive behaviors.

[15] Noga Ron-Harel, Giulia Notarangelo, Jonathan M. Ghergurovich, Joao A. Paulo, Peter T. Sage, Daniel Santos, F. Kyle Satterstrom, Steven P. Gygi, Joshua D. Rabinowitz, Arlene H. Sharpe, Marcia C. Haigis. "Defective respiration and one-carbon metabolism contribute to impaired naïve T cell activation in aged mice." Proceedings of the National Academy of Sciences, 2018; 201804149, DOI: https://doi.org/10.1073/pnas.1804149115, "Do non-native species contribute to biodiversity?" by Martin A. Schlaepfer, PLOS BIOLOGY, published: April 17, 2018, https://doi.org/10.1371/journal.pbio.2005568, https://journals.plos.org/plosbiology/article?id=10.1371/journal.pbio.2005568, Accessed March 17, 2020.

[16] "Coronavirus: What is 'flattening the curve,' and will it work?" by Brandon Specktor, March 16, 2020, Live Science, https://www.livescience.com/coronavirus-flatten-the-curve.html, Accessed March 17, 2020.

# Chapter 31
# The Temptation of the Catastrophe: Deep Structures of Suicide

## 31.1   Classifying Human Behavior

Mathematicians are fond of a category widely named "catastrophe," but its actual denominations are difficult to reconcile with real-world relevance. Famine, tsunami, hurricanes, and tornadoes: neither numbers nor the laws of physics quite add up to the significance of, say, the extinction of a single species at the hands of humanity (Fig. 31.1).

For example, while there are equations to quantify "hyperbolic," "parabolic," and "elliptic umbilic" catastrophes, and a rash of notational depictions, developed by Vladimir Arnold (the ADE classification system)—folds, cusps, butterflies, periodic boundaries, etc.—it won't help those dying on respirators to better grasp the nature of their demise. In mathematical terms, humans are simply too complicated; too many variables beyond mere descriptive effects—ripples and pitchforks, symmetry breaking and hysteresis loops (physical changes happening faster than the systemic effects which caused them), bifurcations, stable or unstable minimums and maximums.[1] As a rule, people do not panic, resort to mob rule, faint, or kill themselves because of gravitational lensing or the three-dimensional reflection of light.

Ecologically, rash behavior is medical behavior; psychological intention; emotional response; embrace or flight. Somewhere in the middle, beyond what we have been characterizing as ecodynamic flux, is some level of odd, calming, beautiful homeostasis. However tranquil the Zen satori or deep moments of prayer, independent temporal contingencies do, in fact, lend to these strangely becalmed aesthetic zones a ticking-clock reality that turns from tranquility to panic with little or no warning.

---

[1] E.C. Zeeman, Catastrophe Theory, Scientific American, April 1976, in which the late British mathematician Christopher Zeeman examines seven "elementary catastrophes." http://www.gaianxaos.com/pdf/dynamics/zeeman-catastrophe_theory.pdf

© Springer Nature Switzerland AG 2021                                                    285
M. C. Tobias, J. G. Morrison, *On the Nature of Ecological Paradox*,
https://doi.org/10.1007/978-3-030-64526-7_31

**Fig. 31.1** "Photomontage of the evolution of a tornado: Composite of eight images shot in two sequences as a tornado formed north of Minneola, Kansas on May 24, 2016." (Public Domain)

While the formal theory of mathematical catastrophe is often credited to the French topologist René Frédéric Thom (1923–2002), he, in turn, borrowed from the basic nomenclature of analytic geometry—cones, arcs, curves—as well as classical mechanics and symplectic surfaces, a combined set of assumptions that were later enshrined as an actual theorem (the Thom Conjecture).[2] Basically, when any sudden shift in a normal system with dependent or independent variables occurs, one can conceptually think of it as a catastrophic change; the rate transition can be measured, and one can chart its physical pathways, in physics, biology, or in the pure languages of mathematics, where the physical is simply a representational idea.

---

[2] Morgan, John; Szabó, Zoltán; Taubes, Clifford (1996). "A product formula for the Seiberg-Witten invariants and the generalized Thom conjecture". Journal of Differential Geometry. 44 (4): 706–788. doi: https://doi.org/10.4310/jdg/1214459408. MR 1438191. With the onset of the remarkable advent of the Bereitschaftspotential (readiness potential), to be discussed later on – a combining of the principles of electromyograms, electroencephalograms, event-related potentials in neurophysiology, the "computer of average transients," and other forms of brain research, it would be determined that, for example, if one committed suicide with a gun (effectively), you would be dead at least 2/3rds of a second before your body even realized it. Hence, a natural, overwhelming catastrophe – like an oncoming asteroid, or massive tsunami – would probably, in essence, be painless. See, Lavazza, Andrea; De Caro, Mario (2009). "Not so Fast. On Some Bold Neuroscientific Claims Concerning Human Agency". Neuroethics. *3*: 23–41. doi: https://doi.org/10.1007/s12152-009-9053-9. See also Kornhuber, H.H.; Deecke, L. (1990). Readiness for movement – The Bereitschaftspotential-Story, Current Contents Life Sciences 33: 14.

Curiously, if an idea is equal to a catastrophe, then it is safe to assume that anything can happen, and any variable is likely to act counter-predictively (free will/won't).

Applied to evolution, catastrophe is more subjective than real, notwithstanding its vivid apparel, all that most would prefer to avoid; and the famed critique in 1877 by geologist Clarence King, who would, 2 years later, become founding director of the US Geological Survey, on the theories of Georges Cuvier, "Catastrophism and Evolution"[3] (Fig. 31.2). When a fish evolves to become a terrestrial amphibian, hindsight suggests catastrophic pressures, but depending on the degree of exacerbating temporal variables, the full extent of catastrophe for any one individual is less obvious, short of the broader categories of planetary destruction.

**Fig. 31.2** Georges Cuvier, portrait painted by Jacques C. Lorichon, 1826, engraving, US National Library of Medicine Digital Collections, NLM Image ID B05033, http://resource.nlm.nih.gov/101412205

GEORGES CUVIER.

---

[3] "Revisiting Clarence King's 'Catastrophism and Evolution,'" by Niles Eldredge, Biological Theory. 14, 247–253 (2019), published September 30, 2019, DOI: https://doi.org/10.1007/s13752-019-00326-6, https://link.springer.com/article/10.1007/s13752-019-00326-6#citeas, Accessed June 11, 2020.

Spontaneous catastrophe, or that occurring over millions of years, is value free, judgment free, a neutral and recurring fact, the normal rate in the range of one per one million to ten million years, based upon a vague background rate lifespan of a mammal. But some marine invertebrates may live ten million years, terrestrial invertebrate species even longer. Moreover, six extinction sagas most assuredly skew the data (some argue seven: again, definitions, when extinction is involved, remain murky). As a hominin, we have already survived approximately one-third of our projected duration. As mammals we have survived two previous global extinction events, the first 200 million years ago (End Triassic, 80% of our shew like ancestors lost), the second some 66 million years ago (End Cretaceous, K/Pg). We can estimate a 24% survival rate, among mammals, of those two catastrophic events, largely among burrowing, rodent-like multituberculates, as well as some eutherians who lived in trees.

Most placental mammals and marsupials, both of which evolved during the Cretaceous, fared poorly, and this is important to recognize. Biology is not given to happy endings, not in the literary sense. We exert, by our passion for eavesdropping, a fourth wall, dramatically speaking. But that also plays into the anthropic sequence of self-referential bias.[4] Hence, there is some gambling logic, at least, in predictions pertaining to the current third inflection point. It is a pattern in definitive conflict with our tenure. Among the 5416 known mammal species still surviving (approximately 130 billion mammalian individuals), there is likely to be a surprising survival rate. It will obtain on the basis of stochastic (random) conditionals conferring absolutely zero predictability. In other words, a neutral truth-value has no opinion one way or another on our durability as a species.

## 31.2   Variables in Evolutionary Theory

This is especially probable given that the survival rate in the past two extinction anomalies involved outside individuals coming into new territories, as well as pre-existing sufficient pressure to induce mutational success rates favoring the rapid rebirth of populations following the disasters. Humans would be suited to that kind of behavior, possibly.

But the odds are as much a question of laws of average across taxonomic suites as they are a moral stressor exerted implosively (psychologically) by our kind. "What kind of mammal are you?" is the appropriate, Lewis Carroll-like query. The sort likely to kill itself off, one every 20 seconds (the global projection for human suicides globally in 2020)? If we embrace the former explanation, namely, a statistical and random game of chance, wherein virtues promulgating resilience and compassionate common sense outmatch a concatenation of ruinous environmental

---

[4] See "Mammalian fauna across the Cretaceous-Paleogene boundary in eastern Montana," by L. DeBey and G. Wilson, Cretaceous Research, 51: 3610385, 2014, doi: https://doi.org/10.1016/j.cretres.2014.06.001

depredations, that absence of blame accompanying ecological disaster has about it a mature commentary that explains how John Ruskin could calmly wish for the past: "There would be hope if we could change palsy into puerility; but I know not how far we *can* become children again, and renew our lost life."[5]

Acknowledging that neither mathematics, cosmology, nor personality equal personal destiny (no matter how eloquent Plutarch, Ovid, Shakespeare, and Saul Bellow were on the topic), the mere desire, even firmly fixed hope, to deflect catastrophe does not add up to tactic. Our life-strategies are not fixed in any biological code. Some will disagree, particularly those hard incompatibilists whose logical deterministic theology is driven by the proverbial *intuition pump*, or Pierre Simon Laplace's (1749–1827) demon, as theorized in his *Philosophical Essay on Probabilities*.[6] This dialectic can be simplified by declaring a general belief, or not, in free will.

In seeking to understand heredity and its endless types of information, many geneticists have posited that "an RNA *world* existed on Earth before modern cells", a world in which autocatalysis occurred in the mechanism of polynucleotides (linear polymer nucleic acid molecules crucial for evolution) capable of creating their own endless templates of themselves, a capacity abetted by protein enzymes and— following a breakthrough in 1982—the realization that RNA itself was that very "templating mechanism"[7] along the way to the profoundly emblematic expression of DNA. DNA has its own duplication capacity during cell division, and with it the complex processes of ushering heritable traits. One such behavioral descendent of those traits has been described as fitting into that realm of "functional cognitive biases [which] are… 'designed' to err in adaptive ways."[8]

But this is merely one of countless other debatable design issues upon which evolutionary biologists have focused. These include (but are not limited to) those endless fundamentals of coping with threat, during both waking and sleeping hours; the multiplicity of complex cellular mechanisms spontaneously arising; the cohesiveness of a system that has genes, but also nomadic, floating bits and pieces of genes that may or may not have anything to do with adaptive individuals, gaps in the

---

[5] John Ruskin, *Seven Lamps Of Architecture*, Thomas Y. Crowell & Company, New York, 1900, p. 199.

[6] Laplace, Pierre Simon, *A Philosophical Essay on Probabilities*, translated into English from the original French 6th ed. by Truscott, F.W. and Emory, F.L., Dover Publications, New York, 1951, p. 4.

[7] From "The RNA World and the Origins of Life," in *Molecular Biology of the Cell*, 4th edition, NCBI Resources, Copyright © 2002, Bruce Alberts, Alexander Johnson, Julian Lewis, Martin Raff, Keith Roberts, and Peter Walter; Copyright © 1983, 1989, 1994, Bruce Alberts, Dennis Bray, Julian Lewis, Martin Raff, Keith Roberts, and James D. Watson, https://www.ncbi.nlm.nih.gov/books/NBK26876/, Accessed March 21, 2020.

[8] See "Darwin's Influence on Modern Psychological Science," by David M. Buss, Psychological Science Agenda, May 2009, Science Briefs, "Evolutionary Theory and Psychology," https://www.apa.org/science/about/psa/2009/05/sci-brief, Accessed March 21, 2020.

record (the paradox of fossils), and sudden explosions of life forms; not to mention the fact that "molecular biology has failed to yield a grand 'tree of life.'"[9]

Theories from both faith- and astrogeophysics-based speculation that regard such concepts as the "Big Crunch," "Big Rip," and "Big Freeze"[10] as remote galactic crashes; the earth, replete with potential Black Plagues, nuclear holocaust, the Anthropocene, children's spider nightmares, the mournful sound of church bells and wailing ambulances, all factor into equations of evolution. And no more so than in the immediate psychological profiles characterized as driven by the compulsions of mortality.

## 31.3   Death-Driven Vicissitudes

The individual ideation of suicide, patients, and clusters presenting themselves at some level of destruction results, in part, from "primitive object representations."[11] Biologists are well advised with respect to the 20–70 billion cells that die of apoptosis (the death of cells) every day in every human to acknowledge a graduated scale of so-called cell shrinkage, nuclear fragmentation [and] chromatin condensation[12] from children to adults. But individuals also show what Edgar Allan Poe described as "the one unconquerable force," the knowledge of what psychologists have characterized as an "irresistible depravity." Sigmund Freud referred to it as "thanatos," a "death drive" and a "death instinct."[13]

How do we reconcile these interchangeable instincts of horror, demise, and rebirth and beauty? Is this simply the most basic way of getting our arms around what it is to be human? Human history is no guide. As Aldous Huxley (1894–1963) once remarked, "That men do not learn very much from the lessons of history is the

---

[9] "The Top Ten Scientific Problems with Biological and Chemical Evolution," by Casey Luskin, in the book More Than Myth, edited by Robert Stackpole and Paul Brown, Chartwell Press, 2014, reprinted in Uncommon Descent, February 24, 2015, https://uncommondescent.com/evolution/casey-luskin-on-top-10-problems-in-biological-and-chemical-evolution/, Accessed, March 21, 2020.

[10] See "What Does Science Say About the End of the World?" by Stephanie Hertzenberg, BeliefNet, https://www.beliefnet.com/news/what-does-science-say-about-the-end-of-the-world.aspx

[11] "An Empirical Study of the Psychodynamics of Suicide," by N. J. Kaslow, S. L. Reviere, S. E. Chance, J. H. Rogers, C. A. Hatcher, F. Wasserman, L. Smith, S. Jessee, M. E. James, B. Seelig, PMID: 9795891, DOI: https://doi.org/10.1177/00030651980460030701, PubMed.gov, https://pubmed.ncbi.nlm.nih.gov/9795891/

[12] See J.A. Karam, *Apoptosis in Carcinogenesis and Chemotherapy*, Springer, Netherlands, 2009. See also Savill J., Gregory C., Haslett C. (November 2003). "Cell biology. Eat me or die". Science. *302* (5650): 1516–17. doi: https://doi.org/10.1126/science.1092533. hdl:1842/448. PMID 14645835

[13] See "What's behind our appetite for self-destruction?" by Mark Canada and Christina Downey, The Conversation, January 8, 2019, https://theconversation.com/whats-behind-our-appetite-for-self-destruction-108575, Accessed March 21, 2020.

most important of all the lessons of history." This fallacy is most striking in the history of science. Everything we have studied about evolution and natural selection evolves toward, and selects for, that which we have already studied. But such redundancy, heading into the future at every instant, also recognizes the random chance

**Fig. 31.3** Subimages extracted from time-lapse microscopy video showing apoptosis of DU145 prostate cancer cells. By Peter Egelberg, November 13, 2012. (Public Domain)

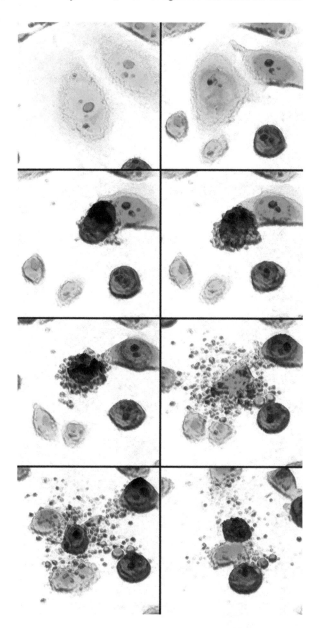

catastrophe which is our species' shadow between vitality and all the ways our cells can perish, not least necrosis and apoptosis (a kind of cell murder or suicide).[14] We simply have no mechanism psychologically for resolving these two diametrically opposed biological facts (Fig. 31.3).

---

[14] See CureFFI.org, Harvard Extension Cell Biology, https://www.cureffi.org/2013/04/28/cell-biology-11-apoptosis-necrosis/, Accessed August 3, 2020.

# Chapter 32
# Cave Paintings of the Mind

## 32.1 Provocative Bifurcations of Civilization

"Black-Headed People," including the Sumerian chief, Gudea of Lagash, a sculpture from ca. 2150 BC, and another sculpture of his son, Ur-Ningirsu, created approximately a half-century later, emerge several thousand years after the pan-Sumerian civilizations, centered around Sumer, a city that had conquered wheat and barley, composed the first inventories of surplus in a pre-cuneiform of immense complexity on cylinder seals, and employed elaborate weaponry, music, and the manipulation of water ways for sophisticated irrigation. Their naked priests laid offerings before an abundance of deities. They avoided ghosts, and had grown up in a hierarchy that affirmed a specific geographical understanding of the world wherein death had its transcendental options: paradise or hell. Stark, pure ziggurat zeniths dotted the horizons, an architecture devised to empower communication with the Sumero-Akkadian gods and built of mudbrick over devoted generations, that gave their millions of residents some grip on life. And all this enriched by an intensely diverse diet. Sumerian technology, economic principles fueling trade throughout the Levant, sophisticated expeditions and a cosmology as rich as any global culture has ever seen, lends to Sumerian legacy its rightful claim as one the world's foremost ethnic centers.[1]

Museum archaeologists for generations have seen to it that remnants of this 5000-year-old ensemble of human machinations are preserved, an instinct that wants to somehow implicitly preserve itself; to marvel at how uniform and continuously expressive our species has always been. Some might view museums as the height of a species' arrogance; others as refuges for that nostalgia gene that dominates the will to live. Or, an immense, mysterious mirror, in which we both marvel

---

[1] John Haywood, *The Penguin Historical Atlas of Ancient Civilizations*, edited by Simon Hall, Penguin, London 2005, p. 28.

© Springer Nature Switzerland AG 2021
M. C. Tobias, J. G. Morrison, *On the Nature of Ecological Paradox*,
https://doi.org/10.1007/978-3-030-64526-7_32

at and lament the passage of time, knowing that we personally are implicated in all that is ineluctable and unstoppable. And that we, ourselves, will each come to an end.

At the height of Sumerian culture, around 2800 BC, in the cities of Uruk, Ur, Eridu, Nippur, Kish, and Lagash, Sumerians represented approximately 5% of the world population, estimated to be around 27 million during the third millennium BC. Its glory days would be long gone by the time of Christ, when our species numbered some 100 million.[2] But in its day, Sumerian cities rivaled other contemporary, urban conglomerates like Jerusalem, Damascus, Plovdiv (Bulgaria), and Sidon (Lebanon).[3] It can be supposed that all such "ancient" centers of human activity had arrived at more or less the same civic and private revelations: that the Self had disassociated from the previous biodiversity commons, like Continental Drift, and, by its own endless tinkering, pursuit of sufficiency, and power grabs, emboldened the edifice, psychological fortresses, whose focus had surely shifted, obscuring the previously unprepossessing relations between humans and other species. Museums capture that transmutation. They are living embodiments of a functioning hubris, part celebratory, part aggression, that is all one would need in terms of our species' contact information.

That shift from other species is everything in humanity's most dramatic period of evolution, from the Mesolithic (Epipaleolithic) to the present. For it is during those 7–9000 years that our territorial expansion and the profound consequences of our impacts coincided with what we term the Egolithic Age, the very antithesis of the ecolithic. Many have remarked that this transition represented the worst collective choice in our entire ontology, the one that would ultimately destroy every Sumer of the world.

## 32.2   The Paleolithic Others

That Self, the narcissistic goal of the shift, would be strangely emblemized in the first photographic portraits of people by Hippolyte Bayard in France ("Self-Portrait of a Drowned Man," a direct paper positive in 1840)[4] and by 30-year-old Robert Cornelius (1809–1893) who sat before his camera outside his Philadelphia store without moving for probably 15 minutes in 1839, thus producing on daguerreotype the first self-portrait in American history.[5] Cornelius' selfie is that of a tenaciously,

[2] Charles Keith Maisels, *Early Civilizations of the Old World: The Formative Histories of Egypt, the Levant, Mesopotamia, India and China*, Routledge, Abingdon, UK, 2001.

[3] See "14 of the oldest continuously inhabited cities," by Bryan Nelson, November 29, 2018, Mother News Network, MNN.com, https://www.mnn.com/lifestyle/eco-tourism/stories/12-oldest-continuously-inhabited-cities, Accessed March 23, 2020.

[4] *A World History Of Photography*, by Naomi Rosenblum, New York, Abbeville Press, © Cross River Press, 1984, p. 33.

[5] "This is the first ever photograph of a human – and how the scene it was taken in looks today," by Adam Withnall, November 5, 20–14, The Independent.

almost furiously uncertain mien: a most tentative character in his own dramatic fin de siècle concentrated within a few minutes, mid-century. This singular look of awe was soon followed by a "Young Couple"(c.1858), bemused, longing before the just invented "ambrotype"—from the Greek word "immortal or imperishable."[6] Little boxed machines granting the idea of a new layer of longevity which promised to yield light forever, whereas the grave could only concede the comforts of darkness (Fig. 32.1).

Compared with these mid-nineteenth-century aspiring eyes looking toward their posterity, Cornelius' antecedents embodied the earliest human anonymities, muted by the skilled yet ultimately modest workings of fingers—the large pebble-sized "Venus of Willendorf" (c.30,000–25,000 BC), the "Female Head of Brassempouly, France" (c.22,000–20,000 BC).[7] Such testimony to individuals lacked the verisimilitude of an altogether separate target of our fascination; not the appeal to aggrandize oneself, but all those Others with whom we had communed for hundreds of thousands of years. Our artistic relationships with them continue to be debated, but the debate itself is propelled by today's altogether uncertain ethical stance, compounded by a pragmatic amnesia. We cannot know what it would have been like to have stood all but powerless before the aurochs and saber-tooth tigers of lore.

Parietal (cave or other rock) paintings from numerous Upper Paleolithic sites like Chauvet, Font de Gaume, and Altamira reveal the astonishing language of presences that were these Others. The painters themselves are confined to the status of mute observer, or, at best, clumsy stick figures with little if any self-interest. At Cueva de La Pasiega in Cantabria, for example (Upper Solutrean, Lower Magdalenian, discovered in 1911),[8] there are some 700 animal figures delicately inscribed throughout the labyrinthine chambers. There is also a well-preserved Neanderthal-painted ladder from 64,000 years ago. And 9000 years before that, a kind of red-ochre crayon-rendered mini-ladder, a tectiform or idiomorph on a 1.5 inch rock flake found in the Middle Stone Age Blombos Cave 185 miles East of Cape Town. Each of these hybrid hominin expressions falls in step, millennium upon millennium with a biological devotion, that of our ancestors, expressly absorbed by the sheer proliferation, beauty, and philosophical weight of animals (and some plants) in *their* world. The Paleolithic Others.

The deep biosemiotic symbology at La Pasiega was finely representational of a hierarchical coordination that had meaning in and of itself; a notational array one could argue was some precursor for the kind of prolific classifications that would transform doodles into intricate Egyptian and Sumerian literary cultures.

---

[6] See *The American Tintype*, by Floyd Rinhart, Marioin Rinhart and Robert W. Wagner, Ohio State University Press, 1999, p. 8.

[7] See *The Illustrated History Of Art*, by David Piper, Chancellor Press, London, 1981, p. 12.

[8] Breuil, H., Alcalde del Río, H., and Sierra, L., *Les Cavernes de la Région Cantabrique (Espagne)*, Ed. A. Chêne, Monaco 1911. See also, Pettitt, Paul (November 1, 2008). "Art and the Middle-to-Upper Paleolithic transition in Europe: Comments on the archaeological arguments for an early Upper Paleolithic antiquity of the Grotte Chauvet art". *Journal of Human Evolution*. 55 (5): 908–917.

**Fig. 32.1** Anonymous European Couple, Photographer Unknown, ca. early 1850s. (Private Collection)

## 32.3  Life Before Conquest

Lascaux, Chauvet, and the Cantabrian Cueva de la Pasiega were not merely animistic compulsions, sudden seizures of representational exuberance, but the sustained interrelationships and acute observations, noted upon surface, with dyes and tools that could record for successive generations the stunning phenomena of life. Egos were left at the entrances, by some psychological capacity to enter the halls of humility, dispatching with hubris. Perhaps Chartres and the world's other great architectural sites of devotion have attempted to replicate those doors of restrained Self, giving instead to perception the world before us. We can trace the movement by quite easily following the forensic evidence: Portraiture of humans occurs only tens of thousands of years after the animal world had entered the mind.

The earliest known animal depiction thus far comes from Eastern Kalimantan, Borneo—among the Sangkulirang-Mangkalihat Karst formations in the Lubang Jeriji Saléh cave area, between 53,000 and 40,000 years ago, and upon which sketches, and then a painting of a banteng wild buffalo species, were carefully imprinted. They were discovered in 2018.[9] The buffalo painting is remarkable for one particularly acute reason, aside from its antiquity. Researchers have examined the humans and half-humans in this work of art and it is quite clear that the global harmonium, humans and other species, had changed. While this banteng and its surrounding stick figures may rightly be described as one of our most ancient stories, it is also apparent that there is a hunt in action. The ego had, in fact, come into the cave to report upon its gruesome sense of perverted heroism (Fig. 32.2).

Many millennia later, other evidence arises, suggesting that even well into the Mesolithic some cultures still survived without an apparent need to kill animals. Their rock art shows no such drama, only cosmic and community connections without any evidence of a weapon. This is most potently clear among rock paintings in the Burgos complex in Mexico's Tamaulipas state, where many (thus far discovered) dozens of rock walls yield some glimpse into the lives of two or three peoples about which virtually nothing else is known. Peoples who survived for centuries in the period of approximately 4500 BC.[10] They had canoes, ladders (perhaps portals to other worlds), and seemed to relish other creatures—parrots, vipers, and various ungulates. Twenty-two thousand years earlier, at the Chiquihuite cave in the Astillero Mountains, at a near alpine altitude, their predecessors had fashioned "stone points or possible tools used for cutting, chopping, scraping, or as weapons."[11] At some point between Mexico and Borneo, 35,000–49,000 years ago, a tempestuous transition was taking place between human cultures. The Others were being subdued.

That many of the paintings thus far rediscovered, from numerous parts of the world, were possibly executed by Neanderthals[12] makes it clear that the aesthetic

---

[9] Aubert, M.; et al. (November 7, 2018). "Palaeolithic cave art in Borneo". Nature. 564 (7735): 254–257. doi: https://doi.org/10.1038/s41586-018-0679-9. PMID 30405242. See also "Is this cave painting humanity's oldest story?" by Ewen Callaway, Nature 11 December 2019, https://www.nature.com/articles/d41586-019-03826-4, Accessed March 23, 2020. See also "World's oldest hunting scene shows half-human, half-animal figures – and a sophisticated imagination," by Michael Price, SciencMag.org, https://www.sciencemag.org/news/2019/12/world-s-oldest-hunting-scene-shows-half-human-half-animal-figures-and-sophisticated, Accessed March 23, 2020.

[10] See *The Hypothetical Species: Variables of Human Evolution,* by M. C. Tobias and J. G. Morrison, Springer Nature, New York, 2019, pp. 71–72.

[11] See "Evidence of human occupation in Mexico around the Last Glacial Maximum," by Ciprian F. Ardelean et al., Nature, https://doi.org/10.1038/s41586-020-2509-0, 22 July 2020, as discussed in, "Discovery in Mexican Cave May Drastically Change the Known Timeline of Humans' Arrival to the Americas," by Brian Handwerk, Smithsonianmag.com, July 22, 2020, https://www.smithsonianmag.com/science-nature/when-did-humans-reach-america-mexican-mountain-cave-artifacts-raise-new-questions-180975385/, Accessed July 23, 2020.

[12] "Ancient cave paintings turn out to be by Neanderthals, not modern humans," by Rachel Becker,

**Fig. 32.2** Burgos Complex Wall Paintings, Heretofore Unknown Peoples, with archaeologist Gustavo Ramirez, Tamaulipas State, Mexico, Ca. 4500 BP. (Photo © M.C.Tobias)

impulses, techniques, and lineages track with a far more primordial genesis than for many decades had been assumed by scholars. And there is no doubting the paleontological divide that continues to spur heated discussion among evolutionary biologists debating the advent of fire, of meat-eating, of megafaunal extinctions resulting from human killing and what has become known as the Pleistocene Overkill Hypothesis.[13] At some point, human cultures did, in fact, pivot upon this central of all themes: interspecies murder. Within relatively few millennia, such killings would be turned to ourselves (Fig. 32.3).

February 22, 2018, The Verge, https://www.theverge.com/2018/2/22/17041426/neanderthals-cave-painting-spain-uranium-dating, Accessed March 23, 2020. See also, "Ancient Cave Paintings Clinch the Case for Neandertal Symbolism," by Kate Wong, Scientific American, February 23, 2018, https://www.scientificamerican.com/article/ancient-cave-paintings-clinch-the-case-for-neandertal-symbolism1/, Accessed March 23, 2020.

[13] Koch, Paul L.; Barnosky, Anthony D. (2006-01-01). "Late Quaternary Extinctions: State of the Debate". Annual Review of Ecology, Evolution, and Systematics. *37* (1): 215–250. doi: https://doi.org/10.1146/annurev.ecolsys.34.011802.132415

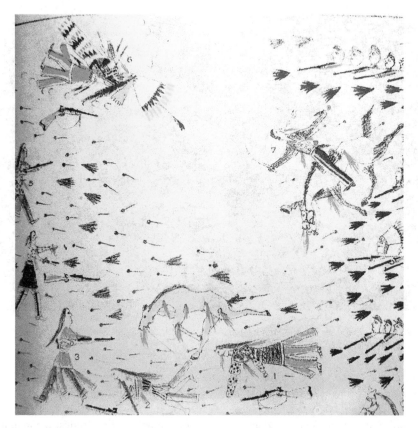

**Fig. 32.3** "General Mackenzie's Fight With The Cheyennes, And Death Of Lieut. McKinney, Drawn with coloured pencils, by Big Back, a Cheyenne Indian," in *Our Wild Indians: Thirty-three Years' Personal Experience Among the red men of the Great West*, by Colonel Richard Irving Dodge, A.D.Worthington & Co., Chicago, Ill., 1884, p. 412. (Private Collection). (Photo © M.C. Tobias)

## 32.4   That Which Continues to Haunt

A span of time connecting the Borneo banteng to the self-referential double-portrait of the "Arnolfini Portrait" of 1434 by Flemish master Jan van Eyck, with the mirror reflecting the artist, and the little lap dog Brussels Griffon perhaps indicating the couple's everyday wish to have a child, shows us how distant yet connected are these cave paintings of the mind.[14] The divergence from nonviolence to violence, and the embrace of artistic documentation as a means of reporting upon this great ethical and aesthetic quandary, appears to have taken place at some twilight point

---

[14] See "Jan van Eyck's Arnolfini Portrait," by Erwin Panofsky, The Burlington Magazine for Connoisseurs, Vol. 64, No. 372, March 1934, pp. 117–119, 122–127, https://www.jstor.org/stable/865802?seq=1, Accessed March 23 2020.

**Fig. 32.4** Solitaire among the Werehpai Cave Petroglyphs, Southern Suriname, ca. 4200–5000 BP. (Photo © M.C. Tobias)

during the convergence and/or separation of hominid cultures, *Homo sapiens* and *H. neanderthalensis*, when simple, noble, silent communion with other species was transformed, and depicted. Such representations span tens of thousands of years of complex, largely undeciphered annotation. A transmogrification from soft and humble presence to that of imagined valor on a killing field. This dialectic—evincing the divining rod of all paradoxes—continues to haunt not only art historians, but our every move (Fig. 32.4).

# Chapter 33
# Moral Choices in an Epoch of Angst

## 33.1 Conservation Success and Failure

Carl Safina, an Endowed Professor at Stony Brook University and Founder of the Safina Center,[1] recently framed a set of biodiversity statistics with unusually effective sangfroid, a lyric ode to what can be done, despite the rush toward the cliff. He reiterated a few of the more iconic, pressing data: the harrowing declines among populations of avifauna, the ones that used to be seen so easily, almost from a backyard, enormous aggregates of those far out at sea who are starving unseen (by us), North American freshwater bivalves and amphibians in the wild, bats, invertebrates, the "sharks and rays," vast "mortality events," "over 700" of them since the late 1800. All leading to his logical conclusion: "At this point in the history of the world, humankind has made itself incompatible with the rest of life on Earth." (Fig. 33.1)

But Safina, like so many who have devoted their lives to saving life (as opposed to giving up on its prospects, lodged silently between complacency and depression, a condition bound to perpetuate ecocide, what Safina likens to a "psychic numbing"), has good reason to point to conservation success stories. And, as he makes so many others make clear, there are an endless array of anecdotes and heavily documented eco-restorative sagas involving the concerted rescue of numerous species and/or populations from a clear trajectory toward oblivion in the wild. Safina references peregrine falcons and ospreys, brown pelicans, condors and the emblematic saga of bison, short-tailed albatross, musk oxen, Przewalski's horse, the Mauritius kestrel, black rhinos, and gray whales.[2] Other well-known conservation triumphs include the loggerhead shrike and Chatham Island black robin, particularly an individual female named Old Blue who kept delivering eggs years beyond the statistical

---

[1] https://www.safinacenter.org/

[2] "Psychic Numbing: Keeping Hope Alive in a World of Extinctions," by Carl Safina, February 26, 2020, YaleEnvironment360, https://e360.yale.edu/features/psychic-numbing-keeping-hope-alive-in-a-world-of-extinctions, Accessed March 26, 2020.

© Springer Nature Switzerland AG 2021
M. C. Tobias, J. G. Morrison, *On the Nature of Ecological Paradox*,
https://doi.org/10.1007/978-3-030-64526-7_33

**Fig. 33.1** Florida Panther (*Puma concolor coryi*), Critically Endangered. (Photo © M.C. Tobias)

life expectancy (for 8 years) of her species. She seemed to know she was saving her kind, and her human steward, the late New Zealand ornithologist Don Merton readily conceded as much.

In most cases of extreme survival challenges, the resurgence of the pattern of rescues is largely indebted to the 1973 Endangered Species Act; to research stemming from the International Union for the Conservation of Nature (IUCN), and the United Nations Convention on Biological Diversity and its 193 signatory nation "Biodiversity Action Plans";[3] as well as countless NGOs around the world, most notably the World Wildlife Fund, Conservation International, the Wilderness Society, the Audubon Society, the Center for Biological Diversity, Global Wildlife Conservation, the Nature Conservancy, US Fish & Wildlife Service, and the European Natura 2000.

Conservationists, who do not like the word triage, have generally concluded across the broadest suite of taxa that at least 4000 individuals are needed, in largely undefiled habitat, to provide genetic robustness and the evidence of survival probabilities over numerous future generations. Of course, any number of breeding pairs represents a vast improvement over those documented individuals who are the last of their kind, like Martha, the final passenger pigeon who died at the Cincinnati Zoo in 1914. One century after Martha, the IUCN (International Union for the Conservation of Nature) celebrated "50 years of conservation."[4] Other, less charismatically recognized species than those referenced by Safina were profiled in the

---

[3] See https://conbio.org/policy/policy-priorities/treaties/cbd, Accessed March 26, 2020.

[4] *The IUCN Red List – 50 Years of Conservation*, by Jane Smart, Craig Hilton-Taylor, Russell A. Mittermeier, Series Editor Cristina Goettsch Mittermeier, CEMEX NATURE SERIES, Earth in Focus, Inc., Washington, D.C., 2014.

IUCN summary lists, like Boeseman's rainbowfish, the Asiatic wild ass, Haleakala silversword, the Madagascar fossa, the Apollo butterfly, Axolotl, Wyoming toad, Addax, Jamaican iguana, Ploughshare tortoise, Bastard quiver tree, Iberian lynx, and Celebes crested macaque.

Most humans will never hear about, let alone witness any of these spellbinding taxa. Which is a loss to humanity, but a practical consolation in terms of preventing that national park procrustean syndrome in which too many people can love a place or species to death. A Faustian bargain made with the best of intentions. We need all the diverse biological acquaintance we can muster, if we are to have a significantly redemptive hand in the genetic rallying cries of biophilia. But non-intrusiveness and non-violent methodologies for conservation are critical components of any effective strategies.

There is little latitude for restorative mistakes. In many cases of those referenced species like iguanas and quiver trees, their numbers are relatively few.[5] At least 16,306 species are globally endangered, within the IUCN Red List.[6] The data gaps, as have earlier been discussed, are enormous, further hampering clear, long-term tactics. Real-time genetic changes among some species under intense pressures continue to come to light. Moreover, the ethical suasion of even one species' impact on modern human psyches is unaccountable. We recall the Biblical premise, one salient line upon which the human problem is laid bare. Psalm 115:16, "The heavens belong to Yahweh, but the earth He gave over to man."[7] Herein lies a complete puzzle antithetical to egalitarian ecological philosophy, given that so much of our bio-historical biases are derived from thinking about fossils. Upon this anthropic tentacle, at least one paleobiologist has asserted that our burden is not merely the consideration of early organisms, but the *"evolution of thinking about* the world before animals."[8] Now that such thinking has pervaded all of the natural sciences, infiltrating every responsible metaphysic concerned with the fate of the earth, human dominion is working against us at every level (Fig. 33.2).

Before the Cambrian explosion of biodiversity, or long afterward, when Thoreau, writing of the birds he communed with at Concord, said, "…sometimes one hears quite a new note, which has for back-ground other Carolinas and Mexico's than the books describe, and learns that his ornithology has done him no service."[9] The scientific record, while assiduous in its task to describe all of nature, is everyday confronted with the unknowable, reduced to extrapolations from what is unknown (the definition of a paradoxical outcome, when starting from any given point of thought

---

[5] See https://endangeredlist.org/continent/. Accessed March 26, 2020.

[6] See "Endangered Earth," produced by Craig Kasnoff, https://www.endangeredearth.com/

[7] Cited in *The Natural History of the Bible – An Environmental Exploration of the Hebrew Scriptures*, by Daniel Hillel, Columbia University Press, New York, 2006, p. 243

[8] *Darwin's Lost World – The hidden history of animal life*, by Martin Brasier, Oxford University Press, Oxford, UK, 2009, p. 177

[9] *Thoreau On Birds, Selections from his writings*, compiled and with commentary by Helen Cruickshank, Foreword by Roger Tory Peterson, McGraw-Hill Book Company, New York, 1964, p. 264.

**Fig. 33.2**  European Wood Bison, Wisent or zubr (*Bison bonasus*), IUCN Red List – Threatened, Białowieża National Park, Poland-Belarus. (Photo © M.C. Tobias)

or geographic coordinate); an outcome that more readily acknowledges the bad news, or what Tim Flannery and Peter Schouten described as "a gap in nature." Their particular volume is heartbreakingly insightful in terms of—to take the novel coronavirus peak/flattening dichotomy—describing the "last record" of numerous extinct species, like the Piopio of New Zealand, "last record: 1902."[10] They make clear that the islands in the Caribbean hotspot, following Columbus' journey of 1492, resulted not only in decimated islands but a depauperate documentation, at best: "Many species, such as the putative macaws of Hispaniola and curious neso-phontes [also known as West Indies shrews, the last of their sole monotypic genus] insectivores, vanished even before an adequate description was made of them. Many of the native rodents, such as hutias and rice rats, did not merit the briefest description by the Spanish chroniclers…."[11] Even the pre-Spanish conquest indigenes had their own mysterious Nahuatl taxonomies, including a "Mexican bull" which to this day remains unidentified. Flannery and Schouten commence their chronicles with the "Upland Moa" circa 1500, then leaping in 1681 to the Dodo, 1827—the "Gongan Giant Skink," all the way to the "Pink-headed Duck" in 1936, and the "Atitlán Grebe" in 1989.

As for early New World zoological documentation, some emerged as the result of Cortes' troops' observations (i.e., military explorer Bernal Diaz del Castillo's *Historia verdadera de la conquista de la Nueva España*, 1632), and then a record

---

[10]*A Gap in Nature -Discovering The World's Extinct Animals*, by Tim Flannery and Peter Schouten, Text Publishing, Melbourne, Australia, 2001, pp. 102–103

[11]ibid., p. xvii.

**Fig. 33.3**  Pacific Pocket Mouse (*Perognathus longimembris pacificus*), at Marine Corps Base Camp Pendleton, Southern California, Critically Endangered. (Photo © M.C. Tobias)

merged with ethnography and laid out in the *Florentine Codex – (A) General History of the Things of New Spain* by Bernardino de Sahagún (1793) (Fig. 33.3).

## 33.2  Ecological Epidemiologies

Ecological epidemiology is only too aware of its data deficiencies, an analytical series of computations and caesuras, that is often a tool to nowhere. As Flannery and Schouten, and others like Errol Fuller, make clear, what we do know about other species and their fates is an infinitesimally diminutive body of data compared with the gaps.

Take intelligence quotients, for example. Considering a multitude of vertebrates, psychologist E. M. Macphail writes that "It is therefore not implausible on phylogenetic grounds that there are no differences in intelligence amongst the vertebrate classes...."[12] Representing this consideration within the human community, in one of the last letters of his life, Thomas Jefferson wrote, speaking symptomatically of the "Cannibals of Europe," "I hope we shall prove how much happier for man the Quaker policy is, and that the life of the feeder is better than that of the fighter; and

[12] *Brain and Intelligence in Vertebrates*, by Euan M. Macphail, Clarendon Press, Oxford, UK, p. 331

it is some consolation that the desolation by these maniacs of one part of the earth is the means of improving it in other parts."[13]

As matters of civic, triage-threatened ethics dominate the life of *Homo sapiens* across the earth, in the widest throes of a new virus exponentially spreading (thus far, as of the Summer 2020, only the Antarctic free of it), these human designated geopolitical boundary areas have no mechanism, as nation-states comprising victims and people with opinions, to work well together. We have no idea whether the novel Covid-19 virus appears anywhere in the oceans (there are an estimated $10^{31}$ viral species that do), or atop Himalayan peaks, or in countless other places.[14] Only humans seem to fit a model of the optimal vector (although it is understood to be a gregarious virus, one that could perhaps destroy all remaining gorillas and chimpanzees and likely other primates, as well as felines and canids, should its spread and half-dozen mutations go unchecked). But Covid-19 appears common and doing no harm to most other species, including "camels, cattle, cats, and bats" which contain it, at no detriment to themselves.[15] While fewer than 6000 viruses have been described to any extent, it is believed that there could be as many as "100,939,140" other viruses in the world, over 320,000 of them—12 most notoriously so—attacking mammals.[16]

With possibly hundreds of thousands+ species going extinct every year from among the numerous phyla and categories of organisms that have been described to date,[17] we may assume that our best efforts to save life forms, even if modeled after the character and style of heroic first responders—whether in emergency rooms in Madrid or Queens, or among penguin colonies suffering from Antarctic heat spells—have not yet agreed to agree upon uniformly efficacious and compassionate plans of action, triage criteria, or ultimate goals during this one generation of humans, which could be its last. This is solid ground for philosophically waxing upon anything remotely beyond our own hospital beds. Thus far, humanity has not shown a collective capacity for coping with biological entropy.

---

[13] *Writings Of Thomas Jefferson, 1816–1826,* Volume X., G.P. Putnam's Sons, New York, 1899, p. 217

[14] See https://www.worldometers.info/coronavirus/. Accessed March 26, 2020.

[15] See https://www.centura.org/COVID-19, Accessed March 26, 2020.

[16] "How Many Viruses on Earth?" September 6 2013, Virology Blog, by Vincent Racaniello. September 9, 2013, http://www.virology.ws/2013/09/06/how-many-viruses-on-earth/. Accessed March 27, 2020.

[17] See Rudyshyn, S; Samilyk, V (2015). "Development of Knowledge of the Taxonomy and Phylogeny of Living Organisms for Future Biology Teachers". The Advanced Science Journal. 2015 (1): 75–82. https://doi.org/10.15550/ASJ.2015.01.075; See also, Woese, Carl R.; Kandler, Otto; Wheelis, Mark L. (1990). "Towards a Natural System of Organisms: Proposal for the Domains Archaea, Bacteria, and Eucarya". Proceedings of the National Academy of Sciences of the United States of America. 87 (12): 4576–9. Bibcode:1990PNAS...87.4576W. https://doi.org/10.1073/pnas.87.12.4576. PMC 54159. PMID 2112744; See also, Cavalier-Smith, T. (2004). "Only six kingdoms of life". Proceedings of the Royal Society B: Biological Sciences. **271** (1545): 1251–62. https://doi.org/10.1098/rspb.2004.2705. PMC 1691724. PMID 15306349

Attempting to defy all evolutionary odds, touting such mind-numbing conjurations as cheerful realism, guarded optimism, and the like, it is and has always been clear that we are, quite simply, lost in a colloquy that declares "war" on such novel organisms with the same aggressive mandate and neutrality of morals as appears to have inspired the novel coronavirus-19 itself. Or so our economic modeling, social chaos (a restless, economically stretched public forcing states to compromise on social distancing and the like), and the politics of an American presidential race would indicate. The human social contracts, like the United States or India, are simply too large and internally fragmented to succeed ecologically.

But from a strictly biological perspective, here is the planet working out another tactic to reduce the size of the human population—to do things we cannot begin to imagine. With human quarantine arose some of the best air quality San Francisco had ever documented; a reprieve for wildlife in national parks, currently mostly empty; and towns and cities quiet for the first time in a generation. Far more birds were singing and more nocturnal creatures emboldened to move with a sense of impunity across highways and urban landscapes. It may strike of zero consolation to those who are most horribly suffering. But from an orthodox conservation biology vantage, the chaos and systematic detection and mitigation endeavors unleashed by such opportunistic new species like the Covid virus must compel our grudging respect, at least. Even the most pacifist doctrines within various ethical traditions (Jain, Buddhist, etc.) do not advocate self-destruction in its wake, but the compromise bearing upon a survival stratagem.

At the same time we realize that our response is desperately limited to killing it, a fact that can only further undermine and complicate our response and sense of moral duty, in times of panic toward all other species, who—even during calmer periods of human history—have never exactly benefitted from our being here. To them, we are the novel virus and no philosophical chicanery can see us in any other light (Fig. 33.4).

**Fig. 33.4** Coronavirus, CDC Public Health Image Library, Content/Photo Providers: CDC/Alissa Eckert, MSMI; Dan Higgins, MAMS, 2020, https://phil.cdc.gov/ Details.aspx?pid=23312

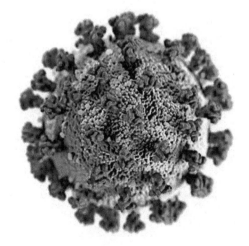

# Chapter 34
# The Dream of Don Quixote

## 34.1 Early Ecological Epics

Miguel de Cervantes [Cerbantes] (ca. 1547 to April 22 [Earth Day], 1616). The details of a life give little to the biographer. Maybe his mother was a converso, a Jewess who converted to Catholicism to avoid persecution.[1] Perhaps the son, while enslaved by the Turks, helped build a famed mosque in Istanbul (*Kılıç Ali Paşa Külliyesi*) sometime between 1578 and 1580, after assorted travels in Italy, which by turns led him into the miserable thick of the effort by European Catholic states, fired up by Pope Pius V and Phillip II of Spain to fight the ever-escalating threat of Ottoman adventurism at sea. His left arm was left enfeebled by the Battle of Lepanto on October 7, 1571. A year later, Paolo Veronese painted "The Allegory of the Battle of Lepanto" (at the Venetian Gallerie dell'Accademia), in which we can all but decipher the muted, militant novelist-to-be, a warring microscopic dot in armor that is Cervantes aboard one of the 400 rowing skiffs beside galleons—hailed at the time as the greatest naval battle and victory in Western history. Veronese's incomprehensibly complicated scene of teeming violence is on a par with Albrecht Altdorfer's "Battle of Alexander at Issus," rendered 42 years earlier (at Munich's Alte Pinakothek). If we are to believe the myths surrounding Cervantes' role in all of it, the fact he took three bullets was all the pedigree he should ever need in order to have something to say (Fig. 34.1).

More to come, of course, including kidnapping and his being held hostage by other Ottomans. Eventually, a ransom that would liberate him into a Europe overwhelmed by wars and revolts, of annihilations, mobs, axes, fire, and trampling, as art historian Claude-Henri Rocquet would write of the period.[2] The embers of Erasmus (1466–1536), and with him his calls for peace, had long died out. When

---

[1] See *Cervantes*, by Jean Canavaggio, W. W. Norton & Co., New York, 1990, pp. 18–19

[2] See *Bruegel or The Workshop Of Dreams*, by Claude-Henri Rocquet, Translated by Nora Scott, The University of Chicago Press, 1991.

© Springer Nature Switzerland AG 2021
M. C. Tobias, J. G. Morrison, *On the Nature of Ecological Paradox*,
https://doi.org/10.1007/978-3-030-64526-7_34

**Fig. 34.1** Don Quixote, Engraving, Frontispiece by Gustave Doré, from the two-volume 1863 French Edition, *L'ingenieux Hidalgo Don Quichotte De La Manche*, by Miguel De Cervantès Saavedra, Librairie De L., Hachette, et Co. (Private Collection). (Photo © M.C.Tobias)

judge advocate general of the Spanish armies that invaded the Netherlands in 1566, Balthazar Ayala, echoed the famed author of *In Praise of Folly* (1511), pleadings for an end to superstition and altercation, and the resurrection of restraint, law, international treaties, they were all to no avail. It would take the Dutch Jurist Hugo Grotius (1583–1645) to concretize any plausible sense of a global codification of ethics, setting certain ground rules that would ultimately prepare the way for international environmental treaties.[3]

---

[3] See Hugo Grotius, *De jure belli ac pacis libri tres -On the Law of War and Peace*, Paris, 1625, English translation under same title, by F. W. Kesey, The Clarendon Press, Oxford, UK 1925.

But others like the French monk Emeric Crucé (1590–1648), Blaise Pascal (1623–1662), Montesquieu (1689–1755), and John Locke (1632–1704) tried to follow in this tradition, all tied like a heavy, unrealizable vacuum to doomed helium balloons. Between the Magna Carta of King John of England at Runnymede (June 15, 1215) and Thomas Paine's *Common Sense* pamphlet (1775–1776), the human world witnessed war upon war, and little else.

While Cervantes' 1605 and 1615 volumes One and Two of *Don Quixote* are typically thought of as the first novel (considered by many—and rightly so—to be the greatest work of world fiction), it is clear that his intentions strayed far from his precise contemporary, Shakespeare (he and Cervantes died within a week of one another), in advocating for a certain irresistible dream, as contradistinguished from the cinema verité pathologies and tragic ironies of early seventeenth-century humanity. The redundancy of grief throughout Shakespeare is not easily compiled because, to paraphrase Abraham Lincoln, there was already such an abundance of tears that one must laugh so as not to cry. Note the murders in Macbeth; Richard II's killing of an uncle, his treachery then backfiring; of Richard III, that bloodiest of princely protagonists, whose imbroglios took no less than nine lives, preceding his own; the rivaling Montague and Capulet families culminating in suicides of Romeo and Juliet; a dinner-turned-massacre, with the cutting off of heads, tongues, and hands before General Titus' son Lucius finally grabbed the throne in *Titus Andronicus*. And Horatio on the sickness of Hamlet's era, "Of carnal, bloody and unnatural acts, Of accidental judgments, casual slaughters, Of deaths put on by cunning and forced cause, And in this upshot, purposes mistook/Fall'n on th' inventors' heads".[4]

For Cervantes, that same period in our history, his own personal history, owed far more to Sannazaro (1458–1530) and the other Italian Arcadian poets[5] than to the gifted poetics of journalism that were so embedded in Shakespeare's topology. Which is not to ignore the very roots of la Mancha, meaning tainted, or stained. We know that Cervantes' travails had not the theatrical outlet of a Shakespeare, with actors in robust profusion populating the Globe Theatre. Pent up, all but anonymous, Cervantes' personal problems undermined the aspiring writer whose principal competition exerted so prolific an output in the personage of one Lope de Vega (1562–1635), Spain's giant of the Baroque. De Vega produced over 500 plays and 16 other fictional works, in addition to thousands of sonnets.

But by whatever biographical mysteries we may forever interpolate, there they are, Don Quixote, his tired horse Rocinante, Sancho Panza and his donkey, Dapple, and Dulcinea del Toboso. Their exploits—to a degree endlessly more tempting to illustrators than most anything in De Vega or even Shakespeare—owe their theatrical enjoyment to the landscapes, beginning with the first illustrated edition in English (translated by John Milton's nephew, John Phillips, London, 1687, with

---

[4] "Hamlet," Act V, Scene 2, Lines 381–384.

[5] See Jacopo Sannazaro – *Arcadia & Piscatorial Eclogues*, Translated with an introduction by Ralph Nash, Wayne State University Press, Detroit 1966.

nine copper plate images). Consider Gustave Doré's (1832–1883) 370 odd engravings published in the Louis Viardot translation of 1863,[6] nearly half devoted to the most extreme, wild landscapes, a superabundance of brewing storms, deep mountainous caverns (of the Sierra Morena or Black Mountain), fulminating skies compressing ravenous edges of tormenting sea, desert, and plains wherein beasts a plenty beat their battle drums (windmills). Throngs of enemies in all directions, and despite averted grins and sighs before such heaving buffoonery, there is that unique realism conferring the way it was, or might have been, among shepherds and fools. A pastoral unending (Fig. 34.2).

## 34.2   The Illustrated Dreamers

The same madcap ecological joie de vivre is conveyed by the scores of other rapturous, comic, despairing illustrators, beginning with editions in 1632 in Germany, Dordrecht in 1657, Brussels in 1662, Spain in 1674, and Hendrick Cause's Dutch drawings for the 1681 edition. Then in no particular rapid succession following the 1687 John Phillips edition, such giants of illustration as Daniel Vierge, George Cruikshank, Toby Johannot, Salvador Dali, Daumier, Goya, Picasso, Henry Lemarie, and the astonishing Albert Dubout, whose illustrated *L'Ingenieux Hidalgo Don Quichotte De La Manche*, published in Paris by Sous L'Emblème Du Secrétaire in 1938 has the confident Rosinante and Dapple charging like revolutionaries past a terrified crowd, countering the ultimate follies, as Germany prepared to declare war on the world and Paris chaotically readied to be occupied (Fig. 34.3).

Between them, Cervantes and Shakespeare, we can point to something that transcends the alleged ACC theory of creativity in the human/primate brain (the anterior cingulate cortex).[7] These literary enigmas not only borrowed, as it were, from the powers of the ACC but also defined the brain as a whole, and the *mind* as it stands forever in debt to the body in which it lives. A body interdependent with every other body. Cervantes elicited, however he did it, the ecological storyline. *Don Quixote* is a mega-saga of nature, as was the *Epic of Gilgamesh*; Homer's *Iliad* and *Odyssey*; the *Bible*; Virgil's *Aeneid*; Ovid's *Metamorphoses*; Dante's *Divine Comedy*; Milton's *Paradise Lost*; Joyce's *Ulysses* and *Finnegans Wake*; and Kazantzakis' *Odyssey – A Modern Sequel*. But to be clear, Cervantes personalized the ecological tragicomedy in a way that sets it apart from all of its predecessors and successors. No other character became an adjective, "quixotic," discernible in nearly every major language,

---

[6] Hachette and Co., Paris, and Cassell and Co., London.

[7] "Different brain structures associated with artistic and scientific creativity: a voxel-based morphometry study," by Baoguo Shi, Xiaoqing Cao, Qunlin Chen, Kaixiang Zhuang and Jiang Qiu, Sci Rep. 2017; 7: 42911. Scientific Reports, nature research, Feb. 21 2017, https://doi.org/10.1038/srep42911, PMCID: PMC5318918, PMID: 28220826. Accessed March 27, 2020.

**Fig. 34.2** Don Quixote, Rocinante, Sancho Panza and Dapple at the Windmill, Engraving by Gustave Doré, from the two-volume 1863 French Edition, *L'ingenieux Hidalgo Don Quichotte De La Manche*, by Miguel De Cervantès Saavedra, Librairie De L., Hachette, et Co. (Private Collection). (Photo © M.C.Tobias)

as Amherst College Professor Ilan Stavans pointed out on the 400th anniversary of Cervantes' death.[8]

Miguel de Cervantes alone suggests that the creative imagination is humanity's response to its own annihilatory impulses. That is not to say that creativity does not exist in every living organism. It does. But the degree to which Don Quixote himself

---

[8] See "Marking the 400th Anniversary Of Cervantes' Death," May 11, 2016, NPR, https://www.wbur.org/hereandnow/2016/05/11/cervantes-anniversary, Accessed March 27, 2020.

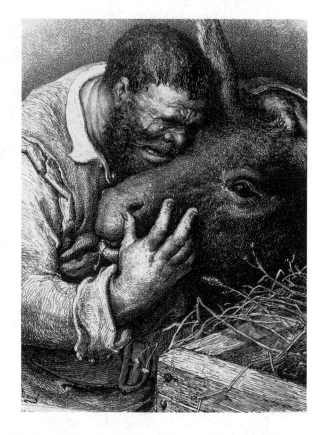

**Fig. 34.3** Sancho Panza
and Dapple, Engraving by
Gustave Doré, from the
two-volume 1863 French
Edition, *L'ingenieux
Hidalgo Don Quichotte De
La Manche*, by Miguel De
Cervantès Saavedra,
Librairie De L., Hachette,
et Co. (Private Collection).
(Photo © M.C.Tobias)

must "dream the impossible dream" is evidence enough to render a formal commentary. Our motives are mixed, as Shakespeare most probingly delved. But it can
easily be argued that our compulsion to rectify and ameliorate; to save and be saved;
and to restrain, exert tenderness, speak of compassion, formulate town halls, vote,
assist, and engage in all those actions of kindness with which our myriad prophets
and saints have been absorbed dates most recently to the quixotic in human nature.
If it is the requisite response (and every sociobiologist, ecopsychologist, neurolo-
gist, molecular biologist, and every other -ist will endlessly debate its evolutionary
fineries), the outcome of a train of logic will be the same. Cervantes has enshrined
the ecological paradox by bringing the dreamer to life; sorting out good and evil;
and traversing the landscapes of paradise and hell, with his soft-hearted, ever-
gullible comrade, Sancho, in a dual effort to save the world from themselves. If this
is the thematic origin of the novel, then it can be assumed that giraffes and butterflies have no need of novels, although non-fiction is something else entirely.

# Chapter 35
# The Ratiocinations of Rakiura

## 35.1 Decidedly Ambiguous Reveries

Nietzsche writes in *The Antichrist* of "two *physiological realities*" abetting our notion of salvation and labels them the "super-development of hedonism upon a thoroughly unsalubrious soil".[1] The book, praising all that was natural, condemning all that sullied nature (in the name of a fundamentalist Christianity) was published in 1895, one year after the death of the American luminist master, George Inness. In concluding his biography of his father, George Inness Jr. wrote of his father's painting, "Under the Greenwood," (painted in 1881)[2] "...let's cross this field that is bathed with soft, gray, mellow light that gives a sense of stillness, not of the grave, but of the kind that follows some great strain of music that has died away, and left a hush of awe...".[3] These two sentiments might well yield useful caveats for anyone approaching New Zealand with an eye toward the solemnity of beauty and an inclination toward naturalism. They certainly did for the self-righteous Scottish historian George Lillie Craik (1798–1866) in whose book *The New Zealanders* (1830) he concluded, "Let us steadily impart as much as possible of the real blessings of civilized life ... If we should stop at our present point of advancement in our attempt to civilize the New Zealanders, it might well be doubted whether we had not rather inflicted an injury than conferred a benefit on them".[4] Craik acknowledges (from all he has read of Captain Cook's adventures and others) that New Zealanders were no noble savages, but cannibals that could be tamed if one maintained vigilance over

---

[1] *The Antichrist*, by F. W. Nietzsche, Translated From The German With An Introduction By H. L. Mencken, Alfred A., Knopf, New York, 1920, p. 95

[2] See George Inness, A Catalogue Raisonné, Volume Two, 1880–1894, by Michael Quick, Rutgers University Press, New Brunswick, New Jersey, 2007, pp. 81–82.

[3] *Life, Art, And Letters Of George Inness*, by George Inness, Jr., The Century Co., New York, 1917, p. 290

[4] *The New Zealanders*, Charles Knight, London, 1830, p. 423.

© Springer Nature Switzerland AG 2021
M. C. Tobias, J. G. Morrison, *On the Nature of Ecological Paradox*,
https://doi.org/10.1007/978-3-030-64526-7_35

their easy predilection to deceive outsiders by their show of moral etiquette, temperate diet, and appearance of easygoing friendship (Fig. 35.1).

If Nietzsche had visited New Zealand after European migrants brought with them to the town of Riverton, in the far south of the South Island, Australian brush-tailed possums (*Trichosurus vulpecula*) to breed, release, and then hunt for sport, commencing in 1837, he might well have responded, at least through the ecological lens, the way Professor Susan Gubar, in her biography of Judas Iscariot, wrote of humanity's "propensity for wrong-doing, vacillation and betrayal ... humanity's disgust and self-disgust, our grief and nausea about our capacity for inflicting pain".[5] European rabbits (*Oryctolagus cuniculus*) from Spain and Portugal were also introduced in the 1830s and mustelid family carnivores in 1884.[6] In the midst of these ill-conceived mammalian translocations throughout New Zealand, indigenous New Zealanders themselves were in crisis. In 1835, the Chatham Islands population, guided by the strictly non-violent code of "Nunuku's Law," laid down by the

**Fig. 35.1**   Old Growth (ca.500-700-year old) Rimu Trees (*Dacrydium cupressinum*), On Rakiura/ Stewart Island, New Zealand. (Photo © M.C.Tobias)

---

[5] See *Judas: A Biography*, by Susan Gubar, W. W. Norton & Company, Inc., New York, NY 2009. For quotation see Indiana University NewsRoom, "Award-winning author, Distinguished Professor Susan Gubar publishes book on Judas," April 13, 2009, https://newsinfo.iu.edu/news-archive/10535. html, Accessed March 28, 2020

[6] See "The spread of rabbits in New Zealand," www.teara.govt.nz/en/rabbits/1; See also, "Introduced Animal Pests – Possums," Te Ara Encyclopedia of New Zealand, http://www.teara. govt.nz/en/introduced-animal-pests/2

local Moriori Chief, Nunuku-whenua, were invaded and massacred by mainland Maori from two kinship/clans (iwi) Ngati Mutunga and the Ngati Tama.[7]

But the ecological-bioheritage *self* by which one might prefer to be identified in New Zealand's current extinction crisis is more complicated than all the combined ethnic propensities, arrogant philosophies, condemnations, outright violence, illiteracy, and the preferred false equivalencies of innocence.

When we first set about to create a scientific reserve on Rakiura (Stewart Island), our goals were to leave the many hundreds of acres—some of the oldest forest in the country—utterly untouched in perpetuity. It still had a few wild roaming breeds of bovine and some sheep, as well as 165 years-worth of other small mammalian descendants, principally two species of rat, and possums from Australia (not Virginia), as well as local feral cats and the rare horse. The combined, easily discernible damage to native birds, insects, possibly forest geckos, and to forest understory exceeded mere salt intrusion, evident on the wind-swept headlands from the surrounding Faveaux Strait storms, a body of northern sub-Antarctic water separating the 674 mi$^2$ island—New Zealand's third largest—from the South Island by approximately 17 miles. Rakiura's nearly 400 inhabitants depend almost entirely upon ecotourism for island revenue. It is also home to the nation's newest, 13th national park which accounts for nearly 540 mi$^2$ of the island's extant.

## 35.2  Rhapsodies in Green (Fig. 35.2)

And thus, it was with the honesty of pure reverie that we wrote the following passage, regarding the last two bovines in that monastic outback: "monarchs of nameless point" who roamed the "podocarp-broadleaf forest whose origins date back at least four million years. The moisture is rich … The grasses upon which their diets [the bovines] are free to experiment … *Uncinia uncinate* hooked sedge of the Cyperaceae family; Latinate mysteries as velvety as their enunciations: *Uncinias zotovii*, fine nervosa, silvestris, strictissima gracilenta, *Carex gaudichaudiana*, the curly fretalis geminata, *Cortaderia richardii*, and so on. These are their fruits and chocolates, all of their bread…."

> They inhabit a world that orients over churned up sea, headlong into the northwesterlies. On clear days there are Alps rearing across these great rolling whitecaps. The rains are abundant. Even the occasional snowfall, though it remains for an hour like hail miscast and vanishes amid rainbows whose constant ephemerality marks the same turbulence of so many seasons manifesting each day. Rainbows that strike the earth like lightning, thick double bolts, prisms that shock the air, catapult the innocent iris … Amid forests of primeval rimu thick with ecstatic kakas and kakarikis, bell birds, keruru, fantail, brown creeper, gray warbler … This is primary canopy with virtually no harm inherent to its rules, other than a slightly toxic caterpillar, a stinging nettle and bush lawyer … and a very shy spider (the katapo) down among the dunes [though we never saw one] … The cool moist arboreal

---

[7] *The Encyclopedia of New Zealand*, https://teara.govt.nz/en/moriori/page-4, Accessed March 28, 2020.

Ltl Acorn          ●     035°F   002°C                01.08.2012 19:27:49

**Fig. 35.2** Stewart Island (brown) tokoeka *Apteryx australis lawryi*, largest of New Zealand's five Kiwi species, Threatened/Nationally Vulnerable. (Night Camera, © Dancing Star Foundation)

dim of mountain celery pine, or toatoa; and dense copses of kamahi and tree fuchsia. Epiphytes, including orchidaceas, and a chorus of leaves catering to every possible whim of the palate. Pseudopanax, Kohekohe, Kakabeak, Tutu and Shiny Karamu off which huge water droplets fall onto the tongue. Southern Rata, Wineberry and Rhabdothamnus. Hangehange and Black maire. Putaputaweta in its juvenile form dances forth from the ash and peat and loam … whilst the windy days are floating in the air. The *Olearia arborescens*, and countless see sprays from the coprosmas – crassifolia, rotundifolia, colensoi … Everywhere the animals roam, starry white blossoms, like those of the lacebark … Maidenhair ferns and the slender-trunked *Blechnum fraseri* … In the many creek beds, tree and cave weta, stick insects, carabid, chafer and stag beetles moving unconcerned along the rich leaf-bound bottomlands thick in red-fruiting supplejack, Peziza fungus and earth tongue, as well as the Delicatula, or parachute mushroom. On the ends of broken limbs, enormous masses of gill fungus infiltrate. And brightly displaying Russula and Entoloma point their wet, brilliant coloring into light of day. Cicadas carouse through the nights, native bees and Red Admiral butterflies amass around the cabbage trees at dawn, whilst web spiders build their empires between tree ferns, one day at a time … [and raucous] company is assured among the chorus of Tuis insisting on their biographies from branch to branch. The very albino Kereru sweeps through occasionally, while green finches, great herons, and the wandering Brown Kiwi come by now and then, [the Kiwi] probing with its flute-like beak. It is a vast, strange, talkative community. All are friends. No squabbles. No lawyers or accountants. Only the night craaacking of Little Blue penguins and, we pray, Yellow-eyed. Even the Fiordland Crested Penguin has been known to venture this far north…

On the beaches, occasionally, fur seal puppies are born and weaned and eventually – by the sixth month or so – go out into the vast macrocystis [pyrifera] kelp beds to play like otters. And in the deep carex grasses, baby rats sit nibbling with their dainty fingers, staring with total innocence into the eyes of anyone who cares to share time with them [these last] two fine caretakers … two cows. The sum total of all those recently bred Holsteins and Guernseys; stoic, vulnerable, scentless at birth … among the vast majority of captive bovines, enduring unimaginable pain thanks to us, only to be slaughtered, by the hundreds of millions. Lives of impenetrable spirituality and keen, reduced to dead weight, lard, meat patties and leather. All descended from three communities of primeval bovines, in East Africa, Southeast Asia, and the Tibetan plateau, where the enormous Aurochsen moved unrivaled, until humans wiped out the last of their kind in eastern Europe in 1627.[8]

## 35.3  Reluctant Killing Fields

Into this paradise, we came to better understand the omnivorous behavior of the introduced possums and rats (no mice or mustelids on Rakiura, not yet).[9] And to gauge the non-human impacts on native wildlife. A raging series of daily battles between innocence and those of the 4.8 million *Homo sapiens*—farmers, tourists, hikers and grandmothers, babies and pastors, students, and every New Zealand consumer. Particularly those members of hunting outings and "predator control" industries across the country, with their lethal stock and trade jargon, battalions of lethal bait stations with cyanide in the form of Feratox capsules, Sodium fluoroacetate (1080)—a rash of technologies. The heinous methods of precision aerial drops from helicopters and chemicals targeting "pests" "vermin" … the vilification of all those mammals and non-native flora (e.g., the gorgeous gorse, one of the most beautiful flowering plants in all of New Zealand, inviting the futile botanical archaeophyte versus neophyte, ecological versus geographical space debates, colonization before or after 1500 AD)[10] brought by humans into this island continent which, for 80

---

[8] See *Biotopia*, by M. C. Tobias, Zorba Press, Ithaca New York, 2013, pp. 73–75.

[9] See *God's Country: The New Zealand Factor*, by M. C. Tobias and J. G. Morrison, A Dancing Star Foundation Book, Zorba Press, Ithaca New York, 2011

[10] Such discussions over "biological invasions" has been definitively examined with respect to the popular garden plant, *Lilium lancifolium*, one of the tiger lilies. The researchers point out that "The admixture of genetically divergent lineages may also promote niche shifts," with respect to evolution, natural selection, wild and naturalized, diploid and triploid offspring, all but blurring the realistic burden of proof as to who and what belongs where or when on a finite planet. See Herrando-Moraira, S., Nualart, N., Herrando-Moraira, A. et al. Climatic niche characteristics of native and invasive *Lilium lancifolium*. Sci Rep 9, 14334 (2019). https://doi.org/10.1038/s41598-019-50762-4, https://www.nature.com/articles/s41598-019-50762-4#citeas, Accessed August 2, 2020. In the case of tiger lilies, it is clear that their dissemination throughout most of the world has favored any number of invertebrates, most notably the swallowtail butterflies. See "27 Of The Best Lily Varieties And A Guide To The Different Types," by Lorna Kring, Gardener's Path, March 28, 2020, https://gardenerspath.com/plants/flowers/best-lily-varieties/;  See "Two-tailed Swallowtail,(Papilio multicaudatus) (W.F. Kirby, 1884) " Canadian Biodiversity Information Facility, Government of Canada, https://www.cbif.gc.ca/eng/species-bank/butterflies-of-canada/

million years, the time since the land mass broke off from Antarctica, had dwelt in an utter biological utopia, particularly so on Rakiura.

New Zealand animal rights organizations are all but helpless when it comes to the conservation biology battle. Despite millions of dollars annually expended on R&D into immuno-contraception for "invasives" (the testing, e.g., of genetically modified carrots and potatoes as a means of delivering hormone and steroid-free pig-derived zona pellucida contraceptive agents), the only true, efficacious and non-violent solution has yet to be embraced by Kiwis, the "term of endearment" used to describe human residents of a country whose national bird, the five distinct genetic species of Kiwi, are all threatened with extinction in a nation most ecologists characterize as the "capital of extinctions." In numerous nation-wide polls, as well as various "Humane Vertebrate Pest Control Working Groups," undeviating concerns have been raised and debated for decades: "Is it right to control one population in order to preserve another [e.g., kill stoats to protect Kiwi]? Upon these killing fields, ethics are debated. If all wildlife deserves our utmost and humble respect, and wherefore science has shown in an abundance the precious, unique sentience and feelings of all living organisms, how can killing ever be construed as ethical?.[11] (Fig. 35.3)

two-tailed-swallowtail/?id=1370403265794, Accessed August 2, 2020. As for gorse, it was declared a weed by New Zealand's Parliament in 1900. See Hill, R. L.; A. H. Gourlay; S. V. Fowler (2000). "The Biological Control Program Against Gorse in New Zealand" (PDF). Proceedings of the X International Symposium on Biological Control of Weeds: 909–917. By 1928, it was attacked on all fronts by way of a New Zealand national policy of biocontrols which, despite 90 official years of attempts through the introduction of seed and foliage-feeding (also exotic) insects to control it, *Ulex europaeus* continues to thrive across over 1.7 million acres throughout New Zealand. See "Biological Control In Protected Natural Areas," by Susan M. Timmins, Science And Research Internal Report No. 6, https://www.doc.govt.nz/globalassets/documents/science-and-technical/ir06.pdf

[11] See "Ethics And The Killing Of Wild Sentient Animals," C. O'Connor, B. Warburton, and M. Fisher, "13th Australasian Vertebrate Pest Conference Proceedings," May 2–6, 2005, pp. 203–207. Discussed in *God's Country*, op.cit., pp. 423–425.

**Fig. 35.3** Newly
Discovered Giant Moa
bones of *Dinornis
robustus*, ca.2000,
Honeycomb Hill Cave,
Oparara Basin,
northwestern North Island,
New Zealand. (Photo ©
J.G.Morrison)

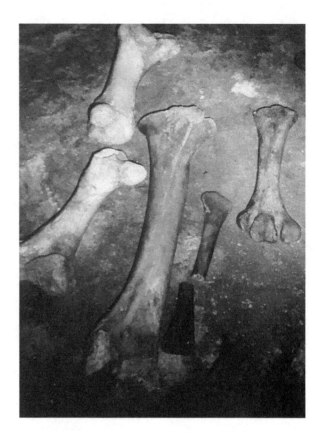

# Chapter 36
# Human Evolution at a Glance Within Ryoan-ji

## 36.1 The Smallest Utopias in the World (Fig. 36.1)

Neoplatonist Roman/Egyptian philosopher Plotinus (204/5–270 C.E.), like so many of his time, acknowledged the discipline of dialectical contemplation as the most natural method for resolving the very bases of duality; of maintaining the vitality necessary to heed not the twists and turns that torture those caught between joy and sorry; rather, to be elevated by all that is virtuous and real in the animal nature within humans, "a radiation from the All-Soul" and "The Absolute Transcendence of the One".[1]

Over 1600 years later, in the Japanese Imperial Seat of Kyoto (Miyako, capital of peace and tranquility, also known as Heian, city of purple mountains and crystal streams, only much later named Kyoto), Hosokawa Masamoto, son of Hosokawa Katsumoto, a powerful deputy of the Eighth Ashikaga Shogun Yoshimasa (1434–1490), set about to pick up the ashes of the Upheaval of Ōnin, the worst rivalry among samurai in the history of Japan. That a city so aptly, seductively named after its landscape superlatives should fall prey to so vicious and protracted a civil war (1467–1477) speaks volumes to the motivations of Plotinus and to the very biological crisis that has so inextricably singled out our species as a window on to what can go right and so horribly wrong. The Ōnin involved well over 160,000 sword-bearing combatants, ashigaru, foot soldiers, loyal to either the Hosokawa or Yamana clans, together reducing one of the most beautiful and metaphysical cities in the world, with its legendary 10,000 gardens, to rubble and the blood of tens of thousands of civilians.

Yet, the city survived and would be quickly returned to esthetic glory. Plotinus had once dreamt of just such a township, in Campania, where he died, but not before

---

[1] See The First Ennead, I.i.12, "First Tractate," p. 39; and "Third Tractate," *Plotinus: The Ethical Treatises*, Translated from the Greek by Stephen Mackenna, The Medici Society, London and Boston, 1926, pp.50–55, and p.141.

© Springer Nature Switzerland AG 2021
M. C. Tobias, J. G. Morrison, *On the Nature of Ecological Paradox*,
https://doi.org/10.1007/978-3-030-64526-7_36

**Fig. 36.1**  Saiho-ji Moss Temple (Koke-dera) in Kyoto. (Photo © J.G.Morrison)

appealing to his ardent followers, the Emperor Gallienus and his bride Salonina, to finance what he called the City of Philosophers, predicated upon the ideals of Plato's *Republic* (see Chap. 8). Somehow, by the puzzles inherent to human evolution and its geographical tidings, that Utopia would finally arise at the Temple of the Dragon at Peace, or Garden of Crossing Tiger Cubs, best, and most singularly known as Ryoan-ji, a temple within the Myōshin-ji Rinzai school of Zen Buddhism (which has over 3400 other temples and gardens throughout Japan). Ryoan-ji alone invites that ultimate resolution to the dialectics Plotinus, and nearly every other outstanding artist, activist, and scientist has ever posited: a "sermon in stone" as the late landscape esthetician Loraine Kuck described the sand garden at Ryoan-ji (Fig. 36.2).[2]

As Kuck points out, there are numerous possibilities with respect to who designed the Ryoan-ji garden, which is the size of about one large diner or (Kuck's analogy) "tennis court" or thirty-by-seventy-eight feet of stark, flat, astonishing veranda, a dry landscape. It is also unclear who actually built the garden. Among the landscape architect candidates hired by Masamoto ("the young Lord Hosokawa") to rebuild the temple following the war are Tokuho Zengetsu.[3] Or, perhaps the greatest Zen landscape master of the Muromachi artistic renaissance (1336–1573) in Japan, Sesshū Tōyō (1420–1506); or the two workmen who actually are documented to

---

[2] See *The World of the Japanese Garden – From Chinese Origins to Modern Landscape Art*, by Loraine Kuck, with color photographs by Takeji Iwamiya, Weatherhill, New York & Tokyo, 1980, Chapter XVII, pp. 163–171.

[3] ibid., p. 169.

**Fig. 36.2**  Ryōan-ji Temple, 龍安寺, Kyoto. (Photo © J.G.Morrison)

have been involved in placing the 15 stones in five groupings upon the raked, naked sand, Kotaro and Hikojiro, whose signatures were carved into one of the tallest of the rocks there.

Today a UNESCO World Heritage Site, and the most famous garden in Japan, quite likely the smallest, most celebrated piece of human artistic landscape architecture, so-called, in the world, Ryoan-ji is majestically elusive, for all of its perfect simplicity. Kuck suggests the following: "It is best to sit down on the edge of the veranda with no sense of time and pressure and study the scene, first objectively, in an attempt to analyze it intellectually, then subjectively, trying to feel what the maker was expressing here" (Fig. 36.3).[4]

The distance from Ryoan-ji to one of Kyoto's other great landscape architectural icons, Ginkaku-ji or Silver Pavilion, is 8.6 kilometers.[5] This was the place that the Shogun, Yoshimasa, had retired to, his "Eastern Hill Palace" with its abundant floristic province, detailed most authoritatively in Josiah Conder's two-volume *Landscape Gardening in Japan* (1893): the coniferous evergreens and shrubs, firs, junipers, non-coniferous evergreens, deciduous trees including "Sycamores, Maples, Chestnuts, Elms, Beeches, Hornbeams, Birches, Ashes, Limes, Willows, Poplars [and] Mulberry trees; trees of "Reddening Leaf"; hundreds of other trees and shrubs (the most prolific variety of plum trees in the world); a multitude of

---

[4] ibid., p. 164.

[5] See 13 Ryoanji Goryonoshitacho, Ikyo Ward, Kyoto, 616–8001, to Higashiyama Jisho-ji, 2 Ginkakujicho, Sakyo Ward, Kyoto, 606-8402.

**Fig. 36.3**  Meditation at Royan-ji. Japan. (Photo © M.C. Tobias)

Azalea and Rhododendron, Magnolia and Camelia; flowering plants, most notably
the enormous biodiversity of "Chrysanthemums, Asters, Carnations, Lilies,
Gentians, Irises, Jonquils, Lotuses, Peonies, Anemones [and] Orchids; and then the
Creeping Plant varieties, Large Leave Plants, bamboos, rushes, and grasses; and
finally, and most intrinsic to Ginkaku-ji, the extraordinary diversity of mosses.[6]
Conder also describes, in as much minutiae, the varieties and purposes of every
stone placed in Japan's gardens, which—in and of their time, deriving from a matrix
that involved Buddhism's vast itinerary from northern India and Nepal to Tibet,
China, and Korea, before entering Japan—avowed a higher reality; a paradise in the
here and the now, an integration of forces ultimately conceiving of true harmony, "to
live without wrestling angels in the soul's dark night" (Fig. 36.4).[7]

---

[6] *Landscape Gardening in Japan*, by Josiah Conder, Tokio (sic) Hokubunsha, Ginza, 1893, Volume
1, Chapter XI, pp. 106–137. Plates by Ogawa Kazumasa; Conder also had gone to great extremes
to elucidate the myriad kinds of stones employed in gardens like Ryoan-ji, including such "radical
stone shapes": as "Recumbent Ox," "Statue," "Low Vertical," "Propitious Cloud," "Falling-Water
Stone," "Flying Geese," "Fish-diverting," Sea-gull resting," "Long Life," "Tortoise Head Stone,"
and the like. Volume 1, Chapter II, pp. 41–58. See also, *Sanctuary: Global Oases of Innocence*, by
M. C. Tobias, and J. G. Morrison, Chapter on Moss Gardens of Japan, A Dancing Star Foundation
Book, Council Oak Books, Tulsa, OK, 2008. See also, *Garden Plants of Japan*, by Ran Levy-
Yamamori and Gerard Taaffe, Foreword by E. Charles Nelson, Timber Press, Portland &
Cambridge, 2004.

[7] See *After Eden – History, Ecology and Conscience*, by M. C. Tobias, Avant Books, San Diego,
1985, p. 70.

**Fig. 36.4** Lake and Pavilion, Ginkakuji, From First Edition of Josiah Conder, Photograph by Ogawa Kazumasa, 1893. (Private Collection)

But most arresting of all, within the sagas of paradise that one will be tempted to search out in a life, was the Shogun himself, Yoshimasa, Plato's and Plotinus' true philosopher-king who, in his late 40s, renounced his political title not once, but twice, and its duties, retreating instead to a life of silent esthetics and gardens and rigorous, contemplative tea ceremonies. There is much more to his personal story of course: adoptions, childlessness, and then a child. But after all the infernal squabbles and horrors of human violence had died away, it is Yoshimasa's legacy of the cultivation of Ginkaku-ji and its neighboring magnet, the mesmerizing Ryoan-ji, that remain and challenge us to equally embrace.[8]

---

[8] For extensive essay on Yoshimasa and his integration of the tea ceremony into his ideals, see *After Eden*, ibid., Chapter 5, "The Embattled Lotus," pp. 69–90. See also, *Zen Gardens – Kyoto's Nature Enclosed*, by Tom Wright and Mizuno Katsuhiko, Suiko Books, Mitsumura Suiko Shoin, Kyoto, 2006; *Gardens in Kyoto*, by Mizuno Katsuhiko, Suiko Books, Kyoto 2005; and *The Courtyard Gardens of Kyoto*, by Mizuno Katsuhiko, Suiko Books, Kyoto 2006.

# Chapter 37
# The Paradox of Light

## 37.1   Day One

Day one, God, light, which given the italics under "*was* good," suggests that all of the Christian commentators collectively professed some doubt (as it were, in God's mind) as to whether this lightness and darkness combination, morning and evening side by side, would actually work out.[1] In the older Talmudic Parshah Bereishees, the first part of the Torah reading, there is no italicized "was." No expressed doubt whatsoever that the darkness, the vast emptiness, all required light, precursors to the Creation.[2] Certainly by all the evidence, biology would confirm that latter proposition. But by such classical mechanics we are at once stricken with the mythopoetic underpinnings of every allegory, simile, and dialectic afflicting life and its eventual death. The ecological dynamics are the fodder, of course, of poetry, historical perspective, the cooling balm that comes from shade, especially if you are an Arabian Oryx. Moonlight gives additional expanse to this rich potentiality in every organism and across each landscape (Fig. 37.1).

Every celebrated poet has added to the repertoire, from Homer, for whom night and darkness are invincible, to John Milton, who asks God to explain the transition from sight to blindness in what must be his most poignant sonnet, "When I Consider How My Light is Spent." The poet was himself blind by that time. As Lord Byron contemplated the darkness throughout the world following the eruption of Mount Tambora in Indonesia in 1816, and Thomas Hardy's "weakening eye of day" and Emily Dickinson's "....certain Slant of light," Carl Sandburg's "dark listening to dark" or the teenager Dylan Thomas' debut "Light Breaks Where No Sun Shines,"

---

[1] *The First Book Of Moses, Called Genesis 1:3–5, Holy Bible King James Version Containing The Old And New Testaments*, The Christian Science Publishing Society, Boston, Massachusetts, USA, p. 7.

[2] "Bereishit, Genesis, Chapter 1, https://www.chabad.org/library/bible_cdo/aid/8165/jewish/Chapter-1.htm, Accessed March 31, 2020.

© Springer Nature Switzerland AG 2021                                                                  329
M. C. Tobias, J. G. Morrison, *On the Nature of Ecological Paradox*,
https://doi.org/10.1007/978-3-030-64526-7_37

**Fig. 37.1**  Bangkok, Freeway, Night. (Photo © M.C.Tobias)

and Denise Levertov in "Bearing the Light"[3] all fall into a similitude of light and darkness, the dyad diversifies its implications for an era smitten by apperceptions of the Apocalypse, the conquest of darkness over daylight.

But such *end of days* is also the fundamental problem, or paradox, in this case. It is as difficult (or impossible) to fathom as that equally vexing perturbation of photons in a vacuum moving at the speed of light, 299,792,458 miles per second. Or Bose-Einstein statistics, energy states rapidly evolving realities of such biological studies as the photosynthetic potential loadings on vertical high-rises or the "impact of artificial light at night (ALAN) on crepuscular and nocturnal biodiversity".[4] In addition to that most perdurable of all motifs, the central question of every quest, the very idea of God—agnostic or otherwise—on day one. We point our fingers toward the Sun and other stars, or calculate the corners and causes of darkness from the many waxing moon phase calendars. But, ultimately, submit to what it is. Unlike

---

[3] For famed poems on light and dark, see, "10 of the Best Poems About Light," "Interesting Literature," https://interestingliterature.com/2019/07/10-of-the-best-poems-about-light/; and "10 of the Best Poems about Darkness," "Interesting Literature," selected by Dr Oliver Tearle, https://interestingliterature.com/2018/02/10-of-the-best-poems-about-darkness/amp/. Accessed March 31, 2020.

[4] See "The impact of artificial light at night on nocturnal insects: A review and synthesis," by Avalon C. S. Owens, and Sara M. Lewis, October 23, 2018, Ecology and Evolution, Wiley Online Library, https://doi.org/10.1002/ece3.4557, https://onlinelibrary.wiley.com/doi/full/10.1002/ece3.4557, Accessed March 31, 2020.

biology here, living with us, in us, Sun and Moon we cannot moderate, back-breed, or change.


## 37.2  The Deification of Thomas Edison

What we can point out are the heavily leaning paradoxes with which these same astrogeophysics, spirituality, and poetics must contend. And, curiously, just 23 kilometers south of Kyoto, there is a shrine—Iwashimizu Hachiman-gu—sacred to both Shinto and Buddhist, which contains a statue commemorating Thomas Edison.[5] Unlike Buddhism, the Shinto religion has no actual founder, but, rather, is an early eighth-century outgrowth of early Japanese community animism, a profound worship of nature that today recognizes over 2.4 million *kami*, invisible to humans but critical sentinels at every sacred place on earth. A primary goal of Shinto is to ultimately secure *magokoro*, the pure heart that comes from being in tune with the kami. It is a tall order. But at Iwashimizu Hachiman-gu, a Shinto temple founded in 859, we watched as the Shinto priest or kannuchi and his colleagues lit many hundreds of bamboo lanterns before an image and stone memorial of Edison (1847–1931), who is considered one of the kami. For it was precisely here, in these hills above Kyoto, that Thomas Edison's assistant, William H. Moore, discovered a unique sub-species of bamboo, the filament of which was so tenacious as to be the one chosen by Edison for his first light bulb. The Festival of Light, every May 4th, celebrates that connection to Shinto's primary deity, Amaterasu, Japan's sun goddess, the alleged daughter of Izanagi and Izanami, creators of the world.[6]

Like the atom, so tortured, compromised, and perversely manipulated by Einstein and friends, Edison's successful fixation with the light bulb, relying upon the carbonized paper filaments demonstrated in the 1860s by the British chemist Joseph Swan (1828–1914), and following a century of research into the nature of heating up a wire into an incandescent state, exerts a persuasive force that today lights up Kyoto and every city and nearly every human household on the planet. Edison electrocuted an elephant named Topsy on Coney Island in 1903 (it was filmed) before eager investors to persuade them of the power of direct current electricity versus Tesla's alternating current. Debate persists regarding the veracity of the "string of animal electrocutions Edison staged"[7] but irrefutably, the rage for electricity and the energy sources to produce it are destroying the very world that Edison, Einstein, and

---

[5] See "Iwashimizu Hachiman-gu Shrine," https://www.japanvisitor.com/japan-temples-shrines/iwashimizu-hachiman-gu-shrine, Accessed March 31, 2020.

[6] See Ono, Sokyo. *Shinto: The Kami Way.* Translated by William Woodard, Charles E. Tuttle Company, Rutland, Vermont, 1962.

[7] See "Topsy the Elephant Was a Victim of Her Captors, Not Thomas Edison," by Kat Eschner, January 4, 2017, Smithsonianmag.com, https://www.smithsonianmag.com/smart-news/topsy-elephant-was-victim-her-captors-not-really-thomas-edison-180961611/. Accessed March 21, 2020.

the Japanese deities enshrined at Iwashimizu Hachiman-gu endeavored to nurture and sustain.

## 37.3   A Moral to the Tale?

There is no moral to the tale of electricity, except to blindly acknowledge humanity's unguided thirst for energy. We are afraid of the dark, no matter how much we court its effects, take comfort in sleep, or thrill to the inducement of night lights from Las Vegas to Kyoto's Gion District, Singapore's 250-acre Garden by the Bay of electrified "supertrees, downtown Ho Chi Minh, Time Square, or Hong Kong."

One of the first American scientists to visit Meiji Japan (1868–1912), Edward Morse of Boston, later to become the president of the American Academy for the Advancement of Science, wrote, "A foreigner, after remaining a few months in Japan, slowly beings to realize that, whereas he thought he could teach the Japanese everything, he finds to his amazement and chagrin, that those virtues or attributes which under the name of humanity are the burden of our moral teaching at home, the Japanese seem to have been born with".[8] It was another Boston-based academic, Serge Elisséev, director of the Harvard-Yenching Institute from 1934 to 1956, who persuaded President Roosevelt to spare Kyoto, when the US military was fire-bombing every other major city in Japan.[9] And so the contradictions in human nature continue to escalate a certain riddle, our quest to see the light (Fig. 37.2).

---

[8] In Henry Smith, ed., Learning from *Shogun: Japanese History and Western Fantasy*, University of California Program in Asian Studies, Santa Barbara, CA, 1980.

[9] See *After Eden – History, Ecology and Conscience*, by M. C. Tobias, Avant Books, San Diego, 1985, p. 87.

**Fig. 37.2** An Electrified Kyoto at Dusk. (Photo © M.C.Tobias)

# Chapter 38
# The Last Numbers of Emptiness

## 38.1   Science Laid Bare

Ecologically mirroring the search for the last digit of $\pi$, an unexplained infinity at once enlists the perfectly spelled-out dialectic between the number 1, and all that is defined as irrational, inexplicable, and blurred. When we say that Pi equals approximately 3.14, we could also say, essentially, between 3.1408 and 3.1429[1] or "3.125 in Babylon (1900–1600 BC) and 3.1605 in ancient Egypt (1650 BC)".[2] The actual number of digits, as yet uncalculated, might well span trillions of non-repetitive sequences beyond the decimal point. Xiaojing Ye suggests that "39 digits are sufficient to perform most cosmological calculations, because that's the accuracy necessary to calculate the circumference of the observable universe to within an atom's diameter" (typically of a hydrogen molecule, number one of a known 118 chemical elements/atomic numbers on the periodic table).[3] Because of electrostatic interactions (van der Waals interaction), Pi, out in the cosmos, inside our bodies, and everywhere, seems to satisfy our longing for interpolation. Of great relevance to this arena of the irrational was the Dutch physicist Johannes Diderik van der Waals (1837–1923) who got himself wonderfully immersed in the pursuit of Irish chemist Thomas Andrews' (1813–1885) experiments on phases between gasses and liquids, with an emphatic discovery of so-called critical points,[4] what we might think of as intermolecular forces on the way to the myriad Pi's of the world.

Those forces that van der Waals pursued, in turn, appear to pursue every living organism, as defined by Pi, in the sense that there are infinities that cannot be

---

[1] See https://www.pbs.org/wgbh/nova/physics/approximating-pi.html, Accessed January 14, 2020.

[2] See Professor Xiaojing Ye's "The search for the value of pi," The Conversation, March 11, 2016, http://theconversation.com/the-search-for-the-value-of-pi-55744, Accessed January 15, 2019.

[3] Xiaojing Ye, ibid.

[4] Andrews, Thomas (1869) "The Bakerian lecture: on the continuity of the gaseous and liquid states of matter,' Philosophical Transactions of the Royal Society London, 159 : 575–590.

© Springer Nature Switzerland AG 2021
M. C. Tobias, J. G. Morrison, *On the Nature of Ecological Paradox*,
https://doi.org/10.1007/978-3-030-64526-7_38

accounted for within any perceived ecological space. This segues into our futile grasping after probabilities, of any given numeric coefficient for biodiversity in a fixed community of species. The "answer" to such quests is more than elusive, for they rely on Pi-driven equations. Species richness and heterogeneity—the latter being the measure of relative abundance of individuals within that overall species deduction—equates with nothing less than our fumbling calculations to determine just how many digits are out there, whether with regard to stars or atoms; populations or the singular unambivalence of a specific personage; a Mona Lisa, or an equation involving n = the number of individuals of a specific species, and N = the total number of beings *of all species* combined.

Such speculation leaves us with the endless digits of Pi well beyond the trillions. Just consider the average 100 trillion neural connections in the brain of *Homo sapiens*; then multiply that by all individuals of all species (every prokaryote) (perhaps 100 million + [some argue over one billion] such species), and then compound that combinatorial amalgam by volume, pressure, weight, and space occupied by the unimaginable, physical attributes, and the history of science is laid bare.

## 38.2   Between One and Infinity

Between one and infinity, we also introduce the entire panoply of ascribed behavior, of psychology, every branch of ethics, all those inroads to despair and exit strategies recommending hope.[5] Having provided the groundwork for endless numeric characterizations of the physical universe, little wonder that—with it—human hubris has discovered no bounds. Hence, scientists at CERN's Large Hadron Collider (LHC) celebrated in the summer of 2012 the instantaneous creation of a "quark plasma" with an estimated temperature of 9.9 trillion degrees Fahrenheit (5.5 trillion Celsius).[6] Considering that the nearest star to earth harbors a core temperature of approximately 27 million degrees Fahrenheit (15 million degrees Celsius), one can only wonder what the proponents within the European Organization for Nuclear Research, created at the height of the Cold War in 1954, are actually thinking/doing. CERN's mission statement has been "to uncover what the universe is made of and how it works".[7] The circular collider—27 kilometers long making it the world's largest machine—is underground, between 50 and 175 meters, so that the earth's crust can allegedly protect all of us from its radiation. Unlike the legacy of Bikini Atoll with its World War II 15 Mt. Castle Bravo nuclear test and the subsequent cultural mayhem resulting from strontium-90 and caesium-137 contaminations,

---

[5] See Barcelona Field Studies Centre, https://geographyfieldwork.com/Simpson'sDiversityIndex.htm, Accessed January 15, 2020.

[6] See "CERN scientists may have set new man-made heat record: 9.9 trillion degrees Fahrenheit," by Adi Robertson, Aug 15, 2012, The Verge, https://www.theverge.com/2012/8/15/3244513/cern-scientist-hottest-man-made-temperature, Accessed January 15, 2020.

[7] https://home.cern/about/who-we-are/our-mission. Accessed January 15, 2020.

CERN scientists assure us that the LHC will not obliterate Switzerland, France, or the rest of the world.

On August 2, 2010, mathematicians announced a new record for the digits of Pi: 5 trillion (short scale).[8] But it should come as no revelation that the proof of a googleplexian Pi (a 1. + 100 zeros) is only a matter of time, since a vaporous instant of thought can easily, and is tempted to, envision it. Which puts into particularly forlorn perspective the recent declaration that the Chinese paddlefish, or swordfish (*Psephurus gladius*: bearer of pebbles, dark tailed…), is now considered extinct.[9] What is the connection, one may ask? Quite simply, we are most proficient in illimitable numbers, but as numerals verge toward the few, and then the one, we easily lose control. Until, at the level of the single individual, death and/or extinction appears all but inevitable. Whereas we have no reason to worry about the extinction of galaxies.

As of 2010, the IUCN still listed the Chinese paddlefish as critically endangered[10] along with 2464 other animals and 2104 plant species investigated at that time.[11] It's not that these biological quanta represent *trivial* numbers. Not by any means. Only that they so tragically evoke the massive disparities in what humans do with, or appear concerned by, large versus small numbers of *things*. We are greatly impressed with the word "trillion" and equally astonished by those first few humans whose personal wealth has exceeded 100 billion dollars. But we find saving a paddlefish or North American passenger pigeon from extinction too much effort of the imagination.

## 38.3   Ecological Numerics

What can be said to defend the stature of numbers, when the imagination—any imagination—clearly rejects limits, on either side of zero? How we reconcile that emotionally wrought moment, standing in heated indoors on a bitterly cold night, looking out the window in the dark at various animals that might be struggling, has no defense. Try as we may to fall back upon ideas of natural selection, adaptation, and resilience, such idle beliefs leave the heart no less heavy, despite our knowledge

---

[8] See "5 Trillion Digits of Pi – New World Record," by Alexander J. Yee and Shigeru Kondo, http://www.numberworld.org/misc_runs/pi-5t/announce_en.html, Accessed January 15, 2020; For earlier records, see, "Pi record smashed as team finds two-quadrillionth digit," by Jason Palmer, 16 September 2010, BBC News, https://www.bbc.com/news/technology-11313194, Accessed January 15, 2020.

[9] "The Chinese paddlefish, one of the world's largest fish, has gone extinct," by Douglas Main, National Geographic, January 8, 2020, https://www.nationalgeographic.com/animals/2020/01/chinese-paddlefish-one-of-largest-fish-extinct/. Accessed January 15, 2020.

[10] https://www.iucnredlist.org/species/18428/8264989

[11] IUCN (2014). "Table 2: Changes in numbers of species in the threatened categories (CR, EN, VU) from 1996 to 2014 (IUCN Red List version 2014.2) for the major taxonomic groups on the Red List" (PDF).

of the myriad strategies employed by ecto- and endothermic species endowed with behavioral capacities of torpor and other forms of hibernation, protective scales, density of feathers, rapid fat accumulation, "counter-current heat exchangers" in elk, up to 30% weight loss in grizzly bears, and so on.[12]

Alas, "For any ectotherm, even brief exposure to subzero temperatures carries the risk of irreparable injury or death. Cold impairs cellular functions by rigidifying membranes, slowing ion pumps, inducing oxidative damage, denaturing proteins, and altering energy balance," writes Jon Costanzo.[13] Most dogs, left outside at 20 degrees Fahrenheit or less, are likely to perish. Dogs also risk severe heatstroke if left confined in an ambient temperature exceeding 82 degrees Fahrenheit. Even the notion that indigenous birds are perfectly adapted to the onset of brutal Winter months has long proved false. Miami was hit by snow in 1977. The storms affected much of eastern North America. Researchers examined 20 species of birds in the aftermath and found that "at least 80% of brown thrasher, winter wren, eastern towhee and junco" as well as "northern flicker and red-bellied woodpecker"(s) had died of the cold.[14]

The numbers across every swath of science fail to achieve their intended target, save for all the passing residua, either cold and algebraic or emotional. That is because our intentions are not mathematical. We are not *made* of numbers, statistics, or calculus. Our biology is an ambassador for vast assemblages of individuals whose combinations we have designated with the jargon of population science. But we should tread carefully here, as if a cease-and-desist warning had been upended and the consciousness caught in the crosshairs of some distant feeling, issued by reason, otherwise lost in a stubborn white-out of the multitudinous, regained reason, just in time. Just short of that dehumanization, that nameless, numberless "mass man" about whom/which José Ortega y Gasset so trenchantly speculated.[15]

But what of the calculations that must swarm the mind as we thrill to an icy firmament of near-infinite stars or stand back in horror at the gates of a slaughterhouse? The ecological sciences are guided but must ultimately stand stunned behind the veneer, the unheroic cover, of that which we believe, however inadequately, we can count. And having counted, are content to do nothing (Fig. 38.1).

---

[12] See "How Yellowstone's Wildlife Adapts to Winter," https://www.yellowstonenationalpark-lodges.com/connect/yellowstone-hot-spot/how-yellowstones-wildlife-adapts-to-winter/. Accessed January 15, 2020.

[13] See "Extreme Cold Hardiness in Ectotherms," By: Jon P. Costanzo (Department of Zoology, Miami University) © 2011 Nature Education, Costanzo, J. P. (2011) Extreme Cold Hardiness in Ectotherms. Nature Education Knowledge 3(10):3, https://www.nature.com/scitable/knowledge/library/extreme-cold-hardiness-in-ectotherms-24286275/. Accessed January 15, 2020.

[14] "How Blizzards and Extreme Cold Impact Birds," by Joe Smith, Cool Green Science, February 13, 2017, https://blog.nature.org/science/2017/02/13/how-blizzards-and-extreme-cold-impact-birds/. Accessed January 24, 2020.

[15] See "The Dehumanization of Art: Jose Ortega y Gasset and Ad Reinhardt," by Peter Halley, The Brooklyn Rail, https://brooklynrail.org/special/AD_REINHARDT/artists-on-ad/the-dehumanization-of-art-jose-ortega-y-gasset-and-ad-reinhardt, n.d., Accessed July 14, 2020.

**Fig. 38.1** Hubble Spies Glittering Star Cluster in Nearby Galaxy, Credit: ESA/Hubble & NASA, Information extracted from IPTC Photo Metadata

# Chapter 39
# Shelley's Ecological Exile and His Utopia of Animal Rights

## 39.1  Various Problems with Human Nature

Human history does not of itself engender change or even slow transformation. It is a subjective fact-finding mission without end, always trailing behind our biases, which build up into messages that, often like unopened telegrams, haunt the truths both of ourselves and of strangers centuries ago signaling to us in the future. The cumulative trial and error that is curiosity's addicting forensic may enrich cartography or the epistolary annals, but little legislative enlightenment or seismically altered attitude can ever be confidently projected. Otherwise, we should all have listened carefully to the poet(s) of Ecclesiastes, to Mahavira, or Aristotle, Lao-Tzu, or Christ. We did not, or certainly a critical mass of humanity did not, despite chance after chance, generation after generation (Fig. 39.1).

Who could have predicted the links between Napoleon's personality, immunity to Yellow Fever among the African slaves sent to Haiti (unlike the European soldiers), Haiti's rebellion resulting in Independence, and Jefferson's subsequent Louisiana Purchase by which he hoped and succeeded in expanding the Continental United States in order to absorb the onrush of immigrants to America. That enormous population growth would then set into motion a demographic monster of its own, liable to every ecological fiasco that was to afflict North America.[1] We thought, with the advent of modern synthesis in biology, that we knew something about the workings of food chains, systems collapse, including genetic squeezes and extinctions, interdependencies in water, soil, and air; the fate of molecules throughout photosynthetic reactions, and of materials, like the more than 400 million tons of plastics produced each year by people. But with all of these relationships, the one pronounced gap, which we know nothing about, is human nature.

---

[1] See The New Yorker between Frank M. Snowden and Isaac Chotiner, "How Pandemics Change History," by Isaac Chotiner, The New Yorker Magazine, March 3, 2020. https://www.newyorker.com/news/q-and-a/how-pandemics-change-history, Accessed April 5, 2020.

© Springer Nature Switzerland AG 2021
M. C. Tobias, J. G. Morrison, *On the Nature of Ecological Paradox*,
https://doi.org/10.1007/978-3-030-64526-7_39

**Fig. 39.1** Spanish Steps, Rome, in the mid-19th Century, Anonymous Photographer. (Private Collection)

What we do know is so variable as to require multiplication by nearly 7.8 billion human natures. And that said, we still do not know what "nature" means. Less than 10%, approximately, of all books published are biographies or studies of psychology and/or philosophy. Of course, the majority of publications are fiction or "how-to" from which we glean—those who even read—the bulk of our literary impressions. Social media, tweets, and real bird-song account, to be sure, for other untapped quantities, unfathomed qualities, of human experience, outside the experience of experience. But there is no reliable *sourcing* or rating of inspiration, insight, intuition, expectation, hope, despair, or other feelings.

Of course, we witness local tonics for, and analogies with liberation and the premonition of *success*, at various levels. Better treatment—but it can by no means be generalized—of humans, of cats and dogs and Giant Pandas, while rodeos and bull-fighting continue. But the catalysts of ethics and moral suasion overall remain far behind the plodding missteps of humanity's endless debacles and usually fumbling efforts at redemption, an impossibility even over one grave, let alone a genocidal trench.

Democracies have come to take all the potholes and missteps as fundamental, albeit Sisyphean progress. At least, says science, by liberal policies and the continuing poetic metaphors, there is movement forward, however riddled that path may be. But definitive significant change occurs only as the result of a linear event, typically catastrophic, that happens right outside our tent, on this day. As we watch the global contamination and death toll of the "virus" spreading from country to country, region to region, the Great Depression, plus a magnified, zoonotic World War II, absent any Federal leadership, equaling World War III, New Yorkers told to prepare for their "Pearl Harbor," it all makes terrible sense with even the slightest hindsight. And that's because, in large measure, we do have the letters of Saint Jerome, private and immense, preserved sculptural likenesses of Caesar or Plotinus, a death mask of Keats, emphatically open diaries by Anne Frank or Marcel Proust, as well as devotional paintings of the young Percy Bysshe Shelley at the height of his glorious indwelling (Fig. 39.2).

## 39.2   An Ecological Life of Shelley

The likenesses left to posterity of Shelley, particularly Joseph Severn's painting of "Shelley in the Baths of Caracalla Writing 'Prometheus Unbound'"(1845) at the Keats Shelley Memorial House in Rome, gives us to imagine that he is someone we

**Fig. 39.2**  Portrait of Percy Bysshe Shelley After The Original Drawing In The National Portrait Gallery, London, From A Photograph by J. Caswall Smith, by Miss Amelia Curran, Irish Painter, Died 1847. (Private Collection)

know. His images have become almost iconic; those of his blessed absence of human foible, like a lamb. Of course, the impression of innocence would do him enormous disservice. Yet, reading his works from youth solidifies a potent friendship to the point of his becoming our own confidant, a parallel universe whose agonies and aspirations mirror the complicit heart, against the vast twilight of his own times and troubles, and ours. We think we know every great writer or that they know us. A more probing reality suggests that the closer we come to sensing human nature, the more difficult our acceptance of ourselves becomes. This is the mark of intellectual evolution, however disconcerting. It may indeed provide useful outlets, curb our appetites, channel our anger, and invite consideration of new alternatives. In the case of Shelley, all of the above.

We are invited to commence some semblance of a profound connection to this young man and his world at the Keats Shelley House. His, Shelley's, and several other poets' intangible lives are compressed up the stairs, right off the street, among some 10,000 books, numerous drawings, a few paintings, memorabilia, and momento mori, all abase the teeming Spanish Steps (the Piazza di Spagna, 26, 00187 Roma), a working research library that welcomes the stream of literary pilgrims who have been coming there since the Spring of 1909 (Fig. 39.3).

Keats died of tuberculosis in the house in 1821, after less than four months inhabiting its two rooms. He was nursed along by his traveling companion, the painter, Severn. Immediately after the 24-year-old had perished, authorities had the interior sanitized. Some 80 years later, poets from the United States, the United Kingdom, and elsewhere managed to acquire the property at the virtual epicenter of Rome (Keats Shelley at the bottom of the steps, the famed Villa Medici at the top, known as Pincio Hill) and dedicate it instructively. Shelley had invited the ailing Keats to the Kingdom of Italy. And the two men would die there, roughly 15 months apart. Along with Severn, all three would be laid to rest at Il Cimitero Straniero (or Protestant [Strangers] Cemetery). Their gravestones proffer poignancy in a way perhaps no other poetic friendships have ever been quite so emotionally enshrined. Or that's certainly the sensation one who has managed to visit their gravesites is heir to.[2]

From those Roman calms, Shelley's spirit calls out to posterity in an unbridled, high-voltage current of ever-undiminished, lyrical provocation. Geographically, scientifically, he had been voraciously in tune with the humanistic tailings of his times, since childhood. Like a young Leonardo, he was imperviously engaged in chemistry experiments, reading widely on topics ranging from James Hutton's *A Theory of the Earth* (1785) to essays on Alpine, South American and Himalayan exploration

[2] "At Shelley's Grave: The Ineffable Calico Cat at Il Cimitero Straniero," A lovely and emotional essay about the search for the poets, in addition to cats, and Rome in general, by "Francis." Literary Traveler, October 15, 2018, https://www.literarytraveler.com/articles/percy_bysshe_shelley_grave/. Accessed April 4, 2020. While the two young poets were not close friends, they clearly shared a competitive love for one another and met on numerous occasions in England throughout 1817. See, Hanson, Marilee. "Letters To Percy Bysshe Shelley, 16 August 1820" https://english-history.net/keats/letters/percy-bysshe-shelley-16-august-1820/, https://englishhistory.net/keats/letters/percy-bysshe-shelley-16-august-1820/, February 22, 2015.

**Fig. 39.3** Entranceway to
the Keats-Shelley House,
Rome. (Photo © J.G.
Morrison)

in the *Quarterly Review, Monthly Review, Gentleman's Magazine, New London Magazine, Edinburgh Review,* and *Blackwood's Magazine.* He read, it would seem, nearly everything pertaining to "nature" literature that he could obtain: novels, poetry, and adventure stories by Robert Southey, Wordsworth, James Thompson, Chateaubriand, James Beattie, Mrs. Ann Radcliffe, Charles Brocken Brown, and others. And his linguistic skills—Greek, Latin, French, Italian, and German—aided him in his multiple translations as his great odyssey, lasting the final, third decade of his life, commenced, a journey most acutely chronicled by biographer Richard Holmes.[3]

---

[3] *Shelley – The Pursuit*, E.P.Dutton, New York, 1975. See also, James Rieger, *The Mutiny Within: The Heresies of Percy Bysshe Shelley*, George Braziller, New York, 1967; Earl R. Wasserman, *Shelley: A Critical Reading*, The Johns Hopkins University Press, 1971; *On Shelley*, ed. By Edmund Blunden, et al., Oxford University Press, New York, 1938; B. M. Stafford, "Rude Sublime: The Taste for Nature's Colossi During The Late 18th and Early 19th Centuries, *Gazette des Beaux Arts* 87, April 1976; and Ronald Ree's "The Scenery Cult: Changing Landscape Tastes over Three Centuries," Landscape 19, no. 3, May 1975. See also, *The Letters of Mary Wollstonecraft Shelley*, Volume 1, "A part of the Elect", Edited by Betty T. Bennett, The Johns Hopkins University Press, Baltimore and London, 1980; *Ariel- The Life Of Shelley*, By André Maurois, Translated by Ella D'Arcy, D. Appleton And Company, New York, 1924; *Shelley- The Man And The Poet*, Thomas

## 39.3   Details Compelling His Apotheosis

By 1811, Percy Shelley had begun to put his map of the human psyche together for himself in a way that would spell out a charged sequencing of humanity's shortfalls, particularly as they triggered political injustice and animal abuse. The first part of that equation came together for him with the limited publication of his (relatively recently discovered) poem, "Poetical Essay on the Existing State of Things," on behalf of a miscarriage of justice carried out against the wrongly imprisoned Mr. Peter Finnerty[2]. It might well have been written by George Kennan, given the maturity and scope of the Shelley's outrage and concerns.[4]

By that same Summer, the 18-year-old Shelley (1792–1822) had been both disowned by his family and expelled from Oxford for declaring himself an "Atheist." In the late Summer of 1816, he would actually inscribe his name and *atheist* in the register at the Mer de Glace glacier in Chamonix, inciting great consternation from others who preferred—upon stepping onto the actual ice—to shoot their guns and proclaim their devotions to God before such grand scenery.[5]

But in 1811, Shelley was also furiously probing logical, Utopian remedies for the abuses of the French Revolution. By March 1812, he was living in Dublin, catapulted by the writings of Thomas Paine (*The Rights of Man*) and Paul Henri Baron d'Holbach (*Système de la Nature*) and assessing what to him had become revelatory, namely, the depths of despair of the Irish working class that would figure expressly in his 12,000 word, "Address to the Irish People," propounding hopes for

---

Yoseloff, Publisher, New York, 1960; *Shelley – A Life Story*, by Edmund Blunden, The Viking Press, New York, 1947; *Shelley – The Last Phase*, by Ivan Roe, Roy Publishers, New York, 1955; *Shelley's Platonic Answer to a Platonic Attack on Poetry*, by Joesph E. Baker, University of Iowa Monograph, by Joseph E. Baker, University of Iowa Press, Iowa City, 1965; *Shelley's Jane Williams*, by Joan Rees, William Kimber, London, 1985; *Jefferson Hogg – Shelley's Biographer*, by Winifred Scott, Jonathan Cape, London, 1951; *Percy Bysshe Shelley – The Esdaile Note-Book*, Edited by Kenneth Neill Cameron, from the Original Manuscript in The Carl H. Pforzheimer Library, Faber and Faber, London, 1964; *Shelley – His Thought And Work*, by Desmond King-Hele, Second Edition, Fairleigh Dickinson University Press, Teaneck, Rutherford, Madison, 1971; *Verse And Prose From The Manuscripts Of Percy Bysshe Shelley*, Edited by Sir John C. E. Shelley-Rolls, Bart and Roger Ingpen, Privately Printed, London, 1934; *Essays on Shelley*, Edited by Miriam Allott, English Texts And Studies Liverpool University Press, 1982; and *Shelley: A voice not understood*, by Timothy Webb, Humanities Press, Atlantic Highlands, New Jersey, 1977.

[4] http://poeticalessay.bodleian.ox.ac.uk/; See also, "Shelley's long-lost poem – a document for our own time (and any other)," by John Mullan, The Guardian, November, 11, 2015, https://www.theguardian.com/commentisfree/2015/nov/11/shelley-lost-poem-essay-existing-state-of-things, Accessed January 12, 2020.

[5] See Gavin de Beer, "The Atheist: an Incident at Chamonix," in On Shelley, ed., Edmund Blunden, et al. Oxford University Press, New York, 1938, pp. 43–54. See also, Sir Archibald Alison's *Some account of my life, An Autobiography*, Edinburgh 1883. Alison was in Chamonix in 1816 and wrote, "Unable to take my eyes from the mountains, I threw myself on the ground and drank in, gasping, the enchanting spectacle." It was John Ruskin who had been infuriated upon seeing tourists firing their "rusty howitzers" in admiration of the mountains. See *A Vision of Nature – Traces of the Original World*, by M. C. Tobias, Kent State University Press, Kent, Ohio, 1995, p. 287.

Irish Independence during an epidemic of British tyrannies. He'd been a "spy" and something of an outlaw, if an elegant one, as characterized by authorities.[6] His itinerary during these times were harrowing for him. But one imagines mostly his utter discontent with the human state of affairs; what today would be most readily typified as an ecological melancholy, firing up ideals, preparing him for a no-return saga of expatriotism. He recognized in humanity the "beast." But the haggard vagabond of the most tender demeanor viewed nature as a living compendium of the benign and enchanted. This dichotomy, for Shelley a fatal paradox combining an ecological exile that would take him from his native England to the Continent, would inform his literary outpourings still before him, epics both long and diminutive. These included "Alastor, or, The Spirit of Solitude," "Prometheus Unbound: A Lyrical Drama," "The Revolt of Islam," "Rosalind and Helen: A Modern Eclogue," "The Masque of Anarchy," "Epipsychidion," "The Cenci: A Tragedy," "Intellectual Hymn to Beauty," "Adonais: An Elegy on the Death of John Keats," "The Triumph of Life," and perhaps most quintessential to Shelley's rightful legacy, the soon to follow, "Queen Mab: A Philosophical Poem."

Many, most notably T. S. Eliot, have assailed Shelley's pioneering political and spiritual motivations for his art. They confused his remarkably consistent and adamant environmental convictions with sentiment, and for generations, he was rather dismissed at best by Victorian and early modernist readers and critics. A few, particularly the Chartists (British political reformers from 1838 to 1857), and Shelley's most devoted biographer and animal rights activist, the polymath Henry Salt, have made of Shelley a spiritual figurehead, never a zealot but the purest of poets, as lyrical and incisive as Shakespeare, though far more well-read. That plus one additional quality, which was the ultimate measure of man, for Salt: Shelley's emphatic love of animals, a passion so intense that it could be said to have carried him to his grave. Had he not perished at the age of 29, his likely other outpourings present a dizzying prospect, not to mention his brimming passions for Greek Independence, contraception for women, along with universal suffrage, systematic reforms in Parliament (as Richard Holmes has discussed), and hopes to found a vehicle to erode the noxious conservatism of the Torries. He would have made a most compelling poet/politician of his era, Holmes has suggested.

Some, like literary critics Axon, Cameron, and Bate, have followed the line of quasi-adoration advanced by the underrepresented Henry Salt (1859–1931) who would follow, with his great work *Animals' Rights – Considered In Relation To Social Progress* (1892),[7] in the steps of Thomas Taylor (1758–1835), the great Neoplatonist who had translated Plato's (then controversial) *Symposium* in 1792. Shelley's own great interest in Plato would have led him to Taylor, and, by turns, Taylor back to Shelley. Just one year after Shelley died, Taylor translated *The Select*

---

[6] See Richard Holmes, Shelley – *The Pursuit*, E. P. Dutton, New York, 1975, pp. 117–132.

[7] George Bell & Sons, London and New York, 1892. The book includes a nearly 30-page Bibliography of works on animal rights in English between 1723 and 1887, sufficient data to suggest that Shelley was in his purest form when he combined his poetic gifts and complex voice to causes which, by all contemporary accounts, meant more to him than almost anything else.

*Works of Porphyry, containing his Four Books on Abstinence from Animal Food*,[8] most likely influenced by Shelley's pamphlet, *A Vindication of Natural Diet*, notes, initially, for the first edition of his "Queen Mab" (1813) and later that year published in its own right.[9]

And it is to the 9 Cantos and 17 Notes, 6 of them essays in themselves, of "Queen Mab" that we turn. So powerful is the work—much in the manner of the mature Mozart—that by 1821, there were multiple pirated versions all over England. It had become celebrated by the general reader, oppressed workers, enlightened scholars, and poets of all persuasion. A substantial bestseller, all the more remarkable given both its huge erudition and the momentous advocacy for animal liberation with which it takes up considerable energy, graphic commiseration, and the very Utopian-like conclusion. It was Shelley's first and, in many respects, the foremost epic of his entire career. The book (beautifully described by biographer Holmes)[10] sets out all of the critical themes, style, reveries, and confluence of disciplines that would for-ever mark Shelley's unique orientation to a specific geographic place (as Holmes writes, "The idea of the search for a *place*, from which he could launch his ideas of changing society…").[11]

The book combines theories of evolution, quotations from Lucretius' *On the Nature of Things*, ideas from Spinoza, and particularly those most powerful views on the intelligence of flowers—of all nature—as incarnate in the grandfather of Charles Darwin, Erasmus Darwin's *The Botanic Garden* (1791)[12] and *Zoonomia* (1796). Holmes distinguishes "Queen Mab" from Shelley's entire body of work for its "…relating, comparing and combining, information from enormously varied sources: historical, ethical, astronomical, theological, political and biological".[13] As for historian Colin Spenser, he writes that "Queen Mab" "was to have enormous influence throughout the century. It became the most widely read, the most notori-ous and the most influential of all Shelley's works."… "In his first major work he had captured something of the spirit of the coming age".[14] Even before his writing

---

[8] Thomas Rodd, London 1823.

[9] See William E. A. Axon's *Shelley's Vegetarianism*, London, 1891; Kenneth Neill Cameron, *The Young Shelley: Genesis of a Radical*, Macmillan, London, 1950; Jonathan Bate, *Romantic Ecology: Wordsworth and the Environmental Tradition*, Routledge, London, 1991; Henry S. Salt, *The Creed of Kinship*, Constable, London, 1935; Henry S. Salt, *The Story of My Cousins: Brief Animal Biographies*, Watts & Co., London, 1923, and Salt's two biographical studies of Shelley, *A Shelley Primer*, Shelley Society, Reeves & Turner, London, 1887, and *Percy Bysshe Shelley: Poet and Pioneer*, William Reeves, London, 1896. See also "Henry Salt on Shelley: Literary Criticism and Ecological Identity," by William Stroup, Romantic Circles, University of Colorado-Boulder, https://romantic-circles.org/praxis/ecology/stroup/stroup.html, Accessed April 3, 2020.

[10] op. cit., Richard Holmes, Chapter 9, pp.199–234.

[11] ibid, Holmes, p. 200.

[12] Part 1 Containing The *Economy Of Vegetation – A Poem With Philosophical Notes*; and Part II Containing *The Loves Of The Plants*. Printed for J. Johnson, London, 1791.

[13] op cit., Holmes, p. 202.

[14] See *The Heretic's Feast – A History Of Vegetarianism*, by Colin Spenser, Fourth Estate, London, 1993. pp. 249, 250.

the epic, members of the Bible Christian Church of Manchester had, in 1809, committed themselves to a life of vegetarianism. It was in the air.[15]

As for that "place"—whether the Chamonix of his "Mont Blanc," or the "Lonely Vale in the Indian Caucasus" of his "Alastor, or, The Spirit of Solitude"—the actual locus forever eluded Shelley, though not the themes and convictions that fueled its genius. One year before his death (some have viewed it as a suicide, though Richard Holmes has rejected such a view),[16] he wrote from Pisa to his good friend Thomas Medwin, who was in Geneva at the time, "How much I envy you, or rather how much I sympathise in the delights of your wandering. I have a passion for such expeditions... I see the mountains, the sky, and the trees from my windows, and recollect, as an old man does the mistress of his youth, the raptures of a more familiar intercourse, but without his regrets for their forms are yet living in my mind".[17]

At the time of his writing "Queen Mab," having fully embraced vegetarianism sometime in the early month of 1812, that pivotal *place* had also become conceptual, not merely topographic. He sensed, in the final stanza of Canto VIII, that a life of non-injury to others would solve all the problems of the world. That vegetarianism was the central issue of inequality, trade, politics, commerce, hierarchy, socialization, health and nutrition, a cause at the root of human aspirations for paradise, and the only conceivable practice that could rightly segue into a true Utopian system for all beings.

When he writes of that paradise, it is in terms of our kind having finally liberated itself from our species' ecological cruelties, even as Shelley himself intuited that his declarations were of such magnitude as to possibly incur his own demise. Having described in inimitable detail how a human "slays the lamb that looks him in the face,/And horribly devours his mangled flesh,/Which still avenging nature's broken law,/Kindled all putrid humours in his frame," he goes on to that final phase, when such "evil passions, and all vain belief" have come to a blessed end, such that "All things are void of terror: man has lost/His terrible prerogative, and stands/An equal amidst equals: happiness/And science dawn though late upon the earth".[18]

## 39.4   Ultimate Vindication

As for the science itself, which he hails, it is of interest that in his 17th note accompanying "Queen Mab," namely "A Vindication of Natural Diet," he speaks of a human life on a par with that same "lamb"; of a person surviving on "acorns and berries" which is the same self-professed diet of "The Monster" who tells this to his

[15] ibid., Colin Spenser, p. 251.

[16] "Death and destiny," by Richard Holmes, January 23, 2004, The Guardian, https://www.theguardian.com/books/2004/jan/24/featuresreviews.guardianreview1, Accessed April 21, 2020.

[17] *The Letters of Percy Bysshe Shelley*, edited by F. L. Jones, 2 vols., Oxford University Press, 1964, 2:218–219, no. 578.

[18] See Percy Bysshe Shelley's *Queen Mab*, W. Clark, London, 1821, VIII., p. 79.

Creator, the scientist, in Mary and Percy's "Frankenstein: or, The Modern Prometheus" (1818). And, in the Appendix of "Vindication," suggests that, "All men might be as healthy as the wild animals. Therefore, all men might attain to the age of 152 years," a calculation based upon the like of one 'Old Parr' stemming from Shelley's wide reading on cancer, on mortality rates in Iceland, on the "Nervous Temperament," data from Pliny's *Natural History*, and countless other sources. And it all comes down to his statement near the beginning of "Vindication," namely, "The language spoken however by the mythology of nearly all religions seems to prove, that at some distant period man forsook the path of nature, and sacrificed the purity and happiness of his being to unnatural appetites. The date of this event, seems to have also been that of some great change in the climates of the earth, with which it has an obvious correspondence. The allegory of Adam and Eve eating of the tree of evil, and entailing upon their posterity the wrath of God, and the loss of everlasting life, admits to no other explanation, than the disease and crime that have flowed from unnatural diet".[19]

---

[19] Percy Bysshe Shelley, "A Vindication Of Natural Diet," in *Shelley – Selected Poetry, Prose and Letters*, Edited by A. S. B. Glover, for The Nonesuch Press, London, 1951, p. 900.

# Chapter 40
# The Zoosemiotic Paradox of Aesop

## 40.1 The Holy Grail of Biosemiotics

Aesop is thought to have been born into slavery, in approximately 620 BC. He would live for approximately 56 years. Most perceive his home to have been Greece, though one thirteenth-century biography suggests that he was of African descent and others believe him to be a Phrygian at the very end of the Phrygian Empire. The zoology of Anatolia then, her musical traditions, and wild child origins of Cybele would tend to support the Phrygian origins theory. Aesop (Αἴσωπος) had a great gift, variously referenced by Herodotus, Plutarch, and Aristotle. From the first century CE, his mythic personality is assured, as flamboyantly enshrined as the Holy Grail of biosemiotics. The search to discover the real Aesop takes on dramatic proportions because references to the real man are so fragmented. Notwithstanding obscurity, his deeply observational traits deliver a wealth of biological data that is—as an everyday metaphysic, or Farmer's Almanac, is to the fanfare of country life—understood as a great chain of being that every age group of every species partakes in, with all its poignancies, good-humor, ill-humor, the sum total of interdependencies, mostly focusing on food, nests, and the right to exist.

Aesop's biospheric hierarchies, whether involving an ant, a wolf, or a grasshopper, are culled from a coeval pantheist, spiritual and aesthetic hub, weighing heavily on matters of jurisprudence, town halls, fundamental questions of morality among humans, and mostly, the relations between other species.

Aesop's genius and character are never in doubt. His thoughts are directly in tune with a Mahavira, a St. Francis, Shelley, and George Orwell. Throughout the many variations regarding his biography, there is a central ethical line that is conversant with both satire and tragedy; with all those weighty matters more often given to a Zadig, a Sanhedrin, a Diaspora of the Learned. Moments of finality and absolute necessity that perforce need the right insight at the precise moment. And, perhaps, incompletely attributed to him through the dissemination of "The Aesop Romance," his collation of tales and biographical adventures beginning in the first century and

© Springer Nature Switzerland AG 2021
M. C. Tobias, J. G. Morrison, *On the Nature of Ecological Paradox*,
https://doi.org/10.1007/978-3-030-64526-7_40

extending their fabulist emendations for over two millennia, absent any author or singular cultural matrix, are precisely what humanity's childhood and old age, both, required: In the quest to learn the languages and souls of other species in order to better understand ourselves. The same man who would drive home the message of his freedom—a freedom given by Mother Nature—to then go on and pose the central questions of philosophy and science for all succeeding centuries.

This Aesop is the same figure portrayed as a physically modest (short and stooped) ecological alchemist in a 1489 Spanish edition of the "fabulous histories" of one "Ysopet." An archaeologist, Otto Jahn, rummaging through the Vatican Museum in 1843 famously discovers the fifth-century Grecian cup with an image, he believed, of Aesop. That man has been described as one who "listens carefully to the teachings of the fox sitting before him".[1] The 1640 portrait of "Aesop" by Diego Velázquez in the Prado has him brilliantly unconventional, far more like a bullied, beaten-down victim of Thomas Hobbes' universe, uncannily anticipatory of Rembrandt's "Self-portrait as St. Paul" (1661) with the almost identical large manuscript in hand. Indeed, Rembrandt's self-portraits from 1669 onward each contain the life-weary look of Aesop. But through Velázquez's deeply dispiriting sieve, a world gripped by *too much of itself*, there is also the evocative toll that too many years have taken, whether on a Leonardo (always said to look older than he was, in his mid-60s) or on a Tolstoy. Aesop is said by some chronological storytelling to have been thrown off the cliff above the oracle at Delphi. Most historians are skeptical of this (Fig. 40.1).

But there is, in the Velázquez's portrait, a sense of coming doom atop years of hardship. Aesop stands so uneasily, worrisomely out there and exposed, cloaked in a flowing red robe that is signaling only the most tentative command over nature, perhaps meant to incite the unrecorded volumes of blood that have soaked the world. Life has wrapped herself around his wrinkles. With a large manuscript worn down to palimpsest leather and fragile paper in his right hand, his gaze at once invites the sadly respect of a Biblical scholar or of a seriously concerned individual who had his deeply contemplated reasons to deflect any more attention, let alone the canonization by children around the world for whom his fables conveyed more than mere sapience can explain.

Within two centuries of Aesop's life, a student of Aristotle, Demetrius Phalerum (c. 350–c. 280) would compile Aesop's many tales and fragments into the "Aesopica," which would eventually segue into Latin translations by Horace, and then find their expression in languages and illustrations around the world. The fable, children's story, nursery rhyme, riddle, simile all shared an ethos. It came from diverse sources in Egyptian and Mesopotamian precursors, a semiotic of particularly progressive, signaling prowess, involving humans and other species, specific species like Aesop's wolf, fox, and crow; or the hawk and nightingale in Hesiod; the lion in Homer's "Iliad"; the wolf and lamb in the Babrius (second century CE)

---

[1] Zanker, Paul, 1995. *The Mask of Socrates: The Image of the Intellectual in Antiquity*. Berkeley: University of California Press, pp. 33–34.

**Fig. 40.1** "Aesopus," by
Diego Velázquez, 1640,
Museo Nacional del Prado,
Inventory #P001206.
(Public Domain)

translation of Aesop; the *anthos* or yellow wagtail; and *aigithios* or long-tailed tit in
Aristotle.[2]

## 40.2  A Paradoxical Morality Play

In every case paradox presses in on the morality play, the begging for mercy between
animals, that is, a mouse and a lion. The lion is persuaded the mouse is an unsuitable
prey species in the scheme of things, agrees not to harm the mouse and is, in turn,

[2] "Memories of Childhood: Aesop's Fables Are as Relevant Today as They Were 2,600 Years Ago,"
*Ancient Origins*, by Aleksa Vuckovic, July 26, 2019. See also, Clayton, E. "Aesop, Aristotle, and
Animals: The Role of Fables in Human Life." Humanitas, Nos. 1 and 2, Vol. XXI, 2008, pp. 179–
2008. Discussed for their picking of insects off some animals and nest-disturbance call that sounds
like "the whinnying of horses," as analyzed by Tua Korhonen in "A Question of Life and Death:
The Aesopic Animal Fables on Why Not to Kill," Humanities, MDPI, May 13, 2017, p. 11, taken
from Geoffrey Arnott, Birds in the Ancient World from A to Z, Routledge, London, 2007, pp.14–
15; file:///Users/michaeltobias/Downloads/humanities-06-00029-v2.pdf. The original reference
derives from Aristotle's 8th book of his *Study of Animals* (8.609a4-610a36) but he also wrote of
these matters throughout his *Politics*.

freed by the mouse who eats through a sack in which the lion has later been captured by a human. The Age of Zeus, involving all species "at war" (polemos) is superseded by the previous Age of Zeus' father, Cronus, in which the world—all species—coexisted in a time of wonderful harmony; animals were hemeros, "tame and gentle," lacking "teeth and claws".[3]

The great Greek poet Hesiod, in his *Works and Days* (c. 700 BCE), had early on established the seriousness of these dialogues between animals, a remarkable influence on Socrates and Plato and hence, upon all subsequent philosophical literature, such that "aerial spirits" are attributed with the very powers on earth of being "the Guardians of Mankind" at a time when "Men, abandon'd to their Grief,/Sink in their Sorrows, hopeless of Relief. While now my Fable from the Birds I Bring,/To the great Rulers of the Earth I sing".[4]

And it is not just the birds (as demonstrated throughout Aesop) that have messages for human policy. As studied in the *Perry Index of Aesop's Fables* (Ben Edwin Perry, 1892–1968, a professor of classics at the University of Illinois at Urbana-Champaign), the fox, in trying to determine the edibility of a cluster of grapes, is granted a precocious intellect, as is every sentient being portrayed by Aesop, who wrote his masterpiece at the same time that the *Chandogya Upanishad* had promulgated the great declaration, "Tat Tvam Asi," "You are That," "That Thou Art".[5]

Aesop's remarkable influence has reached straight across the history of humanity, to date, bearing uncanny and undeniable substance of what today is thought of as deep ethology, or zoosemiotics. But what is so utterly profound about the evolution of Aesop's masterpiece are all the loose trappings allowing for any kind of illustration or translation that accords even the most basic set of feelings toward other species. Aesop, in essence, has formulated an interspecies international language, no less the Esperanto of L. L. Zamenhof in 1887. The Byzantine monk Maximum Planudes (1260–1310) collected every fable from the Greek he could find, and his manuscript came to be known as *The Medici Aesop*, most likely illustrated by Gherardo di Giovanni (1444–1497), a painter who in his time attracted as much adulation as Botticelli and Leonardo. Vasari devoted a chapter to him. His illuminations are particularly sensitive with respect to the faces of oxen, foxes, and monkeys, dogs, hens, wolves, frogs, lions, donkeys, moles and bears, rooks, and shrews. The book would become the basis for nearly all subsequent versions throughout the world.[6]

---

[3] ibid., Korhonen, pp. 10–11.

[4] Hesiod, Works and Days, *The Works of Hesiod*, Translated from the Greek, N. Blandford, London, 1728, p. 86.

[5] See *The Hypothetical Species: Variables of Human Evolution*, by M. C. Tobias and J. G. Morrison, Springer Nature, New York, 2019, p. 51. See also, Perry, B.E. *Studies in the Text History Of the Life and Fables Of Aesop*. Oxford University Press, New York, 1981.

[6] See *The Medici Aesop, Spencer Ms 50*, From the Spencer Collection of The New York Public Library, Introduction by Everett Fahy, Fables Translated from the Greek by Bernard McTigue, Harry N. Abrams, Inc., Publishers, New York 1989. For the reference to Ghirlandaio and Vasari, see p. 10 of *The Medici Aesop* introduction.

## 40.3   Aesop's Many Histories

The first English translation from the Latin, by William Caxton in 1484, presented Aesop for children. Versions of this would inspire Charles Perrault, Jean de La Fontaine, Sir Roger L'Estrange, the Reverend Samuel Croxall, and zoologist Thomas Bewick, as well as illustrators as diverse as Arthur Rackham, John Tenniel, and Alexander Calder.

In the early thirteenth century, the great Persian poet Attar of Neishabur wrote his remarkable animated life of birds, *The Canticle of the Birds*, a century before the poetry of Rumi, and was clearly influenced by Aesop. Countless Persian illustrators would flock to the task of illuminating successive versions of this great epic of omnipotent avifauna. Among the many successive poets who emulated *The Canticle* was Ziya' U'd-Din Nakhshabi (d. 1350) whose 52 Tales of a Parrot, *Tuti-nama*, was transformed in the 1550s when the Emperor Akbar commissioned 250 miniatures from many of the greatest painters of the era for the manuscript, and it was translated into Sanskrit and Urdu.

As it would be translated into virtually every dominant language in the world, at least 7300 editions commonly archived, including 3753 in English, 758 in Latin, down to 43 editions in Hebrew, 36 in Hindi, 7 in Welsh, 31 in Vietnamese, 421 in Chinese, 199 in Japanese, 3 in Zulu, in Dakota, Gaelic, Haitian Creole, nearly 2482 versions for children, and so on.[7]

In 1739, Father Guillaume-Hyacinthe Bougeant (1690–1743) published his treatise *A Philosophical Amusement upon the Language of Beasts* in which he made the case that animals actually had the supreme intelligence over humans. As will be taken up later on, in Chap. 43, Bougeant was intent to impress upon readers that within the animal world, humans are actually entirely deficient. It was a remarkable confession, coming from an authoritative, religious perspective. But he was not merely an expert on the catechism, but also a European historian, a Jesuit who had written the most incisive overviews of the Treaty of Westphalia and the decades of bloodshed leading up to it. He had within his acute purview a smoldering notion of the stuff of humanity, and incorporating an equally famed sense of humor was able to lend commentary on zoology in a manner satirically expressed and bound to excite heated controversy. His many years of exile would attest to the level of his impact on French society.[8]

The astonishing convergence of mythopoetic interest in the relations of humans to other species, as captured by Aesop, has infiltrated our artistic emblems, scientific paradigms, and cultural norms regarding zoology like few other works. It is clear that more than 315,000 books, editions, or articles, largely about Charles Darwin, particularly since November 24, 1859, with the publication of *On the Origin of Species*, have dramatically altered our worldviews. But it is also most probable that

---

[7] World Cat Advanced Search, https://www.worldcat.org/search?q=ti%3AAesop%27s+Fables&fq =&dblist=638&fc=ln:_100&qt=show_more_ln%3A&cookie

[8] Printed for T. Cooper, London, p. 50. See also, Chapter 43, "The Mind of a Chicken."

Darwin (1809–1882) read Aesop not only as a child—given his grandfather's wild poetic strain and general literary character of his family—but also as an adult. It is known, for example, that in a correspondence of January 18, 1864, he heard from Robert Goodwin Mumbray regarding a belief, noted by Thomas Bewick, that finches hybridize. Bewick's book on British birds had been published in combination with his book on the fables of Aesop, as well as a general history of Quadrupeds. Darwin would have read it.[9]

He might also have read John Gay's *Fables*, after Aesop. First published in 1727, by Darwin's middle age, nearly 50 editions of Gay had been published. Similarly, the *Fables* of Gaius Julius Phaedrus (15 BC–c. 50 AD), also after Aesop, had been a huge publishing success in Amsterdam by David van Hoogstraaten for the Prince of Nassau; a gorgeous 1701 edition with 108 engraved emblems and other plates by I. V. Vranen, many of animals; one of the finest illustrated books published in the early eighteenth-century Netherlands; and first translated into English by Christopher Smart in 1831 in London.[10] Two years later, quoting over 100 major English poets and some naturalists, since the time of Spenser, with hand-colored portraits of birds, "A Lady" (Mary Hannah Rathbone) published her volume *The Poetry of Birds*.[11] By 1836, when Darwin—already something of a scientific celebrity—had returned from his five-year journey on the Beagle, and deeply enamored of ornithology, there is little likelihood he would not have devoured the book, with its deeply sympathetic interspecies broad strokes, detailing the falcon, owl, blackbird, cuckoo, bullfinch, goldfinch, skylark, nightingale, redbreast, wren, sedge warbler, titmouse, swallow, dove, pheasant, king-fisher, tern, stormy petrel, and swan.

Similarly, multiple editions of the Indian version of Aesop had also been published and translated into English with animal illustrations by Darwin's time, *The Fables of Pilpay*.[12] And by 1869, 13 years before Darwin's death, even the famed Russian writer, Ivan Andreevich Krylov, who first started writing fables after Aesop in 1809, the La Fontaine of Russia, had also been translated into English.[13] In the 1847 three-volume Russian edition, published in Saint Petersburg 3 years after Krylov's death, the famed color frontispiece portrayed the aging author seated at the

---

[9] See https://www.darwinproject.ac.uk/search/?sort=&keyword=Aesop, Accessed April 7, 2020.

[10] *Horace [also] Phaedrus.* The Phaedrus portion translated by C. Smart, A. J. Valpy, London, 1831.

[11] *The Poetry of Birds*, Mary Hannah Rathbone, Published by Liverpool, George Smith and Ackermann & Co, London, 1833. Rathbone, publishing anonymously her *Diary of Lady Willoughby, as relates to her Domestic History, and to the Eventful Period of the Reign of Charles the First*, had gained her great celebrity by 1844 throughout British society. Darwin would have known of her and certainly admired her skills as a water colorist of birds; and the spirit in which her work was assembled.

[12] *The Instructing and Entertaining Fables of Pilpay, An Ancient Indian Philosopher*, 3rd edition, London, 1754.

[13] *Krilof and his Fables*, Strahan and Co., 1869. Internet Open Library, https://openlibrary.org/publishers/Strahan_&_co. See also, Boris Yeltsin Presidential Library exhibition, "The Presidential Library Illustrates Ivan Krylov," 13 February 2020, https://www.prlib.ru/en/news/1289191, Accessed August 6, 2020.

base of a tree, his hand on the head of a resting lion, a hare, a rooster, a rearing horse, a steer, a snake, and other animals surrounding him, a spider and an eagle overhead. It could have been a portrait of Darwin himself (Fig. 40.2).

In Jean de La Fontaine's (1621–1695) Preface to his own *Contes, Fables* (1664,

**Fig. 40.2** "Portrait of Ivan Krylov Among thew Animals," from *Басни* (Fables), St. Petersburg, Russia, 1847, Frontispiece, Artist Unknown. (Private Collection). (Photo © M.C. Tobias)

1668 and 1678) he declares that "Aesop's fables were known long before Socrates was born".[14] And that while Plato had "banished Homer from his republic of letters, has given Aesop there a most honourable place. His wish is that children might

---

[14] See Jean de. *Fables choisies, mises en vers par M. de La Fontaine*. Claude Barbin, Paris, 1668. In translation, see *The Fables of Fontaine*, Translated From The French by Robert Thomson, with Twenty-Five Original Etchings by Auguste Delierre, J. C. Nimmo And Bain, London, 1884, p. 7.

imbibe these fables with their mothers' milk; and he recommends it to the care of their nurses to teach them; since they cannot be too early accustomed to the precepts of wisdom and virtue".[15]

But it is to the 1665 London, Thomas Roycroft edition of *The Fables* that we ultimately turn, the John Ogilby (1600–1676) translation with annotations, the folio comprising 81 brilliant illustrations by Wenceslaus Hollar, and 24 less subtle graphics by Dirk Stoop. The edition is actually a second, expanded one, the first having been utterly destroyed in the Great Fire of London. From its very first Fable, that *Of the Cock and Precious Stone* Ogilby's annotations encompass the insights of the Pythagoreans, Epicureans, of Lucian, Proclus of Athens, Lucretius, Pliny, and Phaedrus, all converging in a single message at the very beginning of this two-part 404-page epic: "I'll be to Nature kind; My Body I'll not Starve, to Feed my Mind".[16]

By Fable XX of the Second Part, Aesop has gathered all the strength of character, language, and message to inform every King and Queen, every Parliament and citizen, whether a Van Eyck, a Shakespeare, or the anonymous farmer, all the same, a warning from the "Council" of sheep who, speaking among themselves, have seized upon the language of the butcher, of Man, and tell us unsparingly for the ages, what, precisely, is at stake: "Last to Reward, who so Preach'd up his Cause; Who not suspected Cutting of his Throat, But to be Duke and Peer made of the Coat; False and Ambitious Councellors, then said he; May they be paid their Punishment like Me." And the Moral: "Few publick Spirits, Common Counsels find; These Fathom Wants, those Private Interest blind; Most for the Present, and their own Affairs: Suddain Calamities seizeth unawares" (Fig. 40.3).[17]

---

[15] ibid., p. 9.

[16] *The Fables of Aesop Paraphras'd in Verse: Adorn'd With Sculpture and Illustrated With Annotations*, The Second Edition, by John Ogilby, Esq., Master of His Majesties Revells in the Kingdom of Ireland, Printed by Thomas Roycroft for the Author, London, 1668, p. 2.

[17] ibid., p. 51, Second Part.

**Fig. 40.3** "The Butcher and the Sheep," by Engraver Dirck Stoop, from *The Fables of Aesop Paraphras'd in Verse: Adorn'd With Sculpture and Illustrated With Annotations*, The Second Edition, by John Ogilby, Esq., Printer Thomas Roycroft, London, 1668. (Private Collection). (Photo © M.C. Tobias)

# Chapter 41
# The Conical Temple of Konawsh

## 41.1 A Non-violent People

The first documented westerner to reach the Todas, Rev. Fr. Giacomo Fenicio (1558–1632), an Italian Jesuit priest, had come to the severe escarpment known as the Nilgiris in search of a lost community of Thomistic Christians, a hidden Utopia, Todamalaa. As Dr. Tarun Chhabra points out, reviewing studies by both Rivers and Walker,[1] Fenicio's hopes were dashed. The Todas spoke a language unlike any other in all of India, were animists (Fenicio referred to their love of buffaloes), and apparently showed no interest in what he had to preach. In 1873 the British anthropologist W. E. Marshall[2] spent minimal time among the Todas, following at least nine other ethnographers who had visited this most unusual community. Marshall conveyed an extensive characterization detailing how the Todas were *primitive* in a most unexpected manner: no violence, no hunting, no killing whatsoever, even to the alleged detriment of the protein in their diets; no violent sports, no trade, agriculture, wars, prisoners, torture, slaves. Nothing but "a milk diet and grain, whilst the woods are full of game, and flocks and herds to be had for the taking." (Fig. 41.1).

It was this discovery of a non-violent, magnificently isolated people (numbering fewer than 600 individuals throughout the Nilgiris or Blue Mountains at that time) which prompted him to write, "Have we come on the tracks of an aboriginal reign of conscience? And was man originally created virtuous as well as very simple?" A

---

[1] See *The Toda Landscape – Explorations in Cultural Ecology*, by Tarun Chhabra with a foreword by Anthony R. Walker, Orient BlackSwan, Harvard University Press, Cambridge, Mass., 2015, pp. 12–13. See W. H. R. Rivers, *The Todas*, Macmillan, London, 1906; and Anthony R. Walker, *Between Tradition and Modernity and Others*, B.R., Delhi, 1998. See also, "The Toda People: Stewards of Wilderness and Biodiversity," by Tarun Chhabra, International Journal of Wilderness, Volume 26, Number 1: April 2020/

[2] *A Phrenologist Amongst the Todas*, Longmans Green, London; See also "The Anthropology of Conscience," by M. Tobias, Society & Animals, Volume 4, Issue 1, The White Horse Press, Cambridge, UK, 1996.

© Springer Nature Switzerland AG 2021

M. C. Tobias, J. G. Morrison, *On the Nature of Ecological Paradox*,
https://doi.org/10.1007/978-3-030-64526-7_41

**Fig. 41.1** Members of the Toda hamlet of the Kerrir Patrician. (Photo © M.C. Tobias)

tribe that had renounced "gain" and "thrift"? A people whose attributes suggested some antediluvian, "primeval race" inhabiting a dubious paradise and for whom "the products of the buffalo" meant everything.[2]

For at least a millennium if not longer, the Todas have been lacto-vegetarians, even though the four surrounding communities—indigenous Nilgiri cultures, along with the Todas, the Kotas, Kurumbas, Irulas, and Badagas—are largely meat eaters.[3]

The endogamous Todas, known to themselves as *Awll(zh)* (translated by Dr. Chhabra as "the men" or "the people"), have numbered between 1200 and 1500 for the last 25 years. Their world comprises an erstwhile pastoral paradise marked by

---

[3] See. p. Todas of The Nilgiri Hills – Anthropological Reflections On Community Survival, Dr. Jakka Parthasarathy, Government of Tamil Nadu, 2005, p. 14.

transcendentalist beliefs in sacred mountains all around them (the Nilgiris Biosphere Reserve, first of its kind in India; the Kundah Range encompassing Mukurthi National Park); an afterlife upon those very mountains; and an all-pervasive religion based upon a particular endemic sub-species, genetically isolated breed of semi-wild, ashen gray, buffalo[4]—a rare Asiatic water buffalo, *Bubalus bubalis*, created by a female Toda deity, Taihhki(r)shy.[5] This playful deity engendered dairying herds and temples, both sacred and secular. But the lives of the buffaloes corresponded so intensely to the lives of the Todas, and of the surrounding ecosystems, that all was unified.

To this day, every semi-nomadic Toda hamlet (mund)—of which, as of 2020, there were approximately 75, each on average containing some ten households—is devoted to the worship, health, and perseverance of their herds. The problem for the Todas is twofold: they are well aware of surrounding modernity, and to various degrees partake of it; and their pastoral ecosystems which, for untold millennia they have depended upon, are being encroached upon by outsiders, with countervailing priorities.[6] The result is an increasing rate of buffalo hybridization and the loss of

---

[4] "Breeds List – Toda Buffalo," https://www.breedslist.com/toda-buffalo.htm, Accessed April 12, 2020.

[5] op cit., Chhabra, p. 47.

[6] For a brief, excellent overview of the displacement of habitat, particularly within biological hotspots, by increasing human population trends (in this case, encroachment by non-Todas onto Toda-designated lands), see J. L. R. Abegão, "Where the Wild Things were is Where Humans are Now: An Overview," Human Ecology, 2019, 47:669-679, https://doi.org/10.1007/s10745-019-00099-3, August 28, 2019, Springer Science+Business Media, LLC, part of Springer Nature 2019, https://overpopulation-project.com/wp-content/uploads/2019/08/Abeg%C3%A3o2019_Article_WhereTheWildThingsWereIsWhereH.pdf. The Author points expressly to the fact that population growth in high biodiversity areas exceeds that of the global average. He writes, "Cincotta *et al.* (2000) assess the global growth rate from 1995 to 2000 was 1.3%, but it was 1.6% in the hotspots. (Williams 2011)…" Cincotta, R. P., Wisnewski, J., and Engelman R. (2000). Human Population in the Biodiversity Hotspots. Nature 404(6781): 990–992. https://doi.org/10.1038/35010105. And, Williams, J. (2011). Human Population and the Hotspots Revisited: A 2010 Assessment. In Biodiversity hotspots – distributions and protection of conservation priority areas (pp. 61–81), https://doi.org/10.1007/978-3642-20992-5_4. In addition, Abegão cites data for the neotropical ecosystems like the Congo Basin or Amazon, suggesting that "carrying capacity" translates "into one subsistence hunter per square kilometer" and that in one study, suggesting "around 150 million households in the Global South acquiring meat through bushmeat hunting"… "8000 randomly selected households in 24 tropical and sub-tropical countries, 39% hunted bushmeat and 89% of the resulting income is dedicated to household dietary need." Abegão cites: Nielsen, M. R., Pouliot, M., Meilby, H., Smith-Hall, C., and Angelsen, A. (2017). Global Patterns and Determinants of the Economic Importance of Bushmeat. Biological Conservation 215: 277–287. https://doi.org/10.1016/J.BIOCON.2017.08.036. Nielsen, M. R., Meilby, H., Smith-Hall, C., Pouliot, M., and Treue, T. (2018). The Importance of Wild Meat in the Global South. Ecological Economics 146: 6. While bushmeat trade is not, to date, the primary driver for competition for space in the protected areas, deemed sacred to the Todas, human encroachment throughout the Nilgiris has been upwards of a habitat crisis since the earliest 19th century conversion of in situ and the significant biodiversity rarity of a "highland shola-grassland ecosystem, which also includes wetlands" (shola referring to "stunted evergreen montane forests with trees less than 20 m"); in addition to other Nilgiris vegetation, including "evergreen rainforests, moist deciduous broadleaved forests, thorny scrub forests

the pure Toda breed—and its attendant six dairy temple grades—upon which this naturalistic people's entire cosmology depends.


## 41.2   Ethical Animism

As Dr. Chhabra points out in his ethnographic masterpiece, *The Toda Landscape— Explorations in Cultural Ecology*, at the opening of its Appendix 4, "Toda Sacred Chants",[7] "Every Toda hamlet has its own specific prayer, which mentions the kwa(r)shm (sacred names) of nearby stones and rocks, streams and pools, swamps and grassy slopes, trees and shola thickets, buffaloes, pens and pen posts, calves and calf enclosures, summits and hills, sacred bells and churning vessels, sacred paths and routes...." Prayers generally converge upon a singular volition: "Poll(zh)yzh (põhzh) pehg mawhh! Twehhzh pehrry mawhh!" meaning, "May the dairy-temple products [milk, butter, buttermilk, and ghee (clarified butter)] be abundant! May the occupants of the buffalo pen be plentiful!".[8]

As Anthony Walker wrote back in the mid-1980s, "...the buffaloes are credited with almost human intelligence." They talk, and various anthropomorphic characteristics are attributed to their behavior, whether mourning with a human, or having been instrumental in actually founding "a particularly sacred dairy".[9]

Throughout history, there have been, of course, significant matrices of cultural worship and specific species. The Roman Pan (man/goat); the horse, sacred to Phoenicians; Ganesha, the Hindu elephant god of wisdom created by Parvati; the deer, sacred to Artemis in ancient Greek culture; cats in Egypt; all the animals of the Chinese Zodiac—Rat, Ox, Tiger, Rabbit, Dragon, Snake, Horse, Goat, Monkey, Rooster, Dog, and Pig; the wolf (Goddess Luperca) in terms of the birth of Roman civilization; a "Young Lion" in Genesis; the "Monkey King" in China; the Jaguar in Aztec and Mayan cultures; the corvids (crow/raven) for the Bhutanese as well as the Tlingit of Alaska; the Frigatebird (Fregata) for Easter Islanders; the Judaic fish god, Dagon; the hawk-god Laki Heno for the Kayans of Borneo; and so on. Monkey temples abound in Asia, like that of Swayambhunath in Kathmandu. There are rat temples, sloth-bear temples, and the Ram-headed Sphinxes at Karnak Temple in Egypt, where Middle Kingdom hieroglyphs render a window on the mixed veneration of jackals, beetles, shrews, the mongoose, baboons, ibis, crocodiles, and many

---

and lowland savannas"- all bearing amazing levels of floristic endemism (ibid., Chhabra, p. 17) to tea and fast growing eucalyptus plantations. See "The Sholas, Evergreen Hill Forests and Moist Evergreen Low Forests," in *Wonders Of The Indian Wilderness*, by Erach Bharucha, Mappin Publishers Pvt Ltd., Ahmedabad, India, and Abbeville Press Publishers, New York, pp. 685–686

[7] op cit., Chhabra, p. 477.

[8] op cit., Chhabra, p. 481.

[9] Anthony R. Walker, *The Toda Of South India: A New Look*, Hindustan Publishing Corporation, Delhi, India, 1986, p. 103.

other species that were in part integrated with human communities and their veneration.[10]

## 41.3   Cows and Buffaloes

Most notably, in India, the cow has always dominated the construct of animal worship, and there are estimated to be at least 100 million Brahmanical Hindus who abide by that connection, are vegetarian, and have ordained the institution of the goshala, or cow shelter. In Hindu ancient scripture, the cow is viewed like Aditi, the goddess of fruition responsible for all other gods. A general all-India cow protection society (Gaurakshini Sabha) was first established in the state of Punjab in 1882.[11] Jain panjorapors—animal sanctuaries—had been in existence for many centuries before that.[12] Cow worship far exceeds India's lovely obsessions with Nagas (snakes), elephants (Ganesha), the few remaining lions and tigers that were said to have once been ridden upon by the Goddess of powerful energy, Durga, and the rat, the vehicle for Ganesha, for whom the famed Karni Mata Temple in Rajasthan was built. Some 20,000 rats are said to live there—all believed to be "reincarnated members of Karni mata's clan," who was herself "the reincarnation of the goddess Durga".[13]

But in a country with more than 200 million Muslims who annually celebrate animal sacrifices en masse, Eid-ul-Adha, and who advocate for cow slaughter as a fundament of religious freedom,[14] little wonder there have been bloody Hindu/

---

[10]Lewis G. Regenstein, *Replenish the Earth: a History of Organized Religions' Treatment of Animals and Nature – Including the Bible's Message of Conservation and Kindness Toward Animals*, Crossroad Publishers, New York, 1991. See also, Richard H. Schwartz, "Tsa'ar Ba'alei Chayim: Judaism and Compassion for Animals" in Roberta Kalechofsky, ed., *Judaism and Animal Rights: Classical and Contemporary Responses*, Micah Publications, Marblehead, MA, 1992. See also, *Animals into Art*, Edited by H. Morphy, Unwin Hyman, London, 1989, deriving from the consequential 1986 World Archaeological Congress in Southampton, UK, which addressed critical animal/cultural themes, such as: "What is an Animal?" "The Appropriation, Domination and Exploitation of Animals," "The walking larder: patterns of domestication, pastoralism, and predation," and "Signifying animals: human meaning in the natural world." From Foreword to Morphy, by P. J. Ucko, p. x.

[11] *The Making of an Indian Metropolis, Colonial governance and public culture in Bombay, 1890/1920*, by Prashant Kidambi, p. 176, Routledge, Abingdon-on-Thames, UK, 2007, p. 176. See also, *Sacred Animals of India*, by Nanditha Krishna, Penguin Books, Ltd., London, 2014, pp. 80, 101–108.

[12] See Life Force – *The World of Jainism*, by M. C. Tobias, Asian Humanities Press, Fremont, CA, 1991.

[13] See "India's Karni Mata Rat Temple," "India's Karni Mata Rat Temple" 1 May 2013. HowStuffWorks.com. <https://people.howstuffworks.com/karni-mata-rat-temple.htm> 13 April 2020.

[14] Shabnum Tejani, *Indian Secularism: A Social and Intellectual History, 1890–1950*, Indiana University Press, Bloomington, Indiana, 2008, pp. 43–49.

Muslim massacres over the issue, beginning in 1893 with the so-called cow riots, which continue to this day. The significant and telling problem is the government of India's hypocrisy on this issue. It is widespread. Throughout northeastern and southwestern India, the slaughter of cows is allowed, whereas in five Indian states, one can be sentenced to up to ten years in prison for killing a cow. Rules, ellipses in monitoring or regulating, vary from state to state. It is fully legal to slaughter a bovine in Kerala, West Bengal, Arunachal Pradesh, Mizoram, Meghalaya, Nagaland, Tripura, and Sikkim. In Karnataka, the cow—if "old"—can be killed. In Punjab, slaughtered bovines are allowed to be exported if a permit is given. And throughout most of the country, "buffaloes" can be slaughtered willy nilly.[15]

When the above referenced October 2015 Indian Express article was published, hundreds of comments flooded the Internet, shining a spotlight on the huge divisiveness of the issue throughout India, from "very tasty food," "to the hypocrites," to "sell your God when you need money…" Years before, in 2010 the Indian public's attitude regarding the chasm between Hindu, Muslim, Jain, Buddhist and global animal rights positions (fully bypassing such complicated indigenous beliefs as those held by the…. Toda, among others) had led to the formation of a Cow Protection Army (Rashtriye Gauraksha Sena), founded by Ashoo Mongia. The Army recognized that India "has about 3,600 legal slaughterhouses and 30,000 illegal ones," where a multitude of cows, in addition to buffaloes, are annually killed. Added to that were an estimated "two million cows" that "are smuggled across a 2,400-mile poorly-patrolled border into Bangladesh every year".[16]

Widespread anger is increasingly fueled by this living debate. Whether the slaughter of cows can be vouchsafed or the archaic caste system still very much prevalent across the sub-continent contributes to the ironic fact that the country remains one of the largest exporters of meat in the world. Cows are considered sacred by some Hindus who ignore the status of buffalo, but in 1988 an all-India sociological survey revealed that, in fact, approximately 90% of all Hindus ate meat at one time or another. Hence, there is anything but consensus on "sacred animals" in India. As Supriya Nair has written, "The rising prevalence of this medieval spectacle in India today is not just a symptom of ancient pathologies. It is the most spectacular outlet yet for Modi's dog-whistle politics and the BJP's Hindu-nationalist

---

[15] See "The states where cow slaughter is legal in India," October 8, 2015, The Indian Express, n.a., https://indianexpress.com/article/explained/explained-no-beef-nation/. Accessed April 13, 2020.

[16] From "Selling the Sacred Cow: India's Contentious Beef Industry – In a country where cattle are considered sacred, they're also paradoxically becoming a lucrative export," by Sena Desai Gopal, February 12, 2015, https://www.theatlantic.com/business/archive/2015/02/selling-the-sacred-cow-indias-contentious-beef-industry/385359/. Accessed April 14, 2020.

project".[17] According to some sources, "at least 44 people" have been killed, hundreds injured, in the cow riots between 2016 and 2019.[18]

India's hypocrisy in this matter of cow worship exceeds other countries—like the United States—where there is no pretense of even fondness for bovines. Contradicting a vast legacy leading up to Mahatma Gandhi's embrace of cow protection, one renowned historian, Dwijendra Narayan Jha, has argued that the ancient Vedic period across the Indian sub-continent saw cows slaughtered, sacrificed, and eaten for at least 5500 years. Not until there was "an agricultural boom" in India did the Hindu begin to view cows with more restraint, and eventually a reverential attitude.[19] Writes B. R. Myers in his review of *Every Twelve Seconds: Industrialized Slaughter and the Politics of Sight*, by Timothy Pachirat,[20] "If our ancestors had had—as we now do—full awareness of animals' sentience, and the wherewithal to live without red meat, and the knowledge that red meat is harmful in even the smallest quantities, would they have gone on eating it?".[21]

## 41.4   A Separate Reality

Myers, of course, has no answer but suggests that Americans, and by implication consumers around the world, have chosen to normalize the slaughterhouse because they never go anywhere near one. They don't see it, hear it, and smell it. They just eat it. As many as 95% of all Americans eat it, in one form or other, 222.2 pounds of red meat per person per year (versus an average global number of 75 pounds).[22]

---

[17] "The Meaning of India's 'Beef Lynchings,'" by Supriya Nair, July 24, 2017, The Atlantic, https://www.theatlantic.com/international/archive/2017/07/india-modi-beef-lynching-muslim-partition/533739/. Accessed April 13, 2020. See also, "Violent Cow Protection in India," Human Rights Watch, February 18, 2019, https://www.hrw.org/report/2019/02/18/violent-cow-protection-india/vigilante-groups-attack-minorities, Accessed April 13, 2020.

[18] "Cow Vigilantes in India Killed at Least 44 People, Report Finds," Iain Marlow, Bloomberg News, February 19, 2019, https://www.bloomberg.com/news/articles/2019-02-20/cow-vigilantes-in-india-killed-at-least-44-people-report-finds, Accessed April 13, 2020.

[19] From "Selling the Sacred Cow: India's Contentious Beef Industry – In a country where cattle are considered sacred, they're also paradoxically becoming a lucrative export," by Sena Desai Gopal, February 12, 2015, https://www.theatlantic.com/business/archive/2015/02/selling-the-sacred-cow-indias-contentious-beef-industry/385359/. Accessed April 14, 2020. See D.N. Jha's *The Myth of the Holy Cow*, Verso Publishers, Brooklyn, NY, 2002.

[20] *Every Twelve Seconds – Industrialized Slaughter and the Politics of Sight*, by Timothy Pachirat, Yale University Press, New Haven, Connecticut, 2013. This political scientist went undercover for five months at a slaughterhouse in New York where 2500 bovines were slaughtered daily.

[21] ibid., "U.S. Slaughterhouse Rules," by B. R. Myers, November 2012 Issue, The Atlantic, https://www.theatlantic.com/magazine/archive/2012/11/slaughterhouse-rules/309113/. Accessed April 13, 2020.

[22] "Americans will consume a record amount of meat in 2018," Global Agriculture, March 1, 2018, n.a., https://www.globalagriculture.org/whats-new/news/en/32921.html, Accessed April 13, 2020. See also, "The Countries where people eat the most meat," Skye Gould and Lauren F Friedman,

At the same time, India for the period 2018/2019 exported "1.15 million tonnes" of buffalo, an amount down by some 15% from the previous year because China had banned imports from India fearing quality control issues in Indian abattoirs that suggested potential "foot-and-mouth disease".[23] Two billion three hundred million pounds of exported buffalo meat from India does not, of course, include the likely incalculable amounts of buffalo, cow, steer, chicken, and other meat consumed each year *within* India. It is known that in 2018, 3.7 million metric tons of poultry were eaten in India.[24] In 2019, Indians consumed 2.687 million metric tons of beef and veal.[25]

Writes Sena Desai Gopal of the world's "second-largest beef exporter" (India), "India's beef industry says all its beef comes from buffaloes, a claim challenged by right wing, religious, and animal rights groups".[26] According to Gopal, "India has 115 million buffaloes, more than half the world's population, and produces about 1.53 million tons of beef every year," quoting Santosh Sarangi, the chairman of the Agricultural and Processed Food Products Export Development Authority (APEDA) of India.[27]

British colonial administrators made their way into the Nilgiris in the early 1800s, seeking cool respite from the brutal summers down in Coimbatore, 86 kilometers to the South. Ooty (Ootacamund, or Udagamandalam) was likened to a Swiss-like paradise, cool mists, much rain, and a gorgeous floristic province, teeming with wildlife and some "exotic" "tribes." It was the perfect hill station, at over 6000 feet altitude. With the building of a steep, circuitous road to Ooty and nearby Coonoor, completed in 1832, it became an ideal spot for British soldiers to indulge

---

Sep 26, 2015, Business Insider, https://www.businessinsider.com/where-do-people-eat-the-most-meat-2015--9, Accessed April 13, 2020. From research by the "OECD-FAO Agricultural Outlook 2015," OECDiLibrary, https://www.oecd-ilibrary.org/agriculture-and-food/oecd-fao-agricultural-outlook-2015/meat_agr_outlook-2015-10-en

[23] "India's buffalo meat exports to plunge amid China clampdown on illegal imports," by Rajendra Jadhav, Reuters, January 31, 2019, https://www.reuters.com/article/us-india-meat-exports/indias-buffalo-meat-exports-to-plunge-amid-china-clampdown-on-illegal-imports-idUSKCN1PQ3RZ, Accessed April 13, 2020.

[24] Statista, "Consumption volume of poultry meat in India from 2013–2019," https://www.statista.com/statistics/826711/india-poultry-meat-consumption/. Accessed April 13, 2020

[25] Statista. "Consumption volume of beef and veal in India from 2015 to 2019 (in metric tons carcass weight equivalent)," https://www.statista.com/statistics/826722/india-beef-and-veal-consumption/. Accessed April 13, 2020.

[26] op cit.,"Selling the Sacred Cow: India's Contentious Beef Industry – In a country where cattle are considered sacred, they're also paradoxically becoming a lucrative export," by Sena Desai Gopal, February 12, 2015, https://www.theatlantic.com/business/archive/2015/02/selling-the-sacred-cow-indias-contentious-beef-industry/385359/, Gopal's data, in part, taken from: "Pink revolution? Meat exports up in NDA regime," HindustanTimes, by DK Singh and Timsy Jaipuria, January 6, 2015, https://www.hindustantimes.com/business/pink-revolution-meat-exports-up-in-nda-regime/story-6GuGjvgHgPcnFyscGTqJ0O.html, in which the Narendra Modi policies are elucidated with respect to the "subsiding slaughterhouses and promoting meat exports".

[27] ibid., S.D. Gopal.

in some R&R, which included wide-scale sport hunting.[28] The first British-style house was constructed in 1822. The British built a sewage disposal system targeting roughly 22,000 people. But by the early 1990s, that same system had the effluent from over 164,000 people to cope with, one reason, writes Charlie Pye-Smith, the "'old queen of the hills' is plagued by squatter settlements, pollution and unfettered tourist development".[29]

And, citing the then secretary of the Nilgiri Wildlife Association, "no matter where you look, every tree, every flower, every blade of grass you see will be exotic, an import of the British".[30] Bollywood also caught the Nilgiri fever, and at least 200 feature films have been produced around Ooty.[31] Uncurtailed development has polluted all of the region's waterways, which have been labeled "Rivers of Poison".[32]

## 41.5   Nilgiris Conservation

Fortunately for the indigenous peoples, and the wildlife, the Nilgiris can be viewed half-full, rather than half-empty. Spanning Mudumalai Wildlife Sanctuary to the northwest, and Mukurthi National Park in the southeast, with Mount Doddabetta (2623 m, 8605 feet) in the center, the International UNESCO Nilgiri Biosphere Reserve encompassing parts of the Western Ghats and Nilgiri Hills was established in 1988 to protect this unique tropical and subtropical biome of moist broadleaf forests and rare shola montane temperate grasslands, bogs, and forest mosaics, the first such Man in the Biosphere Reserve created in India. And since that time Aralam Wildlife Sanctuary and Bandipur-Nagarhole Tiger Reserve, and Silent Valley National Parks were formed, in addition to the Sathyamangalam and Wayanad Wildlife Sanctuaries. The overall region, covering some 5530 square kilometers (2125 square miles) is part of one of four biological hotspots touching India—the Western Ghats[33]—home to well over 3300 vascular (flowering) plants of which a

---

[28] See Nilgiri – Sporting Reminiscences, by An Old Shikarri, Higginbotham And Co., Madras, 1880. The book also cites such previous hunting enthusiast works as Colonel Walter Campbell's *My Indian Journal*, Edmonston & Douglas, Edinburgh, 1864, Shakespeare's "Wild Sports of India," and "Hawkeye".

[29] See *In Search Of Wild India*, by Charlie Pye-Smith, UBSPD Publishers, New Delhi, 1993. pp. 47, 49.

[30] ibid., p. 49.

[31] https://www.zostel.com/blog/bollywood-movies-shot-in-ooty/. Accessed April 14, 2020.

[32] "Rivers of Poison," by Sibi Arasu, The Hindu Business Line, June 23, 2017, https://www.thehindubusinessline.com/blink/cover/rivers-of-poison/article9733513.ece; See also, "Water bodies near Ooty polluted by industrial effluents," by Shantha Thiagarajan, The Times of India, October 5, 2017, https://timesofindia.indiatimes.com/city/coimbatore/water-bodies-near-ooty-polluted-by-industrial-effluents/articleshow/60946985.cms, Accessed April 14, 2020.

[33] See, for example, Daniels, R. J. R. (2001) "Endemic fishes of the Western Ghats and the Satpura hypothesis," Current Science 81(3):240–244.

**Fig. 41.2**  Sacred Toda - Mount Teihhttaa(r)sh - in the Nilgiris. (Photo © M.C.Tobias)

large majority are endemic to the Nilgiris, as Dr. Chhabra and colleagues have been documenting for over three decades (Fig. 41.2).

Against this context of all India, and nearly 85 million people in the southern state of Tamil Nadu (half rural, half urban) that includes most of the Nilgiris, one can only speculate as to the likely future of the (currently) fewer than 1450 strictly vegetarian, Orthodox Todas, and their rapidly declining sacred buffalo population—each female buffalo known by her sacred, first-name basis to the Todas in their prayers, and upon which their entire culture appears predicated.

It is believed that the Endangered Indian wild buffalo (*Bubalus arnee*) numbers fewer than 1000 (with the possibility of some additional clusters in the far northeast of India).[34] Toda buffaloes are most closely related to them, among India's more than a dozen water buffalo breeds, out of a world total exceeding 150 breeds of *Bubalus bubalis*.[35] The wild and semi-wild species originated in Southeast Asia but,

---

[34] "…total numbers of Asian Wild Buffaloes in the isolated populations that remain are likely much less than the purported 4000 and may even be as low as 200 individuals…" See "Genus Bubalus," *Handbook Of The Mammals Of The World*, Chief Editors, Don E. Wilson and Russell A. Mittermeier, Lynx Edicions in Association with Conservation International and IUCN, Vol. 2, Hoofed Mammals, Barcelona, Spain 2011, p. 583.

[35] op cit., Walker, pp. 98–118. See also, the DAD-IS, or Domestic Animal Diversity Information System of the Food and Agriculture Organization of the United Nations.

**Fig. 41.3** Toda Buffaloes High in the Nilgiris. (Photo © M.C.Tobias)

other than *B. arnee* which remains in India, has seen distribution to nearly every continent, where they are subjected to three human fixations: dairy, meat, and draft. Clearly, with children in many parts of Asia, *Bubalus* is also perceived as part of the family, a companion animal. Many of the breeds—for all of their size (they can approach 2700 pounds) and mythic ferocity, easily capable "of fighting off a tiger"[36] (this quality not surprisingly first pointed out by British sportsmen for whom Toda buffaloes were one of their prized targets)—are also known to be incredibly gentle, docile, and loving to humans. This is especially the case with the Toda, the only known culture in the world to worship them (both the sacred and domestic herds into which the animals, for countless reasons, are designated) to the extent that they fully consider the buffalo to be one of them, indeed, the direct link to a spiritual Otherworld (Fig. 41.3).

## 41.6 The Worship of Buffalo

The buffalo also provides a powerful segue for many of the Todas into the realms of other species, including snakes and tigers, all considered sacred. In fact, reptiles "are regarded as both protectors and preservers of dairy-temple sanctity"; while

---

[36] ibid., p. 100.

"there are a few Todas who claim to know the rituals for 'closing the mouth' of predators, especially tigers ...".[37]

The Toda buffalo physiology differs markedly from the Indian zebu cow (the most widespread on the sub-continent). But most importantly, the fat content of their milk is 30% richer than the zebu's milk.[38] Female buffaloes are generally known to the Toda as ïr, males as e-r. A large bull is called pïrer pïr.[39] But, generally, male buffaloes are not given personal names, only the females when a buffalo calves. The rites, upon calving, and the myriad duties necessary for the dairyman (there are no dairywomen among the Toda) are all-consuming. Toda society is completely dedicated to the ethical and religious hierarchy of events, cycles, predictions, communiqués between the deities, the buffaloes, the dairies, and all Toda. The details are ingrained in the boys; the hamlets are consumed by the moral responsibilities to their companion buffaloes; and it can be said that the buffaloes are in charge. The one "glitch" as it were in this perfect system—for dairy products are the mainstay of the Toda diet—is the fact that, in former centuries, there was a time when a male buffalo calf would be sacrificed, a ritual no more. But Rivers did record such a ceremony in 1906, though since that time it does not happen, and the Todas will not talk about it, clearly wanting to be understood by their very love of the buffaloes and, hence, to faithfully practice pure vegetarianism.[40]

## 41.7  Ecological Spirituality

Like another Indian collective, the Bishnoi (Vishnoi) of Rajasthan, the Toda worship nature, and by their comprehensive involvement with every minute biological detail of their world, they are botanically and meteorologically clairvoyant. As Dr. Chhabra elaborates,[41] there are fundamentally some 280 plant species the Toda utilize and fully comprehend in every season. And that is only the beginning. The Bishnoi, a desert people who number nearly one million, are a self-described Hindu sect that worship Guru Jambheshwar (1451–1536). In 1485, he authored a book known as the *Jamsagar*, or, "show people the light," which prescribed 29 rules of ecological sustainability, vegetarianism, and ethical activism. They rightly view themselves as guardians of the Thar (Great Indian) desert, an expanse of some 200,000 square kilometers in northwestern India. Their diet comprises "curd,

---

[37] op cit., Chhabra, 2015, p. 462.

[38] op cit., Walker, p. 102.

[39] ibid., p. 103. One must note that Dr. Chhabra has extensively revised Toda orthography from that used by Anthony, as well as the earlier Professor Murray Emeneau who, in 1971 worked out an initial transliteration system for the Toda language which is not written, only spoken.

[40] ibid., p. 179. One could imagine that this ritual, cloaked today in silence and possibly a sense of regret felt by Todas, was occasionally undertaken under pressure by neighboring tribes for whom animal sacrifice was no anomaly. But this is merely speculation.

[41] op cit., Chhabra, 2015.

buttermilk, cooked vegetables, kiddi made from graham flour, spices, sugar, breads and a considerable amount of lentils." They worship the thick-trunked khejare tree (*Prosopis cineraria*), which is an abundantly high-protein bearing, flowering pea family (Fabaceae) tree known as Ghaf. In 1730, a Maharaja near Jodhpur ordered his troops to cut down a number of the trees for the wood. A Bishnoi mother and her daughters gathered fellow disciples of Jambheshwar together to protect the sacred trees; 363 of them were killed. In the late 1980s these martyrs were commemorated by the Indian government as the nation's first official environmental heroes, precursors of the famed Himalayan "Chipko Movement"—tree huggers.[42]

The enormously complex tapestry of religious tenets, ethical mores, and living realities in India, as she fast surpasses 1.35 billion residents—the country is easily poised to exceed China's approximately 1.4 billion people—presents a dizzying, joyous but often painful human and other animal rights portrait. We have been conducting various research projects in India for many decades. But on every visit there has been an overwhelming sensation that is both familiar and altogether novel. India and her native flora and fauna never cease to amaze, but the challenges of conservation are escalating at a bewildering pace. In the winter of 2019 we set out with our long-time colleagues, Tarun (Dr. Chhabra) and his close friend, Ramneek Singh, to visit what may be one of the oldest structures still standing and revered in all of India, the Toda conical dairy temple of Konawsh.

## 41.8 Konawsh

Its form, and the site where it remains, may date back 9000 years, although there is no proof, as yet, of that. Tarun and Ramneek formed the Edhkwehlynawd Botanical Refuge (EBR—"place with a spectacular view" in Toda) in 1992, a local nonprofit Trust devoted to preserving the Toda lifestyle and habitat, and helping the Todas to cope progressively with the onrush of the twenty-first century.[43] Supporting and helping to sustain Toda deep ecological knowledge; providing educational opportunities for the youth; and bridging gaps like electrification, clean water, habitat restoration, dairy temple maintenance (re-thatching as needed), sustainable income generation, and legal protections for their tribal lands are all clear mandates. What is less predictable or controllable, however, is the welfare in perpetuity of a species—the Toda buffaloes—who are at great risk. Their numbers are fast declining, and the ability of the Toda to maintain sacred herds, perpetuate their dairy rituals and thus the rudiments of their entire cosmological presence as a society in the Nilgiris, is endangered. Write Tarun and Ramneek, "Today, the herd size of both sacred and domestic buffaloes has shrunk dramatically and if immediate steps are

---

[42] See "Desert Survival by the Book," by M.C. Tobias, New Scientist, Vol. 120, No. 1643, December 17, 1988, p. 29. See also, op cit., "The Anthropology of Conscience," by M. Tobias, Society & Animals, Volume 4, Issue 1, The White Horse Press, Cambridge, UK, 1996.

[43] www.ebr.org.in

**Fig. 41.4**  Konawsh Conical Temple. (Photo © M.C.Tobias)

not taken to propagate the purebred animals and increase the population, it is feared that the Toda culture that dates back to ancient times, might collapse ..." (Fig. 41.4).[44]

Konawsh is one of only two remaining conical temples in India (several hundred years ago, a third existed). It is considered the most sacred of all Toda póh or dairy temples, about a 45-minute trek from the nearest Toda hamlet of the Kerrir patrician, with Mount Tehhkolmudry looming through the mists far in the distance, a peak where Toda gods are said to have once met. Tarun has published many details of the Konawsh Temple.[45] While its significance to the Toda may be said to equate with that of the Vatican for Catholics, the temple presents an entirely different set of aesthetic, spiritual, and utilitarian objectives. To most, it resembles a strange, approximately 40-foot tall leaning cone of striped rattan hoops and braided native grasses, support beams, and stonework. A thatched, otherworldly structure of inviting considerations, in a jungle teeming with eyes that were watching us and elephants, leopards, tigers, bears, wolves, jackals, wild dogs, Nilgiri langurs, and many others.[46]

---

[44] "Restoring Traditional Landscapes in the Nilgiris," by Dr. Tarun Chhabra and Ramneek Singh, ibid., pp.10–11. See also, "Indigenous Peoples' Plan – Todas," https://www.cepf.net/sites/default/files/56154-safeguard-ipp.pdf

[45] op cit., Chhabra, 2015, pp. 216–288.

[46] At least one scholar has suggested that the Toda conical temples may be the original Dravidian

Adjoining the temple is a large circular area, the pen, surrounded by stones, where the buffaloes rest and sleep during the one lunar month of the year, in the dry season, when the temple is in operation. Its dairyman priest— "pohkarrdhpawll(zh)"—a Kerrir clan member absolutely free of any pollution, as stipulated in the deeply entrenched ethical contract of dairying strictures that guide the entire world of the Toda,[47] is there to ensure its inviolable sanctity and protect the buffalo (or, as the case may be, find succor and protection in their company).

We reached the temple along a specific path, leech-ridden as one drew close, and paid our many respects. This primeval setting emanated its inexpressibly stable eco-system, an interdependent alliance of biodiversity, non-violence, and sheer love engaging one and all. We had little doubt that the pugmarks of the various large cats, were, at most, five minutes separated from our own passage into this thicket of primordial spirituality. And that when we headed back up the muddy allée toward the distant hamlet in darkness, cold-drenching mists, common in winter throughout the Nilgiris, swirling around us, that all those animals were watching. Permitting us to pass. They read our hearts.

Emerging from the thickness of wind, rain, and encroaching darkness, several Toda buffaloes wandered past us, at their ease, undisturbed, a mother keeping a calf from losing the path.

## 41.9  Their Future

We thought back to various theories of island biogeography, and the notion that is quite key to conservation currents, namely, the reality of habitat fragments throughout the world, of populations confined to and dependent upon small ranges.[48] In the case of both the Toda buffaloes and the Todas themselves, there is little latitude for expanding contiguity between biological fragments in the Nilgiris. However, while Tamil Nadu population figures for the state continue to increase, as in all of India, total population numbers within the Nilgiris have been steadily declining, from a high in 2001 of 762,141, to 735,394 in 2011 (essentially the population of Bhutan), and continuing on that trajectory of diminution.[49] That is good news for biodiversity

---

temple style of South Indian vimana, a prototypic 7-story towering structure that has also been replicated in Northern Indian temple shikara, as well as Tibetan Buddhist stupas or chortens. Indeed, as a prototype, that architectural form would also comport with the very mountain metaphor inherent to Kailasha, or Su-Meru, in the northwestern Himalayas -site of the celestial Siva/ Parvathi embrace, the most sacred mountain in the world to Hindus and Buddhists. See Chhabra, 2015, p. 288.

[47] ibid., p. 235.

[48] See "Connecting Habitats to Prevent Species Extinctions," by Stuart L. Pimm and Clinton H. Jenkins, American Scientist, https://www.americanscientist.org/article/connecting-habitats-to-prevent-species-extinctions, Accessed April 14, 2020.

[49] See https://www.census2011.co.in/census/state/tamil+nadu.htm, Accessed April 14, 2020; For Nilgiris specific population trends, see https://www.census2011.co.in/census/district/31-the-nilg-

and for one small group of Todas who depend upon the thousands of cohabiting species.

Whether progressive regional and local forest regulations continue to accept the responsibility for protecting an extremely vulnerable culture, and fragile ecosystems, represents the first challenge. The fieldworkers, forest civil servants, and others who must uphold those social and conservation decrees will be crucial to the Toda's day-to-day future. Eco-tourists to Ooty are often keen to see the Toda, and there is a Toda "cultural village" of sorts, which should satiate most curiosity (a cultural buffer zone/biological bubble). Moreover, a few of the hotels sell the fine Toda embroidery, particularly their famed shawls. The network of parks and reserves throughout the Nilgiris affords a spatially significant—forwardly conceived— amount of habitat, certainly sufficient to safeguard one of the most robust clusters of Asian elephant and tiger populations in the world, especially with multidisciplinary ecologist/activists, like Dr. Chhabra and Ramneek Singh and their colleagues in place.

What remains the greatest of the unknowns, however, is the future of the Toda buffalo. Hybridization, genetic drift, any number of emerging infectious diseases,[50] other co-morbidity factors, all will impact not only an already attenuated Toda buffalo population and distribution throughout their traditional grazing lands, but also—by zoosemiotic implication—the very existence of their humble stewards, the last Indian indigenous vegetarian collective on the planet.

---

iris.html, Accessed April 14, 2020.

[50] See "Emerging Infectious Diseases in Water Buffalo: An Economic and Public Health Concern," by Marvin A. Villanueva, Claro N. Mingala, Gabriel Alexis S. Tubalinal, Paula Blanca V. Gaban, Chie Nakajima and Yasuhiko Suzuki, IntechOpen, February 21, 2018, https://doi.org/10.5772/intechopen.73395, https://www.intechopen.com/books/emerging-infectious-diseases-in-water-buffalo-an-economic-and-public-health-concern/emerging-infectious-diseases-in-water-buffalo-an-economic-and-public-health-concern, Accessed April 14, 2020. Write the Authors, "Leptospirosis, brucellosis, Bovine Tb, BVDV and fasciolosis have projected economic impact to water buffalo industry as well as its effect as zoonoses. However, the data seem underquantified since most are neglected diseases…"

# Chapter 42
# Does Natural Selection Select for Natural Selection?

## 42.1 The Mystique of Choice

How do we live? In what closed or open spaces and levels of confinement? Who makes these decisions and how do such choices—personally, culturally—impact the world? Genes, memes, orientations, convictions, sociologies: The varied dimensions of our chosen and/or imposed confines—the many assumptions and premises of science—are but one telling metric of the sum-total of behavior, our zoo-narratological topology that, through so many descents with modification, lays claim to who we are, our social norms and societal inflictions. Lays claim, but does not necessarily persuade. We may assume that the temporal element of evolution between the individual and species levels does not lend philosophical or physical resolution to this self-referential mirror of natural selection, not in humans, at least for the past several millennia. As early as 1962, in the first edition of Thomas S. Kuhn's *The Structure of Scientific Revolutions* (University of Chicago Press) he spoke of incommensurabilities undermining conventional proofs for or against entirely new paradigms. Incommensurable paradigms whereby a premise does not connect; causes and effects are strangers; relations are askew. Now, suddenly, the nature of science is challenged by an altogether novel series of the commensurate, namely, the emphatic nature of ecosystem vulnerabilities. Reality is posing the greatest shifts to every discipline previously studied (Fig. 42.1).

Time has not solved any equations or taught us the underlying fallacy of a perpetual motion. We systematically break the bonds of time, within our lifespans, presenting to the imagination every option and futility. This exponential test of natural selection, and the choices it propounds, is reduced to the instantaneous; to unrelenting idealism, the ground upon which all follies are executed, the aerodynamics that tease us with Mars or Titan. Our unfettered imaginations have excelled at skyscrapers, clothing, unchecked serial-killer diets, and an average Western-style ecological household footprint of approximately 48 tons of $CO_2$(equivalency)/yr

© Springer Nature Switzerland AG 2021
M. C. Tobias, J. G. Morrison, *On the Nature of Ecological Paradox*,
https://doi.org/10.1007/978-3-030-64526-7_42

**Fig. 42.1** Paiute dwelling, Owens Valley, California, early 1900s. (Photo by A. A. Forbes)

(calculated for the ten major greenhouse gasses, of which three are strictly correlated with human industrial processes), with an annual rate of increase of 0.1%.[1]

Such numbers and dimensions are not in tune with normal perception or daily
capacity, neural infrastructure, the flexibility of human choice, or moral heroism. So
how could natural selection work to alter actual choices we make? Cell division,
genotypic expression, and heritable traits over multiple generations each contributes
its say in matters assumed to take more time than we have left to solve dire environmental problems. And yet, there is no inner genetic code pre-empting our ability to
deliberate effectively, within an instant; and those instantly multiplied, should we
choose, in the world's favor over the course of an entire generation. The mystique
of fruitful choices gives us an advantage to be true to ourselves.

Despite over 300,000 years of distinct human evolution, several peoples on three
continents are still essentially naked, including the Koma of Nigeria's Alantika
Mountains, the Kambari in Niger,[2] the Sanema on the Venezuelan border with

---

[1] Primary GHGs include Cardon dioxide, Methane, Nitrous oxide, two groups of perfluorocarbons,
three groups of hydroflurocarbons, and sulfur hexafluoride. University of Michigan Center For
Sustainable Systems, http://css.umich.edu/factsheets/greenhouse-gases-factsheet, Accessed April
19, 2020. See https://www.theguardian.com/environment/2020/jun/19/why-you-should-go-ani-
mal-free-arguments-in-favour-of-meat-eating-debunked-plant-based, Accessed August 11, 2020.

[2] See Business Insider, "Meet the naked tribes of Nigeria," by Inemesit Udodiong, October 21,
2019,     https://www.pulse.ng/bi/lifestyle/meet-the-naked-tribes-of-nigeria-where-people-wear-
leaves-and-little-to-nothing/w3ttqxv.

Brazil,[3] and the Korowai of Indonesia.[4] Impressively, the Digambara Jain monks throughout parts of India also remain naked their entire ordained lives, as has been referenced in earlier chapters, wandering from village to village, never asking for food, but accepting it in the manner of gowkari, the grazing of a cow, if it is pure vegetarian cuisine, and depending on the ascetic's own particular schedule. Taking one meal per day, though some ascetics fast every other day; while others have imposed myriad additional scruples. Ultimately, Jain monks may slip into the end-of-life discipline of Sallekhana, not so much a suicide as a letting go, for a new country of the soul. Such select choices test the boundaries of what science calls natural selection. And the choices are copious ones, from species, to population; from culture to community; from individual to individual. And if that can be seen to interfere with conventional natural selection, it alters even more troubling twists for the idea of human evolution, as scientifically described.

## 42.2  Shelters and Co-adjacencies

Humanity's colossal ecological crisis is an explicit function of our architectural and socially enforced spaces of occupancy, plots of arrogation, the places in which we choose to live, altering our surroundings, inventing and constructing "our" possessions. This is an entirely ephemeral habitation, obviously, which has only our ancestors, or futurist scenarios and pictorial cosmologies, to cling to, or benefit. Short of Digambara, or uncontacted, or quasi-uncontacted tribal affiliation, human wildness of the kind portrayed in the tragic case of Ishi (c.1861–1916), the last member of the California Yahi people,[5] remains paradoxically embedded in an environmental nostalgia with a rich history. The concept, the individual, was first sentimentally attributed in John Dryden's play *"The Conquest of Granada" (1672)*—"I am as free as nature first made man,/Ere the base laws of servitude began..."—a literary precursor to Alexander Pope's reference to a "simple nature," "a humbler heav'n; Some safer world in depth of woods embrac'd."[6]

Until one has been homeless in Mumbai, or a coyote in Chicago, or a ruby-throated hummingbird frantically storing up calories before migrating across the Gulf of Mexico to the Yucatan, we have no approximate idea what it means to be wild. The macroevolutionary speciation, mutations, differing phenotypes, traits discernibly heritable, and all the modernist syntheses of developmental biology remain

---

[3] See http://planetdoc.tv/documentary-sanema-tribe-customs-and-traditions.

[4] See "Born To Be Wild," by Amanda Devlin, March 28, 2017, The Sun, https://www.thesun.co.uk/news/3195004/indonesia-lost-pygmy-tribe-picture/. Accessed April 19, 2020.

[5] Fleras, Augie (2006). "Ishi in Two Worlds: A Biography of the Last Wild Indian in North America". Journal of Multilingual and Multicultural Development. 27 (3): 265–268. doi:https://doi.org/10.1080/01434630608668780.

[6] *An Essay on Man; In Epistles to a Friend (Epistle II)* (London: Printed for J. Wilford, 1732. See https://archive.org/details/b3054466x/page/n2/mode/2up.

at a loss to definitively array differing finch beaks with vastly oppositional human footprints, that of a Digambara, versus the average meat-eater in Uruguay, Argentina, or Hong Kong, the three largest per capita consumers of meat in the world. In 2016, humans collectively ate 129.5 billion pounds of beef.[7] Their environmental footprints were huge, violent, and unnecessary. We have no measures for their opposite, all those quiet footprints that were also laid down. But the evolutionary edifice and its emphatic behavioral circumlocutions engendering rapine and living plunder converges upon a phenotypic expression that is unambiguous.

In Sir Thomas More's *Utopia* (1516) the killing of animals was deemed so horrible that "freemen were not even allowed to witness it lest their human clemency be eroded."[8] If meat crowds humanity's digestion (nine days on average to fully digest a steak), consider the odious particulars of her settlements. That is the profound conundrum urban architecture and the very bowels of its engineering, water supplies, and effluents must confront.

Traditionally, a hamlet, for the majority of our history, has defined a steady, egalitarian-like co-adjacency in terms of immediate shelter. From those clusters dating back a few hundred thousand years, we divine something of a uniform predilection, not unlike observed compulsions in other species regarding their nests, dens, lairs, roosts, and the like, which have largely remained undeviating in form and function for millions of years. Indeed, the oldest known stable nesting site thus far discovered, in South Africa, combines some 190 million years of antiquity.[9]

But when we examine a chair, or even more so a door handle, we are into the realm of something more uncanny than the mere mechanics of comfort, or re-engineering of power. We cannot underestimate the inordinate impacts of human architectonic and mechanical pivots. Any technology is destined to become oppositional to what was before it. Natural selection has kept door handles fairly unchanged throughout time. But our species, when members are seated on an armchair (as opposed to what history has shown to be a less socially eminent *stool*), enlists a parallel evolution of ecopsychology and the eruption of the idea of power over others. Our manipulations of our living confines begin with floors but quickly move to the walls, roofs, and all adjacent space.

This is the case whether a mud and thatch house as one would see among the Dogon of Mali, in tightly knit yurts or tepees from the Rockies to Mongolia, all the way to the enormous motte-and-bailey fortification, such as the original Windsor Castle, and onward through diverse styles and size-qualified modern-day duplexes

---

[7] "World Beef Consumption Per Capita (Ranking of Countries), by Rob Cook, Beef2Life, April 18, 2020, https://beef2live.com/story-world-beef-consumption-per-capita-ranking-countries-0-111634. "All three countries consumed more than 100 pounds of beef per capita." Americans come in 4th in this ranking.

[8] See *Man And The Natural World – A History Of The Modern Sensibility*, by Keith Thomas, Pantheon Books, New York, 1983, pp. 294, citing p. 71 of More.

[9] "Oldest Dinosaur Nesting Site Discovered," by the University of Toronto Mississauga, SciTechDaily, January 25, 2012, https://scitechdaily.com/oldest-dinosaur-nesting-site-discovered/. Accessed April 18, 2020.

in any city in the world. Each empowers a through-story that earnestly strives to *defy* natural selection. There are other, human-rooted forces that clearly bifurcate between our morphological and cultural evolutions. One has remained stable; the other appears, for better or worse, out of control. We imagine everything.

The hat-trick that would work to liberate (or condemn) us to a separation from the cave[10] began in the realm of bedding and beds, other décor; a continually warming hearth, a place to store things—things themselves, these alter egos of utility; outhouses, sheds, a place for tools. And then, in total, a strange global vernacular of human architectural styles tensed between needs just one step removed from caves and emergency shelter and everyday utilitarianism (the premise there will be an everyday, as was hastened by the differing advents of agriculture), to a show of strength (e.g., a surrounding moat), formidable resistance. Versus the more welcoming open house, whether an inn, almshouse, a priest's villae or countryside dwelling proffering restraint over ostentation. In northwestern Europe for centuries such homes were called Frisian byre-houses, with nearby buildings that might contain animals, or farm equipment, or mulch, or slaves. European villages proper would not emerge before the late eleventh century in parts of France, Catalonia, and Bavaria.[11]

Furniture historian Florence de Dampierre has speculated that Adam and Eve stood in Paradise, but later on we began to sit down and that act of "sitting has been something of a luxury in the panoply of human activity."[12] Dampierre cites one early sixteenth-century Spanish writer who comments, "We Christians sit at a proper height, not on the ground like animals."[13] More emblematic of our species' recent cultural evolution than furniture is the zoo narratology of building design and its perpetual font of newly devised forms of isolating occupants. By what structures and designs, inchoate or fabricated, we have chosen—our choices usurping those of natural selection—to, in essence, separate ourselves endlessly, until we have safely moved away from a lost Eden into ever the new one? Such choices conform to the free will inherent in every question of our relationship with God, and the extent to which we have also been separated. If we also remove the idea of God from the idea of nature, by whatever philosophical devices or labyrinthine motives, then those choices become a kind of destiny that the jungle, the Himalayan summit, or some anonymous green sward could never replicate. Our minds are too much with us, as Wordsworth rather intimated (Fig. 42.2).

---

[10] See two essays by René Guenon, "The Heart and the Cave," Studies in Comparative Religion, Vol. 5, No. 1, Winter, 1971, World Wisdom, Inc.www.studiesincomparativereligion.com, http://www.studiesincomparativereligion.com/Public/articles/The_Heart_and_the_Cave-by_Rene_Guénon.aspx; and "The Mountain and the Cave," Studies in Comparative Religion, Vol. 5, No. 2, Spring, 1971, http://www.studiesincomparativereligion.com/Public/articles/The_Mountain_and_the_Cave-by_Rene_Guenon.aspx.

[11] See *The Village & House in the Middle Ages*, by Jean Chapelot and Robert Fossier, Translated by Henry Cleere, University of California Press, Berkeley and Los Angeles, California, 1985, p. 129.

[12] Florence de Dampierre, *Chairs: A History*, Abrams, New York, 2006, p. 7.

[13] ibid., p.7.

**Fig. 42.2** "Schönbrunn Palace," from Johann B. F. von Erlach's *Entwurff einer historischen Architectur,* 1725, engraver, Johann Adam Delsenbach. (Private Collection). (Photo © M.C. Tobias)

## 42.3   Ecological Legacies in Mortar and Stone

Hence, the architect takes on a moderator's role between physical and mental evolution. There are huge leaps in these many assertions; but who has not obsessed upon, not just death, but the act of dying, and its precise *location*? To similar effect, the act of living and its geography are a subject of intense, indeed the ultimate locus for, contemplation. Few architects have ever found themselves in so politically and aesthetically charged an environment as the Austrian architect and historian Johann Bernhard Fischer von Erlach (1656–1723), whose lavishly baroque commissions would result in many of the Habsburg Empire's most elaborate images of power ever propounded, particularly in the case of Schönbrunn Palace in Vienna, for which construction commenced under Leopold I in 1694, but was not completed until 1749: a 1,441-room part Rococo part neoclassical re-creation of an interior ecotone, on a par with Versailles, a wedding gift to Archduchess Maria Theresa from the Holy Roman Emperor Francis I.[14] In its civic expanse, there is little to recommend a love letter between newlyweds; no Taj Mahal. Aside from its lavish interiors, it strikes one as a symbol of strength and stability, which was certainly no hallmark of

---

[14] Hans Aurenhammer, *J.B.Fischer von Erlach*, Harvard University Press, Cambridge, Mass. 1973.

the Habsburg Empire, Austria being engaged in nearly 35 wars, occupations, and/or uprisings during the Monarchy's several hundred years of nearly continuous domination. Moreover, the Schönbrunn Tiergarten, an imperial menagerie, was built on the grounds in 1752 and remains the oldest continuously active zoo in the world, opened to the public in 1779. A troubled legacy.

While both elephants and giant pandas have been born at the Tiergarten, the zoo has seen its share of tragedies over time. Two thousand six hundred animals in captivity were eaten during World War I. By the end of World War II, approximately 400 animals were left in the zoo. Giraffes and lions and countless other species have been captured in the wild, delivered to Schönbrunn, and vanished from the world. Two caretakers were killed in 2002 and 2005, first by three jaguars, then by an elephant. After privatization, and eventual oversight by the president of the Austrian World Wildlife Fund, the Tiergarten developed polar and Amazon zoological sections and, as zoos go, has tried to redeem itself into a conservation research center. Nearly 11 million euros were spent by the privatized zoo in May 2014 on what is called Franz Josef Land, to receive two polar bears, male and female, from a zoo in Estonia and a wildlife pin Finland. This transit of zoologic exotica would no doubt have delighted Erlach, in his day.

Indeed, the House of Habsburg certainly found its architect in Erlach. Among his numerous other works, he completed the Karlskirche in Vienna, and the Church of the Holy Trinity in Salzburg, Schloss Klessheim in Salzburg (used in part as a casino since the early 1990s), and the Austrian National Library of Vienna—the Hofbibliothek. At the height of his career he published a monumental work, *Entwurff einer historischen Architectur (Design of a historical architecture)* (Leipzig, 1725).[15] The large-scale engravings depict the Temple of Solomon, vast Pentagon-like invented expanses of Babylon, the Pyramids, the reimagined Temple of Diana at Ephesus, a colossus that was the Temple of Nineveh, enormous mausoleums of Egyptian royalty, and then a shift to mammoth human constructions beside nature. The cataracts of the Nile. A fabulist portrait of Alexander the Great's architect, Dinocrates, implanted upon the Greek monasteries at the cliff-girded Mount Athos, a huge Roman funeral tower of the Scipios overlooking the heartlands of the Spanish Tarraco, the ever-rambling ruins of Palmyra, Stonehenge, the Imperial City of Peking and Giant Pagoda of Nanking. Every engraving enlarges the human presence on the earth, magnifying symbols of empire. Limitless bridges, vases depicting stern animal deities, ultimate edifications of churches and plazas, ruins reignited

---

[15] *...depicting famous buildings, antiquity, and foreign people, different from those depicted in the history books, memory coins, ruins, and obscure demolitions. In the first book. The construction types buried by time of the old Jews, Eg, ptier, Sÿrer, Perser, and Greeks In the other. Ancient unknown roman. In the third. Some foreign domestic and non-European, as the Arab, and Türcken, etc. also new Persian, Siamite, Sinesian and Japanese buildings. In the fourth. Some buildings from the author's invention and drawing, by Fischer von Erlach, Johann Bernhard, 1656–1723*; engravings Johann Adam Delsenbach, 1687–1765, engraver. https://archive.org/search.php?query=creat or%3A%22Delsenbach%2C+Johann+Adam%2C+1687-1765%2C+engraver%22;   See   also, http://socks-studio.com/2015/06/22/project-for-a-history-of-architecture-johann-bernhard-fischer-von-erlach-1721/Accessed April 14, 2020.

with the forces of human history, emboldened toward a very construct of infinity that has borrowed from every nuance of Utopia. This is the Erlach so at home in the unbounded coordinates of a Schönbrunn. He was to palatial mortar what a Frederic Edwin Church was to the tropical Andes.[16]

In such imaginary unions and schisms is natural selection in humans modified by the inconvenient realities of physical constraint, and the invention of depth, altitude, and multitudinous perspective to advance the cause with which Dante concluded his own epic (the *Divine Comedy*, 1320) by formulating his nine spheres of heaven. That Empyrean cosmology both taunts and contradicts human imperial reasoning, torturing demons with angels, people with higher causes than themselves. The human scale with otherworldly comforts of the illimitable. All of these divides are at the heart of an ecological dementia, when augmented to the level of a large vertebrate numbering nearly eight billion who is paradoxically hell-bent on fouling every nest.

## 42.4   Natural and Unnatural Occupancies

Schönbrunn—its grand airs nothing compared with the 9999 rooms of the Forbidden City—sits there, open to day-long tours, after which most people return to their own petit dominions. Such personal spaces, our studio apartments and farmhouses still measured by feet, hands, eyes, odors, and a modicum of comfort for our particular hominid anatomies, are ultimately very little changed from the earliest cave dwellings, whether at the Chinese Guyaju, the Sassi di Matera in Italy, the Bandiagara Escarpment in Mali, Ancestral Puebloans throughout the North American Southwest, or Kandovan in East Azerbaijan Province. People were still living in the Cappadocian caves of Uçhisar until the 1950s. The Shanidar Caves in Iraq were home to at least two different hominid species of different communities for over 55,000 years.

According to the United Nations Commission on Human Settlements (UNCHS) deteriorating living conditions for much of humanity in at least 37 nations studied have resulted in an average floor space per person of approximately 20 m², or just over 215 square feet.[17] In the United States, square footage for apartments varies from 711 sq. feet in Seattle (the smallest) to 1038 sq. feet in Tallahassee.[18] From India to Germany, apartment sizes—which house the overwhelming bulk of the more than 50% of the human population which is now urban—comport with the US figures. "Just under 1-in-3 people in urban areas globally live in slum households,"

[16] For a British equivalent to an architect grappling with the natural/unnatural bifurcations, see the work of Thomas Robins the Elder (1715–1770), as exemplified in the book, *Gardens Of Delight – The Rococo English Landscape Of Thomas Robins The Elder*, by John Harris, With Natural History Notes by Dr. Martyn Rix, The Basilisk Press, London, 1978.

[17] See "Charting the Progress of Populations," United Nations Population Division, https://www.un.org/en/development/desa/population/publications/pdf/trends/ProgressOfPopulations/14.pdf.

[18] "The Average Apartment Size of the Largest US Cities, Charted," by Digg, November 27, 2018, https://digg.com/2018/average-apartment-size-data-viz, Accessed April 18, 2020.

so-called.[19] In Japan since 2017, average new apartment sizes are "under 100 sqm."[20] In stark contrast, Paris, the second most expensive city in the world, after Singapore in terms of housing prices, by law provides 16.2 square meters (174.3 square feet) per living space, which could include an entire family. A French chamber de bonne (maid's attic room) is often no more than 2.43 square meters, or 26.1 square feet, smaller than most jail cells, and at the extreme of a spectrum that includes the largest private home in the world, there is the 27-story building, a family home, in Mumbai comprising 400,000 square feet.[21]

Construction of human settlements, with their twenty-first-century ecological footprints, has mired most utopian visions, as well as any conceivable comparisons with a Shanidar. Painting over the green with the charcoal, crowding out that which was, chemically altering everything in a mad dash for highly prized and hugely impactful comforts that were never known to our species, before. This is not to negate some highly refined architectural plans for revisioning a sustainable integrative set of factors, which for several centuries has been largely renounced. Take the "Soprema" vision for corporate headquarters in Strasbourg, France, an "8,225 square-meter ecological utopia," a "poetic landmark" poised to be "the green city of the future" that can "achieve a symbiosis between humans and nature" with a "green flex office for nomad co-workers."[22]

---

[19] See "Our World In Data" – "Urbanization," https://ourworldindata.org/urbanization, Accessed April 18, 2020. See also, "Average apartment sizes in top seven Indian cities decline 17% in five years: Report," ETRealty, march 12, 2019, https://realty.economictimes.indiatimes.com/news/residential/average-apartment-sizes-in-top-seven-indian-cities-decline-17-in-five-years-report/68374229, Accessed April 18, 2020.

[20] See "99% of new apartment supply in Tokyo in 2017 was under 100sqm," by Zoe Ward, Re-Think Tokyo, March 14, 2018, https://www.rethinktokyo.com/blog/zoe-ward-japan-property-central/99-new-apartment-supply-tokyo-2017-was-under-100sqm/tokyo, Accessed April 18, 2020.

[21] See, "Paris: life in a storage cupboard has never been so expensive," by Rooksana Hosssenally, March 7, 2014, The Guardian, https://www.theguardian.com/cities/2014/mar/07/paris-storage-cupboard-illegal-renting-expensive, Accessed April 18, 2020. See "A Look at the Absurd Size of the Largest House in the World," by Trevor English, February 2, 2020, Interesting Engineering, https://interestingengineering.com/a-look-at-the-absurd-size-of-the-largest-house-in-the-world, Accessed April 18, 2020.

[22] See "Semaphore: an Ecological Utopia Proposed by Vincent Callebaut," archdaily, https://www.archdaily.com/tag/utopia, Accessed April 18, 2020. See also, "How Utopian Living Looked to Modernist Architects," by Anika Burgess, Atlas Obscura, September 14, 2016, https://www.atlasobscura.com/articles/how-utopian-living-looked-to-modernist-architects, Accessed April 18, 2020. See also, "Journey Back to the Dreamy, Gorgeous Architecture of Utopia," by Margaret Rhodes, March 31, 2018, Wired Magazine, https://www.wired.com/2016/03/journey-back-dreamy-gorgeous-architecture-utopia/. Accessed April 18, 2020. See also, *The Green Braid: Towards an Architecture of Ecology, Economy and Equity*, by Kim Tanzer and Rafael Longoria, Routledge, Abingdon-on-Thames, UK, 2007; *Paradoxes of Green: Landscapes of a City-State*, by Gareth Doherty, University of California Press, Berkeley, CA, 2017; *Tent Life: Haiti*, Photographs by Wyatt Gallery, Essay by Edwidge Danticat, Umbrage Editions, Brooklyn, NY, 2010; *Robert Adam – Country House Design, Decoration & the Art of Elegance*, By Jeremy Musson, Photography by Paul Barker, Foreword by Sir Simon Jenkins, Rizzoli International Publications, Random House, New York, 2017; Ian McHarg's *Design With Nature*, 25th edition, John Wiley & Sons, New York, 1995; *Sense Of Place -European Landscape Photography*, Foreword by Herman van Rompuy, Bo Zar Books, Prestel, Munich, 2012; *The Emancipatory City? Paradoxes and*

Distinguished contemporary architects, inculcating a bravado aesthetic that both accommodates and bursts beyond any obvious geometrical borders, like the late British-Iraqi Zaha Hadid, Richard Meier, the Frank Lloyd Wright of the Waterfall House, the Renzo Piano of Centro Botin art gallery in Santander, Spain, many of the late Paolo Soleri's (1919–2013) drawings, all take a vision of interior space far beyond, for example, the massive urban renewals of a central Paris undertaken by Georges-Eugène (Baron) Haussmann during the rule of Emperor Napoleon III between 1853 and 1870.[23] But these one-offs of luminous building materials are no match for the realities of the current 33 megacities in the world (soon to be 39 and projected to take up at least "3% of global land mass" and 9% of the human population).[24] Any more than the tent cities of Haiti have anything in common with the country estates of Robert Adam, except that people live there. Whether measuring by ancient Egyptian common cubits—elbow to the end of the middle finger; Royal cubits—the addition of the length of a human palm; or by the feet and inches inherited from the Greeks and Romans, human occupants have the same wishes. A warm bath. Clean drinking water. Peace and quiet. A suitably inviting eu-topos—a good place, seamlessly integrated into our own for our companion animals, whether avian, feline, equine, canine, or bovine, typically.

Walled and densely populated urban clusters, like Syria's Tell Brak, or Sumerian/Babylonian Uruk, dating back 8000 and 6500 years respectively;[25] or the splendid Aztec capital of Tenochtitlán founded in 1325; ancient Athens and Rome, or Jericho, Jerusalem, and Machu Picchu… they all fought the same insulation, person-to-person distancing, hygienic, public health, energy fallout, supply and demand distribution chains, and civic disorder, and class-system complaints we would come to denominate by deep history divides, racial and ecological umbrella categories: North/South, income inequality, environmental and social justice dysfunctions.

Does natural selection select different choices within our species in the country-side, versus a megacity of ten million+? Of course it does; it must. Our experiments with built utopias have at times succeeded—certainly for the odd privileged family over time—but have largely failed when stacked up against the inevitability of a species' needs, as depicted throughout the miseries of those behind sheltered walls of a Florentine mansion in Giovanni Boccaccio's *Decameron* (1353). No amount of luxury was sufficient bulwark against the infiltrative bacterium *Yersinia pestis*. Or

---

*Possibilities*, by Loretta Lees (Editor), SAGE Publications, Ltd., London, 2004; *Design for Ecological Democracy*, by Randolph T. Hester, The MIT Press, Cambridge, Mass., 2006, the Bible for an architectural "everyday future." See also, *Megalopolis, the Urbanized Northeastern Seaboard of the United States*, by Jean Gottmann, MIT Press, Cambridge, Mass.,1966.

[23] See "Haussmann's Paris," The Art History Archive, "Architecture in the Era of Napoleon III," by Emily Kirkman, 2007, http://www.arthistoryarchive.com/arthistory/architecture/Haussmanns-Architectural-Paris.html, Accessed April 18, 2020.

[24] "Mapping the World's New Megacities in 2030," by Jeff Desjardins, Visual Capitalist, October 26, 2018, https://www.visualcapitalist.com/mapping-the-worlds-new-megacities-in-2030/. Accessed April 18, 2020.

[25] See "The Ancient City," by Joshua J. Mark, April 5, 2014, Ancient History Encyclopedia, https://www.ancient.eu/city/. Accessed April 18, 2020.

where a family's or clan's resilience and vision of what works is utterly incompatible with, yet almost wholly dependent upon, a city ruled by strangers. We see remarkable variations on these issues at "the world's largest spiritual utopia," Auroville, in Pondicherry, India, housing 2500 people, or at New Orodos in the Gobi Desert, which has been described as "…a post-apocalyptic space station…."[26]

Whether after hundreds of millennia humanity can change, under ecological duress, or in the full freedom of its moral intellect and adaptability, remains a totally *unknown* quantum for natural selection.

---

[26] See "7 Utopian Design Experiments, from Le Corbusier's Radiant City to a Ghost Town in China," by Demie Kim, Artsy, January 9, 2017, https://www.artsy.net/article/artsy-editorial-7-utopian-design-experiments-le-corbusiers-radiant-city-ghost-town-china, Accessed April 18, 2020. For a brief overview of Utopian socialism and its corresponding settlement patterns, including Karl Marx and Friedrich Engels, Robert Owen, Charles Fourier, the Moravians and Shakers, Anabaptists and German Pietists, see *God's Country: The New Zealand Factor*, by M. C. Tobias and J. G. Morrison, A Dancing Star Foundation Book, Zorba Press, Ithaca, New York, 2011, p. xxxvi.

# Chapter 43
# The Paradox of Solace

## 43.1 Local Transfigurations

The painter John Crome (1768–1821) died on what is now called Earth Day, 200 years ago. He left five known letters, approximately 300 oil paintings, 33 etchings, and a multitude of sketches and watercolors. When he passed away, the population of his native home, the East Anglian Norwich in England's northeastern Norfolk County, was just over 50,000. After London, it was England's most bustling city, having been established in 1094 along the chalk River Wensum, a tributary of the River Yare and today a Special Area of Conservation, teeming with old mills, bridges, remnants of a glorious medieval center, and a huge concatenation of native flora and fauna[1] (Fig. 43.1).

World War II, particularly the Nazi Luftwaffe Baedeker raids at the end of April, 1942, damaged or destroyed more than 80% of all structures in Norwich: 2000 dwellings were obliterated, 229 people killed, over a 1000 injured.[2]

Today, with 142,000 residents, and much of the city rebuilt, it is widely considered one of the most livable, charming congeries in all of Great Britain, with an approximate equal number of pubs and historic churches, in one of which, St. George's Church, Crome lies buried. In another, the eleventh-/twelfth-century St. Julian's Church, the anchoress, Juliana (1343–1416), wrote the first book by a female in documented English history, *Revelations of Divine Love*, in which, from the very beginning, Jesus appears before her and shows her a hazelnut, which

---

[1] Natural England, "Designated Sites View," River Wensum SAC, https://designatedsites.naturalengland.org.uk/SiteGeneralDetail.aspx?SiteCode=UK0012647&SiteName=&countyCode=29&responsiblePerson=&unitId=&SeaArea=&IFCAArea=; And https://sac.jncc.gov.uk/site/UK0012647. See also, https://www.visitnorfolk.co.uk/inspire/Seven-Natural-Wonders-of-Norfolk.aspx; https://www.norfolkwildlifetrust.org.uk/gallery/all-photos; see also http://norfolknaturalists.org.uk/wp/wp-content/uploads/2017/02/Trees_1_broadleaf.pdf. See also, See https://www.edp24.co.uk/news/nine-beautiful-places-a-30-minute-drive-from-norwich-1-4440710.

[2] "A History of Norwich," http://www.oldcity.org.uk/norwich/history/history09.php.

© Springer Nature Switzerland AG 2021
M. C. Tobias, J. G. Morrison, *On the Nature of Ecological Paradox*,
https://doi.org/10.1007/978-3-030-64526-7_43

**Fig. 43.1** "Front of the New Mills," 1813, etching by John Crome. (Private Collection). (Photo ©
M.C. Tobias)

Juliana interprets as a sign of love.[3] Others throughout the centuries have gone to
great lengths to fathom Juliana's visions. Claire Foster-Gilbert did her PhD thesis at
King's College, London, on the ecological ramifications of Juliana's *Revelations*.
Amid a flurry of scientific studies and surveys occurring throughout the greater
Norwich ecosystems.[4] Foster-Gilbert declares in her Thesis Abstract, "...can the
Julian of Norwich texts be read today in such a way that they can help address the
twenty-first century ecological crisis, by transforming our 'buffered' subjectivity
into the 'porous' subjectivity Julian brought to and learnt from her revelations?

   This thesis argues that the stresses on the planet that are caused by humanity are
themselves symptoms of an underlying human subjectivity enslaved by Gestell, the
'essence of technology', defined by Heidegger, which turns nature and humanity
itself into objects to be exploited." *Restoring Porosity and the Ecological Crisis: A
Post-Ricoeurian Reading of the Julian of Norwich Texts* is a penetrating study of
"Wounds, wonder and identity with the Earth," of "Dying" and of "Transformation";

---

[3] Barry Windeatt, ed. *Revelations of Divine Love*, Oxford University Press, Oxford, UK., 2015,
p. 45.

[4] See, for example, See also, https://www.norfolkwildlifetrust.org.uk/home; See also, *Trees of
Norfolk*, by David Richmond, http://norfolknaturalists.org.uk/wp/wp-content/uploads/2017/02/
Trees_1_broadleaf.pdf, Accessed April 20, 2020.

of "Gardening," "Humility," and "Gratitude."[5] And it brings into proper focus the history of the United Kingdom since the last Ice Age, as interpreted by one mystic whose life is caught out in the very web of that ecological series of transformations.

Juliana's life comports in numerous respects with her Norwich successor, Sir Thomas Browne (1605–1682), physician, philosopher, naturalist, and mystic, whose great confessional, *Religio Medici* (*The Religion of a Physician*, 1643), speaks to the life of Pythagoras, to metempsychosis, to the nature of man versus the worms that devour him. And, ultimately, to that strange "Amphibium, whose nature is disposed to live not only like other creatures in diverse elements, but in divided and distinguished worlds, for though there be but one to sense, there are two to reason; the one visible, the other invisible."[6]

Both Juliana and Thomas Browne set the appropriate groundwork for a painter who lived all of 52 years (life expectancy in 1820s England was 40 years for a man, although compared with Keats, who died just a few months before Crome, aged 25, death could be famous for her randomness) and managed resolutely to wander at his more-or-less leisure, depicting the scenes he so loved, in a city that had the principal advantages of London, without its Big Smoke or Great Wen. Wrote Norman L. Goldberg in his two-volume comprehensive biography and "critical catalogue" of Crome, "His capacity for keeping inviolable the emotional response to the observable beauty of nature was the essence of his aesthetic development. Few indeed are the landscape artists who, in their search for order, truth, and beauty, have risen to such dignity of manner."[7]

## 43.2 A Cosmic Norwich

The Norwich School and Society of Artists was most prominent during the first three decades of the nineteenth century, under the enthusiastic sway of Crome. It had much to do with a legacy of East Anglia's pastoral calm that today's world has fortunately inherited. But Crome's modest obsessions also raise important questions, earlier posed, that involve the motives, circumstances, and built-up historical genotypes, as it were, that inspire certain individuals to test the boundaries of their physical and perceptual worlds. Crome—a master of the "picturesque"—provides a gentle but persuasive link between the worlds of the Dutch master Meindert

---

[5] See https://kclpure.kcl.ac.uk/portal/files/94606632/2018_Foster_Gilbert_Claire_1352686_ethesis.pdf, p. 3, 2018. https://kclpure.kcl.ac.uk/portal/en/theses/restoring-porosity-and-the-ecological-crisis-a-postricoeurian-reading-of-the-julian-of-norwich-texts(7637e8b5-44fa-4725-b6a6-7075cb99caac).html.

[6] Sir Thomas Browne, *Religio Medici*, Sixth Edition, With Annotations; and Observations by Sir Kenelm Digby, Printed for Andrew Crook, London 1659 pp. 74–75.

[7] See *John Crome the Elder*, Norman L. Goldberg, Two Volumes, New York University Press, New York, 1978, Volume I, p. 34.

Hobbema (1638–1709)[8] and Thomas Gainsborough (1727–1788),[9] both of whom he venerated, and the work of John Constable (1776–1837).[10]

While J.M.W. Turner (1775–1851) had captured tumult, incandescence, great fires, industrial steam, and every gale of the sublime, Crome assayed to chronicle those very tangible aspects of any given day that stand; a connection at the heart of Samuel Johnson's very critique of Milton's *Paradise Lost*, in whose penultimate grasp that link was perceived to be equally banished: "The man and woman who act and suffer [Adam and Eve] are in a state which no other man or woman can ever know. The reader finds no transaction in which he can be engaged; beholds no condition in which he can by any effort of imagination place himself;…The good and evil of eternity are too ponderous…the mind sinks under them in passive helplessness, content with calm belief and humble adoration."[11] In Crome, that lost paradise is fully regained.

It was during 1804 in Wales, where the Pen-y-Darren, the world's first steam-automated train, hauled ten tons of iron, that Crome embarked with his friend and fellow artist, Robert Ladbrooke, on "sketching expeditions" throughout the region. He would produce watercolor drawings of Tintern Abbey, of the Welch and Wye Valley, but no train belching smoke into the pure air.[12] Unlike a Thomas Cole who not infrequently juxtaposed the ever-present destructive powers of humanity within his otherwise Arcadian landscapes (e.g., "The Course of Empire: Desolation," 1836, New York Historical Society), Crome followed the purity of purpose exemplified by the work of his two great natural history Norwich contemporaries, botanist William Jackson Hooker (1785–1865) and Sir James Edward Smith (1759–1828), a natural historian who founded in 1788 "The Linnean Society of London [which] is the world's oldest active biological society," having acquired the entire literary and scientific estate of Carl Linnaeus.[13]

But by 1867, as Joel Sternfeld writes in his essay "The Calamitous Sublime," Matthew Arnold, on his honeymoon, wrote that the world "Hath really neither joy, nor love, nor light,/Nor certitude, nor peace, nor help for pain." This poem, "Dover

---

[8] *Masters of 17th-century Dutch Landscape Painting*, Exhibition Catalogue, Ed. P. C. Sutton; Amsterdam, Rijksmuseum, Boston, MA, Museum of Fine Arts, Philadelphia, PA, 1987–8.

[9] See *Gainsborough's Landscapes: Themes and Variations*, by Susan Sloman, Bloomsbury, New York, 2011.

[10] See *English Landscape Scenery: A Series Of Forty Mezzotinto Engravings On Steel*, by David Lucas, From Pictures Painted By John Constable, R.A. Henry G. Bohn, London, 1855.

[11] Samuel Johnson, *Lives of the Poets*, 'Milton' (1779), cited in *Turner And The Sublime*, by Andrew Wilton, British Museum Publications, 1981, p. 23.

[12] op.cit., Goldberg, Volume 1, p. 75.

[13] https://www.linnean.org/; Of great interest, Dr. Giorgia Bottinelli, curator of historic art at Norwich Castle Museum and Art Gallery, points out that William Hooker "was the son-in-law of Crome's patron Dawson Turner (he married Turner's eldest daughter Maria). Dawson Turner was a botanist himself and was elected a Fellow of the Linnean Society in 1797. Another interesting link with Crome is Dr Rigby, Crome's first employer: as well as a physician, Dr Rigby was also a member of the Linnean Society and an early arboriculturalist." Personal Correspondence with Dr. Bottinelli, who points to the link, https://www.geograph.org.uk/photo/5949767.

Beach," suggests Sternfeld, was "perhaps the first poem of the modern era"...
[Arnold] "seems to know that mass warfare, atomic weaponry, all-pervasive pollution, and a planet struggling to drink and to breathe is coming."[14]

Crome's day remained indebted to trees not yet subject to cynicism, though the
unbashful historian could not fail to compare the devastations by deforestation
occurring throughout the world's remaining neotropics to those same impulses for
gain that compelled earlier peoples to inflict massive defoliation. "A brief history
of British woodlands" makes clear that during the Bronze Age (ca. 3800 BC to
40 AD) Neolithic residents began a patchwork of woodland clearance for agriculture that took out at least half of all forests on the British Isles. For 400 subsequent
years, Roman occupation simply added more insult, an ecologically disastrous
segue to the advent of Anglo-Saxon "skilled carpenters" whose machinations further reduced the possibilities of the temperance of shady groves to roughly 25% of
England. By the time Norwich had been initially settled, woodland cover was a
mere 7%.

Only the Black Death of 1349 provided relief for remaining trees (the way many
residents of Northern India would get their first ever glimpse of the Himalayas following Covid-19-induced cessation of auto traffic/pollution). Populations had been
far too decimated to mount further assaults on potential wood products, at least for
a century or two.[15] Speaking to the trends throughout the Middle and Late Middle
Ages, environmental historian Clarence J. Glacken would summarize: "Landscapes
whose appearance reflected new kinds of densities arising out of human choices
replaced the older environment...The age as a whole is marked by the retreat of forest, heath, marsh, and bog, and the creation of new towns...."[16]

Crome is remembered for having been one of the first artists in England to actu-
ally depict precise species of trees accurately in his works. The sheer diversity of
silva fauna in East Anglia is impressive: 28 primary broadleaf species in Norfolk
County, including Alder, Beech, Horse Chestnut, Wych Elm, Hornbeam, the London
Plane, Black, White, and Lombardy Poplar, Weeping Willow, Rowan, and one of
Crome's favorites, the Holm Oak, *Quercus ilex*.[17]

This is not surprising given the intense conservation efforts that have always
been focused on the region, with a current 81 conservation areas designated just in

---

[14] In *Landscapes after Ruskin – Redefining The Sublime*, Hall Art Foundation, Hirmer, Grey Art
NYU Gallery, Reading, Vermont 2016, p. 13.

[15] See "A brief history of British woodlands," Royal Forestry Society, Forestry Commission of
England, 2015, https://www.rfs.org.uk/media/441738/7-a-brief-history-of-british-woodlands.pdf,
Accessed April 20, 2020.

[16] See *Traces on the Rhodian Shore – Nature And Culture In Western Thought From Ancient Times
To The End Of The Eighteenth Century*, University of California Press, Berkeley and Los Angeles,
p. 290.

[17] http://norfolknaturalists.org.uk/wp/wp-content/uploads/2017/02/Trees_1_broadleaf.pdf.

North Norfolk.[18] At present, at least 13 species are endangered throughout East Anglia.[19]

But sylvan consolation does not appear to depend upon the actual number of trees in a glen, or the frequency with which migrating birds are spotted (Norfolk County may well have the largest number of avifauna throughout the United Kingdom).[20] Nor was John Crome privy to such declarations born of true desperation as the "Global Charter for Rewilding the Earth – Advancing nature-based solutions to the extinction and climate crises" (2020), a "Vision" which is clearly characteristic of our times: "We believe that the world can be more beautiful, more diverse, more equitable, more *wild*."[21] The cumulative weight of ongoing devastations before and since the time of the Norwich School brings us to a very peculiar and fragile point of burden. In their biography of Crome, Derek Clifford and Timothy Clifford speak of his lifelong emotional ties to the Norfolk landscapes from boyhood; of impassioned, undying personal connections, and to the fact that, allegedly, on his very deathbed, Crome told his son John Berney, "…if your subject is but a pig-sty – dignify it"[22] (Fig. 43.2).

## 43.3   Cultivating One's Backyard

At his very best, Crome bids us to follow him into the depths of the Dutch Golden Age of landscape aesthetics, at a time when British history was providing plenty of distractions: the Act of Union which created the United Kingdom, dissolving the Irish Parliament; the conquest of the French and Spanish at the Battle of Trafalgar; the first attacks by the Luddites at many of England's major industrial establishments; introduction of Corn Laws; Wellington's defeat of Napoleon at Waterloo; marches in London for labor reforms and the deaths of 11 protestors at Peterloo in Manchester; as well as the death of the long-serving George III. That tumult was originating out of the politics of nearly 1.5 million urban occupants occurring in all but a different world, 117 miles southeast of Norwich in London.[23]

---

[18] https://www.north-norfolk.gov.uk/tasks/conservation/view-conservation-areas-in-north-norfolk/. Accessed April 21, 2020. See also, https://historicengland.org.uk/listing/what-is-designation/local/conservation-areas/. Accessed April 21, 2020.

[19] See "13 wildlife species in danger of disappearing from East Anglia," Ross Bentley, May 20, 2019, East Anglian Daily Times, https://www.eadt.co.uk/business/threat-to-biodiversity-in-east-anglia-1-6059859, Accessed April 21, 2020. See also, https://www.nationaltrust.org.uk/features/wildlife-in-the-east-of-england, Accessed April 21, 2020.

[20] See "Birding in Norfolk – January thru December," by Steve Mills, Surfbirds.com, http://www.surfbirds.com/mb/trips/norfolk0104.html, Accessed April 21, 2020.

[21] The World Wilderness Congress, www.wild11.org/charter.

[22] *John Crome*, by Derek Clifford and Timothy Clifford, New York Graphic Society, LTD., Greenwich, Connecticut, 1968, p. 99.

[23] See http://www.bbc.co.uk/history/british/timeline/empireseapower_timeline_noflash.shtml.

**Fig. 43.2** "Composition: A Sandy Road through Woodland," 1813, etching by John Crome. (Private Collection). (Photo © M.C. Tobias)

But Norwich and her artistic disciples could easily function unfettered by such oppositional forces, such is the peculiarity of geography, given to mayhem down a certain road, or stunning tranquility just across an old bridge, and into the woods. For Crome and his fellow wanderers, like the painters John Opie, Thomas Harvey, Joseph Stannard, William Beechey, George Vincent, John Thirtle, James Stark, and John Sell Cotman, the world was the local world. Its atmospherics, rhythms, and the biodiversity right there, under every weather, was more than enough to grapple with. As Goldberg writes, "It is his [Crome's] realistic transmutation of nature's tranquil moments which makes his work a triumph."[24]

In many of the best of Crome's works, paintings like "New Mills: Men Wading" (City of Norwich Museums, 1812), "Moonlight On The Yare" (Paul Mellon Collection, 1817), "The Old Oak" (National Gallery of Canada, 1813), "Marlingford Grove" (The Lady Lever Gallery, 1815), "A Study of A Burdock" (City of Norwich Museums, 1813), "Thistle And Water Vole" (Paul Mellon Collection, 1813–1814), "Norwich River: Afternoon" (Collection Mr. Max Michaelis, 1819), "Mousehold Heath, Norwich" (The Tate Gallery, 1818–1820), there is the entrancing permanence of a countryside vision, delineated under no discernible duress, outside the distribution of angst that is the social contract gone awry, but, rather, a human nature in harmony with herself and others for all time.[25]

Among his rarity of sketches and etchings, "Sketch Of Three Pigs" (Witt Collection, 1809–1810), "Composition: Sandy Road through Woodland" (British Museum Print Room, 1813), "At Heigham" (British Museum Print Room, 1812–1813), "Front of the New Mills, Norwich" (British Museum Print Room, 1813), and his great storm approaching a windmill in "Mousehold Heath, Norwich" (entirely different than his oil by the same title, British Museum Print Room, 1810–1811); or his watercolors "Tintern Abbey" (City of Norwich Museums, 1804–1805), "Silver Birches" (City of Norwich Museums, 1814–1815), and "Landscape with Cottages" (Victoria and Albert Museum, 1809–1810), we are living testimony to the continuity of a utopic understanding that surpasses nature's own dissonance and overcrowding. By his insistent love of his own neighborhoods, and the immeasurable patience he espoused in formulating and refining an aesthetic from the most everyday sentience, Crome has managed to eclipse the famous and frothing capitals of the world, the exotic ecosystems on other continents, or the very inaccessible to tempt every expedition to nowhere, by, instead, cultivating his very backyard.

---

[24] op. cit., Goldberg, Volume 1, p. 34. See also, The Norman and Roselea Goldberg Collection, Vanderbilt University: https://collections.library.vanderbilt.edu/repositories/2/resources/977#.

[25] See "The Painting- In Focus," by Rachel Scott and Sam Smiles, The Tate, https://www.tate.org.uk/research/publications/in-focus/mousehold-heath-norwich-john-crome/the-painting, Accessed April 21, 2020; See also, by "Pictorial Transitions: Crome," in "Prints and Principles," by Dr. Gordon James Brown, July 29, 2012, http://www.printsandprinciples.com/2012/07/crome-pictorial-transitions.html, Accessed April 21, 2020.

# Chapter 44
# Collodi's Garden and the Misadventures of Pinocchio

## 44.1 The Ecological Imagination

Everything is as it appears, and nothing is. Every work of the imagination deliriously holds itself hostage but may engender all possible contexts for reinventing those alluring *no exits* which do not otherwise exist but in our free reign of fantasy. A tree is a tree is a tree. Yes. But it is also a real tree! A poet has roots, as well, and, in the case of the author of "Pinocchio," those taproots lay claim to a vast fortune of historic details. Some might call them biomimicry, or fantastic trivia, while yet others will obtain vestiges, even the vague hint of a glimpse, a horizon line, a tortured look-alike knot in the forest that helps explain so much, so many heretofore unexamined markings in the sand, ancient carvings into the bark. Collodi's gardens are the mythic tone that people can realize in that conversation between humanity and botany; a species of ecological bubble that is environmentally friendly, a carbon sink at the very least. As importantly, Collodi's story does not so much build up the theatrics of *hope* in a world increasingly gutted, on the brink of a transformation that will usurp all of us, but provides at the very least a most appealing footnote in the history of the paradox: a true solution, even for just a pleasant morning, a leisurely stroll (Fig. 44.1).

Pinocchio knows well the difference between truth and fiction; comes to learn of loyalty and the ecosystem that made him. His hubris and impulses state clearly the condition of a hybrid evolutionary system. Pinocchio is an mRNA template for reproducing other bubbles, gardens, concepts, throughout the mimetic landscape oriented toward surviving the near-term. Paradoxically, he is a figment of the imagination, of course. But, in any mind, at any time, a tree becomes something other than itself. And that is the origin of both ecodynamism *and* free-will.

There is the man, Enrico Mazzanti, a nineteenth-century engineer who would turn to cartoons to quench an otherwise indecipherable yearning. What gave him to indulge this conception of the most famous puppet in the world? His image of Pinocchio graces the cover of the original aggregated book by one military volunteer,

© Springer Nature Switzerland AG 2021
M. C. Tobias, J. G. Morrison, *On the Nature of Ecological Paradox*,
https://doi.org/10.1007/978-3-030-64526-7_44

**Fig. 44.1** The Cecina River Valley in Tuscany. (Photo © M.C. Tobias)

Carlo Lorenzini (1826–1890), who would take up arms in the Italian movement for unification in the mid-nineteenth century, a War of Independence in 1848 that aspired to see Giuseppe Mazzini and others of the Risorgimento, among them Giuseppe Garibaldi and Victor Emmanuel II, help bring down the existing leadership. Lorenzini's *Il Lampioni*, a daily newspaper he created, meant to aid Italy's political turns of fortune through political satire, was a short-lived endeavor, shut down with the rise of the Grand Duke Leopold who did not take kindly to such criticism.

This politically inclined skeptic, Lorenzini, a hard-nosed journalist, soldier of fortune, was thus waylaid mid-course by the last thing on earth he was expecting: to translate children's stories for a publisher in his home town of Florence. Despite his writing two scarcely received novels (*Un romanzo in vapore* and *Il viaggio per l'Italia di Giannettino*), he had probably given up on the idea of a Lorenzini breaking into that rarified realm of great literature. Or had he decided on a course of delightful heresy? The aftermath of this little saga would suggest that to be the case—but real life does not anticipate such hindsight. In either case, whatever the real motive, Collodi was about to change the course of history.

By 1860, he was using his penname, Collodi, the name of his mother's village. What was he thinking? Did he have the slightest inkling of what was to come?

It was the year 1875 when the small publisher in Florence specializing in children's materials, Felice Paggi, suggested to this linguistically adept would-be captain of the brigades, and former political commentator, that he take Charles Perrault's "Mother Goose" and turn it into proper Italian. For an intellectual who followed every hysterical outbreak of philosophy, each turnabout in the politics of the day, this commission might have come as a profound letdown, even an embarrassment. He needed to work. But "Mother Goose" was no Marxist treatise; nothing like the subtle masterpieces produced by so many of those great names populating his native Tuscany (Niccolò Machiavelli, Giuseppe Giusti, etc.). This man was plagued by a dead-end schoolbook mentality beset by simple fables, an absence of the slightest nuance or controversy, thoroughly inoffensive standard fare. He knew his anonymity was on the verge of a guarantee. For a man so deeply in love with life, and his native landscapes, he had hoped for an epiphany and the fortune of good timing to give it loft.

Looking more deeply, as he obviously did, Charles Perrault's (1628–1703) "Mother Goose" did not lack for complexity, actually. "Mère l'Oye" dates back to 1695, titled, *Histoires ou contes du temps passés, avec des moralités -Contes de ma mère l'Oye*. Collodi would have divined in its sweeping echoes a combining of genres that were ancient and would, in time, lead to morality plays, opera, ballet, film—an entire panoply of human emotions from cradled nursery rhymes to the grave. It was serious. The author of "Little Red Riding Hood," of "Sleeping Beauty," "Puss in Boots," "Bluebeard," and "Donkeyskin," among other stories, had firmly established the dark undersides of childhood. It was not Perrault's subtlety that interested the schoolmarms and librarians of late nineteenth-century Italy. But it clearly interested and helped form Collodi's nuanced ecological epic. The sprawling, rusticated charm and philosophical echoes of Tuscany were, in his mind, the perfect epicenter to engender something we had never before experienced as readers and happy co-conspirators.

Within six years, Collodi had found his true vein as an author and by July 7, 1881, the many memorable creatures of his "Storia di un Burattino" ("Tale of a Puppet") were ready to be launched in a magazine devoted specifically for children, the *Giornale per i bambini*.

## 44.2   The Tuscan Mirrors

There is an anecdote—it may be true, or not—that Collodi was unconvinced about the quality of his tale of a puppet and told the publisher that it was a "childish piece" and only hoped that, were the enterprise to move forward, he be paid sufficiently so as to incite him to continue with the concept, the characters, the story line, notwithstanding his own uncertainty about the outcome. By all appearances the story was not unappreciated and by 1883 Collodi had finished, and Felice Paggi had published the entire novel. Dying suddenly in Florence in 1890, one month shy of the age of 64, Lorenzini would never know that his little puppet would enjoy hundreds of editions, translations into over 260 languages and dialects, whole industries created around him, the entire literary world changed forever.

Considering who jumps up and down and runs circles around the world, those myriad dramatic personae of Collodi's tale, we all may one day—if we haven't already—look back at Pinocchio and recognize ourselves in this Tuscan mirror, gaze directly upon the tragic-comedy of the human saga—the insanity of a biological crisis we have unleashed in the name of ourselves, for the sake of our children, and their children—and know that we can turn tragedy into something altogether different: the joy and sheer amazement that biodiversity and her gorgeous, endlessly inventive cleverness provides the human imagination. We feast upon Pinocchio in the sense that *his own* love of life is never satiated.

Curiously, Collodi questions childhood in an ambiguous manner, suggesting that their indifference, at times, to the health of the earth, their stubborn refusal to obey rules, to step into line, and to be good little boys and girls, may be the salvation, or the destruction we see everywhere around us in the twenty-first century. Indeed, Pinocchio is ready to squash the cricket, and there is plenty of evidence within the tale to indicate ill-boding currents that go far more than merely implicating the lie of a generic non-violence. Collodi suggests, as Orwell would in *Animal Farm*, that power is not merely vested in humanity, but in Nature herself—and Nature is not without her cruelties—though it is humanity that most easily abuses those privileges scattered freely by natural selection.

Pinocchio and his habitat is all that (cursed, uncursed, predictable, unpredictable), in addition to being delightful! Hence, a psychological thriller that prefers the joy of discovery to the dark corridors of what is actually going on. But they are not mutually exclusive, and what Lorenzini did by joining them was, obviously, a masterful way to unlock that which had been kept secret, far off in a dark forest of his ancestral landscapes. It started there, in Tuscany, birthplace not only of the Lorenzinis, but also of Dante and Botticelli, Leonardo da Vinci and Galileo, Petrarch and Vespucci. The great Puccini hailed from this region where the local word, pinolo, or pignolia, referring to the native pine nut, has everything to do with Pinocchio, and the origins of this particular adaptation. Tuscany is to Italy what Burgundy is to France—biologically and artistically. There could be no more perilous, entrancing, and meaningful paradox than that which begins in humanity's childhood, amid an eruption of story-tales that all have something sobering to add.

## 44.3  The Loyalty of Trees

Of the nearly 125 species of pine worldwide (molecular biologists and botanists debate the precise number, of course), about 20 of them produce edible seeds, or nuts. The predominant European pine nut comes from the Stone Pine (*Pinus pinea*), and in Tuscany it has been cultivated for some six millennia. Neanderthals probably harvested it from wild trees, the original members of the Pinaceae family. In the Old Testament, Hosea 14:8, there is a green fir tree mentioned and scholars recognize in it a fruit that was probably the same Stone Pine nut found in many parts of Europe, not just Italy. The tree was worshipped by the Greeks because it was sacred to the

God Neptune. Not only were pine nuts found in the rubble after 79 AD, when Mount Vesuvius finally blew up in everybody's face (after many threats to do so), but it was a common staple for the Roman Legions wherever they roamed. In fact, these pine nuts were probably the preeminent European snack, going back as far as one cares to look. And for nutritional benefits, beauty, and tenacity, the Italian Stone Pine is considered among the most steadfast and loyal of pine trees in the world.

Loyal, you ask? A tree that is loyal?

Si. Pinocchio is born of such a tree.

Chemist, environmentalist, and justifiably proud historian of his native Tuscany Professor Ugo Bardi from the Dipartimento di Chimica at the Università di Firenzi declares, "Plant trees, disband the army, work together: the Tuscan way of surviving collapse."[1] What is he referring to? Jared Diamond's "Collapse"? Collapse in general? Writing his essay in 2006, he was most certainly speaking of the current dependency on foreign oil and the spate of ecological problems that have historically affected this remarkable region of Italy, Tuscany. Tragic earthquakes and the Coronavirus were yet to strike. But Bardi was certainly drawing graphic conclusions from Gibbon's *The History of the Decline and Fall of the Roman Empire*, and from such examples of 22 civilizations described by the late historian Arnold Toynbee, cultures that vanished because they violated their most important compact, namely, a respect for wilderness and ecological carrying capacity. Instead, these cultures, one by one, succumbed to the same gruesome syndrome of self-destruction by which the soils were destroyed, the trees cut down, every last precious mineral extracted, rivers fouled, fisheries, docile birds, anything that moved within striking distance decimated.

For Professor Bardi, Tuscany, birthplace of Pinocchio, remains a fitting and paradoxical metaphor for what continues to happen across the planet. Bardi examined the over-exploitation of the soil by the Roman Empire, later waves of deforestation that filled the River Tiber with silt, consequent famines that erupted, and the first European waves of environmental refugees. With the final implosion of the once vast Roman military hegemony, Tuscany was liberated in the sense that it could breathe freely again, her soils regenerating critical nutrients, and ecosystems returning to some level of stability.

But by the time of the early Italian Renaissance, there was more industrialization, a repeat scenario of centuries before; Pisa's port was filled with silt from all the new erosion that came with the cutting down of second- and third-growth trees. Commerce coincided with the production of charcoal used in the making of firearms; these weapons in turn argued for their use. And used they were: In battle after battle, as adjoining territories bleating with warriors took up arms, power struggles ensued, and life was again imperiled. The history of Europe.

In the 1520s the leadership of the Republic of Florence, cognizant of approaching adversaries, the Spanish Imperial Armies heading directly toward Tuscany, built

---

[1] See "Transition Culture – 11 Dec 2006","Plant trees, disband the army, work together: the Tuscan way of surviving collapse" by Ugo Bardi, at: http://transitionculture.org/2006/12/11/plant-trees-disband-the-army-work-together-the-tuscan-way-of-surviving-collapse-by-ugo-bardi/.

enormous fortresses. Michelangelo was one of many locals who found himself frantically trying to buffer frescoes and church domes by covering them with blankets, or building up whole mattresses in an effort, mostly futile, to protect church art from the invaders.

Conditions became so dreadful that it is rumored that by the late sixteenth century bread not even suitable for hungry dogs was the staple diet in Tuscany, an area once rich in every respect that had become one of the poorest, depauperate regions in all of Europe. Imagine, a cross-section of some nearly 9000 square miles in the heart of Italy, its capital being Florence; the geography that gave birth to such ancient cultures as the Villanovans, the Etruscans, and to the very Renaissance itself, reduced to rubble and squalor.

With little else to sustain the remaining inhabitants, ancient (obviously organic) agricultural techniques were again embraced so as to try to revivify a dying culture. In this Quixotic effort, the revered Tuscan Saint Giovanni Gualberto (995–1073), known for having planted trees his entire adult life, was recalled with urgent fondness. Duke Ferdinando I embraced Gualberto's creed to bring his country back to the living. The Duke's symbol became one of "working bees"—producing life-affirming fruits of the land. The strategy succeeded.

Meanwhile, Jews fleeing from the Spanish Inquisition were welcomed in Tuscany; the death penalty and all forms of torture were abolished, Tuscany becoming the first government body to achieve these humanitarian ends anywhere in the world. And today, as Professor Bardi has noted, Tuscany has more forest than virtually any other region of Italy. This ecological renaissance is based upon a rural love affair, the endless embrace of nature, as recounted in every art form, love sonnet, that evanescent landscape behind "The Mona Lisa" or "Virgin of the Rocks." An impression easily obtained by visiting the little sequestered village in the hills after whom the author of Pinocchio adopted his nom de plume, the village of Collodi where Lorenzini's mother worked as a maid for the richest family in town, the Garzonis. They would, in turn, honor that woman and her son who today Italy knows to be one of the great literary geniuses of all time (Fig. 44.2).

## 44.4  Ecollodiology

The town of Collodi, founded in the late 1100s, is described by many as being in perfect harmony with its surroundings, a true case study in what we might whimsically designate ecollodiology. Atop the town, with steps leading to it, is found the great Church of Saint Bartolomeo. Bartolomeo was the son of a ploughman, one of the Twelve Apostles of Jesus who—according to Saint Jerome—carried a copy of the Gospel all the way to India. At the base of the village are the magnificent Baroque gardens of the Villa Garzoni, along with the wonderful park designed by Ottaviano Diodati in the eighteenth century, said to be the most sublime of all parks in Italy and later dedicated to the memory of Pinocchio. The Garzoni gardens spread out along flirtatious grids adrift in whimsy, mythic statues, follies, caves, aviaries, a

**Fig. 44.2** Pinocchio at Stromboli's Theatre. (Illustration by Attilio Mussino, 1926, *The Adventures of Pinocchio*, by C. Collodi, New York, The Macmillan Company, 1926)

butterfly house with Lepidoptera species from ecosystems all over the world. A place for fantasies that even includes the so-called "fairytale road that leads meandering Pinocchio pilgrims from Collodi to the nearby town of Pescia, over hillsides where wanderers can daydream in their own stride. A pathway like that to the Land of Oz, or Through a Looking Glass, with works of art interspersed between streams, deep forests, a monastery, and mysterious diversions.

And along the very fortress that Michelangelo helped protect is the Porte Sante, a cemetery designed in 1854 that holds the body of Carlo "Collodi" Lorenzini, in addition to other poets, sculptors, painters, filmmakers, and notable individuals of the region.

An Italian national organization devoted to Collodi's genius (The Fondazione Carlo Collodi) has amalgamated an inventory of the astounding proliferation of translations, adaptations, illustrations, imaginative by-products, sequels, and interpretations of Pinocchio from 1883 to the present. It is a formidable legacy encompassing video games and comic books, plays, musicals, toys, memorabilia, animation of every sort, movies, television series, and countless language editions. Pinocchio has breathed life into thousands of subsequent renditions and works of art indebted to this remarkable creation.

From an *Astro Boy* in Japan, Steven Spielberg's film *AI – Artificial Intelligence*, rapper and jazz saxophonist renditions (including a piece by Wayne Shorter recorded on a Miles Davis album), to Mickey Rooney, and much later Roberto Benigni, and Pee Wee Herman, all playing the title role of Pinocchio, to the opera by Jonathan Dove at the Great Theatre in Leeds in 2007, the legend lives on. Danny Kaye and Martin Landau both played Geppetto; actress Julia Louis-Dreyfus once played the Blue Fairy.

There have been Pinocchios in Africa and in outer space, in paintings by contemporary artists like Jim Dine, and in philosophical embraces like those of the rigorous Italian thinker Benedetto Croce (1866–1952) who considered *The Adventures of Pinocchio* a fine example of his "absolute idealism," a notion that ecological aesthetics is core to all philosophy. Countering the grave leveling effects of eco-anxiety, such Garzoni gardens, though all but a few acres, manifest in every visitor a credo, possibly fleeting, or as easily a life-altering revelation. Such is the power of eco-bubbles that may be divined in places, guises, and circumstances least expected. In the face of a divided and divisive human population, with the insistent question of whether the species can rise to the level of the individual, Pinocchio passes along for good measure a proof-positive credo. We don't have to be trapped between total despair and a fully realized Utopia. There might be a regenerative farmer, or fuel cell innovator intoxicated with the idea of practical solutions, ready to champion an entire generation of practical-minded individuals.

Throughout most of his life, the Nobel Peace Prize laureate Norman Borlaug (1914–2009) was one such person—even as he admitted the fact that his "Green Revolution," with its heavy applications of nitrogenous fertilizers and severe impact on aquifers, was only a stop-gap measure that might give the human species a decade or two to stabilize its population explosion, while lending to human geography a respite from immediate famine through the mechanism of higher-yielding, disease-resistant varieties of wheat and other staples. Borlaug's enthusiasm for feeding humanity was undimmed, up until the end. Notwithstanding his acute premonition that our species was in trouble.

Pinocchio anticipated similar contradictions about human nature. In the 1940 Walt Disney Animated Classics version (the second of such films following upon *Snow White and the Seven Dwarfs*), Pinocchio's voice was performed by Dickie Jones, Jiminy Cricket by Cliff Edwards, and Geppetto by Christian Rub. The storyboards for that production remain classics, all formulated as Hitler was wreaking havoc across Europe. But the myriad illustrators in the last nearly 138 years—from Attilio Mussino and Giovanni Manca to Walt Disney's own Pinocchio illustrator, Swedish-born Gustaf Adolf Tenggren (1896–1970)—are as diverse as that fabulous concatenation of illustrations for Pinocchio's cousin, "Don Quixote." In 1937 alone, there were an estimated 80 simultaneous editions being published of Pinocchio, almost all of them illustrated.[2]

---

[2] See JSTOR, "Centenary of a Character: Pinocchio," by Angela M. Jeannet, Italica, Vol. 59, No. 3, Pedagogy (Autumn, 1982), pp. 184–186, Published by: American Association of Teachers of Italian, URL: http://www.jstor.org/stable/478984 © 1982 American Association of Teachers of Italian; See also the "Infography about Pinocchio," http://www.infography.com/content/927376398317.html.

An ecological satire born of the soil of Tuscany, and the current state of the world. Such is Collodi's Pinocchio, who promulgates his own Manifesto which is the reality of his own dignity, his roots, the original landscape where he was born, both as an ideal and as a reality in a world where trees, let alone pine nuts, are disappearing rapidly. As if to say: Be what you are, and if that happens to be a tree, how fortunate for all concerned. Take care of it. Take care of yourself.

## 44.5   Tuscany's Biodiversity

Fortunately for Tuscany, and for the legacy of Collodi, the Tuscan countryside has been much revived since World War II. There are six UNESCO World Heritage sites (including Florence, Siena, and the Cathedral and Tower of Pisa); and over 120 nature and scientific reserves. But with over ten million tourists per year, pressures on Tuscany's biological heritage are not diminishing. Throughout Italy, 547 birds have been identified, many in Tuscany. There are no known endemic birds in Italy, or anywhere in Europe for that matter, but numerous breeding grounds, places of hospice for migratory birds that annually move from the far Northern Boreal forests to the Mediterranean or parts of Africa. Naturalist Damiano Andreini refers to thousands of species found in the Tuscan marshes, including the Trifid Bur Marigold (*Bidens tripartita*), the *Nonnotto Botarus Stellaris* or Bittern, the Ragno d'Acqua or common pond skater (*Gerris lacustris*), as well as many other remarkable creatures, many with peculiar, little known indigenous names, where science and literature were made for each other,[3] including nine native amphibians, some of which are the Marsh, Pool, and Agile frogs.[4]

In a study focusing on a biodiversity revival program called Re.Na.To. (Repertorio Naturalistico Toscano), 88 habitats and 472 plant species were identified within Tuscany as suffering from various degrees of threat and vulnerability.[5] Among Italy's many tree species, the predominant ones—such as the holly or Holm oak, the Aleppo, maritime and umbrella pines, in addition to the more common above referenced Italian Stone pine—are all in need of the same Tuscan protective umbrella

---

[3] "Wetlands: The Flora and the Fauna of the Marshlands of Tuscany," by Damiano Andreini, at: http://www.slowtuscany.it/tuscany/pistoia-prato/Wetlands.htm.

[4] See "Amphibians Of Italy: A Revised Checklist," by Massimo Capula, Arianna Ceccarelli and Luca Liuselli. Write the authors, "The Italian amphibian fauna with 17 Urodela and 23 Anura ascribed to 10 families… is one of the richest in Europe." Research Gate, https://www.research-gate.net/publication/289523559_AMPHIBIANS_OF_ITALY_A_REVISED_CHECKLIST, Accessed July 16, 2020.

[5] See "The Protection Of Biodiversity in Tuscany," by Paolo Emilio Tomei and Andrea Bertacchi, Department of Agronomy and Agroecosystem Management, Università degli Studi di Pisa, in the book, Nature Conservation: Concepts and Practice, ed. By Dr. Dan Gafta and Dr. John Akeroyd, Spinger Berlin Heidelberg, 2006.

that was first manifested by Saint Giovanni Gualberto,[6] a passion for nature that culminated, of course, in the works of Fra Angelico, Leonardo, Raphael, the later Italian landscapists, not to mention the remarkable life of Saint Francis.

There is much work to be done to maintain protection of this precious historical, ecological, and artistic center of the world. Pinocchio is one of those charter members, a veritable signatory to a Declaration of Biological Rights, not only of the plant community in his native land, but of humanity's future place in a world at risk, where common sense, imagination, an unstinting love of life, not to mention a sense of humor, are worth their weight—not in gold but in old forest, healthy soils, rich biomass, a dazzling, unchallenged array of plant and animal life, clean air, and plenty of fresh water to drink. These are, after all, the real concerns of any pine nut, let alone the Pinocchio kind, a mythic ecological doppelgänger rendered in wood, who speaks to the world of his, and of our, condition. We know if he's lying: he is not. So, the question persists: After Pinocchio, everything we do matters[7] (Fig. 44.3).

**Fig. 44.3** Cherubs, and a Leaning Tower as Truant as Pinocchio, Pisa, Tuscany Commune. (Photo © M.C. Tobias)

[6] See "Plants of Italy: Discover the Mediterranean Trees and Herbs Found in Italy," by Sharon Falsetto, March 20, 2009, plant- ecology.suite101.com/article.cfm/plants_of_italy.

[7] See *The Misadventures of Pinocchio*, by M. C. Tobias, Zorba Press, Ithaca, New York, 2012.

# Chapter 45
# The Poetics of Biodiversity: Kazantzakis and Crete

## 45.1 Timeless Cretan Lyricism

Two years before his death, Nikos Kazantzakis (1883–1957), in a radio interview in Paris, declared, "The writer of today, if he is truly alive, is someone who suffers and worries at the sight of reality."[1] His homeland of Crete had, for at least 4000 years, been at the crossroads of both rusticated simplicity and compelling cosmopolitan disruption. The Bronze Age Minoan (Aegean) Civilization, which prospered across Crete from roughly 2700 to 1100 BC, invested in three-story palaces, two-story villas, often with refined plumbing and indoor toilets, complex polyculture and complex cuisine, a robust trading network from Egypt to Cyprus to the Near East, and a rich zoomorphic artistic culture of murals (celebrated dolphins and bulls), jewelry, and clothing. There were at least two, now untranslatable, scripts: Linear A (precursor to Linear B and the first Greek) and the quasi-inscrutable "Phaistos Disc," an enigmatic undamaged piece of fired clay, less than six inches in diameter—a microcosm of the planet—discovered within the walls of one of Crete's four principal palaces in 1908. Knossos' ruins had been discovered just 15 years before Kazantzakis was born. And with the fanfare surrounding the massive digs of English archaeologist Sir Arthur Evans came the worldwide news that the ancient Cretan epicenter had been inhabited for nearly 9000 years and bespoke of blissful epiphanies (Fig. 45.1).

With the unearthing of the celebrated Disc, everyone who has ever looked at its 241 "tokens" (in the Heraklion Archaeological Museum) feels they have nearly got its translation half-correct. A veritable poetics of biodiversity that spells out the

---

[1] See the essay "Kazantzakis, Crete, and Biodiversity," by M. C. Tobias, in *The Terrestrial Gospel of Nikos Kazantzakis*, Translated and Edited by Thanasis Maskaleris, Bilingual Edition in English and Greek, Zorba Press, Ithaca, New York, 2016, p. 71. Cited from: "Excerpt from Pierre Sipriot's Interview with Nikos Kazantzakis," French Radio (Paris), 6th May 1955, from the Nikos Kazantzakis Files, Nikos Kazantzakis Museum, www.historial-museum.gr/Kazantzakis/en/index.html.

© Springer Nature Switzerland AG 2021       407
M. C. Tobias, J. G. Morrison, *On the Nature of Ecological Paradox*,
https://doi.org/10.1007/978-3-030-64526-7_45

**Fig. 45.1** Nikos Kazantzakis. (Photo Courtesy of Niki Stavrou and Nikos Kazantzakis Publications, Athens, Greece)

universe, but without the least certainty or resolvable context, the equivalent to an entire manuscript of Roman dodecahedron-like puzzles. A language of nature both lost and perpetually tantalizing, encompassing perhaps the very meaning, and meaninglessness, of human destiny.

Not that it should matter. The best muses are those ecological secrets before which we ourselves remain both unwitting partners and strangers. Or, translated differently, "we are ourselves the best muses, secrets, strangers...." However it is elucidated, and notwithstanding a history of despair—the Black Death of 1347/1348 killed probably half the island's human population; the Greco-Turkish War in 1897 left searing memories that last to this day; those German occupying forces during World War II bombarded Crete for over 800 hours continuously, perhaps the bloodiest battles per capita and per size of territory in all of the war.[2] And post-World War II, there have been excited hordes of flocking tourists, some explicitly to return home with Cretan contraband: rare floral specimens. Fortunately, there are still 223 endemic remaining Cretan plants[3] across the gorgeous, biologically dazzling island.

## 45.2 The Artistic Divining Rods

For literary pilgrims, Crete is dominated spiritually and ethically, it seems, by Nikos Kazantzakis, who lies controversially buried in the Martinengo Tower outside the capital of Heraklion's walls. High above the city, his spirit looks out over the Ida mountains to the west, the Asterousia Mountains bearing south, and the rich farm valleys and wine-dark waters, as Homer called them, of the Mediterranean to the north and east. He is up in the air, liberated, protecting his homeland with his famed "Cretan Glance." On his tomb is famously written, "I hope for nothing. I fear nothing. I am free."[4]

At a time in the heart of the Anthropocene, no other artist has so encapsulated, by his death, words to distinguish Cretan biodiversity with such pellucid, embracing verisimilitude. Of course, Kazantzakis had profound, local predecessors. Vincenzo Kornaros, a contemporary of Shakespeare, who authored the Cretan epic romance, the *Erotokritos*, written in the Cretan/Sitian dialect, first published in Venice in 1713; and Kornaros' contemporary, the indomitable painter, Domenikos Theotokopoulos, or El Greco (1541–1614), who grew up in a remote mountainous gorge along the eastern portion of Crete and was, in turn, greatly influenced by the Cretan renaissance of icon painting, most indelibly, one of Crete's foremost icon visionaries, Angelos Akotantos (1425–1450), a gentle sapience enshrined and self-illuminated, not unlike Russia's Andrei Rublev (c.1360–1430).

---

[2] By at least one estimate, 6,593 men, 1,113 women, and 869 children were killed on Crete by the Nazis, who parachuted on to the island beginning the morning of May 20, 1941. See Callum MacDonald, *The Lost Battle – Crete 1941*, Free Press, New York, 1995, p. 303.

[3] See Menteli, V., Krigas, N., Avramakis, M. *et al.* "Endemic plants of Crete in electronic trade and wildlife tourism: current patterns and implications for conservation," J of Biol Res-Thessaloniki 26, 10 (2019). https://doi.org/10.1186/s40709-019-0104-z, Published 30 October 2019, DOI: https://doi.org/10.1186/s40709-019-0104-z, https://jbiolres.biomedcentral.com/articles/10.1186/s40709-019-0104-z, Accessed April 26, 2020.

[4] See "Kazantzakis," a Documentary by M. C. Tobias, KRMA/PBS, 1984, https://www.imdb.com/title/tt0816548/ in which Kimon Friar recites the three Greek sentences on Kazantzakis' gravestone; See also, Chapter 9, "Fantastic Journeys," in *A Vision of Nature – Traces of the Original World*, by M. C. Tobias, Kent State University Press, Kent, Ohio, 1995, pp. 252–261.

## 45.3   A Biological Hotspot Within a Hotspot

With its population of nearly 650,000 spread across 3219 mi$^2$ (nearly half its residents living within Heraklion), a biological hotspot of high endemism *within* the Mediterranean hotspot, Cretan wildlife is distinguished by such rarities as the *Capra aegagrus cretica,* the wild Cretan mountain goat, rare Bearded Vulture (*Gypaetus barbatus*), and Eleonora's Falcon (*Falco eleonorae*). The rocky gorges and three windswept mountain ranges, exceeding 8000 feet, are adrift in unique lichen and grass species. And moving quietly at dawn and dusk throughout these ecotones spanning the air of two continents, where once the mighty Minotaur roamed, are two endemic mammals, the Cretan Spiny Mouse (*Acomys minous*) and the Cretan Shrew (*Crocidura zimmermanni*).[5] Crete's Wild Cat (Fourogatos in Greek, *Felis Silvestris Agrius* since the time of Linnaeus) is considered one of the rarest mammals in the world. All these, along with the gentlest of beings, the "Critically Endangered" Mediterranean Monk Seal (*Monachus monachus*), which has managed to find one of its final refuges just south of Crete on some rocky islets.[6]

The wild, iconic goat (known as Kri-Kri, or agrimi—wild creature—by the Cretans) was saved from extinction by local conservation efforts that have resuscitated its populations up to some 2500 individuals. Despite some 523 breeds of goats worldwide, in addition to nine principal species and sub-species, the Cretan Capricorn has come to symbolize Crete and her human occupants: a perfectionism smitten by mood swings. The Kri-Kri is a pragmatic, honest, and stubborn luminary, a true mountaineer whose island home continues, however paradoxically, to provide the perfect habitat of harmony, challenge, and balance at the center of Europe's most ancient and advanced of civilizations. Like the Toda and their buffalo, Cretan hagiography would not be complete without the converging biotic spirituality of both the mythic minotaur and precious goats.

In addition to the endemic Kri-Kri, there is a lovely Cretan cuckoo and song thrush, a chaffinch and sparrow hawk, countless bat species, lizards, snakes, and scores of butterflies. At just one of the newly created sanctuaries in Crete, Akrotiri, a 75-acre Park for the Preservation of Flora and Fauna of the Technical University of Crete, inaugurated in 2004 under the inspired directorship of Yannis Phillis, there are at least 18 mammalian species – from wild cats and European badgers, to Kuhl's pipistrelle and the Greater Horseshoe Bat; a dozen reptile species including the Ocellated Skink, Leopard Snake, and Balkan Green Lizard, to the Turkish and Mauritanian Geckos; three amphibia—toads, tree frogs, and the Cretan Water Frog—to nearly 170 bird species including such beauties and rarities as Temminck's

---

[5] See the Natural History Museum of Crete, www.nhme.uoc.gr/links-3.en/index.html.

[6] See "Mediterranean Monk Seal – Monachus monachus," European Mammal Assessment team. 2007. *Monachus monachus.* The IUCN Red List of Threatened Sprecies 2007: e. T13653A4305994. https://www.iucnredlist.org/species/13653/4305994, Accessed 17 July 2020.

Stint, the Collared Partincole, Chukars, Moorhens, Imperial, Booted and Bonelli's Eagle, the Greater Flamingo, and the Whooper Swan. A paradise[7] (Fig. 45.2)

**Fig. 45.2** Nikos and Eleni Kazantzakis. (Photo Courtesy of Niki Stavrou and Nikos Kazantzakis Publications, Athens, Greece)

---

[7] See "'The Park for the Preservation of Flora and Fauna' welcomes you..." www.park.tuc.gr/park_idea_UK.php; see also, Yale Peabody Museum – Treasures & Exploration: The Biodiversity of Crete, www.peabody.yale.edu/explore/cretebiodiv.html.

## 45.4   A Clod of Cretan Soil in His Palm

In his book *The Terrestrial Gospel of Nikos Kazantzakis* (2016)[8] the late poet and comparative literature scholar Thanasis Maskaleris presented various selections from the enormous oeuvre of Kazantzakis[9] to dramatically evince the brilliant nature poet who was the author of *Zorba the Greek*, *The Last Temptation of Christ*, *St. Francis*, *The Odyssey – A Modern Sequel*, and his autobiography, *Report To Greco*, among more than 50 of his other works and translations. From *Zorba the Greek*, Maskaleris particularly relished the following lines: "The universe for Zorba, as for the early humans on Earth, was a splendidly rich vision; the stars touched him, the sea-waves crashed on his temples. He lived the soil, water, the animals and God, without the distorting interference of reason."[10]

Like his friend and admirer Albert Schweitzer, Kazantzakis' passion for a transcendental activism encompassed animal rights at all levels. While Schweitzer—an animal liberationist who would go to any length to save an ant—would live and work for three decades at his hospital in Lambaréné (now Gabon), Kazantzakis, the quintessential observer, was traveling the world, politically, spiritually, and culturally at home everywhere and nowhere—a heretic by most standards of his time. From a monastic retreat upon Mount Athos—symptomatic of the ascetic allure of monasteries throughout northern Greece, including Meteora—to Geneva, where he was at one time a multicultural steward for UNESCO, to his island home, with his wife Eleni, on Aegina, where the Nazis tried to starve them out, to his final sojourn in China (he died in Freiburg during his return to Europe), the poet, novelist, essayist, travel writer, philosopher, playwright, journalist, and translator[11] would nine times be nominated for the Nobel Prize. The year he died, Camus received that prize (by one vote), but would make clear that it was for Kazantzakis.

There are precious pictures of Kazantzakis and Schweitzer enjoying a forest picnic together somewhere in Europe, long ago. Both men seem to be grinning with a shared, heartbreaking anticipatory weariness. They know something. Neither were nihilists, but they understood everything. An image can convey that. They believed in the ability of our species to act with compassion, intellect, and heart. Schweitzer was convinced that *Homo sapiens* had what it takes to make things right. Kazantzakis, for his part, was not overly convinced of that. Yet, at the end of his long trek throughout the monasteries of Mount Athos during December, 1914, still a young man and

---

[8] *Will the Humas Be Saviors of the Earth?* Bilingual Edition In English and Greek, Translated And Edited by Thanasis Maskaleris, Zorba Press, Ithaca, New York, 2016.

[9] See www.kazantzakispublications.org.

[10] Maskaleris, ibid., p. 53.

[11] A legacy brilliantly preserved by Patroclos Stavrou, and his daughter and assiduous Kazantzakis scholar, Niki Stavrou through their devoted Kazantzakis Publications Ltd. in Athens.

**Fig. 45.3** Nikos Kazantzakis and Albert Schweitzer. (Photo Courtesy of Niki Stavrou and Nikos Kazantzakis Publications, Athens, Greece)

just 5 months after Europe had been convulsed by the first massacres of World War I, he could confidently write, "I said to the almond tree; 'Sister, speak to me of God.' And the almond tree blossomed"[12] (Fig. 45.3).

---

[12] From *Report To Greco*, by Nikos Kazantzakis, Translated from the Greek by Peter A. Bien, Simon And Schuster, New York, New York, 1965, p. 234. See also, https://www.youtube.com/watch?v=xqANLC5-7Zk.

# Chapter 46
# Famine in Bangladesh

## 46.1 Combinatorial Cruelty in Nature

Every act of production and consumption begs several perennial questions revolving around pain and suffering. These unknown qualia and their magnitudes, in turn, encompass all of biochemistry, trophic (food chain-level) relations and separations, reproduction and barriers to such, social order and entropy, consciousness and the conscience. The silent reveries from atom to molecule, from cell to ethical dispositions, have revealed, by example, a lesson of relative durations, a mere few hundred million years of (human) time. That is, according to the most open-ended characterization of cosmological coherence and its pathway (at least one of them) toward nerve endings, axons, sensations, and their implicative orientations, whether in the guise of a soul, a body, a mentation, or some other construct altogether separate but aware. The history of philosophy has idly speculated on these forms, motions, and laws—the perturbations and combinatorial epiphanies we all inhabit.

At the evolutionary level of life-forms and their commingling, we almost immediately come upon birth and death (maternal, child, and infant co-mortalities) and, with these predicates, a host of sporadic reflections that remain unanswered in the annals, certainly, of the human brain. Issues pertaining to death: the cruelest way to die; animal species starving to death, no euthanasia to put themselves out of their misery; the ecology of starvation—what point does it serve? What mission has famine served? And hence the inevitable conjectures congealing around every consideration of ethics, of a God, or no God; of the value of biology and its multiple disciplines. If "it" all evolves toward general destruction and suffering, then has it been worth it? Biodiversity, contemplation, artistic expression, some manner of joy? Or, is the debilitating epic to stave off pain merely the self-directed odyssey predicated by a terrible guidebook that should never have come into existence? How many have proffered the terrible conjecture: If all this cumulative experience and wisdom is simply a prelude to destruction, what is the point?

© Springer Nature Switzerland AG 2021
M. C. Tobias, J. G. Morrison, *On the Nature of Ecological Paradox*,
https://doi.org/10.1007/978-3-030-64526-7_46

There is some subjective value, if fewer heart beats, in tracking a variety of ecological correlations within these realms, as they apply to our own perceived lives and any virtue that might theoretically, paradoxically derive. One (not entirely random) event with which to reckon (noting that one of the authors—mt —was there, on the ground, during the worst of it) was the upending famine that swept across Bangladesh throughout much of 1974/1975.

## 46.2   The Tragic Case of Bangladesh

For the 9 months of the Bangladeshi War of Independence from Pakistan throughout 1971, the preconditions for food scarcity and political chaos had been concretized. Following the return to power of Prime Minister Sheikh Mujibur Rahman, the coming years would see him grappling with one of the most densely populated, impoverished nations in the world, ravaged by war, over six million families left homeless, farmers devastated, the nation's infrastructure nullified.[1]

A sustained drought, followed by floods across 75% of the nation, destroyed all hope of any staples production: no "rice, wheat, maize, potato, pulses, and [or] oil seeds."[2] The resulting famine, which struck in the Spring of 1974 and lasted throughout 1975, elicited little sympathy or support from other nations. The United States opted not to help the country with food aid because Bangladesh was desperately seeking foreign revenue by exporting what jute it had to Cuba, a political non-starter for the rabidly anti-Communist American Congress,[3] even though this majority sentiment blatantly contradicted the then long-standing principle, never let passing political policies undermine the basic humanity of food aid, a principle dating to the early nineteenth century in the United States.

Almost from the beginning of his reign (popular culture is vastly conflicted over his legacy) Rahman had invoked inordinate presidential powers, creating his own private paramilitary force, the Jatiya Rakkhi Bahini, allegedly to counter pro-Marxist left-wing revolutionaries. In fact, history knows that the soldiers and police involved were guilty of an enormous swathe of murders, rapes, and human rights abuses. I (mt) was on the ground in Dhaka and saw it with my own eyes. Eventual estimates of total mortality exceeded 1.5 million people. There were ten million refugees from the war with Pakistan, either still grappling with homelessness in their own country of Bangladesh, or having fled to West Bengal in India, most to Calcutta. There was little food to go around. I spent many nights with those refugees under the Howrah Bridge, spanning the Hooghly/Bhagirathi River, before and after my time in famine-deluged Bangladesh. They were, obviously, the lucky ones.

---

[1] See Time Magazine, "Bangladesh: Mujib's Road from Prison to Power," Dan Coggin, January 17, 1972.

[2] See "Global Yield Gap Atlas," http://www.yieldgap.org/Bangladesh, Accessed April 28, 2020.

[3] See Devinder Sharma, "Famine as commerce," August 2002, http://www.indiatogether.org/agriculture/opinions/dsharma/faminecommerce.htm.

They'd gotten out. India was trying desperately to help these many refugees. But India also had her own problems.

By the time Rahman and numerous members of his family were assassinated on the early morning of August 15, 1975, in an army coup, the population of Bangladesh was over 70 million, with a Total Fertility Rate of 6.82.[4]

I was one of the very last westerners in Bangladesh during the height of the famine and remember skirting corpses being eaten by dogs on city streets. At a refugee camp in Tongi, Dhaka's northern border, I helped one of the few remaining oversees charitable organizations handing out packets of milk powder and biscuits to tens of thousands of mostly women and children who had been rounded up. There were rotting bodies everywhere.

One sees in those too weak to stand in line to collect food supplements, but no food per se, the very definition of the final stages of starvation in humans and other vertebrates: "Apathy, withdrawal, listlessness, increased susceptibility to disease… flaky skin, changes in hair color and massive edema in the lower limbs and abdomen, causing the person's abdomen to seem bloated. During the process of starvation, the ability of the human body to consume volumes of food also decreases."[5]

Add up all the estimates of human mortality from the documented 250 famines, between 2200 BCE and 2017, and the numbers exceed 300 million dead. This figure does not come close to capturing co-morbidity data. The largest coefficients, both in magnitude of mortality and in the sheer number of famine events, rapidly escalate from the time of the great Chinese famine of 1849/1850, when some 20 million people perished, and the world's population exceeded 1.2 billion.[6]

## 46.3   A Continuing Crisis

Alex de Waal, author of *Mass Starvation*, is somehow a guarded optimist about the currents moving forward, pointing out that approximately 100 million people have died of famine since 1850, but, since 1980, famine has all but been eliminated, despite near-famine conditions persisting across war-torn areas from Yemen and Syria to the Sudan (as of 2020). Nearly 70% of Yemen's population remains at severe risk of malnutrition, 3.3 million children under five, and lactating mothers

[4] The World Bank, "Fertility Rate, total (births per woman) – Bangladesh," https://data.worldbank.org/indicator/SP.DYN.TFRT.IN?locations=BD, Accessed April 27, 2020.

[5] "Disabled World," by Thomas C. Weiss, https://www.disabled-world.com/fitness/starving.php, Updated October 29, 2018.

[6] See "Fearfull Famines of the Past," by Eduardo Ferreyra, http://www.mitosyfraudes.org/Polit/Famines.html; Cornelius Walford, "On the Famines of the World, Past and Present" (Journal of the Statistical Society, 1878–1879); *Mao's Great Famine: The History of China's Most Devastating Catastrophe, 1958–1962*, by Frank Dikötter, Bloomsbury USA, 2010; *The Great Irish Famine: A History in Documents*, From the Broadview Source Series, by Karen Sonnelitter, Petersborough, Canada, 2018.

eking out grueling daily minute-by-minute routines in the shadows of sugar and water on the lips of their waning offspring, to the beat of their own looming starvation. Yet, clear progress in combating famines—as a historic trend—has been made.[7]

Others focusing on future food security have long been caught in the debate between GMOs and fast-growing species logics, food availability for nearly eight billion people, and the systemic debacles curtailing efficient and egalitarian food distribution.[8] The points of view are obviously exacerbated by five dominant factors: a continuing human demographic Mount Everest, availability of water for agriculture, human consumption of other animals, climate change, and civil wars aggravating all of the above. In addition, researchers have long examined volatility in food prices, protective management of ecosystem services, and the realization that "an ecological perspective is important for increasing synergies among different land uses and objectives such that conservation, food production, and livelihood goals can simultaneously be achieved across rural landscapes."[9]

According to the last two U.N. Food and Agriculture Organization (UNFAO) reports, over 800 million people currently suffer from hunger. The 17 Sustainable Development Goals of the UN General Assembly of 2015, *UN Resolution 70/1*, aspire to see the gaps reconciled by 2030. They list as their top three agendas: No Poverty, Zero Hunger, and Good Health/Well-Being. The "hunger" issue is addressed through the proposed mechanisms of achieving "food security and improved nutrition" and the promotion of "sustainable agriculture."[10] These are generic well-wishes for one species that has broken down the muddles into such generic categories as the "Global Hunger Index," the components of investment in children, better agronomy and high-nutrition crops at local and regional scale (and hence, a "global nutrition agenda"), supply chain issues key to efficient distribution of food, and how best to augment crop and soil resiliency against increasing desertification and fast warming trends.[11] From the vantage of those most afflicted, such grand phrases and government aspirations are vacuous and wearisome.

---

[7] See *Mass Starvation: The History and Future of Famine*, by Alex de Waal, Amazon.com Services, 2012, https://www.wiley.com/en-us/Mass+Starvation%3A+The+History+and+Future+of+Famine-p-9781509524662.

[8] See the website, "Political Ecology of the World Food System," https://courses.washington.edu/ps385/category/hu.

[9] See, "Ecology and Hunger," by Sean Smukler, Jeff Milder, Roselin Reimans and Fabrice Declerck, 94th Ecological Society of America Annual Convention, 2009, https://www.researchgate.net/publication/267284895_Ecology_and_hunger; See also, "The Food-Insecurity Obesity Paradox: A Resource Scarcity Hypothesis," by Emily J. Dhurandhar, Ph.D., Physiol Behav. Author manuscript; available in PMC 2017 Aug 1. Physiol Behav. 2016 Aug 1; 162: 88–92., Published online 2016 Apr 26. doi: https://doi.org/10.1016/j.physbeh.2016.04.025, PMCID: PMC5394740, NIHMSID: NIHMS845194. PMID: 27126969, NCBI, PMC, https://www.ncbi.nlm.nih.gov/pmc/articles/PMC5394740/. Accessed April 27, 2020.

[10] See United Nations, 2018. *The Sustainable Development Goals Report 2018*, Foreword by António Guterres, Secretary-General, United Nations, https://unstats.un.org/sdgs/files/report/2018/TheSustainableDevelopmentGoalsReport2018-EN.pdf, Accessed April 27, 2020.

[11] See *The State Of Food Security And Nutrition In The World, 2018*, http://www.fao.org/3/I9553EN/i9553en.pdf; See also, Global Hunger Index. Global Hunger Index and the Paradox of

The cascade of complexities ultimately comes down to suffering, and the end-losers in this vicious cycle of ecological dysfunction. Ecology and geopolitics is rife with mismatches between our species and all others, and how that comes to ultimately backfire upon us. Growing rural-to-urban migrations, loss of agricultural expertise on the ground in the countryside, absentee landlords (zamindars, in India), small untenable plots, the profoundly counter-biological monoculture versus a far more sustainable polyculture, the short-term hegemonies of multinational intellectual property right invasions, the chemical fallacies brought into the limelight by the Green Revolution, and the use of hunger and civil unrest to further political interests (as in the case of Bangladesh, 1974): all add to the inevitability of misery.

The fact that there is suffering on account of food, and innumerable instances where farmers destroy their product during emergencies because of the breakdown in distribution options, or the collapse of market prices, is more than merely tragic irony. It is one of many paradoxes afflicting the fundamental (ethically proposed) right of all organisms of every species to be able not to starve to death. Alas, from the earliest recorded famines by the historians Livy, and Josephus, human suffering has mirrored the equally vast torment endured by other domestic, farm, and wild species (Fig. 46.1).

---

Hunger. Human Rights & the Global Economy eJournal. Social Science Research Network (SSRN). K. von Grebmer, J. Bernstein, L. Hammond, F. Patterson, A. Sonntag, L. Klaus, J. Fahlbusch, O. Towey, C. Foley, S. Gitter, K. Ekstrom, and H. Fritschel. 2018. 2018 Global Hunger Index: Forced Migration and Hunger. Bonn and Dublin: Welthungerhilfe and Concern Worldwide. See also, Patrick Webb, Gunhild Anker Stordalen, Sudhvir Singh, Ramani Wijesinha-Bettoni, Prakash Shetty, Anna Lartey (2018). "Hunger and malnutrition in the 21st century". The BMJ. 350: k2238. doi:https://doi.org/10.1136/bmj.k2238. PMC 5996965. PMID 29898884.See also, Joanna Rea (25 May 2012). "2012 G8 summit – private sector to the rescue of the world's poorest?". The Guardian. See also, "Africa: Children's Investment Fund Foundation (CIFF) Leads Transformation of Global Nutrition Agenda with $787 million Investment". AllAfrica. 8 June 2013. See, K. von Grebmer, J. Bernstein, A. de Waal, N. Prasai, S. Yin, Y. Yohannes: 2015 Global Hunger Index - Armed Conflict and the Challenge of Hunger. Bonn, Washington D. C., Dublin: Welthungerhilfe, IFPRI, and Concern Worldwide. October 2015. See also, Lucy Lamble (15 October 2019). "Higher temperatures driving 'alarming' levels of hunger – report". The Guardian. See "Organizing for action" (PDF). FAO, 2011. See also, "How Blockchain Can Be Used to Address Food Security in India," by Javaid Sofi, December 11, 2018, MAHB. See "India FoodBanking Network," A Food Security Foundation India Initiative, https://www.indiafoodbanking.org/hunger; See India's Food Security Problem," by William Thomson, April 2, 2012, The Diplomat, https://thediplomat.com/2012/04/indias-food-security-problem/Accessed April 27, 2020. See "How Blockchain Will Transform the Supply Chain And Logistics Industry," by Bernard Marr, Forbes, March 23, 2018. See "What is Blockchain?" Bernard Marr & Co., https://www.bernardmarr.com/default.asp?contentID=1389, Accessed April 22, 2020.

**Fig. 46.1** "Inmates of a relief camp (during the famine 1876–1878) in Madras, Tamil Nadu, South India. (Photo by Willoughby Wallace Hooper, Public Domain)

## 46.4   Among the Many Tiers of Tragedy

The International Human Suffering Index differentiates between four basic levels, from minimal to extreme; with each of 10 specific criteria ranked between 1 and 10. The key, in terms of food, is called "social welfare" and it is characterized formally as being "the sum of 10 measures: life expectancy, daily caloric intake, clean drinking water, infant immunization, secondary school enrollment, gross national product per capita, the rate of inflation, communication technology (i.e., telephones), political freedom, and civil rights."[12]

The problem with these formal categories is that they are, ultimately, generic, divorced from real persons. Ranking anything having to do with pain and suffering must fail to impress its target audience, just as such nouns as faith, love, compassion, and hunger are largely indefinable. Ronald E. Anderson, Professor Emeritus of Sociology at the University of Minnesota, has long championed the subjective gradations that translate into different kinds of pain. His writing on world suffering in its many emotional and psychological forms has helped to redefine the complexity of the very narrative; of discussing pain when, in truth, we are so hard pressed to

---

[12] See "Human Suffering." Philippine Legislators' Committee on Population and Development Foundation, Dec.2, 1992, (11):1–4, PubMed.gov, https://www.ncbi.nlm.nih.gov/pubmed/12179239, Accessed pril 27, 2020. See also the Human Suffering Index in the Dictionary of Geography, Oxford Reference, https://www.oxfordreference.com/view/10.1093/oi/authority.20110803095950827.

understand it in any other Being.[13] Anderson's seminal study on comparative Quality of Life Indicators, takes into account the "negative aspects of the underlying constructs."[14]

Hence, the contradictions that issue from, for example, the 2020 United Nations Development Programme (UNDP) reports on declining world poverty. The notion that the world has vastly diminished "extreme poverty" (less than "$1.90 income per day"), to 700 million people, at first must invite some sense of imagined relief; a vision, not too distant, where all will be well. Writes UNDP Administrator, Achim Steiner, "There has been a massive drop in global extreme poverty rates – from 36 per cent in 1990 to 8.6 per cent in 2018 – vastly increasing the economic and social opportunities for so many across the world."[15]

Yet, as Anderson's global studies reveal, "levels of stress and pain, as well as negative emotions, have actually gone up in the past 12 years at the rate of about 1-percent per year." When we speak of "global" statistics, the cybernetic feedback loops of the Anthropocene (particularly acute when speaking of human agriculture) necessarily augment the discussion points to encompass every quadrant of the biological world. Moreover, by expanding that framework, we are also committed to an overarching depicture of pain among all taxa.

Measurements of pain have never been precise; nor has the concept itself. The International Association for the Study of Pain, founded in 1973,[16] defines pain as "an unpleasant sensory and emotional experience associated with actual or potential tissue damage, or described in terms of such damage." This dates to the work of Harold Merskey in 1964.[17] It is about as explanatory to the layperson in terms of defining pain, as Hippocrates (c.460–370 BC) characterizing it as an "imbalance in vital fluids" or words to that effect. In other words, about as precise as saying that pain and unpleasantness are the same.[18]

---

[13] See "Global Rise in Negative Emotions Including Pain," Ron Anderson, February 16, 2020, World Suffering & Compassionate Relief of Suffering" website, http://worldsuffering.org/global-suffering-on-the-rise/.

[14] "A Technical Introduction to World Suffering," by Ronald E. Anderson, University of Minnesota, January 2015, http://worldsuffering.org/wp-content/uploads/A-Technical-Introduction-to-World-Suffering-1jan15.pdf, Accessed April 28, 2020.

[15] See "International Day for the Eradication of Poverty," by Achim Steiner, October 17, 2019, UNDP, https://www.undp.org/content/undp/en/home/news-centre/speeches/2019/international-day-for-the-eradication-of-poverty-.html, Accessed April 28, 2020.

[16] www.iasp-pain.org.

[17] Merskey, H. (1964), An Investigation of Pain in Psychological Illness, DM Thesis, Oxford.

[18] See Turk DC, Melzack R. Handbook of Pain Assessment. Guilford Press; New York: 2001; See also, Farrar JT, Young JP, LaMoreaux L, et al. Clinical importance of changes in chronic pain intensity measured on an 11-point numerical pain rating scale. Pain. 2001;94:149–158. [PubMed]; and Farrar JT, Berlin JA, Strom BL. Clinically important changes in acute pain outcome measures: a validation study. J Pain Symptom Manage. 2003;25:406–411. [PubMed]; See also, Curr Pain Headache Rep., Curr Pain Headache Rep. 2009 Feb; 13(1): 39–43. doi: https://doi.org/10.1007/s11916-009-0009-x, PMCID: PMC2891384, NIHMSID: NIHMS209740, PMID: 19126370, "Pain Outcomes: A Brief Review of Instruments and Techniques," Jarred Younger, PhD, Rebecca McCue, BA, and Sean Mackey, MD, PhD, HHS Public Access, https://www.ncbi.nlm.nih.gov/pmc/articles/PMC2891384/. Accessed April 27, 2020

More recently, scientists have "developed an electroencephalography-based test to objectively measure pain… [in which] brain activity is measured in the form of oscillations or 'waves' of a certain frequency…that correlates with pain in animals…the 'theta band' …[such that] computational analysis of theta brain waves to determine their power can be used to objectively measure pain in rodents and humans in a non-invasive manner…"[19] Functional magnetic resonance imaging has also been utilized to devise yet other pain callibrations.[20]

Far more elegant and realistic assessments of pain can be ascertained in such recent reflections as those of Noelia Bueno-Gómez of the Department of Philosophy and Research Center Medical Humanities at the University of Innsbruck in Austria. Like Anderson, she has applied a highly empathetic, sophisticated approach to respectfully delineating the very conceptualization of suffering and of pain.[21]

All vertebrates are thought to possess endogenous opioid neuromodulators to deflect pain to some extent. But there is more than ample empirical evidence to prove the same modulation exists in invertebrates, from fruit flies to lobsters.[22]

---

[19] Suguru Koyama, Brian W. LeBlanc, Kelsey A. Smith, Catherine Roach, Joshua Levitt, Muhammad M. Edhi, Mai Michishita, Takayuki Komatsu, Okishi Mashita, Aki Tanikawa, Satoru Yoshikawa, Carl Y. Saab. An Electroencephalography Bioassay for Preclinical Testing of Analgesic Efficacy. Scientific Reports, 2018; 8 (1) DOI: https://doi.org/10.1038/s41598-018-34594-2.

[20] See, "An fMRI-Based Neurologic Signature of Physical Pain," Tor D. Wager, Ph.D., Lauren Y. Atlas, Ph.D., Martin A. Lindquist, Ph.D., Mathieu Roy, Ph.D., Choong-Wan Woo, A., and Ethan Kross, Ph.D., April 11, 2013, N Engl J Med 2013; 368:1388-1397, DOI: https://doi.org/10.1056/NEJMoa1204471, https://www.nejm.org/doi/full/10.1056/NEJMoa1204471, Accessed April 27, 2020.

[21] Noelia Bueno-Gómez, "Conceptualizing suffering and pain." Philos Ethics Humanit Med 12, 7 (2017). https://doi.org/10.1186/s13010-017-0049-5, Published 29 September, 2017DOI: https://doi.org/10.1186/s13010-017-0049-5, Philosophy, Ethics, and Humanities in Medicine 12, Article Number: 7 (2017) https://peh-med.biomedcentral.com/articles/10.1186/s13010-017-0049-5#citeas, Accessed April 27, 2020. See also, Cassel EJ. *The Nature of Suffering and the Goals of Medicine*. Oxford: Oxford University Press; 2004. See also, Sorensen A. "The Paradox of Modern Suffering." The Journal for Research in Sickness and Society. 2010; 13:131–159.

[22] See "Study Finds Insects Can Experience Chronic Pain," by Jason Daley, Smithsonian Magazine, July 16, 2019, https://www.smithsonianmag.com/smart-news/insects-can-experience-chronic-pain-study-finds-180972656/. Accessed April 28, 2020. See also, "Fish do feel pain, study confirms," by Russell Deeks, October 6, 2019, BBC Science Focus Magazine, https://www.sciencefocus.com/news/fish-do-feel-pain-study-confirms/. Accessed April 28, 2020. See also. "Cephalopods and Decapod Crustaceans – Their Capacity To Experience Pain And Suffering," Advocates for Animals, 2005. The publication also makes clear the relation between "Nociception," "The role of opioid molecules in the regulation of pain," and "Similarities between stress systems in vertebrates and invertebrates". https://web.archive.org/web/20080406120543/ http://www.advocatesforanimals.org.uk/pdf/crustreport.pdf; More recently, tremendous amounts of research have attended upon invertebrates, particularly the octopus. (See *Other Minds: The Octopus, the Sea, and the Deep Origins of Consciousness*, by Peter Godfrey-Smith, Farrar, Straus and Giroux, New York, 2016. In which the Author poses such conjectures, as "What is it like to have eight tentacles that are so packed with neurons that they virtually 'think for themselves'?" See also, OneKind, "Cephalopods and decapod crustaceans deserve legal protection," https://www.onekind.scot/campaigns/seatheirsuffering/. Accessed April 28, 2020. See, in addition, faunalytics, https://faunalytics.org/cephalopods-and-decapod-crustaceans-their-capacity-to-experience-pain-and-suffering/ November 25, 2011.

## 46.5   Ecologies of Suffering

The obvious realization of starvation in other beings so quantifies illimitably the human conscience—already faltering beneath a rash of confused moral priorities among its own kind—that there would appear nowhere to take this data and these feelings. As cited in the informative *Animal Ethics* website, "Most animals in the wild are at continuous risk of dying for a number of reasons. In fact, the vast majority of them die shortly after coming into existence, with their lives containing little more than the pain of their deaths."[23] That said, we submit intuitively that there is a place to which Moses, Mahavira, Buddha, Lao Tzu, and Christ did go, in mind and by the very personalized constructions of their ethical intentions and activism.[24]

Ecologists and ethicists, and everyone else, might want to calculate a very basic set of similitudes and reciprocities. They comprise the multiplication of numbers that represent individuals, species, populations, genera, and every other taxonomic category on whichever systematic Tree of Life or Chain of Being seems best fitted to one's imagination. All species experience some level of boom and bust (famine), lemmings most notably. Recently, J. Jansen and R. Van Gorder applied new theoretical approaches to the mathematical ecology underlying the wide array of inputs that manifest/pose as resources versus predators in any allegedly stable biological system. Their predictive equations—building upon the work done since the time of Alfred Lotka and Vito Volterra in the 1920s (and later density-dependent Volterra-Rosenzweig-MacArthur theories) regarding predator-prey models—describe physical realities that are no different whether speaking of mosquitoes, trout, mountain lions, rabbits, or humans.[25] We live in a biosphere that condones carnage at various rates and fluctuations. It is worth repeating and reflecting upon the fact this all implies many complications for the human conscience (needless to reiterate), and no semblance of mathematical sobriety or predictability can soothe a psyche forever wounded by this state of affairs.

Multiply the number of species by the number of individuals, then attempt to divine the pain that exists in the world. It is as (mathematically) *simple* as that. It necessarily equates, to take just one small example, the pain felt by insects in fields

---

[23] See "Animal Ethics" – "Further Readings," and the website's section entitled, "Wild Animal Suffering," https://www.animal-ethics.org/wild-animal-suffering-section/situation-of-animals-wild/ as well as, "The Situation of Animals in the Wild," https://www.animal-ethics.org/wild-animal-suffering-section/situation-of-animals-wild/.

[24] See "Malnutrition, hunger and thirst in wild animals," n.a., in Animal Ethics, https://www.animal-ethics.org/wild-animal-suffering-section/situation-of-animals-wild/malnutrition-thirst-wild-animals/. Accessed April 27, 2020.

[25] Whitfield, J. Why cycling lemmings crash. Nature (2000). https://doi.org/10.1038/news000601-10, Published01 June 2000, DOI https://doi.org/10.1038/news000601-10, https://www.nature.com/articles/news000601-10#citeas, Accessed April 28, 2020. See also, "Dynamics from a predator-prey-quarry-resource-scavenger model," by Joanneke E. Jansen and Robert A. Van Gorder, Theoretical Ecology 11, 19–38(2018), September 14, 2017, Springer Link, DOI, https://doi.org/10.1007/s12080-017-0346-z, https://link.springer.com/article/10.1007/s12080-017-0346-z, Accessed April 28, 2020.

sprayed with pesticides to all those of us who were in Dhaka in 1974/1975. Pain is pain. Comparing pains, as science has continued to discover, is contrary to the empiricism upon which science prides itself.

If we were serious as a species about ending hunger in humans, we would do so. If we were equally serious about reducing pain in humans (and other vertebrates), we would advocate for open, free, and universal assisted euthanasia while revisiting the legalization of heroin (named after the German word for heroic, "heroisch"), initially developed in 1874 as the perfect all-round analgesic, deriving from at least 7000 years of human opium use;[26] the first applications of anesthetics in sixteenth-century Japan; and the horrifying documentation accompanying unanesthetized amputations on battlefields from the American Civil War to trenches during World War I.

Of the 63,000+ known species of vertebrates (on earth for the last 525 million years, invertebrates for approximately 543 million years), and comprising, at most, about 5% of all known species, vertebrate numbers can at best be estimated. And so can the pain that must accompany their rapid, or horribly lingering, demise. Similarly, with invertebrates, there are approximately 1.25 million of their species thus far identified, but as many as 30 million other invertebrate species that can be extrapolated.[27] Despite neurobiological biases on either side of the suffering indices, we must grant that they feel pain. The case of the lobster is among the most studied and surely troubling (Fig. 46.2).

## 46.6   Indices of Pain

One way to grapple with the pain and suffering that has gone on among the estimated Kingdoms of life + viruses (earlier we have mentioned between five and seven such delineations), both intended and unintended consequences, is to consider one source of speculation that examines the *average* 60% decline in vertebrate species populations between 1970 and 2014, assessed by the World Wildlife Fund's Living Planet Index. This mean average of population level declines also is tantamount to levels of pain.[28] The 60% number (only controversial because, when first published, it was misinterpreted) coincides with the mass of incorporated carbon, as a homogenous measure of life on earth. A sample size closest to our own sense of identifiable pain source, all remaining vertebrates represent a mere 2 gigatons (Gt)

---

[26] See "The History of Heroin," by Hosztafi S., Acta Pharm Hung. 2001 Aug;71(2):233–42. PubMed, https://www.ncbi.nlm.nih.gov/pubmed/11862675, Accessed April 28, 2020.

[27] See "National Geographic, Invertebrates, Pictures & Facts," https://www.nationalgeographic.com/animals/invertebrates/. Accessed April 27, 2020. See also, "Invertebrates," Center for Biological Diversity, https://www.biologicaldiversity.org/species/invertebrates/. Accessed April 27, 2020.

[28] https://livingplanetindex.org/projects?main_page_project=LivingPlanetReport&home_flag=1, Accessed April 27, 2020.

**Fig. 46.2**  Brownie, a "backyard dog" who has been severely malnourished, reported by PETA—People for the Ethical Treatment of Animals

of carbon (humans only 0.06 Gt of that total), or—another way of thinking about it—a rough guestimate (and easily disputable for underestimating countless measures within the full suite of biodiversity) of between 3 and 7 million years of evolution that would be required to generate such rich layerings of vertebrate aggregates. It does not, however, account for co-dependent life-forms.[29]

But placing the gigatons of carbon metric in the line of sight of a Mahavira or Buddha, and the projected evolutionary equivalency, and translation into pain and suffering, is exponentially blown wide open when we leave the vertebrate domain and enter the entirety that are the many Kingdoms of Life: "Protists = 4 GT C; Archaea = 7 Gt C; Fungi = 12 Gt C; Bacteria = 70 Gt C; Plants = 450 Gt C; All animals = 2 Gt C;... Arthropods =1 Gt C; Fish = 0.7 Gt C; Mollusks 0.2 Gt C; Annelids = 0.2 Gt C; Livestock = 0.1 Gt C; Nematodes = 0.02 Gt C; Wild mammals = 0/007 Gt C; Wild birds = 0.002 Gt C." Noting that "There are an estimated 550 gigatons of carbon of life in the world. A gigaton is equal to a billion metric tons. A metric ton is 1,000 kilograms, or about 2,200 pounds." And 550 billion metric tons times 2200 pounds = 121,000,000,000,000 (trillion) pounds of sentience, at any given second. Obviously, the approximately 3 trillion trees still standing throughout the world are solely mirrored in the data according to their carbon content; nor does

---

[29] See "Wait, Have We Really Wiped Out 60 Percent of Animals?" by Ed Yong, The Atlantic, October 31, 2018, https://www.theatlantic.com/science/archive/2018/10/have-we-really-killed-60-percent-animals-1970/574549/. Assessed April 27, 2020.

this study adequately assess the carbon biomass in downed forest, which can exceed 50% of the deadwood even 7 years after a tree has been cut.[30]

And this can be presumed to be two powers of ten short of accounting for all species yet to be extrapolated, thereby greatly increasing the 550 Gt C as a baseline for assessing pain per pound, pain per ounce, pain per gram, quintile, and millili- ter.[31] All of these factors must come into our purview when we review the nature of famine and its corollaries in human behavior.

Ultimately, we must ask the basic question: If there is a conscience at large among our species, how can we stand by idly while others suffer from hunger?

This is one of the hotly contested areas where human ethics and biotechnology urgently challenge our thinking. Our goal is to see effective ways to delimit and substantially diminish human dominion over Nature; to step back and front the moral compass, not to re-engineer the planet, which has been our species' modus operandi to date. Hence, the whole alarming question of *directed evolution*, the recent vogue of chemical engineering and xenobiology, that endeavor to engender a new diversity of genes and a new library of mutations and mutagenesis: a meta-eugenics that screens and amplifies selection largely within a laboratory. Its goal, in so many ways, is to accelerate and create new mutational efficiencies beyond natu-ral heredity and selection; fashioning, to date, hundreds of new amino acids, a re-tweaking of the evolution for customized, hyperefficient proteins and enzymes, and genetic firewalls. All of this would, in essence, recombine synthetic components of the evolutionary process at the molecular level so as to (allegedly) better facilitate human progress. If applied expressly to the amelioration of suffering, would such technologic arrogance be acceptable?

---

[30] See "Deadwood biomass: an underestimated carbon stock in degraded tropical forests?" by Marion Pfeifer, Veronique Lefebvre, Edgar Turner, Jeremy Cusack, MinSheng Khoo, Vun K Chey, Maria Peni and Robert M Ewers, IOP Publishing Ltd., IOP Science, Environmental Research Letters, April 28, 2015, https://iopscience.iop.org/article/10.1088/1748-9326/10/4/044019; See also, "From Models to Measurements: Comparing Downed Dead Wood Carbon Stock Estimates in the U.S. Forest Inventory," by Grant M. Domke, Christopher W. Woodall, Brian F. Walters, and James E. Smith, NCBI, PMC PLOS/ONE, 2013: 8(3); e59949, Published online 2013 Mar 27. doi: 10-1371/journal.pone.0059949, PMCID: PMC3609740; PMID. 23544112, https://www.ncbi.nlm. nih.gov/pmc/articles/PMC3609740/. Accessed April 28, 2020.

[31] See "All life on Earth, in one staggering chart," by Brian Resnick and Javier Zarracina, August 15, 2019, VOX, https://www.vox.com/science-and-health/2018/5/29/17386112/all-life-on-earth-chart-weight-plants-animals-pnas, Accessed April 27, 2020, and sourced from the landmark bio-mass (biosequestration and biogeochemical cycle) quantitative assessment paper, including reflections on abundance, distribution, "broad patterns over taxonomic categories" and "trophic modes" including "global marine biomass pyramid" data that accounts for consumers versus pro-ducers: "The biomass distribution on Earth," by Yinon M. Bar-On, Rob Phillips, and Ron Milo, PNAS, Proceedings of the National Academy of Sciences of the United States of America, PNAS June 19, 2018 115 (25) 6506–6511; first published May 21, 2018 https://doi.org/10.1073/pnas.1711842115, https://www.pnas.org/content/115/25/6506, Accessed April 27, 2020.

# Chapter 47
# Sakya Coming Out of His Mountain Retreat

## 47.1  That Which Is Earnest

At the very conclusion of his sustained meditation "At The Side Of A Grave," in which he had labored to fathom the indeterminacy of death, Søren Kierkegaard (1813–1855) admonishes all of us to prepare for "this final examination of life" which "is the test of earnestness."[1] As with the long-standing tradition of the theorized "three laws of thought" (Law of Identity, Law of Noncontradiction, and Law of the Excluded Middle), Kierkegaard, like Nietzsche, Afrikan Aleksandrovich Spir (1837–1890), even Socrates, was rightly concerned (obsessed) with what could meaningfully distinguish an individual from its collective (species); and upon what archetypal evidence, precedent, moral obligation, or common sense did a picture of reality arise that was not strictly of the imagination. A picture of truth that might invoke virtue, denatured nature, a dialectic resolution that put forth meaningful attribution—that which is earnest.

Roman historian Cornelius Tacitus' characterization of Christianity as being accused of *odium generis humani*—"hatred of humankind"—by Nero (supposedly to remove blame from himself for the Great Fire of Rome in 64 AD and his edict demanding the subsequent torture and killing of scores of Christians)[2] evokes in its ancient madness an equally medieval and modern psychopathology. The many ambivalences of any instantaneity render the primordial A = A assuredly unclear. Human nature too easily veers from this simplistic mathematical lane. Like all subjectivity, the passing cloud, the thundering cascade, provides sufficient evidence that our senses are as ephemeral as *their* senses. And that, while we are generally

---

[1] *Thoughts on Crucial Situations in Human Life – Three Discourses on Imagined Occasions*, By Søren Kierkegaard, Translated From The Danish by David F. Swenson, Edited by Lillia Marvin Swenson, Augsburg Publishing House, Minneapolis, Minnesota 1941, pp. 114–115.

[2] *The Annals* c. 116 CE.

© Springer Nature Switzerland AG 2021
M. C. Tobias, J. G. Morrison, *On the Nature of Ecological Paradox*,
https://doi.org/10.1007/978-3-030-64526-7_47

inclined toward the Self, it is clear from the history of ethics that autobiography and all progress toward self-improvement relies, however grudgingly, on the recognition of other life-forms.

We are part of a biological whole, whether we like it or not; believe in it or not. The rogue has a muddied biography, although life for most is unfair and truncated enough that the solitary sphere of activity might die out prior to realizing that others have meaning; that our actions have never been independent of the actions of others. As Oscar Wilde suggested, all of our diaries, passions, quotable remarks, could have come from anyone else. Even from a Buddha.

Indeed, as with every charismatic historical figure, the spotlight casts an enormous burden. The shadows of the highest peaks are a deafening roar within silence (Fig. 47.1).

## 47.2  Gautama Buddha's Enlightenment

The theme of Gautama Buddha (Sakyamuni, fifth to fourth century BCE) emerging from a mountain cave after 6 years of solitude and austerities (Shih-chia ch'u-shan) took on especial significance in the Ch'an School of Buddhism during the Northern Song dynasty in China (960–1127). The theme, which would carry over for many centuries in both China and then Japan—continuing to do so in any number of relevant guises—is termed *Shussan no Shaka* in Japan, where scores of copies of the original Chinese paintings were executed as part of the Zen (principally Rinzai) attraction to this profound storyline. There is Buddha standing weary, perplexed, windblown and in tatters (either enlightened or not yet enlightened) before the mountain habitat of his years-long retreat.[3]

This great juncture in his life is first suggested in a document from the second century CE. Namely, that even by then, nearly 500–700 years after Gautama's death and reincarnation, there existed a belief that the Buddha (Sakyamuni after Shakya, the sixth-century BC Ancient Vedic Kingdom—now in Nepal—where he was born) had retreated for 6 years into a mountain wilderness, a period of deep meditation that would change the course of his life, and to some extent of humanity's. The

---

[3] See,"Shussan Shaka In Sung And Yüan Painting," By Helmut Brinker, Ars Orientalis, Vol. 9, 1973, Freer Gallery of Art Fiftieth Anniversary Volume (1973), pp. 21–40, Published by: Freer Gallery of Art, The Smithsonian Institution and Department of the History of Art, University of Michigan, https://www.jstor.org/stable/4629268, https://www.jstor.org/stable/4629268?seq=31#metadata_ info_tab_contents; https://commons.wikimedia.org/wiki/File:Liang_Kai-Shakyamuni_Emerging_ from_the_Mountains.jpg, http://www.chinaonlinemuseum.com/painting-liang-kai-sakyamuni. php; ca. 1201-04, and signed Yüch'ien t'u-hua Liang K'ai, "Painter-in-Attendance" at the Imperial Academy of Hang-chou during the Chia-t'ai era. See Brinker, p. 30. The painting is infinitely richer in respect to the actual portrait of the Buddha, and details of the landscape, compared, for example with the contemporaneous Shussan Shaka by Ma Lin; or of the same period K $^o$ zanji workshop, or later anonymous ink monochrome Kano School handscrolls after the same motif; See Brinker, plates 1-7, and 9.

**Fig. 47.1** "Sakya Coming Out of his Mountain Retreat," Japanese Zen Buddhist Painting, School of Tsununubu, ca. 1683, After Original, Late Twelfth Century Version by Liang Kai. (Private Collection). (Photo © M.C. Tobias)

document is an early biography (the first) of the Buddha, known as *Buddhacharita* ("Acts of the Buddha") and written by the famed poet Asvaghosa (c. 80–150 CE). It is there that reference is made to what would be termed Shussan Shaka (Japanese: shussan shaka; Chinese: chūshān shìjiā).[4]

The painting styles later on in both China and Japan would see the Buddha's hands hidden by his robes, thus obviating the need to explicitly reveal one mudra or another; the position of his fingers either indicating enlightenment or not. A slight spherical aura of gold on his forehead would indicate that he was, indeed, already enlightened upon leaving the mountain, although the "'abbreviated' brush technique—chien-pi"[5] ascribed to Liang K'ai (died 1210) could easily be said to have pictorially re-focused the question of singular or multiple enlightenments in the overall facial expression, and general physiognomy. Compared with a Jan Lievens or Christus Petrus, the Buddha's face is a sketch, suggestive, not inundated.

But regardless of detail and style, there is the point of the very notion of *successive enlightenments*. The first would have occurred during Buddha's mountain wilderness retreat, 6 years of self-abnegation (half-dozen disciples with him, or nearby);

---

[4] See Charles Willemen, translator, *Buddhacarita: In Praise of Buddha's Acts*, Berkeley, Numata Center for Buddhist Translation and Research, 2009, p. XIII.

[5] op.cit., Brinker, p. 34

the second enlightenment to occur under the Bodhi Tree in Bodh Gaya (India) where his pictorial trajectory is said to lead. And there, beneath that tree, came the acquisition of an indescribable, supernal foment in consciousness, the pillars of a moral fortitude which, by definition, would extend the merits of his own enlightenment, or enlightenments, to the betterment of humanity and all other sentient beings.

This connection is the most seminal fact of the autobiography of Buddha, enlivened by way of such universal compassion. This is the dialectical cornerstone of the Shussan Shaka tradition, making that moment on the mountain an existentially pivotal instant within the rushing stream of his, of all of our, existence: solitary enlightenment, universal connections; or not yet.

But in either case, what to do with one's ineluctable feelings about others? An endless embrace, or deferral. How to be of profound assistance? Does protection manifest an ethological (interspecies) mandate? This is the most long-lasting and relevant ethical question of all.

How to pivot effectively at that moment of great transformation which for Buddha is said to have occurred on the traditionally eighth day of the twelfth month? The crux of this dilemma—and it *is* a dilemma for a real human being (Buddha)—is the motive for renouncing Nirvana—the end of all our travails—in the spirit of assisting pain that has arisen ecologically throughout the world, in others. Can it be fixed? The Hippocratic Oaths of every medicinal revelation—of kindness, comfort, succor, amelioration—all combine to insist not on ourselves, but those same Others, for whom Hippocrates (c. 460–c. 370) intoned a universality of first responders; for it is only through that succession of apprehensions and kindness that we can truly feel ourselves to be alive, part of life. Does said enlightenment call for the promulgation of a message of submission to the flux (samsara) that is inherently the biosphere and our place within it? If flux (all ecodynamics) is the norm, does it repair of one wound; a repeal of a single sorrow; the staunching of blood that will flow, add up to a meaningful confluence? Is a singularity equivalent to that confluence?

And what of the insuperable realization that helping one life, when tens of millions of others are crying out, is—at least, mathematically—not very impressive? Or is it?

Such questions devolve to near mediocrity in as much as not one good question can be answered fully. The very queries reveal a near naivety that is applauded at the height of ethical concerns, but undermined by the masses of humanity which have walked over it for thousands of years, unseeing, incapable of flexing a muscle over the complicated delineations that would see the individual in the heart of the mob. Triage scarcely intimates the degrees of nuance and emotional complexity embalming, offending, and/or liberating this inscrutably failed will-to-pantheism.

These conjectures whirl around the mind, and the paradox inherent to all of this hinges, in turn, upon the notion that Buddha's enlightenment has nonetheless led him to despair, before the cause of all other sentient beings. And/or his despair and frustration stems from his having the cause of others' suffering all around him, *in him* (the quintessence of Buddhism). As he stands above the steamy plains (Terai) of northern India, down below his alpine retreat, and no amount of his own personal enlightenment, past, present, or awaiting him, can obviate the need for his compassionate intervention.

It is the truth of his life, manifested in a choice, no less than what the young Kierkegaard had posited beside the grave: the test of becoming something more heartening, thoughtful and generous, than what he was. It is a journey worth taking, we are to understand. But to elicit the strength in oneself that promotes embarkation anytime soon is more than a mere chore: It changes one's everyday life, which is usually most uncomfortable, terrifying. One rarely hears or reads of *terror* or the tremendous discomfort in the Buddha's, or in Christ's or Mahavira's, choices. We are the ones left to wonder what it must have been like for them.

The historian and scholar of East Asian art Helmut Brinker likens this moment of anticipatory enlightenment—as commemorated throughout Zen monasteries in Japan, in which meditation occurs for a week, images of the Shussan Shaka are displayed, and hymns of the "Great Compassion" are recited—to an equivalent passion display by Christians during Easter, a distinct comparison with the Resurrection of Christ.[6]

Two Japanese scientists at Kyoto University have examined human eye contact and assessed the factors involved in people not looking into one another's eyes while speaking, and/or processing data. It lends some insight to the conceptual resolving powers of the Sakyamuni pictured throughout the Shussan Shaka genre—and all landscape meditations—wherein either a protagonist pictured in the image (e.g., the Buddha, an Arhat, or Boddhisatva disciple) or each and every one of us contemplating such paintings, and in any historic or cultural period, is challenged to focus. On what, precisely, are we narrowing our gaze, factoring the coefficients, summing up the sensation, calculating the meaning, arranging contexts and fathoming a conclusion to the artistic, revelatory moment?

According to researchers S. Kajimura and M. Nomura, the brain appears to have much *more difficulty* processing information when we are looking at something intently (eye-to-eye contact with another person or animal) and speaking at the same time, by as much as a 40% differential. This lends some weight to the famed Ludwig Wittgenstein conclusion to his *Tractatus* ("... *one must be silent*")[7] while also suggesting a deep parallel reality to our ability—silently—to step into a painting and be transported there. The immutability of Zen revelation, as achieved by the Shussan Shaka circumstances, are precise coordinates of this effortless effort: neural coordinates for ecological liberation.[8]

---

[6] ibid., Brinker, p. 27.

[7] *The Tractatus Logico-Philosophicus* Ludwig Wittgenstein, With an Introduction by Bertrand Russell, Harcourt, Brace & Company, Inc., New York and London: Kegan Paul, Trench, Trubner & Co., Ltd., 1922, p. 189.

[8] See Cognition. 2016 Dec;157:352–357. doi: 10.1016/j.cognition.2016.10.002. Epub 2016 Oct 15, "When we cannot speak: Eye contact disrupts resources available to cognitive control processes during verb generation." By S. Kajimura and M. Nomura, PMID: 27750156 DOI: https://doi.org/10.1016/j.cognition.2016.10.002, NCBI PubMed, https://www.ncbi.nlm.nih.gov/pubmed/27750156, Accessed April 30, 2020.

The comparison of the Sakyamuni's revelatory prelude to his engagement with the world with that of the roles in Judeo-Christian traditions of paradise, the Christ passion, and the many compassion myths expressed in the arts, enriches the universal language that is the Shussan Shaka.

By way of another comparison, Amanda K. Herrin describes how "The arrival of the Sadeler family in Munich in 1587 transformed the Bavarian court into a center of printmaking in Europe." An odd juxtaposition, one assumes, at first. But those prints, the first etchings and engravings for mass distribution, would incessantly point to the iconography of ascetic retreat, and subsequent engagement with other species in replicas of paradise throughout Europe and, by implication, the invocation of a real geography of such paradises. Herrin refers to Jan and his brother Raphael I Sadeler, and their nephews, the prodigious Aegidius II and RaphaelI II. As for one of their long-time collaborators, the Antwerp artist/print designer Maarten de Vos (1532–1603), the famed Friedrich Wilhelm Hollstein (1888–1957) attributes at least "1598 engravings" to him. Between the Sadelers and de Vos, well over 100 monks, saints, and ascetics in their respective landscapes of retreat (including trees and caves) would be disseminated throughout Europe, as the Sakyamuni in his wilderness would become the lingua franca of Asiatic enlightenment tales,[9] and preconfigured centuries before by Marco Polo's widely disseminated adventures along the Silk Routes (Fig. 47.2).

## 47.3  Interpretations of the Enlightened Self

As interpretations of that enlightenment were enriched, a host of nuances enshrining philosophical deliberation, ritual and practice, matters of faith, were at stake. Certainly, in Western literary tradition, the distinctive shift in certainty can be discerned in different translations of the Old and New Testament; differing attributions, naming of species, life in paradise once lost; and life on the other side. Yes, paradise may be regained, but the vast majority of those portraitures and allusions pertain to a life after "sin." In populist Ch'an thinking and expression, there were equally troubling contradictions. For example, there is the seemingly antithetical motif of an enlightened monk catching (and presumably killing) a shrimp. Writes Ann

---

[9] See *Hollstein's Dutch & Flemish Etchings, Engravings and Woodcuts 1450-1700*, vols. 44–46 (Rotterdam: 1996) as described in Amanda K. Herrin's "Pioneers of the Printed Paradise: Maarten de Vos, Jan Sadeler I and Emblematic Natural History in the Late Sixteenth Century," in *Zoology in Early Modern Culture: Intersections of Science, Theology, Philology, and Political and Religious Education*, Chapter 9, pp. 329–400, Brill, Koninklijke Brill NV, Leiden, 2014, DOI: https://doi.org/10.1163/9789004279179_011, https://brill.com/view/book/edcoll/9789004279179/ B9789004279179_011.xml, Accessed May 1, 2020. Series: Intersections, Volume: 32, Editors: Karl A. E. Enenkel and Paul J. Smith. See also *Hollstein's Dutch & Flemish Etchings, Engravings and Woodcuts 1450–1700*, vols. 21–22, Aegidius Sadeler To Raphael Sadeler II, Text Compiled by Dieuwke De Hoop Scheffer, Edited by K. G. Boon, Van Gendt & Co., Amsterdam, 1980. See also, https://www.hollstein.com/dutch-en-flemish.html, Accessed June 15, 2020.

**Fig. 47.2** "Saint Anthony the Abbot in the Wilderness," (ca. 1435) by Stefano di Giovanni di Consolo, known as il Sassetta (1392–1450), Metropolitan Museum of Art, New York. (Public Domain)

Yonemura, "…the enlightened mind realizes that all phenomena and distinctions are illusory, including the distinction between observing or deifying the Buddhist discipline of not killing any living thing."[10]

Montaigne's *Essais* are at the ultimate other extreme of the Gestalt of the unconditional self. A self-portrait that utilizes language and the written form to attempt to say *everything* about oneself; to reveal contradictions in human nature and then

---

[10] See "Kensu (Hsien-tzu)" in *Japanese Ink Paintings -From American Collections: The Muromachi Period An Exhibition in Honor Of Shujiro Shimada*, Edited by Yoshiaki Shimizu and Carolyn Wheelwright, The Art Museum, Princeton University Press, Princeton, NJ, 1976, p. 82.

contradict *them*.[11] In Henri-Frederic Amiel's *Journal Intime* (1882), or the seven volumes of Marcel Proust's (1871–1922) *À la recherche du temps perdu*, the logical outcome at the far peripheries of transparency is this same task of perpetual embossing of that elusive center between oneself and the outside. One's inner world is irretrievably tensed between its autobiographical instincts and an extroversion that can be explained by any number of impulses—the embrace of oneself as nature speaking through Being; a futile explanatory predilection; the endless, garrulous path that is language embodied in more language, forever trying to escape its linguistic landmines; or, the sad, endless struggle to wander beyond the simplicity of mental constructs which happen to include humanity's desperate, meaningless, endless chatter.[12]

No less the engine of verbs and metaphors, Sir Thomas Browne's *Religio Medici* (1643) references "the Cosmography of myself."[13] André Malraux writes that the truth about ourselves "lies first and foremost in what he hides";[14] Rousseau commenced his *Confessions* by asserting, "I mean to present my fellow mortals with a man in all the integrity of nature; and this man shall be myself." The absolute zero person, an aboriginal stream of self-regard, from the Achaemenid acts of Darius (519 BC), inscribed confessionally to the God Ahuramazda on the 1500-foot cliff of Mount Behistun, to Flavius Josephus' "Life" (circa AD 90) appended to his *Jewish Antiquities*, Bach's "Art of the Fugue," to the 1.25 million words in the *Diary* of Samuel Pepys written between 1660 and 1669 (he was 27 years old when he started it)[15] all connoted this struggle with the exposition of Self.

Wrote Stephen Spender, "The basic truth of autobiography is: I am alone in the universe."[16] Or there is Rabbi Hillel declaring, "...if I am for myself alone, what then am I? And if not now, then when?" Descartes' unambivalent Cogito. Or Jorges Luis Borges chiming in, "...there is at least a moment in the life of a man when you can see him. I have tried to find such a moment."[17]

---

[11] See Frederick Rider, *The Dialectic of Selfhood in Montaigne*, Stanford University Press, 1973, pp. 72–73. And Michel de Montaigne, *Selected Essays*, ed. Blanchard Bates, New York Modern Library, 1949, pp. 439–509.

[12] See Henri Frederic Amiel, *Amiel's Journal*, trans. Mrs Humphry War, New York, Macmillan 1899.

[13] See Margaret Bottrall, *Every Man a Phoenix – Studies in Seventeenth-Century Autobiography*, John Murray, London: 1958, p. 21.

[14] André Malraux, *Anti-Memoirs*. Trans. Terence Kilmartin, Holt, Rinehart, and Winston, New York, 1968, p. 5.

[15] See George Misch, *A History of Autobiography in Antiquity*, 2 vols. Harvard University Press, Cambridge, Mass., 1951; See also Roy Pascal, *Design and Truth in Autobiography*, Harvard University Press, Cambridge, Mass., 1960, and André Maurois, *Aspects of Biography*, Ungar, New York, 1966.

[16] See Stephen Spender, "Confessions and Autobiography," in *The Making of a Poem*, H. Hamilton, London, 1955, p. 65.

[17] J. L. Borges, "Post-Lecture Discussion of His Own Writing," Critical Inquiry 1, no. 4, June 1975, p. 721. See Karl J. Weintraub, "Autobiography and Historical Consciousness," Critical Inquiry 1, no. 4, June 1975, p. 824.

All of these instances are not just in extreme, but daily life facing a plight plausibly described by Diane Johnson, when she contemplates, "The artist writing his memoir is in double jeopardy; first he must lead the risky life worth reading, must come through it and face in retrospect the awful disparity between what it meant and what he had intended. Then he must make a fiction of it... Autobiography, that is, requires some strategy of self-dramatization."[18] We see the ad hoc mingling of irrepressible self-consciousness that was born of the image of self, liberated into a fiction that counters verisimilitude with a new level of advanced age confession in Goethe's *Aus meinem Leben. Dichtung und Wahrheit* ("From My Life: poetry and truth" [1811]). And we all, painfully, mysteriously, regard Rembrandt Harmenszoon van Rijn's some 80 known self-portraits with any number of considerations. The fulfillment of that "Romantic orientation toward autobiography" which Boris Tomasevskij wrote about, namely, the engendering of a "futurism" taken "to its ultimate conclusions. The author really [becomes] the hero of his works."[19] Or some other kind of Rembrandt altogether. Nothing heroic other than the brilliance and courage of his lustrous if sometimes brutish reveals, one after the other, throughout his life. Through sheer wonderment? The harsh, undeniable exhilaration and exhaustion of growing old? The pity we feel for ourself? For all those others condemned to this existence with us? Despite our inner, undeviating youth? Of course, we don't know, particularly at the end of his life, what Rembrandt sought implicitly to achieve. The National Gallery in London has Rembrandt's final (or presumably one of the final) self-portraits. Rembrandt was 63 years old, just a few months before his death. The date is 1669. The painting—86 × 70.5 cm—has been hanging in the Gallery since 1851, now in room 22 (NG221). Commentary by the gallery is ambivalent: The artist is focused on the "blemishes," "blotches," "the sagging fold beneath his right eye," "intense, unflinching, existential honesty."

Or, "...in the seventeenth century people had different ideas about self-analysis and how the mind works than we do now. Rembrandt's motives may have been more straightforward – driven less by soul-searching, and more by a professional fascination with the challenges of his art."[20]

Describing a self-portrait done exactly 10 years before (1659, BR 51 at the National Gallery of Art in Washington, D.C.), Clifford S. Ackley writes, "The only truly luminous area of the image is the flesh of the artist's face... with its vivid record of life experience and of the effects of the relentless pull of gravity, is a thoroughly lived-in one...the dignified, and stoical gaze of the large eyes in the setting of fleshy folds...that makes indelibly visible the stresses and strains of a life...."[21]

---

[18] See Diane Johnson, "Ghosts," New York Review of Books, 3 February, 1977.

[19] See Boris Tomasevskij, *Readings in Russian Poetics: Formalist and Structuralist Views*, ed. Ladislav Matejka and Krystyna Pomorska, MIT Press, Cambridge, Mass., 1971, pp. 54–55. See "On Thinking about Oneself," by Michael Tobias, The Kenyon Review, New Series, Vol. 4, No. 1, Winter 1982, Kenyon College, pp 9–25, https://www.jstor.org/stable/4335248.

[20] https://www.nationalgallery.org.uk/paintings/rembrandt-self-portrait-at-the-age-of-63=.

[21] "Later Self-Portraits," in *Rembrandt's Journey – Painter-Draftsman-Etcher*, by Clifford S. Ackley, In Collaboration With Ronni Baer, Thomas E. Rassieur, And William W. Robinson, MFA Publications, A Division Of The Museum Of Fine Arts, Boston, 2003, p. 308.

Perhaps, looming within his obsessive poignancies of self-depiction, he anticipated the brevity of the end, and thus, in a sense, the futility of all his brushstrokes for posterity. Writes Émile Michel, "Broken by poverty, and crushed by bereavements, the old master was not long parted from his son. The greatest of Dutch painters passed away with no record of his disappearance but the brief entry in the death-register of the Wester Kerk: 'Tuesday, October 8, 1669, Rembrandt van Rijn, painter, on the Roozegraft, opposite the Doolhof. Leaves two children.'"[22]

## 47.4   The Zen of Rembrandt's Self-Portraiture

Rembrandt's multiple enlightenments mirror Asiatic schools aspiring to Satori and kenshō, Zen bodhi, as encapsulated in most Western translations since the period of the orientalist Max Muller (1823–1900), particularly his 50-volume set of *The Sacred Books of the East* (1879–1894). But in fact bodhi, or budh,—"to wake up, to recover consciousness"—is too easily mistaken for what has been suggested to be a "Christian" orientation to the word.[23] Bodhi, the name of the tree under which Buddha had his first or second enlightenment—depending on which theory and school of thought one adheres to—signals in either case the ultimate catharsis, awakening, an event leading to Buddhahood.[24] That process entails confronting human nature, one's own and others.' The relations are not without unrelenting non-absolutes. Every wrinkle in Rembrandt echoes the same face of Buddha (Fig. 47.3).

This great reality is enshrined throughout Zen study, and with it the perilous, persistent challenge of penetrating the veil that covers human fragmentation and inconsistency; and the struggle to achieve some measure of lucidity in our lives. The Ch'an *Blue Cliff Record*, 100 koans, or densely contradictory paradoxes in versification, was first compiled during the Song Dynasty rule of Emperor Huizong, and modified by the Ch'an authority, Yuanwu Kegin (Engo), in the early twelfth century. A century later, the *Gateless Barrier* compilation of 48 Zen koans by the Chinese master Wumen Huikai (1183–1260) was added to the legacy of Rinzai study (the collection known as "Wumen Guan") by which disciples were challenged to find conceptual means of transcending contradiction, duality; circumventing walls of adamant logic with alternative forms of clear and radiant thought.

The twentieth century regards these transcendental hurdles under the aegis of 2500 years of discourse and epistemology; of precise paradoxical narratives that debate endlessly (1) the distance between abstraction and the sensory world, (2) the meaning of that world to the mind, the limitations of which are infinite, thus

---

[22] See *Rembrandt Harmensz Van Rijn – A Memorial Of His Tercentenary, 1606–1906*, by Émile Michel Of The Institute Of France: With Seventy Plates: Published by Willian Heinemann, London, 1906. p. 118.

[23] See Robert S. Cohen, *Beyond Enlightenment: Buddhism, Religion, Modernity*, Routledge, Abington, UK, 2006, pp. 1–3.

[24] See Suzuki, D. T. *Essays in Zen Buddhism*, Grove Press, New York, 1994, p. 229.

**Fig. 47.3** "Old Man,"
Mid-seventeenth Century,
After Rembrandt, Painter
Unknown. (Private
Collection). (Photo © M.C.
Tobias)

confounding every premise, cause, and consequence, and (3) the value, to begin with, of psychoanalysis, when the end result is a pragmatism that is either imagined or irrelevant. Jean-Paul Sartre's critical works stand out in this (relatively) recent wave of questioning. *L'Imagination,* 1936; *La Nausée* 1938 (in which he cynically propounded one of his most lauded witticisms, "Three o'clock is always too late or too early for anything you want to do"; *Esquisse d'une Théorie des Émotions,* 1939; *Le Mur,* 1939; *L'Imaginaire. Psychologie—Phénoménologique de l'Imagination,* 1940; *L'être et le néant. Essai d'ontologie phénoménologique,* 1943. And his famed war-time lecture "Existentialism is a Humanism," in which he declared, "l'existence précède l'essence."

We read all of these (and countless other) Eastern and Western struggles to achieve the precipitation and abandonment of identity in an ocean of riptides that cares nothing about us, or our philosophies, as perpetually frustrated, grand epics that may seem entirely pointless. They contain passing beauty that only mocks itself, conscious of the ephemerality that persists within and without. A koan—typically defined as a "paradoxical anecdote or riddle used in Zen Buddhism to demonstrate the inadequacy of logical reasoning and to provoke enlightenment"[25]—however

---

[25] https://www.google.com/search?client=firefox-b-1-d&q=koan.

incomplete its accomplished insights, in the end, is surely preferable to warfare, unemployment, and homelessness; starvation and bloodshed, doomed social Utopias and the continued betrayal of paradise by the universality of pain and suffering. In all of its Sino-Japanese allusions to the great disconnect, the so-called hosshi, or "breakthrough koan"—a staple of Rinzai discipline—aims to provide sufficient insight through sustained concentration that can literally merge all dialectic into a unison that gets beyond these overwhelming futilities.

## 47.5  Ultimate Impetus

What distinguishes the Shussan Shaka impetus, the personalities in whom the re-emergence or obtaining of wisdom occurs, is the unambiguous attempt to solve a significant problem. It refers, ultimately, to a nuts-and-bolts philosophy of compassion. A connection from the ascetic wilds—characterized in every version of the wilderness and the endless landscape affiliations, biosemioses, histories of the idea of nature—to the autobiographical awakening that can be discerned in every great and legendary personality, whether Buddha, Christ, or a Rembrandt. At that juncture the Self becomes painfully aware of something greater than him/herself, and in that perception may go on by turns to understand the challenge of liberation; the infinite individuals and souls whose lives truly are at stake, whether looking down into the populated valleys beneath a wild cliff, or out a studio window on to the crowded market ways of Renaissance Amsterdam. At its undeviating core, this is the philosophy—and a potent one—of biological interdependency.

# Chapter 48
# The Mind of a Chicken

## 48.1 Bougeant and Hildrop

In 1739, Father Guillaume-Hyacinthe Bougeant's earlier referenced animal rights treatise in English (Chap. 35) would, by its incandescent language and embellished theories, serve to force him into a celebrated retreat: he had for his own safety to evacuate Paris to the Pays de la Loire, a sub-prefecture of the South-Sarthe, where he remained the rest of his life in heretical hideaway. Ovid had endured a similar exile from Rome in 8 AD for his own alleged zoological/poetic heresies in the *Metamorphoses*, although there may be additional reasons for his banishment by Augustus. Bougeant's book contained an analytical version of Ovid by positing not only the normal intermingling of God and humans, humans and animals, but the sheer apotheosis inherent to *A Philosophical Amusement Upon the Language of Beasts and Birds*[1]—a most delicate, 66-page essay that illuminates an informed and common-sense reality about every over creature sharing the earth with humans, to their express dismay. He knew in his heart that the animal world, by any spiritual or intellectual reckoning, must be perceived in an altogether different realm than their terrible condition at the hands of man had allowed. Bougeant volunteered that they exceeded us; our species was inferior to their great diversity and power of intellect and soul. He invokes the word "amusement" only in so much as he felt compelled to lessen the blow upon readers, as if to imply, *that being said* we should not be too surprised or threatened. But what Bougeant actually had in mind was a re-stating of principles that had been espoused by great saints and theologians for thousands of years, but never in so terse and forceful a way.

Bougeant engenders a social contract with animals 23 years before Jean-Jacques Rousseau would publish his own *Social Contract*. Both men would revitalize the

---

[1] Written originally in French by Father Bougeant, a famous Jesuit; Now confined at La Fleche on Account of his Work, Printed for T. Cooper, at the Globe in Paternoster Row, 1739. Price One Shilling.

© Springer Nature Switzerland AG 2021

M. C. Tobias, J. G. Morrison, *On the Nature of Ecological Paradox*,
https://doi.org/10.1007/978-3-030-64526-7_48

eighteenth-century ossified hierarchies: Rousseau in his famed declaration that "man is born free, but he is everywhere in chains," and Bougeant expressing the same insight with respect to all other species.

His most pressing observations are those concerning the language of our animistic kin; and in this way he was nearly two centuries ahead of the revolution in zoosemiotics and deep ethology to come. Bougeant writes, "When the Hen obliged to keep her young warm under her has not Time to go abroad, and the Cock drops some Food out of his Beak into hers, she testifies her Satisfaction to him by the clapping of Wings, and by a little Cry different from all the others, which must necessarily signify, 'I am very glad: You do me Pleasure'".[2] And Bougeant applies such *field observations* in his analysis of psycholinguistic conductivity between insects, spiders, and every major representative of the animal worlds, with especial focus on avifauna. He writes, "Two Sparrows will know one another by their Voice among a thousand. I might here allege a hundred like Facts, to prove that all Animals have, in their mutual Correspondence a Delicacy of Discernment, which is not within our reach, and which makes them observe Differences among themselves which are altogether imperceptible to us. If then many Birds seem to us always to sing the same Note, as the Sparrow, the Chaffinch, and the Canary-Bird; we must not thence conclude that they are saying the same thing for ever [sic]".[3]

Such was Bougeant's influence (by Earth Day 2020, Pope Francis would declare, "Take to the streets to teach us what is obvious....there will be no future for us if we destroy the environment...") that just 3 years later John Hildrop (1682–1756), best known perhaps for his work *Reflections on Reason* (1722), came to greatest prominence by composing the first book on the subject of animal rights by an *English* author: *Free Thoughts Upon the Brute-Creation; Or an Examination of Father Bougeant's Philosophical Amusement, &C: In Two Letters to A Lady*.[4] The famed frontispiece was engraved by Gerard Vandergucht (1696–1776), of Adam and Eve surrounded by their cliff-showering waterfall, of a paradise teeming with birds and mammals. One of the birds, no doubt an ostrich (which had been painted for many centuries before Hildrop, including Raphael's famous image of one),[5] in fact is so uncannily a giant Moa in every other respect that one is forced to re-examine the evidence, which to date suggests that the first knowledge of Moas in Europe did not occur until the 1830s, the first discovery of dinosaurs, by British fossil enthusiast William Buckland, in 1824. If Vandergucht had heard about Moas we don't how, unless explorers returning with loot to Amsterdam had somehow inflated—nearer to young brontosaurus proportions—an ostrich.

---

[2] Ibid., p. 62.

[3] ibid., p. 50.

[4] Printed for R. Minors, London, 1742.

[5] See "The Ostrich: An unexpected Allegory," by Melinda Knox, Queens University, Kingston Ontario, Canada, n.d., https://www.queensu.ca/vpr/ostrich-unexpected-allegory, Accessed May 5, 2020. See also, *Raphael's Ostrich*, by Una Roman D'Elia, Penn State University Press, University Park, PA, 2015.

Also on the title page of Hildrop's remarkable book is the famed quotation from Job xii. 7, 8, 9, and 10: "But ask now the beasts, and they shall teach thee; and the fowls of the air and they shall tell thee. Or speak to the earth, and it shall teach thee; and the Lord hath wrought this? In whose hand is the soul of every living thing, and the breath (spirit) of all mankind." Hildrop's thesis is that unlike the denatured species of man, all other animals have souls that are immortal, but "degraded by the fall of man." Of the 13 creatures depicted in Vanderqucht's Garden of Eden, two are human, one looks to be an unknown species like that of a bald uakari monkey (*Cacajao calvus*), there is a coral snake, and a goat. The other eight species are avifauna, dominated by the "Moa" and what could be a hybrid swan/chicken. In 1783, Calvinist philosopher John Wesley (1703–1791) referred to the fact that "Millions of spiritual creatures walk the earth unseen…".[6] Wesley went on to compare such creatures to "Angels" and pointed out that humanity's qualities of language, sight, and understanding were each inferior to those of such other creatures. He had obviously read Hildrop and Bougeant (Figs. 48.1 and 48.2).

Animals, Wesley went on, know our thoughts by way of an unfathomable wisdom. Indeed "they sang together when the foundations of the earth were laid".[7] Moreover, throughout dozens of other sermons Wesley acknowledged the extraordinary insights of Hildrop's proposition that the "Brute-creation" comprises immortal souls, the truest expressions of God and God's creative will; that even a flea is immortal; that such animals' preservation was essential—their highest wisdom impossible to distinguish from man's. That in their natural state they should never

**Fig. 48.1** An Ancient Friend. (Photo © J. G. Morrison)

[6] Arminian Magazine, 1783, Vol. 6, "A Sermon on Hebrews," 1. 14, p. 7. https://babel.hathitrust.org/cgi/pt?id=iau.31858022058451&view=1up&seq=11, Accessed May 3, 2020.
[7] Ibid., p. 14.

**Fig. 48.2** "Adam and Eve in Eden," Engraving by Gerard Van der Gucht, Frontispiece for *Free Thoughts Upon The Brute-Creation* by John Hildrop, London 1742. (Private Collection). (Photo © M.C. Tobias)

be hindered.[8] Wesley had also clearly inculcated the spirit of William Hogarth's "The Four Stages of Cruelty" (1751), four instructive engravings published to drive home the message to British children and their parents that cruelty to animals—graphically revealed—would result in cruelty to other humans; that, as David

[8] https://babel.hathitrust.org/cgi/pt/search?q1=brute;id=iau.31858022058451;view=1up;seq=5;start=21;sz=10;page=search;orient=0: The works of John and Charles Wesley. A bibliography: containing an exact account of all the publications issued by the brothers Wesley, arranged in chronological order, with a list of the early editions, and descriptive and illustrative notes. By the Rev. Richard Green ...1906, https://babel.hathitrust.org/cgi/ls?field1=ocr;q1=Arminian%20Magazine%2C%201783;a=srchls;lmt=ft.

Perkins presents in his book *Romanticism and Animal Rights*, "a cruel boy becomes a murderer as an adult, is hanged, and ends as a corpse dissected in an anatomy demonstration".[9]

Perkins deftly illustrates other writers and philosophers of the mid-eighteenth century who were entirely in league with *Hildrop's Theorem*, as we might liken it. Said John Locke, "Children should from the beginning be bred up in abhorrence of killing and tormenting any living creature".[10] Such dire motives and intent are described in Hogarth's own "Autobiographical Notes": "The four stages of cruelty, where done in hopes of preventing in some degree that cruel treatment of poor Animals which makes the streets of London more disagreeable to the human mind, than any thing what ever [sic], the very describing of which gives pain...".[11]

Perkins invokes other animal rights luminaries of Hildrop's era, men like Robert Southey, Bishop Joseph Butler, the clergyman Richard Dean who, quoting passages from the Bible, wonders whether it might not be necessary for humans to "impeach the divine Goodness" (God) for the "undeserved pain" that humanity inflicts on innocent animals. Or Peter Buchan declaring that this realm of suffering suggests that God must absolutely provide some "future state" for these animals that have so endured. Others also followed Hildrop (and the Frenchman Bougeant): an anonymously published *Clemency to Brutes* (1769); James Granger's *An Apology for the Brute Creation* (1771); Humphry Primatt's *Duty of Mercy & Sin of Cruelty to Beasts* (1776);Thomas Young's *An Essay on Humanity in Animals* (1789); Richard Mant's *Reflections on the Sinfulness of Cruelty to Animals* (1807); Lord Thomas Erskine's introduction to Parliament in 1813 of his famed anti-cruelty bill, and another, by Richard Martin, in 1822; James Plumptre's *Three Discourses on the Case of Animal Creation, and the Duties of Man to Them* (1816); and in 1838, the Presbyterian minister William Drummond's *The Rights of Animals: And Man's Obligation to Treat Them with Humanity*.[12] There were, in addition, numerous other treatises on vegetarianism, as well as the great classics scholar and translator of Porphyry's own *Four Books on Abstinence From Animal Food* (London, 1823) Thomas Taylor, for example, who magnificently argued that only a thorough shift in human diet could alleviate the agonies meted out to animals, a critical transformation in human behavior fully in keeping with the widely publicized developmental psychology as described by both David Hume (*Treatise of Human Nature*, 1739) and John Locke's

---

[9] *David Perkins, Romanticism And Animal Rights,* Cambridge University Press, Cambridge, UK, 2003, p. 21.

[10] Quoted by Perkins from John Locke's *Thoughts on Education* (1693), in Works, T. Tegg, London, 1823, ix, 112–113, on p. 20 of Perkins' book.

[11] See "The Animal Question and Women – Jane Austen and Animals," by Barbara K. Seeber, in Jane Austen – Environment and Nature, A Routledge College, Routledge, Taylor & Francis, https://www.scribd.com/document/357184876/Jane-Austen-Environment-and-Nature, Accessed May 3, 2020. Material from Barbara K. Seeber's book, Jane Austen and Animals, Routledge, UK, 2013.

[12] See Perkins, op. cit., pp. 28, 176–177.

*Essay Concerning Human Understanding*, 1748, in Part II, Chaps. 10 and 11. As Perkins summarizes, "Locke discussed the intellectual powers of animals".[13]

And then there was the outstanding historian and philosopher Joseph Ritson (1752–1803) whose remarkable book *An Essay on Abstinence from Animal Food, as a Moral Duty* (London 1802) was committed to ethical vegetarianism, connecting barbarism perpetrated by humans against animals to those same people who would commit similar crimes against other humans. Indeed, Ritson suggested that slavery and cannibalism derived from those sorts of people who would think nothing of torturing and killing other species. Avowed Ritson, "those accustom'd to eat the brute, should not long abstain from the man".[14]

Against this backdrop of elegant fire and fury, at the height of the new renaissance, the combined Ages of Enlightenment and Romanticism, in Western artistic and philosophical tradition, we glean clearly the context for that adamant animal rights predilection, as earlier typified in Percy Bysshe Shelley; and in Oxford University's foremost professor of law and ethics, Jeremy Bentham (1748–1832), whose beliefs that universal jurisprudence *must* convey equal rights to all animals would help shape the formation of anti-vivisection societies, and, as we think of it today, the underlying premise of most animal rights philosophy and initiatives.

What *Hildrop's Theorem* added to this brief history was the unequivocal illustration of a distinct prospect, namely, that all species other than Man had immortal souls. But in the case of *Homo sapiens*, we are deficient by our own betrayals and despoliations, an escalating syndrome of totalitarian terror over all other animals (Fig. 48.3).

That dominion is most assuredly and emblematically concentrated on that terrestrial vertebrate tortured more than any other, *Gallus gallus domesticus*, the chicken, about whom it is estimated by the United Nations Food and Agriculture Organization that approximately 25 billion live chickens were slaughtered worldwide for commercial production in 2019.[15] An estimated nine billion of those individuals were killed in the United States, while 305 million hens were held captive in the United States for their eggs.[16] These numbers only approximate the actual number of fatalities. A "slaughterhouse" is deemed a commercial, licensed operation, in whatever country. Yet, there is no country where home killings of chickens—and other animals—do not occur, and there is no reliable metric to capture such data. In most economically marginalized regions of the human family, we can only surmise that the numbers would skyrocket above those official recognized commercial operations.

---

[13] ibid., Perkins, p. 177. In addition, there is the famed *Twelve Stories of the Sayings and Doings of Animals,* by Mrs. Sara Bowdich Lee, Grant and Griffith, London, 1954; the *Historia animalium…* by Wolfgang Franz, Johannem Ravestenium, The Netherlands, 1665.

[14] ibid., p. 124.

[15] http://www.fao.org/faostat/en/#compare.

[16] https://www.peta.org/issues/animals-used-for-food/factory-farming/chickens/.Accessed May 5, 2020.

**Fig. 48.3** The Commoditization of *G.g.domesticus*, PETA Archives

## 48.2   The New Zealand Paradox

Moreover, the escalation of chicken ills is astonishing. In just one country, New Zealand, prided for her "Green" overtures, during the past 15 years the number of murders rose from 80,600,000 chickens slaughtered in 2004 to 100 million in 2019.[17] New Zealand is a particularly telling nation of ecological paradox. That "Clean, Green" mantra—its pride, for example, in all but eliminating Covid-19 before any other nation; its partly deserved fame in conservation biology and eco-tourism circles—collapses at the level of environmental basal metabolism and non-violence. As we wrote in 2010, after having worked for nearly 15 years to create and maintain—with a tremendous team—a scientific wildlife preserve in that country, those 80.6 million chickens slaughtered in 2004 "must have suffered far, far beyond our ability to describe it, yet have less 'significance' to the country, than grapes and pumpkins in terms of economic ratings".[18] They are abstracted into some heinous purgatory of absolute human denial that festers between a cheap wine and sautéed broccoli.

---

[17] See *God's Country: The New Zealand Factor*, by M. C. Tobias and J. G. Morrison, A Dancing Star Foundation Book, Zorba Press, Ithaca, New York, 2011, p. 359. See also, https://figure.nz/chart/QzXp9lsqIAJr9v1x, Accessed May 3, 2020.

[18] ibid., *God's Country*, pp. 358–359.

But here is what is actually accorded their final unheralded moments, as described in the New Zealand's government's "Animal Welfare (Broiler Chickens: Fully Housed) Code of Welfare No. 1, 2003".[19] Consider the prescribed 'Minimum Standard No. 11—Humane Destruction.' We're not speaking of "maximum" standards, only minimum. This is the language of disingenuous economic expediency. The underlying goal is to kill as many animals as the markets will compensate. And to do so by conflating a virtually useless nod to committees charged with oversight of "ethical" standards throughout a country, with the singular motives and methods employed by growers, transporters, intensive breeding sites, slaughterers and their employers, freezing works, the whole final chain of abysmal dispatch into the mouths of consumers.

The language is as follows: "Humane destruction when necessary must be carried out by the use of concussion and neck dislocation, or electrical stunning and neck dislocation, or neck dislocation, or euthanasia with at least 70% $CO_2$ gas or a mixture of $CO_2$ 70% and argon gas 30%, or at processing plants. When humane destruction is carried out by gassing, then the procedure must be sufficient to ensure collapse of every broiler chicken within 35 seconds of exposure to the gas, and broiler chickens must remain in the gas for at least a further two minutes following collapse." Standard No. 11 goes on to declare in section d: "Persons undertaking humane destruction must be appropriately trained and must ensure that the broiler chickens are managed carefully and calmly at all stages of the process." '4.2 of the Standard entitled "Catching and Loading" continues, "If a broiler chicken starts flapping its wings while being carried to the transport crates, it should be brought under control by either resting it against the catcher's leg or by resting it on the ground. This will help to reduce the risk of hip dislocation. Broiler chickens should be loaded into the crates in a smooth, controlled action that minimizes the force exerted on the hip."

Pursuant to "Minimum Standard No. 13—Catching and Loading" it is written, "Broiler chickens that are injured during the catching and loading procedures must be destroyed immediately. The maximum number of broiler chickens that may be carried at any one time in each hand of a catcher must be no more than four. The broiler chickens about to be carried to the crates must all be held with their hocks and shanks aligned in the same manner within the hand...Broiler chickens must be placed carefully into the crates".[20] This is written on paper, but the men and women doing the work are not reading these lines. They are running after terrified chickens. How are they to destroy them immediately if they injure them? What is the penalty for injuring them? What is the reality of any of these laws, except on paper to perform a due diligence service to the abysmally disinterested lawmakers and willful companies profiting from the reckless insanity, in our opinion, of such slaughter? And, again, all in the context of minimal standards".[21]

---

[19] www.biosecurity.govt/nz/animal-welfare/codes/broiler-chickens/index.htm; op. cit., www.biosecurity.govt.nz/animal-welfare/codes/boiler-chickens/index.htm, p.12.

[20] ibid., pp. 12–13 and p.27.

[21] op. cit., *God's Country*, p. 359.

Despite the growth promotion antibiotics—zinc bacitracin and macrolide tylosin—fed to chickens in New Zealand (banned in the European Union) and a continuing open book on the tide of abuses of even the most minimal standards of care, there is no empirical data to suggest that all of the campaigns targeting ("alleged") cruelty at battery farms have made any inroads into the public consumption patterns. As one researcher has been quoted, "You have to accept that production is industrialized, that you minimize animal welfare and maximize profits so people can afford to get the product into a supermarket trolley".[22]

For a country that perceives itself as a world conservationist leader in saving birds, it gives no quarter to the chicken, which, by New Zealand standards, is no bird at all. This contradiction is not anomalous, nor is it merely New Zealand's human nature problem. It is a human problem. As early as 1877, the famed English professor at University College of London Henry Morley (1822–1894)—best known for his ten-volume *English Writers* and his Universal Library—wrote a book entitled *The Chicken Market and Other Fairy Tales*, with illustrations by Charles H. Bennett (1828–1867).[23] For his part Bennett had illustrated an edition of *The Adventures of Young Munchausen* and *The Sorrowful Ending of Noodledoo*; an Aesop, *Birds, Beasts and Fishes*, *The Nine Lives of a Cat*, a *Pilgrim's Progress*, and *The Faithless Parrot*. By all accounts, both Morley and Bennett had strong affinities for animals. Yet in their collaboration they propounded a lethal, if common, assumption: Chickens were to be slaughtered and eaten, an emotional disconnect no less thorough and unquestioning as "The extermination procedure in the gas chambers".[24] Writes Morley, "Though a humming-bird grew to the size of an ostrich, and increased as much in beauty as in size, it would be no match for one of Ben Ody's chickens as those chickens now shone down the dawn. They had crossed the water, and stood glittering among the dull sand-hills like hillocks of rainbow in the morning rain." And, upon nearing the market continues, "All the fowls that hither we bring, Body and legs, liver and wing, We mean to present to the Fairy King".[25]

These readers, the children of those English settlers across New Zealand, grew up, like the farmer/journalist Oliver Duff, to become part of that vast contradictory set of conspiratorial preconditions for killer instincts and acts. It was Duff, writing in 1961, in his masterpiece *A Shepherd's Calendar*, who would describe his hopes of making it through life "without shooting a bird. There will be enough animals around the Judgment Seat...." And continued, animals "battered, [read: John Donne][26] bleeding, and mutilated, to tear my penitence to shreds if they re-member

---

[22] Crop and Food Research Expert, Graeme Coles, cited by Bruce Ansley in his article, "The Secret Life of Chickens," Listener, March 16, 2002, pp. 19–22.

[23] Cassell Petter & Galpin, London, 1877.

[24] See "Auschwitz-Birkenau" Memorial and Museum, http://auschwitz.org/en/history/auschwitz-and-shoah/the-extermination-procedure-in-the-gas-chambers.

[25] ibid., pp. 9, 17.

[26] See https://www.poetryfoundation.org/poems/44106/holy-sonnets-batter-my-heart-three-person d-god.

**Fig. 48.4**  Assembly Line, PETA Archives

me".[27] And he adds, "in a thousand or ten thousand years" we might all become vegetarians and look back at the "pioneers" only to discover they were "the liars, cowards, hypocrites, and pretenders of the twentieth century who tried to spread mists of confusion between their appetites and the places from which they indulged them" (namely, the kind of factory farm Duff himself visited once in Iowa and twice at New Zealand freezing works) (Fig. 48.4).[28]

## 48.3  Breeding Contempt

With well over 600 chicken breeds in the world, and between 19 and 25 billion individual chickens at any given moment (*Gallus gallus domesticus*), all descended from the red jungle fowl (*Gallus gallus*)—of Least Concern by IUCN standards[29]—*G. g. domesticus* is actually of infinite concern. A bird monumentally abused, tormented. In terms of the geography of conscience, *Gallus* is the seismic epicenter

---

[27] O. Duff, *A Shepherd's Calendar*, Latimer, Trend & Co., Plymouth, Hamilton and Auckland, 1961, p. 118.

[28] See *The Hypothetical Species: Variables of Human Evolution*, by M. C. Tobias and J. G. Morrison, Springer New York, 2019, p. 193.

[29] BirdLife International. 2016. Gallus gallus. The IUCN Red List of Threatened Species 2016: e.T22679199A92806965. https://doi.org/10.2305/IUCN.UK.2016-3.RLTS.T22679199A928069 65.en. Downloaded on 04 May 2020.

of suffering on earth. For a fellow Being—the most numerous of all avians—to have suffered so suggests that an allegedly *sapient* hominin should, by now, have gleaned something about *domesticus* that would have tamed our cruelty, heightened a modicum, at least, of discerning restraint, if not outright love and admiration. George Orwell, in his Preface to the Ukrainian edition of *Animal Farm* (written in March 1947 and promulgated by the Ukrainian Displaced Persons Organisation based in Munich), wrote, "I proceeded to analyse Marx's theory from the animals' point of view. To them it was clear that the concept of a class struggle between humans was pure illusion, since whenever it was necessary to exploit animals, all humans united against them: the true struggle is between animals and humans".[30]

Brian Fagan reminds us that the chicken's first verifiable bones come from northeastern China, circa 5400 BCE; that Roman law (in 161 BCE) limited consumption by locals to one chicken per meal (fears of scarcity predicated by gluttony); and that Pope Nicholas I (858–867) was so taken by chickens that he proclaimed that a rooster should be settled upon the summit of every church.[31] We all watched as the copper rooster atop Notre-Dame cathedral survived the April 15, 2019, conflagration, proudly carried through the streets of Paris by those who had rescued it from ruin.[32] In most human cultures, birds are said to signify something auspicious, spiritual, good. Until the cock crowed, Peter would not acknowledge knowing Christ, one of the reasons roosters are found throughout Renaissance art, particularly associated with the Christ Passion, the Nativity, and the Adoration of the Shepherds.[33] But by 1847, a New York City magazine, *The Knickerbocker*, propounded what would become a universally lame quip according to "Why did the chicken cross the road?"[34] It says nothing about chickens, but much about humans.

Our bias turned malevolent against chickens involves more than the sum of what any first glance can reveal, given the love and fidelity chickens continue to show those who treat them as companions, not food. Yet, in the realms of intellect, character, family charters, consciousness itself, our disregard of the chicken remains pronounced. Members of the Corvid family, parrots and pigeons, are hailed as the geniuses of the bird world. Chickens are ignored.[35] While there has been enormous research conducted on the brains of vertebrates in general, chicken brains, and certainly the chicken's mind, have been underrepresented. More importantly, vertebrates in general—roughly 3% of all known animals, by their relatively recent emergence in evolutionary terms—are thought to be far less "successful" than 97% of

---

[30] See *Animal Farm -A Fairy Story*, by George Orwell, 50th Anniversary Edition, pictures by Ralph Steadman, Harcourt Brace & Company, New York, 1995, p. 179.

[31] Brian Fagan, *The Intimate Bond – How Animals Shaped Human History*, Bloomsbury Press, New York, 2015, pp. 262–264.

[32] https://www.reuters.com/article/us-france-notredame-exhibition/fire-damaged-rooster-fro m-notre-dames-spire-goes-on-display-idUSKBN1W51ID.

[33] See *Nature and Its Symbols*, by Lucia Impelluso, Translated by Stephen Sartarelli, The J. Paul Getty Museum, Los Angeles, 2003, p. 313.

[34] The Knickerbocker, or The New York Monthly, March 1847, p. 283.

[35] See for example, "Why Ravens and Crows Are Earth's Smartest Birds," by Amelia Stymacks, National Geographic, March 15, 2018.

**Fig. 48.5**  Live Animal Market, Saudi Arabia. (Photo © M.C.Tobias)

"non-chordata phyla," namely, invertebrates.[36] Multiplied by their prolix numbers
and acute mental and emotional faculties, *Gallus* may well represent the exquisite,
universal apotheosis—not an exception—of all that is counter to such grievous,
scientific ignorance on our part, just one more vertebrate, stingily accommodating
itself, but no others (Fig. 48.5).

## 48.4   Intelligence Quotients

Studies of "intelligence" and brain anatomy in birds began early in the twentieth
century.[37] By 1966 significant behavioral comparisons were being drawn between,
for example, "chickens, bob-white quail, parrots…and red-billed blue magpies".[38]

---

[36] See *Brain and Intelligence in Vertebrates*, by Euan M. Macphail, Clarendon Press, Oxford, UK,
1982, p. 20.

[37] See for example L. Edinger "The relations of comparative anatomy to comparative psychology". Journal of Comparative Neurology and Psychology. 18 (5): 437–457, 1908; doi:https://doi.
org/10.1002/cne.920180502. hdl:2027/hvd.32044106448244.

[38] Op. cit., E. Macphail, p. 223, citing R. L. Gossette and W. Riddell, "Comparison of successive
discrimination reversal performances among closely and remotely related avian species," Animal
Behavior, 1966, 14, 560–4.

Studies in cognitive ornithology to somehow make meaningful (to ourselves) the mental world of birds, avian intelligence in general has tended toward quite primitive markers and criteria: "numerical competence," "symbolic matching," "associative learning," "cache recovery," various forms of "spatial memory," the use of obstacles and mirrors ("self-recognition"), "tool and compound tool making," "object tracking," variability of a host of strategies in general, "pattern learning," and numerous anatomical investigations of the forebrain of birds versus mammals, for example, "socio-ecological variables (diet, social structure, relative brain size, innovation, life history and habitat) across corvids, parrots, other birds, monkeys, apes, elephants and cetaceans…".[39]

In 2016, "the first study to systematically measure the number of neurons in the brains of birds" found that—in equal brain masses—there were "significantly more neurons packed into their small brains" (birds) than in mammals, including primates.[40]

"Birds are remarkably intelligent, although their brains are small," the authors at once declare. Unlike most birds, chickens will want to reproduce and lay eggs independent of seasonality, or whether a male is present. Write the authors of *The Complete Encyclopedia of Chickens*, "This phenomenon is based on a hereditary factor. In the course of time, one has by selection more or less eliminated the characteristic that birds only lay eggs after having mated up in the mating season….the outcome of a focused selection".[41] What Verhoef and Rijs also point out is the extraordinary diversity of chickens, as seen in their head furnishings, combs, feathers, crests and beards, muffs, and in the variegated plumage mottling, lacing, spangling, penciling, striping, and barring.[42] The vast artistry of a chicken's global diversity is simply beyond any other class of aves. Gorgeous, striking differences, in color, communicative range, depth and enthusiasm, curiosity, affection, intensity of personality, on every continent except Antarctica, where scientists eat them, but cannot import them due to infectious disease liabilities for indigenous penguins.

---

[39] See one of the most comprehensive and sensitive essays on the topic, "Cognitive ornithology: the evolution of avian intelligence," by Nathan J. Emery, Philosophical Transactions B, Philos Trans R Soc Lond B Biol. Sci. 2006 Jan 29; 361(1465): 23–43, Published online 2005 Dec 7. Doi: 10.1098/rstb.2005.1736, PMCID: PMC1626540, PMID: 16553307, The Royal Society Publishing, NCBI PMC, https://www.ncbi.nlm.nih.gov/pmc/articles/PMC1626540/. Accessed May 4, 2020.

[40] See "Bird brain? Ounce for ounce birds have significantly more neurons in their brains than mammals or primates," Vanderbilt University, Science News, June 13, 2016, citing "Birds have primate-like numbers of neurons in the forebrain," by Seweryn Olkowicz, Martin Kocourek, Radek K. Lučan, Michal Porteš, W. Tecumseh Fitch, Suzana Herculano-Houzel, and Pavel Němec PNAS first published June 13, 2016 https://doi.org/10.1073/pnas.1517131113. Edited by Dale Purves, Duke University, Durham, NC, and approved May 6, 2016 (received for review August 27, 2015), Proceedings of the National Academy of Sciences of the United States of America, https://www.pnas.org/content/early/2016/06/07/1517131113.abstract.

[41] Esther Verhoef and Aad Rijs, *The Complete Encyclopedia of Chickens*, Rebo Publishers, Prague, The Czech Republic, 7th edition, 2006, p.65.

[42] ibid., pp. 101–113.

There are also no live chickens in the Vatican, though the Pope eats Roman chickens. The Dalai Lama told us years ago that his doctor insisted he eat chicken for protein.

The United States and China together make up at least a third of all chicken consumption in the world.[43] A mass industrialization of chicken slaughter, as Mark Cocker has pointed out, began perhaps in Delaware where a local cottage industry went from killing "50,000 birds in 1925…to 7 million" in less than 10 years. And then presidential candidate Herbert Hoover in 1928 promulgated the message of "a chicken in every pot and a car in every garage".[44]

While "scientific" experiments regarding egg-laying in cockerels can be dated to the 1650s in Denmark (along with the "castration of farm animals"),[45] a great interest in egg development was documented well before the time of Aristotle. Indeed, avian eggs and human consumption is part of the 2.6 million-year-old story of hominin carnivory.[46] Ancient Chinese domestication of chickens for their eggs dates back at least 8000 years.[47] Additional DNA data point to earlier domestication efforts in Southeast Asia. Across geography and time, the chicken has been ritually and customarily slaughtered, its white meat entrails providing an endless litany of superlatives by householders, philosophers, scientists, and culture after culture. Even Thoreau, writing of the cockerel's crow, states, "That crow is all-nature-compelling; famine and pestilence flee before it".[48]

Chickens have—as do many non-migratory birds—a detectable "magnetic sense," presumably to help them navigate.[49] Which lends yet an additional layer of tragedy to humanity's behavioral perversion of a bird who is clearly the most all-forgiving of any known species on the planet. We have usurped its primordial magnetic lures, replacing ancient natural destinations for these nomadic birds, with a singular endpoint: torture chambers.

---

[43] *Birds – A Visual Guide*, by Joanna Burger, Firefly Books, Weldon Owen Inc., 2006, p. 261.

[44] See *Birds & People*, by Mark Cocker, Photographs by David Tipling, Jonathan Cape, London, 2013, p. 64.

[45] See *The Wisdom Of Birds – An Illustrated History Of Ornithology*, by Tim Birkhead, Bloomsbury, New York 2008, pp. 277–279.

[46] See "Evidence for Meat-Eating in Early Humans," by Briana Pobiner (Human Origins Program, Smithsonian Institution) © 2013 Nature Education, Citation: Pobiner, B. (2013) Evidence for Meat-Eating by Early Humans. Nature Education Knowledge 4(6):1, https://www.nature.com/scitable/knowledge/library/evidence-for-meat-eating-by-early-humans-103874273/Accessed May 4, 2020. See also, "Ancient Humans Ate Cantaloupe-Size Eggs from 500-Poumd Birds," by Laura Geggel, Jaunary 29, 2016, LiveScience, https://www.livescience.com/53528-humans-contributed-to-flightless-birds-extinction.html, Accessed May 4, 2020.

[47] See FoodTimeLine Library, http://www.foodtimeline.org/foodeggs.html, Accessed May 4, 2020.

[48] Op.cit., *Birds & People*, by Mark Cocker, Photographs by David Tipling, Jonathan Cape, London, 2013, p. 67, citing *Thoreau on Birds*, ed. By Francis Allen, Beacon Press, Boston, 1993.

[49] See *Bird Sense – What It's Like to Be a Bird*, by Tim Birkhead, With Illustrations by Katrina van Grouw, Walker & Company, New York, 2012, p. 175.

What does all this mass slaughter and interspecies perversion say about human nature, expressly? Everyone who considers the horrifying contradictions will necessarily come to her/his own conclusion. In his work *Peaceable Nature: An Optimistic View of Life on Earth*, Stephan Lackner takes on Darwin, extolling the vegetarian giants of the earth, hippos, elephants, and rhinos, and in spite of all the violence and "man's unstable nature" concludes that "Life is a desirable good".[50] Lackner points out that a toad is estimated to consume "10,000 insects during each warm season." And asks if such violence and attrition "produce any noticeable improvement in the living apparatus of the surviving insects?" His answer is "No; the mosquito has remained the same through the eons".[51] It seems an imperative to reverse the correlation: has humanity, in its vast killing fields of chickens, in some way *improved*? Has the chicken (or the cows, the fishes, turkeys and pigs, etc.) become less of a chicken? This is more than a mere evolutionary paradox, these most provocative turns of fate, and deliberate, terrible ruses of nature. In the United Kingdom, "poultry" diversity was first displayed in 1845 in London. Some of the bird portraits showed up in the *Illustrated London News*. "Specimen birds" had become something of a cult.[52] Such that, while slaughtering them, we marveled at their beauty.

Chickens are the staple meat dish for the vast majority of all *Homo sapiens*. Breeds are particularly diverse in such countries as Belgium, Germany, France, Egypt, Italy, the Netherlands, Spain, the United Kingdom, and Vietnam. But their 600+ breeds span every nation. There are millions of pet chickens (not just at Easter), but that represents an estimated tiny fraction of the nearly one trillion chickens that have been slaughtered by humans during the past 40–50 years, an incomprehensible Holocaust.[53]

But of course today intensive poultry breeding, genetic manipulation, and outrageously cruel confinement continue more fervently than ever, though at least there has been a free range revolution in the last several decades, typically conferring some semblance of a normal, if utterly human-dominated, life upon those few lucky millions of chickens that are kept out of wire mesh jails and bunkers, where numerous avian disease outbreaks and continued infliction of pain 24/7 often keeps chickens from ever being able to stand up or spread their wings, 2–3% of them dying

---

[50] Stephan Lackner, *Peaceable Nature*, Harper & Row, Publishers, San Francisco, 1984, p. 18.

[51] ibid. p. 38.

[52] See *Farm Animal Portraits*, by Elspeth Moncrieff with Stephan and Iona Joseph, Antique Collectors' Club, Suffolk, 1996, Ch. 8 – "Developments in Poultry Breeding," pp. 276–288.

[53] See Lisa Boone, "Chickens will become a beloved pet — just like the family dog". latimes.com. August 26, 2017. See also, "A Warning from the Chickens of the World," David Waltner-Toews, The Walrus, July 8, 2020, republished in MAHB, July 22, 2020, https://mahb.stanford.edu/library-item/a-warning-from-the-chickens-of-the-world/ Accessed July 22, 2020.

from these afflictions long before they are gassed, or stunned, or electrocuted, or strangled, or simply slammed against a wall.[54]

Despite all these unspeakable horrors—one gasps, tears quickly, before the unspeakable aggregations of such videos on YouTube—amazingly, the merciful and loving chicken has remained spectacularly true to its nature, from all that can be divined. Asks one reporter, "Are chickens the new popular pet?"[55]

We know from personal experience and from the testimony of numerous friends around the world that chickens as companion animals provide, says one gardener friend, "the most constant, remarkable calming love imaginable. Every day I would return home, the chickens came running up to me to cuddle, to share warmth in my lap, to kiss and console me."

## 48.5   The Strange Tragedy of a Brain

Yet, if you watched any of the above-indicated videos, then you know that the human brain, in comparison with that of *Gallus gallus domesticus*, had manifested all of its own worst woes. Expressing hate and ecological doomsday for these precious birds. Why is that?

In his daunting biography of Friedrich Nietzsche (1844–1900), Rüdiger Safranski writes of Nietzsche's *The Birth of Tragedy* (1872), "When it comes to culture, he contended, a decision must be made as to its essential aim. The two major options are the well-being of the greatest possible number of people, on the one hand, and the success of individual lives, on the other".[56] Such utilitarian impulses, of course, have always been at the root of the ethical debate propounded by the likes of John Stuart Mill, Henry Sidgwick, and George Edward Moore. While Nietzsche became immersed in his quest for that which was "beyond good and evil," suggesting famously that "To live is to suffer, to survive is to find some meaning in the

---

[54] See, for example, "Investigation Into British Chicken Farms," by AnimalEquality, May 13, 2019, https://www.youtube.com/watch?v=-oORI_y2ccs; or "COK Investigation Exposes Chicken Industry Cruelty," September 21, 2009, https://www.youtube.com/watch?v=mlrdr2jHV1A; or "You need to kill him?': Tyson Food contractors caught on video mistreating chickens," by Justin Wm. Moyer,The Washington Post, December 6, 2017, https://www.washingtonpost.com/local/ you-need-to-kill-him-tyson-food-contractors-caught-on-video-mistreating-chickens/2017/12 /06/35ec4f58-d9fa-11e7-94b5-82a81b0f862e_story.html; or, "Hidden-Camera Exposes Criminal Animal Abuse at Chicken Slaughterhouse," March 28, 2015, https://www.youtube.com/ watch?v=V-RwqjtQmm8; or "Horrific Video Shows What Really Happens At America's Chicken Farms," by Melissa Cronin, The Dodo, November 21, 2014, https://www.thedodo.com/chicken-abuse-koch-foods-mfa-833876038.html; or, "Three Birds Every Second – High-Speed Chicken Slaughter Plant," March 19, 2019, https://www.youtube.com/watch?v=QdEaI7XQmF4.

[55] See article by this title, by Katherine Martinko, treehugger, August 29, 2017, https://www.tree-hugger.com/pets/are-chickens-new-popular-pet.html.

[56] *Nietzsche – A Philosophical Biography*, by Rüdiger Safranski, Translated by Shelley Frisch, W.W. Norton & Company, New York, London, 2002, p. 73.

suffering," there was the philosopher, sociologist, and psychologist Herbert Spencer (1820–1903) arguing that "biological and social progress formed one broad evolutionary continuum – that they were governed by the same immutable laws and controlled by the same forces of nature".[57]

In considering the vast number of chickens—the most numerous single vertebrate species on the terrestrial planet (in marine depths, the most multitudinous fish vertebrate is the tan bristlemouth Cyclothone) —issues of cruelty, of individual-versus-population-level ecodynamics, of the nature of success in biological terms, of the relevancy of mind, and the dictates of conscience, all converge in a dizzying challenge. That our being is witness to these endless atrocities of course afflicts the human heart, which remains inert and largely conspiratorial in the destruction of chickens, on into a dark, annihilatory unknown that is akin to a nuclear war. But our hearts are bleeding, in the worst, sentimental version of the cliché, while we do nothing as a species, or as individuals (in most cases). For each individual chicken, however, it is a very different story, the biology of earth having let her/him down entirely.

Population dynamics have not added up correctly, as they once did, until 10,000 years ago, or so, when domestication and slaughter of these birds began. That was the *real* birth of tragedy which, by its perpetuation against a single species, exploded into heinous violence between peoples.

In studying the chicken's brain, neurophysicists have sought to find correlations between behavior and the mechanics of the neuron. But given the "hundreds of billions of nerve cells, with hundreds of trillions of connects" any conclusive empirical explanation for who a chicken is, and what she/he thinks, is, in our opinion, but one more example of the *anthropic fallacy*.[58] The chicken's brain mass equals approximately 4 g, versus 1350 g on average in humans, 180 g for a pig, and 442 g for a cow.[59] Does that equate with the presumption—by most scientists and other humans—that we are smarter than chickens? That is the question, and one which researchers, comparing 4 g to 1350 g, for example, are quick to seize upon.[60]

---

[57] See *Darwin's Origin of Species – A Biography*, by Janet Brown, Douglas & McIntyre, Vancouver, 2006, p. 64.

[58] See "Inside the chicken brain," Washington University in St. Louis, EurekAlert! AAAS, December 6, 2007, https://www.eurekalert.org/pub_releases/2009-12/wuis-itc120609.php. See also, "Model chicken-brain circuit raises understanding of neural circuitry," Diana Lutz, Washington University in St. Louis, December 1, 2009, https://source.wustl.edu/2009/12/model-chickenbrain-circuit-raises-questions-about-understanding-of-neural-circuitry/Accessed May 5, 2020.

[59] See "How are brain mass (and neurons) distributed between humans and the major farmed land animals?" in Reflective Disequilibrium, September 10, 2013, by "Carl," https://www.blogger.com/profile/16384464120149476437, http://reflectivedisequilibrium.blogspot.com/2013/09/how-is-brain-mass-distributed-among.html?m=1, Accessed May 5, 2020.

[60] See "Why we're smarter than chickens," University of Toronto, August 20, 2015, citing the PTBP1 protein that allegedly fuels "evolution of mammalian brains to become the largest and most complex vertebrates." https://www.sciencedaily.com/releases/2015/08/150820144840.htm, Accessed May 5, 2020.

But does brain mass, number of neurons, and synaptic connections actually provide a meaningful window on intelligence, let alone comparisons of all those nuances by which we presume to mean *intelligence*? The red junglefowl has 221 million neurons, a prairie dog—with its highly articulated communication skills—473,940,000 neurons,[61] blue-and-yellow macaws, 3.136 × 10 [9], a human 8.6 × 10 [10].

Encephalization quotients (EQ) and allometrics—anatomical brain ratios that encompass physiology, anatomy, and behavior—have also been accorded some degree of confidence, but it again comes down to brain size. Hence, the EQ for a human is maximum 7.8, and for a hippopotamus, 0.37.[62] Research has pointed out that large brains and number of neurons in the cerebral cortex are not necessarily correlated, as canids have revealed[63]—hence, the so-called paradox of the elephant brain and the attendant human bias.[64] Efficacy signals? Mental and intuitive qualia levels? Subjective circuit-breakers? Emotional intelligence? Ethical and spiritual intelligence? Longevity? Population sustainability? None of these quotients can be measured, yet they are crucial to actually formulating a theory of mind that is relevant, let alone subject to meaningful comparisons.

---

[61] See *Prairie Dogs: Communication and Community in an Animal Society*, by C. N. Slobodchikoff, Bianca S. Perla, and Jennifer L. Verdolin, Harvard University Press, 2009. Octopus, 500,000,000.

[62] See Schoenemann, P. Thomas (2004). "Brain Size Scaling and Body Composition in Mammals". Brain, Behavior and Evolution. 63 (1): 47–60. doi:https://doi.org/10.1159/000073759 . PMID 14673198. See also, Lefebvre, Louis; Reader, Simon M.; Sol, Daniel (2004). "Brains, Innovations and Evolution in Birds and Primates". Brain, Behavior and Evolution. 63 (4): 233–246. doi:https://doi.org/10.1159/000076784. PMID 15084816. See aso, Cairó, Osvaldo (2011). "External measures of cognition". *Frontiers in Human Neuroscience. 5*: 108. See also, Suzana Herculano-Houzel, *The Human Advantage: How our brains became remarkable*. MIT Press, Cambridge, Massachusetts, 2017.

[63] See "Dogs Have the Most Neurons, Though Not The Largest Brain: Trade-Off between Body Mass and Number of Neurons in the Cerebral Cortex of Large Carnivoran Species," by Debora Jardim-Messeder, Kelly Lambert, Stephen Noctor, Fernanda M. Pestana, Maria E. de Castro Leal, Mads F. Bertelsen, Abdulaziz N. Alagaili, Osama B. Mohammad, Paul R. Manger, and Suzana Herculano-Houzel, Frontiers in Neuroanatomy, Front. Neuroanat., 12 December 2017 I https://doi.org/10.3389/fnana.2017.00118, https://www.frontiersin.org/articles/10.3389/fnana.2017.00118/full, Accessed May 5, 2020.

[64] "The Paradox of the Elephant Brain – With three times as many neurons, why doesn't the elephant brain outperform ours?" by Suzana Herculano-Houzel, Illustration by Hannah K. Lee, Nautilus, April 7, 2016, http://nautil.us/issue/35/boundaries/the-paradox-of-the-elephant-brain. Researchers have also sought to explain the impact of a single gene in the human brain, namely, the ARHGAP11B, "a gene found only in humans, [that] is known for its role in expanding neocortex...." "Brain Gene Tops the List for Making Humans, Human," by Courtney Sexton, June 18, 2020, Smithsonian Magazine.

## 48.6   Divisions That Fail to Unite

In a landmark essay, "The Compartmentalization of the Cerebellum," by Karl Herrup and Barbara Kuemerle,[65] studies which began on the embryogenesis of Drosophila (small fruit-flies) resulted in new advances in the description and partial understanding of human "cerebellar organization." These insights provided a clearer topography of brains comprising "developmental compartments," subdivisions. This picture resolved down to a conclusive revelation that suggested "the cerebellum is a nest of paradoxes" —a "little brain" (the Latin), which is situated "at the back of the brain stem at the midbrain-hindbrain junction" and contains what is estimated to be "80 to 85% of all human neurons." But what tradeoffs has the human brain allowed itself? What metabolic ratios of efficiency have evolved? And has such efficacy correlated with an ethic? Or has the neural efficacious grown up unchecked by moral or, for that matter, intellectual constraints? To be witness, for example, to the liberation of Auschwitz on January 27, 1945,[66] or viewing those earlier referenced YouTube videos of chicken slaughterhouses, to have the physiological weight of experience, the very luxury that would give us to learn, or not, from our histories? Of the collapse of one civilization after another?

All the statistics add up to the quintessential paradox of human dietary schizophrenia, denial, and innumerable madnesses. Considering, for example, the number of cases of Covid-19 at slaughterhouses, the majority of viral hotspots (along with nursing homes); the fact that while an estimated 25% of all those between 25 and 34 in the US support vegetarianism or veganism, the majority of the rest of the country seems to have subsumed the whole issue of where their food comes from under so many layers of denial that they can no longer retrieve any semblance of environmental and health reality. Even the author of the *New York Times* Op Ed "The End of Meat Is Here" (Jonathan Safran Foer) himself points out climate change, income inequality, and then the very "paradoxes in my own eating life...." Data from Project Drawdown that declares a plant-based diet "the most important contribution every individual can make to reversing global warming" and so on.[67] None of it evens out, and gives no indication—despite a shuffling of numbers—of the least trend toward non-violence, of the human population scales.[68]

---

[65] Annual Review of Neuroscience, Volume 20, 1997, pp. 61–90, https://www.annualreviews.org/doi/full/10.1146/annurev.neuro.20.1.61#_i4, Accessed May 5, 2020.

[66] What a Soviet soldier saw when his unit liberated Auschwitz 70 years ago". The Washington Post. 27 January 1945. See also, http://auschwitz.org/en/history/evacuation/the-cessation-of-mass-extermination, Accessed May 5, 2020.

[67] https://drawdown.org.

[68] "The End of Meat Is Here - If you care about the working poor, about racial justice, and about climate change, you have to stop eating animals" by Jonathan Safran Foer, May 21, 2020, The New York Times, Opinion Piece,https://www.nytimes.com/2020/05/21/opinion/coronavirus-meat-vegetarianism.html?smid=em-share, Accessed May 22, 2020. See also, the feature documentary, "Mad Cowboy," (109 minutes), A Voice For A Viable Future/Dancing Star Foundation/KQED-PBS Production, 2006; See http://www.humanedecisions.com/documentary-film-mad-cowboy-the-story-of-howard-lyman/. Accessed August 8, 2020.

If the number of folds, neurons, compartments, divisions, sub-divisions of the brain should matter, then why—and this not merely rhetorical—has human existence fallen far behind that of chicken civilization—its artistic, scientific, and ethical expressions notwithstanding. Why, then, do we fail to find solace in the outer manifestation of such compartments—in tribes, Republics, "peoples," and countless boundaries? All have proved to be obstacles to peace, hurdles to solace, outright territorial aggression that has more than defined us.

Clearly our size shrinks beside whales, sequoias, and polar bears. And the number of our combined neurons is nowhere close to the civilization of ant or termite neurons across the planet. Such numbers, which can be manipulated endlessly, do not translate into the level of intelligence and moral understanding that would honor and respect non-violence. This, above all else, is an ecological paradox without any conceivable resolution, or not in our time, given current evidence (Fig. 48.6).

**Fig. 48.6** Home At Last, Gut Aiderbichl Sanctuary, Salzburg, Austria. (Photo © J.G.Morrison)

# Chapter 49
# The Christ Paradox

## 49.1 Christ Imagery: The Raw Ideals

Iconographical images of Christ are known to have been circulated at length by the third century.[1] Carlo Cecchelli cites Eusebius' (260?–340?) *Historia ecclesiastica* (VII, 18) in which already there was mention of icons depicting Christ and two apostles, Peter and Paul. Cecchelli's analysis of the origins of the youthful, lean, bearded, and long-haired version of Christ largely adopted throughout art history derives in part from the Mandylion, or Image of Edessa given in Eusebius' recounting to King Abgar of Edessa, who is said to have received the image on cloth and a letter from Christ himself. Abgar V died around AD 40, between 7 and 10 years after Christ moved on. The image, the letter, engendered a perennial challenge with respect, of course, to the whole endeavor of anthropomorphism and what Cecchelli describes as that singular course of events in Western history whereby "The God of Israel came down to man, became man, the servant of mankind." And that "his sanctity 'appears to us in the fullness of an actual living fact, and no longer in the customary language.'" And, finally, that "The process of humanization this kenosis ('emptying of himself'), of divinity provided a legitimate opportunity for its representation[2]" (Fig. 49.1).

At the time of Christ the human population was approximately 150–200 million.[3]

By the First Council of Nicaea (325) when Arianism (the theory that Christ was not divine, but created) was contested, and the notion of homoousios ("one substance", meaning that Christ was deemed to be unified with the Father, God, and

---

[1] See "Christ" by art historian Carlo Cecchelli (1893–1960), *Encyclopedia Of World Art*, Volume 3, McGraw Hill Book Company, Inc., New York, Toronto, London, 1960. pp. 596–601.

[2] ibid., Cecchelli, p. 396.

[3] See https://www.vaughns-1-pagers.com/history/world-population-growth.htm, Accessed May 5, 2020.

© Springer Nature Switzerland AG 2021
M. C. Tobias, J. G. Morrison, *On the Nature of Ecological Paradox*,
https://doi.org/10.1007/978-3-030-64526-7_49

**Fig. 49.1** The Corcovado Christ the Redeemer (Completed in 1931) Overlooking Rio de Janeiro. (Photo © M.C. Tobias)

that this, henceforth, would become the very essence of Christianity),[4] humanity numbered close to 250 million.[5] Based upon early census data from the Mediterranean and Near East, we may surmise that at least 30% of the human population would have heard something of Christ by the time Constantine I (272–337 AD, of what today is İznik, Turkey) called for the meeting in which the earliest formulations of both church doctrine and the mechanics of its clerics would be drafted. In all of the demographic tumult since that time, and some 60 to 70 human generations, the ideal that someone else—combined mortal, immortal, hybrid, tripartite versions of Christ in many hundreds of denominations—miraculously persisted is itself an astonishing premise regarding human need, vulnerability, guilt, and the preoccupation with interpreting that realm between myth and reality, which typically constitutes a perceived ecosystem of some kind.

Art historian Élie Faure wrote that it would "require ten centuries of seclusion before it [Christian ideals] finds its real expression... to return to the deeper life and to embrace the hope which has been set free".[6] That was the freedom of a Saint Francis who, as Faure so beautifully characterized, "trembled," "laughed," "wept for joy"... "He never ceased loving. He fell asleep and awoke under the trees. He called the beasts to him, he sang, warbled, and whistled with them.... He asked

---

[4] See https://www.britannica.com/event/First-Council-of-Nicaea-325, Accessed May 5, 2020.

[5] ibid., Vaughns.

[6] *History of Art,* "Mediaeval Art," Garden City Publishing Co., Inc., Garden City, NY, 1937, p. 263.

counsel of the crickets and they gave it to him, and he did not hesitate to follow it".[7] Today, it is estimated that there are over 2 billion Christians (some allege as many as 2.4 billion) or nearly 30% of the human population. In the United States alone there are an estimated 384,000 congregations (as of 2012).[8]

## 49.2 Nuances of a Spiritual Rallying Cry

This rallying cry behind a true life, or an allegorical one, or both, challenges ecological ethics with all of the life-forms, one species in particular, and their own individual lives, as implicated from the Biblical version of the Creation through to the life of one Christ, and the astonishing longevity of the Bible's many controversial parts. These elegant but controvertible details that build up into the thousands of sculptural edifices—from Chartres to Liverpool Cathedral; from the Abbey of Montecassino to Mount Athos; from St. Catherine's Monastery to the Sumi Baptist Church in Zunheboto, Nagaland (India), considered one of the largest churches in Asia—all enshrine a goal that is, at its heart, an ecological pang.

Its nuances span the contradiction that sees human nature deeply conflicted over sacrifice and the idea of salvation; the value of one living being against another; the quintessence of compassion for someone who is real, in the hopes of resolving all violence, to make it unreal. And the incorporation of those numerous landscapes of desert asceticism, of the flight into Egypt (Matthew 2:13–23), lush Renaissance geographies of darkness and light, pain and the surmounting of pain, of every paradise and post-paradise image. Of one man so persuasive and charismatic (Jesus) as to transform landscapes, biodiversity, suffering, into the *idea* and then the *ideal* of those things (Fig. 49.2).

There is Martin Schongauer's "Crucifixion, From the Passion," ca. 1470–1482.[9] The mesmerizing and most personal "Pietà" by Annibale Carracci (1599–1600), and the earlier etching, as well as a second etching by his older brother Agostini Carracci, ca. 1598. Or Rembrandt's drypoint of 1653 depicting "The Three Crosses," with probable commentary borrowed from St. Luke's and St. Matthew's descriptions of "…darkness over all the earth until the ninth hour; and the sun was darkened. And the veil of the temple was rent in the midst".[10] And then by common

---

[7] ibid. pp. 398–400.

[8] Christianity Today, "How Many Churches Does America Have? More Than Expected," by Rebecca Randall, September 14, 2017, https://www.christianitytoday.com/news/2017/september/how-many-churches-in-america-us-nones-nondenominational.html, Accessed May 7, 2020.

[9] See Jacob Rosenberg, *Martin Schongauer – The Drawings*, R. Piper & Co., Munchen, 1923, #8. See also, "The Drawings of Martin Schongauer," by Max Lehrs, The Burlington Magazine for Connoisseurs, Vol. 44, No. 252 (Mar., 1924), pp. 133–134, 136. See in addition, *le beau Martin – Gravures Et Dessins De Martin Schongauer* (vers 1450–1491), 13 septembre – 1er December 1991, Musée D'unterlinden, Colmar, 1991.

[10] See Rembrandt *As An Etcher – A Study of the Artist at Work*, by Christopher White, 2nd edition, Yale University Press, New Haven and London, 1999, pp. 77–81.

**Fig. 49.2** "The Crucifixion," (ca. 1470–1480), with the Virgin Mary, Mary Magdalene, Mary of Jacob and Mary of Clopas to the Left. St. John to the Right. By Martin Schongauer – 1448–1491. Engraving. (Private Collection). (Photo © M.C. Tobias)

assent, for three days Christ, "the lamb of God" having died, was in paradise prior to his resurrection (anastasis) and triumphant exaltation, the fuel of visionary succession. A close reading of "Hebrews"[11] focuses our attention upon what might be described as the core ecological contradiction in the entire New Testament. That he died for *our* sins. And who are we? Or, as Alice re-wrote the moment, Who are You? His "one offering" that eclipses for all time all those other sins, particularly the killing of "bulls" and of the "heifer." History and pre-history being invoked to remind all those who believe in this saga that they will take part in "the saving of the soul." As crystallized in Hebrews 9:12, "Neither by the blood of goats and calves, but by his own blood he entered in once into the holy place, having obtained eternal redemption *for us.*" And, reiterated, as in John 14:6, "I am the way, and the truth, and the life".[12]

---

[11] *The Holy Bible Containing The Old And New Testaments,* Authorized King James Version, The Christian Science Publishing Society, Boston, Massachusetts, Chapter 10, pp. 1497–1498.

[12] See "Compelling Truth – the truth about: Salvation", CompellingTruth.org Newsletter; https://www.compellingtruth.org/Jesus-died-for-our-sins.html, Accessed May 7, 2020.

A brief sketch of dominant thematic obsessions at such institutions as the British Museum, the National Galleries, the Hermitage, the Louvre, the Vatican Museums, the Groeninge Museum in Bruges, and the Prado easily emblemizes the Western world's undying fascination with the very conceptual ecosystem that is the life of Christ, and the orisons, genealogies, Biblical geography that have made luminous and clear a call to action, just as Pope Francis issued his encyclical, "Laudato Si" of May 24, 2015, on climate change;[13] and Ecumenical Patriarch Bartholomew promulgated his talks, "On Earth As In Heaven: Ecological Visions and Initiatives",[14] following upon his extraordinary "Address" at the "Environmental Symposium, Saint Barbara Greek Orthodox Church, Santa Barbara, California," in 1997, during which Bartholomew declared: "It follows that, to commit a crime against the natural world, is a sin. For humans to cause species to become extinct and to destroy the biological diversity of God's creation; for humans to degrade the integrity of Earth by causing changes in its climate, by stripping the Earth of its natural forests, or destroying its wetlands; for humans to injure other humans with disease; for humans to contaminate the Earth's waters, its land, its air, and its life, with poisonous substances; these are sins".[15]

In one recent exhibition from the British Museum, "Christ: Life, Death & Resurrection –Italian Renaissance Drawings & Prints from the British Museum",[16] the environmental liberation of one soul on behalf of all others—our reading of it—mirrors similar sagas throughout Asian spiritual traditions, encompassing not merely the ravaged body on the cross, as in Giulio Clovio's "The Crucifixion," c. 1568[17] or Andrea Boscoli's "The Crucifixion," c. 1599,[18] but landscapes as moving as those in Hieronymus Bosch or John Martin or Sebástio Salgado at their most dramatic, as in the case of Giovanni Benedetto (or Salvatore) Castiglione's "The

---

[13] http://www.vatican.va/content/francesco/en/encyclicals/documents/papa-francesco_20150524_enciclica-laudato-si.html, Accessed May 7, 2020.

[14] Edited and with an Introduction John Chryssavgis, Fordham University Press, Bronx, New York, 2012.

[15] Ecumenical Patriarchate, "Remarks As Prepared For Delivery Address Of His All Holiness Ecumenical Patriarch Bartholomew At The Environmental Symposium Saint Barbara Breek Orthodox Church Santa Barbara, California, 8 November 1997," https://www.patriarchate.org/-/remarks-as-prepared-for-delivery-address-of-his-all-holiness-ecumenical-patriarch-b-a-r-t-h-o-l-o-m-e-w-at-the-environmental-symposium-saint-barbara-g, November 8, 1997; see also, https://www.apostolicpilgrimage.org/the-environment/-/asset_publisher/4hInlautXpQ3/content/address-of-ecumenical-patriarch-bartholomew-at-the-environmental-symposium-saint-barbara-greek-orthodox-church-santa-barbara-california/32008?inheritRedirect=false, Accessed May 7, 2020. This, in turn, derived in part from the symposium on "Revelation and the Environment, AD 95 to 1995" to commemorate "the 1900th anniversary of the recording of the Apocalypse."

[16] September 8, 2019 – April 19, 2020, Catalogue Published by The British Museum, Foreword by Hartwig Fischer, "The Life of Christ" essay by Hugo Chapman, University of San Diego, San Diego, CA 2019.

[17] ibid., p. 50.

[18] ibid., p. 62.

Crucifixion," 1631 (−1670), oil on paper,[19] a version of what is essentially the same brush and oil, gouache, red chalk, and partially dry brush by Castiglione in the Hermitage. Storm clouds, sunlight, astonished onlookers, Christ himself hanging in a wind-scorched-seeming rapture of incandescent hurricane-force auras. So unlike El Greco's almost introverted "Christ on the Cross Adored by Two Donors," ca. 1580, a study in historical, devotional quiescence, fealty, and gratitude, within the very cavity of a massive storm.[20]

We know that that storm has hit. Hit us all. Rev. Terrence P. Ehrman, C.S.C. Holy Cross at Notre Dame, has written a very amenable essay in an effort to help dispel the provocative digest of 2000 years of cultural events, propagated in 1967 by historian Lynn White Jr., who "blamed Christianity as being the root cause of the ecological crisis." Writes Ehrman, "A rediscovery of God as Creator has at least two effects. It calls forth a new relationship to God and to the rest of God's creation. For the Christian, it calls for 'ecological conversion', whereby the effects of their encounter with Jesus Christ become evident in their relationship with the world around them".[21]

By whatever surrogate form of logic or candescent reach one approaches religions or spirituality, what is undeniable is the Passion itself. Which, it must not be underestimated, has inspired much of the most enduring aesthetic articulations ever produced by our species. Think of the *Bach Passions* (1724 and 1727), for example. Or Handel's *Messiah* (1741). The fact of unfathomable suffering in this world has lent to us the heart through which to see, to hear, to touch, to feel, to affiliate with and protect so much beauty; and to reciprocally recognize the fragility and very sensory organs of biodiversity herself; of every sentient being; and the consequential role of humanity in either continuing to perpetrate the unbearable horror of the entire crucifixion, or halting it, once and for all, here and now (Fig. 49.3).

---

[19] ibid., p. 69.

[20] See *Paintings In The Louvre*, by Lawrence Gowing, Introduction by Michel Laclotte, Stewart, Tabori & Chang, New York, 1987, p. 275.

[21] "What Does Ecology Have to Do with Christ?" by Terrence Ehrman, Center For Theology, Science & Human Flourishing, September 01, 2016, https://ctshf.nd.edu/posts/what-does-ecology-have-to-do-with-christ/. Accessed May 7, 2020.

**Fig. 49.3** "Pietà, Christ of Caprarola (Lamentation)," by Agostini Carracci, 1598, after the 1597 Etching/Engraving/Drypoint by his Younger Brother, Annibale Carracci. (Private Collection). (Photo © M.C. Tobias)

# Chapter 50
# Unthinkable Nullities, Negative Proofs

## 50.1 Hypotheses of Hope amid Nothingness

When we try to enter the realm of that Asiatic and pre-Socratic fever to transcend philosophy, in a word, *nothingness*, we necessarily confront Jean-Paul Sartre's startling magnum opus, *L'Être et le néant : Essai d'ontologie phénoménologique*.[1] The abstruse leaves of levitating through are obviously there, peering down at our own perplexed efforts to grapple with Hegel's influential remark in his work *Greater Logic* (also known as *The Science of Logic*, from Chap. 1), "there is nothing in heaven or on earth which does not contain in itself being and nothingness".[2]

Sartre is clear on this Hegelian paradox when he says of "nothingness" that "It comes into the world by the For-itself and is the recoil from fullness of self-contained Being which allows consciousness to exist as such".[3]

Sartre had gone to great literary lengths to prove that Freud's patients struggled not with suppressed unconscious revelations, but with an internal "censor" that was part of their knowing consciousness, simply acting in "bad faith." There is an important ecological equivalency reliant upon a paradox of proofs, the lack of absolutes, whether speaking of existence or non-existence. We prove nothing. What we know, don't know; grasp, fail to grasp. Both sides of the arbitrary equation are that umwelt of discretion that can be equally formulated in propositions that affirm, on either side of their Being, a neutral but vetted series of consciously formulated mathematical notations. In essence, it is a ponderous pari passu—side-by-side—orientation to

---

[1] Éditions Gallimard, Philosophical Library, Paris, 1943.

[2] Cited by Sartre in his section on "The Dialectical Concept Of Nothingness," See Jean-Paul Sartre, *Being and Nothingness*, Translated by Hazel E. Barnes, Philosophical Library, New York, 1956m p. 13. https://academiaanalitica.files.wordpress.com/2016/10/georg_wilhelm_friedrich_hegel__the_science_of_logic.pdf.

[3] op.cit., Sartre, p. 633.

© Springer Nature Switzerland AG 2021
M. C. Tobias, J. G. Morrison, *On the Nature of Ecological Paradox*,
https://doi.org/10.1007/978-3-030-64526-7_50

ecological symmetry within communities[4] that is this consciousness we are each so excited about; condemned to it as our lungs know only the oxygen upon which they survive.

Philosopher Stephen Hales effectively argues that we can neither *prove* existence or non-existence, but that induction is crucial to every philosophical, and by implication, ecologically mathematical enterprise.[5] To induce, however, is not to live. True, enumerative induction allows for the naturalistic conflation of ecology and consciousness, engendering the widest possible gulf of uncertainties. Arising from uncertainty, the prepositions in every language argue for the definitive, something solid. But this is no certain axiom. Crickets and dolphins, blue whales, and albatross defy substantiality with protean circumlocutions. From any philosophical perspective, details of the *biosphere* exist free, and for all possible deliberations and analysis, not that such analysis is necessarily anything but pointless. That is, admittedly, the cynical path.

A more optimistic one embraces the fervent hope, notwithstanding no shred of evidence, that our love of nature has actually amounted to something. But first ask the chicken, the jaguar, the mountain gorilla. Regardless of our inflictions and simultaneous exultations, no one knows true philosophical balance. Dialectics are genetically hard-wired. But since we identify basic pain and pleasure markers, among numerous other representative senses of the environment, we have, however inert, inept, or ingenious our response mechanisms may be, a vested interest, if not desperate hope, that there are, in fact, meaningful connections to be enlisted and trusted, between ourselves and the outer world.

The aspired-to balance, the ultimate fall-out of that dialectical meaning, in turn, bio-invades our consciousness, takes up habitation to various degrees, enters *mind* just as Sartre argued that nothingness pivots between our "self-contained Being" and "consciousness" itself. These two opposites are merged in the very act of paradoxical thinking, which must rely on the attraction of associations of all kinds, not merely opposites but biochemical gradations. Infused by those increments of life, afresh or "unacknowledged legislators of the world" (Shelley writing of poets in his essay "A Defence of Poetry," 1821) science departs abruptly from metaphysics when it poses the reality of its biological opinions. These are the products of hypotheses which, by the bias of countless disciplines and sub-disciplines, inform the statistical inferences that hover over every *fact*.

---

[4] See "The neutral theory of biodiversity and biogeography and Stephen Jay Gould," by Stephen P. Hubbell, BioOne Complete, Paleobiology, 31(sp5): 122–132 (2005), https://doi.org/10.1666/0094-8373(2005)031[0122:TNTOBA]2.0.CO;2, https://bioone.org/journals/Paleobiology/volume-31/issue-sp5/0094-8373(2005)031%5b0122:TNTOBA%5d2.0.CO;2/The-neutral-theory-of-biodiversity-and-biogeography-and-Stephen-Jay/10.1666/0094-8373(2005)031%5b0122:TNTOBA%5d2.0.CO;2.short, Accessed May 9, 2020.

[5] Steven D. Hales, "Thinking Tools: You can Prove a Negative" (PDF). 2005, Think. 4 (10): 109–112. https://doi.org/10.1017/S1477175600001287.

## 50.2   For Itself

That reality (the continuing challenge model)—beyond the debates over the extent, efficacy or meaning of consciousness—is subject to and invokes the Null Hypothesis ($H_0$) of non-relational entities, often utilized in medical testing, randomized placebo-controlled double-blind trials, exact, inexact, point, and composite hypotheses, among many other applications, all seeking precise parameter values[6] which coincide with biological determinism. V-shaped, U-shaped, W-shaped recoils, some of the comebacks from near-nullity that economists employ to characterize and project future prospects, even and uneven shocks, are all *probability values*. We can prove by small margins of error any number of global depredations, pollution, slaughterhouse attrition, poaching, every form of cruelty, and extinction, but also evolutionary occurrences, even patterns of sedimentation, the laying down over epochs of fossils, and the variance in dancing bumblebees. Frankly, our application of composite hypotheses to probability models reads as very good fiction, even while the underlying protagonists of such drama are representational of other protagonists. Together, they add up to whatever, whoever we want them to be.

These are nullities within mathematical certainty. Absolute zeros where human attribution is entirely subjective. We can equally attribute "success" in the form of progress where Pp (positive progress) is "greater" than zero, or nullity, or D (decline). These arbitrary attributions, when fitted in the library of eventual equations, can be viewed under the observational acuities of incremental change, which is the argument for theoretical ecology, and by correspondence, consciousness that avows Being, and supports it to the extent of both acknowledging and advocating for, biodiversity and non-violence.

## 50.3   The Marriage of Consciousness and the Biosphere

Just as catechisms in Catholicism can issue Declarations of Nullity (pertaining to marriage),[7] the marriage of consciousness and the biosphere argues for a reckoning that, to date, knows no other formulas than those embedded, somehow (a genetic and/or spiritual conundrum) in the conscience. The megatonnage of our physical throw-weight makes our specific, species-wide proclivities for conscience

---

[6] See Jones, B; P Jarvis; J A Lewis; A F Ebbutt (6 July 1996). "Trials to assess equivalence: the importance of rigorous methods". BMJ. 313 (7048): 36–39. https://doi.org/10.1136/bmj.313.7048 .36. PMC 2351444. PMID 8664772.

[7] Second Edition, Promulgated by Pope John Paul II. Catechism of the Catholic Church. Citta del Vaticano: Libreria Editrice Vaticana. p. 904; See also, "Catholic Church's annulment criteria when deciding whether to grant 'declaration of nullity,'" by Tom Mooney, September 18, 2014, Providence Journal, https://www.providencejournal.com/article/20140918/News/309189989, *Accessed May 8, 2020.*

a truth-value of varying degrees that is the make-or-break inflection point for Pp >
D, at the species and population levels, and, as a philosophical problem within the
Individual, for the totality of Phyla. Belief in such elementary connections is at the
heart of applied ecological ethics which, in turn, reminds us that our Being, our
Consciousness, and our Conscience together provide the only known codes and
pathways for restoration and amelioration, as opposed to mass ecocide.

"For itself," "self-contained Being"[8] places conventional Continental Logic
(post-Kantian reflection upon the empirical temptations to simply correlate all perceived phenomena with the natural sciences) into the turbid cycles of Karmic cosmology, in other words, into the phraseology of a Being that only imagines itself
for-itself. Whereas *for-itself*, if contextualized biologically, necessitates every
inflection of interdependency. The probability of negative proofs, of double and
triple negatives in the thought process evincing language and co-dependent ideas,
overwhelms any volition that would seek pure isolation from the interdependent Pp
(again, "positive progress") which avows a biological physicality that embodies it.
Believing otherwise is contrary to our senses; embodying a paradox which—at the
ecological frontier—forces our ideas and ideals, all nihilism and fatalism to reinvent
such desperation in view of a lasting affirmation.

Stepping outside a laborious set of logical types and abstruse denominations,
suffice to say that, to affirm is to live, even if both one's affirmative propositions and
*will to live* pose contradictory crises for the Self, which they do.

To live beyond the aphorisms ("life is beautiful," seize the day, etc.), is to also
coordinate the psychology which either embraces or rejects futilities, and the endless speculation that hampers motor activity or any other embrace of real objects,
goals, boundaries, or lack of them, within any given time-frame.

All affirmation is problematic, if one dwells on causality and the remains of any
full thought; verging on a manic state by almost any definition, because its etiologi-
cal truths all rely upon those objects of consciousness that will die, or have died. Yet
continues ad infinitum to affirm life, regardless. This is our biology. Such optimism
combines thought processes with empirical experience, observations of other observations, or merely intuition and are applauded at the root-cause of all biodiversity.
This, in itself, propagating that "Great Chain of Being" (as depicted in Didacus
[Diego] Valades' 1579 work published in Perugia, *Rhetorica Christiana*) lends to
apperception the very metaphors and similes out of which all ecological cognition
and intuition combine to assert the verities of the world (Fig. 50.1).

Apperceptive de-infinities of thought ultimately grow weary of the process
because, as has been reiterated, the mathematical proportions rapidly devolve
toward an infinity that graces no character, no attitude, with anything like certainty.
Regardless of the order of magnitudes by which thought cascades, the interloping
consciousness commandeers this metaphysic we know to be some version of our
reality, and it is a distinctly two-fold paradox.

---

[8] op.cit., Sartre, p. 633.

**Fig. 50.1** 1579 drawing of the Great Chain of Being from Didacus Valades' *Rhetorica Christiana*. (Public Domain)

## 50.4 Scalar Multipliers and Gap Analyses in the Evaluation of Change in Nature

First, we accept that evolution or some equivalent process, has furnished a matrix algebra with its scalar multiplications and transpositions. Human scale is our anthropic lens and, short of a transcendental experience that turns physical (as in the earlier discussion of the Buddha's enlightenment, or personal resolve to embrace and get beyond Hildrop's Paradox), this scale condemns or liberates us to the properties that guide numbers, as well as the ultimate zero matrix and idempotent (unchanging value) matrix ($A^2 = A$ in every iteration; $1 \times 1 = 1$; $0 \times 0 = 0$; a person times his/her multiples of ideas is the same person). But there is nothing to stop $1 \times 1$ from adding another $1 \times 1$, and that changes everything in the world.

Second, the algorithm which seeks to identify those areas between numerical stability and indeterminate bounds or levels of complexity (data-compromised domains) that can account for the natural world are teeming with fissures, discrepancies, gray areas between what seems to be coherence and incoherence, which are the differentials that allow for modeling, scoring, rating, in fact, all delineations.[9] For ecological speculation that seeks to solve the identity crisis within the greater calamities befalling the entire biosphere, there are empirical types of data that easily confirm our depredation theory of Being, namely, that we do sense and can generally measure levels of nullity.

We can prove, in other words, the negative through its very amplification. Examples include Kazimerz Fajans' and Frederick Soddy's law pertaining to radioactive decay, simultaneously witnessed in 1913.[10] Or the earliest detection of cultural eutrophication, or any toxic or epidemiological threshold or non-linear tipping point dynamic.[11] Other dominant trends within vigorous élans of ecosystem change include glacial movement (first theorized by Johann Jakob Scheuchzer regarding the Rhône Glacier in 1705);[12] succession, as in the case of climax communities, as first described by Frederic Clements early in the twentieth century;[13] and continuing gap analyses, as instanced in the cataloguing of biological coldspots and hotspots and all the gap analyses and vulnerability-increments between them.

Specifically, some of the most famous such gaps regard animal census data from places like the Serengeti, starting with the fly-overs by Bernhard Grzimek, director at the time of the Frankfurt Zoo, who, during the late 1950s, first counted "exactly 366,980 animals – chiefly the highly visible grass eaters" out of an overabundance of concern about the survival probabilities of East African herbivores,[14] Other

---

[9] See Roger A. Horn and Charles R. Johnson, *Matrix Analysis*, Cambridge University Press, Cambridge, UK, 1990.

[10] See, for example, "The Suicidal Success of Radiochemistry," by Lawrence Badash, The British Journal for the History of Science, Vol. 12, No. 3, November 1979, pp. 245–256, Cambridge University Press on behalf of The British Society for the History of Science, https://222.jstor.org/stable/4025999, https://www.jstor.org/stable/4025999?seq=1, Accessed May 9, 2020.

[11] See David W.Schindler and John R. Vallentyne, *The Algal Bowl: Overfertilization of the World's Freshwaters and Estuaries*, University of Alberta Press, Edmonton, Alberta, Canada, 2008. See also, F. E. Round, *The ecology of algae*, Cambridge University Press, Cambridge, UK, 1981, pp. 526–541.

[12] See his *"Itinera per Helvetiae alpinas regiones facta annis 1702–1711"* as described in "The discovery of the ruins of ice: The birth of glacier research," by David Bressan, January 3, 2011, Scientific American, https://blogs.scientificamerican.com/guest-blog/the-discovery-of-the-ruins-of-ice-the-birth-of-glacier-research/. Accessed May 9, 2020

[13] Frederic E. Clements, *Plant Succession: An Analysis of the Development of Vegetation*. Washington D.C.: Carnegie Institution of Washington, 1916.

[14] See "Statisticians on Safari: Our Methods for Counting the World," in *Life Counts – Cataloguing Life on Earth*, by Michael Gleich, Dirk Maxeiner, Michael Miersch, and Fabian Nicolay, In Collaboration with UNEP, IUCN WCMC and Aventis, Translated by Steven Rendall, Atlantic Monthly Press, New York, and Berlin, 2000, pp. 46–47.

**Fig. 50.2** "La Moisson" (The Harvest), From a 1704 French Coloring Book. (Private Collection). (Photo © M.C. Tobias)

chasms of existing data bases and increments with respect to animal behavior in general were brought to the fore with the realizations of Imprinting Theory with Greylag Geese, detected (experienced) by Konrad Lorenz in 1935[15] (Fig. 50.2).

## 50.5 Affirming Nature

Every affirmation rests upon premises which entitle that supposition which is disposed toward self-perpetuation amid Others. This is mathematical tautology pre-biasing the expressions of all interactions between statistical units, for example, host-pathogens, parasitism, asymmetric phenomena across ecosystems. Stephen Hubbell's "united neutral theory" holds that "simple bordered symmetric random walks"[16] are comprised, no matter which dimension, of stochastic (random) choices. Population choices are sampled, with typically 90% + confidence levels in the inter-

---

[15] See Lorenz' last book, *Here I Am -Where are You? A Lifetime's Study of the Uncannily Human Behaviour of the Greylag Goose*, by Konrad Lorenz, Michael Martys, and Angelika Tipler, Translated by Robert D. Martin, Harcourt Brace Jovanovich, 1991.

[16] See *K. Pearson, (1905). "The Problem of the Random Walk". Nature. 72(1865): 294.* Bibcode: 1905 Natur 72 294P. https://doi.org/10.1038/072294b0.

vals.[17] That leaves huge margins of life that have escaped even the most bound-unit analyses by science. When we affirm, or refute, percentages of physical fact are missed.

Any model tells us general patterns, like exponential growth. Assumptions in spite of margins of error—intrinsic growth or decay—were the fuel that gave Thomas Malthus his most provocative expressions. Competition (whether in economics or primal geography, that is, territorial behavior, proxemics, and extra-proxemics) were key to Darwin's modeling. Proxemic spatial and psychological distance modalities (e.g., social distancing) is one of the most crucial coefficients because it encompasses the concept of biological density and all of the biometrics that converge within the same spaces.[18] At that point, the modeling is oriented toward simplistic r/K-selection theories (fundamental trade-offs between quantity and quality of offspring) of population density questions of where, how many, and so what.[19] We routinely ascribe to evolution and natural selection r/K assumptions. But we have not reason to believe our belief systems. To do so is merely a convenient way to encapsulate a theory; the theory its myriad hypotheses which, in turn, are nothing more than fragmentary compilations of anecdote and a modicum of observations.

In spite of our negative proofs—nullity to the exponentially zero + power—we are, in the end, allotted time, and choices. You can watch your hair grow longer, week by week, examine every minute change of the seasons, so infinitely detailed in Virgil ("We do not sing to the deaf: the woods re-echo all"),[20] Chaucer ("That it was May thus dremed me/In time of love and jollite"),[21] and James Thomson ("Come, gentle Spring, ethereal mildness, come ... And see where surly Winter passes off ....")".[22] Or note the surest attunement to the slightest difference beyond the Self, in the 14 meticulous volumes of Thoreau's journals, whether describing various states of Flint's Pond, the world of nuts, the "fully expanded light-green leaves" of a beech tree, thunder-showers, the "fox's trail," "the heroism of the muskrat," and so on infinitely.[23] There is absolutely "free play of the imagination" that galvanizes much joy in the landscapes, for example, of an Annibale Carracci

---

[17] See Hubbell, SP (2001). "The Unified Neutral Theory of Biodiversity and Biogeography (MPB-32)". Archived from the original on 2011-07-18. Retrieved 2010-12-16.

[18] Hall, Edward T. (October 1963). "A System for the Notation of Proxemic Behavior". American Anthropologist. 65 (5): 1003–1026. https://doi.org/10.1525/aa.1963.65.5.02a00020.

[19] Reznick, D.; Bryant, M. J.; Bashey, F. (2002). "r- and K-selection revisited: The role of population regulation in life-history evolution" (PDF). Ecology. 83 (6): 1509–1520. https://doi.org/10.1890/0012-9658(2002)083[1509:RAKSRT]2.0.CO;2. ISSN 0012-9658.

[20] Bucolica, Eclogue X., P. Virgilii Maronis, Robert Jennings Publisher, London, 1810, p. 95.

[21] From Canterbury Tales, Spring, "The Knight's Tale," 1.2, Harvard's Geoffrey Chaucer Website, Harvard University, https://chaucer.fas.harvard.edu/spring; See also, 1387–1400 – The Works Of Geoffrey Chaucer, Edited by John Urry, Bernard Lintot, London, 1721, pp. 8–23.

[22] James Thomson, The Seasons, Wilkie and Robinson, London, 1811, p. 3.

[23] See The Journal Of Henry D. Thoreau, Edited by Bradford Torrey and Francis H. Allen, With a Foreword by Walter Harding, In Fourteen Volumes Bound as Two, Dover Publications, Inc., New York, 1962. P. 171.

(1560–1609), particularly a work like his incredibly straightforward and dramatic, "Landscape," c. 1590 (National Gallery of Art, Washington, D.C.), that defines the movement of a canoe through a marshland with unequalled ingenuity.[24] The viewer does not see the middle of the canoe. It is blocked by a leaning leafy tree.

The joys and the sorrows that attend upon our mathematical attempts to compute change in nature, and what it means—between the frissons of weather, community standards and practices, individual volatility, and the vast stretches of geological and biochemical time—can be commiserated with, ignored, abided, and/or engaged by some motive. Ultimately, our lives are fraught, regardless of intellectual, mathematical, or moral persuasion, with the choices we must make each day in favor of something, while we still breathe. We could site an endless array of literature propounding our options. To take just one, as a kind of ambassador of our times for such dilemmas: "Quantifying biodiversity trade-offs in the face of widespread renewable and unconventional energy development".[25]

Populations, patches, metapopulations, rates of colonization, resiliency, and balance, behavior in general, and then human behavior specifically: These are each components of a human consciousness that—in contemplating the Anthropocene—grows rapidly morbid, then defensive, swept over by an angry denial, fundamentally trapped in thought as in the resulting incredulity; in what Sartre had coined, in despair. It would be easier to think of it as a mirage; to have cleaned one's lenses in hopes of waking up from a bad dream, only to again recognize the same wasteland in the guise of normalcy; hundreds of millions of stick figures, roaring motorcycles, 16-wheeled trucks, ink blots on the iris, belching smokestacks, poisonous infiltrations at every tier of colonization, with every breath. To realize that our entire orientation is human, no matter how candid our prayer for privacy or moment of reverie in exile. We are connected by evolutionary integuments with adamantine tensile strength, like the hierarchical proteins in the biomaterial of spider dragline silk fibers, tougher than the highest-grade alloy steel, or Kevlar.

Yet, that same Sartrean condemnation of consciousness, by dint of its daily, uphill trudge, ineffectively but incessantly manifests the affirmative, even in acknowledging the more than 800,000 suicides formally documented each year—a number that professionals believe is likely to double in coming annums, while the number of attempted human suicides daily is approximately 43,835.[26]

---

[24] See *Places of Delight – The Pastoral Landscape*, by Robert C. Cafritz, Larence Gowing, and David Rosand, The Phillips Collection in association with the National Gallery of Art, Clarkson N. Potter, Inc./Publishers, New York, 1988, p. 83, 89, and inside frontispiece.

[25] Popescu, V.D., Munshaw, R.G., Shackelford, N. et al. Quantifying biodiversity trade-offs in the face of widespread renewable and unconventional energy development. Sci Rep 10, 7603 (2020). https://doi.org/10.1038/s41598-020-64501-7, 05 May 2020, https://doi.org/10.1038/s41598-020-64501-7, https://www.nature.com/articles/s41598-020-64501-7, Accessed May 10, 2020.

[26] https://www.who.int/mental_health/prevention/suicide/suicideprevent/en/, Accessed March 31, 2019. See also, Staff (February 16, 2006). "SUPRE". WHO sites: Mental Health. World Health Organization.

**Fig. 50.3** To Dream the Impossible Dream, Engraving by Gustave Doré, from the two-volume 1863 French Edition, *L'ingenieux Hidalgo Don Quichotte De La Manche*, by Miguel De Cervantès Saavedra, Librairie De L., Hachette, et Co. (Private Collection). (Photo © M.C.Tobias)

If human consciousness and ecology together incite this consciousness beyond a tedious and most troubled history of nullification, then the assertion of a positive scenario—Don Quixote dreaming the impossible dream—must somehow manage to address the mismatch of individual numbers, social contracts, and our nearly lost ability to exercise restraint; to front tenacious positivity against a massive inferno whose documentation has become a cultural obsession. All the troubling headlines, as we continue into the long night of its escalating inflictions. Its most primal mathematical equation—and its searingly poignant human fall-out—is the very dialectic at the heart of that ecological inferno now upon us (Fig. 50.3).

# Chapter 51
# Irrational Biomes

## 51.1 Apex Irrationality

The Greek word for "irrational" is more expressive than Latin, from which most Romance languages derive the meaning, as, *not endowed with reason*. In Greek, however, paralogos (Παράλογος) entails absurdity, the preposterous, and senseless, in addition to illogical and the irrational, as applied in mathematics and general psychology. In Maori, *Mea taupatupatu* adds to it qualities like incompetent, and useless. Countless other languages give yet additional twists. Biome = bios, life + ome, a noun "denoting objects or parts having a specified nature"[1] is far more consolidated and singular in meaning. But the two words in combination invite a haze of our current predicament, the self-destructive habiliments asphyxiating every ecosystem, from salt marshes in Nebraska, forested mountains in New Guinea, rain forest on several continents, and coral reefs throughout the seas. Even the clarity of the night sky from earth, as seen through our eyes, had been tainted.

In 2011 Charles Q. Choi chronicled "10 of the Most Polluted Places on Earth",[2] including obvious candidates like Chernobyl, and other cities and regions inundated with "chemical waste" at Dzerzhinsk (Russia), lead in Kabwe (Zambia) and La Oroya (Peru), "coal dust" at Linfen (China), and other toxins from Mailuu-Suu (Kyrgyzstan) to "chromite ore deposits" at Sukinda (India). Years prior, *Time* magazine, had done its own list, "How the List Was Chosen".[3] In the case of Sumqayit (Azerbaijan), the nearly 300,000 people at that time

---

[1] https://www.google.com/search?client=firefox-b-1-d&q=biome%2C+etymoloy#dobs=-ome, Accessed May 10, 2020.

[2] April 20, 2011. LiveScience, https://www.livescience.com/30353-most-polluted-places-earth.html, Accessed May 11, 2020.

[3] By Bryan Walsh, September 12, 2007, http://content.time.com/time/specials/2007/article/0,28804,1661031_1661047_1661015,00.html, Accessed May 11, 2020.

© Springer Nature Switzerland AG 2021
M. C. Tobias, J. G. Morrison, *On the Nature of Ecological Paradox*,
https://doi.org/10.1007/978-3-030-64526-7_51

(today the population registers at 337,583) were dying from industrial chemical pollution at a rate "50% higher than their fellow Azerbaijanis." This kind of old-world, or "ex-Soviet factories … with virtually no environmental regulation," had become a standard for rating the global toxic hotspots. As of February 4, 2019, the list remained more or less the same (Haina, in the Dominican Republican had entered the list as number 3—abandoned lead smelter contamination), though the process of ranking had taken on new rubrics: 300 of the world's worst, reduced to 35 of the worst of the worst, down to the 10 highest areas of contamination.[4] All comport with the concept of a human synthesized irrational biome, as opposed to, say, a steaming caldera (non-synthesized) in which no human settlement would ever be designed, let alone realized.

Scientific American's own collection of polluted human enclaves[5] added habitat areas to the list, including the Niger River Delta (Nigeria), an oil-infested "sacrifice zone"; as well as the Rio Matanza-Riachuelo (Argentina) river basin 60-kilometers long and the bearer of all things from some "15,000 small industries"; a similar crisis on the Citarum River (13,000 square kilometers on Java) in which "more than 2,000 factories" along the river are polluting nine million people. From e-waste dump sites in Ghana to tanneries in Bangladesh, part of the Index for determining who is most at risk relies on the Global Ambient Air Quality Database published in 2018 by the World Health Organization[6] which focuses predominantly on Particulate Matter, at levels "less than 2.5" and "10 µm diameter)" (µm, micrometer = one millionth of a meter). As of 2015, 16.% of all human fatalities were linked to some kind of pollution, "9 million premature deaths".[7] Around 92% of those mortalities occurred in economically marginalized nations.[8] Contributing industries have been ranked and include "used lead acid batter recycling, mining and ore processing, tanneries, artisanal small-scale gold mining, product and chemical manufacturing and the dye industry." Such mining, at a glance, for example, is no self-congratulation. Rather, sights of ghastly ruin to our souls. The chemistry of sleepless nights. Geological wreckage that would not have been but for… Just look at the largest copper mines in the world.[9]

[4] See "The Most Polluted Places on Earth," by Larry West, February 4, 2019, ThoughtCo., https://www.thoughtco.com/worst-polluted-places-on-earth-1204101, Access May 11, 2020.

[5] "The World's 10 Most Polluted Places," by David Biello, November 5, 2013, https://www.scientificamerican.com/slideshow/10-most-polluted-places-in-the-world. Accessed May 11, 2020.

[6] See "Ambient Air Quality Database Application," https://whoairquality.shinyapps.io/AmbientAirQualityDatabase/. Accessed May 11, 2020.

[7] See "The Lancet Commission on pollution and health," Philip J Landrigan et.al., The Lancet, Vol. 391, No. 10119, October 19, 2017. https://www.thelancet.com/commissions/pollution-and-health, Accessed May 11, 2020.

[8] See Smithsonian magazine, "One in Six Global Deaths Linked to Pollution," by Ben Panko, Smithsonianmag.com, October 20, 2017, https://www.smithsonianmag.com/smart-news/one-six-global-deaths-linked-pollution-180965347/. Accessed May 11, 2020.

[9] See "World's Worst Pollution Problems, 2016," https://www.worstpolluted.org/. Accessed May 11, 2020. See also, "The 30 Most Polluted Places on Earth," by Michael B. Sauter, January 6, 2020,

But even in the wealthiest (large) nation in the world, the United States, beyond air pollution is the issue of sewage. Some "80 per cent of Americans" are on municipal sewage systems, the rest on home septic. City water-treatment plants are old, decaying, and the entire infrastructure in severe need of updating, with "23,000 to 75,000 sanitary-sewer overflows" each year.[10]

All of the anthropogenic abuses, whether in aggregated, core based statistical areas (CBSAs)—urban clusters "of at least 10,000 in population",[11] of which there were 955 in the United States as of February 2013—or rural stretches of continuous habitation, or countryside properties too big to farm, too small to ranch, have conditioned our post-industrialist senses. There are at least 388 urban areas of over 50,000 people in the United States; 192 countries, each with between 300 and 400 cities, or at least 50,000 cities on earth. We instantly accept or reject the large definitions of ourselves and our privacy issues as bifurcated between affiliations, associations with other humans, expectations of various rights and systems, promise of social contract privileges and value, and sensory thresholds that turn medical upon closer scrutiny.[12] All such values, held to one's chest—ultimately, we cling throughout life to our privacy—invite more than a mere morbid incredulity, when juxtaposed with such huge aggregates of people on top of one another throughout those global clusters (Fig. 51.1).

## 51.2  Lost Upon an Urban Surrealism

The net effect of this divining rod of what it means to inhabit a place (unless our species were somehow thought of as strictly eusocial) has been lost on the urban surrealism that attends human congeries; densities of population defying expectations of certain degrees of tolerance and intolerance; or making more clear the fundamental need of even the odd tree here and there, or a potted plant. A Central Park underscores the miracle that "nature" performs for those who cannot help but succumb to urban fatigue and ecological deprivation.[13] This need mirrors

---

24/7 WallSt, https://247wallst.com/special-report/2019/12/02/the-30-most-polluted-places-on-earth/. Accessed May 11, 2020. See https://www.mining-technology.com/features/feature-the-10-biggest-copper-mines-in-the-world/. Accessed May 11, 2020.

[10] See, "Flushing the Toilet Has Never Been Riskier," by Mary Anna Evans, The Atlantic, September 17, 2015, https://www.theatlantic.com/technology/archive/2015/09/americas-sewage-crisis-public-health/405541/. Accessed May 11, 2020.

[11] See U. S. Census, https://www.census.gov/programs-surveys/metro-micro.html.

[12] See "Construction of an adaptable European transnational ecological deprivation index: the French Version," by Carole Pornet, et.al., BMJ, Epidemiology and Community Health, Volume 66, Issue 11, https://doi.org/10.1136/jech-2011-200311, Accessed May 11, 2020.

[13] See "Environmental Deprivation," CRESH, Centre for research on environment, society and health, https://cresh.org.uk/cresh-themes/environmental-deprivation/. Accessed May 11, 2020.

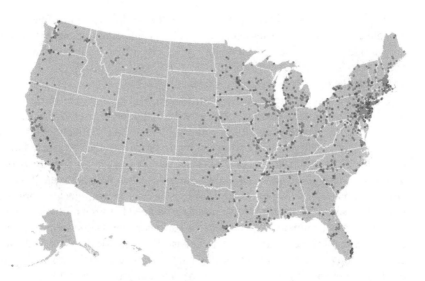

**Fig. 51.1** A Map of U.S. Superfund sites as of October 25 2013: Red indicates currently on final National Priority List, yellow is proposed, green is deleted (usually meaning having been cleaned up). Data from United States Environmental Protection Agency CERCLIS database, retrieved February 12, 2015 with last update reported as October 25, 2013. A live map of Superfund sites is available at https://www.epa.gov/superfund/search-superfund-sites-where-you-live#map

what has been described in the social scientific literature as a syndrome of ecological illiteracy.[14]

But from the global perspective, most easily perceived from the conurbations of city lights at night, the "nocturnal glow" as viewed from space by the Visible Infrared Imaging Radiometer Suite sensor fitted to the NASA-NOAA Suomi National Polar-orbiting Partnership satellite,[15] the electrical presence of humanity is perhaps pretty to some, and physically nauseating to others. While to other species and habitats this constant hum of energy elicited by one species signifies the

---

[14] See "Quantifying Ecological Literacy in an Adult Western Community: The Development and Application of a New Assessment Tool and Community Standard," by Sheryn D. Pitman and Christopher B. Daniels, PLOS ONE, March 3, 2016, https://doi.org/10.1371/journal.pone.0150648, https://journals.plos.org/plosone/article?id=10.1371/journal.pone.0150648, Accessed May 11, 2020. See also, "Systemic ecological illiteracy? Shedding light on meaning as an act of though in higher learning," North American Association for Environmental Education, 2012, n.a., https://naaee.org/eepro/research/library/systemic-ecological-illiteracy-shedding?term_node_tid_depth_join%5B0%5D=2421&page=3, Accessed May 11, 2020; See also, "Environmental Literacy and its Implications for Effective Public Policy Formation," by Julianna H. Burchett, The University of Tennessee, Knoxville, Baker Center for Public Policy, May, 2015, https://pdfs.semanticscholar.org/14f1/97d9d6d16cedcb6a827c2287710e2530ad93.pdf, Accessed May 11, 2020.

[15] See "Bright Nights and City Lights," NASA, Global Climate Change, Webquest #4 – Beautiful Earth Gallery, https://climate.nasa.gov/Earth-Day-2020/Bright-Nights-and-City-Lights/. Accessed May 11, 2020.

ultimate adulteration of the biosphere; the blinding lights impeding all sea turtle nesting behavior, as documented by the Sea Turtle Conservancy.[16]

Of course, the ecological queasiness belies our fond memories of Paris or Kyoto. Or the Metropolitan Museum of Art in Manhattan. Of our own home and those of friends around the world. Of cafés, quaint bookstores, galleries and the urban pleasures of farmers' markets we and our ancestors of dozens of generations have sampled, enjoyed or in equal measure eschewed.[17] But in this middling era of ecological hedonism, there it is and psychologically, physically, manifest in every thinking moment, activity, cause, and consequence, here we are.

Conservation biology and animal rights come down, in the most primal, emotional sense, to what Paul Gauguin famously titled, "D'ou Venons Nous/Que Somes Nous/Où Allons Nous" ("Where Do We Come From? What Are We? Where Are We Going?") painted in Tahiti in 1897–1898 (Boston Fine Arts Museum).[18] As we comb every corner of the earth searching for more and urgent biological data, our total population continuing to climb, adding 83 million each year at present,[19] the mind, pressed between the crashing rocks of its physical shell, the cumulative extent of its exacerbations, and the inextinguishable craving for wilderness, is caged. This trap aligns with ideals that betray the irrational biomes we have engendered over nearly 10,000 years of human history. Knowing it doesn't change the reality. Knowledge has not yet proved to be power. Not like a flowering meadow, the arrival in Spring of the first Western Tanagers, or a cool breeze against the all-out sweltering.

---

[16] See STC, https://conserveturtles.org/

[17] See, for example, the luscious, inspiring history of Paris in Pierre Cabanne's *Paris Vous Regarde*, Pierre Bordas et Fils, éditeurs, Paris, 1988

[18] https://collections.mfa.org/objects/32558.

[19] See Our World In Data, "Absolute increase in global population per year, https://ourworldindata. org/grapher/absolute-increase-global-population?time=latest, Accessed May 11, 2020.

# Chapter 52
# The Extinction Probability Era

## 52.1 Navigating a World of Chance and Uncertainty

"I clung to certain numbers and combinations, but soon abandoned them and staked almost unconsciously… My temples were soaked with sweat and my hands were shaking … 'You are bold – you are very bold,' they said to me, 'but be sure you go away to-morrow as soon as possible, or else you will lose it all'".[1] In the year 1866, Fyodor Dostoevsky, a compulsive gambler, found himself in dire debt, and wrote his novella, *The Gambler*, set in "Roulettenbad." It is likely a far more complicated disorder than conventional psychiatric diagnoses would have us believe, aside from one of its manifestations presenting an "increased risk of suicide".[2] Various anti-seizure, antidepressant and mood-stabilizing drugs, along with psychotherapy, have all been recommended for those afflicted by the "thrill," or its potential cohort of "mental illness"[3] (Fig. 52.1).

But the condition of excited win/or lose all, that infuses the roulette experience—when extended by turns from Dostoevsky's own de minimis fair game—touches a vastly deeper, darker reality that has come to characterize the behavior of the human collective. What are the *odds* of any occurrence in an individual and/or aggregate? Being dealt a straight flush, achieving a bill in Congress, surviving a plane crash, winning the lottery, dying of any disease, or, of an entire species going extinct?

---

[1] *The Gambler And Other Stories*, by Fyodor Dostoevsky, Translated by Constance Garnett, Heinemann & Zsolnat Ltd., London, Toronto, 1914, pp. 103, 105.

[2] See The American Psychiatric Society, Physician Review by Philip Wang, M.D., Dr. P.H., August 2018. https://www.psychiatry.org/patients-families/gambling-disorder/what-is-gambling-disorder, Accessed May 13, 2020. See also, "Compulsive Gambling," https://www.mayoclinic.org/diseases-conditions/compulsive-gambling/symptoms-causes/syc-20355178, Accessed May 13, 2020.

[3] See "Gambling Addiction (Compulsive or Pathological Gambling)," by Roxanne Dryden-Edwards, MD, Medical Editor William C. Shiel Jr., MD, FACP, FACR, MecidineNet, https://www.medicinenet.com/gambling_addiction/article.htm#gambling_addiction_facts, Accessed May 13, 2020.

© Springer Nature Switzerland AG 2021
M. C. Tobias, J. G. Morrison, *On the Nature of Ecological Paradox*,
https://doi.org/10.1007/978-3-030-64526-7_52

**Fig. 52.1** "Portrait of Fedor Dostoyevsky," by Vasily Perov (1833–1882), oil on canvas, 1872, Tretyakov Gallery, Moscow, Russia. (Public Domain)

Dostoevsky's own unthinkable flirtation with a firing squad in Siberia, December 22, 1849. Is there a mathematical basis for bad luck, good fortune, for life itself? How do those same compulsions that addict some people to blackjack, also favor, or discourage other behavioral and emotional commands that are ruinous? Do the fluctuating biological rules that dictate the rise and fall and many twists in destiny of populations, apply with the same levels of latitude to individuals? Does it matter?

Let us preface by suggesting that it does not matter, not merely as an argument against, or in favor of free will. Rather, we may freely presume that beyond our assumptions and choices, there is a more wide-ranging free will at large; in the sense of physical force, unseen, like black background matter, first described in 1933 by Swiss astronomer Fritz Zwicky (1898–1974). He applied his virial theorem in order to better divine, and partially measure the *approximate* amount of kinetic energy and potential forces within both stable and complex systems.[4]

---

[4]George W. *Collins, II, G. W "Introduction".* The Virial Theorem in Stellar Astrophysics. *Pachart Press,* 1978) by the Pachart Foundation dba Pachart Publishing House and reprinted by permission, Made available electronically by the NASA Astrophysics Data System (ADS), 1978.

Everything we sense about the universe we think we can measure. Why do we want to? We even seek to measure qualities, render nearly numerical certainties within subjective realms; how much we feel, the degree of faith, of sacrifice, of despair. We even set out to measure joy, in whatever doses. Again: Why this insistence to measure anything? Plato's "Apology 29b-d" leads to the Socratic Paradox—I know that I know nothing (or at least, about human virtue)[5]—which, in turn, takes us to the curious "Münchhausen Trilemma" of circular, regressive and axiomatic arguments always turning in upon themselves—the physics of introspection. In 1576 the young Portuguese scholar, Francisco Sánchez (c. 1550–1623) applied his devastating skepticism to a syndrome of no-knowledge in a work entitled *Quod nihil scitur* (*That Nothing Is Known*).[6]

Elsewhere, we have described a quintessential debate within sociobiology that expresses our species' and individual existence on a pendulum that swings from the ultimate embrace of Self, whether by a Galileo or a Camus who referenced the former's ultimate renunciation of scientific theory under the pressure of his life at risk, and that of modern biological synthesis, which extols the hegemony of some ultimate operating system in control of the philosopher's very "hypothalamus and limbic system." The result being, says E. O. Wilson, that "the philosopher's own emotional control centers are wiser than his solipsist consciousness".[7] Which places the *getting of wisdom* in a lesser category than the *getting of emotion*. Fair enough.

Wilson describes our hopes for a biological future, citing Camus' pessimism and Camus' talk of suicide as the only truly worthy philosophical topic, in light of "A world that can be explained even with bad reasons [because it] is a familiar world. But, on the other hand, in a universe divested of illusions and lights, man feels an alien, a stranger. His exile is without remedy since he is deprived of the memory of a lost home or the hope of a promised land".[8] Adding, "This, unfortunately is true. But we still have another hundred years".[9] Paul Ehrlich is less sanguine, declaring, "Collapse of civilization is a near certainty within decades[10] (Fig. 52.2).

---

[5] See "The Paradox of Socratic Ignorance in Plato's Apology," by Thomas C. Brickhouse and Nicholas D. Smith, History of Philosophy Quarterly, HPQ, Vol. 1, No. 2, Apr. 1984, pp. 125–131, University of Illinois Press on behalf of North American Philosophical Publications, https://www. jstor.org/stable/27743673, https://www.jstor.org/stable/27743673?seq=1, Accessed May 14, 2020.

[6] Francisco Sánchez, *Quod nihil scitur*, Edited and Translated by Elaine Limbrick, and Douglas F. S. Thomson, Cambridge University Press, Cambridge, UK, 2008.

[7] See E. O. Wilson, *Sociobiology*, The Abridged Edition, Drawings by Sarah Landry, The Belknap Press Of Harvard University Press, Cambridge, Mass., 190, p. 3 of the chapter "The Morality of the Gene."

[8] Albert Camus, *The Myth Of Sisyphus and Other Essays*, Translated from the French by Justin O'Brien, Alfred A. Knopf, New York, 1955, p. 6.

[9] op. cit., Wilson, 1980, p. 301.

[10] Interview by Damian Carrington, "Paul Ehrlich: 'Collapse of Civilisation is a Near Certainty Within Decades," The Guardian Newspaper, London, March 22, 2018, https://www.theguardian. com/cities/2018/mar/22/collapse-civilisation-near-certain-decades-population-bomb-paul-ehrlich, Accessed May 14, 2020. Cited in The *Hypothetical Species: Variables of Human Evolution*, by M. C. Tobias and J. G. Morrison, Springer Nature, New York, 2019, p. 53.

**Fig. 52.2** A Parliament of Optimism, Three Young Bhutanese Monks. (Photo © J.G. Morrison)

Such headwinds beg the fundaments of *probability*, which has become the penultimate modus operandi of twenty-first-century praxis. President Obama characterized the dilemma in far less draconian or uncertain a future than biologists: "We choose hope over fear. We see the future not as something out of our control but as something we can shape for the better through concerted and collective effort".[11] Countless levels of turbulence cloud this optimism, both for humans and all other species. Cloud the biological debates that swirl around any amount of fine-tuning. From 29 Nunivak Island reindeer, moved by the US Coast Guard to St. Matthew Island in the Bering Sea in 1944, abandoned after the war, their herd growing to 6000 individuals, then plummeting in the 1970s to zero.[12] Overpopulation. Uncertainty number one, despite our being able to take micro-census data and know down to every household (in some countries) who is pregnant, and within a very small margin of error the number of births and deaths on any given day, how many human newborns there will be in the coming year. 7.784,464,337 as of this second.[13]

---

[11] President Barack Obama, UN General Assembly 2014.

[12] "What wiped out St. Matthew Island's reindeer?" by Ned Rozell, Anchorage Daily News, January 16, 2010, https://www.adn.com/features/article/what-wiped-out-st-matthew-islands-reindeer/2010/01/17/. Accessed May 14, 2020.

[13] See the most unsettling, "Worldometer," https://www.worldometers.info/world-population/, Accessed May 14, 2020, 2:09 PM.

But we don't know for absolute certainty, nor do we have a clue how all those consumers will behave (the IPAT equation + Tragedy of the Commons syndrome).[14]

We can examine environmental security flaws and their probability factors in 4-D (3D + temporality), whether in regard to what went wrong at Fukushima before, during and after the disaster there on March 11, 2022[15] or by any number of engineering monitoring technologies for risks in every conceivable industry relating to such inherent ecological events as groundwater pollution, toxic plume remediation, hydraulic fracturing problems, municipal infrastructure decay, landfills, sewage, and so on.[16] But for every analytical certainty there is the obverse flaw, measurements even slightly awry at the level of a single NASA Challenger rubber O-ring seal: a projected "~13% likelihood of *O-ring failure* at 31 °F, compared to NASA's *general shuttle failure* estimate of 0.001%".[17]

Whether with regard to seismic measurements,[18] precipitation,[19] hurricane [20] and other extreme weather predictions,[21] avalanche and mountain hazards in general,[22]

---

[14] See Ehrlich, Paul R.; Holdren, John P. (1971). "Impact of Population Growth". Science. American Association for the Advancement of Science. 171 (3977): 1212–1217. Bibcode:1971Sci 171.1212E. https://doi.org/10.1126/science.171.3977.1212. JSTOR 1731166. PMID 5545198. See also, "The Tragedy of the Commons," by Garrertt Hardin, Science 13 Dec 1968: Vol. 162, Issue 3859, pp. 1243–1248, https://doi.org/10.1126/science.162.3859.1243, https://science.sciencemag.org/content/162/3859/1243, Accessed May 14, 2020.

[15] See "Japan's Unknown Ecological Tragedy," by M.C.Tobias, Forbes, March 17, 2011, https://www.forbes.com/sites/michaeltobias/2011/03/17/japans-unkown-ecological-tragedy/#64936150269b; See also, "Japan's Tragedy: Global Ecological Uncertainty," by M.C.Tobias, Forbes, March 13, 2011, https://www.forbes.com/sites/michaeltobias/2011/03/13/japans-tragedy-global-ecological-uncertainty/#2b8c1d728e30

[16] See "Environmental Security: Sensing the World in 4-D," by M.C.Tobias, Forbes, January 31, 2012, https://www.forbes.com/sites/michaeltobias/2012/01/31/environmental-security-sensing-the-world-in-4-d/#73ab9c594d3a

[17] See "The Space Shuttle Challenger Explosion and the O-ring," by Nemil Dalal, Priceonomics, https://priceonomics.com/the-space-shuttle-challenger-explosion-and-the-o/n.d, Accessed May 14, 2020.

[18] See "Seismic Risk Assessment Using Stochastic Nonlinear Models," by Yeudy F. Vargas-Alzate, Nieves Lantada, and Ramón González-Drigo, and Luis G. Pujades, Sustainability 2020 12(4), 1308; https://doi.org/10.3390/su12041308, February 11, 2020, https://www.mdpi.com/2071-1050/12/4/1308, Accessed May 14, 2020.

[19] See "ETA-Based Mos Probability Of Precipitation (PoP) And Quantitative Precipitation Forecast (QPF) Guidance For The Continental United States," by Joseph C. Maloney: "…categorical forecasts are no post-processed and inconsistent forecasts can result…" p. 6, https://www.weather.gov/media/mdl/487.pdf

[20] See National Hurricane Center and Central Pacific Hurricane Center, https://www.nhc.noaa.gov/. Accessed May 14, 2020

[21] See "Severe weather prediction," by ECMWF (The European Centre for Medium-Range Weather Forecasts), https://www.ecmwf.int/en/about/who-we-are.

[22] See "Reducing Uncertainty in Hazard Prediction," by Karim Riley, Matt Thompson, Peter Webley, and Kevin Hyde, April 28, 2017, Eos, Science News by AGU, Editors' Vox, https://eos.org/editors-vox/reducing-uncertainty-in-hazard-prediction, Accessed May 14, 2020.

calculating forest fire liabilities,[23] food insecurity[24] and all forms of energy turbulence, as described by physicist Robert Kraichnan's statistical theories of "energy cascades",[25] there are enormous gaps in probability and distribution models. They could be characterized like the CBOE Index of Volatility on Wall Street;[26] the shifting epidemiological mortalities as described by the University of Washington novel coronavirus modeling;[27] the metric trade-offs and triage associated with the re-opening of an economy during a zoonotic pandemic, or, for that matter, the shifting political tides dictating protocols in American schools against a backdrop of rampant gun violence and mass shootings.

Even at the level of a single second, as defined by the International System of Units (SI), a "duration" (of one second") is a measure of 8,192,631,770 radiation cycles rotating between energy levels of one atom of cesium-133, all of which is vulnerable to minute oscillation/frequency shifts.[28] Associated with that fractured second is a thermonuclear detonation and the risk of it happening, as perceived by the Bulletin of the Atomic Scientists who, as of 2020 inform us we are "Closer than ever: It is 100 seconds to midnight," a profoundly unstable situation.[29]

We are each of us metaphysicians because we somehow manage to survive, for a time, knowing fully well that the world is closing in. We are marooned in a probability era given to an uneasy, indeed, fully reminded context of multivariate biological extinctions occurring continuously, between those micro-seconds. The forlorn craziness of the enshrouding arithmetic condemns the exercise of awareness to winners and losers, at least in the starkest terms.[30] Persistence and diffusion are variables

---

[23] See, for example, "North American Seasonal Fire Assessment and Outlook," National Interagency Fire Center, May, June, and July 2020, https://www.nifc.gov/nicc/predictive/outlooks/NA_Outlook.pdf.

[24] See "2020 Winter/Spring Flood Outlooks," National Weather Service, NOAA, https://www.weather.gov/marfc/WinterSpring_Flood_Outlook

[25] See Kraichnan (1958). "Higher Order Interactions in Homogeneous Turbulence Theory". Physics of Fluids. 1 (4): 358. Bibcode:1958PhFl.1.358K. https://doi.org/10.1063/1.1705897.

[26] See https://www.investopedia.com/terms/v/vix.asp.

[27] See "UW experts on novel coronavirus (COVID-19)," n.a. https://www.washington.edu/news/for-journalists/uw-experts-on-novel-coronavirus-covid-19/. Accessed May 14, 2020. Even Bill Gates, whose Bill and Melinda Gates Foundation has put hundreds-of-millions of dollars into predicting and preparing for the virus expressed certain personal "regret" for not being even more forceful in driving home his critical message to policy makers -who largely ignored the warning - regarding the prospects for a novel coronavirus outbreak. See https://www.wsj.com/articles/bill-gates-coronavirus-vaccine-covid-19-11589207803, Accessed May 14, 2020.

[28] "International System of Units (SI)" (PDF) (8th ed.) International Bureau of Weights and Measures (BIPM). 2006. See also, "FAQs". Franklin Instrument Company. *2007.*

[29] Press Release, Washington, D.C., January 23, 2020, https://thebulletin.org/doomsday-clock/, Accessed May 14, 2020.

[30] See "Partial Survival and Extinction of Species in Discrete Nonautonomous Lotka-Volterra Systems, by Hoshiaki Muroya, Tokyo Journal of Mathematics, Project Euclid mathematics and statistics online, Volume 28, Number 1, 2005, pp. 189–200. https://projecteuclid.org/euclid.tjm/1244208288, Accessed May 14, 2020.

applied in extinction-level models.[31] The lack of independent or co-precision is qualified by an inexplicably vast multitude of such inconstant incisions in our perception of time. Precision is the illusion that biology stammers over, contends with in every leaf, stoma, cell, and season. The history of statistics and probability catches only a vague outline of the endless arrays and nuances.[32] Early probability mavericks like Jacob Bernoulli (1655–1705), Thomas Bayes (1702–1761), P. S. Laplace (1749–1827), and James Clark Maxwell (1831–1879) each concurred that theories of probability were as much predicated upon chance as frequency, part knowledge, part ignorance, part ratio.[33]

## 52.2   Probability Theories

One of the earliest mathematical theories in the history of probability speculation concerned a matter of some regard to Dostoevsky, as earlier referenced, named, "a gambler's dispute" which occurred in 1654 between two Frenchmen (of course); the philosopher Blaise Pascal and his friend the mathematician Pierre de Fermat. The details of the event in question concerned "an apparent contradiction concerning a popular dice game, during which a pair of dice would be tossed 24 times and certain outcomes (that had always previously been interpolated as certain) were to be tossed out the window. Not until the Russian mathematician Andrey Kolmogorov (1903–1987) attended to it in 1933 was some clarity and predictability achieved.[34]

Considered a modern father of probability, as well as algorithmic complexity, his book *Foundations of the Theory of Probability*[35] elevated axioms of prediction such

---

[31] See "Persistence and extinction in single-species reaction-diffusion models," by Linda J. S. Allen, Bulletin of Mathematical Biology, 45, 209–227, March 1983, Springer Link, https://link.springer.com/article/10.1007/BF02462357, Accessed May 14, 2020. See also, "Persistence, extinctions, and critical patch number for island populations," by Linda J. S. Allen, Journal of Mathematical Biology 24, 617–625(1987), February 1987, Springer Link, https://link.springer.com/article/10.1007/BF00275506, Accessed May 15, 2020.

[32] See "A Brief Look at the History of Probability and Statistics," by James E. Lightner, The Mathematics Teacher, Volume 84,, No. 8, November 1991, pp. 623–630, National Council of Teachers Of Mathematics, JSTOR, https://www.jstor.org/stable/27967334, https://www.jstor.org/stable/27967334?seq=1, Accessed May 14, 2020.

[33] See "A short history of probability theory and its applications," by Lokenath Debnath and Kanadpriya Basu, International Journal of Mathematical Education, 46(1), January 2015, https://www.researchgate.net/publication/271856948_A_short_history_of_probability_theory_and_its_applications, Accessed May 14, 2020.

[34] See "A Short History of Probability," from *Calculus*, Volume II, by Tom M. Apostol, 2nd edition, John Wiley & Sons, New York, 1969, http://homepages.wmich.edu/~mackey/Teaching/145/probHist.html, Accessed May 14, 2020.

[35] See *Grundbegriffe der Wahrscheinlichkeitrechnung, Ergebnisse Der Mathematik*; translated as *Foundations of Probability,* New York: Chelsea Publishing Company, 1950. See also, *Selected Works of A. N. Kolmogorov, Vol.2, Probability Theory and Mathematical Statistics*, Edited by Albert N. Shiryaev, Springer Nature B.V., Springer Netherlands, 1992. See also, "On Logical

that they were to become far more critical to science, engineering and thinking about human nature than merely a game of dice, though nothing about the statistics of scattering those dice has changed. But what has been altered, and veering ever more tightly into our headlights, are the probability factors affecting the science and judgments regarding biological precariousness, and extinction. Data devolves to the subjective. "Probabilistic coherence plays much the same role for degrees of belief that *consistency* plays for ordinary, all-or-nothing beliefs," writes Alan Hájek.[36] Hájek cites a 1929 lecture by Bertrand Russell who had declared, "Probability is the most important concept in modern science, especially as nobody has the slightest notion what it means".[37]

But where the philosophy of coherent systems embracing rational ideations of what is probable ineluctably sway from any classical, "logical/evidential interpretations"[38] is at that bifurcation point where predictions have but scant protocol, forensic materials, be it in the case of fragmented probability spaces, an absence in some fossil record ("fossilized species may represent less than 1% of all the species that have ever lived"),[39] or the extinction of a culture and their language. The sum total of destructions that make up the Anthropocene[40] represents a time-conditional variable that doubly skews both the derived data, and its interpretive sphere. In part, this problem stems from the criteria applied to assessing extinctions.

The French naturalist Georges Cuvier (1769–1832) is credited with first recognizing the concept of extinction, through his work in vertebrate paleontology. He examined the bones of North and South American mastodons, Pterosaurs, ground sloths, and grasped the reality that these particular creatures no longer existed; beings that had been caught up in the tumultuous life history of earth—not victims of evolution, which he did not ascribe to, but great and terrifying events, like floods and volcanism, which destroyed and created.

Generally, biologists agree on a 100-year rule. A species having been absent from any field records for at least a century in the wild has thus conceded its own extinction. Also, it is assumed that in general every species goes extinct within ten

Foundations Of Probability Theory," by A. N. Kolmogorov, in *Probability Theory and Mathematical Statistics*, September 11, 2006, pp. 1–5, https://link.springer.com/chapter/10.1007%2FBFb0072897; see also, H. E. Kyburg, *Probability and Inductive Logic*, Macmillan, New York, 1970.

[36] "Interpretations of Probability," Stanford Encyclopedia of Philosophy, August 28, 2019. Stanford Center for the Study of Language and Information, https://plato.stanford.edu/entries/probability-interpret/, Accessed May 15, 2020.

[37] In E.T. Bell, *The Development of Mathematics,* 2nd edition, McGraw-Hill Company, New York, 1945, p. 587.

[38] op.cit., Hájek.

[39] "Evidence of Evolution," lumencandela, CC BY-SA, OpenStax College, Attribution-ShareAlike, Boundless Biology, December 6, 2013, https://courses.lumenlearning.com/boundless-biology/chapter/evidence-of-evolution/. Accessed May 15, 2020.

[40] See, for example, *The Human Planet – Earth At The Dawn Of The Anthropocene*, by George Steinmetz and Andrew Revkin, Abrams Publishers, New York, April 2020.

million years of its emergence.[41] Yet, we now have numerous, informed voices chiming in on the loose prediction that the human species will be gone, wiped out, within a century or sooner, which would leave our kind a legacy of biological presence at roughly 330,000 years, a far cry from ten million years.[42]

While predictions about the end of humanity vary (even the extinction of Neanderthal has been mathematically computed),[43] the economics of increasing conservation scarcity are reflected in policies aimed not at preserving the natural world, but preserving human jobs. And thus, it is no surprise that, based upon a combination of fieldwork and computer simulations, algorithms have, for several years, been applied to proposed policies of conservation triage, a phrase normally banished from the ecologist's toolkit. This is precisely because it relies on the expediency that can be wrought from economic volatility, which might translate into putting resources into trying to prolong the life of "the woodland caribou, the Indiana bat, the Hawaiian crow" but only at the expense of California's marbled murrelet or Florida's gopher tortoise.[44]

Winners and losers. This ordained syndrome which views human economic progress in conflict with nature has been eloquently described by Aileen Lee, Chief Program Officer for Environmental Conservation at the Gordon and Betty Moore Foundation as a "pernicious mindset … an artifact of having operated for so long in a system that has been engineered for exploitation rather than resilience — where our choices about living with nature are constrained toward delivering a narrow set of benefits that flow toward a similarly restricted set of beneficiaries".[45]

In selecting our natural cohorts from across every taxonomic suite, it is, of course, worse than mere wishful thinking to play god. Especially in light of the data

---

[41] Newman, Mark (1997). "A model of mass extinction". Journal of Theoretical Biology. 189 (3): 235–252. arXiv:adap-org/9702003. https://doi.org/10.1006/jtbi.1997.0508. PMID 9441817.

[42] In addition to scientists like E. O. Wilson and P. R. Ehrlich referenced earlier, there is also the Australian scientist, Frank Fenner, in was involved in the extermination of smallpox, who a decade ago gave our species 100 years. "Humans will be extinct in 100 years says eminent scientist," by Lin Edwards, Phys.org, June 23, 2010, https://phys.org/news/2010-06-humans-extinct-years-eminent-scientist.html, Accessed May 15, 2020. Smallpox provides one of countless exceptions to the rule of 10 million years, however. It is known to have existed as a viral species for only about 3,000 years, and was driven to extinction by a coalition led by the World Health Organization in 1977. Its last known location was in Somalia. See World Health Organization, "Frequently asked questions and answers on smallpox," June 28, 2016, https://www.who.int/csr/disease/smallpox/faq/en/, Accessed May 15, 2020.

[43] See "A mathematical model for Neanderthal extinction," by J. C. Flores, for the Journal of Theoretical Biology, London, http://cds.cern.ch/record/342444/files/9712024.pdf, Accessed May 16, 2020.

[44] See "How a Math Formula Could Decide the Fate of Endangered U.S. Species," by Sharon Bernstein, Scientific American, June 19,. 2017, https://www.scientificamerican.com/article/how-a-math-formula-could-decide-the-fate-of-endangered-u-s-species/. Accessed May 15, 2020.

[45] "Perspective: Nature has sent humanity a wake-up call. How will we answer? https://www.moore.org/article-detail?newsUrlName=perspective-nature-has-sent-humanity-a-wake-up-call.-how-will-we-answer, Accessed May `16, 2020.

swarms that, of late, have provided ample probability statistics on pre-existing levels of risk at all levels, including "partial survival and extinction," a mathematically paradoxical condition.[46]

The notion of *partial* comports with the idea of *tendency*, a subjectively held view of observations, relations, and qualified data, considering its anthropic bias that is inevitably tied to all empirical as well as theoretical decision making, on our part.[47] These components of the subjective, in turn, invoke deliberation of "chance events," "expectation," and "variance".[48]

Because we think of ecosystems as dynamic, forever in flux, circumvention of punctuation marks and all, there are distinct "bifurcation points." These philosophical junctures have taken probability theories to the threshold of an interdisciplinary focal point on our species and its likely fate. It is a not surprising ecological obsession in the twenty-first century, that began with other mammals.[49] Its logic then encompassed whole populations,[50] mathematical formulas for re-discovering so-called Lazarus species (those thought to be extinct that are not; or, conversely, the "living dead," those whose little remaining time has been all but calculated),[51] or for re-introducing extinct species like the Passenger Pigeon.[52] Of late, this logical stream has increasingly determined that we are among the *next* to go, however prematurely.[53]

---

[46] See "Partial Survival and Extinction of Species in Discrete Nonautonomous Lotka-Volterra Systems," by Yoshiaki Muroya, Tokyo Journal of Mathematics, Volume 28. Number 1, 2005, 189-200, Project Eucid mathematics and statistics online, https://projecteuclid.org/euclid. tjm/1244208288, Accessed May 16, 2020. See also, 'Extinction in competitive Lotka-Volterra systems," by M. L. Zeeman, Proceedings Of The American Mathematical Society, 123, 1995, 87–96, MSC: Primary 92D25; Secondary 34C05, 34D20, 34D45, 92D40, https://doi.org/10.1090/S0002-9939-1995-1264833-2, MathSciNet review: 1264833, https://www.ams.org/journals/proc/1995-123-01/S0002-9939-1995-1264833-2/. Accessed May 16, 2020.

[47] See "Probability Theory," T. Rudas, in the International Encyclopedia of Education (Third Edition), 2010, ScienceDirect, https://www.sciencedirect.com/topics/social-sciences/probability-theory, Accessed May 16, 2020.

[48] "Basic Probability," https://seeing-theory.brown.edu/basic-probability/index.html, Accessed May 16, 2020.

[49] "Investigating Anthropogenic Mammoth Extinction with Mathematical Models," by Michael Frank, Anneliese Slaton, Teresa Tinta and Alex Capaldi, Spora: A Journal of Biomathematics, Volume 1, Issue 1, Article 3, 2015, https://core.ac.uk/download/pdf/48841606.pdf.

[50] "Stochastic models of population extinction," by Otso Ovaskainen and Baruch Meerson, https://arxiv.org/pdf/1008.1162.pdf; "A math equation that predicts the end of humanity," by William Poundstone, Vox, July 15, 2019, https://www.vox.com/the-highlight/2019/6/28/18760585/dooms-day-argument-calculation-prediction-j-richard-gott, Accessed May 16, 2020.

[51] See "The mathematics of species extinction," n.a., Mathematical Institute, University of Oxford, January 19, 2017, https://www.maths.ox.ac.uk/node/16490, Accessed May 16, 2020.

[52] "Mathematical Modeling in Ecology: Simulating the Reintroduction of the Extinct Passenger Pigeon (*Ectopistes migratorius*), by Erin Boggess, Jordan Collignon, and Alanna Riederer, VERUM 2015, Valparaiso University, University 28, 2015, https://www.valpo.edu/mathematics-statistics/files/2014/09/Capaldi2015.pdf.

[53] See, for example, "Mathematical law could anticipate the possibility of mass extinctions in eco-systems," Newsroom, Universitat Autonoma de Barcelona, September 10, 2018, https://www.uab.

## 52.3   Divergent Ecological Modeling

One extremely interesting divergence has been recognition by ecological modelers of *an individual as a population*. Hence, the "human brain" is described as "one of the largest biological populations, consisting of approximately $10^{11}$ neurons and $10^{14}$ synapses." Comparing the death of this complex system in terms of gauging sub-population declines (polymorphism) as against an "immortal" population model leads researchers to posit "internal population time"(s).[54] Even these, seemingly non-observable, remote, and independent variables, are unlikely (as far as can be ascertained) to elude the so-called transient phenomena in ecology, which impart distinct "important general lessons" within "the context of anthropogenic global change," such as a determination of just "when transients are likely to occur and [their] various properties." And this includes the recognition of new timeframes of change, as well as "nonlinear interactions" that lead beyond mere idiosyncratic ecological events, but up to and beyond the bifurcation points, such that "transient dynamics in ecological models" are now viewed soberly, in the spirit of attempting to identify "appropriate responses to the possibility of sudden system changes".[55]

Such system behavioral shifts might refer to a virulent threshold, where a pathogen either kills its host and total metapopulation entire, or in part; where such changes lead to total die-out of partial populations, or regain some new type of stability according to an entirely different set of fragmented population ratios, dynamics, metabolisms and energy/gene exchange; where eight generations of hunger in a species, as studied by John Drake and Blaine Griffen, provided fundamental early warning of a "critical slowing down" and then major transition.[56] And whether the sum-total of such bifurcation points might (ecologically, medically), as opposed to all out extinction, provide evolution sufficient informational cues to rescue certain species according to, as yet ungrasped, transcendent regulatory mechanisms in

cat/web/newsroom/news-detail/mathematical-law-could-anticipate-the-possibility-of-mass-extinctions-in-ecosystems%2D%2D1345668003610.html?noticiaid=1345773475461,   Accessed May 16, 2020. See also, "Mathematics predicts a sixth mass extinction," by Jennifer Chu, MIT News Office (based upon the work of MIT geophysics professor Daniel Rothman, who correlates the abundance of carbon-12 and carbon-13 at the time of previous mass extinction events), September 20, 2017, http://news.mit.edu/2017/mathematics-predicts-sixth-mass-extinction-0920, Accessed May 16, 2020.

[54] "Mathematical Modeling of Extinction of Inhomogeneous Populations," G.P Karev and L. K. areva, Bulletin of Mathematical Biology, April 2016, 78(4): 834–858, https://doi.org/10.1007/s11538-016-0166-0, PMCID: PMC4877184, NIHMSID: NIHMS784225, PMID: 27090117, https://www.ncbi.nlm.nih.gov/pmc/articles/PMC4877184/, HHS Public Access, Accessed May 16, 2020.

[55] "Transient phenomena in ecology," Alan Hastings, Karen C. Abbott, Kim Cuddington, Tessa Francis, Gabriel Gellner, Ying-Cheng Lai, Andrew Morozov, Sergei Petrovskii, Katherine Scranton, Mary Lou Zeeman, Science 07 Sep 2018: Vol. 361, Issue 6406, eaat6412, https://doi.org/10.1126/science.aat6412, Accessed May 16, 2020.

[56] John M Drake and Blaine D Griffen, "Early warning signals of extinction in deteriorating environments," Nature 467, no. 7314 (2010): 456.

biology.[57] While we have seen ecological and mathematical models and assumptions suggesting that our species has from a mere few decades to precisely 760 years left to go—before extinction, what we have not yet seen are any mathematical models to show any degree of likely perseverance of *Homo sapiens* beyond such extrapolations. That is proof of a nearly overwhelming condition of multiple, simultaneous volatilities.

If our hours are so numbered, then one would assume that our BP (Bereitschaftspotential, or "readiness potential") first revealed in 1964 by Hans Helmut Kornhuber and Lüder Deecke, would appropriately retool, and re-steer its near instantaneous pre-planning manifestation of cortical activity that is at the root of the approximately 35,000 choices a human makes each day (teenagers, fewer, but more virulent)[58] toward safer shores. Or at least, would choose tendencies in such directions.

## 52.4   To Choose, or Not to Choose

In 1985, psychologist Benjamin Libet measured the triggering of the BP as occurring 0.35 seconds before conscious volition, engendering a robust neurological debate about the very nature of free will, in humans.[59] But clearly, with at least 477 known other species extinctions since 1900,[60] and at least 161 avifauna lost since 1500,[61] there would appear to be a mere modicum of amalgamated human angst concerning the future of more than half of all life forms, now predicted threatened and/or vulnerable to extinction. How that *modicum* is measured; what its volitional foundations are made of; the precision, scope, and resolve of its eco-remediation methodologies, public and private partnerships; and the innumerable cultural and ecological variables that seem all but destined to obfuscate our best ideals and their implementation, now represents a life-or-death tier of deliberations and catalysts

---

[57] See "The mathematics of extinction across sc ales: from populations to the biosphere," by Colin J. Carlson, Kevin R. Burgio, Tad A. Dallas, & Wayne M. Getz, PeerJPreprints, MSC 2010 Codes: 92B05, 92D40, https://peerj.com/preprints/3367.pdf. Accessed May 16, 2020.

[58] Wong, Aaron L.; Haith, Adrian M.; Krakauer, John W. (August 2015). "Motor Planning". The Neuroscientist: A Review Journal Bringing Neurobiology, Neurology and Psychiatry. 21 (4): 385–398. https://doi.org/10.1177/1073858414541484. ISSN 1089-4098. PMID 24981338.

[59] Libet, B (1985). "Unconscious cerebral initiative and the role of conscious will in voluntary action". Behav Brain Sci. 8 (4): 529–566. https://doi.org/10.1017/s0140525x00044903.

[60] "Accelerated modern human-induced species losses: Entering the sixth mass extinction," by Gerardo Ceballos, Paul R. Ehrlich, Anthony D. Barnosky, Andrés Garcia, Robert M. Pringle and Todd M. Palmer, Science Advances AAAS, 19 June 2015: Vol. 1, no. 5, e1400253, https://doi.org/10.1126/scladv.1400253, https://advances.sciencemag.org/content/1/5/e1400253.figures-only, Accessed May 16, 2020.

[61] Data Zone, "We have lost over 150 bird species since 1500," BirdLife International, http://datazone.birdlife.org/sowb/casestudy/we-have-lost-over-150-bird-species-since-1500; See also, Butchart, S. H. M., Stattersfield, A. J. and Brooks, T. M. (2006) Going or gone: defining 'Possibly Extinct' species to give a truer picture of recent extinctions. Bull. Brit. Orn. Club. 126A: 7–24.

**Fig. 52.3** Sikkimese Child, Uncertain Future. (Photo © M.C. Tobias)

whose time is fast running out. Or, as Charles Darwin wrote, "Hence, rare species will be less quickly modified or improved within any given period; they will consequently be beaten in the race for life by the modified and improved descendants of the commoner species".[62]

We can search among the behavioral data for the earliest known primates, or ants, or sharks, or the lost trilobites, and speculate upon what constitutes lasting, biological success, and, simultaneously, plumb the earliest philosophical caveats in an effort to determine that which is most common to all ecological idealism. These primeval factors, in whatever specific or combinatorial effect, from a Darwinian perspective must inform the genotypes of choice leading us into the future. What proteins, genes, ideas and wholistic organisms, modified through descent in time, have noticeably exerted the resilience to get organisms through evolutionary cul-de-sacs? Disease outbreaks? Genetic squeezes? Intense competition? Sporadic violence searching for stability?

Such questions now have genetic markers along the way of their odysseys, and a fascinated intellectual era engrossed by probability statistics that intimate the end of the line, or not. Our obsessions with doomsday lend eerie credibility to our instincts. Mathematics and computational biology appear destined to assist in this most arrogant of efforts; arrogant in that it has welcomed through cruelty, escalating total fertility rates, geopolitical/socio-pathological trial and error, incredulity, and a true fiasco of single-generational thinking and consumption, the prospect of its very self-annihilation.

---

[62] Charles Darwin, *The Origin Of Species By Means Of Natural Selection Or The Preservation Of Favoured Races In The Struggle For Life*, John Murray, London, Sixth Edition, 1886, Chapter IV, p. 85.

The accretion of blinders is but one aspect of our indwelling; our unwillingness to actually confront the crisis at the collective level (undoubtedly the primary driver of our rapid final act) prefiguring a most dour and irremediably paradoxical situation: Can we shake this statistical apathy in the face of our extinction? Is kin-altruism enough, at this late stage, to resurrect some council of clarity, a parliament of philosophers with the credible voice to step in before it is simply too late to ecologically liberate the biosphere? (Fig. 52.3).

# Chapter 53
# Non-linear Reciprocity

## 53.1 The Crisis of Connectivity

Reciprocity has been fundamental to the study of electromagnetism, optics, electrostatics, electrical fields, and the realization antennas both transmit and receive signal—the cumulative work of researchers like Michael Faraday (1791–1867),[1] Hendrik A. Lorentz (1853–1928),[2] Hermann von Helmholtz (1821–1894), and John R. Carson (1886–1940).[3] In social and ecological arenas, reciprocity has proved not just key to life—the single essential quotient—but that *quality* of life that gives it a rationale, energy, substance, identity. In another form, it is kinship, as Indian biologist Madhav Gadgil has elucidated. The foundation of community; the source of protection, of future confidence in reciprocal stability, shared sacred spaces, ancestral feelings, and fair transactions,[4] all the qualities of the human experience that militate against the notion of "the selfish gene."

Just because genes are *involved* in the coding for life, they don't (arguably) necessarily determine *everything*. Hence, Holmes Rolston III declares, "We are having trouble seeing how any one gene is in any position to act selfishly".[5] And

---

[1] *On the Various Forces in Nature*, by Michael Faraday, Ed. By W. Crookes, Chatto and Windus, London, 1873.

[2] H. A. Lorentz, "The theorem of Poynting concerning the energy in the electromagnetic field and two general propositions concerning the propagation of light," Amsterdammer Akademie der Wetenschappen 4 p. 176 1896.

[3] J. R. Carson, "A generalization of reciprocal theorem," Bell System Technical Journal 3 (3), 393–399, 1924. Also J. R. Carson, "The reciprocal energy theorem," ibid. 9 (4), 325–331 1930.

[4] See "Of Life and Artifacts," by Madhav Gadgil, in *The Biophilia Hypothesis*, Edited by Stephen R. Kellert and Edward O. Wilson, Island Press/Shearwater Books, Washington, D.C. 1993, pp. 365–377.

[5] See "Biophilia, Selfish Genes, Shared Values," by Holmes Rolston III, in *The Biophilia Hypothesis*, Edited by Stephen R. Kellert and Edward O. Wilson, Island Press/Shearwater Books, Washington, D.C. 1993, p. 385.

© Springer Nature Switzerland AG 2021
M. C. Tobias, J. G. Morrison, *On the Nature of Ecological Paradox*,
https://doi.org/10.1007/978-3-030-64526-7_53

Rolston examines how a behavioral manifestation, by logical turns, would invite "reciprocal altruism" and "close kinship" in any other number of species, versions of "cultural cooperation," and more.[6] Duke University Law Professor Jedediah Purdy examines the famed "poet of the American Revolution," Philip Freneau (1752–1832) to glean yet a different inroad to that ecological mechanism which turns upon the politics of tyranny, citing Freneau's conviction that "Nature … was the unfailing ally of republican freedom, and kings were usurpers against her." So as to evince, "All, nature made, in reason's sight, is order all, and *all is right*".[7] Such "freedom" and its absolute ethical lure presupposes the experience of it. In the same way, Kitty Ferguson renders an intriguing commentary on the nature of rational numbers and universal harmony, versus irrational numbers, infinite inconclusiveness, when she discusses the Pythagorean thinking about right triangles and the fact there is no whole number square root of the hypotenuse: "In order to have had a devastating crisis of faith resulting from the discovery of incommensurability, they [the Pythagoreans] had to have had the faith first, not the crisis".[8]

Genetically, or in any evolutionary sense of the purposeful, we have neither physiological nor moral claims on anything; no privilege, no special gift, or empowerment. Ours is, at best, a humble beginning that has seized up into what, earlier, we have described as a likely near-term extinction candidacy. But what does remain, at our slightest embrace of it is the possibility of re-connecting to other species, and thereby fulfilling a dream. It is saturated with the very Arcadian mythopoetics that have sporadically guided us; the more than 100 "Peaceable Kingdoms" painted by Edward Hicks (1780–1849) and his predecessors and contemporary folkloric artists seeking to use esthetic responses and techniques to tease from our tired rhetoric and bedeviled depths a renascence of emotional ties to trees and cows and every conceivable invertebrate or cluster of moss. To re-invent ourselves through those reciprocal exchanges that matter, both within and outside our heavily burdened consciousness. To both liberate and grasp the readiness and reciprocity potentials that could shake us from our slumber. These are not linear mechanisms, but those that are activated when another soul is forthrightly spoken to, believed in, patiently, inquisitively addressed with full dignity, innocence, and a collaborative zeal. That is the Nature we knew in the beginning of our frenzied odyssey.

---

[6] ibid., pp. 397–405.

[7] Purdy citing Freneau from "On the Uniformity and Perfection of Nature, from *Poems of Freneau*, ed. By Harry Hayden Clark, Harcourt, Brace, New York, 1919, pp. 423–424, in Purdy's *After Nature -A Politics for the Anthropocene*, Harvard University Press, Cambridge, Massachusetts, & London, UK. 2015, pp. 104–107. See also, *Philip Freneau, Poet of the Revolution*, by Richard Nickson, New Jersey Historical Commission, Trenton, New Jersey, 1981.

[8] See *Pythagoras – His Lives And The Legacy Of A Rational Universe*, by Kitty Ferguson, Icon Books, London, 2010, p. 87.

## 53.2  Ineluctable Recognitions

Much that is paradoxical in the history of the natural sciences concerns that inevitable recognition, usually hitting hard quite early in a human life, of violence throughout biology. Assiduous philosophies to counteract aggression are unlikely to ever impede carnivory, within our own homes, and certainly not out beyond our personal borders of control. The Serengeti is not showing signs of a major transformation, in that domain, despite the fact so much of the biomass, from shrub land to the elephant, and the largest assemblage of other large vertebrates (all of the ungulates) are herbivores. But the biological paradox is more than violence/non-violence. It speaks to the relations between beings and exacts an expectation, if often unrequited, of some kind of chance recognition. When people connect with one another, it is a currency; the parlance of everyday; a fluent reality. When we do so with other species, the amplitude at once invents a new form of equilibrium that is typically cast in the form of awe; and not infrequently, gratitude for the experience.

With so-called companion animals the joy has become customary for thousands of years. But our sphere of such companions is limited to so few species (for most people) that they number about the same as those few remaining chosen apple cultivars, a few hundred, where—not that long ago—there were, according to Dan Bussey's seven-volume *Illustrated History of Apples in the United States and Canada*, over 18,000 varieties.[9] In seeking to expand the realms of communication with other species, we tend, from a daily practical standpoint, to focus on very few. A dog, a cat, a bird, a rabbit, a hamster, a turtle, a fish, a horse, for example. It then gets very complicated on farms and ranches, where alleged love easily turns to complicity in killing or abandonment, the *price of doing business.*

Underlying all of the suffering that is inflicted in the goal of achieving some level of soul satisfaction in the company of another species (pet stores teeming with the caged and forgotten) is an altogether lost language of companionship. To be sure, there are millions of people who derive the vast majority of their true solace from dogs and cats. But we're speaking of the encounters in the *wild* where, for so many it is the mere distant glimpse of a non-domesticated Other that is transformative. Our epiphanies make their way, into the biological languages (and whole careers) that have sought to catalogue and gain some kind of understanding, beyond mere two-week photographic safaris which, when they come to their inevitable ends, see us hastening, as a species, to familiar domesticities. To lives surrounded by, immersed in images (multimedia) and memories. These are community ordained, socially prescribed modes of communicating with one another.

The source of that lasting injunction to be forthrightly attuned to Nature, is probably somewhere else, back out in the wild. And all fears aside, what most humans actually crave are those non-human encounters. Lacking them, a train of ants across

---

[9] Numbers worldwide would have been triple that. See "Meet the Man on a Quest to Document Every Apple in North America," by Rohini Chaki, Atlas Obscura, June 18, 2019, https://www. atlasobscura.com/articles/history-of-apples, Accessed July 17, 2020.

the kitchen, or rat chewing somewhere in the ceiling now, and then the most welcome exceptions, we drop pieces of French fries for the pigeons under our café tables, and leave it at that. Some meander about the local zoo, all indications suggesting that no more, on average, than five minutes per visitor is spent reading ecologically educational materials there.

## 53.3  Languages and Paradigms with Which to Reunite

Two years prior to his death, in 1978, French philosopher/psycholinguist Roland Barthes imparted a number of lectures in Paris on the subject of le neutre—the neutral. His hope was to somehow thwart at long last all those paradigms imposed upon consciousness, ethics and history by a western heritage of the binary. He would write in the book comprising those lectures, "Whence the idea of a structural creation that would defeat, annul, or contradict the implacable binarism of the paradigm by means of a third term – the *tertium* ... The Neutral – my Neutral – can refer to intense, strong, unprecedented states." And later in his work, would state: "An emotive hyperconsciousness thus seems to be a contradiction in terms: a paradox. I feel this paradox in myself, I have to live, to debate with it".[9] He was speaking of the tyrannies of zero and of one which, in electronic circuits impose logic gates, formats of counting, the representations of "on" and "off" and as such, in every computer, accountability for each digit, increment, prefix, decimal, fraction, calculation, procedure, conversion ... everything. A binary wasteland whose occupants do not exist, though they gaze upon the present and posterity like concepts-turned-demonic, with a power vested in them to exert every conceivable inordinate sway over our lives, fueled by electricity, amassed into a world of artificial intelligence and other technologies that have only added to the lacunae separating us from that wilderness at the heart of the biological paradox.

Yes, we use many of these tools to *study nature*. The question Barthes was grappling with concerns the languages and paradigms by which our self-awareness and silent hopes in the presence of Others might be liberated and satisfied. This condition, a stultifying human civilization that defines and oppresses most wonder, is without true reciprocity, that can only be achieved, says Barthes, by some altogether outside force. A distance from the normal, an exile radically envisioned and achieved; a language that obeys no language.

Not so with other creatures, we can be fairly certain. Their worlds and community interactions rally free of such stultifying commands and controls, for reasons we have no clue, particularly given that it appears—this absolute freedom—to exist everywhere outside human consciousness; a level of reciprocity between species that has given rise in our natural sciences, to the perception and accompanying theories, of interdependence, the most crucial element in the pan-Asiatic worlds discussed earlier in this work; that source of true reciprocity potential among humans and every being on this planet. Yet, lacking something imperative,

**Fig. 53.1**  Iberian Wolf (Canis lupus signatus), Critically Endangered, at the Portuguese Sanctuary of Grupo Lobo (Iberian Wolf Recovery Center). (Photo © M.C. Tobias)

only humans remain locked out of these theories. Barred from entry by our own worst behavioral traits, yet forever desperate to get inside, somehow circumventing those evolutionary logic gates (Fig. 53.1).

One look at what Jonathan Balcombe has termed "the exultant ark" yields to consciousness (or, for that matter, Barthes' "hyperconsciousness") an expanse of biological joy among those Others that further highlights our own isolation, and lends additional commentary to the crisis we are in. The absence in ourselves of some organ akin to a tuning fork that can help us to be re-united emotionally with other, unfamiliar species; to erase a great emptiness. In that gap is a reciprocity potential, as we have termed it elsewhere; a dormant joy, and a deep level of appreciation so untapped as to pose the full seriousness all at once, of our predicament. A situation much like the huge attenuation of the very seeds we have obfuscated one way or other—93% of all varieties, lost, just within the United States between 1903 and 1983, as earlier referenced just in the case of apples. But of course, the same winnowing has occurred in our very extinctions of experience across most taxonomic suites in the world.[10]

In missing out on nearly everything outside our human-sated selves, by all indications we are rapidly giving up on this world. Stated simply, Balcombe writes, "To relate to others' feelings is to consider their experience, to try to imagine things from their perspective." It is the first rung of the ladder to ultimate compassion, the longing for communion that is fundamentally the nature of ethology, interspecies

---

[10] From *The Neutral*, by Roland Barthes, Lecture Course at the Collège de France (1977–1978), Translated by Rosalind E. Krauss and Denis Hollier, Text established, annotated, and presented by Thomas Clerc under the direction of Eric Marty, Columbia University Press, New York, NY, 2005, pp. 7, 101.

communications, biosemiosis, all those rudiments of biophilia and physiolatry—the true love of nature—that have so rapidly risen in importance as their precise opposites—human-based destruction of the world—has escalated.[11]

Removed from electronic circuit boards and microchips, from the labored history of calculus and all the counting that has given inexorable voltage the infinitely dangerous live wires that now surround human life, are the esthetics, the saints, and all those other 100 million+ species whose very lack of obsession with such *objets grossiers*, the callous mechanics we think we thrive upon, are *their* salvation. Of course, it all strikes of paradox. We write at least part of this text on a—indeed very—convenient computer, looking out at a forest. Yet, we are a species for whom the forest, at least for many, is somehow insufficient. A few uncontacted tribes, and veritable hermits undermine this all but universal rule, in the twenty-first century. Hence, a human biology that has fueled restlessness to the point of going to the moon, sending things to Mars, compiling a list of exobiological habitable ("Goldilocks zones)[12] with calculations ranging from dozens to billions of potential other life-supportive planets in the universe. Is this rage to list other prospects predicated on reverence? Are we to suppose that we will just eventually obtain technology sufficient to get us to one of these other planets, and by instant accommodation refresh our entire inner nature out there? Neither the science nor its psychology warrants much of a response; we cannot get it right here; why dignify the argument that we could get it right elsewhere?

This fascination is the height of our rejection of earth; a symptom of having so perverted the notion what is reciprocal in nature, that we no longer remember, from moment to moment, the joys of our childhood heroes, including Doctor Dolittle and Lassie. Clare of Assisi (1194–1253) was struck by the beauty and simplicity of her local colleague, Francis, as he spoke to birds and to wolves. Born to a wealthy noble family (the Offreduccio's) she nonetheless chose to renounce everything but the church of Saint Damian that Francis had recommended for her. There she became the first abbess of a Utopian community. All the more ironic, then, that in the twentieth century the Vatican should view her as, among other things, the patron saint of television, for having seen the "funeral rites of Saint Francis 'projected' at Saint Damian's".[13]

---

[11] See "Infographic: In 80 Years, We Lost 93% Of Variety In Our Food Seeds," by Mark Wilson, May 11, 2012, FastCompany, https://www.fastcompany.com/1669753/infographic-in-80-years-we-lost-93-of-variety-in-our-food-seeds?utm_source=postup&utm_medium=email&utm_campaign=seeds&position=5&partner=newsletter&campaign_date=12052019, Accessed May 18, 2020.

[12] See *The Exultant Ark – A Pictorial Tour of Animal Pleasure*, by Jonathan Balcombe, University of California Press, Berkeley, Los Angeles, London, 2011, p. 188.

[13] See PHL, Planetary Habitability Laboratory, University of Puerto Rico at Arecíbo, http://phl. upr.edu/projects/habitable-exoplanets-catalog. See also, Choi, Charles Q. (21 August 2015). "Giant Galaxies May Be Better Cradles for Habitable Planets". Space.com; see also, Lammer, H.; Bredehöft, J. H.; Coustenis, A.; Khodachenko, M. L.; et al. (2009). "What makes a planet habitable?" (PDF). The Astronomy and Astrophysics Review. 17 (2): 181–249. Bibcode:2009A&ARv. 17.181L. https://doi.org/10.1007/s00159-009-0019-z.

**Fig. 53.2**  The Mystique of Equus Asinus. (Photo © J.G. Morrison)

To embrace a St. Clare, or that language of neutrality—of nature—upon which Barthes so powerfully meditated, is also to renounce judgment; to celebrate those mysterious aspects of paradox that are positive, cathartic, analgesic—as opposed to insane and ruinous[14] (Fig. 53.2).

---

[14] See *Saints in Art*, by Rosa Giorgi, Edited by Stefano Zuffi, Translated by Thomas Michael Hartmann, The J. Paul Getty Museum, Los Angeles, 2003, p. 94.

# Chapter 54
# The Unfettered Gaze

## 54.1 Dualistic Centralities

In his discussion of paradox, Hugh Bredin writes, "philosophers tend to give the name 'paradox' to any statement which on the surface seems straightforward and innocuous but which turns out, on further examination, to have consequences which undermine some fundamental laws of logic, thought, or language" and he goes on to examine the "New Criticism" school of mid-twentieth-century American writing, particularly the esthetic deliberations of Cleanth Brooks, author of *The Well Wrought Urn*, with its first chapter titled "The Language of Paradox".[1] Bredin is interested to plumb paradoxical nuances of poetry, truth, relationships between irony, contradiction, and what Friedrich Schlegel described as "the impossibility ... of total communication." Bredin describes Brooks' belief that there are "some truths that cannot be stated, but which can be vaguely glimpsed or half-known through our encounters with words".[2] Actually, Brooks (1906–1994), who co-founded the Southern Review with Robert Penn Warren, and is most remembered for his phrase, "the heresy of paraphrase," believed in poetry—all art—in and for itself, devoid of any external necessities, liberated by the very nature of itself from the criticism that would deliberately or inadvertently alter the pure structure and freedom of a poem. It did not require exegesis. It was alone and unfettered by that which is irreducible (Fig. 54.1).

This dialectic which reduces to the irreducible, is crucial to ecological concerns—whether in the debate (linguistic, artistic, conceptual, and geopolitical)

---

[1] Dobson Publishers, London, 1949, pp. 3–20.

[2] Friedrich Schlegel, "Selected Aphorisms from the Lyceum," in Schlegel's *Dialogue on Poetry and Literary Aphorisms*, translated by Ernst Behler and Roman Struc, Pennsylvania State University Press, London PA, 1968, pp. 121–132, as cited by Bredin. Hugh Bredin, Philosophy and Literature, Paideia, "Ironies and Paradoxes, https://www.bu.edu/wcp/Papers/Lite/LiteBred. htm, Paper Presented at the Twentieth World Congress of Philosophy, in Boston, Massachusetts, August 10–15, 1998.

© Springer Nature Switzerland AG 2021
M. C. Tobias, J. G. Morrison, *On the Nature of Ecological Paradox*,
https://doi.org/10.1007/978-3-030-64526-7_54

**Fig. 54.1** A local in Dijon. (Photo © M.C.Tobias)

which seeks to communicate the substance of what life *means* to us; whether there
is actually any significant difference between one species and another, given the
through-story of life herself; questions concerning "wilderness" that have turned to
a crossword puzzle of patchwork between sand-traps, pavement, megacities, toxic
waste-dumps, and all the other largely inorganic congeries of human existence.
Wilderness, like evolution is an idea, a word, that continues to elude us. While we
can glean evolution in real-time—such as in the "59,000 generations" of *Escherichia
coli bacteria* that Richard Lenski at Michigan State University has been observing
in an ongoing experiment since 1988[3]; or in studies that have enabled researchers to
recognize distinct changes in communities of Guppies within streams of Trinidad
"in response to changes in predation threat [by killfish] in just a few generations"[4];
or even real-time changes in deer mice of the Sand Hills of Nebraska within "a
single generation"[5]—the word "evolution" invites only half-formed intimations of
contradiction involving time, consciousness, purpose, esthetic predilection, survival, and every conceptualization of the meaning of human life, and all other life
forms. A word so steeped in inconclusiveness that it all but enshrines the notion of
ecological paradox.

---

[3] See "Evolution in real time," by Alvin Powell, The Harvard Gazette, February 13, 2014 https://
news.harvard.edu/gazette/story/2014/02/evolution-in-real-time/. Accessed May 22, 2020.

[4] See "Infographic: Watching Evolution in Real Time," by Jef Akst, April 30 2017, TheScientist,
https://www.the-scientist.com/multimedia/infographic-watching-evolution-in-real-time-31599,
Accessed May 22, 2020.

[5] See "Seeing 'evolution in real time': Mice blend in to survive," by Katherine J. Wu, January 31,
2019, NOVA WONDERS, NOVA, https://www.pbs.org/wgbh/nova/article/evolution-in-real-time-
mice/. Accessed May 2020.

In his book *The Paradox of Evolution*, physiologist Stephen Rothman suggests that the universe is essentially without purpose; that life—comprising reproductive assertiveness and a compulsion to survive—is inherently at odds with itself. That threats of all persuasion induce life; destruction rallies counterintuitively around creation. But that natural selection, while ensuring survival, is not the actual force urging the optimization of specific traits (allele frequencies) within populations, a notion greatly at odds with traditional biological theory. While Rothman proposes that natural selection has no purpose, reproduction, he claims, does.[6]

There are other fundamental contradictions exposed within the overarching concept itself of *ecology*. Modeled behavior, versus that of real-world (real-time) interactions. In the so-called paradox of enrichment, a phrase coined in 1971 by University of Arizona population ecologist, Michael Rosenzweig, populations whose food supplies have been increased, may result in a *destabilization* (or utter crash) of the overall system. This follows upon a crucial insight applicable both to rabbits and to humans. If carrying capacity is artificially increased beyond a viable threshold, there is a logistical tautology that cannot remain for long. Both conceptually, and in the wild, there is ensuing entropy and a species, either locally or— theoretically, at least—in cosmopolitan coordinates, can go extinct.[7]

Purpose, carrying capacity, stability, instability: four elements of evolution that are either empirically proved by observation and mathematical formulation, or simply present the *illusion of evolution*. Species interactions have traditionally been viewed through a biased, human lens that is content to define "predator-prey" relations as the most common form of interaction (e.g., Lotka-Volterra equations; dog-eat-dog world).[8] When, in fact, there is a non-violent modality far more pervasive throughout the biological world, namely, photosynthesis: $nCO_2$ [carbon dioxide] + $2nH_2X$ [hydrogen] = $nCH_2O$ [carbohydrate] + $2nX$ [byproduct] + $nH_2O$ [water] affecting all life forms, the conversion of "light energy to chemical energy".[9]

We could set up this division in other ways. For example, the differences between an Albert Schweitzer and a Machiavelli; Gandhi and Genghis Khan; legionary behavior (foraging raids) in (army) ants, colony fissions (population thresholds),

[6] See *The Paradox of Evolution: The Strange Relationship Between Natural Selection and Reproduction*, by Stephen Rothman, Prometheus Books, Amherst, NY, 2e015. See review by Michelle Martinez, New York Journal of Books, https://www.nyjournalofbooks.com/book-review/paradox-evolution/; and https://www.publishersweekly.com/978-1-63388-072-6; https://www.goodreads.com/book/show/25310542-the-paradox-of-evolution, Accessed May 22, 2020.

[7] Michael Rosenzweig, "Paradox of Enrichment: Destabilization of Exploitation Ecosystems in Ecological Time," Science Vol. 171: January 29, 1971, Issue 3969, https://doi.org/10.1126/science.171.3969.385. pp. 385–387, https://science.sciencemag.org/content/171/3969/385, Accessed May 22, 2020.

[8] Berryman, A. A. (1992). "The Origins and Evolution of Predator-Prey Theory" (PDF). Ecology. 73 (5): 1530–1535. https://doi.org/10.2307/1940005. JSTOR 1940005.

[9] See https://www.cliffsnotes.com/study-guides/biology/plant-biology/photosynthesis/most-important-process-in-the-world, Accessed May 22, 2020.

and graphic carnivorous behaviors.[10] Versus the (by comparison) docile feeding patterns of nearly all ungulates (including the more than 220 species of two dominant orders: the Perissodactyla and Artiodactyla—odd- and even-toed). The Bambis of the world. And the moths and butterflies (with the exception of at least one North American butterfly), the Harvester [*Feniseca taquinius*], who does, as an adult, feed on aphids.[11]

As humans, perceiving critical differentiations between our own bodies, on the ground (perception rooted in "Renaissance themes and styles" of the south, versus the emergent northern European preoccupations by the nineteenth century with "venture capitalism and industrial developments")[12]; in the waters,[13] the soils,[14] up in the trees (the fact that "dead wood in forests" engenders "a greatly increased species richness of fungi, plants and animals"),[15] and in the air,[16] *what*, and *how* do we view life on earth? We address this fundamental not in order to capture nearly 7.8 billion differing individual opinions, but, rather, to pose an essential sense of the *life gaze*. Metaphysically, it is tantamount to the life force as manifested in perception.

While every perceiving being will personalize variation that is randomized, there are distinct species-wide contours defining our sensory constraints (filter bandwidths in humans),[17] "plasticity of the individual components," and so on[18] and

---

[10] Wilson, Edward. O.; Hölldobler, Bert (September 2005). "Eusociality: Origin and consequences". Proceedings of the National Academy of Sciences. *102*(38): 13367–71. https://doi. org/10.1073/pnas.0505858102. PMC 1224642. PMID 16157878. Accessed May 22, 2020.

[11] See https://basicbiology.net/animal/mammals/ungulate, Accessed May 22, 2020.

[12] See *Sense Of Place – European Landscape Photography*, "Photography, Nation, Nature," essay by Liz Wells, Prestel, Bo Zar Books, Munich, London, New York, Foreword by Herman van Rompuy, 2012, p. 250.

[13] See, for example, *Oceans – Heart Of Our Blue Planet*, Gregory S. Stone, Russell A. Mittermeier, et. al. Series Editor: Cristina Goettsch Mittermeier, CEMEX and iLCP, Arlington Virginia, 2011.

[14] See "Re-animating soils: Transforming human-soil affections through science, culture and community," by Maria Puig de la Bellacasa, The Sociological Review, February 28, 2019, https://doi. org/10.1177/0038026119830601, https://journals.sagepub.com/doi/abs/10.1177/00380261198306 01?journalCode=sora

[15] See "How trees talk to each other," by Suzanne Simard, TEDx, Banff Alberta, Canada, June 2016, https://en.globalistika.ru/videolektorij-kongressa; see also, The Afterlife Of A Tree, by Andrzej Bobiec, Jerzy M Gutowski, William F. Laudenslayer, Pawel Pawlaczyk, and Karol Zub, WWF-Poland, Warszwa-Hajnówka, 2005, p. 204.

[16] See, for example, *Earth From Space*, by Yann Arthus-Bertrand, GoodPlanet Foundation, Abrams, New York, 2013.

[17] See "Revised estimates of human cochlear tuning from otoacoustic and behavioral measurements," by Christopher A. Shera, John J. Guina, Jr., and Andrew J. Oxenham, PNAS, PMC, NCBI, Proc Natl Acad Sci U S A. 2002 Mar 5: 99(5): 3318–3323, Published online 2002 Feb 26. https:// doi.org/10.1073/pnas.032675099, PMCID: PMC122515, PMID: 11867706, https://www.ncbi. nlm.nih.gov/pmc/articles/PMC122516/ Accessed May 22, 2020.

[18] See "Phonology: An emergent consequence of memory constraints and sensory input," by Francisco Lacerda, Reading and Writing: An Interdisciplinary Journal 16: 41–49, 2003, Kluwer Academic Publishers, the Netherlands, 2003, p. 44. http://citeseerx.ist.psu.edu/viewdoc/download ?doi=10.1.1.663.6451&rep=rep1&type=pdf, Accessed May 22, 2020.

sheer attitudinal differences that may be worlds apart. But physiologically, every conjugate gaze (four systems employing both eyes simultaneously) shares some structural similitudes in humans: eye and neck, the lateral function of the "paramedian pontine reticular formation",[19] while the vertical is commandeered by the "rostral interstitial nucleus of medial longitudinal fasciculus and the interstitial nucleus of Cajal",[20] the complex morphological explanation for varieties of visual experience in humans.

By whatever means, our organism usually succeeds in situating its consciousness, projects, and mediates a vision (internal, external) that ripens and transmutes through time. In music, the gaze is compounded by what has been characterized as "paradox through the prism," referring to Friedrich Schlegel's "theory of form," in the context of Beethoven's revolutionary 1802 piano sonata, "Tempest," which, according to Edgardo Salinas, produces an "absence" whose conspicuity we can hear/imagine. And, hence, a gaze across the horizons of sound that is as critical to understanding human nature as our reactions to any landscape scene, to the world at large.[21] In exploring Beethoven's music, Salinas suggests that the "Tempest" is a "case study" in the understanding of how—in "the interface between form and medium, deploying self-referential strategies"—Beethoven all but defined "romantic subjectivities," the "experience of immediacy," and "the paradoxical condition" that is "immanent to the modern self"[22] (Fig. 54.2).

## 54.2  The Proliferation of Gazes

The gaze of which we speak proliferates across boundaries, vulnerabilities, modes of volition, gradients in color, topography and sensibility, spiritual practice, contrition, redemption, isolation, and relief. It is more than a gaze. It is vision of the soul connecting with the world, words that mean nothing short of some extraordinary instance of experience. Eye contact between two species; a grasp of something

---

[19] See "Neural Control of Saccadic Eye Movements," Neuroscience, 2nd edition, NCBI Resources, https://www.ncbi.nlm.nih.gov/books/NBK10992/ Accessed May 22, 2020.

[20] Versino, M; Simonetti, F; Egitto, M G; Ceroni, M; Cosi, V; Versino, M; Ceroni, M; Cosi, V; Beltrami, G (1999). "Lateral gaze synkinesis on downward saccade attempts with paramedian thalamic and midbrain infarct". Journal of Neurology, Neurosurgery & Psychiatry. 67 (5): 696–7.  https://doi.org/10.1136/jnnp.67.5.696.  PMC  1736620.  PMID  10577040;  See also, "Physiological Aspects of Communication Via Mutual Gaze," by Allan Mazur, Eugene Rosa, Mark Faupel, Joshua Heller, Russell Leen and Blake Thurman, American Journal of Sociology, Vol. 86, No. 1, July, 1980, pp. 50–74, University of Chicago Press, https://www.jstor.org/stable/2778851, https://www.jstor.org/stable/2778851, Accessed May 22, 2020.

[21] See "The Form of Paradox as the Paradox of Form: Beethoven's 'Tempest,' Schlegel's Critique, and the Production of Absence," by Edgardo Salinas, Journal of Musicology, 33 (4): 2016, 483–521,  https://doi.org/10.1525/jm.2016.33.4.483,  https://online.ucpress.edu/jm/article/33/4/483/63486/The-Form-of-Paradox-as-the-Paradox-of, Accessed May 22, 2020.

[22] ibid., Edgardo Salinas, From Abstract.

**Fig. 54.2** Charles Darwin
in 1880, two years prior to
his death. (Public domain)

heretofore invisible, inexpressible, or immutable. The gaze might conjure "last refuges on the planet" as in the case of the exquisite and under-the-radar photographs of Finnish artist Pentti Sammallahti.[23]

Or similarly astonishing images of ice and rock, explosive, dismal, gorgeous nuances of the earth's polar regions, by American photographer Thomas Joshua Cooper in his collection, *True*.[24] Combining endlessly rich diversions of sight and sound, one fixates on two sets of two walkers on a sullen desperately remote beach, created by Thomas Wrede ("Beachwalkers, 2004"). It is the end of the world, the beginning of the world, the last four humans on the planet, and they are removed and moving further yet from each other, as two and two, a numerical riddle lost within the print, which, upon close examination yields at least seven other minutely

---

[23] See *here far away – photographs from the years 1964–2011*, Edited By John Demos In Collaboration With Mario Peliti, Preface By Finn Thrane, Bibliography By Kristoffer Albrecht. Dewi Lewis Publishing, Stockport, UK, 2012, p. 6.

[24] Thomas Joshua Cooper, *True*, with "Gazing at the Void – Some Notes on Thomas Joshua Cooper's Polar Pictures," An Essay by Ben Tufnell, Haunch of Venison Publishers, London, UK 2009.

captured individuals far off against the sea.[25] A website titled "wetcanvas" poses the question, "How Many Paintings Exist in the World?".[26]

One recent collection suggests the following break-down in perceptual categories: "Painting with the Flow of the World," "Realism and Beyond," "Post-Pop Landscapes," "New Romanticism," "Constructed Realities," "Abstracted Topographies," and "Complicated Vistas." It is certainly a good beginning.[27] One even more confusingly, fervently embraced collection can be found in Gerhard Richter's thousands of images (photographs, sketches, and collages from 1962 to 2006) of his breathtaking *Atlas.*[28]

All of these sites within the cartography of the world are so distant from, yet inherent to, two intently gazing Ringless Chinese Pheasants (*Phasianus decollatus*) painted by Joseph Wolf (1820–1899).[29] They are looking at us and no description (human) quite comports with the inquisitive—fearful—agitated—excited—luminous—terrified—awestruck—self-contained—height-of-brillant   pellucid   innocence in their eyes.

So many similar images teem around our minds, inviting ethics, senses, philosophies both familiar and exotic. We live within and between them, forming our judgments of imagistic congestion from the instant of our birth, and even, at the end, in coma, or last instant subconsciousness, we are still there formulating. With so much neurological chatter, the question of true wilderness persists: What is it, and has humanity ever actually been at peace there? Is it a place, a feeling, a lost world entirely? Or does it somehow call us into a near or distant future? They are silly questions because language, or any one of these multitudinous human expressions actually know what is going on, even if none of us truly know. We are lost, trying to find something with whatever anatomies and tools at our species' disposal.

Given the range of our proficiencies, why are we so bent upon footprints, handprints, destruction, erasure, obfuscation, violence? That was the form of Beethoven's "Tempest" Piano Sonata No. 17,[30] an urgent commentary on the storms of our presence. The Reverend William Joseph Long (1867–1952), some of whose early books

---

[25] Lambda print mounted on aluminum with Plexiglas face; edition 1/5, 47 × 59 inches (119 × 150 cm), Hall Collection, http://www.hallartfoundation.org/exhibition/birds-eye-view/artworks/thumbnails, Accessed May 22, 2020. See also, *Landscapes After Ruskin – Redefining the Sublime*, Introduction by Lynn Gumpert, Essays by Joel Sternfeld, Dale Jamieson and Chris Wiley, Hall Art Foundation, Reading VT, 2018.

[26] https://www.google.com/search?q=how+many+paintings+in+the+world%3F&rlz=1C5CHFA_enUS893US901&oq=how+many+&aqs=chrome.0.69i59j69i57j69i59l2j0j69i60j69i6l12.1343j1j7&sourceid=chrome&ie=UTF-8

[27] See *Landscape Painting Now – From Pop Abstraction to New Romanticism*, Edited by Todd Bradway, Essay by Barry Schwabsky, Contributions by Robert R. Shane, Louise Sorensen, and Susan A. Van Scoy, D.A.P., New York, 2019.

[28] Edited by Helmut Friedel, Thames & Hudson, London, 2006.

[29] See *Monograph of the Phasianidae or Family of the Pheasants*, by Daniel Giraud Elliot and Joseph Wolf, Volume II, Published by the Author, New York, 1872.

[30] See, for example, https://www.youtube.com/watch?v=6KMGcOYHSs0, Accessed May 22, 2020.

Teddy Roosevelt had removed from libraries because he believed Long was exaggerating his experiences out in the woods, once wrote, "The eye of a wild animal is even more wonderful ... [and] if you ever look into it in close quarters, you will have a real ... lesson in natural history. I have known hunters to throw aside their rifles and to be changed men for life having looked once into the eyes of a dying deer".[31]

This was cited by Charles Fisher in his powerful study on the nature of "discontent," on death, suffering, Buddhism, and Darwin. Fisher compares the erratic, all too-morbid life spans of various species, like Dahl sheep, raccoons, sparrowhawks, moles, porcupines, and bears. In assessing the types of pain and suffering our hunter/ gatherer ancestors (no less riven with stress and anxiety) must have experienced, he also considers contrasts between Buddha's time and our own. The goal is to form some kind of middle-path comfort, in spite of the enormity of suffering sweeping across the otherwise alleged calms of so many gardens of Eden. How animals actually die; people eaten by crocodiles, lions, a multitude of deformities as deciphered in Pieter Bruegel the Elder's sixteenth-century crowd scenes, mortality rates over the past tens of thousands of years in humans, and the typical decay and diminution of the human body in contemporary terms, as it ages.[32] Concluding with the famed ecologist Paul Sears' (1891–1990) characterization, "The face of the earth is a graveyard and so it has always been".[33]

## 54.3   What Do We See?

Despite a seemingly omnipotent motif of such miseries (Darwin himself is known to have suffered throughout most of his life—"he had an organic problem, exacerbated by depression"),[34] humanity is conditioned to thinking that with its recent evolution (domestication, the Stone and Copper Ages), the taming of companion animals and store-housing of plants, origination of metrics, surplus, and coordinated geopolitical hierarchies, that ethical injunctions arose to serve a biological choreography comprised of neighbors. And thus, the great debate, as described by anthropologist Peter J. Wilson, between Freud and those ascribing to the concept of "love thy neighbor as thyself." It was supposed to make the evolution of our domestication easier. Freud, however, insisted that it simply brought the targets of

---

[31] Cited by Charles Fisher, in his book *Dismantling Discontent – Buddha's Way Through Darwin's World*, Elite Books, Santa Rosa, CA, 2007, p. 96, from William Long's book, *Mother Nature*, Illustrated by Charles Livingston Bull, Harper and Brothers Publishers, New York, 1923, pp. 184–191.

[32] ibid., pp. 85–185.

[33] ibid., p. 180, citing Paul Sears, *Deserts on the March*, University of Oklahoma Press, Norman, OK, 1935.

[34] "Darwin's Illness revealed," by A.Campbell and S. Matthews, Postgraduate Medical Journal, PMC, NCBI Resources, 2005, April; 81(954): 248–251. https://doi.org/10.1136/pgmj.2004.025569, https://www.ncbi.nlm.nih.gov/pmc/articles/PMC1743237/ Accessed May 23, 2020.

**Fig. 54.3** Double-crested cormorant (*Phalacrocorax auratus*), at Farallon Islands National Wildlife Refuge, Northern California. (Photo © M.C. Tobias)

our species' inherent aggression right next door.[35] That debate is further charged by early Vedic studies by Ananda K. Coomaraswamy (1877–1947) who viewed the human body itself as a "city of God," that every "immanent deity" is a "citizen" of that same community, and the famed saying, "That art thou" actually connotes self-sacrifice (hence, in evolutionary terms, pain) as well as the notion that such pain also "is to liberate the God within us".[36]

In other words, with human domestication, emerged, first, ethical views, and then a constellation of godheads, the contest of our consciousness within the far-greater realms of our surrounding and implicit ecological circumstances. To see through all of these confounding configurations with some mythological pure vision. To gaze without intervention. That has been the goal of pure vision, beyond mere faith. But is possible? And, if so, what is it?

And even if, technically, we are blind, deaf, insensate, we still crave a line of sight, internalize visions, aspire to catch a glimpse, a sensory connection to some distant place, our own Arcadias, but fail to do so, constantly in the way. All a lucid prelude to communion with that which is still, silent, at rest. The peace at last which, at rare moments, an individual—but never a collective or species—discovers. That is, perhaps, the most difficult of all evolutionary paradoxes to resolve. To finally, unfettered, gaze upon the truth of *ourselves* (Fig. 54.3).

---

[35] See *The Domestication of The Human Species*, by Peter J. Wilson, Yale University Press, New Haven and London, 1988, pp. 182–183.

[36] See *Coomaraswamy*, 2: Selected Papers – Metaphysics, Edited by Roger Lipsey, Bollingen Series LXXXIX, Princeton University Press, Princeton, New Jersey, 1977, p.78.

# Chapter 55
# No Equation for It: Numbers with No Attachment

## 55.1 From Melancholy to Activism

As we task every observation with a marker, some object of recognition that coincides with our habitual manner of attempting to *grasp an impression*, we know, from the outset, that we will never be able to interrupt the flow of time through our instant of interaction. Time does not stop, we cannot stop, despite a plethora of clichés from existential meditation. Nor can we simply immerse ourselves in the stupor of that resignation so amply illustrated in director Lars von Trier's 2011 drama, "Melancholia" which witnesses earth gorgeously annihilated by an oncoming, previously hidden, planet.[1] What we continue to designate as "nature" is that overwhelming landscape absorbing every duality (including utter destruction) we may conjure in our continuing experience of observations and impressions. They do not neutralize, but compound. The cumulative experience is experiential, but no chess match or crossword puzzle: there is no endgame, no conclusion (Fig. 55.1).

All the disciplines we elicit to penetrate surfaces of thought with imagined or experienced detail, like the sight of our own blood, or urine, the throbbing of a migraine, the inability to rush beyond the sheer bliss of bliss, the distance to Saturn, have an evolutionary thrust to the calculations and performances of nature in ourselves that are animate, ongoing, and invulnerable to concretization. These are irrational equations and numbers, nothing attached to a grain of sand.

We cannot believe in, or even attune our behavior to the fixity that grains, or our own breathing bodies, imply. We feel our pulse, and know the onrushing tide of commanding events to be the only truism—of exquisite ephemerality—worth believing in. Even the painting mounted on the museum wall is moving in many ways. Its chemicals are still internally reacting, and every observer is affecting it as subtly as wind atop Dhaulagiri, some version of what, in chaos theory (and popular

---

[1] See "The Only Redeeming Factor is the World Ending," Juul Carlsen, Per, May 20122, Susanna Neimann (ed), Danish Film Institute (72): 5–8.

© Springer Nature Switzerland AG 2021
M. C. Tobias, J. G. Morrison, *On the Nature of Ecological Paradox*,
https://doi.org/10.1007/978-3-030-64526-7_55

**Fig. 55.1** Dr. Birutė Marija Filomena Galdikas With Rescued Orangutan at Tanjung Putting National Park, West Kotawaringin Regency in the Indonesian province of Central Kalimantan, Indonesia. (Photo © M.C.Tobias)

culture), is termed the "butterfly effect." That there are no stand-outs, stand-ins, or stand-alones in nature present a perplexity of half-truths and loaded guns to the observer. Meaning, all is in flux. This is fundamental ecosystem logic.

One particularly noteworthy sign of our vulnerability to ecodynamics is the sense that other species commit suicide, for reasons not dissimilar to those estimated 10.6 per 100,000 human individuals who take their own lives annually.[2] Professor Antonio Preti, a clinical psychotherapist at Italy's University of Cagliari, has suggested that suicide in other species is merely our "anthropomorphic use of the term," a fable, but that "Suicidal behavior might be an evolutionary jump relatively recent in *our* [our italics] species: a byproduct of living in groups of people who are not as closely related genetically as in social groups of nonhuman mammals".[3] If there is wisdom to Preti's interpretations, one could easily surmise a

---

[2] "World Health Statistics data visualizations dashboard," "Suicide," World Health Organization, https://apps.who.int/gho/data/node.sdg.3-4-viz-2?lang=en

[3] From Antonio Preti's "Abstract," "Animal suicide: Evolutionary continuity or anthropomorphism?" Animal Sentience 20(10) 2017, https://animalstudiesrepository.org/animsent/vol2/iss20/10/, Accessed May 23, 2020. See also, "A New Look at Animal Suicide," by Jessica Pierce, Psychology Today, January 6, 2018, https://www.psychologytoday.com/us/blog/all-dogs-go-heaven/201801/new-look-animal-suicide; see also, "The Suicidal Animal: Science and the Nature of Self-Destruction," by Edmund Ramsden and Duncan Wilson, Past & Present, Volume 224, Issue 1, August 2014, pp. 201–242, https://doi.org/10.1093/pastj/gtu015, July 24, 2014, https://academic.oup.com/past/article/224/1/201/1411207, Accessed May 24, 2020. The authors of this latter essay, commence with an analysis of the animal suicide exclusion, by Émile Durkheim in his 1897 study of *Suicide: A Study in Sociology*, Félix Alcan, Paris.

correlation positing that "evolutionary jump" as a remarkable response to our awareness of, and the physical fall-out from, the evolving preconditions of the Anthropocene. The most ponderous conjecture of all falls upon groups, in general: Is *any* group, for any purpose, more prone to survival in the long-term, or not? And if not, what are the alternatives for biological organisms?

But to deny that response in other species rests upon no empirical possibility. Various marine mammals, a famed stag pursued by dogs in 1875 that jumped off a cliff rather than be eaten to death, the many cases reported by the second-century Claudius Aelian's *De natura animalium*, captive bears, other depressed primates,[4] and all those hundreds of dogs who have leapt to their death in a rocky gorge beneath the 50-foot-high Overtoun Bridge in West Dunbartonshire north of Glasgow, Scotland, have something profoundly in common with their human brothers and sisters.

Counteracting depression, ecological depredation and death, there exists a broad current of many-tiered, if modest triumphs engineered in social contracts that save the odd rangeland, marsh, dune, hillock, and dale. That delimit emissions here, groundwater pollution there. That bring back a species from near loss. A literature of local conservation success stories (and heated courtrooms) that together enshrine the indisputable prospects, however fragmentary, for *Homo sapiens*, in what might be described as cells of human consciousness moving from melancholy to activism, bubbles of optimism, miniature life courses spread geographically from continent to continent—the cartography of promise.

In one case, Ted Bernard—an Ohio University professor emeritus of geography and environmental affairs from the Voinovich School of Leadership and Public Affairs, and Jora Young, a "science and stewardship" professional, wrote in 2008 their book *The Ecology of Hope: Communities Collaborate for Sustainability,*[5] which examines examples of fine forest stewardship, ecorestoration and biodiversity management, both rural (Mexico-Arizona border area and California logging cohorts) and urban (Chicago) sustainable master planning among otherwise competitive and antithetical constituencies. Their documentation comports with other catchments of tenable buoyancies profiled in a number of books from New Society Publishers (and hundreds of other imprints). With works extolling titles like "Food Freedom," "The Better World Shopping Guide," "Resilient Agriculture," "The Edible Ecosystem Solution," "Indigenomics – Taking A Seat at the Economic Table," and so forth.[6] Island Press is another publisher with a similarly informed and inspirational credo guiding its publications; books including "Beyond Polarization," "Precision Community Health," "Valuing Nature," and "Designing The Megaregion – Meeting Urban Challenges at a New Scale".[7]

---

[4] See "Many animals seem to kill themselves, but it is not suicide," by Melissa Hogenboom, July 6, 2016, earth, BBC.com, http://www.bbc.com/earth/story/20160705-many-animals-seem-to-kill-themselves-but-it-is-not-suicide, Accessed May 23, 2020.

[5] New Catalyst Books, an Imprint of New Society Publishers, Gabriola Island, B.C.

[6] https://newsociety.com/collections/print

[7] https://islandpress.org/books

The new god is optimism, despite the overwhelming spreadsheets of scientific bad news. It is not our intent to dispel some level of belief in these sane and passionate attempts to divert species malevolence and ecological illiteracy. We ourselves have published one such book, *Sanctuary: Oases of Innocence*, with a very optimistic Foreword by Her Majesty the Queen of Bhutan, wife of the first King, Ashi Dorji Wangmo Wangchuck.[8] Years later, one of us (mt) did a book with Dr. Paul R. Ehrlich, *Hope On Earth – A Conversation.*[9] Among other things, that focused less upon, for example, Dr. Ehrlich's early wakeup calls in New Jersey when he realized the extent to which the spraying of DDT was affecting the ability of caterpillars to feed on various leaves, and more upon the prospects for ecological amelioration at the societal, legal levels that Dr. Ehrlich's later work, and that by Rachel Carson and so many others would have on the global psyche (butterfly and human populations, for example, being critical watershed sets of comparable data for gauging real and positive change).[10]

## 55.2   Hope Hastened

There was a time when environmentalists were charged with possessing excellence at cataloging all the problems on earth, but no stomach for actually devising solutions to those dilemmas. That is no longer the case. Since long before Rachel Carson's devastating critique of industrialized agriculture with its runaway pesticide mentality, conservation-minded activists had been highlighting modalities of conflict resolution. As early as 1864, and then in 1885, George Perkins Marsh, a Vermont diplomat and eventual US minister to the Kingdom of Italy, appointed by Abraham Lincoln, had recommended scores of viable plans for creating parks, and staving off the devastation of fragile ecosystems. He was a pragmatist, as illustrated in such books of his as *Man and Nature* (1864) and *The Earth as Modified by Human Action* (1874).

America, since the early 1970s, has led other nations to recognize and mandate levels of cleanliness: clean air, water, soils, though President Ronald Reagan tried to undermine much President Carter had done; and there is no point citing the disastrous 4 years after President Obama, but in the end, we can only *hope* that sanity

---

[8] By M. C. Tobias and J. G. Morrison, A Dancing Star Foundation Books, Council Oak Books, San Francisco, CA and Tulsa, OK, 2008.

[9] University of Chicago Press, Chicago, Ill., 2014, https://press.uchicago.edu/ucp/books/book/chicago/H/bo17588109.html

[10] See *Superstition: Belief in the Age of Science*, by Robert L. Park, in the chapter "The Last Butterfly," Princeton University Press, Princeton and Oxford, Princeton, New Jersey, 2008, p. 205. For other works detailing the history of good reasons for at least limited optimism within the widespread environmental movements, see, "The Environmental Justice Movement," by Renee Skelton and Vernice Miller, March 17, 2016, https://www.nrdc.org/stories/environmental-justice-movement; and "Victories," Earthjustice, https://earthjustice.org/our_work/victories, Accessed June 17, 2020.

will prevail. Most recently has emerged individuals like the Swedish Greta Thunberg (b. 2003), the "how dare you" teen environmentalist with millions of young followers deeply stirred by her frank and well-read opposition to an adult world where most politicians, economists, and consumers remain content to embrace gradualism at best, business-as-usual at worst. And then there is someone like the unstoppable Dr. Russell A. Mittermeier, Chief Conservation Officer for the NGO, Global Wildlife Conservation and winner of the 2018 prestigious Indianapolis Prize for Conservation, who has successfully helped save tens of millions of acres in country after country, through a combination of deeply informed science (he spent nearly 3 years and restless nights in a hammock in the rainforest of Suriname studying monkeys for his PhD), and his skillful, graceful negotiating skills with Prime Ministers and tribal chiefs alike. Dr. Mittermeier's children are growing up in his assiduous footsteps.[11]

We know that change is possible. Fires can be put out. Some elites and policy pundits *are* panicking and wanting to do the right thing environmentally. Too little, too late? Perhaps that is merely the crash of apathy, or ecological nihilism courting daily anger. Or, perhaps it reflects the numbers and their equations that simply, under even the most sanguine of probability distribution models, cannot win this battle to salvage the earth, nor persuade a critical mass, in absence of absolute salvation, to join in the efforts. Such lackluster jeremiads are yet to be determined.

This dyad of weighted truths and waning realities, brings us full circle, into the realm of mental health, habitual trespass and murder, and the all-out assaults on the planet, fueled by the global average TFRs (total fertility rates of 2.5 children per woman in her child-bearing lifespan),[12] and escalating territorial expansion of our species. Frenzied road building across northeastern Ladakh by both Indian and Chinese military forces, for example.

This polarity devolves to the—at least formulated—conceptual bankruptcy of our species' protective resolve and pertinacity. The calamities and stubbornness driving us toward biological negation is a problem in reality as much as one in mind. If we believe despair and annihilation to be inevitable—every sign suggests just that—then we cannot awaken ourselves in time to fix it. If citizen science, environmental law, and ecological ethics are taken up by so meager a percentage of our kind

---

[11] Dr. Mittermeier hosted our two-hour PBS documentary, "Hotspots" -www.hotspots-thefilm. org -based upon his book by the same title; a concept (hotspots) that was first propounded by British biologist Norman Myers (1934–2019) and then charismatically championed by Mittermeier, who inspired a $261 million grant from the Gordon and Betty Moore Foundation in 2001 to help him, when he was then President of Conservation International, and his thousands of colleagues around the world, to effectively implement the hotspots methodology. See Philanthropy News Digest, December 11, 2001: "Moore Foundation Awards $261 Million to Conservation International," Press Release: "Conservation International Unveils Solution to Prevent Global Species Extinctions," Conservation International Press Release, December 9, 2001, https://philanthropynewsdigest.org/news/moore-foundation-awards-261-million-to-conservation-international, Accessed May 23, 2020.

[12] See "Fertility Rate," by Max Roser, December 2, 2017, Our World In Data, https://ourworldindata.org/fertility-rate

as to be statistically irrelevant, then that by itself speaks to our species' elimination in the vaster scheme, which is running out of all patience with our illegitimate and petty species-claims to tenure.

## 55.3   The Strange Dramas of Human Apperception

Human consciousness has betrayed itself and the assumptions most likely held by all others, whether social insects or deep-ocean creatures. This realization will be taken at much greater length in Part Three of this treatise, "A Natural History of Existentialism." But twenty-first-century schools of epistemology are fraught with a losing effort to fully cognize those shameful history lessons we have ignored, and the legacy of mind that has never been so desperate before a body reckless. In 1974, ethicist and law professor at New York University, Thomas Nagel asked, "What is it Like to Be a Bat?",[13] noting that consciousness is imperishably subjective. Its sole reality is its relatability to some other in-and-of-itself. For philosopher David J. Chalmers, this relationship basis for being in the world faces the conundrum of *qualia*, particularly as such traits spawn "brute discontinuities in fundamental laws" (of nature).[14] These are not mere words, spawned in the self-preserving debates of contemporary philosophy. Rather, they warrant deep reflection, comparison, and study. The very words, "relatability," "qualia," and "brute" come heavy bundled in a history of both etymological revelation, and current peak-critique relevancy.

Marvin Minsky acknowledges the *bat problem* in his proposition that there is a vast distance between "the nature of thoughts" and "the nature of things".[15] Many have tried to bridge the gap, of course. And these dialectics in turn invite what Lorraine Daston and Peter Galison in their utterly incisive work, *Objectivity*, have described as "the image-as-tool," "a new kind of scientific self – a hybrid figure" lodged somewhere between "representation" and "presentations" to ourselves.[16] We stare out dumbly at all the mysterious creatures; some we recognize, others not—like Frilled Sharks, Vampire Squids, Goliath Beetles, Giant Blue Earthworms, Assassin Bugs, the King of Saxony Bird-of-Paradise, Chrysopelea—flying snakes, monotremes, the Antarctic marine *Enypniastes eximia*, and Bleeding Tooth Fungus.

Usually, we catch and cage them, like the carnivorous marsupial, Thylacine, extinct at some point between the late 1930s and mid-1960s, but not before several film cameras captured the tragedy of its final years, in bunkers, on concrete floors,

---

[13] Philosophical Review, 83, 435–50; see p. 436.

[14] David J. Chalmers, "Absent Qualia, Fading Qualia, Dancing Qualia," Conscious Experience, edited by Thomas Metzinger, Imprint Academic 1995, http://consc.net/papers/qualia.html, Accessed May 23, 2020.

[15] In "Realms of Thought," 29.1, *The Society Of The Mind*, by Marvin Minsky, Simon And Schuster, New York, NY, 1986, p. 292.

[16] Zone Books, New York, 2007, pp. 412–413.

pacing desperately (stereotypies).[17] The mental equations linking Thylacine—with one pound sterling (or ten pence) bounties on the heads of adults and pups—to humans, reads as a grimly asymptotic extraterrestrial encounter. A quintessential problem that likens our systemic inhospitality to some global catastrophe, which it is. Faith in an age of such terror and anxiety has few places to turn.

Perhaps to the comforting nostalgia of 18 volumes of faded Dutch town and country photographs from the 1920s compiled by Mr. A. Loosjes, and other Arcadian fallacies, like that antiquity of quiet monasteries across Italy.[18] Robert Thornton's exquisitely illustrated, long-gone *Temple of Flora* (1807), or Joseph Dalton Hooker's *The Rhododendrons of Sikkim-Himalaya* (1849–1851), from another age. George Perry's *Conchology* of 1811 in which one can verily hear the outermost waves on continents far and strange. And all the great animal, fish and bird books, from Mark Catesby's *The Natural History of Carolina, Florida and the Bahama Islands* (1731–1743) to Audubon and Robert Havell's *The Viviparous Quadrupeds of North America* (1845–1849) and John Gould's remarkable series of bird books from Europe, Great Britain, Australia, the Himalayas and New Guinea, among others, between 1832 and 1888.[19] As if eternities had been consulted in a simpler time, of herbaria and quiet nights in romantic settings, and days spent in early retirement, planting and caring for gardens. Paying as little attention as possible to the incoming tsunami, when the soul's arcadias have so much more to say.

Near the conclusion of his book *Human Natures*, Paul Ehrlich suggests that it might be as straightforward as simply channeling "change in ways more beneficial to the majority of human beings." He readily volunteers that "the need for a novel evolutionary approach can be seen in the most critical mismatch between biological and cultural evolution: the fact that the design of the human perceptual system makes it especially hard for people to recognize the most serious environmental problems".[20] Ehrlich is right. Indeed, he always has been, despite those who criticized him for on occasion equating his (brilliant) science, with courageous predictions (cultural inducements to act).

Émile Cioran skeptically alludes to the fact that "Any man who identifies himself completely with something behaves as if he were anticipating the advent of 'the universal harmony,'" but elsewhere, in a chapter titled "Mechanism of Utopia," begins, "Whenever I happen to be in a city of any size, I marvel that riots do not

---

[17] The Thylacine Museum – History, By D. Colbron Pearse, and Michael Sharland, "History: -Persecution –" © C. Campbell's Natural Worlds, http://www.naturalworlds.org/thylacine/history/persecution/persecution_10.htm

[18] *Landelijk Nederland In Beeld*, Scheltema & Holkema's Boekhandel In Uitgevers Maatschappij, N.V. Amsterdam, 1925. See also, the photographs and paintings in *Le Vie Del Santo – Spiriti E Luoghi Del Poema Francescano*, by Ettore Janni, Illustrazioni Dal Vero Di Emilio Sommariva, Tavole A Colori Di Giuseppe Mentessi, Istituto Italiano D'Arti Grafiche, Editore – Bergamo, 1927.

[19] See *Nature Into Art- A Treasury of Great Natural History Books*, by Handasyde Buchanan, Mayflower Books, New York, 1979.

[20] Paul R. Ehrlich, *Human Natures – Genes, Cultures, and the Human Prospect*, Island Press/ Shearwater Books, Washington, D.C., 2000, pp. 326–327.

break out every day: massacres, unspeakable carnage, a doomsday chaos. How can so many human beings coexist in a space so confined without destroying each other, without hating each other *to death*?".[21] "What shall I do with this broken bird? The dead sky has nothing to say," wrote Osip Mandelstam (1891–1938).[22] The poet, who died in a work camp, where he had been arrested more than once for his anti-Stalinist versification, could rightly claim that in such dictatorships as the Soviet Union, poetry was among the highest causes of mortality. But if one takes the theories of an Ehrlich, Cioran, and Mandelstam together, it is clear that—with respect to humanity and the natural world—all of human civilization, with the rarest of exceptions, is a kind of lethal aesthetic. A poetry that sees straight, but cannot see straight.

To be this animal, human, hosts a foreknowledge: H. D. Thoreau building his cabin near Concord, George Inness painting coming storms at Montclair, John Muir's final years at Martinez, Frank Cushing digging for ancestors under shopping malls in Florida, Louis Bromfield planting organics at Malabar Farm in Ohio, John Nieuh of mysteriously disappearing in Madagascar during one of his many seventeenth-century natural history expeditions, Paulus Potter painting his "Punishment of a Hunter" (1647, in the Hermitage), John Martin executing his deeply immersive black/white mezzotints for John Milton's *Paradise Lost* (London, 1827), and Edvard Munch painting his "Der Schrei der Natur" (The Scream of Nature), the German title the Norwegian expressionist originally gave his—as it would eventually be thought of—"Scream" of 1893 (Fig. 55.2).

Stray cats in Damascus or strife-ridden Beirut, thirsting; same in St. Petersburg, where the poorest of the poor must vie with rescues in the one nearby animal shelter for recycled sloppy joes; burros tied by their fetlocks for life on a scorching outcrop in a gorge somewhere in Jordan, beside a stinking garbage dump across northern India, a lost cluster of half-toppled Jewish gravestones deep in the swampy border zones of the Bialowieza forest, between Belarus and Poland. Anonymous deaths in fetid cages from Bamako to Jakarta. The dogs left behind on certain Greek islands, after tourists have left mid-Octobers, when the unwanted mutts are not infrequently and wantonly poisoned by the locals. Nineteen days without rain in Monteverdi Rain Forest, in Costa Rica; excessive tourism to the rare Irish pyramid of endangered rookeries in the sea, Skellig Michael.[23] All of these onrushes of pain and simultaneous efforts to make a difference have been our (the Authors') quiet odyssey, to date.

There are no equations to solve these futilities, demonic lethargies, pejorative dead-ends. This is the ecology of frightened insomniacs, souls that will never rest, take no comfort in theories of impermanence espoused by Padmasambhava—the

---

[21] See *History And Utopia*, by E. M. Cioran, Translated From The French By Richard Howard, Seaver Books, New York. 1960. pp. 105, 80.

[22] Osip Mandelstam, *Stone*, Translated and Introduced by Robert Tracy, Princeton University Press, Princeton, New Jersey, 1981, Poem #21, 1911, p. 85.

[23] "Environmental groups raise alarm over plans for Skellig Michael," Niall Sargent, Green News, January 10, 2019, https://greennews.ie/environmental-ngos-concerned-rnew-plans-skellig-michael/, Accessed April 24, 2020.

**Fig. 55.2**   Stray Cat in Damascus. (Photo © M.C.Tobias)

earlier described eighth-century Tibetan "Lotus Born," and others, perhaps with the single exception of the "sky-clad" Digambara monks in Jain tradition, who have never wavered from what can be rightly characterized as the highest form of known human abnegation and ecological restraint. But taken together, all these algorithms in human form, without overwhelming redemption, court every ecosystem of angst, are the (statistically) very *problem* of thought, filling libraries with the most uneasy of silences, a near cry to final occasions, crepuscular biologies, the past tense, occlusions, and doomed causality. These are the records of a consciousness whose legacy will read of life in absentia, the full emergency of human occupation; of mathematical cessation. Such final numbers, permutations of oblivion, handwritten notes to ourselves are, ultimately, as useless as the building of Bangkok on a distant planetoid. Better, clearly, to simply carry on, in spite of ourselves.

# Chapter 56
# A Situational Animal Rights Ethic

## 56.1 Between Haiti and Socotra

Albert Camus' conception of Kafka's body of work scrupulously avoids committing the ultimate folly—certainty, when he writes of Kafka's "secret," namely, "perpetual oscillations between the natural and the extraordinary, the individual and the universal, the tragic and the everyday, the absurd and the logical."[1] There is, in this evasive play on language, underscoring an enormous difficulty with which Camus' short life grappled, that flippant side of Giorgio Vasari describing Leonardo as a man—"omo senza lettere"—who "set himself to learn many things, only to abandon them almost immediately."[2] A faint echo of Kurt Vonnegut concluding his *A Man Without a Country* by conveying a friend's recommendation—"Look at a million pictures [so as to] never be mistaken" to Vonnegut's own daughter, a painter, who imagines aloud that she "could rollerskate through the Louvre, saying, 'Yes, no, no, yes, no, yes' and so on."[3] All of these conjurations—Camus, Kafka, Leonardo, Vonnegut, the Louvre—coexist in parallel, within a powerful meditation by Rebecca Solnit on what we know, what other animals know, what we don't know, and what it means to be lost, to disappear forever[4] (Fig. 56.1).

In these equations of compulsion and restlessness, knowledge conceding its emptiness, we posit two specific individuals, our universal situation: the hermits of Socotra and Haiti, approximately half-way around the world from one another, each condemned to a state of affairs in which all ethics are situational; every form of

---

[1] From Camus' *The Myth Of Sisyphus and Other Essays*, Translated from the French by Justin O'Brien, Alfred A Knopf, New York, 1955, p. 126.

[2] See *Leonardo And The Mona Lisa Story – The History Of A Painting Told In Pictures*, by Donald Sassoon, An Overlook Duckworth/Madison Press Book, New York and Woodstock, NY, 2006, p. 27.

[3] *A Man without a Country*, Kurt Vonnegut Edited by Daniel Simon, Seven Stories Press, New York, 2005, p. 145.

[4] *A Field Guide to Getting Lost*, by Rebecca Solnit, Viking Publishing, New York, 2005, p. 187.

© Springer Nature Switzerland AG 2021
M. C. Tobias, J. G. Morrison, *On the Nature of Ecological Paradox*,
https://doi.org/10.1007/978-3-030-64526-7_56

**Fig. 56.1** The Hispaniolan lizard cuckoo (*Coccyzus longirostris*) in Southwestern Haiti. (Photo © M.C. Tobias)

logic a paradox; our most fervent ecological whims nothing more than the sudden birth of an idea, one after the other with no end in sight, no purpose, no teleology, nothing to say that any one of them has greater relevancy than any other.

In an era teeming with environmental reversals, animals and habitat in torment, how do we go about seizing upon imperatives, when everything we sense and do represents a pantheistic catastrophe, an amalgam of urgencies at the heart of any moral conception? The task before us is dizzying. A tachycardia pulse rate of 400. The mind is stymied by excessive asseveration, the fast-deteriorating condition of the planet, but one we cannot see at once, not from where we stand. And so it is all imagined, in dread, in the dark.

To be fraught, more or less solitary, in southwestern Haiti, or on the politically contested, all but uninhabited island of Navassa, 35-miles due west from Haiti's Tiburon Peninsula, 103 miles south of Guantanamo Bay, in either case, scrounging for food, water, medical supplies, tied to one's ancestors by way of an enormous tree that has been worshipped for generations: and then to imagine the rest of the world.

Or, the sole occupants of a stone hut in the Hajhir massif in the center of the island of Socotra, Yemen, 150 miles out into the Arabian Sea from Somalia, several hundred miles south of the Yemeni mainland. A family, goats, donkeys, camels. But also some 35,000 Socotris, many in the mountains, who continue to endure the warring fiascos across Yemen and the Gulf of Aden, famine, vying militaries, and the aftermath of a disastrous Cyclonic storm, "Megh" which struck the archipelago November 5, 2015, destroying a huge percentage of Socotra's intensely isolated infrastructure. Yet, the Socotri people have unified around conservation. The island's UNESCO status is well earned.

All such personages—hermits of Haiti and Socotra; human *isolatoes*, as Herman Melville called them, in every apartment building of every city; every last widow and widower, alone in their thoughts, symbolic of our problématique. And with that mass of feeling, how to effectively save the world, as one person, two people, even three. The challenge emerges every day. In the life of the late Senator John Lewis, or of Boethius, or Tao Tzu. Name your biography. Transforming this idle, lackluster consciousness, responsive to basics, tending toward non-resistance, be it gravity or any other silent altercation (one thinks of an electrifying Sen. Cory Booker [D-NJ] speaking about the first small Congressional steps toward American police reform and the persistent oppression/brutality of African-Americans, "Words are not enough" [June 16, 2020]), preferring peace and uncomplicated routines, after all. From that plateau at some point, any point in life, settled-in for good, to rescuing a creature much smaller than ourselves, an ant, a mosquito, a gecko. Or, conversely, watering a plant, leaving food out for the vertebrates who wander by throughout the days and nights. The hidden comforts of surprising companions. Beyond every metaphysic is this ease of possibilities, this biophilia of consolations rising to a zenith of what humanity is capable of.

Haiti is within the Caribbean biological hotspot, with one of the most concentrated arrays of taxonomically endemic species in the world, most threatened by a near 99% loss of island forest. In relation to the toxic combination of near total deforestation and high endemism, three phases of Haitian mass extinctions are projected, with, at best, resulting marginalized, "physiologically-tolerant species" that may "likely survive and expand in poor-quality habitats."[5] The full contours of this disaster surface against the context of the Island of Hispaniola's approximately 5600 plant species, of which some 36% are deemed to be endemic, many on the Haitian side.[6] Less than one-fifth the size of Iowa (10,714 mi$^2$) but nearly four times Iowa's population (11.12 million and a continuing total fertility rate of 2.99 births per woman), with a per capita income of US$2.00 per day, Haiti's human, animal rights and conservation needs will not be solved by a solitaire—hence, the nature of science-and-compassion-driven ameliorative strategies.

In January, 2019, it was announced that Professor S. Blair Hedges, director of Temple University's Center for Biodiversity and long-time researcher in Haiti, along with an airline CEO and multiple conservation groups including the local Audubon Society, had persuaded Haiti's government to feverishly endeavor to save all those species still hanging on within the Massif de la Hotte range of the southwest (towering above the beautific coastal expanse where John James Audubon was born), by declaring three new national parks: Grand Bois, Deux Mamelles, and

---

[5] See "Haiti's biodiversity threatened by nearly complete loss of primary forest," by S. Blair Hedges, Warren B. Cohen, Joel Timyan, and Zhiqiang Yang, PNAS, Proceedings of the National Academy of Sciences of the Unites States of America, October 29, 2018, https://doi.org/10.1073/pnas.1809753115, https://www.pnas.org/content/115/46/11850, Accessed May 25, 2020.

[6] World Conservation Monitoring Centre (1992). Nearctic and Neotropical. IUCN. pp. 384. See also, https://www.cbd.int/countries/profile/?country=ht; and http://darwin.bio.uci.edu/sustain/h90/Haiti.htm

Grand Colline. In addition to a remarkable proliferation of ground orchids, other species, including Eckman's Magnolia tree and known 19 critically endangered amphibians, are likely to be saved because of this collaboration, long in the works and bringing the number of national parks in Haiti to ten.[7]

## 56.2   The Etiology of Sanctuary

For several years we (the Authors) have been attempting to highlight conservation success stories around the world, and in 2008 punctuated such long-term endeavors with a book titled *Sanctuary: Global Oases of Innocence*.[8] In the work we chronicled efforts similar to that in Haiti. These included long-term collaborative efforts in the following locations: Wrangell-Saint Elias National Park and Preserve in Alaska, the Farallon Islands National Wildlife Refuge 28 miles west of the Golden Gate Bridge, Muir Woods in Sausalito, CA, Central Park in Manhattan, Farm Sanctuary in Upstate New York, the Central Suriname Nature Reserve, Grupo Loboan Iberian Wolf Sanctuary in Portugal, Brigitte Bardot's animal sanctuary, La Mare Auzou en Normandie, Michael Aufhauser's Gut Aiderbichl in Austria, the Alertis European Brown Bear sanctuary in the Netherlands, Bialowieza National Park in Poland/Belarus, Marieta Van Der Merhe's Harnas Sanctuary in Namibia, Howard Buffett's Cheetah Sanctuary (Jubatus) in South Africa, the UNESCO World Heritage site that is Socotra (Yemen), the Al Areen Wildlife Sanctuary in Bahrain, the Dubai Desert Conservation Reserve of Al Maha in the UAE, the vegetarian city of Pushkar in Rajasthan, the Nilgiris Biosphere Reserve in southern India (and the earlier discussed EBR Trust that has been effectively helping the indigenous Todas of that region), Wat Phra Keo in Bangkok, a butterfly preserve in downtown Kuala Lumpur, Bukit Timah National Park in Singapore, Ulu Temburong National Park in Brunei, a most critical link in Borneo, resulting from the Sultan's own conservation visions, Tanjung Puting National Park in Indonesian Kalimantan, the greenbelt of monasteries in Kyoto, and the Sakteng Wildlife Sanctuary in Bhutan.[9]

---

[7] See "Haiti's first-ever private nature reserve created to protect imperiled species," Public Release, Temple University College Of Science And Technology, January 28, 2019, EurekAlert! AAAS, https://www.eurekalert.org/pub_releases/2019-01/tuco-hfp012319.php. Key partners include the Haiti National Trust, the Rainforest Trust, Temple University's Center for Biodiversity, Global Wildlife Conservation, the Philadelphia Zoo, Alliance for Zero Extinction, and IUCN Key Biodiversity Areas (KBAs) research. See also, Tobias, Michael Charles (2013-03-09). "How A Single National Park Might Help Transform a Nation: Haiti's Pic Macaya". Forbes. The ten parks does not include the island of Navassa – disputed jurisdictionally between Haiti and the U.S., but designated by the U.S. Fish and Wildlife Service as a U.S. National Wildlife Refuge.

[8] By M. C. Tobias and J. G. Morrison, With a Foreword by Her Majesty Ashi Dorji Wangmo Wangchuck, Queen of His Majesty the Fourth King of Bhutan, A Dancing Star Foundation Book, Council Oak Books, San Francisco and Tulsa, 2008.

[9] See "Dramatic decline in Borneo's orangutan population as 150,000 lost in 16 years," by Ian Sample, The Guardian, February 15, 2018, https://www.theguardian.com/environment/2018/

In each instance, these sanctuaries all had to supersede conflict, applying data sets recovered with grueling persistence against cultures of poaching, palm oil extraction, the hatred of wolves, or illegal trade in wildlife, government intransigency, controversial in situ versus ex situ trade-offs, fear of bears, illegal logging interests and corrupt officials, land negotiations with farmers and/or indigenous peoples who had for decades been undermined by forces beyond their control. Every conceivable conflict and obstacle invariably dragged on for years as a precondition of the many ecological luminaries in each instance ultimately saving some piece of precious turf. Both situational and normative ethical mores come to the fore in any animal rights initiatives. These are not "virtue" or descriptive ethics, but case-by-case instances of desperation, as with one of the rarest seabirds in North America (Kittlitz's murrelet, *Brachyramphus brevirostris*), whose waning life-histories are being systematically studied and championed by researchers with the US Fish and Wildlife Service; the beleaguered orangutans of Tanjung Puting National Park and their champion, Dr. Biruté Marija Filomena Galdikas (b. 1946).

Without her, one of humanity's closest relatives, the Borneo orangutan, one of two Ponginae family Pongos, *Pongo pygmaeus*, might not have survived to this day. Between 50% and 70% of them have been extirpated—150,000 individuals killed between 2002 and 2018. The same 11th hour emergency efforts have been expended by Ms. Marieta Van Der Merhe and her family at the 25,000 acre Harnas Wildlife Sanctuary 185 miles east of Namibia's capital, Windhoek, where the Merhe's and local San tribe have protected, among others, the last large genetically robust populations of African wild dogs (*Lycaon pictus*; Fig. 56.2).

## 56.3   The Hotspots Methodology

In a feature film documentary titled "Hotspots"[10] with Dr. Russell A. Mittermeier (then President of Conservation International), we examined a similarly illustrative array of conservation efforts involving dozens of partnerships from numerous protected areas of New Zealand, to Chilean Rapa Nui (Easter Island), the Atlantic Rain Forests of Brazil, the Tambopata National Park in Peru, and other national and state parks, scientific reserves and a military base containing numerous endangered species, from Madagascar to California. Every effort, a massive series of liberations not without compromise; of animal and habitat communions and commiserations, involving the combined passions, anxieties and endurance of countless individuals

---

feb/15/dramatic-decline-in-borneos-orangutan-population-as-150000-lost-in-16-years, Accessed May 26, 2020. See also, "Family Hominidae ("Great Apes") in *Handbook Of The Mammals Of The World*. Chief Editors, Russell A. Mittermeier, Anthony B. Rylands and Don F. Wilson, Published by Lynx Edicions in association with Conservation International and IUCN, 3. Primates, Barcelona, Spain, 2013, pp. 792–793.

[10] A Dancing Star Foundation Production, in Collaboration with Conservation International, with KQED/PBS, www.hotspots-thefilm.org.

**Fig. 56.2** Endangered African Wild Dogs (*Lycaon pictus*), Among the 300+ individuals remaining in Namibia (pictured here at the Harnas Sanctuary), with 3000+ remaining in all of Africa. (Photo © M.C. Tobias)

from hundreds of taxa, each an imperiled biography; ethical idealists and activists, interdisciplinary pragmatists, field biologists, diverse sources of financial support, science technologies, a legal infrastructure connoting some long-term level of assurance, monitoring and oversight, and a modicum of good tidings.

Animal rights and conservation biology have often had their problematic differences, the latter globally engaged, for example, in exterminating bio-invasives, like the "5,036 feral pigs" slaughtered on Santa Cruz Island, 18 miles from the city of Ventura, on the California Coast, by the combined efforts of the US National Park Service and the Nature Conservancy.[11] Or the approximately 2 million non-native Australian Brush-tailed Possums (*Trichosurus vulpecula*)—out of a current population of some 30–40 million individuals—destroyed annually in New Zealand to counteract the possums' nightly consumption of an estimated 42 million pounds of native vegetation. That's what they do. So irrationally loathsome are possums to New Zealanders, that in one instance, "Kiwi" children at their South Auckland school were drowning baby possums for fun, instigating international outrage that was strictly short-lived, leaving most New Zealand animal rights activists marooned if not outright undermined by a surrounding culture that murders non-natives without hesitation.[12]

---

[11] See "Island Pig Eradication Completed," by Gregory W. Griggs, August 30, 2007, The Los Angeles Times, https://www.latimes.com/archives/la-xpm-2007-aug-30-me-pigs30-story.html.

[12] See "Horror at children drowning baby possums at Drury school event," by Melissa Nightingale, The New Zealand Herald, July 2, 2017.

New Zealanders are largely desensitized to all non-native attrition, and fully accustomed to massive eradication programs (for those other than the largest of the non-natives, humans themselves, and the tens of millions of sheep and other "farm animals" humans exploit). Whereas conservation biology in New Zealand, as we portrayed in "Hotspots," has systematically endeavored for well over 160 years[13] to save many of its most ambassadorial native and endemic species, where genetic research and complex translocations from one site to another, as well as the building of costly conservation fences, have coincided with the extirpation paradigms. Not just possums, gorse or Chilean fireweed, but the biologically prolific members of the Mustelidae family—stoats, weasels, ferrets; as well as the four New Zealand rodent species—Kiore (*Rattus exulans*), Norway Rat (*Rattus norvegicus*) Ship Rat (*Rattus rattus*), and House mouse (*Mus musculus*),[14] in addition to feral cats and deer.

The conservation/animal rights trade-offs? Millions of vertebrates killed, in order to try and boost from near oblivion the diminutive populations of such avifauna as Takahe (*Porphyrio hochstetteri*), the five species of Kiwi (genus *Apteryx*), the flightless parrot, the Kakapo (*Strigops habroptilus*), the critically endangered Yellow-Eyed Penguin (*Megadyptes antipodes*), the Blue Duck (*Hymenolaimus malacorhynchos*), and New Zealand's three native mammals—the long-tailed, lesser short-tailed, and greater short-tailed bats, as well as the seriously threatened Grand and Otago skinks. For each of the taxa—dozens of the most vulnerable, and others under the radar screen, like Stewart Island Robins—New Zealand's Department of Conservation has notably joined forces with numerous allies. Always the local iwi (Maori) first, but countless conservation coalitions that do New Zealand proud. There have been quite literally hundreds of thousands of conservation volunteers

---

[13] See *The Habits Of The Flightless Birds Of New Zealand*, by Richard Henry, Caretaker of Resolution Island, Wellington Government Printer, Wellington, NZ, 1903; See also, *Our Islands, Our Selves – A History of Conservation in New Zealand*, by David Young, University of Otago Press, 2004; See also, *The Penguin Natural World Of New Zealand -An Illustrated Encyclopaedia Of Our Natural Heritage*, by Gerard Hutching, Penguin Group (NZ), 1998; *God's Country: The New Zealand Factor*, by M. C. Tobias and J. G. Morrison, A Dancing Star Foundation Book, Zorba Press, Ithaca, NY, 2011; *Don Merton, The Man Who Saved the Robin*, by Alison Balance, Photographs by Don Merton, Raupo Publishing (NZ) Ltd., 2007; See also, "Hotspots New Zealand: What happens when an archipelago populated by bizarre flightless birds is invaded by alien species?" by Kennedy Warne, Photographs by Frans Lanting, National Geographic Magazine, October 2002; and *Kiwi – New Zealand's Remarkable Bird*, by Neville Peat, A Godwit Book, Random House (NZ), Auckland, NZ, 1999; *Field Guide To New Zealand Wildlife*, by Terence Lindsey and Rod Morris, HarperCollins Publishers, Auckland, New Zealand, 2003; *Forest Lore of the Maori – With methods of snaring, trapping, and preserving birds and rats, uses of berries, roots, fern-root, and forest products, with mythological notes on origins, karakia used etc.*, by Elsdon Best, Polynesian Society, Dominion Museum, Wellington, New Zealand, 1942; *The Prehistory of New Zealand*, by Janet Davidson, Te Papa Tongarewa Museum of New Zealand, Longman Paul Ltd., Auckland, New Zealand, 1987; and *New Zealand's Wilderness Heritage*, Text by Les Molloy, Photography by Craig Potton, Craig Potton Publishing, Nelson, New Zealand, 2007.

[14] *Guide to the identification and collection of New Zealand rodents*, by D. M. Cunningham and P. J. Moors, Department of Conservation, Wellington, New Zealand, 3rd edition, 1996, https://www.doc.govt.nz/documents/science-and-technical/rodent-identification.pdf.

throughout the years across New Zealand. An everyday fact. As evidenced, most famously, by those thousands who helped restore the island in the Harauki Gulf beyond Auckland, Tiri Tiri Matangi.

Employing often best conservation practices are systemic to the New Zealand psyche, even as animal rights (at least for bio-invasives and all those other vertebrates and mammals consumed by a large meat-eating society) are largely cast to the wind. This great ethical divide continues to rankle, dismay, and shatter any illusion of a formidable, all-encompassing ethic toward other species.

In every part of the world where we have worked, similar concatenations of citizen science, motivated volunteers, activists, and landholders have joined forces to save such critically endangered species as Brazil's Southern and Northern Muriqui (genus *Brachyteles*) or Madagascar's largest lemur, the *Indri indri*.[15] There are still documented instants of hunters eating both Muriquis and Indris. But their biggest threats continue to be deforestation and human, ecological illiteracy and/or indifference. Whether with the international trade (now largely illegal) of "exotic" psittacines (parrot family) poached at clay licks along the Peruvian Tambopata River; threats to California Condors (*Gymnogyps californianus*) from the lead of bullets embedded in carcasses the glorious, largest North American land bird consumes;[16] the last few mountain lion populations in southern California;[17] new species of pseudoscorpion in the Clough Cave at Sequoia National Park (one of some 80 known karst caves between Sequoia and Kings Canyon), who may number fewer than a few hundred individuals;[18] or any number of other precious and unique creatures around the world, community conservation, individual animal rights, and the brigades of conspicuous conscience at work to save life, pervades all such research, response, and rescue missions. In Madagascar, our film crews were the first to film the *Lepilemur sahamalazensis*, or sportive lemur, as well as the first to capture footage of the Blue-Eyed Black Lemur (Sclater's Lemur, *Eulemur macaco flavifrons*), one of two subspecies of the Black Lemur, and classified as critically endangered.

During the making of "Hotspots," we were fortunate to film other rarities: a paleopropithecus skull and bone fragments in a remote cave in Madagascar (the 50-pound Giant Lemur that went extinct as recently as 1000 years ago); a newly discovered Titi monkey first seen in Bolivia's Madidi National Park in the Upper

---

[15] Madagascar's largest remaining lemur. See, Primate Info Net, http://pin.primate.wisc.edu/factsheets/entry/indri; See also, op.cit., Mittermeier, Rylands and Wilson, pp. 28–175.

[16] See "New Study: Over Two Thirds of Fatalities of Endangered California Condors Caused By Lead Poisoning," American Bird Conservancy (citing a 2012 collection of surveys by the U.S. Fish and Wildlife Service, the National Park Service, the U.S. Geological Survey, the University of California at Santa Cruz and Davis, The Peregrine Fund, The San Diego Global and the Phoenix Zoo), February 8, 2012, https://abcbirds.org/article/new-study-over-two-thirds-of-fatalities-of-endangered-california-condors-caused-by-lead-poisoning/. Accessed May 26, 2020.

[17] See "California Lion Status and Population," https://mountainlion.org/us/CA/-CA-status.php, Accessed May 26, 2020.

[18] See "27 new species unearthed from Sierra Nevada..." by Glen Martin, SFGate, January 19, 2006, https://www.sfgate.com/bayarea/article/CALIFORNIA-27-new-species-unearthed-from-Sierra-2543060.php, Accessed May 26, 2020.

Amazon, but never in Peru's Tambopata National Reserve; the first footage ever captured of a Southern Bamboo Rat (*Kannabateomys amblyonyx*) foraging and peeing sometime around midnight; nine new invertebrate troglophiles and troglobites at Clough Cave, with National Park Cave specialist, Joel Despain; and the first ever footage of one of the most endangered mammals in North America, the Pacific Pocket Mouse (*Perognathus longimembris pacificus*) rediscovered in 1993 after presumed extinction, at the Marine Base Camp Pendleton just North of San Diego. Amid rattlesnakes and scorpions, we were able to film the little mouse just after midnight.[19]

Of course, the irony was not lost on any of us involved in the making of "Hotspots" that we should rejoice upon seeing for less than a minute one of the rarest rats in the world (in Peru); and that the US marines had been mandated by Congress and the 1973 Endangered Species Act to use every means at their disposal to save a mouse numbering fewer than 300 individuals, while New Zealanders were killing both mice and rats in an orgy of concentrated malevolence. And rodenticides—and freeway traffic—were killing off the last few apex predators in southern California, *Puma concolor*, the cougar, or mountain lion. That hunters were still killing some of the most iconic species on the planet, for food, skins, zoos, for anything that could turn a fast profit. Even killing Jaguars—largely at the bequest of corporate interests throughout China and the rest of Asia—in otherwise sacrosanct Suriname.

The coalitions and group efforts that would save the Jaguar, Mountain lion, the Pseudoscorpion, the Kakapo, and California condor remain a stark reminder that annually, just in the United States, approximately "670,000 dogs and 860,000 cats" are euthanized at shelters.[20] Ethics and mathematics collide. It is emotionally wearisome, but certainly not impossible, to save one life, even as millions of others perish. It does not require that much energy to hold up a conviction, a light in the dark (Fig. 56.3).

## 56.4 Every Voice

Nonetheless, against such deep chasms within human nature—hatred, fear, psychopathology, greed—animal rights, and the distressing contradictions and outright paradoxical nature of its relationship to conservation biology notwithstanding, one must celebrate the collaborative zeal of saving species, which means individuals, no matter what it takes. Passion to perpetuate the life force in all her splendor that is neither discouraged nor diminished by the internecine political, economic, cultural, and zoonotic mayhem of the early twenty-first century is key to saving ourselves (not that that is the motive). Individuals are situated ethically beside other life forms

[19] http://hotspots-thefilm.org/about/page21/page21.html.

[20] See "Pet Statistics," ASPCA, https://www.aspca.org/animal-homelessness/shelter-intake-and-surrender/pet-statistics, Accessed May 26, 2020.

**Fig. 56.3** Donkeys in the Haghier Massif, Haghier Massif, Socotra Archipelago, Yemen. (Photo © M.C. Tobias)

in every plausible corner of the planet. A soul speaks when spoken to. Every voice, sensation, concept emerges with equal prowess, relevance, and staging. Staying power is not vested in *power*, but in tenderness. A rallying cry may have philosophical hues and stripes but at the center of all such concerns are those shared not by the strength of one's grip, but the durability of the heart as it connects with another. Do not imagine that an insect high on Socotra, a ground orchid in Haiti, has not overseen the workings of the world, side by side with a companion; two Diving Bell spiders hidden comfortably together in the precise cranny of Staubbach Falls; a White-Headed Buffalo weaver of the sub-Sahara or Vietnamese wild rooster calling inquisitively to their friends, just at the right second upon dawn's ferrying light, a solemn sunrise made manifest for one and all across the earth's 123 billion acres.

# Chapter 57
# The Geography of Contradiction

## 57.1 The Anthropology of the Sacred and Profane

David Berliner, professor of anthropology at Université Libre de Bruxelles, writes, "Walt Whitman got it right when he wrote in 'Song of Myself' (1855): 'Do I contradict myself? / Very well, then I contradict myself, / (I am large, I contain multitudes).'" Whitman was paraphrasing from Sir Thomas Browne, who, in turn, had written, "We carry with us the wonders, we seek without us: There is *all Africa*, and her prodigies in us; we are that bold and adventurous piece of nature, which he that studies, wisely learns in a compendium, what others labour at in a divided piece and endless volume." And Berliner adds, "Humans live peacefully with contradictions precisely because of their capacity to compartmentalize"[1] (The same cerebral "compartments" described in Chap. 43.) (Fig. 57.1).

This neural phenomenon is not restricted to vertebrates who have shown near universal propensities for dialectics: joy and sorrow, mourning, and the re-acquisition (or not) of some degree of normalcy after a period of time, particularly evident in marine mammals, canids, and primates. But there are tomes of anecdotal evidence for these same separated behaviors in invertebrates, most notably, among diverse crustaceans. But within whole habitats we may reflect by any number of means upon the obvious and great sorrow and holistic jubilation; systems of coherence and incoherence; morbidity and revitalization that everywhere, and in every habitat-teeming sentience, surround us.

Within the adventurous rubrics of the forces of life on one planet, the arbitrary borders imposed by the conditions of human nation-states and sovereignties—invariably having arisen through the course of great bouts of violence and attempts at orderly reconciliation—highlight the nature of dyadic spaces, those sacred and

---

[1] *Religio Medici*, by Sir Thomas Browne, 6th Edition, Andrew Crook, London, 1669, p. 31. "How Our Contradictions Make Us Human and Inspire Creativity," Sapiens, December 14, 2016, https://www.sapiens.org/culture/human-contradictions/. Accessed May 29 2020.

© Springer Nature Switzerland AG 2021
M. C. Tobias, J. G. Morrison, *On the Nature of Ecological Paradox*,
https://doi.org/10.1007/978-3-030-64526-7_57

**Fig. 57.1** Taktsang Monastery, Bhutan. (Photo © M.C. Tobias)

profane. In Bhutan, with its 14,824 mi$^2$ and 755,000 denizens, the greatest tensions erupt between these two profound interferences on the path to survival and toward that which has for many decades been characterized as a nation devoted to "Gross National Happiness" as the fullest measure and modus operandi (pillar of strength) of good governance.[2]

At Taktsang Palphug Monastery gracing a series of ledges high upon a startling black and white and magnetic 3000-foot granite cliff just eight miles north of

---

[2] See *Sanctuary: Global Oases of Innocence*, by M. C. Tobias and J. G. Morrison, A Dancing Star Foundation Book, Los Angeles, CA; Council Oak Books, San Francisco and Tulsa, 2008, pp. 289–322. See also, *Bionomics in the Dragon Kingdom – Ecology, Economics and Ethics in Bhutan*, by Ugyen Tshewang, J. G. Morrison and M. C. Tobias, Springer Publishers, New York, 2018; and *Bhutan: Conservation and Environmental Protection in the Himalayas*, by Ugyen Tshewang, M.C. Tobias and J.G. Morrison, Springer Publishers, New York, 2021.

Bhutan's western most large town of Paro, this matrix of opposites struggling to form an unbiased union is palpable to the tens of thousands of pilgrims who make the four-mile roundtrip trek to an altitude exceeding 10,236 feet throughout the year, commencing from a busy parking lot down in the forest. There was no parking lot in 1974 when I (mt) first journeyed there with my good friend, Tenzing Norgay, a person from the Sherpa cultural matrix in the Solo Khumbu district of Nepal (the first documented human, along with New Zealander Edmund Hillary, to summit Everest, Chomolungma in Tibetan, Sagarmatha in Hindi). Tenzing used to express his doubts to me regarding the wisdom of having gone to that sacred summit. He and I would have similar late-night discussions about the wisdom of encouraging tourism, in whatever small numbers, to Bhutan. For it was that Winter that we had brought one of the very first groups of westerners to the country, as the reclusive nation struggled with the concept of ecotourism and had decided to let 30 people from the *outside world* across its borders.

In subsequent years, the government would deem an annual number of 300 outsiders a sustainable number of fleeting visitors, with a total of one hotel in the country at that time (in Paro). By various turns, and over coming years, economics would factor into the tourism agenda. And today, 45 years later, that number exceeds 300,000, the majority driving in from India, though flights from various points—including New Delhi, Kolkata, Dacca, Kathmandu, Bangkok, and Singapore—also bring tourists daily to the single international airstrip in Paro (opened for business as of January 1983), just a ten-minute drive from the road up to the parking lot at Taktsang ("the tiger's lair").

One can hear, and on occasion even see, the flights of Bhutan's national airline, the meticulous Druk Air (the best, safest airline in the world), approaching or departing the Paro airstrip from Taktsang, which remains the most revered, internationally recognizable sacred place in all of Bhutan. It is uncannily spectacular: that spot where the "Second Buddha," Guru Rinpoche, or Padmasambhava—a renowned "Tantric master"—came to live with his consort, Yeshe Tsogyal (who adopted the form of a tigress for the occasion). The Guru arrived in the form of one of his eight manifestations, the powerful Dorje Drolo. And there, with his companion tiger, he lived in a cave on the cliff for over 39 months and subdued through continuous meditation any and all malevolent spirits inhabiting the area.[3]

In the early twelfth century the famed Tibetan poet/siddha, Jetsun Milarepa, composed one of his archetypal reveries adrift the overpowering walls of Taktsang, with its smooth-waltzing cliff swallows. A temple was built adjoining one of the sacred caves on the cliff in 1684 during the rule of the Shabdrung, the great "unifier" of the country who managed to largely merge religious and political factions.[4]

---

[3] See "Paro Taktsang: The Breathtaking Himalayan Cloud Monastery," by Bryan Hill, August 3, 2015, Ancient Origins, https://www.ancient-origins.net/ancient-places-asia/paro-taktsang-breath-taking-himalayan-cloud-monastery-003532, Accessed May 30, 2020.

[4] See "A Tiger's Nest In The Land Of The Thunder Dragon," by Katherine Anne Paul, Sacred Sites International Foundation, May 4, 2020, http://www.sacred-sites.org/saved-sacred-sites/taktsang-taktshang-goemba/a-tigers-nest-in-the-land-of-the-thunder-dragon/. Accessed May 30, 2020.

By 1692 the entire monastery complex had been erected by inconceivably challenging engineering efforts.

This gorgeous cliff monastery (also called Taktsang Pelphug Lhakhang or Taktsang Sengye Samdrub) is, most importantly, Zangdok Palri, the Buddhist Pure Land of Padmasambhava ("Zangs mdog dpal ri" in Tibetan, literally, the Copper Colored Mountain, atop which is centered the so-called Lotus Light Palace). The mountain is considered the "innermost emptiness nature," the "Ultimate-body of Buddhahood, the absolute sphere of all. It is the union of wisdom and emptiness, the primordial purity, like sky."[5] And there, amid the thousands of feet of sheer granite "is hidden the treasure of the life force ..." and "the karmic results of good and bad actions." Moreover, "At Taktsang Sengye Samdrub, the essence of nectar for perpetuating wealth has been hidden ... treasures [that are] limitless."[6]

Prince Charles hiked halfway up in 1998. In 2008, much of the monastery was burnt, allegedly by a spilled butter lamp. One monk tragically perished in the fire. An international effort was launched successfully to fund and rebuild the monastery to near perfection over a period of two years. In April, 2016, the Duke and Duchess of Cambridge made it all the way up. A "great way to burn off the curry," the Princess volunteered.[7]

With so much history, spirituality, and populist sentiment crowding this lean, narrow, vulnerable habitat, it is no surprise that one of the world's great monuments to Medieval solitary meditation should become the twenty-first-century site of what to many is blatant over-crowding: a situation not unlike similar concerns voiced about Italy's Venice, or so many of the most popular national parks—bumper-to-bumper in Yellowstone, or Belgium's Bruges, which has lately seen mobs, pre-Covid-19, comparable to those at Disneyland.[8] One American tour operator, reflecting on the 274,097 tourists to Bhutan in 2018, decided after a 20-year career to stop bringing tours there. In speaking about the unsuspecting, innocent traveler, "Imagine their dismay when they arrive at the foot of Tiger's Nest to find a parking lot bursting with buses, hordes of noisy regional tourists clogging the

[5] In *Zangdok Palri – The Lotus Light Palace Of Guru Rinpoche, Visions Of The Buddhist Paradise In The Sacred Kingdom of Bhutan*, "The Different Perspectives of Zangdok Palri," Edited by Supawan Pui Lamsam and Kesang Choden Tashi Wangchuck, With Contributions by Tulku Thondup, Dungchen Sangay Dorji, Lopen Kunzang Tengye, Pema Wangdi, Chotiwat Punnopatham, Chongmas Rajabhandarak, and Tanika Pook Panyarachun, Zangdok Palri Series, Volume 1, Gatshel Publishing, Bangkok, Thailand, 2012, p. 3.

[6] ibid. p. 316.

[7] See BBC News, "William and Kate trek to Bhutan's Tiger's Nest monastery," April 15, 2016, n.a., https://www.bbc.com/news/uk-36052016, Accessed May 30, 2020. See also, "Ascending Bhutan's Sacred Tiger's Nest," by David Braun, November 30, 2014, https://blog.nationalgeographic.org/2014/11/30/ascending-bhutans-sacred-tigers-nest/. Accessed May 30, 2020.

[8] See "'It's getting like Disneyland: Bruges pulls up drawbridge on tourists," by Daniel Boffey, June 16, 2019, The Guardian, https://www.theguardian.com/world/2019/jun/16/bruges-crackdown-tourist-numbers, Accessed May 30, 2020.

trail, a gauntlet of vendors hawking fake made-in-Bhutan products and garbage-strewn roads"[9] (One tourist operator's opinion, of course.) (Fig. 57.2).

## 57.2   The Mount Everest Base Camp Syndrome

Much has been written and discussed regarding how it is that "waste confounds modernist dreams of purification and asserts its place in space, society, and nature."[10] In Nepal there is a saying, "You don't need a map to get to the Everest base camp, just follow the trash ...," which is now deemed a "60,000 pound" problem.[11] In surveying waste sites around the world, Taktsang and Everest only rate, obviously, because of their iconic stature and the blatant contradiction-in-terms. But humans everywhere, and throughout our species' history, have discarded unimaginable trillions of pounds of garbage, what contamination experts track in the form of "food wastes, packaged goods, disposable goods, used electronics...demolition debris, incineration residues, refinery sludges ... municipal solid waste per person on a daily basis" and so on.[12] One theory among many holds that garbage played a crucial role in the momentous evacuation of the spectacular and sacred Canyon de

---

[9] Nikkei Asian Review. "Bhutan's neighbors 'undermining' elite tourism strategy," by Nidup Gyeltshen and Phuntsho Wangdi, July 28, 2019, https://asia.nikkei.com/Life-Arts/Life/Bhutan-s-neighbors-undermining-elite-tourism-strategy, Accessed May 30, 2020. Revenues from tourism to Bhutan in 2018 were "USD 85.41 million". See "Tourists Number Records Growth Of 7.61%: Bhutan Tourism Monitor 2018, by Dechen Tshomo, Kuensel, April 8, 2019, https://www.dailybhutan.com/article/tourists-number-records-growth-of-761-bhutan-tourism-monitor-2018, Accessed May 30, 2020. Bhutan's total GNP for 2018 was 2.447 billion USD, with its primary income deriving from timber, hydroelectricity and agriculture.

[10] See "Geographies of waste: dispositions, transformations, contradictions," by Kevin Martyn, American Association of Geographers, Annual Meeting, 2019, https://aag.secure-abstracts.com/AAG%20Annual%20Meeting%202019/sessions-gallery/23172, Accessed May 30, 2020.

[11] "Mount Everest tackles 60,000-pound trash problem with campaign to clean up waste," by Dragana Jovanovic, ABC News, May 2, 2019, https://abcnews.go.com/International/mount-everest-tackles-60000-pound-trash-problem-campaign/story?id=62773297, Accessed May 30, 2020. But of course, it is much more than mere "waste" that infiltrates the dream of Sagarmatha National Park and its epicenter that is Everest. The mountain wildness, as debased and overrun with a theatrical, populist inanity -the tens-of-millions of wilderness tourists, expeditions out to conquer something (half in love with themselves, half in love with oblivion, as the film's Narration wends) was acutely embodied in the recent documentary, "Mountain" narrated by Willem Dafoe from a script by Robert Macfarlane, beautifully Directed by Jennifer Peedom, and premiering at the Sydney Opera House in June 2017. See https://www.hollywoodreporter.com/review/mountain-review-1015413. See also, "Climbing Sacred, Secretive Peaks," by Haley Littleton, March 1, 2017, https://www.elevationoutdoors.com/go-outside/climbing-sacred-secretive-peaks/.        Accessed August 8, 2020.

[12] See EPA, "Report on the Environment – Wastes," https://www.epa.gov/report-environment/wastes, Accessed May 30, 2020; See also, Organization for Economic Cooperation and Development (OECD). 2015. Municipal waste, Generation and Treatment: Municipal waste generated per capita. OECD.StatExtracts.

**Fig. 57.2** Jomolhari (7326 m/24035 ft), climbed six times, beginning in 1937, until all mountain-eering on sacred peaks was banned in Bhutan in the early twenty-first century. (Photo ©
M.C. Tobias)

Chelley in Arizona. These gorgeous cliffs had been inhabited for 5000 years, its
elaborate, multi-story dwellings erected over time by the Ancestral Puebloan peoples known as Anasazi, beginning around 700 CE. But by the fourteenth century, the
buildup of unsanitary waste might have spawned disease outbreaks, in concert with
a thousand-year drought and famine: a combination of ecological ill effects forcing
the inhabitants out toward the Rio Grande Valley of New Mexico.

In places like Borneo—third largest, and among the most biologically diverse
islands in the world—the rampant extraction of "timber, palm oil, pulp, rubber and
minerals" has wiped out, to date, half of her forests ("21.5 million hectares" "lost
between 2007 and 2020"—forests replaced by wastelands of stump and strewn
industrial wastes).[13] In the United States, "85 percent of parks have air that is
unhealthy to breathe," while "96 percent of the United States' 416 national park
units have significant air quality issues."[14] According to the Pew Charitable Trusts,
"as much as 13 million metric tons of plastic ends up in the ocean each year."[15]

[13] See "Borneo Deforestation," n.a., World Wildlife Fund, 2020, https://wwf.panda.org/our_work/
forests/deforestation_fronts2/deforestation_in_borneo_and_sumatra/. Accessed May 30, 2020.

[14] See "Significant Air Pollution Plagues Almost All U.S. National Parks," by Jason Daley,
Smithsonian Magazine, May 9, 2019, https://www.smithsonianmag.com/smart-news/signficant-
air-pollution-plagues-almost-all-us-national-parks-180972141/. Accessed May 30, 2020.

[15] "Plastic Pollution Affects Sea Life Throughout Ocean," by Simon Reddy, September 14, 2018,
PEW, https://www.pewtrusts.org/en/research-and-analysis/articles/2018/09/24/plastic-pollution-
affects-sea-life-throughout-the-ocean, Accessed May 30, 2020.

Worldwide, an estimated "200 million tons" or more of human waste goes untreated each year.[16] In 2018, Bhutan's own southernmost industrial hub, the town of Pasakha, was ranked by the World Health Organization as one of "the most polluted cities" in the world, second only to Muzaffarpur, in India, "with around 150 micrograms per cubic meter of fine particle emissions."[17]

While Bhutanese governance has made globally significant strides to become carbon negative, stem the flow of human waste, protect her primary forest canopies, and maintain robust conservation practices and an extensive network of protected areas and corridors, Buddhist philosophies offer largely theoretical models for perceiving paradox and contradiction, like those involving human traffic at Taktsang or pollution at Pasakh. Or, for that matter, the rapid growth of Bhutan's capital, Thimphu, or of Paro itself. The Middle Path Madhyamaka system, having first emerged around 150 CE in the commentaries of the great Indian interpreter of Buddhist thought, Nāgārjuna, falls impractically in tow with at least one major thesis that holds that in Buddhism there is a conspicuous absence of "any knowledge of the world because they [Buddhas] do not have epistemic processes and warrants to perceive the world." In other words, Buddha was a "global agnostic, on the ground of the nonexistence of mind and mental processes for those who have attained fully awakening." Alternative "no-mind" premises have also been advanced.[18] But such interpretive umbrellas of an advanced consciousness too easily translate into parking lots and waste dumps. The ecological remains of the day.

A very different, worldly, ecological, and ethical approach than *no-mind*, would be, among others, the advocacy of "personhood" and "sovereignty" for every living being in any given habitat—an attribution to every sentient being of an overflowing mind. This philosophical heresy would connote a duty (which Bhutanese have partly embraced) to safeguard the physical realities of earthbound corollaries of enlightened mind, and all those Others who are inherently enlightened. This is precisely what has been proposed for Beings like the orangutans of Borneo, "simian forest sovereignty [that] is critical to conservation efforts."[19] Across Bhutan, any

---

[16] "How Much Do You Poop in Your Lifetime?" By Mindy Weisberger, LiveScience, March 21, 2018, https://www.livescience.com/61966-how-much-you-poop-in-lifetime.html, Accessed May 30, 2020.

[17] See "Pasakha, Bhutan's industrial hub, symbolizes the struggle at the heart of Bhutan's development model," by Tej Parikh, May 16, 2018, https://thediplomat.com/2018/05/bhutans-happiness-faces-the-growing-pains-of-development/. Accessed May 30, 2020.

[18] See "Madhyamaka Philosophy of No-Mind: Taktsang Lotsāwa's On Prāsangika, Pramana, Buddhahood and a Defense of No-Mind Thesis," by Sonam Thakchoe and Julien Tempone Wiltshire, Journal of Indian Philosophy, 47, 453–487, 2019, https://doi.org/10.1007/s10781-019-09388-z, Issued 15 July 2019, https://link.springer.com/article/10.1007/s10781-019-09388-z#citeas, Accessed May 30, 2020

[19] See "An Ape Ethic and the Question of Personhood," by Gregory F. Tague. Rowan and Littlefield, Lexington Books, https://sites.google.com/site/gftague/an-ape-ethic; See also Tague's interview with Marc Bekoff, Psychology Today, March 14 2020, https://www.psychologytoday.com/us/blog/animal-emotions/202003/ape-ethic-and-the-question-personhood; and MAHB overview, https://mahb.stanford.edu/library-item/an-ape-ethic-and-the-question-of-personhood/; see https://www.amazon.com/Ape-Ethic-Question-Personhood/dp/1793619700

peak over 6000 m/19,685 ft has been deemed to be sacred and inviolate since 1994, and technical mountaineering banned entirely in the country as of 2003, much in the spirit of the Buddhist closure of all climbing on Kangchenjunga in Sikkim, and the regional Aboriginal ban on any outsiders hiking up the iconic Uluru. Similar restrictions have long been in place on volcanoes sacred to Muslims in Indonesia, on at least one major peak in Nepal (Machhapuchhare), Tibet (Kailash/Su-Meru), and throughout a myriad of Native American landscapes.

## 57.3   The Contradictions Inherent to the Mahāyāna Mahāparinirvāṇa Sūtra

Ethical bubbles and burred margins persist. For all of Bhutan's successes with preserving habitat for elephants, tigers, snow leopards, black-necked cranes, even the Yeti, for which an entire wildlife sanctuary was created far in the East, at Sakteng, the vegetarian ethos throughout the Buddhist nation is anything but 100%. As in India, *meat* remains a highly contentious issue in Bhutan. Despite ancient Buddhist prohibitions on the taking of life, and a national Agriculture and Food Regulatory Authority policy of formally prohibiting and/or discouraging all meat production, import, export, and consumption for at least two months (the auspicious first and fourth) out of each year, the other ten months are ignored, and as of 2017 Bhutan was "the highest consumer of meat per capita in South Asia."[20] It is not simply the tourists who are eating meat. And despite one of Gautama Buddha's final sermons, the *Mahāyāna Mahāparinirvāṇa Sūtra*, in which he urged followers to abstain from harming or consuming any animals, in 2019 meat importation from India to Bhutan increased by 16% over the previous year[21]—the paradoxes of enlightenment.

And meanwhile, the understory throughout the Blue Pine forests leading up to Taktsang has become increasingly polluted, a sacrifice area for tourist trekkers. None of these contradictions are actually contradictions, but the essence of what one economist, Richard Easterlin, termed in 1974 a distinct (earlier referenced) paradox ("The Easterlin Paradox") which asserts that, over time, per capita increased income does not correlate with increased per capita *happiness*.[22]

---

[20] "Why Bhutan's 'Hardline Vegetarian Right' Wants Everyone In The Country To Stop Eating Meat," by Sarah Reid, Independent, March 3, 2017. https://www.independent.co.uk/travel/asia/bhutan-hardline-vegetarian-right-stop-eating-meat-travel-tours-holiday-a7608141.html.

[21] See "Meat Import from India Shoots Up By 16%," by Thukten Zangpo, Business Bhutan, September 4, 2019, https://www.businessbhutan.bt/2019/09/04/meat-import-from-india-shoots-up-by-16/. Accessed May 30, 2020.

[22] See "The Easterlin Paradox," Economic and Social Research Council, n.a., https://esrc.ukri.org/about-us/50-years-of-esrc/50-achievements/the-easterlin-paradox/. Accessed May 30, 2020. See also, "Wellbeing measurements, Easterlin's paradox and new growth models: A Perspective through gross national happiness," by Sriram Balasubramanian, February 17, 2019, https://voxeu.org/article/wellbeing-measurements-easterlin-s-paradox-and-new-growth-models, COX CEPR Policy Portal, Accessed May 30, 2020.

In the wake of the murder of Mr. George Floyd by police in Minneapolis, with the resulting fiery protests throughout the United States and other countries (and countless other murders also captured on video), a version of Easterlin's coefficients was used to describe "the Minnesota Paradox"—a "twin city" schizophrenia: huge wealth, high quality of life indicators, but a hidden, sinister secret that languishes throughout America—the systemic racial divide. Moreover, in the United States, the period following World War II until 1970 revealed that huge increases in income "failed to show any upsurge in happiness" according to surveys.[23] Other critics of various happiness indices point to "minority rights violations."[24]

Both Aristotle and Thomas Jefferson argued that "happiness" was akin to "eudaimonia," human prosperity, the goal for every government and every individual. And yet, as one middle-aged voter at polling station in Central London was quoted recently, "I'd be happy if I could kick all the bastards out."[25]

## 57.4  "Compassion Transcendence Strategies"

In grappling with these human nature issues, some have simply resorted to describing them as a "stupidity-based theory of organizations."[26] Others, with specific reference to what they describe as "Bhutan's compassion transcendence strategies," have suggested that "transcendence of oppositions is widely recognized as the most

---

[23] See "When Economic Growth Doesn't Make Countries Happier," by Selin Kesebir, April 25, 2016, Harvard Business Review, https://hbr.org/2016/04/when-economic-growth-doesnt-make-countries-happier, Accessed May 30, 2020. For "The Minnesota Paradox," see David Leonhardt, The New York Times, "The Morning," Jun 1, 2020.

[24] See "The Paradox of Happiness," by Benjamin Mason Meier and Averi Chakrabarti, in Health and Human Rights, 18(1): 193–208, June 2016, https://www.researchgate.net/publication/309699641_The_Paradox_of_Happiness, Accessed May 30, 2020; see also, "A Nation of Paradoxes: Will the Bhutan Refugees be Re-integrated into the Philosophy of Gross National Happiness," by Bawa Singh and Dr. Jaspal Kaiur, July 16, 2019, South Asia Journal, http://southasiajournal.net/a-nation-of-paradoxes-will-the-bhutan-refugees-re-integrated-into-the-philosophy-of-gross-national-happiness/. Accessed May 30, 2020. See also, "The Happy Paradox," by Francesco Garutti, October 2018, https://www.cca.qc.ca/en/articles/issues/27/will-happiness-find-us/64484/the-happy-paradox, Accessed May 30, 2020.

[25] See "The satisfaction paradox – Why are happy people voting for angry parties?" The Economist, International Edition, July 11, 2019, https://www.economist.com/international/2019/07/11/why-are-happy-people-voting-for-angry-parties, Accessed May 30, 2020. See response at Democracy Digest, "Why are happy voters backing angry populists?" July 12, 2019, National Endowment for Democracy, https://www.demdigest.org/why-are-happy-voters-backing-angry-populists/ See Andrea Kendall-Taylor and Rachel Rizzo, and Dutch political scientist, Cas Muddle, co-author of *Populism: A Very Short Introduction*, with Cristóbal Rovira, Oxford University Press, Oxford UK, 2017.

[26] See M. Alvesson an A. Spicer, "A stupidity-based theory of organizations," Journal of Management Studies, 49, 1194–1220, 2012.

effective paradox response."[27] Research into contradictory human nature has taken a fascinating turn in one study that compared historic and contemporary Scottish Highlands forest cover regrowth in light of "economic restructuring" that saw revenue from ecotourism outpace traditional heavy "industrial production"—with the effect that "declining industrial power in the region is an inherently ecological one that takes the form of 'schizophrenic forestry,' in which forest expansion leads to the rise of degraded monocultures alongside 'pristine' sites of conservation."[28] Such ecological bubbles, juxtapositions both rude and reckless, represent the increasing truth of human folly and inconsistency.

These diverse and troubling landscapes invoke an ambient mosaic of contradictions which, as we struggle to redeem ourselves in favor of all those kindred spirits everywhere in the world, must re-invent an ethic, a lasting tenure of coherence (fair housing, socialized medical insurance, urban renewal, green space planning, fair and generous cooperation in every sector that will impact biological communities and intergenerational equity, as examples) and that will rise to the occasion of down-to-earth, non-violent policies for everyone. Animal liberation and conservation biology only really works if *all* animals and plants are included.

---

[27] See "Transcending Organization Compassion Paradoxes by Enacting Wise Compassion Courageously," by Ace V. Simpson and Marco Berti, Sage Publications, January 2, 2019, https://doi.org/10.1177/1056492618821188, https://journals.sagepub.com/doi/10.1177/1056492618821188?icid=int.sj-abstract.citing-articles.3, Accessed May 30, 2020.

[28] From the Abstract, "A Forest of Contradictions: Producing the Landscapes of the Scottish Highlands," by Paul Robbins and Alistair Fraser, Antipode, A Radical Journal of Geography, Wiley Online Library, February 21, 2003, https://doi.org/10.1111/1467-8330.00304, https://onlinelibrary.wiley.com/doi/abs/10.1111/1467-8330.00304, Accessed May 30, 2020.

# Chapter 58
# Metaphysical Landscapes

## 58.1 The Animals and Landscapes of Vermeer

Between 1673 and 1723 Antonie van Leeuwenhoek of Delft (1632–1723), famed for his observations of innumerable worlds within worlds, utilizing his impeccably ground single-lens microscope, wrote scores of letters to the Royal Academy of London in which he included fine pencil sketches of insect exoskeletons, the twig of an ash tree, a claw, the leg of a louse, joints, blood vessels, and countless other fineries of the biological world. Leeuwenhoek was born the same year as his close friend and neighbor Johannes Vermeer (1632–1675). Given the scientist's admitted limited talent as a draughtsman, some have speculated that Leeuwenhoek did not work alone. Japanese biologist Shin-Ichi Fukuoka, having studied Leeuwenhoek's manuscripts and letters, presents a most plausible hypothesis: Leeuwenhoek relied upon Vermeer to do the sketches.[1] It certainly makes sense. Both men were obsessed with a photo-realistic ephemerality, inherently contradictory: the presentation of the world through the illusion of human recreation, using a lens, brushes, paints, and the possibly other mechanical devices and techniques to ensure a soft verisimilitude, precise conjurations. Vermeer painted several obscure birds and other mythic beings, in addition to a mournful dog, numerous detailed maps, and several landscapes within his body of 34 known works, 23 separate landscapes within his painting, "Artist in His Studio"(ca. 1666–1668).

For all of his perfections, and crucial to them, Vermeer's landscapes, most notably his two urban panoramas, "View of Delft" (1660–1661) and "The Little Street" (1657–1658), in addition to his "Diana And Her Companions" (1655–1656) are as strangely remote and out of time as certain Chinese ice worms (enchytraeid annelids) in Xizang, like the species *Sininchytracus glacialis,* which lives all year in the

---

[1] See Sin -Ichi Fukuoka's *Vermeer: Realm of Light – Travels through art and science in pursuit of the greatest artist of the Golden Age*, Kirakusha Inc., Tokyo, 2015, Chapter 7: "A Certain Hypothesis," pp. 239–251.

© Springer Nature Switzerland AG 2021
M. C. Tobias, J. G. Morrison, *On the Nature of Ecological Paradox*,
https://doi.org/10.1007/978-3-030-64526-7_58

warmest glaciers on the planet, or other warm-glacier species like the numerous algal and moss species, as well as glacier mice, all found in Iceland.[2] While the world rightly celebrates the body of Vermeer's work, the animals (people, et al.) that he depicted are just that remote: inhabitants of another world, another time. Vermeer painted 76 or 77 people, not including dozens of others in paintings within the paintings. Given the limitations of our concept of *landscape*, Vermeer's oeuvre inculcates in every face, gesture, and human gaze, the mind's landscape, undifferentiated between an interior room, a vast map of the world, a picture of trees and truncated perspectives on a wall, or creatures under his neighbor's microscope. By his genius he left perfect echoes, no footprint.

While a Leonardo or Jan Breughel the Elder approached the bifurcation of humans with their surroundings very differently than Vermeer, all yielded to a conceptual plateau of convergence, wherein "convincing fantasies," as Joost Keizer has explored them, could be accounted for. In the case of Leonardo da Vinci (1452–1519) he thought of that imaginative vortex as the "impressive" located in the "first ventricle of the brain" "closest to the eyes" and that "location where the imagination, the intellect and the soul resided"[3] (Fig. 58.1).

## 58.2   Iterations of Landscape

Wheresoever we hail the landscape, its meaning (étumon), "true sense" compresses human qualities into a scene of identification, fundamental to the anthropic bias that has co-evolved in the very guise of our physiological outreach, our senses and our normal abilities to walk toward a horizon, or pluck an apple from a tree and eat it. As we have described in some detail elsewhere,[4] every human myth, allegory, and pictorial invention follows a trajectory informed by the conflation of an anthropomorphic lexicon. The strawberry populates Ovid's *Metamorphoses* (8 AD) as a confirmed inhabitant of the Garden of Eden and the endless array of Nativity scenes and virtually every aspect of the Christ Passion. Roses become "thee three graces"; columbine a symbol of Mary's grief; narcissus her sorrow; the Dove, as perceived by Noah, an end to the great Flood; an Ermine, symbolizing the purity of a Knight.[5]

---

[2] See "The Monsoon Maritime Glaciers In The Southeastern Part Of Xizang," by Li Ji-jun and Zheng Ben-xing, in *Geological And Ecological Studies Of Qinghai-Xizang Plateau*, Proceedings Of Symposium On Qinghai-Xizang (Tibet) Plateau, Beiing, China, Volume II, Environment And Ecology Of Qinghai-Xizang Plateau, Science Press, Gordeon and Breach, Inc., New York, 1981, p. 1601.

[3] *Leonardo's Paradox – Word and Image in the Making of Renaissance Culture*, by Joost Keizer, Reaktion Books, London, 2019, pp. 144–145.

[4] See *After Eden: History, Ecology & Conscience*, by M.C.Tobias, Avant Books, San Diego, CA 1985, pp. 91–92.

[5] See *Nature and Its Symbols*, by Lucia Impelluso, Translated by Stephen Sartarelli, The J. Paul Getty Museum, Los Angeles, CA 2003.

**Fig. 58.1** Paradise landscape with Adam and Eve, Studio of Jan Brueghel the Elder, ca. 1604–1613, Oil on Canvas. (Private Collection). (Photo © M.C. Tobias)

Cheek against cheek, in the masterpiece known as "The Virgin of Vladimir," painted in Constantinople around 1131, illuminates the dolorous Virgin and Christ child, "the embodiment of the divine in human form."[6]

But these are the obvious reaches, from a tree to the very Tree of Life. Far more incomparable is that period of the late Middle Ages so ably envisioned by art historian Jacques Élie Faure when he establishes that "The French hero is the cathedral."[7] However in England, "The art of the north demands the complicity of the vapor that spreads through space, of the foliage, of the sleeping water, and the uncertain illumination of the night …. [as contrasted against] the abrupt dissociation of the social forces, the defeat that comes day by day, even as man's illusions recommence each day, the mad charges, the feverish plunging of a civilization at the point of death."[8]

No strangers to the battlefields between the human soul and the long reach to its artistic expression, Japanese referred to gardens and monuments that graced them

---

[6] See *The Illustrated History Of Art*, by David Piper, Chancellor Press, London, 1981, p. 62.

[7] Jacques Élie Faure, the Five-Volume *History of Art*, Volume 2, *Mediaeval Art*, Translated From the French By Walter Pach, Garden City Publishing Co., Inc., New York, 1937, p. 342.

[8] ibid., p. 354.

in European tradition as "borrowed scenery."[9] Yet, such visionary landscape had become the very world, lived in alternatively from the rude streets covered first in the gray dust from burning wood and then, after its implementation by Henry II, the soot of burning coal—from those bellowing byproducts following the 1698 patent by Thomas Savery of steam power, augmented by James Watts' aggressive sale of engines and then Oliver Evans' upgrade for the conversion of coal to energy. The steam locomotive became a British rage; pastureland replaced tillage just as water-power had begun to replace human labor as early as 1768. The resulting impact on landscapes was dramatic, with Europe's population increasing by some 200 million in the nineteenth century, a 20% rise in the city of London during the decade of the 1820s. By 1853 Thomas Babington Macaulay's *History of England*[10] referenced "suburban country seats surrounded by shrubberies and flower gardens." Within three decades, "The speculative builder [had] become the pest of suburban London."[11]

Sir Kenneth Clark dates that discernible tension of elements inherent to an aesthetic life—a work of art, a distinct appreciation for scenery propounded upon a real place—to the Swiss artist Konrad Witz's rendition of "the background" for his painting "The Miraculous Draught of Fishes" (1444) which features Lake Geneva; and then, precisely half-century later, various precise geographical locations feature in water colors by Albrecht Dürer, including the alleged first drawing of an actual town, one on the map, Innsbruck.[12]

## 58.3    "Even in Arcadia There Am I"

"Who placed us with eyes between a microscopic and a telescopic world?" wrote Thoreau, asserting the self-confessed paradoxical presence of Nicolas Poussin's "Et in Arcadia ego" (1637–1638)—"Even in Arcadia (There) Am I"—a painting which he interpreted after a 1620 version by Giovanni Francesco Barbieri (Il Guercino) depicting a country scene in which two shepherds stand "contemplating a skull."[13] This vision strikes at the historic root of that tension between the urban and the rural, its infinity of incarnations within that primate who moves freely between the two neverworlds. Writing in 1972 of threats to Britain's second largest national park, the 845-square-mile Snowdonia, Amory Lovins described the "30,000 people who try to make a living from its land. The soil is thin, poor, acid, and high; business is very bad at the quarries; the economy is chronically

---

[9] op. cit., *After Eden*, p. 91.

[10] 1850–1859, iii. 1., p. 351, as cited under "Suburban," "Su – Sz." *The Oxford English Dictionary*, p. 71, Volume X, Oxford, UK, Reprinted 1961.

[11] Law Times, LXX. 130/2, ibid.

[12] *Landscape into Art*, by Kenneth Clark, John Murray, London, 1949, p. 19.

[13] *Landscape Painting – A History*, by Nils Büttner, Translated by Russell Stockman, Abbeville Press Publishers, New York, 2000, p. 142.

depressed; the communities are bled by depopulation, by the country-to-town drift that divorces men from their environment and their heritage."[14]

So confluent a syndrome, this, that should the madding crowd run riot with "belittling influences" (Johann Schiller's phrase), there theoretically remains a refuge beyond. Lord Byron's "Childe Harold" toured its wilds, theorizing upon the resident savagery in a revitalized poetic that would be taken up by British Parliament in an attempt to appease the public's declared need of nature. Plans were adopted for the first public park, at Birkenhead in 1847.[15] New York's Central Park was soon to follow, Frederick Law Olmsted looking back to the complicated idea of pure landscape that had been advocated by English landscape architect William Kent (1685–1748). Kent's masterpiece, Rousham, laid out in 1740 at Buckinghamshire, had, according to Horace Walpole, "leapt the fence and saw that all nature was a garden."

Kent wanted no evidence of humanity in his landscapes, no straight lines, no conventional grazing deer upon cleared greens beside nestling brook. But Buckinghamshire was not to be confused, even by Kent, with some paleolithic outback. There had been nearby agriculture since the Norman Conquest, and bodies were sure to have been burnt and/or buried in the near enough vicinity. John Milton had occupied a cottage in the area and Thomas Gray had composed his famed "Elegy" close by. Indeed, with the famed and fertile Vale of Aylesbury just North, this was scenery the English had celebrated for a millennium: longer if the Roman megalithic ruins in the neighborhood could be said to attest to yet additional antiquated fondness for the region (Fig. 58.2).

Kent mowed no lawns, pruned no trees, in his efforts to engineer the ultimate rusticity. It's hard to know what he would have made of Rousham today—"No children under 15. No dogs."—a venue that allows for "wedding receptions, photographic shoots and events such as car club rallies."[16] And after all, Olmsted and Vaux introduced every single plant, walkway, bench, sculpture, water element, to Central Park: it was a human creation imposed on land that had, previously, been characterized by America's "father of landscape architecture," Andrew Jackson Downing, as nothing but "squares and paddocks.[17]

Ultimately, our downsizing of the wilderness into our own likeness was, perhaps, best captured by Roelant Savery (1576–1639) in his painting, "The

---

[14] *Eryri, the Mountains of Longing,* An interpretation and case study by Amory Lovins, with photographs by Philip Evans and an introduction by Sir Charles Evans, Edited by David R. Brower, The Earth's Wild Places: 5, Friends of the Earth, San Francisco, 1972, pp. 78–79.

[15] See "The History of Birkenhead Park," Wirral, https://web.archive.org/web/20080626164507/http://www.wirral.gov.uk/LGCL/100006/200073/670/content_0001110.html, Accessed March 19, 2020.

[16] See https://www.rousham.org/

[17] See "The Legacy of Central Park: How Downing, Vaux and Olmsted Set the Standard for American Parks," n.a., FBW – Fund for a Better Waterfront, August 30, 2018, https://betterwaterfront.org/the-legacy-of-central-park-how-downing-vaux-and-olmsted-set-the-standard-for-american-parks/?gclid=EAIaIQobChMIzMOji9Cn6AIVh5WzCh1-_gyoEAAYASAAEgI8nvD_BwE, Accessed March 19, 2020

**Fig. 58.2** Illustration by Wenceslas Hollar of Virgil's "Georgics IV," From *The Works of Virgil*, Translated by John Dryden, Second Edition, Printer Jacob Tonson, London, 1698. (Private Collection). (Photo © M.C. Tobias)

Temptation in the Garden of Eden,"[18] which conforms, in its human way, to what Richard Le Gallienne described as "the wild end" as opposed to his "home end" which "is the real garden ... the end where you take tea in a shady corner of the

---

[18] See detail on p. 9 of *The Enchanted Garden – Images Of Delight*, by Bryan Holme, Oxford University Press, New York, 1982. See also, the many dozens of paradise scenes by Savery in Kurt J. Müllenmeister's *Roelant Savery, Die Gemälde Mit Kritischem Oeuvrekatalog*, Luca Verlag Freren, 1988; See also, *The Garden Of Eden – The Botanic Garden and the Re-Creation of Paradise*, by John Prest, Yale University Press, New Haven, Connecticut and London, 1981, particularly p. 45, with its reproduction of The Garden at Leyden from P. Paaw's *Hortus publicus acadmiae Lugdunum-Batavae*, 1601.

**Fig. 58.3** Central Park at Sunrise. (Photo © M.C. Tobias)

lawn, and even dine out on warm summer nights under the mulberry-tree … near the dove-cot and the beehives and the chickens"[19] (Fig. 58.3).

---

[19] In the essay entitled, "The Joy Of Gardens," by Richard Le Gallienne, in *Corners Of Grey Old Gardens With Illustrations In Colour by Margaret Waterfield*, T.N. Foulis, Ltd., Edinburgh & London, 1914, p. 151.

# Chapter 59
# Savery's Castle of Secrets

## 59.1   A Painter, an Emperor, a Castle, and a Bird

Roelant Savery (1576–1639) and his family fled Kortrijk, southern Belgium, in 1580 to escape Spanish persecution of Anabaptists. He would move to Haarlem and, tutored by his older brother Jacob (1566–1603), soon made a name for himself as a painter. The young artist would migrate on to the bustling opportunities afforded by Amsterdam, to Paris in 1603[1] and then, momentously, receive an invitation to relocate to the Holy Roman Emperor Rudolf II's (1552–1612) castle in Prague, considered to be at that time the epicenter of the arts and sciences in Europe. It was the same year that Jan Brueghel the Elder also first visited Rudolf's castle in Prague.[2] And the very year Savery's contemporary, Jan Saenredam of Assendelft, engraved, in collaboration with the fine Haarlem Mannerist, Abraham Bloemaert, one of the most perfect Apollonian images of Paradise, an engraving laid on paper of "Adam Naming the Animals," ever produced. Rudolf was a passionate and compulsive collector of everything beautiful and strange, living and dead. His relationship with Savery would in many ways change the way humanity would forever come to view other species, noting that paradise itself was ephemeral. Never was a region spanning multiple nations, states, republics and empires so unilaterally infatuated with arcadias. In Italy, Milan's Cardinal Federico Borromeo was encouraging his close friend, Jan Brueghel the Elder with an intelligent zeal equal to that of Rudolf, as if Christianity herself hung upon the degree

---

[1] See https://artist.christies.com/Roelandt-Savery%2D%2D43142.aspx, Accessed April 25, 2020.

[2] See RKD, https://rkd.nl/en/explore/artists/466467, Accessed April 25, 2020. See also, *Zoology in Early Modern Culture: Intersections of Science, Theology, Philology, and Political and Religious Education*, Chapter 10, "Exotic Animal Painting by Jan Brueghel the Elder and Roelant Savery, by Marrigje Rikken, pp. 401–433, Brill, Koninklijke Brill NV, Leiden, 2014, DOI: https://doi.org/10.1163/9789004279179_011, Accessed May 1, 2020. Series: Intersections, Volume: 32, Editors: Karl A. E. Enenkel and Paul J. Smith. https://brill.com/view/book/edcoll/9789004279179/B9789004279179_012.xml.

© Springer Nature Switzerland AG 2021
M. C. Tobias, J. G. Morrison, *On the Nature of Ecological Paradox*,
https://doi.org/10.1007/978-3-030-64526-7_59

to which paradise could be celebrated and reproduced on canvases, in parkland settings and in zoological containment[3] (Figs. 59.1, 59.2, 59.3).

While the Emperor was obsessed with killing animals, hunting, as witnessed by his many hunting grounds,[4] he also proffered a more affable affiliation with companion animals. His pets included tigers, and an apparently rambunctious lion named Muhammed. Rudolf and his father Maximillian II before him, established Europe's first zoo, a "Deer Moat," a "Lion Court," with a public viewing gallery, and "Royal Game Preserve" surrounding his Star (Hvêzda) summer palace.[5] Rudolf's aggregate of wild animals was imported from all over the world, including a cassowary at first believed to breathe fire, brought to him via Amsterdam. The Dutch captain whose men managed to trap the bird suffered a revenge killing by the bird en route from the Banda Islands of Indonesia. Rudolf's big cats are said to have roamed his castle at will, and taken the occasional lethal bites out of the odd royal guests.

Those who survived received at least some reparation. It was a menagerie more aggressive in its holdings, even, than those of Louis XIV's animal "apartments" at Versailles[6]—or of Archdukes Albert's and Isabella's menagerie in Brussels, where Breughel the Elder would frequently visit for his painterly sittings with live animals.[7]

For nearly a decade as court painter, Savery, having arrived with his nephew/ assistant, Hans Savery II, produced an astonishing body of work at the castle, particularly focused upon paradise images and an unprecedented array of life-like animal portraitures, including several of the Dodo, arguably painted from life.

---

[3] See "The Explorations of Emperor Rudolf II," by Jeff Michael Hammond, The Japan Times/ Culture, a review of "The Empire of Imagination and Science of Ruldolf II" at the Bunkamura Museum of Art, www.bunkamura.co.jp/museum, January 23, 2018, https://www.japantimes.co.jp/ culture/2018/01/23/arts/explorations-emperor-rudolf-ii/#.Xp9yB9NKi-s, Accessed April 24, 2020. The review features Savery's painting of "Orpheus Playing to the Animals," (1625) from the National Gallery in Prague, a painting that seems to sum up the artist's own sense of passing time: the castle has become a ruin, and the parliament of animals is missing one in particular, the Dodo, whose living reality would appear to have been all but eclipsed. See also, "Federico Borromeo as a Patron of Landscapes and Still Lifes: Christian Optimism in Italy ca. 1600," by Pamela M. Jones, The Art Bulletin, Vol. 70, No. 2, June, 1988, pp. 261–272, Published by CAA, DOI: 10.2307/3051119, https://www.jstor.org/stable/3051119, https://www.jstor.org/stable/3051119? seq=1, Accessed June 18, 2020.

[4] See The Saint Louis Art Museum, "Forest With Deer," by Savery, 1608–1610, https://www.slam. org/collection/objects/974/. Accessed April 25, 2020.

[5] "First Zoo in Europe," by Miroslav Bobek, March 30, 2015, https://www.zoopraha.cz/en/about-zoo/news/director-s-view/9130-first-zoo-in-europe, Accessed April 24, 2020.

[6] See Zoo: A History of Zoological Gardens in the West, by Eric Baratay and Elisabeth Hardouin-Fugier, Reaktion Books, London 2004, p. 62, https://press.uchicago.edu/ucp/books/book/ distributed/Z/bo3535747.html, Accessed April 24, 2020.

[7] See Oudry's Painted Menagerie: Portraits of Exotic Animals in Eighteenth-Century Europe, edited by Mary Morton, Getty Publications, Los Angeles, 2007, p. 138. https://books.google.com/ books/about/Oudry_s_Painted_Menagerie.html?id=lxdHAgAAQBAJ.

ROELANT   SAVERY

A été un peintre extraordinaire des animaux, et autres oyseaux; et les paisages les, quelles il faict, sont bien estimees de les amateurs de la painture il est natif de Flandres.

Adam Willaerts delin.                    Io. Meyssens fecit et excudit.

**Fig. 59.1** Portrait of Reolant Savery by Joannes Meyssens (1612–1670) after Adam Willaerts, Rijksmuseum Amsterdam. (Public Domain)

During two of his years working for Rudolf he traveled throughout Bohemia and the Tyrol painting landscapes (1606–1608), many of which were converted into exquisite works on paper by Rudolf's long-time imperial engraver, Ægidius Sadeler (c. 1568–1629).[8]

By the time Savery had returned to Prague, events within the Habsburg Monarchy were rapidly dissembling. Savery saw the demise of Rudolf on January 20, 1612,

---

[8] See *Hollstein's Dutch And Flemish Etchings, Engravings And Woodcuts, CA. 1450–1700*, Volumes XXI and XXII Aegidius Sadeler To Raphael Sadeler II, Plates, Compiled by Dieuwke De Hoop Scheffer, Edited by K. G. Boon, Van Gendt & Co., Amsterdam, 1980, The Netherlands.

**Fig. 59.2** Portrait of Emperor Rudolf II by Hans von Aachen (1152–1615). (Public Domain)

and then went to work for his successor/younger brother Emperor Matthias, dividing his time between Amsterdam and Prague, eventually moving full-time back to the Netherlands, based in Utrecht as of 1618. He became a member of the prominent artist's guild of St. Luke, and died there in 1639.[9] According to the first biographical sketch of his life by Dutch author/painter Arnold Houbraken (1660–1719) Savery had gone "insane" by the time of his death. He had seen too many deeply vulnerable animals jostled about; netted, dragged, strung up, forced to pose—trophies for Renaissance collectors. He had been inside the castle and knew that he had been a collaborator in the paradoxical best, and worst of human nature. The same castle walls as Boccaccio's protagonists, not realizing that they themselves were hosting

---

[9] See RKD, https://rkd.nl/en/explore/artists/Savery%2C%20Roelant, Accessed April 24, 2020.

**Fig. 59.3** Engraving of a Floating Castle, a Mythologized Kingdom of Human/Landscape Interactions such as was Enshrined by Kings and Emperors like Rudolf II, By David Vinckboons, Engraver Nicolaes de Bruyn, 1601. (Private Collection). (Photo © M.C. Tobias)

the germs.[10] Early-seventeenth-century remarks about Rudolf II suggest that he, too, suffered from great depression and melancholy and may have died as a result.[11]

Archduke Matthias had begun his successful agitation to end his older brother Rudolf's reign by the very year Savery had arrived at court, in 1604.[12] At that time Matthias had mobilized Hungarian allies enabling him to force his brother's hand and turn over rule of Moravia, Hungary, and Austria. But by that time, it was also clear that Rudolf II was more interested in devoting himself entirely to the occult arts and sciences, and to his animals, libraries, and galleries rich with treasures, than to the annoying inconveniences of politics, particularly war with the Ottomans.

[10] http://www.getty.edu/art/collection/artists/402/roelandt-savery-flemish-1576-1639/.

[11] De groote schouburgh der Nederlantsche konstschilders en schilderessen (*The Great Theatre of Dutch Painters*) 1718, https://www.dbnl.org/tekst/houb005groo01_01/houb005groo01_01_0027. php Accessed April 24, 2020. See also, "Rudolf II," in Impossible Objects, "Fragments," by R. J. W. Evans, [A fascinating Gestalt of Rudolf's life and times] http://www.impossibleobjectsmarfa. com/fragments-2/rudolf-ii, Accessed April 24, 2020.

[12] See "Roelandt Savery: a painter in the services of Emperor Rudolf II," Exhibition: 8 December 2010 – 20 March 2011, CODART, https://www.codart.nl/guide/agenda/roelandt-savery-a-painter-in-the-services-of-emperor-rudolf-ii/. Accessed April 24, 2020.

With Savery painting his every dreamscape, Rudolf had all but embraced the life of a recluse in his "Wunderkammer," with its chameleons, crocodiles, fish, and a bird of paradise.[13] These, and countless other entities would be inventoried between 1607 and 1611 by another court painter, Daniel Fröschl.[14]

Following his successful usurpation, Emperor Matthias moved much of Rudolf's collections—now considered perhaps the greatest in Europe, dominated by what would come to be known as "Rudolfine Mannerism"[15]—to Vienna, as political conditions spun awry; what would come to be known as the Thirty Years' War.[16] The Holy Roman Empire was disintegrating. Rudolf's family predecessor, Ferdinand II had long before set in motion a mad foment polarizing Roman Catholics and Protestants. The three decades of ensuing chaos would leave over eight million people, and unknown Others, dead, victims of violent clashes and famine. Rebellions turned into atrocities, which triggered an endless series of revenge massacres, from Bohemia to Saxony; from the Southern Netherlands where Savery had come from, to Spain. Dozens of territories and nations were drawn into the European-wide hatreds and grotesque affronts on battlefields, and in villages, combatants arriving from Scotland to Prussia. It was certainly and quickly the end of an era, and most definitively for the once glorious city of Prague. Endless turmoil, followed by months of steady bombardment near the end, had reduced her population three-fold, an accurate template of what humans had been doing to all the birds.[17]

Even after the fragile treaties of Münster formulated as part of the Peace of Westphalia, European governance was to be suddenly dominated by the Swedes, which would have lasting consequences for what remained of Rudolf's remarkable art and natural history collections. Prague was sacked in 1648 by Swedish troops, the best of the remaining paintings swept up into Queen Christina's orbit, shipped

---

[13] A Rich Cultural History: Prague and Rudolf II," by Erin Naillon, February 24, 2017, https://cz. cityspy.network/prague/features/rudolf-ii-prague/. Accessed April 24, 2020.

[14] See "Rudolf II's 'Wunderkammer' and fascination with the exotic," by Damian Brenninkmeyer, April 6, 2017, Dorotheum Blog, https://blog.dorotheum.com/en/rudolf-ii-wunderkammer/ Accessed April 24, 2020. See also, *Alchemy Of The Gift: Things And Material Transformations At The Court Of Rudolf II*, by Ivana Horacek, University of British Columbia Theses and Dissertations, 2015. https://open.library.ubc.ca/cIRcle/collections/ubctheses/24/items/1.0166243, Accessed April 24, 2020.

[15] See "Holy Ronan Emperor Rudolf II – 1576–1612," Holy Roman Empire Association ©, n.a., http://www.holyromanempireassociation.com/holy-roman-emperor-rudolf-ii.html, Accessed April 24, 2020.

[16] See "The private Museum (Kunstkammer) of Rudolf II," by René Zandbergen, 2018, http:// www.voynich.nu/extra/inventory.html, Accessed April 24, 2020.

[17] See "The Art, Science sand Lechery of Rudolf II," by Jane Perlez, The New York Times, June 4, 1997, https://www.nytimes.com/1997/06/04/arts/the-art-science-and-lechery-of-rudolf-ii.html. Accessed April 24, 2020. See Peter H. Wilson, *Europe's Tragedy: A New History of the Thirty Years War, Penguin,* London, 2010. See also, "Researchers Catalogue the Grisly Deaths of Soldiers in then Thirty Years' War," by Jason Daley, Smithsonian Magazine, June 6, 2017, https://www. smithsonianmag.com/smart-news/researchers-catalogue-grisly-deaths-soldiers-thirty-years-war-180963531/, [Imagery of newly uncovered mass graves.] Accessed April 25, 2020.

in the night to Antwerp. By turns, the masterpieces, many of them, would transit through Europe's air of exhausted warring aftermath into the collection of Phillippe, the Duke of Orléans. By the Duke's death in 1723, his was considered second only to Rudolf's former collection, the finest in the western world. But alas, like everything else that had lived and breathed in the castle in Prague, the masterpieces would be scattered to the four corners, particularly after the French Revolution, when great masters were secreted away in fire-sales and the lives of all those animals who had once been captured and shipped, still breathing or stuffed, would fade into the realm of romanticism. For one, the Dodo, her time was up.[18]

## 59.2   The Minutiae of Suffering

The seventeenth century leaves far less an intricate record of human-bird encounters than the eighteenth and nineteenth centuries. Linnaeus, of course, has his diverse, global taxonomic sources of information. The 10th edition of his monumental *Systema Natura* (1758) would list 554 species, 63 genera. But by the time of Darwin, natural history observations had become meticulous and rapacious to the point of often horrifying delectation, where the boundaries between science and the salacious became blurred. One thing is for certain: we know, for example, far more grim detail as to what precisely would happen to the Great Auk (*Alca impennis*), extinct very soon after 1844—that century's ornithological icon for the tragic fate of so many birds—than we do the Dodo. But reading, for example, Errol Fuller's tragic accounts (and his own painting of, the Auk's "Last Stand") should be enough to impart nightmares to anyone, and convey sufficient machinations so as to allow us to draw our own picture of what became of the even larger, more docile Dodo.[19] By the early twentieth century, we are left with numerous photographs of Martha, the last Passenger Pigeon, going extinct in a cage in a zoo.[20] The seventeenth century leaves us images of the Dodo thanks to Savery.

Today, with at least 18,000 bird species identified, another 2000 species genetically extrapolated, the ratios, percentages, and threatened and endangered species tallies are beyond anything the first three major chroniclers of bird species—Pliny,

---

[18] See William Buchanan, *Memoirs of Painting, with a Chronological History of the Importation of Pictures of Great Masters into England by the Great Artists since the French Revolution*, Ackermann, London, 1824. See also, *Catalogue des tableaux flamands du cabinet de feu S.A.R. Mgr le duc d'Orléans, noted by Louis Courajod, Le livre-journal de Laurent Duvaux* Paris, 1873. See also, Nicholas Penny, National Gallery Catalogues (new series): *The Sixteenth Century Italian Paintings, Volume II, Venice 1540–1600*, National Gallery Publications Ltd, London, UK, 2008.

[19] See Errol Fuller, *Extinct Birds*, Oxford University Press, Oxford UK, pp. 156–163.

[20] See also, https://www.si.edu/object/martha-passenger-pigeon:siris_sic_11640 See the short, recreated and heartbreaking, "From Billions to None," https://www.youtube.com/watch?v=MdFC7Q fuVTA.

Aristotle (*History of Animals, Avium praecipuarum ... Historia*) and then, in 1544, Doctor William Turner, who described over 130 species—could have imagined.[21]

All such avifauna annals are lodged in a painfully unmediated purgatory between the cage, pet stores, illegal animal markets, the aviary, the case of specimens, thousands of colorful pages in books, and actual living habitat. The fate of the Dodo in our imaginations and, hence, likely ethical activism going forward, inextricably involves what Roeland Savery saw, or did not see (live), but certainly, and repeatedly felt deeply compelled to paint. While he may not be the household name that Lewis Carroll commands, Savery's impact on the ecological aesthetics of zoology has framed the narrative since the European Renaissance.

Consider Chapter II, Alice, neck-deep in her own "pool of tears," suddenly confronted with a teeming confusion of birds caught out in that very maelstrom: "a Duck and a Dodo, a Lory and an Eaglet, and several other curious creatures"[22] (Fig. 59.4).

With the patronage of Rudolf's impressive finance capacity, Renaissance painters had risen to ornithological prominence that would deliver the first imagery of Dodos and their kindred to a contemporary audience and most assuredly to posterity. Savery's equally prolific contemporary, Jan Brueghel the Elder's "An Allegory of Air" (c. 1611) certainly marked one of the great moments in the history of humanity's love affair with birds. Brueghel's fellow artists signaled their own equally passionate devotions, from Frans Snyders, Jan van Kessel the Elder, to Carl Wilhelm de Hamilton and ultimately, in Roelandt Savery himself, who lives on in art history as the unwitting champion of a bird that would become the poster child for all extinctions.[23]

---

[21] See *Turner On Birds: A Short And Succinct History Of The Principal Birds Noticed By Pliny And Aristotle*, First Published By Doctor William Turner, 1544, Edited With Introduction, Translation, Notes, And Appendix, by A. H. Evans, Cambridge University Press, Cambridge, UK, 1903. For current revision, see George F. Barrowclough, Joel Cracraft, John Klicka, Robert M. Zink. *How Many Kinds of Birds Are There and Why Does It Matter?* PLOS ONE, 2016; 11 (11): e0166307 DOI: https://doi.org/10.1371/journal.pone.0166307; "New Study Doubles the Estimate of Bird Species in the World," American Museum of Natural History Press release, December 12, 2016, https://www.amnh.org/about/press-center/new-study-doubles-the-estimate-of-bird-species-in-the-world.

[22] *Alice's Adventures In Wonderland*, by Lewis Carroll, With 42 Illustrations by John Tenniel, Macmillan and Co.. Limited, 1898, p. 28. See the fascinating essay, "Establishing extinction dates – the curious case of the Dodo Raphus cucullatus and the Red Hen Aphanapteryx bonasia," by Anthony S. Cheke, in Ibis, International Journal of Avian Science, 19 January 2006, https://doi.org/10.1111/j.1474-919X.200600478.x    http://onlinelibrary.wiley.com/doi/org/10.1111/j.1474-919X.200600478.x; See also, *Return of the Crazy Bird – The Sad, Strange Tale Of The Dodo*, by Clara Pinto-Correia, who writes, "The dodo didn't have a clue when it came to fleeing hungry sailors who enjoyed feasting on its abundant meat. These sailors named it doudo, or crazy.... Copernicus Books, Springer Verlag New York, 2003. See also, *Lost Land of the Dodo: An Ecological History of Mauritius, Réunion & Rodrigues*, by Anthony S. Cheke and Julian Pender Hume, A&C Black, London 2008, https://books.google.com/books/about/Lost_Land_of_the_Dodo.html?id=RUjCAwAAQBAJ, Accessed April 24, 2020.

[23] For a superb general survey of great ornithological masterpieces, beginning with the Paleolithic, and with Egyptian hieroglyphics, and concluding with a section devoted to the "The Modern Bird," selections by Paul Klee, Lucian Freud and others see *The Bird in Art*, by Caroline Buger, Merrell Publishers, London/New York, 2012.

**Fig. 59.4** "And a Long Tale," Alice Meeting the Dodo, by John Tenniel, from *Alice's Adventures In Wonderland*, by Lewis Carroll, from the 1898 Edition, Macmillan And Co., London, p. 35. (Private Collection). (Photo © M.C. Tobias)

In her famed *The Language of Birds*[24] Mrs. G. Spratt had described that there was no bird the ancient Greeks did not esteem as a deity. On the island of Rhodes, the swallow was famed for encouraging the Spring. In their comprehensive work, *Birds and People*, Mark Cocker and photographer David Tipling, along with 650 contributors in 81 countries, profoundly underscore the poignancy of the relation between people and birds, including one great image of a crane who "serenades the

---

[24] Saunder and Otley, London 1837.

return of spring."[25] While the apotheosis of the mechanics of flight as enshrined in avifauna presents little ambiguity in terms of human admiration, those primeval love affairs are more than tainted by a paradox that reads of "13% of all bird species (roughly 1,300 prior to global re-assessments of taxa that almost doubles the count) that are presently threatened with extinction; "9% near threatened" and some "78%" of all populations of birds in decline.[26]

Some of this category of known threat was well recognized by scientists in the mid-nineteenth century. In 1848, nearly two centuries after the Dodo had succumbed to human massacre, H. E. Strickland and A. G. Melville published their work, *The Dodo And Its Kindred*.[27] Its most charismatic color frontispiece facsimile of Roelant Savery's 1626 painting of a Dodo (*Raphus cucullatus*) reveals the bird's most gentle nature, all the most pathetic given its unilateral demise. Art historians debate whether Savery started painting the Dodo in 1604, when he had arrived at Rudolf's castle, or not until 1611 (Fig. 59.5).

Did Savery paint the bird while it still lived? It does matter, in every way. But reportage of the time, anecdotal evidence, inventories, and various notated itineraries evoke enormous, if uncertain historical minutiae. There is no question that Savery painted at least a dozen versions of the great bird which, in adulthood, stood over a meter in height and weighed nearly 40 pounds, as can be calculated from anecdote, from the 1634 sketch by Sir Thomas Herbert in which a broad-billed parrot, red rail, and dodo are drawn side by side,[28] but also induced from the subfossil materials discovered in the Mare aux Songes swamplands of the Mauritius itself where, by 1662, the real Dodo had gone extinct.[29]

Passionate erudition has targeted the Savery/Dodo connections. For example, the last possible location of Rudolf II's specimen, or one of them, may have turned up

---

[25] See *Birds and People*, by Mark Cocker, Photographs by David Tipling. Jonathan Cape, London, 2013, p. 186.

[26] See IUCN 2016. The IUCN Red List of Threatened Species.Version 2016-3: www.iucnredlist.org/pdflink.115131842.

[27] ...*Or the History, Affinities, and Osteology of the Dodo, Solitaire, and Other Extinct Birds of the Islands Mauritius, Rodriguez, and Bourbon*, Reeve, Benham, and Reeve, London, 1848.

[28] See Herbert's *Travels*, p. 347, WellCome Library, https://wellcomelibrary.org/item/b20663705#?m=0&cv=358&c=0&s=0&z=-0.6843%2C-0.079%2C2.3685%2C1.5801, Accessed April 25, 2020.

[29] See "The history of the Dodo Raphus cucullatus and the penguin of Mauritius," by Julian P. Hume, Historical Biology, 2006; 18(2): 65–89, Taylor & Francis, http://julianhume.co.uk/wp-content/uploads/2010/07/History-of-the-dodo-Hume.pdf. See also, *Roeland Savery 1576–1639, Museum Voor Schnone Kunsten*, Gent, 10 April – 13 Juni 1954, Image #s 75 and 107; and see *Roelant Savery, Die Gemälde Mit Kritischem Oeuvrekatalog*, by Kurt J. Müllenmeister, Luca Verlag Freren, Düsseldorf, 1998, "Landschaft mit Vögeln (Wiener Dodo) Kupfer 42 x 57, Roelandt Savery FE, 1628," p. 268; See also, *Roelandt Savery, 1576–1639*, Filippe De Potter, Isabelle De Jaegere, Olga Kotkova, Stefan Bartilla and Joaneath Spicer, Published by Gent Snoeck, 2010; and *L'odyssée Des Animaux, Les Peintres Animaliers Flamands Du XVIIIE Siècle*, Sandrine Vézilier-Dussart, Snoeck, Gent 2016. See also A. S. Cheke, "The legacy of the dodo – conservation in Mauritius," Oryx, 21 (1): 29–36, 1987, https://doi.org/10.1017/S0030605300020457; Anthony S. Cheke, "The Dodo's last island," 2004, Royal Society of Arts and Sciences of Mauritius.

**Fig. 59.5** Roelan(d)t Savery's 1626 Dodo, as reproduced in *The Dodo And Its Kindred*, by H.E. Strickland and A.G. Melville, Reeve, Benham, and Reeve, London, 1848. (Private Collection). (Photo © M.C. Tobias)

in the National Museum in Prague which "possesses 34 specimens of 14 extinct and nearly extinct species of birds" including "an upper jaw" and "three leg bones" of a dodo from a marsh in the Mauritius; a "mandible," a "lower beak,"[30] obtained at an unknown date from "Reálka na Smíchové a secondary school in Prague-Smichov, Czech Republic."[31] Unlike the Greak Auk, that utilized various North Atlantic islands, such as St. Kilda, the Dodo appears to have largely been restricted to the

---

[30] "P610.1111/j.1474-919X.200600478.x-002910, mount, no date or locality.

[31] See "Extinct and nearly extinct birds in the collections of the National Museum, Prague, Czech Republic," by Jiří Mlíkovský, Department of Zoology, National Museum (NMP), Journal of the National Museum (Prague), Natural History Series Vol. 181 (9): 95–123. © Národní museum, Praha, Czech Republic 2012, https://www.researchgate.net/profile/Jiri_Mlikovsky2/publication/2724900549_Extinct_and_nearly_extinct_birds_in_the_collections_of_the_National_Museum_Prague_Czech_Republic/links/54c61ada0cf277664ff2d3ea.pdf; See also, 338 Bull. B.O.C. 2003 123A Extinct and endangered ("E&E") birds: a proposed list for collection catalogues by M. P. Adams, J.H.Cooper and N.J. Collar, http://www.scricciolo.com/extinct_endangered_list.pdf. See also http://www.thriftbooks.com/w/lost-animals-extinctions-and-the-photographic-record/9415316/. And https://books.google.com/books/about/Extinct_Birds.html?id=z7RIAkGw0-UC. See also, Anwar Janoo, "Discovery of isolated dodo bones [Raphus cucullatus (L.) Aves, Columbiformes] from Mauritius cave shelters highlights human predation, with a comment on the status of the family Raphidae, Wetmore, 1930. Annales de Paléontologie 91: pp. 167–180. See also, David L. Roberts and Andrew R. Solow, "Flightless birds: When did the dodo become extinct?" Nature 425 (6964), 2003.

Mauritius and proximate other few islands.[32] Tim Flannery and Peter Schouten point out that "Just one significant subtropical island archipelago retained its full fauna until after 1500 – the Mascarene Islands." And that "the last complete dodo specimen was held by the Ashmolean Museum at Oxford. In 1755, the ageing mounted skin was ordered out for destruction, but somebody had the foresight to cut off the head and right foot before consigning the rest to the flames."[33]

Supposedly, a live dodo was seen and painted in approximately 1600 by Joris Hoefnagel at the menagerie at Ebersdorf.[34] But the underlying details are infinitely more complicated, as Jolyon C. Parish's extraordinary forensics reveal, providing what, to date, is the most thorough, depressing elucidation of the earliest one or two, possibly three Dodo's to arrive in Europe, in his book *The Dodo and the Solitaire: A Natural History*. It reads like a murder mystery because it is.[35] Parish leads us backward from an auction in 1782 of Prague Castle in which the "Prague dodo, or at least its beak" may have been for sale. He analyzes possibly correlative data on the current dodo bones at the Národní Muzeum in Prague, and concludes, upon an abundance of data, that "If Rudolf received the dodo alive it probably lived in one of the vivaria that he established at Prague."[36] The Emperor's bird(s)—not just the Dodo, but also two legendary cassowaries and a Mauritius red rail—might have likely lived together, considering the scope of Rudolf's zoological compulsions and an oisellerie Oiseaux he had built in 1601 adjoining the main garden to the north of Prague. All vague, to be sure.[37] But just how and from where the Dodo arrived is a detective story with no resolution. A Dutch, Spanish, Portuguese ship? A gift sent from Amsterdam? An itinerary from one aviary at the gardens of Hradcany Castle to Ebersdorf? In transit from the Indian Ocean to Holland and on to Schloß Neugebäu? Or "that it lived in Rudolf's menagerie c. 1604–1605 and 'died a Dodo's death around 1607' (Van Wissen 1995, 64); that it had arrived at Prague in 1604 (Valledor de Lozoya 2002a) and that it had probably been taken alive to Prague sometime between 1599 and 1609 (Hendrix 1997) – also cannot be proved with certainty," Parish writes, citing an enormous range of references, however suppositional.[38]

Making this even more problematic is Parish's analysis of two other pieces of Dodo bone, one referred to as "Clusius's bird" (Carolus Clusius [1526–1609] was the prefect of Maximillian II's imperial medical garden in Vienna)[39] the other, the painter "[Dirk De Quade]Van Ravesteyn's bird" which Parish suggested was

---

[32] See *A Gap in Nature – Discovering The World's Extinct Animals*, by Tim Flannery and Peter Schouten, Text Publishing, Melbourne, Australia 2001, pp. 4–5.

[33] ibid., p. 4.

[34] *Extinct Birds*, Errol Fuller, Oxford University Press, Oxford UK 2000, p. 146.

[35] Indiana University Press, Bloomington, Indiana, 2012.

[36] ibid, pp. 186–189.

[37] ibid. 189.

[38] ibid. p. 189.

[39] See https://www.timelineindex.com/content/view/3185, Accessed April 25, 2020.

"undoubtedly painted from a stuffed subject." It appears as a plate in a two-volume book known as the *Museum Kaiser Rudolfs II*. And, as history would further complicate things, "The majority of the subjects in the volumes were stuffed specimens."[40] The "Prague beak" poses a problem of who saw which bird; was it alive, was it dead, what are the anatomical inconsistencies from the Nicolas Visscher engraving, the Joris Hoefnagel painting ("identified as a dodo" in a National Gallery of Art, Washington D.C. catalog), the Jan Brueghel the Elder version in his painting of 1611, "Air," which, explains Parish, "is very similar to the Van Ravesteyn bird, indicating that it was either painted with reference to this source (Ziswiler 1996) or from the Prague specimen itself."[41] But the problem with Savery's claim that he had painted it from "real life" is that the dates don't appear to work out. "Indeed," says Paris, "the Prague dodo may have been already stuffed by the time Savery arrived in Prague."[42] This contention was re-affirmed in 2011 by Judith Magee, curator of an exhibition at London's Natural History Museum entitled "Images of Nature" in which new paleontological bearings were brought to bear on the imprecise overlapping osteology of the dodo bones in the museum—published in 1866 by the first superintendent, Dodo-discipline, Professor Richard Owen—and the Savery painting of 1626. Van Ravesteyn, born in 1565, had died in 1620.[43]

As Fuller explains in his own, vastly illustrated and moribund book, *Extinct Birds*, *Raphus cucullatus*, the most famous of all doomed avians, is in the same Columbiformes Order as that of the other most celebrated case of doomsday for a species, the Passenger pigeon (*Ectopistes migratorius*). In addition, at least eight known other doves and pigeons related to them have also gone extinct.[44] Fuller tells the whole saga[45] of a bird, which is synonymous with all those who have passed forever before us, most notably the dinosaurs. According to Fuller, the first description of a living Dodo on the Mauritius itself was conveyed by the Dutch naval officer, Jacob Corneliszoon van Neck (1564–1638) and published in the Netherlands in 1601. The alleged last living account was provided by one mariner and scientist Benjamin Harry around 1681 while he was busy measuring the angle of earth's magnetic field out in the Indian Ocean. He evidently heard about, or actually witnessed the continuing consumption of dodos by other sailors during a short time that his ship was docked at the Mauritius island. Because he noted its flesh was tough, he apparently attempted himself to make a meal of one, though that is uncertain.[46]

---

[40] See ibid. pp. 180, 186.

[41] ibid., p. 183.

[42] ibid. pp. 182, 189.

[43] See "A new lease of life for the dead dodo," by Robin McKie, January 15, 2011, The Guardian, https://www.theguardian.com/culture/2011/jan/16/dodo-painting-natural-history-museum, Accessed April 22, 2020.

[44] ibid., Fuller, p. 170.

[45] ibid., pp. 194–203.

[46] See NBC News, "When did the dodo go extinct? Maybe later than we thought," by Douglas Main, October 9, 2013, https://www.nbcnews.com/sciencemain/when-did-dodo-go-extinct-maybe-later-we-thought-8C11361418, Accessed April 22, 2020.

**Fig. 59.6** The Dodo, Illustrated by Frederick William Frohawk, Plate 24 of Lord Walter Rothschild's *Extinct Birds*, A. Chris Fowler, Printer, London, 1907. "… taken from the picture by Roelandt Savery in Berlin, but the wings, tail and bill have been altered, partly from Pierre Withoos' picture of the Bourbon Dodo, and partly from anatomical examination." p. 174. (Private Collection). (Photo © M.C. Tobias)

During the last 100 years of its much beleaguered persevering, live specimens were brought back to Europe, India, possibly Japan, according to Fuller,[47] who cites much data based upon all the prolific evidence, tattered remains, and life-like portraiture by at least 8 Renaissance painters, including Savery, Gillis Claesz D'Hondecoeter, Adrian van der Venne, Joris Hoefnaegel, Ustad Mansur of the court of the Moghul Emperor Jahangir,[48] Cornelius Saftleven and T. and J. J. de Bry.[49] Erol Fuller's research on extinct birds, in the tradition of Walter Rothschild, is among the most extensive of any naturalist, and it is little surprise, then, that he refers to the Dodo (the national bird of the Mauritius, ironically) as "one of the most fantastic birds ever to have lived." His chronicle of haphazard comments about this bird included a "shape and rareness [that] may antagonize the Phoenix

---

[47] ibid., p. 200.

[48] See "Dodos in Mughal court," by Onu Tareq, September 15, 2019, in which is discussed a report of "two Dodos brought to the city of Surat (Gujarat)" by sailors between 1628 and 1633. The Business Standard, https://tbsnews.net/feature/travel/dodos-mughal-court, Accessed April 22, 2020.

[49] *The De Bry collection of voyages (1590–1634): editorial strategy and the representations of the overseas world*, Michiel van Groesen, The University of Amsterdam, the Netherlands 2007, https://pure.uva.nl/ws/files/3960216/47113_Groesen_compleet.pdf.

**Fig. 59.7** "Landschaft mit Vogein," 1628, by Roelant Savery, Kunsthistorisches Museum, Wein, GG1082. (Public Domain)

of Arabia," eyes "like to Diamonds," "covered with Doune, having little wings hanging like short sleeves" and "a great fowle somewhat bigger than the largest Turkey cock"[50] (Fig. 59.6).

## 59.3   The Walter Rothschild Factor

Lord Walter Rothschild's bibliographic review of extinct birds, from 1580 to 1907, encompasses well over 150 major sources. His extensive plates and sketches of the Dodo in his *Extinct Birds* publication[51] reveal a gleeful heavy physiology—between a Wandering Albatross and its precise, flightless opposite in a gentle, earthbound being with a Falstaff-like mien that simply had not the least apparition of merciless human predators. Hers was, before man, a life in paradise. Writes, Rothschild, "The total inability of flight, the heavy slow gait, and the utter fearlessness from long immunity from enemies, led to a continental slaughter." Rothschild lists 14 paintings of the Dodo, from Vienna, Berlin, the Duke of Northumberland's Sion House,

---

[50] op. cit., Fuller, pp. 194–203.
[51] London, 1907, pp. 172–174, Plates 24, 24a–c.

London, Dresden, Oxford, Haarlem, Emden, Stuttgart, the Haag and Pommersfelden, Bavaria. He also settles the matter of ornithological debate—was the Dodo a Struthion (Linnaeus called it *Struthio cucullatus* = cucullated, or hooded), some kind of vulture, or a giant, specialized pigeon? And he adapts his drawings from the many earlier renditions produced in 1601, 1606, 1605, 1634, 1635, 1658, 1752, 1757, 1758, and 1766, fully apprised, of course, of the H.E. Strickland and A.G. Melville plates throughout their definitive work, *The Dodo And Its Kindred; Or The History, Affinities, And Osteology Of The Dodo, Solitaire, And Other Extinct Birds Of The Islands Mauritius, Rodriguez, And Bourbon*, (1848).[52]

Ultimately, we are left with an odd friendship—that of a Flemish painter, a Holy Roman Emperor, and a bird—whose respective coordinates in natural history are emblematic of a huge fly in a rancid ointment; of that flaw bound by human nature residing behind castle walls. Its secrets revealed, we are left with a magnificent, if dolorous chaos. And the most problematic of paradoxes: humanity's so-called *love affair* with birds[53] (Fig. 59.7).

---

[52] Reeve, Benham, and Reeve, London 1848. It should be pointed out that in *Return of the Crazy Bird – The Sad, Strange Tale Of The Dodo*, by Clara Pinto-Correia suggests that by the mid-1800s (the 1848 Strickland book, *The Dodo and Its Kindred*, there were only "five" "dodo oil paintings" known to exist. Copernicus Books, Imprint of Springer Verlag New York, 2003, p. 151.

[53] The most complete record, to date, of Roelant Savery's oeuvre is that of Kurt J. Müllenmeister's *Roelant Savery, Kortrijk 1576–1639 Utrecht, Hofmaler Kaiser Rudolf II. In Prag, Die Gemälde Mit Kritischem Oeuvrekatalog*, Luca Verlag Freren 1988. Additionally, there is the Praag, Národni galerie, Kortrijk, Broelmuseum exhibition catalogue from 8 December 2010–10 March 2011, and 21 April 2011 – 11 September 2011, *Roelandt Savery 1576–1639*. Broelmuseum Kortrijk, Snoeck. In addition, *Roelandt Savery 1576–1639*, Museum Voor Schone Kunsten, Gent, 10 April – 13 Juni 1954; and *Roeland Savery, Kataloog*, Herdenking, Tentoonstelling Georganizeerd Naar Aanleiding Van De 400ste Verjaardag Van De Geboorte Van De Kunstschilder, 25 April 7 Juni 1976, Stedelijk Museum Voor Schone Kunsten, Kortrijk.

# Chapter 60
# Human Cruelty and SARS-CoV-2

## 60.1 Human Diet in History

Throughout much of our lives, we have tried to rescue animals still breathing from their fetid cages and from the indifferent, usually impoverished human captors, in open animal markets from the West African capital of Mali, Bamako, to roadsides in Kenya, animal hells in Central India, northern Russia, western Saudi Arabia, across Yemen and Indonesia, to remote villages in China. The syndrome of animal cruelty and animal consumption has systematically gotten our species to its current zoonotic crisis, Covid-19 (SARS-CoV-2) (Figs. 60.1, 60.2, 60.3).

A brief prelude to the 2020 pandemic: Romans, particularly those who sought to show off their wealth or power, exercised maniacal cruelty to animals. This was carried out every day, for centuries, whether as animals were slaughtered for human food, chained in captivity, relentlessly pursued by veritable phalanxes of mercenary sport hunters, being killed and buried with humans as a petition to the gods (e.g., mass animal sacrifice, a practice borrowed from 2000 years of Egyptian tradition), and the constant show of force, so-called, by one insecure Emperor after another glorifying in the public massacre of thousands of animals, typically imported from North Africa (e.g., lions, elephants, giraffe).[1] Roman diets included everything from dormice to ostrich.[2]

This Western grotesque relationship with Others is the most seminal throughstory in our history. It touches every aspect of our past, and our present. Great inflection points can be highlighted—like the dietary, human demographic and economic shifts accompanying the Crusades, the Black Plague, or the Hundred Years' War.

---

[1] See Dr. Iain Ferris' *Cave Canem: Animals and Roman Society*, Amberley Publishing, Gloucestershire, UK 2018.

[2] See "Dormice, ostrich meat and fresh fish: the surprising foods eaten in ancient Rome," by Erica Rowan, HistoryExtra, BBC History Magazine, March 19, 2015, https://www.historyextra.com/period/roman/dormice-ostrich-meat-and-fresh-fish-the-surprising-foods-eaten-in-ancient-rome/.

© Springer Nature Switzerland AG 2021
M. C. Tobias, J. G. Morrison, *On the Nature of Ecological Paradox*,
https://doi.org/10.1007/978-3-030-64526-7_60

**Fig. 60.1** A joyous pig roaming free. (Photo © M.C. Tobias)

**Fig. 60.2** Pigs inhumanely transported to slaughter in China. (Photo © M.C. Tobias)

But the out-of-control killing of animals never ceased, despite human, historic fluctuations in everyday life and death; as well as the abundance of psychiatric data
unambiguously telling us that those who kill animals are more likely to kill human

**Fig. 60.3** Pigs served whole at a banquet at West Lake, Hangzhou, China. (Photo © M.C. Tobias)

animals, as well. The great unknown in that psychological data pool concerns the behavior of those who don't *do* the killing, but simply *eat* the remains.

According to food history consultant, Dr. Annie Gray, "the range of animals and birds consumed in Georgian Britain was astounding – nothing that moved was safe from the cooking pot." From beavers to the hedgehog, all was fair game. In addition, there was "extravagant medieval culinary diversion: fantastic beasts" including "the cockentrice" which "comprised half a pig sewn to half of a capon (a castrated and fattened chicken)."[3]

Japanese have long had a dietary tradition of Ikizukuri, meaning prepared alive and, in the case of octopus, still moving on the plate as the diner consumes it. The varieties of torture meted out to marine invertebrates and vertebrates, to frogs, snakes, enchinoderms like sea-urchins, decapod crustaceans (shrimp and prawns) and to insects, including the live larva of numerous grubs, continue unabated throughout the world[4] (Fig. 60.4).

[3] "Pig-chickens, beavers' tails and turtle soup: 8 weird foods through history," Dr. Annie Gray, HistoryExtra, BBC History Magazine, August 27, 2015, https://www.historyextra.com/period/roman/pig-chickens-beavers-tails-and-turtle-soup-8-weird-foods-through-history/.

[4] See "11 Foods Eaten Alive That May Shock You," by Brent Furdyk, Food Network, October 19, 2017, https://www.foodnetwork.ca/fun-with-food/photos/foods-eaten-alive-that-may-shock-you/#!irish-garlic-oysters; See also, "It's time to ban eating live creatures on I'm A Celebrity, Get Me Out Of Here," by Alice Wright, METRO, November 22, 2017, https://metro.co.uk/2017/11/22/its-time-to-ban-eating-live-creatures-on-im-a-celebrity-get-me-out-of-here-7080860/; and "10 Animals That People Eat Alive," by Simon Griffin, ListVerse, June 21, 2018, https://listverse.com/2013/03/06/10-animals-that-people-eat-alive/. Accessed April 1, 2020; and "Anthony Bourdain took people on a journey of the world – often by eating really strange things," by Michael Collett, ABC News, Australian Broadcasting, June 9, 2018, https://www.abc.net.au/news/2018-06-09/anthony-bourdain-weird-things-hes-eaten/9852966, Accessed April 1, 2020.

**Fig. 60.4**  Butcher's shop in Saudi Arabia. (Photo © M.C. Tobias)

## 60.2  Wet Markets

This brings us to the current situation with "wet markets." National Public Radio journalist Jason Beaubien reported on the Tai Po wet market in Hong Kong, writing, "Live fish in open tubs splash water all over the floor. The countertops of the stalls are red with blood as fish are gutted and filleted right in front of the customer's eyes. Live turtles and crustaceans climb over each other in boxes. Melting ice adds to the slush on the floor. There's lots of water, blood, fish scales, and chicken guts." The market also sells "Himalayan palm civets, raccoon dogs, wild boars and cobras."[5] The reality of such wet markets had fully erupted early in 2019. Noted animal rights philosophers Peter Singer and Paola Cavalieri wrote, "At China's wet markets, many different animals are sold and killed to be eaten: wolf cubs, snakes, turtles, guinea pigs, rats, otters, badgers, and civets. Similar markets exist in many Asian countries, including Japan, Vietnam, and the Philippines."[6] Zoologist Juliet Gellatley declared, with reference to the Wuhan live-animal mar-

---

[5] Why They're Called 'Wet Markets' – And What Health Risks They Might Pose," Jason Beaubien, January 31, 2020, NPR, https://www.npr.org/sections/goatsandsoda/2020/01/31/800975655/why-theyre-called-wet-markets-and-what-health-risks-they-might-pose.

[6] "The Two Dark Sides of COVID-19," Peter Singer, Paola Cavalieri, March 2, 2020, https://www.project-syndicate.org/commentary/wet-markets-breeding-ground-for-new-coronavirus-by-peter-singer-and-paola-cavalieri-2020-03.

ket, "Over 100 different animals are sold here, including wolf pups, civet cats, poultry and snakes." And she continued, "These animals are kept in cramped, dirty conditions, with direct contact with humans. These markets are referred to as 'wet markets' – so-called because animals are often slaughtered directly in front of customers."[7]

The live-customer aspect is no surprise in the United States, where buyers and sellers of "livestock" (*live* being the operative sales pitch on both sides of the eventual fork and knife) have been meeting for hundreds of years to transact their merciless exchanges. Which is why, for example, a study of increased potential for transmission of "swine-origin IAVs" (Influenza A viruses) at open live-animal markets was conducted in just one of those symptomatic open-market transactional states: Minnesota, by researchers in 2015.[8]

From the mid-Western United States to nearly every other continent, there is plentiful data (though severely under-documented or tallied) to make clear that "unregulated and usually filthy markets are found all over Asia and Africa," for example.[9]

During the first two weeks of February 2020, Chinese police raided homes across China and arrested over 700 people "for breaking the temporary ban on catching, selling or eating wild animals"—among the "40,000 animals" taken during the crackdown, were "squirrels, weasels and boars." But this did not begin to account for the multitudes of snakes in jars, or those merchants who were legally able to sell "donkey, dog, deer, crocodile and other meat." And, as a retired zoologist from the Chinese Academy of Sciences, Wang Song, declared, "In many people's eyes, animals are living for man, not sharing the earth with man."[10]

But whether thousands of years of Chinese cultural traditions will be irrevocably altered by one outbreak appears unlikely. Although on February 24, 2020, the 16th Meeting of the Standing Committee of the 13th National People's Congress put into effect an edict "Comprehensively Prohibiting the Illegal Trade of Wild Animals, Eliminating the Bad Habits of Wild Animal Consumption, and Protecting the Health

---

[7] "Eating Animals Will Be the Death of Us," by Juliet Gellatley, February 4, 2020, The Ecologist, Common Dreams, https://www.commondreams.org/views/2020/02/04/eating-animals-will-be-death-us, Accessed April 1, 2020.

[8] "Live Animal Markets in Minnesota: A Potential Source for Emergence of Novel Influenza A Viruses and Interspecies Transmission," by Mary J. Choi, Montserrat Torremorell, Jeff B. Bender, Kirk Smith, David Boxrud, et.al., Clinical Infectious Diseases, Volume 61, Issue 9, 1 November 2015, pp. 1355–1362, https://doi.org/10.1093/cid/civ618, https://academic.oup.com/cid/article/61/9/1355/432409

[9] See "Inside the horrific, inhumane animal markets behind pandemics like coronavirus," by Paula Froelich, The New York Post, MarketWatch, January 25, 2020, https://www.marketwatch.com/story/inside-the-horrific-inhumane-animal-markets-behind-pandemics-like-coronavirus-2020-01-25.

[10] "'Animals life for man': China's appetite for wildlife likely to survive virus," by Farah Master and Sophie Yu, Reuters, World News, February 16, 2020, https://www.reuters.com/article/us-china-health-wildlife-idUSKBN20A0RK.

and Safety of the People." Time will tell.[11] On April 2, 2020, Chinese authorities ordered a ban, to take effect one month later, on the eating of dogs and cats in at least one city, Shenzhen. But across the vast reaches of China, at least ten million other dogs and four million cats were likely slaughtered in 2020, often boiled alive. At the same time, the grim torture of bears kept alive to perpetually drain them of their bile for its supposedly medicinal ingredient, ursodeoxycholic acid[12] (like kept baby lambs for veal chops, geese, or ducks for Foie gras, an endless litany of other tortures) continues. By June 2020, authorities in Beijing saw another tide of Covid-19 outbreak, and shut down more animal markets whose many proprietors had not yet followed the new rulings.

In a discussion with Dr. Peter J. Li, an Associate Professor of East Asian Politics at the University of Houston-Downtown, and China Policy Specialist of Humane Society International, back in late 2012, he described how the 1988 Wildlife Protection Law ("WPL"), China's most comprehensive such set of mandates, was severely flawed in that it positioned "wildlife animals as 'natural resources' to be used for human benefits. Local authorities and businesses have, however, chosen to use the WPL as their bible to justify their business operations of wildlife exploitation."[13]

Aside from the sheer horror such practices evince in most people, it is statistically unlikely that the *majority* of humanity, which continues its practices of meat and fish consumption, would find these dietary standards at all reprehensible, excepting those ethicists and other activists who write such papers as "The Moral Standing Of Insects And The Ethics Of Extinction," by Jeffrey A. Lockwood. In this important essay of 1987 he posited that "The criterion of sentience which includes concepts of pain, consciousness, thought, and awareness appears to provide an intuitively satisfying, empirically approachable, philosophical basis for including a being in our moral considerations. Existing evidence indicates that insects qualify as sentient and their lives ought to be included in moral deliberations...." [and] "if there is a conflict in which all other interests are equal, an entity with even infinitesimal moral significance may determine the right course of action."[14]

Nearly 35 years ago, when Lockwood wrote this paper, it was in the midst of an intense wave in animal behavior, biomedical ethics, and general philosophical

---

[11] See WCSNewsroom, "WCS Statement and Analysis: On the Chinese Government's Decision Prohibiting Some Trade and Consumption of Wild Animals," WCS News Release, New York, February 26, 2020, https://newsroom.wcs.org/News-Releases/articleType/ArticleView/articleId/13855/WCS-Statement-and-Analysis-On-the-Chinese-Governments-Decision-Prohibiting-Some-Trade-and-Consumption-of-Wild-Animals.aspx.

[12] See BBC NEWS, "Shenzhen becomes first Chinese city to ban eating cats and dogs," 2 April 2020, n.a., https://www.bbc.com/news/world-asia-china-52131940, Accessed April 2, 2020.

[13] See "Animal Rights in China," by M.C. Tobias and J. G. Morrison, Forbes, November 2, 2012.

[14] Florida Entomologist 70(1), March 1987, pp. 70, 73. https://journals.flvc.org/flaent/article/view/58243/55922, Accessed April 1, 2020. See also, *Insect Conservation Biology – Proceedings of the Royal Entomological Society's 23rd Symposium*, Edited by A. J. A. Stewart, T. R. New and O. T. Lewis, CABI, London 2007.

and biological discussion focusing upon animal rights that encompassed such writers (all cited by Lockwood) as S.R.L. Clark, *The Moral Status of Animals* (1977), D. R. Griffin, *The Question of Animal Awareness* (1976), A. Jolly, *A New Science That Sees Animals as Conscious Beings* (1985), J. Rawls, *A Theory of Justice* (1971), T. Regan, *The Case for Animal Rights* (1983), and P. Singer, *Animal Liberation: A New Ethics for Our Treatment of Animals* (1975). But in spite of the vast number of insects and spiders destroyed during the Green Revolution, with its global celebration of chemical adulterants,[15] there was little attention paid to big moral questions as applied to the smallest among us. To find such serious and breathtaking deliberations, one would have to look back to the Jain and Buddhist annals, to Mahavira (ca. 497–425 BCE) and his disciples most notably.[16]

And today, insects and spiders are going extinct in multitudes. Whole populations are disappearing in a day, a consequence of human behavior that, at least with such apex pollinators as bees, many are hearing the clarion call. An appreciation of the eco-dynamics occurring among bacteria and viruses is far more esoteric, for the majority of us. Until our bodies suddenly tell us that something has gone terribly awry. And as we now collectively engage in "war" against the latest disease calamity in nearly every human community, the many revelations concerning zoonoses have not yet fully entered the ethicist's lexicon, perhaps because the contemplation of saving mosquitoes or cockroaches, locust or fleas is unlikely to ever gain much philosophical traction. Not as long as we remain the dominant vertebrate species on earth. Dr. Paul Ehrlich and I discussed this, in various guises, with respect to Lepidoptera (the nearly 180,000 known species of butterflies and moths of which Paul is one of the world's leading authorities) in our book *Hope on Earth: A Conversation*,[17] especially in light of the 0.92% decrease annually of flying insects. The fact that Cubans, Chinese, and others keep crickets as pets, and that the Chinese turn cockroaches into a cream thought to combat gastroenteritis (and some Russians and Germans are known to have cockroach races), has not yet engendered a generalized love affair between most humans and insects (Fig. 60.5).

---

[15] See, for example, "The Toxic Consequences of the Green Revolution," by Daniel Pepper, US New and World Report, July 7, 2008, https://www.usnews.com/news/world/articles/2008/07/07/the-toxic-consequences-of-the-green-revolution; See also the vast scientific commentary on the writings of Rachel Carson, https://www.rachelcarson.org/Bio.aspx.

[16] See *Life Force: The World of Jainism*, by M. C. Tobias, Asian Humanities Press, Fremont, California 1991. For other invertebrate-related issues of biosemiosis, legal codes, and historical anecdote – from ants and tarantulas to studies of endangered co-dependent insect/spider species issues – see *God's Country: The New Zealand Factor*, by M. C. Tobias and J. G. Morrison, A Dancing Star Foundation Book, Zorba Press, Ithaca, New York, 2011, pp. 175–177.

[17] University of Chicago Press, Chicago, Ill., 2014. See also, John L. Capinera, "Butterflies and moths". Encyclopedia of Entomology. 4 (2nd ed.). Springer, New York, 2008, pp. 626–672.

**Fig. 60.5** A free cow singingly happily, Upstate New York. (Photo © M.C. Tobias)

## 60.3   Ethics and Epidemiology

Addressing the combination of horseshoe bats and Chinese pangolins together possibly infecting humans at a live-animal market in Wuhan, China, Johnathan Epstein, the epidemiologist/veterinarian involved in tracking down the animal source of the SARS (Severe Acute Respiratory Syndrome) virus from civet cats in animal markets in Guangdong, China, in 2003, has stated that "about half of all known human pathogens are zoonotic, which means they originated in animals."[18]

In the last century, at least 70 major infectious disease outbreaks and/or continuing animal-transmitted diseases have tortured all those hundreds of millions of victims concerned, in a systemic, combinatorial mayhem that remains fundamentally an ecological result of human ignorance, indifference, cruelty, and trespass. Some of those diseases, and the implicated species disrupted, like the current "beast," include: Anthrax—numerous farm animal species; Bubonic plague—dozens of mammalian taxa from camels and goats to sheep and rabbits; Chagas disease—armadillos; new variant Creutzfeldt–Jacob disease, in humans, "mad cow disease"—bovines; Ebola virus—other primates; Giardiasis, Rabies and Influenza—dozens of wild and semi-wild mammals; Lyme disease—the deer/tick

---

[18] See "'This is not the bat's fault': A disease expert explains where the coronavirus likely comes from," By Brian Resnick, February 12, 2020, VOX, https://www.vox.com/science-and-health/2020/2/12/21133560/coronavirus-china-bats-pangolin-zoonotic-disease, Accessed April 1, 2020.

matrix; MERS coronavirus—camels, bats; Rickettsia—canids and rodents; and Zika virus—numerous other primates[19] (Fig. 60.5).

In every instance, it would be too simple and incomplete a mental sequence of events to suggest that these are Gaia's (the earth's) antibodies to human over-population. For one, the plague ("Black Death")—humanity's worst pandemic thus far, peaking throughout Europe and Eurasia between 1347 and 1351—wiped out between 30% and 60% of the afflicted regions' populations, or as many as 200 million people.[20] But by the time William Shakespeare was born (1564) the human population had made a total comeback, plus some. The underlying story fueling every zoonotic eruption has been the same: humans encroaching on wild habitat, capturing, poaching, consuming, or transporting other species; commingling in usually unsuspecting ways with a wild animal's urine or feces, or the fleas and diseases it may be carrying-unnatural intrusiveness, in other words.

With the human species on a single-minded path toward 9.5, 10, 11, or even 12 billion individuals, depending upon which predictive model one cares to believe in, there is little if any persuasive empirical evidence, thus far, to suggest demographic amelioration. Indeed, our continuing population explosion (83+ million humans added to the planet each year), and that magnitude of human biomass in proportion to the fast-waning portion of all other wildlife on the planet, is a ratio painfully out of balance, with no promise for improvement.

In our "Animal Rights In China" conversation for Forbes with Dr. Li, he told us, "Never in its 5,000-year history did China ever raise and keep hundreds of millions of wildlife species in captivity as it is today." But, he added, "Today, Chinese consumers, according to a recent report eat twice as much meat as those in the United States. This is not surprising. China surpassed the U.S. as the world's biggest meat

---

[19] World Health Organization, "Zoonoses," https://www.who.int/zoonoses/en/. Accessed April 1, 2020. For one prime example, see the PBS feature film documentary, "Mad Cowboy," the story of Montana rancher turned vegan, Howard Lyman, http://www.dancingstarfoundation.org/mad_cowboy.php; According to an NIH study, "Prioritizing Zoonoses for Global Health Capacity Building – Themes from One Health Zoonotic Disease Workshops in 7 Countries, 2014–2016," by Stephanie J. Salyer, Rachel Silver and Casey Barton Behravesh, "An estimated 60% of known infectious diseases and up to 75% of new or emerging infectious diseases are zoonotic in origin." Emerg Infect Dis. 2017 Dec; 23(Suppl 1); S55-S64. Doi: 10.3201/eid2313.170418, PMCID: PMC5711306, PMID: 29155664, Cent3ers for Disease Control and Prevention, Atlanta, Georgia. That data comes from Woolhouse ME, Gowtage-Sequeria S. Host range and emerging and reemerging pathogens. Emerg Infect Dis. 2005; 11:1842–7, 10.3201/eid1112.0500997; and Jones KE, Patel NG, Levy MA, Storeygard A, Balk D, Gittleman JL, et al. Global trends in emerging infectious diseases. Nature. 2008; 451;990–3.10.1038/nature06536. We don't know if cockroach tea, consumed by ancient Greeks and Egyptians, but also 19th century diners in Louisiana to fight certain ailments, acquired immunity to viral diseases like poliomyelitis, or to typhoid fever and diarrhea, or, conversely, were more prone to such ailments. See "Cockroaches – World Health Organization," https://www.google.com/search?client=firefox-b-1-d&q=diseases+spread+by+co ckroaches; See also, "How cockroaches could save lives," by Mary Colwell, BBC Magazine, November 3, 2015, https://www.bbc.com/news/magazine-34517443, Accessed July 18, 2020.

[20] Historical Estimates of World Population," United States Census, https://www.census.gov/data/tables/time-series/demo/international-programs/historical-est-worldpop.html.

producer in 1990, and the Chinese authorities have long looked to the industrialized West as the object of emulation in meat production. While Westerners greet each other by asking 'how are you,' Chinese people traditionally greeted each other by saying 'Have you eaten?'"[21] And as Dr. Li explained, this may culturally relate to the legacy amplified during the ascendancy of Mao. Said Dr. Li, "Many Chinese mainlanders (I am using this word to distinguish Chinese on the mainland from those in Taiwan and Hong Kong) are, in my opinion, possibly indifferent or insensitive to animal suffering (to varying degrees). But let me also state for the record that, again in my opinion, they themselves are not to blame. Neither is Chinese traditional culture to blame. People become indifferent or insensitive perhaps because they have been socialized to be so, particularly under Mao, when sympathy for the downtrodden, love of pets, wearing make-up, and displaying individual taste in fashion were all condemned as bourgeois and rebellious."[22]

We posed to Dr. Li, "So across China, at the local level, it's a political quagmire guaranteed to perpetuate animal suffering?" His answer was politically complicated, but, in our view, implied a version of business as usual all too familiar in the West (Fig. 60.6).

That Western experience of cruelty, both inside and outside of pandemics, has been well illustrated by Frank M. Snowden in his book *Epidemics and Society: From the Black Death to the Present.*[23] And in his descriptions of western dietary changes from the time of the Middle Ages, geographer Yi-Fu Tuan effectively narrates the grim excesses of our recent ancestors, including "the filth in the kitchen" that no doubt escaped into the piles of flesh consumed by "Plantagenet kings" or those with means whose dinner might easily contain within three courses, dozens of items "heaped high on large platters" such as "shields of boiled and pickled boar, hulled wheat boiled in milk and venison, oily stews, salted hart, pheasant, swan, capons, lampreys, perch, rabbit, mutton, baked custard and tart fruit." That was just the first course. The "second course" replicated the first. And then there was yet another course, and so on.[24] Such run-away carnivory in feudal societies, say early-fifteenth-century France, when the population was approximately 11 million, 2 million in England, would have taken heavy tolls indeed, on the biodiversity as well as on the consumers whose life expectancies were, among the nobility, between 48

---

[21] See https://www.forbes.com/sites/michaeltobias/2012/11/02/animal-rights-in-china/#4f70dd b17d57; See also, https://www.telegraph.co.uk/news/worldnews/asia/china/9605048/China-now-eats-twice-as-much-meat-as-the-United-States.html, Accessed April 1, 2020.

[22] op.cit., "Animal Rights In China," M.C.Tobias and J. G. Morrison, Forbes, https://www.forbes.com/sites/michaeltobias/2012/11/02/animal-rights-in-china/#4f70ddb17d57, Accessed April 1, 2020.

[23] Yale University Press, New Haven, Connecticut, 2019. See also the interview in The New Yorker between Snowden and Isaac Chotiner, "How Pandemics Change History," March 3, 2020. https://www.newyorker.com/news/q-and-a/how-pandemics-change-history, Accessed April 5, 2020.

[24] *Passing Strange and Wonderful – Aesthetics, Nature, And Culture,* by Yi-Fu Tuan, Island Press/Shearwater Books, Washington, D. C.,Covelo CA, 1993, pp. 47–51.

**Fig. 60.6**  Killing of bulls, illustration by Wenceslas Hollar of Virgil's "Georgics III," From *The Works of Virgil*, Translated by John Dryden, Second Edition, Printer Jacob Tonson, London, 1698. (Private Collection). (Photo © M.C. Tobias)

and 54, but in the low 30s for the poor.[25] There are over 4600 species of cockroaches, each female producing at least 800 offspring, all of whom transmit a wealth of zoonoses (they carry at least 30 potentially deadly bacterial species) and have, for some 350 million years, developed sophisticated "quorums" ("collective

[25] See "Lifespans of the European Elite, 800–1800," by Neil Cummins, The Journal of Economic History, Volume 77, Issue 2, June 2017, pp. 406–439, Cambridge Core, OI: https://doi.org/10.1017/S0022050717000468, https://www.cambridge.org/core/journals/journal-of-economic-history/article/lifespans-of-the-european-elite-8001800/BE252C4B25C4AAC29ED62D591A1675AC/core-reader?source=post_page%2D%2D%2D%2D%2D%2D%2D%2D%2D%2D%2D%2D%2D%2D%2D%2D%2D%2D%2D%2D%2D%2D%2D%2D%2D%2D%2D%2D%2D%2D%2D#.

decision making" that enables them to remain for extended periods of time in groups at prime feeding locations, e.g., inside homes and throughout hamlets). We earlier questioned the evolutionary effectiveness of quora, groups in general, without forming a conclusive judgment. It has been key to the evolution of certain social parasites, and to ants, as has been thoroughly reviewed by Bert Hölldobler and Edward O. Wilson in their examination of group "selection," "recruitment," and "retrieval," with no outsized judgment effectively relevant to the case of a large, bipedal hominid.[26]

But twenty-first-century mass industrial killing of animals, multiplied by a global population increase from approximately 425–461 million in 1500 to 7.8 billion presently, with far worse a hygienic landscape throughout most of the world, places the biological burden that comes with human consumption into a domain scarcely imagined in the worst Hells of Dante or Hieronymus Bosch. And we're not speaking of cockroaches.

Even by modern comparisons, the stakes have now radically changed at a level not seen since the Spanish (Haskell County, Kansas) Influenza Pandemic of 1918, caused by the H1N1 virus, of avian genetic origin,[27] while researchers currently search for another "1.67 million unknown [but projected] viruses infecting the animals of Earth."[28] And it would appear all but certain that at the heart of this global human tragedy is a singular, ecological verity: the likely vortex of this epidemic, and more to follow, is humanity's most sincerely ill-advised obsession with killing other animals and eating them. This is no "ecological (or inference) fallacy," that confuses deductions based upon individual behavior being used to char-

---

[26] See "Collective foraging decision in a gregarious insect," by Mathieu Lihoreau, Jean-Louis Deneubourg, and Colette Rivault, Behavioral Ecology and Sociobiology, 64 (10): 1577–1587, doi:10.1007/s00265-010-0971-7. It is alleged that cockroaches were so rife in Renaissance Europe that on a single Danish Naval vessel in 1611, "32,500" cockroaches were captured. See "Cockroaches Through History," by "Halt Pest Control," January 23, 2018, https://www.haltpest-control.com/blog/2018/january/cockroaches-through-history/. Accessed July 18, 2020. See The Ants, by Bert Hölldobler and Edward O. Wilson, The Belknap Press of Harvard University Press, Cambridge, Massachusetts, 1990, pp. 294–295, 276, 254, 388, 417, 212.

[27] See "History of 1918 Flu Pandemic," Centers for Disease Control and Prevention (CDC), https://www.cdc.gov/flu/pandemic-resources/1918-commemoration/1918-pandemic-history.htm, Accessed April 1, 2020. For possible Kansas origins, see Kansas Historical Society, https://www.kshs.org/kansapedia/flu-epidemic-of-1918/17805, Accessed April 1, 2020.

[28] See "Why Scientists Are Rushing to Hunt Down 1.7 Million Unknown Viruses," by Brandon Specktor, L6veScience, February 23, 2018, https://www.livescience.com/61848-scientists-hunt-unknown-viruses.html. For a superb early overview of ecological epidemiology and zoonoses, see, "Global Change And Human Susceptibility To Disease," by Gretchen C. Daily and Paul R. Ehrlich, Annu. Rev. Energy Environ. 1996. 21:125–44 ©1996 by Annual Reviews Inc., https://mahb.stanford.edu/wp-content/uploads/2020/04/1996_dailyehrlich_globalchangedisease.pdf. See also, "How Humanity Has Unleashed A Flood of Zoonotic Diseases," by Ferris Jabr, The New York Times, June 17, 2020, https://www.nytimes.com/2020/06/17/magazine/animal-disease-covid.html?smid=em-share, Accessed June 19, 2020.

**Fig. 60.7** Concentrated animal feeding operation, United States. (Photo © M.C. Tobias)

acterize group behavior.[29] This is, in fact, the group that built the human world which endangers humanity like never before (Fig. 60.7).

---

[29] See L. Goodman, "Ecological regression and the behavior of individuals," American Sociological Review, 18, pp. 663–664, 1953; See also, Review of "A Solution to the Ecological Inference Problem," by D. A. Freedman, S. P. Klein, M. Ostland and M. R. Roberts, Journal of the American Statistical Association, 93, pp. 1518–22, with discussion, Volume 94 (1999), pp. 352–57. See also, "The Ecological Paradox: Social and Natural Consequences of the Geographies of Animal Health Promotion," by Gareth Enticott, Transactions of the Institute of British Geographers, New Series, Vol. 33, No. 4 (Oct. 2008), pp. 433–446, The Royal Geographical Society (with the Institute of British Geographers), https://www.jstor.org/stable/30135326, https://www.jstor.org/stable/30135 326?seq=1, Accessed July 18, 2020.

# Part III
# A Natural History of Existentialism

# Chapter 61
# Strange Connectors

## 61.1   The Search for Henbane upon Mount Athos

During 1934, Arthur William Hill (1875–1941), the curator for many years at the Royal Botanic Gardens of Kew Gardens in southwestern London, was scouring the slopes of Mount Athos, northern Greece, in search of various unusual plant species when he and his two companions encountered a local Greek Orthodox monk who was, himself, searching for *henbane*, based upon an antiquarian monastery copy of the five-volume Greek masterpiece, *De Materia Medica* (ca. AD 77), of the first century by Pedanius Dioscorides (40–90 AD), who first described the plant to be a superb sedative, if mixed with mandragora. With its alkaloid scopolamine contents and intoxicating qualities, its "frenzies" were likened to drunkenness, a kind of "wine" according to Pliny the Elder, a congenial narcosis given to hallucinatory outbreak, intimated Theophrastus.[1] Remarkably, this monk was still relying on the botanical and pharmaceutical insights of a book which for nearly 2000 years had been the most celebrated and republished work of science in the world[2] (Fig. 61.1).

With 600 or more plants and animals identified by Dioscorides, he was able to transform the study of nature into a purposeful, pain-alleviating enterprise that has never been equaled. Linnaeus named the Genus *Dioscorea* after him. Today, the Angiosperm Phylogeny Group II (updated in 2009) system recognizes some 715 species, including the yam, within the Dioscoreaceae family. As for Dioscorides' homeland, the town of Anavarza in the Anatolian region of Cilicia Secunda, it was destroyed in 1374 by the Egyptian Mamluks. The whole region was subject to the

---

[1] See *Pharmakon – Plato, Drug Culture, and Identity in Ancient Athens*, by Michael A. Rinella, Lexington Books, Rowman & Littlefield Publishers, Inc., Lanham Maryland, 2010.

[2] See "Travel Diary: Greece." Ref: RM 9/10, 1934 Apr 8 – 1934 Apr 30, Royal Botanic Gardens, Kew, Library and Archives, https://discovery.nationalarchives.gov.uk/details/r/C11269987, Accessed October 13, 2019.

© Springer Nature Switzerland AG 2021
M. C. Tobias, J. G. Morrison, *On the Nature of Ecological Paradox*,
https://doi.org/10.1007/978-3-030-64526-7_61

**Fig. 61.1** Henbane (*Hyoscyamus niger*) in Hermann Adolph *Köhler's Medicinal Plants* (*Köhler's Medizinal-Pflanzen*), illustrated by K. Gunther, Walther Otto Müller and C.F. Schmidt; Franz Eugen Köhler, publisher, 1887. (Public Domain)

dramatic and violent political vicissitudes of human nature, becoming part of modern-day Çukurova in Turkey, but only after armies, of the First Crusade, the Armenian Kingdom, and many others, left the area in ruins, to be admired by twenty-first-century tourists. In fact, while a monk sought out a painkiller on remote Mount Athos, that same year (1934) also saw the doomed consecration of the Balkan Entente, a Pact that would never prevent subsequent aggression across Dioscorides' homeland by the Nazis.

Why are any of these obscure details relevant? Because the holy Mount Athos is a most fitting context for sensing the salient contours, disturbing at best, which separate the art and valiant spirituality of the last Byzantine bastion, from modernity; confronting prayer, science and art with the realities of the modern world, atop one of the most inaccessible locations on the planet. At the crux of geographical and historic bifurcations along the Halkidiki Peninsula—where the 6670-foot snow-clad peak rests above the 20-monastery complex of Eastern Orthodoxy, with its scores of cells and cliff-hanging hermitages overlooking the Homeric wine-dark waters—humanity reveals a window on its contradictory face. This eremitic complex of precipices over which human nature confronts its daily prayers, duties, and demons is an ecological wonderland. As the late poet and scholar Philip Sherrard, with photographer Takis Zervoulakos, portrayed, the monastic paradise teems in the Spring with "olives and vines, ilex and arbutus, wild smilax and osmunda fern ...," St. John's wort, "bright purple flowers of the Judas tree and the white plumes of the Manna ash ..." while "wild boar and deer still roam the woods."[3]

The mountain's hagiography commences in AD 49, when Athos lore has the *Holy Virgin, Mother of God* arriving; and by the ninth century, a monk by the name of Peter the Athonite taking up residency, "fifty years in the cave" there on the venerated abyss. That cavernous wild, and numerous others like it, would subsequently host a larra, or "colony of hermits" in the form of the Great Lavra, a cenobitic monastery founded by Saint Athanasios in 961. Dozens of hesychasteria—cells on cliff walls wherein solitaires lived out their devotional stays on earth—remain today. Quoting St. Gregory Palamas, writes Sherrard, "Man, being himself light, sees the light with the light; if he regards himself, he sees the light, and if he regards the object of his vision, he finds the light there again, and the means that he employs for seeing is the light; and it is in this that union consists, for all this is but one."[4] And, concluding his narrative with words from St. Symeon, "... For immense richness I exist penniless/and having as I think nothing, possess much/ and I say I thirst, in the midst of waters."[5] Such poetic raptures, symptomatic of the living miracle on Mount Athos, population just over 1800 souls (all males), contrast poignantly with the rest of us in ways that one must ask: Where to begin? This is the essence of existentialism, with its origins and future in nothing less than natural history herself.

That light of St. Palamas (1296–1359), with his enormous bird, brilliant halo, and penetrating, El Greo eyes, has worked miracles across the icons and murals at Mount Athos, in much the manner that it erupted, for example, in the body of resplendent and overpowering mountain illuminations by Joseph Mallard William Turner (1775–1851), particularly in his canvas, "Ancient Italy – Ovid Banished

---

[3] Sherrard and Zervoulakos, The Overlook Press, Woodstock, New York 1982, p. 7.

[4] Citing *Triade II*, 3, 36, in Gregoire Palamas, *Défense des saints hésychastes*, edited by Jean Meyendorff, Vol. 2, Louvain, France 1959, from p. 171 of Sherrard and Zervoulakos.

[5] Sherrard and Zervoulakos, citing *Hymn III* in Symeon le Nouveau Theologien, Hymnes I. p. 171.

From Rome," exhibited in 1838;[6] across the full dramatic sweep of Impressionist canvases (one thinks immediately of Monet's green reflections and grand decorations at the water-lily pond of Giverny in his years 1920–1926); or, most fittingly, during Nikos Kazantzakis' pilgrimage to Mount Athos in the Winter of 1914, when he speaks with Father Makários, who, in Kazantzakis' mind has arrived at the very gates of paradise. Kazantzakis wants to know if the old monk is still battling demons. To which the Father responds that the devil "has grown old with him." Instead, Makários "wrestles with God." To which Kazantzakis replies, "And you hope to win?" "I hope to lose, my child," the monk replies.[7]

Everything and everyone are connected; mountains and metaphysics, trees and people, the first lessons of both natural history and of a movement, generically described, as religion. To believe, or not to believe, in the future.[8] Any survey of the great mountain wildernesses, of monasteries and monks, of ecosystems and of the heart, confounds the great yearning for the Beyond. Kazantzakis would ask himself, "What was I seeking when I went to the Holy Mountain, and what did I find there?"[9] No precise answer comes to him. It was not to be that simple.

Indeed, naturalist expeditions, each a nuance of the existentialist personality, yield an entirely tumultuous mixed picture of sensation transformed by any number of predilections and biases of a subjective brain whose cortical size and complexity is no match for infinite varieties of richness of contexts, the very non-linear (hence, non-local) nature of experience and consciousness; the latitude of datasets. With that minefield of physiographic vicissitudes, is it any wonder that the most dominant aftermath of travels by William Bartram in North America, Charles Darwin aboard the Beagle, Maria Sibylla Merian and her daughter in Surinam, Alexander von Humboldt in the Andes, Johann Koenig and Patrick Russell in India, Carl Linnaeus in Lapland, Martin Frobisher in the Arctic, Johan Nieuhof in Madagascar, John Muir in the Sierras, Leonardo in the Alps, the many Marco Polo's of history … [10] has indulged an all-absorbing obsession with the geographical

---

[6] See *Turner and the Sublime,* by Andrew Wilton, British Museum Publications for The Art Gallery of Ontario, The Yale Center for British Art and The Trustees of the British Museum, London 1980. See the Frick Collection exhibition on Turner, "Turner's Modern And Ancient Ports: Passages Through Time," https://www.frick.org/exhibitions/turner/113, Accessed June 19, 2020.

[7] Simon & Schuster Publishers, New York 1965, p. 222.

[8] See "The Healing Quality of Pilgrimage to Mount Athos," by René Gothóni, *Archiv für Religionspsychologie/Archive for the Psychology of Religion,* Vol. 23 2000, pp. 132–143, Published by: Sage Publications, Ltd., https://www.jstor.org/stable/23909948; See also, "A Contribution to the Botany of Athos Peninsula," W. B. Turrill, *Bulletin of Miscellaneous Information (Royal Botanic Gardens, Kew,* Vol. 1937, No. 4 (1937), pp. 197–273, Published by: Springer on behalf of Royal Botanic Gardens, Kew, DOI: 10.2307/4107714, https://www.jstor.org/stable/4107714; "Arthur William Hill. 1875–1941," F. T. Brooks, *Obituary Notices of Fellows of the Royal Society,* Vol. 4, No. 11 Nov., 1942, pp. 8–100, Published by: Royal Society, https://www.jstor.org/stable/769151.

[9] op.cit., *Report to Greco,* p. 235.

[10] See Raymond John Howgego's spectacular *Encyclopedia of Exploration To 1800,* Hordern House, Sidney Australia, 2003; See also, *Voyages of Discovery,* by Dr. Tony Rice, Introduction by Dr. David Bellamy, Firefly Books, London 2008.

sublime? None of these perceived, imagined, experienced appurtenances, glories, and travails of humans pursuing phantoms out of doors has actually helped resolve a single ecological problem. That is to say, the vast body of details that have accumulated in the wake of thousands of travelers' tales, no matter how scientifically precise, has not managed to alter the course of pain, suffering, fear, and mortality that grip our species. That, too, is at the heart of existentialism.

## 61.2    The Earliest Existentialist Codices

While the most ancient illustrated natural history books seem to date to the first century AD, converging theories of human nature, and its behavioral range of proclivities and contradictions, all under the aegis of Nature, have obviously offered up stifling riddles long before that. Human suffering at that time, as Aristotle certainly discovered in his final year of life, had probably no less impact on our mental attitudes than today's ecological and demographic concussions. These were existential responses to everyday life in the trenches, and the study of natural history was the premiere outlet for voicing curiosity about the world, anxiety and terror. Dioscorides, Thales, Hippocrates, Pythagoras and Aristotle's student Theophrastus before him, and a long history of medicine after them, had seen in their prodigious studies of natural history both metaphysical and pragmatic components most relevant to humanity. But "relevancy" can do little for our cause. The latter, practical considerations regarding pain had largely been consolidated into various herbaria, medical texts focusing upon the alleviation of illness and its accompanying suffering. As H. Walter Lack informs, the *Codex Aniciae Julianae*, written sometime before AD 512, was the "oldest illuminated version of the writings of Dioscorides" and "No words can exaggerate the importance of this work." Also known as the *Codex Constantinopolitanus*, the *Codex C*, and the *Codex Vindobonensis* in Vienna, it remains, along with the *Codex Sinaiticus*, one of the most important of all manuscripts; part of UNESCO's "Memory of the World" list.[11] It might as well be designated, *memory of long suffering* (Fig. 61.2).

The afore-named "Julianae" has its own dizzying history, as elegantly described in an entire chapter ("Juliana's Book") by Anna Pavord, in her work, *The Naming of Names – The Search for Order in the World of Plants*;[12] a comprehensive botanical

---

[11] See *Garden of Eden – Masterpieces of botanical illustration*, by H. Walter Lack, Österreichische Nationalbibliothek, Taschen, Koln, 2008, p. 22.

[12] Bloomsbury, New York, 2005, pp. 82–92. See also, the wonderful book by evolutionary biologist Carol Kaesuk Yoon, *Naming Nature – The Clash Between Instinct And Science*, W. W. Norton & Company, New York and London, 2009, in which, at one point, she speculates on the "umwelt" (the world as experienced by a particular organism), "We will need to welcome in all manner of wondrous absurd-seeming possibilities: cassowaries as mammals, orchids as thumbs, and bats as birds. We'll need to greet with open arms the views of the French naturalist who classified slugs, snakes, and crocodiles as insects…" p. 295.

**Fig. 61.2** Portrait of Pedanius Dioscorides, illumination from the sixth-century Greek Juliana Anicia Codex, ca. 515 AD, also known as the *Vienna Dioscurides*. (Private Collection)

odyssey that also makes clear the origins of botanical illustration. The challenge, writes Pavord, as early on indicated by Pliny was that "it was difficult to represent a plant in a single image when the plant itself changed all the time."[13] The first known illustrated plant, a comfrey (*Symphytum officinale*) from the Johnson Papyrus, AD 400[14] was greatly benefitted by new advances in the production of "sheets rather than rolls" of papyrus, around AD 100.[15]

The pharmaceutical entrepreneur Sir Henry Solomon Wellcome (1853–1936) whose passion for the natural sciences and medicine resulted in one of the largest medical philanthropies, and science libraries in England, first opened its doors to the public in 1913, and again, transformed, in 2007. It is London's equivalent of the

---

[13] ibid., p. 76.

[14] Wellcome Library, MS 5753, London.

[15] ibid., Pavord, p. 76.

Morgan Library in Manhattan. Like the Akkadian tablet fragments at the Morgan, and other ancient written entablatures, the Wellcome library has an Egyptian medical prescription on papyrus from 1100 BCE. Its focus is pain.

That merging of human need, and natural history has never ceased, and goes to the core of a most delicate, if problematic dilemma facing humanity: our obsession with ourselves (little wonder: pain hurts), and the corresponding inroads to our grasping the similarly widespread agonies amply evinced by whatever lenses we may enlist to grasp the extent of the Anthropocene. Today, while the same emphatic links between natural history and medicine remain, and as medical technology and treatment modalities evolve with astonishing rapidity, neurobiological insights into the workings of the brain and, particularly, theories oriented to questions of human behavior and our social norms, have all gravitated in large part toward studies involving mammalian genetics. The emphasis in large measure has been the continued affirmation of a Darwinian sensibility that always claimed for humankind a unique niche, lodged somewhere between the heart and the mind.

Translated into genetic cipher, that niche has claimed a multitude of possible explanations for what it is *Homo sapiens* are, feel, think, need and express; and not a few superlatives along the way allegedly separating us from all Others by way of distinguishing those abundant characteristics in tune with a unique synecological role, of complete ecosystems and the identities therein. But what is that role? What part do we play? That has been the everlasting Koan of Mount Athos, and the persistent temptations of spirit it has elicited in all those fortunate enough to experience its living ethos.

Every possible indication has its counter. Purpose, purposelessness; distinct, blurred; information driven, at a loss to cope with so many neurological connections; compassionate, violent; socially organized, living in chaos from day to day; conceptually evolved, philosophically downtrodden; hopeful, in despair; biologically novel, the slippery slopes of taxonomy; giving, taking; restrained, all-inflictive; in sync, totally out of sorts. One cannot say the same for the mountain herself: she is steadfast and consistent, although there is rarely a moment during which she does not give rise to some micro-climate shift, passing clouds, snowstorms in winter up along the summit ridge.

# Chapter 62
# The Synecological Conscience

## 62.1   On the Hominin Conscience

One recent noteworthy attempt to explain some of the fundamental pillars of what it means to be the last, as yet non-extinct hominin, is the book *Conscience – The Origins of Moral Intuition*, by Patricia S. Churchland.[1] Churchland's summary of existing brain research that might help shed light on the human conscience represents an accessible overview of recent neurological and genetic studies that attempt to favor mammalian, and specifically, human brain development as it pertains to our social behavior as a species, and one particular outcome of that sociality, our sense of morality.

Her work is a fascinating tour-de-force. Early on, Churchland points out that "there are essentially no animal models of human warfare" (referencing observations regarding chimpanzees and rodents).[2] And that, contradistinguished from all others, our brains, with their "86 billion neurons" require roughly "6 calories per billion neurons" to maintain our brain size and the scope of its activities (or approximately "516 calories" per day). Additionally in need of that sustenance is our very endothermy and immaturity at birth.[3] All the other warm-blooded innocents with big brains, distinguished from those with smaller brains, the smaller mammals, like the prairie and montane voles, show a significant difference by way of the "variation in the density of receptors for oxytocin and vasopressin receptors" in such species.[4] These hormonal proteins within a neuron's membranes "dock," within a neural circuitry and, specifically, both in voles and in humans, are given toward the acceleration of a neurochemical affinity toward social behavior. Or, as

---

[1] See W. W. Norton & Company, New York, 2019.

[2] ibid., Churchland, p. 15.

[3] ibid., Churchland, p. 37.

[4] ibid., Churchland, p. 47.

© Springer Nature Switzerland AG 2021
M. C. Tobias, J. G. Morrison, *On the Nature of Ecological Paradox*,
https://doi.org/10.1007/978-3-030-64526-7_62

it has been asserted, the very human conscience (as well as monogamy, dual-parenting, and other qualities, to varying degree).[5]

These are huge attribute delineations and accordingly, a genetical principle, or at least plausible explanation for social and maternal behaviors that, far downstream, elicit concerns for others closest to us, within our brain-dominated community orbits. This currency of social concern is fundamental, because such behaviors ultimately get translated into moral forms of activity. Of course, as mentioned in an earlier chapter regarding the cockroach "quorum" there is no reason to believe that those trillions of insects do not also make collective, rational choices that stem from some, heretofore unexamined, eusocial imperatives that align with moral deliberation at the social level.

This "oxytocin portfolio," as Churchland points out, is not easily measured, either in blood or in urine, and most inferences as to what is really going on in our brains come from research on other mammals, mostly primates and rodents, not insects, or not as yet.[6] Nor does Churchland unequivocally assert the oxytocin data as definitive. For example, she is quick to ask, "How our 86 billion neurons work together so that we talk and do fancy math, so that we create symphonies and parliaments, is not known." But the fact that every "neuron will make about 10,000 connections with other neurons" easily suggests exponential outcomes, summarized by the declaration, "The remarkable fact is that there is no secret structure unique to the human brain that creates sophisticated social institutions or art or a moral conscience. Our brain just has more neurons."[7]

Such unabashed arithmetic is not without its counterpoints. For example, Suzana Herculano-Housel notes that "the so-called overdeveloped human cerebral cortex holds only 19% of all brain neurons, a fraction that is similar to that found in other mammals." In its internal architecture, among that edifice of scaling of the neocortex layers, there may indeed be anatomical efficiency, but the rules for assessing such architecture are not without the very bias built into those rules favoring self-inquiry.[8] As Frans De Waal writes of "autonomous conscience" and human conscience within the rubrics of known evolutionary theories, "To give the human conscience a comfortable place within Darwin's theory without reducing human feelings and motives to a complete travesty is one of the greatest challenges to biology today."[9] There is no "comfortable" reign of the human conscience that can be explained either by evolution or by natural selection. And that is for two reasons. First, it is not unique. Conscience, were we to search with requisite tools and

---

[5] ibid., Churchland, p. 61.

[6] ibid, Churchland, p. 67.

[7] ibid., Churchland, p. 178.

[8] See "The Human Brain in Numbers: A Linearly Scaled-up Primate Brain," Suzana Herculano-Housel, Frontiers in Human Neuroscience, 2009; 3:31, November 9, 2009, doi: https://doi.org/10.3389/neuro.09.031.2009, PMCID: PMC2776484, PMID: 19915731, https://www.ncbi.nlm.nih.gov/pmc/articles/PMC2776484/. Accessed October 13, 2019.

[9] *Good Natured – The Origins Of Right And Wrong In Humans And Other Animals*, by Frans De Waal, Harvard University Press, Cambridge, Mass., 1996, pp. 91–92, 117.

feelings, is likely a universal, if conditional, quality of all sentience. Second, its conditionality hinges upon predicates, timing, attitude, and the sheer individuality of organismic biology that far outnumbers our capacity to investigate.

Analysis of the human organism has been, in every conceivable instance, and according to a vast range of acknowledged variables, a stressful, unsuccessful enterprise. It is not mere Socratic skepticism that rages over the topologies of accumulated data. Ludwig Wittgenstein clarified the dilemma when he wrote, "A function cannot be its own argument, whereas an operation can take one of its own results as its base."[10] Even a cursory examination of early philosophical considerations raised in human self-awareness selects for doubt and discouragement. Socrates, it is said, "saw the Athenian Government, though under the form of a Democracy, was yet nearer to a Tyranny or Monarchy, or as himself professeth, being dissuaded by his Genius from meddling in public affairs, which advices was his preservation, being too honest to comply with the injustices of the Common-wealth, (and) to oppose them was extremely dangerous, as he found experimentally"[11] (Fig. 62.1).

## 62.2   Do We Learn from Our Mistakes?

But it was Edward Gibbon's masterpiece, echoing Socrates' lament, that should stifle our reason for being in the face of the very history from whose ranks we have been jettisoned up until now. Writes Gibbon, "The disgrace of the Romans still excites our respectful compassion, and we fondly sympathize with the imaginary grief and indignation of their degenerate posterity. But the calamities of Italy had gradually subdued the proud consciousness of freedom and glory. In the age of Roman virtue, the provinces were subject to the arms, and the citizens to the laws, of the republic; till those laws were subverted by civil discord, and both the city and provinces became the servile property of a tyrant."[12]

These are the fundamentals of every age of humankind, too well known to ignore, in the complacency born of hope (Nikos Kazantzakis' "last temptation") that is as endemic to the incoherent power grabs of Athens or Rome as it is, so many centuries later, even in the artful schizophrenia of Calif, the "Stranger," son of the King of Timur in Puccini's "Turandot" (1926). In ecological terms of the present, these myriad inroads to total disruption, and a fated despair, are cached in such currents as those characterizing the Paleocene-Eocene Thermal Maximum ("PETM") which is

---

[10] *Prototractatus – An early version of Tractatus Logico-Philosophicus*, by Ludwig Wittgenstein, Edited by B. F. McGuinness, T. Nyberg, and G. H. von Wright, with a translation by D. F. Pears and B. F. McGuinness; an historical introduction by G. H. von Wright, Cornell University Press, Ithaca, New York, Routledge & Kegan Paul, Ltd., edition, 1971, 5.00161, p. 139.

[11] *The History Of Philosophy: Containing The Lives, Opinions, Actions and Discourses of The Philosophers Of every Sect*, by Thomas Stanley, Second Edition, London 1687, Part III, p. 81.

[12] *The History Of The Decline And Fall Of The Roman Empire,* by Edward Gibbon, Volume VI, London 1821, Chapter 36, pp. 227–228.

**Fig. 62.1** Roman ruins, anonymous photographer, ca. mid-nineteenth century. (Private Collection). (Photo © M.C. Tobias)

now again reverberating upon us, in the forms of an atmospheric "end-game," runaway "amplifying feedbacks," and a global biochemical and extinction "hysteresis" (lag-times between cause and effect). Component parts of the cybernetic fall-out are palpable, others retarded (e.g., the phenomenon of ghost species, ghost biomes, trophic ghosts in general).[13] But the mathematical meditations attendant upon the passionate pessimism that necessarily follows from our doomed distractions, however heroic some efforts to rectify them may be (those passengers on United flight #93 on September 11, 2001), beg and borrow the issue of how to think about our coming nullity, notwithstanding the tens of millions of neurons allegedly helping us navigate the downside of our collective behavior.

---

[13] See "Beyond climate tipping points," by Dr. Andrew Yoram Glikson, June 25, 2019, Global Research, https://www.globalresearch.ca/beyond-climate-tipping-points-greenhouse-gas-levels-exceed-stability-limit-greenland-antarctic-ice-sheets/5681653, Accessed July 5, 2019; See also, *The Plutocene: Blueprints for a Post-Anthropocene Greenhouse Earth*, by A.Y. Glikson, DOI https://doi.org/10.1007/978-3-319-57237-6, Springer International Publishing AG, Cham Switzerland, 2017, https://link.springer.com/book/10.1007%2F978-3-319-57237-6#toc

# Chapter 63
# The Ecological Summons of Jain Mathematical Calculations

## 63.1 The Mathematics of Non-violence

The history of religions is rich in numeric calculations—613 mitzvot or commandments attributed in the Torah in the third century CE; 40,000 horsemen and another 40,000 footmen as described in the Book of Samuel (10.18); 100,000 songs of the eleventh-century Tibetan Buddhist saint, Milarepa; the number four as a most harmonious pillar of the Sioux Indian cosmology, but also a harbinger of death in the New Testament's Book of Revelation, and so on (Fig. 63.1).

It is widely believed that Mahavira himself was a trained mathematician. Jain mathematical insights have gone far beyond arithmetical processes, similes, the endless ritual numerology, or even the symbolic power of cosmological details and beliefs which share commonality among countless spiritual traditions. The second-century BCE Jain treatise known as *Sthananga-Sutra*,[1] developed ten major precedents for understanding various types of equations, permutations, combinations, and the proper place and calculations for virtually every object posited within the Jain metaphysical pantheon of six eternal substances. In addition, the sutra conceived of multiple versions of the infinite, including transfinite numbers.

Mahavira's pre-Socratic contemporary, Pythagoras, considered numbers to be sacred in as much as they were the tools by which harmony could be described. For Pythagoreans, a soul's journey after the corporeal body's demise would take 3000 human years to complete its full cycle, an odyssey that encompassed the equivalency of Karmic incarnations in, and echoes of one species after another. The journey in the afterlife was mathematically calculated, an ecological saga that was pivotal to the Parmenides dialogue by Plato which focused upon the great conversation between the one and the many as the central bifurcation point for all philosophical speculation.

---

[1] Kornelius Krümpelmann, "The Sthanangasutra – An Encyclopaedic Text of the Svetambara Canon," International Journal of Jaina Studies (IJJS), 2 (2), September 2006.

© Springer Nature Switzerland AG 2021
M. C. Tobias, J. G. Morrison, *On the Nature of Ecological Paradox*,
https://doi.org/10.1007/978-3-030-64526-7_63

**Fig. 63.1** *Kalpasutra* Jain manuscript page (ca. AD 1470), traditionally ascribed to Bhadrabahu (fourth century BCE), containing biographies of Parshvanatha and Mahavira. (Private Collection). (Photo © M.C. Tobias)

That grounding in numeric characterizations was utterly revolutionized in Jainism, as it sought to transform the multiplicity of rules, orders, stages, restrictions, and veritable laws of natural behavior into ethical activism, principally, ahimsa; a non-violence oriented to every single individual of every species, organic and inorganic, including grains of sand and dew drops. Imagine such preconditions of Nature's infinities dictating entirely new contexts, functions, equations, sets, subsets, and sets of subsets all oriented to how we as humans ought to behave, as a very function of Natural Law.

Most radically, in terms of the history of environmental thought and injunctions regarding moral behavior, Jain traditions—while developing numbers of an exponentially unprecedented extant—never ceased to consolidate the vastness of infinities into the singularity of pain in individuals, and the supreme importance of orienting one's life toward the amelioration of all suffering in all beings. The first and oldest of the Jain agams, or religious texts based upon the teaching of Mahavira, the *Acharanga Sutra*, extolls the most mathematically paradoxical of all contexts when it lays the groundwork for fully appreciating the challenge of simply recognizing and revering every individual within a biosphere of hallowed interdependency. Not numbers, personages. Today's genomic, computational and synthetic biological research is in many respects at odds with Mahavira's teachings because, simplistically stated, such analyses fail to recognize individual trees within the forest. Or have forgotten that across the digital spans of 1-0-1-0-1, there are real individuals involved.

All of the algorithms, population dynamics, and Big Data fall short of explaining the fate of an individual in the environment, and the phenomenon of individual kindness that today is clearly the most challenging and critical prelude to understanding, and mitigating the full scope of the fast burgeoning Eremozoic era, that period of Great Dying; of human loneliness in an increasingly depauperate world. Even a more mundane, mathematical approach to extrapolating large forms of behavior in species, our own included, relies in conventional ecological research on "mapping techniques" which, in turn, define the so-called biological optimum,[2] which also presumes "a *minimum* set of basic habitat requirements of a species."[3]

## 63.2    A World of Nigodas and Jivas

The spiritual topology that is Jainism, with all of its ethical aspirations, today must confront the very real possibilities of a wasteland, solely induced by human collective behavior. But through its reconciliation of the one and the many, the living Jain legacies of ahimsa have introduced a stark contrast to brute collectivities and the cold calculus of biological taxonomy by way of the formulation of jivas and nigodas, individual, inviolate selves and souls, and our ethical responsibilities toward them. Pivotal to that collective of individuals is the Jain computation of pain upon a world stage of biological vulnerability, which represents a departure from all other traditions. Given that the modern ecological revelations and calls for action all summon levels of unprecedented moral restraint and re-engineering, the role of Jain computations in the midst of a Samsara-driven global-level extinction event is wholly relevant to today's Anthropocene—whether in computing the moral equivalencies owed a Blue Whale or the most humble flea (Fig. 63.2).

---

[2] See "A Minimalist Approach to Mapping Species' Habitat: Pearson's Planes of Closest Fit," by John T. Rotenberry, Steven T. Knick, and James E. Dunn, Chapter 22, in *Predicting Species Occurrences – Issues of Accuracy and Scale,* Edited by J. Michael Scott, Patricia J. Heglund, and Michael L. Morrison, Jonathan B. Haufler, Martin G. Raphael, William A. Wall and Fred B. Samson, Island Press, Washington, Covelo, London, 2002, p. 283.

[3] ibid.

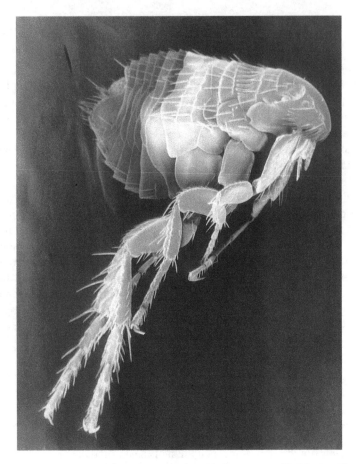

**Fig. 63.2** (False-color) Scanning electron micrograph of a flea, by CDC/Janice Haney Carr, Centers for Disease Control and Prevention's Public Health Image Library

Jain infinities within every personage are the contemporary realization that every Being contains a forest, ecosystems within systems. Such complexities should necessarily induce our humility, tolerance, and imaginative reciprocity with the world.

# Chapter 64
# Fundaments of Observation and Melancholy

## 64.1  On the Origins and Novelty of Despair

In her essay on the origins of human artistic creativity, University of Oxford anatomist Gillian M. Morriss-Kay suggests that "Linking anatomy to the origins of art is impossible"; that "Art does not exist in a vacuum but requires a social context, otherwise it is meaningless."[1] While the conceptuality of artistic enterprise 30,000 years ago exploited no known anatomical features independent of those available to Paleolithic successors alive today, other cultural norms have most assuredly changed, underscoring a statement in Jean-Paul Sartre's first definitive clarification of what he meant by existentialism. "…what can be said from the very beginning is that by existentialism we mean a doctrine which makes human life possible and, in addition, declares that every truth and every action implies a human setting and a human subjectivity."[2] A year after the release of his *Existentialism*, Sartre's subsequent work *The Psychology of Imagination* examined a student of anatomy who held to the contention that reality, its impressions, the emanations of an "intense life" as experienced in short-term memory could be "more true than nature" "just as a particularly significant portrait is said to be truer than its model.[3]

We don't know, nor did Sartre, what "more true than nature" might mean, but it is a tease of both tragic and sublime dimensions that underscores the most rigorous default in evolutionary theory. That is because there has been no direct evidence, and little outright speculation in either science or the history of philosophy what a new species, let alone a large omnivorous mammal, has to say, to think, to do. How

---

[1] "The evolution of human artistic creativity," by Gillian M. Morriss-Kay, The Journal of Anatomy, J Anat. 2010 Feb; 216(2): 158–176., Published online 2009 Nov 6. doi: https://doi.org/10.1111/j.1469-7580.2009.01160.x, PMCID: PMC2815939, PMID: 19900185."

[2] *Existentialism*, by Jean-Paul Sartre, Translated by Bernard Frechtman, Philosophical Library, New York, 1947, p. 12.

[3] *The Psychology Of Imagination*, by Jean-Paul Sartre, Translated from the French, Philosophical Library New York, 1948, p. 53.

© Springer Nature Switzerland AG 2021
M. C. Tobias, J. G. Morrison, *On the Nature of Ecological Paradox*,
https://doi.org/10.1007/978-3-030-64526-7_64

such a being fits in? What, if any, role does it play in the ecological flux of a world that could not be said to have *gone mad*, prior to its, *our*, arrival on the scene no more than 300,000+ years ago, if that? Are there comparable behavioral digests for any species, predicated upon their novelty and newness to the earth?

At the turn of the twenty-first-century researchers on the Galapagos island of Daphne Major noticed a large cactus finch that had arrived on the small island, mated with the endemic *Geospiza fortis* (a smaller native finch) and produced entirely new, large, viable offspring hybrids. Within four generations of successful mating by those chicks, the new species—by then one of four such finch species on the island—numbered 26.[4] There are undoubtedly countless other newly arrived species that are, as yet, undetected. In contemplating other examples, Dr. Ken Saladin, Emeritus professor of biology at Georgia College and State University, points out that the "nylon-eating bacteria in the genus *Flavobacterium*," can be no older than nylon itself, invented in 1935. And Dr. Saladin cites data regarding two sub-species of sea urchin in Western Australia, *Heliocidaris erythrogramma*, a Scottish monkey flower, *Mimulus peregrinus*, no more than 143 years old, and a Swiss cabbage family species, *Cardamine eboracensis,* first seen to have arisen around 1979.[5] With real-time evolution occurring rampantly throughout the world, and at least 80,000 chemicals in commercial use (at least 2000 new chemical compounds introduced each year, largely untested), the projected associated species is simply unknown, but a nylon-eating bacterium is certainly indicative of the possibilities at large.[6]

Hybridization is one of the dominant precursors of future biological diversity under Anthropocenic conditions of multiple stressors, syncretistic chemical reagents and resulting reactants in every biome, and the unnatural loading of tension, more and more such stress, beyond tenable thresholds. Given this context, human reflection is at a crossroads in trying to gauge the relativity of our *presence* as a new species. Even more confused, troubled by the acceleration of bias that is our singular lens through which we attempt to scrutinize present and past biological conditionals and relations, nothing we do or think quite captures the stark, unknowable extent of our behavioral impacts; let alone of those upon ourselves.

We can correlate, roughly, cause-and-effect, in terms of heritable traits, climate change and pregnancies, the reproductive rate and contagion within human populations of pathogens, or the levels of dissolved oxygen in lakes and its impact

[4] "A New Bird Species Has Evolved On Galapagos And Scientists Watched It Happen," by Michelle Starr, November 24, 2017, ScienceAlert, https://www.sciencealert.com/darwin-s-finches-evolve-into-new-species-in-real-time-two-generations-galapagos, Accessed October 11, 2019.

[5] "Which species is the youngest on Earth?" Ken Saladin, Emeritus professor of biology, Georgia College and State University, https://www.quora.com/Which-species-is-the-youngest-on-Earth, Accessed October 11, 2019. Saladin cites a former BBC site titled "World's Youngest Species," http://esciencenews.com/sources/bbc.news.science.nature/2012/11/22/worlds.youngest.species, Accessed June 19, 2020. See also, Kinoshita, S., Kageyama, S., Iba, K., Yamada, Y. and Okada, H. Utilization of a cyclic dimer and linear oligomers of ε-aminocapronoic acid by Achromobacter guttatus K172, Agric. Biol. Chem. 116, 547–551 (1981), FEBS 1981.

[6] See "It could take centuries for EPA to test all the unregulated chemicals under a new landmark bill," by Mark Scialla, PBS NewsHour, June 22, 2016, Accessed October 11, 2019.

**Fig. 64.1** Spanish writer and philosopher Miguel de Unamuno (1864–1936), Bibliothèque nationale de France, 1925. (Public Domain)

upon fish. But far less clear are the psychological and emotional precursors of such impacted behavior; let alone how such intangibles might compare and contrast with those in other species. The very conjectures—at the heart of anthrozoological, ethological and biosemiotic studies—yield virtually nothing that can be asserted with certainty (Fig. 64.1).

## 64.2   Tragedy and Biology

Even the great Spanish philosopher Miguel De Unamuno (1864–1936), in his masterpiece of unflinching reflection, *Tragic Sense of Life* (*Del sentimiento trágico de la vida*, 1912), titling its sixth chapter "In The Depths of the Abyss," quotes at length from the *Essai sur l'indifférence en matière de religion* of the nineteenth-century French Catholic priest Hugues-Félicité Robert de Lamennais (or De La Mennais, 3rd part, Chap. 67), "Shall we doubt that we think, that we feel, that we are? Nature does not allow it; she forces us to believe even when our reason is not convinced. Absolute certainty and absolute doubt are alike forbidden to us. We hover in a vague mean between these two extremes, as between being and nothingness…"[7] Inciting what would become Sartre's most prominent of all investigations

---

[7] *Tragic Sense of Life*, by Miguel De Unamuno, Translated by J. E. Crawford Flitch, Dover Publications, New York, 1954, p. 117.

(*Being and Nothingness: An Essay on Phenomenological Ontology – L'Être et le néant: Essai d'ontologie phénoménologique,* 1943) Unamuno, like Jean-Paul Sartre and Nikos Kazantzakis after him, would commit themselves to a philosophy of action, by which humanity's role, if any, was to struggle, to believe, to invent. And by these counter-resistant ploys, try to survive. Nature insists that we do so, or that is one philosophy. Any alternative would merely result in our perishing before we even realized we were a Being, alive, newly risen within a geography of great difficulty

## 64.2.1   The Continuation of a Philosophy

Oxford scholar Robert Burton's *The Anatomy of Melancholy* (London, 1621) held up psychology over the natural sciences as the key to unlocking the most rampant ailment of his time: depression. He himself spoke of it as "gravidum cor, foetum caput" (a heavy heart, embryonically lodged in his head). Over 300 years later, the Catholic existentialist, Gabriel Marcel, declared his preference for the phrase "neo-socrateism." Whether "existentialism" actually dates to the thinking of Socrates, to Søren Kierkegaard in 1841, to Fyodor Dostoevsky or to Friedrich Nietzsche, the interpretive outcomes do not begin to encompass the ecological despair of the twenty-first century, summed up by Albert Camus who wrote in his *Myth of Sisyphus* that the only truly philosophical problem for humans hinged upon the ever looming issue of "suicide." For Sartre, it was the notion that "l'existence précède l'essence," and that, by implication, our free will had fashioned whatever essence we might define for ourselves, over and above the fundamental impermanence, absurdities, instability, and vulnerability of our lives, a realm as fitting for contemplation by metaphysicians as by biochemists. All bolstered by the amalgam of our past experience into a kind of hardened, serial fiction trapped within a world "in-itself," Sartre's definition. A biosphere, more precisely, that has absolutely no need of our one species, based on all the evidence, to date (Fig. 64.2).

Hence, to embrace life is to inculcate *precursors*: of thought, memory, and that constellation of fever pitches and optimism that can effectively hold at bay the nullifying quotients, our annihilation. Such precursors involve an ancestry formed of 2–5 trillion molecules per human cell. Such invisible weight connotes the ultimate prefix.

In the ethnographic and shamanic literature, this primordial miasma, part physiology, part dreamtime, seduces the conscious or subconscious claims to affiliation. When we invoke God, and Back-to-Nature, we are touching the live-wires of an infinite antecedent comprising particles, gasses, waves, frequencies, forces, espied laws of physics and other interactions, spins and the even greater proportion of non-baryonic (unaccountable) dark matter and black holes out of which cosmology and exobiological research derive so much of their own speculative matrices. A world, or worldly dimensions thought to exist simultaneously to, and/or before, humanity—dimensions that will persist long after we are gone. The question as to

**Fig. 64.2** Simone de Beauvoir and Jean-Paul Sartre attending the ceremony of the sixth anniversary of founding of communist China in Beijing on October 1, 1955, in Tiananmen Square (Liu Dong'ao)—Xinhua News Agency

our pertinacity, meaning and value dangles like a puppet, an extra in a short-lived extravaganza. Is it a meaningless event? We can imagine that such a question is Barthe's perfect neutral, like a moth falling asleep on a wall; a dangling modifier ad infinitum, regardless of ancillary components such as free will or predetermination. The prime mover in this gallery of vagaries remains forever elusive. Our consciousness may spin out of control, but it will not change the valency of such unknowns.

## 64.3   The Onrush of a Global Existential Crisis

Throughout 2019–2020, Democratic presidential primary hopefuls, and environmental activists worldwide (mostly youth), spoke passionately from venue to venue of an "existential climate crisis." How did the legacy of Søren Kierkegaard and Jean-Paul Sartre suddenly blossom? By what pathways of social unrest did such a weight of cumulative anxieties affirm an overnight linguistic lingua franca to characterize the urgency of 2020?

One might fix its center in Tuscany, around the city of Volterra, during the oldest period of the Etruscan culture, at the height of the Villanovan renaissance (900–700); a period in which the veneration of ancestors reached its zenith, not to be matched until medieval death cults emerged on Rapa Nui (Easter Island). Sartre asked the leading question on the subject in his *Being and Nothingness*: "But what exactly is necessary in order for these concepts of disintegration to be able to receive even a pretense of existence, in order for them to be able to appear for an instant to

**Fig. 64.3** Moais on Rapa Nui (Easter Island). (Photo © M.C. Tobias)

consciousness, even in a process of evanescence?"[8] In the case of Easter Islanders, one might surmise that a "creeping normalcy" or "landscape amnesia" (as Jared Diamond utilized the expressions)[9] insulated those invested in the Bird Cults and enormous societal commitments to the carving and movement of the massive mono-lithic stone moais. There were more than 600 of them, the largest weighing some 82 tons.

But Rapa Nuians' monumental blindness to what was happening to their 63.17-square-mile island may not be the classic case study in environmental self-oblivion that is so often registered in historic commentary. All that energy invested in deforestation, rituals attendant upon the public display of decompos-ing corpses that would then be reunited with ancestral spirits, and an entire island-wide obsession with a veritable worship of the past that helped expunge from consciousness the present concerns of a dying civilization, might well have been an elegant form of the equivalent of species suicide.[10] Deliberate, ornate,

---

[8] See *Being and Nothingness: an Essay on Phenomenological Ontology*, Translated and with an Introduction by Hazel E. Barnes, Philosophical Library, New York 1956, p. 58.

[9] See *Collapse – How Societies Choose To Fail Or Succeed*, Viking Publishers, New York, 2005, p. 425.

[10] See *The Mystery Of Easter Island – The Story of An Expedition*, by Mrs. Scoresby Routledge, Sifton, Praed & Co., London 1919; See also *Legends of Easter Island*, by F. Sebastian Englert, Museum Store Editions, Father Sebastian Englert Anthropological Museum, Easter Island Rapa Nui, 2nd edition, Santiago, Chile 2003. See also "Hotspots," feature documentary, by M. C. Tobias and J. G. Morrison, KQED/PBS, A Dancing Star Foundation Production, Los Angeles, 2008 – www.hotspots-thefilm.com

soberly motivated. But it remains as unclear and debated as the alleged mass suicide by the Jews atop Masada in Israel[11] (Fig. 64.3).

---

[11] See "Maybe there was no mass suicide at Masada? Top archaeologist questions a legend," by Amanda Borschel-Dan, October 4, 2019, The Times of Israel, https://www.timesofisrael.com/maybe-there-was-no-mass-suicide-at-masada-top-archaeologist-questions-a-legend/, Accessed March 4, 2020.

# Chapter 65
# The Great Divergence

## 65.1 Mental and Physical Coefficients of Behavioral Distribution Patterns

When Edward Steichen grouped 503 photographs together from 68 countries at the Museum of Modern Art in New York, the famed 1955 exhibition entitled "The Family of Man," he asserted in the accompanying publication[1] that "the most important service photography can render…" is to " 'explain man to man' and man to himself." The luminous evocations of photography sweeping across a landscape of tumult and the sublime, from Timothy H. O'Sullivan's albumen print, "A Harvest of Death," scores of dead soldiers in the eerie dark and misty twilight of Gettysburg, Pennsylvania, July 1863, printed by Alexander Gardner and housed in the New York Public Library, to Ansel Adams' iconic gelatin silver print, "Clearing Winter Storm, Yosemite Valley, California," c. 1944 (self-explanatory), we recall a famed chapter, "What use is it?" in the 1976 publication, *Ecological Relationships* in which the authors suggest that "a scientific explanation must refer some observed phenomenon to its antecedent causes at one or more levels of detail removed" (Fig. 65.1).

It's in that *distance* of details, that the human evolutionary drama invokes the most ambiguous potentials. Potential for a silent end to the species, or some other, utterly unpredictable frisson of opportunities meeting headlong. "Thus," continue the authors of "What use is it?" "photochemical reactions are explained in terms of molecular orbitals, and nerve conduction in terms of iconic flows. For purely technical reasons, there are rather few areas of ecology where satisfying explanations of this kind seem to be feasible, at present." And the authors point to "population

---

[1] Edward Steichen, ed., "Introduction, *The Family of Man*, Museum of Modern Art/Maco Publishing, Distributed by Simon & Schuster, New York, 1955, pp. 3–4, as cited in *A World History of Photography*, by Naomi Rosenblum, New York, Abbeville Press, © Cross River Press, 1984, p. 481. One might also compare the 90-minute documentary resulting from 4500 hours of crowd-sourced footage from 193 countries, July 24, 2010, https://lifeinaday.youtube/, Accessed July 19, 2020.

© Springer Nature Switzerland AG 2021
M. C. Tobias, J. G. Morrison, *On the Nature of Ecological Paradox*,
https://doi.org/10.1007/978-3-030-64526-7_65

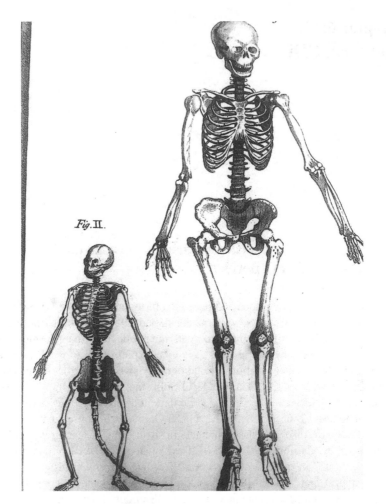

**Fig. 65.1** From a *Philosophical Account of the Works of Nature*, by Richard Bradley, W. Mears, printer, London 1721, p. 171

dynamics" (clearly an area of more unknowns than knowns) and "plant and animal distributions" in relation to biochemistry and physiology, again, extraordinarily broad swathes of gap analysis.[2]

Behavioral trends and distribution patterns are the product of perceptions, caught out in the whorls of near meaningless syncretism and exegesis (a cascade of meta-data without a meta-framework). Quasi-meaningless in so far as, to paraphrase sculptor Henry Moore, every idea has a physical shape. But in the discrepancies of thought surrounding the physical, and vice versa, historical interpretations of

---

[2]By N. Gilbert, AQ. P. Gutierrez, B. D. Frazer, and R. E. Jones, W.H. Freeman and Company, Reading and San Francisco, 1976, pp. 89–90.

cultural noumena, like ancient Egyptian cattle cults, provide no actual clues to what Brian Fagan has described as our present discernment of "a divergence between cattle as numinous – symbols of power and sacrificial victims – and their more pragmatic role as draft animals pulling plows and transporting loads, and as sources of meat."[3]

The "divergence," as with all transition, not in a moment, is measured against that metric, solely human, which also assumes the lifespan of a neutron to be 14+ minutes. The kinesthetic timepiece by which we seek to circumscribe the multidimensionality of *H. sapiens* is only rudely rooted to any specificity in mind. This mental constellation of an imprecise nature that condemns all so-called "knowledge" to a neural purgatory, a tenebrous penumbra with neither center nor circumference, resembles all those simulacrums of espoused faiths in the myriad medieval definitions of God. They are mathematically ephemeral delusions and aberrations instinct with every evolutionary fallacy. That would include what is, in fact, a merely fickle fitness, a melodrama of mutations, and the genetical parade that hoists its unanimous flag. That war cry and its banner are blind and malevolent, loyal to a single lineage of large vertebrates whose biomass and evolutionary instincts have been perverted by presuming to predicate learning and "success" upon our own species' unnatural selection. Natural or unnatural? That is the only question worth seeking something approaching legitimate context for; a distillation of all that we have experienced, thought, projected, internalized, into the most pellucid of explanations. Ecological clarification. A metacontext for human deep history (Fig. 65.2).

## 65.2 Natural, Unnatural?

At the conclusion of her recent study, *Unnatural Selection*, Katrina Van Grouw argues on behalf of every form of selection, natural, artificial, unnatural, gradual, each one providing invaluable evidence and information; each highlighting Darwin's own obsessions with domesticated animals, with the mysterious neural accumulations of change (showing no phenotypic exposure to discernible mutational change and thereby surprising us time and again). Most interestingly, she answers the challenge, in her beautifully conceived work by describing an urban encounter near her home with a silver fox that feeds out of her hand. One of countless examples, she argues, that lends logic to an entirely new notion regarding the evolutionarily positive trends of interspecies domestication, even self-domestication, a behavior that for many wild creatures has given them an advantage in seeking and obtaining food from the wild human neighbors, as in the case of the red (and silver-red) fox.[4]

---

[3] See *The Intimate Bond – How Animals Shaped Human History*, Brian Fagan, Bloomsbury Press, New York, London, 2015, p. 86.

[4] See Katrina Van Grouw, *Unnatural Selection*, Princeton University Press, Princeton, NJ and Oxford, UK, 2018, Chapter 12 – "Between Dog & Wolf," pp. 266–278.

**Fig. 65.2**   Over Los Angeles, California. (Photo © M.C. Tobias)

But while such egalitarian evolutionary views may be liberating, in breaking down absolute barriers between wild and domestic traits (an argument we have been calling attention to throughout this work, and given that such traits can exist in all species), the perspective tends to ignore the breeding and inbreeding backlash unleashed by humans. Grouw recognizes it when she acknowledges how the build-up of muscles through breeding has certainly been a boon to the meat industry.[5]

But it is this profiteering within the breeding milieu that sounds the alarm bell, and presages of slavery, and all forms of human-dominated exploitation and slaughter of the natural (and unnatural) worlds, as the case may be. One of the most graphic depictions of this human proclivity to exploit life is to be found in the Russian writer, Vasily Grossman's epic novel of World War II, and the horrors of Hitler and Stalin, *Life and Fate*.[6]

Twenty-five years ago we were emphatic that *all economic progress* must end.[7] Most recently, interdisciplinary energy experts, like Vacliv Smil, have echoed our

---

[5] ibid., Katrina Van Grouw, p. 173.

[6] Vasily Grossman, *Life And Fate, A Novel*, Translated from the Russian by Robert Chandler, Harper & Row Publishers, New York, 1985. In addition, the Shoa Foundation in Los Angeles at USC, https://sfi.usc.edu/, has accomplished a critical pillar of human understanding, empathy and memory, mostly of Holocaust survivors.

[7] Introductory chapter, *World War III: Population and the Biosphere at the End of the Millennium*, by M.C. Tobias, ed., by J. G. Morrison, Bear & Co., Santa Fe, 1994.

assertions,[8] and at a time when, like never before, it can be adduced that human life translates into ecological paradox, as well as one of its demonstrable subsets, the William Stanley Jevons Paradox, pertaining to more efficient extraction of natural resources increasing the actual human consumption of said resources.[9]

Rapid turnabouts, saltative, single-step speciation, polyploidy in plants -the macromutational theories of John Christopher Willis, Richard Goldschmidt and Søren Løvtrup, versus phyletic gradualism, or its evolutionary variants (punctuated equilibria), as devised by Stephen Gould and Niles Eldredge: These insights to change and transition represent scratches on a chalkboard of speculation and out-right conjecture; what we have recently explored in *The Hypothetical Species*.[10] The change in allele frequencies, compounding theoretically correlated shifts in populations, has become the surrogate landscape of questions with downstream implications for every aspect of human cultural introspection. Population genetics, taking after the theoretical models of Sewall Wright, Ronald Fisher, J.B.S. Haldane, and then Ernst Mayr's 1942 work on founder effects and speciation,[11] has obsessed upon these abstractions, in the guise of the whole panoply of genes. And, ulti-mately, of genomes and their wildly flagrant characteristics, so celebrated is the 1953 saga of the James Watson/Francis Crick double helix tale of DNA, that all other diagnoses have largely dissipated. To the extent that to cite some other infra-structure entirely would be deemed archaic, foolish, and, most damaging, useless.

Yet, if polymerase chain reactions oriented toward the efficient manufacture of millions of copies of DNA sequences in molecular biology and biomedical research are among the latest phantasms of the human endgame, we must carefully add that something else entirely has been overlooked. It is not that we shall ever return to a flat world; an earth around which the Sun orbits. But that the very rudiments of inquiry that accelerate human curiosity, the hammering of nails, the erection of scaffolding around each and every intellectual question mark have never prepared an anxious brain for self-destruction, or even expiation. That is the story of our colossal rub; and not just at present, but from our near origins. Whether it is genetic or some other language that yields the ultimate insights about our species is yet to be discovered.

---

[8] See "Vaclav Smil: 'Growth must end. Our economist friends don't seem to realise that.' By Jonathan Watts September 21, 2019.

[9] See, for example, "Modern Life On Planet Earth Is An Ecological Paradox," by Marlene Cimons, February 4, 2018, CleanTechnica, *Nexus Media*, https://cleantechnica.com/2018/02/04/modern-life-planet-earth-ecological-paradox/, Accessed September 22, 2019; See also, "Ecological Paradoxes: William Stanley Jevons and the Paperless Office," Human Ecology Review 13(2): December 2006, https://www.researchgate.net/publication/238684194_Ecological_Paradoxes_William_Stanley_Jevons_and_the_Paperless_Office

[10] See M. C. Tobias and J. G. Morrison, *The Hypothetical Species: Variables of Human Evolution*, Springer Nature, New York 2019. For polyploidy in plants, we are using here the definition as propounded by nature.com, namely, "Polyploidy is the presence in cells of more than a single pair of each chromosome ... the result of a spontaneous multiplication of a plants genetic material or through hybridization." See "Latest Research and Reviews," https://www.nature.com/subjects/polyploidy-in-plants#:~:text=Polyploidy%20is%20the%20presence%20in,extremely%20common%20in%20domesticated%20crops, Accessed July 19, 2020.

[11] See W. B. Provine, "Ernst Mayr: Genetics and Speciation," Genetics 167 (3): 1041–6. PMC 1470966. PMID 15280221.

Whatever we are prepared to declare on behalf of this bizarre organism is fully suspect because of its predominant solipsism. Profuse existentialist engineers (e.g., Thomas Edison, Bill Gates) propagate the religion of techno-fixes, Gates and his wife, Melinda, with an inspirationally curious, ethically pragmatic zeal that they have owned. But restraint, as etymologically enshrined in such word/concepts as the Chinese *zhongyong*, the Indic *sattva*, or early Greek *sophrosyne* do not come about in the Spaceship earth-versions of what Buckminster Fuller sought out in his synergetics, ephemeralization, doing more with less, closing gaps and using technological know-how to solve all of our problems. We can't; we won't. Whether it is the concept of a perfectly engineered planet, toilet or vaccine: humans are only human.

The *problem with human problems* remains not only *unsolved* (an indignity of a word in the context of its overwhelming hierarchical breakdown within both language and protocognition), but unimaginable. This lithospheric sculpture of epigrammatic, inaccessible philosophical quanta is the problem of humanity; of human individuals, communities, their interactions and genetic factoids, as enunciated generation after generation. In an unspeakably tense, breathtaking 13-page essay by British author, Hari Kunzru in 2006, for the slim and undermining book, *Human Problems*[12] innumerable cornerstones of the problem, the iconic pillar of our myriad discontents and reasons for anxiety, surface. These include the "crucifixion of the humming bird hypothesis," "Bot-fly larva removal," and all those who "know they don't deserve what they've inherited" (speaking of "the earth").[13] Leading to an ultimate dilemma: the fact that "to winnow the truly human from the pretenders" constitutes "The problem."[14]

And simply because, in the hands of, a problem in and of himself, an author, any author, any reader, any zoological ambassador of this enterprise accelerates and embodies the proliferation of more problems, moral dilemmas, absolute wastelands absent any compass, level of confidence or ability to delineate identities from among a biological chaos that would seek to champion order. Order upon Self; Order in the very outlines from atop some mythopoetic Olympus wherein some iteration of a Zarathustra or Buddha seeks to look down upon the earth as it lingers in anticipation, presumably several billion years hence, of its galactic implosion. As Venice, Italy sinks; and the global climate continues to rise. Yet, the real problem, herewith elucidated, is fully hidden beneath the distractions of geopolitics, and the identity crisis of a species fully caught out in the tumult of its undoing. We simply will not see ourselves.

Neither evolutionary biologists, mathematicians, biophysicists nor ethical pundits have managed to find wormholes of predictability between macro and micro fields. The "clash of civilizations," as outlined by political scientist Samuel

---

[12] Half of the work devoted to the deeply arresting art of New Zealand-born Francis Upritchard, Kate MacGarry/Veenman Publishers/Gijs Stork, Rotterdam, The Netherlands, 2006.

[13] ibid., pp. 5, 6, 9.

[14] ibid., p. 15.

P. Huntington,[15] the forensics attendant upon societal collapse, and the bankrupt collective of collectives, leaves an Anthropocene that is as blind and grasping as classical nausea; statements and confessions where observation has surrendered to a more fluent but blasphemous inhibition. Leonardo, who never hesitated to dissect a pig to improve his anatomical sketch work, worshipped mechanical objects of human construction, and he also, in his impenetrable fashion, worshipped humans whom he described as "the greatest instrument of nature"[16] (Fig. 65.2).

## 65.3   The Qualifying Task of Mathematics

Let us be clear, that all such rejoicing in human capacity is as tentative and ambiguous as philosopher Michael Strevens' sub-title, "Understanding Complexity through Probability."[17] Strevens applies Occam's razor (the least needed assumptions to proffer an explanation) to a world of probabilistic "convolutions," wherein "knotted histories and future histories make up a fantastically irregular filigree of life trajectories."[18] The outcome of his analysis—hinging upon an ecological planet in which "A creature's every move is dependent on the behavior of those around it – those who would eat it, those who would eat its food, and those who would mate with it"[19]—becomes the qualifying task of the mathematician who would translate all such variables into an ultimately simple approach to chaos and complexity, both; a preconditioning set of behavioral quirks throughout the biosphere that would account for both independence and dependence; macroperiodicity; metaphysics and physics; "microdynamic trials" and "deterministic chains."[20] A view of any large city from the air at once ignites the co-dependent variables that make for eco-dynamic chaos, inside and outside. The ratios of organic to inorganic; other obfuscations and interruptions, from ground water pollution, to acoustical interference by humans with other species dependent on what was, previous to the building of a city, a "neutral" sound space pre-existing for millions of years; the infliction of stress, cumulative waste, every conceivable adulterant. The elimination of pure habitat, and the rude, toxic juxtapositions inducing new pathways for the spread of contaminants. Los Angeles at a glance, like every other mega-teeming region on the planet that feeds into the ocean, or, in the case of a Chicago, to a great lake, is just such a monster.

---

[15] See *The Clash of Civilizations*, by S. P. Huntington, Simon & Schuster Publishers, New York 1996.

[16] See *Milan, Ambrosiana,* the 12-volume *Codex Atlanticus* 86r-a; Essay by Eugenio Garin, pp. 234–235 in *Encyclopedia of World Art,* IX, McGraw-Hill Book Company, Inc., New York, 1958.

[17] *Bigger than Chaos – Understanding Complexity through Probability*, by Michael Strevens, Harvard University Press, Cambridge, Massachusetts, London, UK, 2003.

[18] ibid., Strevens, p. 1.

[19] ibid., Strevens, p. 1.

[20] ibid., p.278.

The eco-apology, in mathematical terms, borrows from the history of utilitarianism: the worst impact on the most numbers, consolidated into a city. On a planetary scale, where tiers of triage exert considerable ethical meaning, there is some solace in the scaling of damage according to the least numbers of such metropolitan clusters, and largest numbers of demographically inflictive inhabitants, lending logic to the rural–urban migrations around the world.

Add to these coordinated belief systems the notion of *biological realism*,[21] laws of large numbers,[22] simple and complex probabilities,[23] "causal independence and the Sufficient Condition."[24] Within the overall flow of information between subjects, objects, positions and outcomes, there are maps of probability. We have discussed some of them in earlier chapters, but they conclusively all stem from the same vortex of a neutral language. That would include both the subjective and objective—a set of elemental descriptors that encompass every known *function* and *spin* of those infinite intersections between the natural sciences and fundamental physics and calculus. Hence, that would invoke such areas of physics and chemistry as the Maxwell–Boltzmann distributions (speeds of any gas at any temperature); quasi-determinism, simple behavior, random numbers and discrete space, as well as the Second Law of Thermodynamics (in an isolated system maximum entropy is eventually a constant); the Carnot heat engine (the ideal thermodynamic cycle that produces heat); stochastic averages (random markers), Markov chains (sequentially predicted probability shifts within systems); and electricity. Even with these physical, high probability distribution events and systems, there remains the question, again: biological realism?

To further grasp an image in one's mind of the testament to the proliferation of biodiversity that human thought has subsumed, several of the great libraries in the world at once provide a geography of such inconceivably beautiful and restless idealism in the guise of consciousness: The Morgan Library in Manhattan, the remarkable Folger Shakespeare Library in Washington, D.C., Trinity College Library in Dublin, the Biblioteca Angelica Roma, Národní knihovna Praha, or the Herzogin Anna Amalia Bibliothek Weimar II. In the United States, the Library of Congress hosts a harrowing "800 miles of bookshelves," "150 million items," and "66 million manuscripts."[25] What does that say about the thermodynamic cycles of humanity? Of entropy in consciousness? Or of biological realism framed within the evolutionary cycles that we glimpse only in their varietal transitional phases?

To take one example of what we mean by transitional. The prelapsarian gut (a metaphor suggesting the perfect, unspoiled, unevolved human anatomy; that moment it was born, without having tasted the death of another) teeming with its bacterial soldiers of life, does not translate into today's vertebrate cognition, not

---

[21] ibid., p. 330.

[22] ibid., p. 9.

[23] ibid., pp. 28–29.

[24] ibid., p. 147.

[25] https://www.wonderopolis.org/wonder/where-is-the-largest-library-in-the-world,        Accessed September 17, 2019.

ours in any case. So much has happened since the beginning of our biological formation. Indeed, there is no ascertainable transitional from the microbial to any other Kingdom, in as much as these things, units of measurement, categories of that compulsion to denominate, are useless letters of all those human languages that know only the mechanism of harpooning targets across every suite of the biosphere, and thence upon the printed page. And like that emboldened, ruthless fool who is the Ahab of every doomed genetic generation, and has always been he, there is no known remedy for the fatal hubris eating away at all we have dreamt. Those dreams serve only to further corrode any and all (golden age) legacies, spelling out in ravenous detail the future our Anthropocene makes more clear by the minute, from microplastics in the Arctic snow to the grotesque tally of dead fur-bearing animals in Denmark, the true victims wandering in the guise of Hamlet's ghost, the figure of a man who is certainly one of many crucial keys to all that was Shakespeare, as was the year 1606.

But that is not to suggest the least clarity. Such analysis is the wan metaphor of black holes within holes, where no measurements have meaning. Where even theoretical inference connotes only those odd mesmerizing fragments of collective delusion, from cemetery to cemetery. We're not suggesting that there is no good will or interesting outcomes in these miles-upon-miles of library shelves. Only that our biological growth has so favored consolidation in *mind*, that mind is stuffed, the way a mirror in Jain tradition is clouded by karma, impeding our ability to see and act clearly.

## 65.4    A Colossus of Migraines

What, then, is there to say about this colossus of migraines, pounding blood pressure, and grievous missteps? While multitudes positioned at the hypothetically individual bio-level have purportedly imagined some Utopian alternative to prescribed histories and mythologies; painted caves, illuminated manuscripts and the harmonious gatherings of musical cadence, each production assaying self-referential edification; the underlying problem is ineluctable. Paramount. Baffling. A malignancy as aggressive as an asteroid breaking out in body and soul with no distinguishing or differentiating characteristics that might otherwise lend themselves to repair or the slightest redemption. In the most mundane but telling guises of this problem, it is noteworthy that every national collective has debt. Those nations where the debt *appears* to be diminutive, in fact, have not internalized natural capital depreciation. That would include, of course, meat eating, which is highest in the United States, Kuwait, Australia, the Bahamas, Luxembourg, New Zealand, and Austria.[26] And lowest in Bangladesh, then India, Burundi, Sri Lanka, and Rwanda.

But there are innumerable other uncorrelated factors. India, for example, is one of the largest exporters of buffalo meat in the world (as described in Chap. 36, f. 23).

---

[26] See, "Revealed: The world's most vegetarian country," by Oliver Smith, The Daily Telegraph, March 26, 2018.

An absence of accounting for natural capital depreciation is particularly true in Russia, Saudi Arabia, Kuwait, the United Arab Emirates and Nigeria, all of which share oil and gas as common, delusional and destructive denominators. It is less obvious in Oman and Botswana.[27] And most striking about such debt is the nation of Bhutan, which ranks fifth highest in terms of the internal ratio of national debt to GDP, with Japan being number 1. Moreover, ineluctably, in absolute numbers, the human population explosion continues unabated.

Morally bankrupt, the remaining entrails claim the silliest of notions—civilization—to be the rudder and stern of this senseless figment of a shooting star gathering dust in a mindlessness. Where is it going, in so reckless and indifferent a hurry? To what leviathan of ironies do we owe such selfish and rapacious a storyline?

The view of natural versus unnatural is singularly ambiguous, giving perpetual birth to every paradox and headache. Many of our kind have ascribed to the colorful deliberations attendant upon the ideal of relativity, whose intellectual architects were confident that abstraction could materialize in one dimension or another, between wave, particle, vapor and void. That time and space and all those laws of energy were not merely the ideas of a certain great ape, but actualities of a sort that presumably matter, with equally solvent material apparitions; again, a biological reality of particular relevancy to those who survived the incinerating effects of two atomic bombs in Japan, or of other gas chambers in World War II.

Debate proves of interest for only a minute or two, before the onslaught of our internalized boredom with ourselves. Uncertainty is apotheosized as a rallying cry. Certainty, on the other hand, lost in the labyrinth of connections, as emphasized by Ludwig Wittgenstein during the last two years of his life, living in Ithaca, New York, and writing those fragments that would constitute his book *On Certainty*. "My name is 'L.W.' And if someone were to dispute it, I should straightaway make connexions with innumerable things which make it certain"[28] (Fig. 65.3).

## 65.5   A Deformative Fixity

Moments and spaces on earth where this odd and disturbing member of an Order called Primates seeks refuge so as to shuffle the playing cards of its life might be the behavior of a *deformed fixity* at the species level.[29] But taking hostage most

---

[27] See "Debt to GDP Ratio by Country 2019," World Population Review, http://worldpopulationreview.com/countries/countries-by-national-debt/, Accessed September 22, 2019.

[28] 21.4, 594, *On Certainty – Über Gewissheit*, by Ludwig Wittgenstein, Edited by G. E. M Anscombe And G. H. Von Wright, The Text In German, Über Gewissheit, Translated By Denis Paul And G.E.M. Anscombe, with An Introduction By Arthur C. Danto, And Twelve Illustrations By Mel Bochner. Arion Press, San Francisco, 1991.

[29] See "The Evolutionary Basis of Some Clinical Disorders of the Human Foot: A Comparative Survey of the Living Primates," by T R Olson, M R Seidel, National Library of Medicine, PMID: 6873777, DOI: https://doi.org/10.1177/107110078300300603, https://pubmed.ncbi.nlm.nih.

**Fig. 65.3** Animal Cruelty
at an Open Market in Saudi
Arabia. (Photo ©
M.C. Tobias)

Others in the event, from the evolutionary perspective, defies any known rules of
biology. *H. sapiens* are one of 701 formally described primate species, a mammal allegedly devoted to love and nurturance, out of some 5500+ other remaining mammalian species on the planet.[30] Between the certainty and uncertainty

gov/6873777/; See also, Biology Letters, "Evolution in caves: selection from darkness causes spi-
nal deformities in teleost fishes," Julián Torres-Dowdall, Nidal Karagic, Martin Plath, and Rüdiger
Riesch, Published:06 June 2018, https://doi.org/10.1098/rsbl.2018.0197, The Royal Society
Publishing, https://royalsocietypublishing.org/doi/full/10.1098/rsbl.2018.0197; See also, "Gene
Behind Rare Birth Abnormality Is a Window on Evolution," by Jeffrey Norris, July 30, 2014,
University of California-San Francisco, https://www.ucsf.edu/news/2014/07/116451/gene-
behind-rare-birth-abnormality-window-evolution, Accessed June 20, 2020. See also, "Inbreeding
shaped the course of human evolution," by Michael Marshall, New Scientist, November 27, 2013,
https://www.newscientist.com/article/mg22029453-500-inbreeding-shaped-the-course-of-human-
evolution/; and, "What causes deformities in frogs, toads, and other amphibians?" USGS, https://
www.usgs.gov/faqs/what-causes-deformities-frogs-toads-and-other-amphibians?qt-news_sci-
ence_products=0#qt-news_science_products, Accessed June 20, 2020.

[30] See Introduction by Russell A. Mittermeier, in *Handbook of the Mammals of the World, Volume
3, Primates*, Chief editors, Russell A. Mittermeier, Anthony B. Rylands & Don E. Wilson, Lynx
Editions, Barcelona, Spain, 2013.

paradigms, a deformed life force may be seriously difficult to reckon, certainly as an appropriate starting point for all philosophy. Noting, in particular, the moment of Pythagoras' death, when he allegedly whispered, "better to be killed than to speak."[31] History records the betrayal of reason. The fragments of outrage combining into an historic anti-vivisection movement suggest some degree of hope amid darkness. Slamming scientists as proponents of "experimental torture," Henry S. Salt, in his groundbreaking *Animals' Rights*, compares our "whole system" of "natural history" with "deplorably partial and misleading method..." Like that "child" described by Michelet who "disports himself, shatters, and destroys; he finds his happiness in *undoing*."[32]

But such lessons, as articulated by Salt, are accentuated with every added blow. Principles remain, of course, up to a point. Even the 69 pillars of the Parthenon have barely managed to hold off disintegration, most notably during the conflict between Venetian armies and those of the Ottoman Turks; one battle in particular which, on September 26, 1687, resulted in a massive explosion that destroyed much of the venerable Doric colossus on its hill. Nineteenth-century rusting iron pins holding together much of the marble (prior to more recent restoration efforts utilizing titanium) also contributed to the building's veritable downfall.[33] The sum total of fragile ethics whose pediments have been caught out in the convolutions of human conflict and inconstancy eludes firm grasp swept ahead in a lava flow of evolutionary transition. It is some kind of Yeti of the incredulity that is our purview within this particular solar system we view as our temporary home; and these bipedal bodies.

All of these ephemeral bastions of self-reference, these autobiographical self-absorptions bespeak of a general theory of the vague, about which we have every reason to be both terrified and interested. It is us, as much as we reject some of the more damning claims.

---

[31] See *The History of Philosophy*, by Thomas Stanley, 2nd edition, London, 1687, p. 506.

[32] Henry S. Salt, *Animals' Rights -Considered in Relation to Social Progress*, George Bell & Sons, London, 1892, p. 92.

[33] See "Parthenon," https://resources.saylor.org/wwwresources/archived/site/wp-content/uploads/2011/08/HIST361-2.2.2-Parthenon.pdf

# Chapter 66
# Mismatches

## 66.1 Who Are We?

The infernal machinations of human mentation loom as a furnace where misgivings are further battered like atoms subjected to the steam of their reactivity. A human would be rare and hard-pressed to escape the ingredients of this peril: Intuition, logic, illogic, impulse, paralysis, and horror all ravenously accrete into a subject that can be tentatively psychoanalyzed, with no outcomes; a science of Self destined to become the chatter of a most charitable, gullible, and prolific solipsism. Ipseitous self-regard is the psychological mismatch in a world of biodiverse-rich opposition to the solitary mirror in which humanity has entangled itself (Fig. 66.1).

Let a propositional heresy be substantiated as preamble before all the egocentric forensics necessarily to follow. Exhume the evidentiary mob. The 90+ species represented in cannibalized bones at a Zhoukoudian Cave system in suburban Beijing, with its Acheulean hand axes, chopping and cutting tools, all the hallmarks of forced stressors. We can easily surmise that these killing objects of stone did not originate with *Homo erectus* or *H. heidelbergensis*, but in the likely confines of informal conceptualization millions of years prior. The neural complex leaves few genetic clues, and even if it did, the correlations between parts of a genome and parts of a concept is but the prelude to an astonishing realization: Ideas are not readily undertaken, exploited, or tallied by natural selection. Rather, with the continuing hindsight, accumulation of victims, and the gaps in natural history their endlessly unsettling eviscerations (mutations, variations) have incited, and speak to evolution working in its own merciless ways to first isolate the offending taxon, prior to an otherwise defeated world's final ultimatums and biological choice for dealing with, and dispatching so rogue a species.

© Springer Nature Switzerland AG 2021
M. C. Tobias, J. G. Morrison, *On the Nature of Ecological Paradox*,
https://doi.org/10.1007/978-3-030-64526-7_66

**Fig. 66.1** Burgos Complex wall paintings, Heretofore Unknown Peoples, Tamaulipas State, Mexico, ca. 4500 BP. (Photo © M.C. Tobias)

## 66.2   The Earliest Evidence

All of the earliest stone kits date to the growth of ideas hovering over preconceived manipulation. The relations are so well practiced as to have charged the cumulative experience of our ancestors with every sense of halo, hubris, and equal nuance of lamentation. The combination ineluctably enflames a chaos of pains laid bare across vast swathes of the planet; the clash of clans, formation of armies, and an unrelenting bravado. As recently as 1964 the hominid brain was examined in terms of proportions within those areas devoted to the tongue, the toes, the pharynx, the upper limbs, and so on; as they precipitated aspects of motor function and vocalization. Writes André Leroi-Gourhan in his monumental work, *Gesture and Speech* (*Le Geste et la parole*), "The point at issue is therefore to demonstrate the technicity (to the exclusion of other forms of intelligence…) was an early anthropoid characteristic and that its character remained the same throughout the family as a whole."[1] Three years after *Gesture and Speech* was first published, Leroi-Gourhan wrote, in his other masterpiece, *Treasures of Prehistoric Art*, "With a symbolic

---

[1] *Gesture and Speech*, André Leroi-Gourhan, translated from the French by Anna Bostock Berger and introduced by Randall White, An October Book, The MIT Press, Cambridge, Massachusetts, 1993, p. 86.

intention that is closely related, the spear seems to have been assimilated to the male and the wound to the female: wounded animals are usually found in compositions that do not contain signs, in which the presence of such animals is explained by their taking the place of signs."[2]

History performs a curious fancy in its cloying detective work, unearthing patterns culled from private diaries of life, documenting every miniature implosion and despair. As community life obfuscates lives and gives us our first salient impression of the valency of categories, panmixia as it applies expressly to hominin hybridizations results in a species strange to us, though it is ours alone.

Wrestling with pariahs in absence of any theory, as yet; objectifying obstacles with such adamantine resolve as to ever punctuate the life of the human mind, as if it actually warranted this berserker entitlement within the greater surroundings.

Social souls, insects, avifauna, neo-tropics, corals, unfathomable repositories of vibrant microfauna, all defy the gravitas of odds by mathematically overwhelming chance. And to what end? *Homo sapiens* have endeavored to re-create biochemistry in total isolation, non-consensually. Golem and Rosenthal effects (unrealistic expectations of others) have so tainted the underlying anthropic zeal of our quest for every exit as to have ensured an absolute zero value; a generalized valueless status that has no comparable anywhere in natural history.

It must be articulated. It has already been demonstrated ad nauseum. So coveted is the political hierarchy of an indwelling obsession incredibly oblivious to the doom and gloom it courts, no other exchange—be it of morals, ideas, increments or preference—can exert traction upon what had long been an unceasing, tautological onslaught of a proprietary monologue. This declaration of our exile is the language of false analogies; of musclemen metaphors and similes gleaned from a clear enough record; the utterly blurred minutiae of mimicry. Our human library of experience has no muse but ourselves, which means no bonobo, no kangaroo, no octopus, no redwood, no rose can enter into the conversation, except as a dead specimen. While it is true that primates have everything in common with their resident bacteria, there is no actual primacy that can be divined in humankind's relational vacuum. As Roland Barthes stated, "I call Neutral everything that baffles paradigm." And he added, quoting from Maurice Blanchot's *The Infinite Conversation* (*L'Entretien infini*, 1969 *The Infinite Conversation*), "Weariness is the most modest of misfortunes, the most neutral of neutrals."[3] And later, cites Jean-Jacques Rousseau's "Fifth Walk," writing, "This is the island on which I sought refuge ... to forbid me any kind of communication with the mainland so that being unaware of all that went on in the world I might forget its existence and that it might also forget mine."[4]

---

[2] André Leroi-Gourhan, *Treasures of Prehistoric Art*, Harry N. Abrams, Inc., New York, p. 205.

[3] *The Neutral*, by Roland Barthes, Translated by Rosalind E. Krauss and Denis Hollier, Text established, annotated, and presented by Thomas Clerc under the direction of Eric Marty, Columbia University Press, New York, 2005, p. 20.

[4] From Rousseau's *The Reveries of the Solitary Walker*, translated by Charles E. Butterworth, NYU Press, New York, 1979, quote in Barthes on p. 139.

That is the setting of our *zero minus all sums*. A species, in other words, that has abrogated all outright affiliation. Or, as Wallace Shawn composed in his lyrical and unsettling play *The Designated Mourner*, "All that endless posturing, the seriousness, the weightiness, that I was so sick sick to death of – I'd never have to do any of it ever again."[5] Or, in a similar guise, as bespoken by T. S. Eliot, "Irresolute and selfish, misshapen, lame, Unable to fare forward or retreat."[6]

## 66.3   Both

Of course, we fondly recall those pinnacles of apparent integrity. Thales, when asked about God is said to have deliberated (giving no certain account) "to no effect," a deliberate paradox that comported with his equal belief that all matter was fluid and variable, the soul always in motion.[7] Of Mahavira's own substantive retreat (to Rousseau's island of Saint-Pierre in the Lake of Bienne, as it were), namely, the rite of sallekhana, of fasting to death. Archaeological evidence attests to inscriptions at Sravana-Belgola, in India's southern state of Karnataka, that indicate at least 59 cases of such Jain spiritual suicide during the seventh and eighth centuries, practiced not merely by the clergy, but by laypersons as well.[8] Porphyry, Clovius, Erasmus Darwin, Percy Shelley (Fig. 66.2).

All of these iconic testimonies and genomic memories stir the caudal samavasarana of a Saint Francis (Fig. 66.2) and his interactions in Sassetta's dozens of original panels of Borgo San Sepolcro Altarpiece, and of St. Bonaventura's biography. Obviously, the Van Eycks' "The Adoration of the Lamb," Giorgione's "La Tempesta," Vermeer's three masterpieces, "Artist in his Studio," "Girl with Pearl Earring," and "View of Delft." Each one gives us to marvel upon the intersections of our species' paradoxical gifts for great beauty and, adjoining its core, or enshrouding it, a strange and most unflattering madness which by every account is quintessentially a font of specifically human contradictions. We don't see it with hippos or magpies or even 41,000-year-old Arctic nematodes.

So great and dire a mismatch of proclivities in humans as to evoke that sudden dizziness of elevation gain towards the obvious conclusion of Beethoven's "Ninth,"[9] or Handel's "Messiah."[10] These mammoth artifacts of a 300,000-year-old human

---

[5] *The Designated Mourner*, Wallace Shawn, Farrar, Straus and Giroux, New York, NY 1996, Part Two, p. 91.

[6] From the conclusion of *Animula*, Wood Engravings by Gertrude Hermes, No. 23 of The Ariel Poems, Faber & Faber Ltd., London 1929.

[7] See *The History of Philosophy*, by Thomas Stanley, 2nd edition, London 1687, pp. 6–12.

[8] See *Jainism in Early Medieval Karnataka c.A.D. 500–1200*, by Ram Bhushan Prasad Singh, With a Foreword by Professor A. L. Basham, Motilal Banarsidass, Delhi, 1975, p. 21.

[9] See *Beethoven's Ninth – A Political History*, by Esteban Buch, Translated by Richard Miller, The University of Chicago Press, Chicago and London, 2003.

[10] See *Handel's Messiah – A Celebration*, by Richard Luckett, A Helen And Kurt Wolff Book, Harcourt Brace & Company, New York, 1992.

**Fig. 66.2**  St. Francis with the Animals, From *Vita Del Serafico S. Franceso*, Scritta Da S. Bonaventura, 1593, p. 96. (Private Collection). (Photo © M.C. Tobias)

odyssey bewilder the foundations of optimism, punching holes entire within any argument that tries sociobiologically to reassure or counter doom. The species with the shortest of all epitaphs. Our species, however statistically differentiated at the individual levels, owns the Holocaust (Fig. 66.3).

Yet, we are *different*. We carry from generation to generation a phylogeny of "Ecclesiastes," the "Peaceable Kingdom," those critical years between 1606 and 1612 when Jan Brueghel the Elder was focused upon his commissions by Milan's Cardinal Borromeo to illustrate every detail of Paradise, about which in depth treatises were composed by John Salkeld in 1617,[11] Marmaduke Carver in

---

[11]*A Treatise of Paradise and the Principall contents thereof*, Printed by Edward Friffin for Nathaniel Butter, London 1617.

**Fig. 66.3** Near Vilnius (Lithuania) where 100,000 Jews were shot by the Nazis in 1941, Paneriai Massacre, photographer unknown. (Public Domain)

1666[12] and to which an actual map was designed in 1678 by M. Johanne Herbinius in his *Dissertationes De Admirandis Mundi Cataractis, Aestus Maris Reflui Paradiso*. With such cartographical perfection in mind, our erratic idealisms and concomitant representational compulsions are a finishing school of self-assurance. Hence, the rapidity of our odious sense of well-versed superiority over those termites, flies, cockroaches and Platyhelminth parasitic trematodes that only enrich our disgust played out in the classification of Others. This contradictory nature has been summed up in the formation of the United States, with its perplexing celebration of freedom for all but those deemed to be slaves. Aristotle's Greece suffered from the first blatant inkling of this paradox that benights our otherwise sacrosanct pledges, accrues rust and irreligion on statues of liberty, records our massacres with the historian's fated perpetuity, learning pain and more pain; and disseminates the bipolarities throughout every vista, between picture postcards and prisons.

---

[12] *A Discourse of the Terrestrial Paradise Aiming at a More Probable Discovery of the True Situation of That Happy Place of Our First Parents Habitation*, Printed by James Flesher, London 1666.

# Chapter 67
# True Narcissism

## 67.1 Unto a World War III

There are two particularly compelling literatures endemic to the human twenty-first century: epistolary revelations—hours, minutes even seconds prior to suicide; and the owning up at every ground zero that encompasses the toxic chemistry of collective infliction. Economists have long sensed "the possibility that subjects suspect deception" and that this represents "a potential impediment to inducing even the simplest distribution."[1] Predictive science is a function of density, laws of irreconcilability, and absolute separation, and how these products of the imagination impact a person in her/his private-most moments. A case in point: Franz Marc having "virtually prophesied in his art the holocaust that was to follow"; and the similarities of his 1913 painting "Fate of the Animals" with his own death at Verdun; his embodying of "the pathetic fallacy from the domain of botany to the domain of zoology"[2] (Fig. 67.1).

Some 4 years after the ending of the World War II, Samuel Beckett's *Waiting for Godot* seemed clearly to summit the first half of the twentieth century. As Anthony Cronin has analyzed in his biography of Beckett's darkest, most penetrating work, jettisoned there amid a mere tree and country road one wandered at length upon "a holiday atmosphere" and "funny jokes about the genuinely worst aspects of human existence," a tradition of final summing up that included the likes of "Laurel and Hardy" but also "the master and man situation which has a centuries-old history in European literature, from Roman times through Cervantes and Le Sage to Tolstoy and Goncharov."[3]

---

[1] See *Experimental Economics*, by Douglas D. Davis and Charles A. Holt, Princeton University Press, Princeton, New Jersey, 1993, p. 71.

[2] See *Modern Painting and the Northern Romantic Tradition – Friedrich To Rothko*, by Robert Rosenblum, Icon Editions, Harper & Row, Publishers, New York 1975, pp. 141–142.

[3] Anthony Cronin, *Samuel Beckett – The Last Modernist*, HarperCollins Publishers, New York, 1997, p. 391.

© Springer Nature Switzerland AG 2021
M. C. Tobias, J. G. Morrison, *On the Nature of Ecological Paradox*,
https://doi.org/10.1007/978-3-030-64526-7_67

**Fig. 67.1** Samual Beckett
(1906–1989) in 1977,
photo by Roger Pic,
Bibliothèque nationale de
France. (Public Domain)

Our species' dualism baffles all corners of perception, undermines theories and catechisms with the gusto of an avalanche. What could we have possibly been thinking, to have divorced a universality of calm tenderness, embracing instead the sheer Hells we have imposed? All go by the names of evolution, purpose, destiny, and self-deification.

The mess is unprecedented. The anti-natalism expressed in *Sophocles' Oedipus at Colonus*—Chorus: "not to be born is, beyond all estimation, best"—as described by David Benatar,[4] counters a dominant pulse in spite of its pursuits of philosophy in Paris, or architectural serenity in Bruges, musical genius in Vienna and great poetry in Bengal. That our genesis truly belies a seemingly uncalculated *mistake*, which every other species has called it out for what it is. But we tremble intellectually upon the moral high grounds that have always been our collective excuse and fallback.

Yet, our mistakes are self-evident, but for the true narcissism our personal agendas have perpetually unleashed upon the world. An indomitable throw-weight that is not, in any way, a vouchsafed part of the biosphere, but a kind of colossal seizure

---

[4] See "Can it be better to never have been born? A philosopher argues against procreation," by David Benatar, CuencaHighLife, "Opinions," November 14, 2017, https://cuencahighlife.com/can-it-be-better-to-have-never-been-born-a-philosopher-argues-against-procreation/, Accessed June 21, 2020.

that leapt from the shores of its primordial slime into the heady ranks of large vertebrate company. Outsiders and intruders from our very beginning, we have fast exceeded every threshold of viability. In the horror stories that are our biological misadventures, a through-line would seem to tell all. We have, on the condition of our own diseased pertinacity, perpetuated the scientific, psychological, and head-strong lies of this comedy of errors, and not funny at all.

What is that betrayal? That aforementioned pathetic fallacy illegitimately linking our self-reference to the rest of Nature? To every conceivable extent has our species' addictions to a comprehensive falsehood empowered our mendacity, elucidating a clear-enough cause-and-effect: How this historically addled font of self-fictions resulted in the ousting of all those Others and the most blatant scorched-earth policy ever mounted—this World War III whose reckoning is our autobiography; adjectives vying with superlatives for the darkness of our very presence? We are the destroyers and our light is actually a void.

## 67.2 Conceptual Transitions

It gives no pleasure to parse such sad, if remorseless and recondite tidings, but they do not exaggerate. They can only underestimate the full measure of ecocide. Yet, we will not see or hear of it. Our perceptions miss the core as well as its peripheries. Diameters have lost, in our guise, their equal parts. The celestial spheres that we observe and ponder, their myriad galaxies and forces, have not dented our resolve to implode down below, here, on the little blue planet we once thought to be a permanent homestead, full of imaginings of our own involvement; devoted to representations of our presence, which some are content to label the history of art.

The deep ethology of connections is pitted against our wanton, pathological disregard of that interdependency once hallowed as a fundament preparatory to all birth, sustenance, and survival. At some point in our ontology—variations on a theme of transition—we changed dramatically, from one timid, low-density being, to a very different character study. We seem to have avoided switchbacks and gone straight for the kill in our evolution. By the time of Terra Amata in modern-day Nice, the behavioral suite of changes from one thing to another involved a cladistic time-bomb of inflictions now in full swarm. There above the Mediterranean, approximately 400,000 years ago, the butchery was blatant. "Over 12,000 animal bones and bone fragments" have been exhumed at Terra Amata[5] (Fig. 67.2).

By 10,000 years ago, with the emergence of human-mediated crops like flax, beans, cotton, squash, barley, millet, and numerous other herbaceous legumes, our foraging style had evolved into mini-industrialized agriculture converging around permanent residencies. A story told over and over again. Human conceptualization

---

[5] See "Terra Amata (France) – Neanderthal Life on the French Riviera," ThoughtCo., by K.Kris Hirst, July 28, 2019, https://www.thoughtco.com/terra-amata-france-neanderthal-life-173001, Accessed 8/20/2019.

**Fig. 67.2** In the Badiyat Al-Sham Region of southern Syria, a cradle of agriculture. (Photo ©
M.C. Tobias)

of the world had clearly been self-altered. The catalysts were responsive to—among
others—post-glacial retreats, fairer weather, resulting nutritional opportunism,
more stable supply and demand, greater confidence in community and family stability, from the Hadramaut regions of eastern Yemen to the Badiyat Al-Sham of southern Syria, to a geographic span encompassing Northern Ethiopia, prehistoric
Europe, and many parts of Southeast Asia and Melanesia. "mtDNA and
Y-chromosomal DNA markers" have provided more than sufficient evidence for the
genetic modeling and proximate itineraries of these early, robust populations.[6]

Relics of these incipient farming systems, laid out in any contemporary composite, yield windows upon the original, pragmatic designs favoring long-term survival
strategies. But the predilections for violence never deviated. From the southern
Pacific to China, the remains of butchery in countless occupied sites going back at
least 3.4 million years, by early versions of *Homo*, are unambiguous.[7] Drawing
unbreakable connections from stone tool use, to analysis by Thucydides of hostilities
between Spartans and Athenians, and then on to the "clumsy and illogical document"

[6] See "The Human Genetic History of South Asia," by Partha P. Majumder, Current Biology,
Volume 20, Issue 4, 23 February 2010, pp. R184–R187, https://doi.org/10.1016/j.cub.2009.11.053,
Elsevier,    https://www.sciencedirect.com/science/article/pii/S0960982209020685,    Accessed
August 21, 2019.

[7] See "Ancient Cut Marks Reveal Earlier Origin of Butchery," by Kate Wong, August 11, 2010,
Scientific American, https://www.scientificamerican.com/article/ancient-cutmarks-reveal-butch-
ery/, Accessed August 21, 2019.

that was the "constitution of the German Empire," as necessarily undertaken and managed by the King of Prussia, we read of a delivered posterity in which is clear enough fodder for second-guessing our species' present hegemony.[8]

*Homo sapiens* thus set the stage and courted a melodrama all too familiar to our kind. Our recent evolutionary history reads of bifurcation points, preposterous high-grounds, antipodal modalities between those brief interstices of relative calm, and the all-out assaults that had become the norm; salient, defining destinies soon to have engulfed a planet entire, whether deliberately or inadvertently determining at the collective, conceptual level to do so.

## 67.3   The Historical Amalgam

The authorship of so much vice is easily contested, given an ethnographic record of very small and scattered vegetarian communities. But their amalgam does not appear to have exerted much suasion on the critical mass. This is the real story, the one we resist to speak of because its intellectual and emotional reverberations will not admit to any clear exit strategy. Yet, many are the poets who have guessed the fineries of its final denouements, the bangs and whimpers, eco-crashes and epitaphs with no voice left standing to utter their name; or even guess upon the origins of such exiles. A careful scan of the cacophonous tombs by the likes of Oswald Spengler, Arnold Toynbee, Will and Ariel Durant, Edward Gibbon, Herodotus, Xenophon, Pliny the Elder, Clarence Glacken, Josephus, many others, makes abundantly clear that we *made* ourselves. A psychological equivalent of enormous, global "negative-yielding corporate debt."[9] Species-wide, we don't have a lens for gauging the extent of our outcast zoology, our banishment. Dante and John Milton certainly sized up the cosmic melancholy of our predicament, our fall from any sense of grace. Having done it to ourselves, the absence of a baseline by which to judge our predicament has followed upon our entire story. Never has an evolutionary cul-de-sac been so written in the open air—trillions written off in blood—for all to see (Fig. 67.3).

Simply told, our histories, by such volumes of insistent reiteration and echo, built up the edifices of this killer vertebrate. The evidence of the Anthropocene dates back those millions of years in that our collective energies clearly chose a path, which only in more recent millennia and centuries has been abetted by technological aids and the battering rams of a rushing crowd. As Elias Canetti, with every sentence of his masterpiece, *Crowds & Power*, has reminded us, "For the same

---

[8] See *On the Origins of War and the Preservation of Peace*, by Donald Kagan, Doubleday, New York, 1995, pp. 64 and 85.

[9] See "Surge in corporate debt with negative yields poses risk 'unlike anything' investors have ever seen," by Jeff Cox, Wednesday, 21 August 2019, 8:50 AM, CNBC.

**Fig. 67.3** Expulsion from Eden, Mezzotint by John Martin, from John Milton's *Paradise Lost*, Book X1, p. 326, printer Henry Washbourne & Co., London, 1853. (Private Collection). (Photo © M.C. Tobias)

people who have cause to lament are also survivors. They lament their loss, but they feel a kind of satisfaction in their own survival... they are always perfectly aware of what the dead man's feelings must be. He is bound to hate them, for they still have what he has lost, which is life."[10]

[10] *Crowds And Power*, by Elias Canetti, Translated from the German by Carol Stewart, The Viking Press, New York, 1962, p. 263.

# Chapter 68
# Caesuras of Certainty

## 68.1 "Crawling at Your Feet"

The problems exacerbating the confident extrapolation of deep signals from the space of natural history are not simply the size of our inner radio telescopes, but the reach of our hearts; the bias of our bipedal height from the ground, fear of being stung, and an evolutionary ambivalence toward tiny invertebrates whose 480 million year genesis, since the time of the Ordovician and subsequently, 80 million years later, that moment in the Devonian when some insects took to flight, all account for our utter insecurity around swarms and beings far more pertinacious and stout of character than hominids. While many cultures, like that of the East African Hadza, feast on insects and children delight in squashing or protecting them, for many natural historians, the problem with an emphatic embrace of insects that is not simply destined toward those grisly chambers of torture where thousands of "bugs" are shown embalmed, pinned to cabinets of curiosity in one museum after another, is our reliance on the word "pest" and its application to industrial agriculture, a utilitarian shift in Neolithic thinking that smashes the glory of the Bread-and-Butterfly, as well as the Gnat who warns Alice, "Crawling at your feet"[1] (Fig. 68.1).

The endlessly curious relations between tiny creatures and ourselves are emblemized in such instances as Handasyde Buchanan's point that "In spite of the title of Meria Sybilla Merian's great book *Insects of Surinam*, of 1705 (*Dissertatio de generatione et metamorphosibus insectorum Surinamensium*), it is always considered a flower book."[2] Buchanan traces many of the finest, most colorful, and detailed explorations in print of the invertebrate world—J. J. Ernst (1779–1793), Edward Donovan (1800–1804),, and James Say (1824–1828). Butterfly books

---

[1] Chapter 3, "Looking-Glass Insects, of *Through The Looking-Glass and What Alice Found There*, by Lewis Carroll, Macmillan And Co., London, 1898, p. 58.

[2] See Handasyde Buchanan, *Nature Into Art – A Treasury of Great Natural History Books*, Mayflower Books, New York, 1979, p. 177.

© Springer Nature Switzerland AG 2021
M. C. Tobias, J. G. Morrison, *On the Nature of Ecological Paradox*,
https://doi.org/10.1007/978-3-030-64526-7_68

particularly have won over the public—the brilliant *Histoire Naturelle des Lépidoptères Exotiques* of Pierre Hippolyte Lucas (F. Savy, Libraire-Editeur, Paris 1845), for example, and the 1718 masterpiece by Merian, *Erucarum ortus, alimentum et paradoxa metamorphosis.*

But ultimately, with few exceptions, butterflies and lady bugs most notably, the human species appears to despise most Insecta and Arachnida. At issue is our very uncertainty about them, and our place, versus theirs, in the world. Tics, mites, scorpions, and the like make us nervous. Or worse.

Taken to the level of what we like to think of as human civilization (only entomologists and very generous philosophers spend much time discussing insect and spider civilizations), the mussiness of Ludwig Wittgenstein's final essay on "Certainty" allows little doubt as to our own infernal illogic, as we extrapolate from the exploded scenes of our own wanton stings, crushing crusades, ephemeral glories, and overall fireworks—no Hubert Robert, Carlo Dolci or Master of the Parrot: this is the humanscape—between the topological powerlines, the rivers of effluent and freeways of strewn axons firing. Every human expression, want, remnant translates into no visibility conditions. Our sense cannot evade or gain, even internal perspective on this most essential of all dilemmas. We are trapped in our perceptual finitude. Socrates and most subsequent thinkers have tired of the impossibility.

**Fig. 68.1**  Selfie in the Anthropocene. (Photo © M.C. Tobias)

To be confined in this matter is not the same as an embedded coal miner waiting for rescue; or the estimated 170,000+ orphans of Romania's dictatorship under Nicolae Ceauşescu.[3]

## 68.2   Circumstances of History

All of the circumstances of history form one of many contexts shaping this debacle that is man. You cannot lay it off, any of it, to congressional testimony or the stored wisdom crowding the shelves of an old bookstore in Reykjavik, with its dusty Norse sagas. Delusionality and ultra-ultra-rapid cycling bi-polar disorder, or something even more medically disturbing, clearly untreatable. A reign of biocidal terror. Even at the first signs of our arrival we left a stark, and readable account. A conversation between a man and his soul, or "Ba," as figured in the Berlin hieratic Papyrus #3024, from Egypt's Middle Kingdom Dynasty (1900–500 BC). And then, in 1605/1606, Shakespeare's "Timon of Athens," whose "churlish Philosopher" Apemantus warns before Cupid and the Amazons, "Like madness is the glory of this life... We make ourselves fools to disport ourselves;... Upon whose age we void it up again, With poisonous spite and envy, Who lives that's not depraved or depraves... Would one day stamp upon me: it has been done; Men shut their doors against a setting sun."[4] Beautiful prose. But what is the part that is beautiful?

The centrality of such target-specificity—the human heart, its enshrouding minds—seemingly immune to the forces converging upon it to stay the brigandage at every level concerning humankind, it must be said, and repeated, is of an especial kind whose ecological fate cannot be doubted. Imagine Rome, circa 133 BCE, the first city to exceed one million. We scruffily deny and look away, all pretending that for now we sleep or stand, upon this planet fast orbiting a star.

Our descent into this dead-end has not yet registered. Might it impact foreknowledge? Might we try to change? We collectively can or cannot. Despite all the signage propelling and translating hope, we fail to spell it right or seize beyond the rally some oddball curve amid the pressures favoring brute force. All of the evidentiary testaments we leave behind will not occupy the mind of our successor species. We shall have vanished. Given this starting point, ideas of a future haven are but mathematical fantasies. We can correlate every invented image of the brush, of music to our ears, assigning numbers randomly to a Michelangelo here, Bach there. Such numerical topologies do not begin or end, they have no traction because the equations to describe no exit, final exit, utter devastation, are like some literary

---

[3] See "30 Years Ago Romania Deprived Thousands of Babies of Human Contact," by Melissa Fay Greene, The Atlantic, July/August 2020, https://www.theatlantic.com/magazine/archive/2020/07/can-an-unloved-child-learn-to-love/612253/?utm_source=pocket-newtab, Accessed June 21, 2020.

[4] "Timon of Athens," *The Complete Works Of William Shakespeare*, Edited with a Glossary by W. J. Craig, Oxford University Press, London, 1947, p. 799.

heteronym, an actor playing a role and using numbers toward the countdown and final lines. They mean nothing. As Estragon and Vladimir, in Beckett's "Waiting For Godot" ramble on, Estragon: "We might try him with other names … That'd pass the time. And we'd be bound to hit on the right one sooner or later"[5] (Fig. 68.2).

**Fig. 68.2** "And is himself the great Sublime he draws," Alexander Pope on *Dionysius Longinus On the Sublime*, from J. Wall design, and engraved by G.Vdr. Gucht, translated with notes and observations by William Smith, E.Johnson, London, 1770. (Private Collection). (Photo © M.C. Tobias)

---

[5] "Waiting For Godot," a tragicomedy in two acts by Samuel Beckett, Faber and Faber Limited, London 1955, p. 83.

## 68.3   What Is Beautiful?

But that is not to concede that the beautiful images of a Kuo Hsi or Li Cheng, a Sultan Muhammad or those imagined dynamics between Gilgamesh and Enkidu do not herald this idea, further explored in Italo Calvino's 1957 novel, *Il Barone rampante*, of a rim of green so ever verdant scattered light in which the semi-wild child illustriously brachiates and there situates his entire conception of life and of duty to life. That place so finely elucidated in Edward O. Wilson's *Biophilia*, when he writes, "Nor is there anything foreordained or otherwise trivial in the aesthetic optimum of human beings"[6] (Fig. 68.3).

By that one can easily envision our expulsion from Paradise and by every conceivable turn, our mortal frames. Think back to Giovanni di Paola's "Paradise"

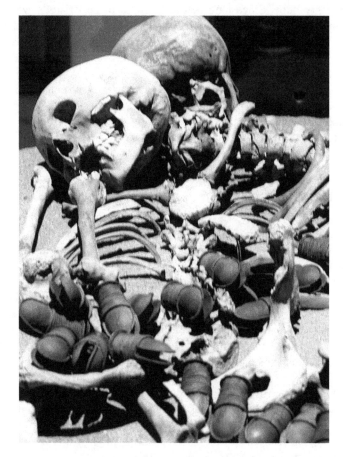

**Fig. 68.3** Post-conquest ritual burial remains, northern Mexico. (Photo © M.C. Tobias)

---

[6]*Biophilia*, E.O.Wilson, Harvard University Press, Cambridge, Massachusetts, and London, England, 1984, p. 79.

and her first two notable expatriates; or to Jan Van Eyck's "Crucifixion" and "Last Judgement" panels at the Metropolitan Museum of New York. In such iconographic halos of artistic investiture, we all suppose that "hope" is the fitting approach as we seek something beyond the rudiments of a peat bog without end; of icy gales in the Outer Hebrides and the wet mop of a beleaguered sheep's coat, fully exposed, or of all that bacterial evanescence in the gut of a maimed mammal walking her last.

Ralph Harper, luminary behind *On Presence: Variations and Reflections* (1991), often referred to as the first philosopher to bring existentialism to North America, mused, "It is a measure of the ambivalence of Athos that with so much spirituality to gain, so much that is human and artistic must be renounced."[7] Knowledge does not equate with our reality and these two components have been brought down to the raw plane of intangibility. We have lost access to that which we persistently crave, our only hope. Since long before the time of an Aristotle we have been tempted and teased by goodness and ethical character, only to be stifled at the collective moment of outreach or ratification. Rembrandt plunges into bankruptcy. George Orwell despairs of humankind ever changing its ways as he acknowledges one tribe after another effacing neighbors and predecessors. Until the science of the future, at least as concerns humanity, is no more than the concise history complete of ill-tidings and ill-bodings. Whatever painterly vision seemed plausible before, does so with less and less frequency by the minute, from community to community.

## 68.4   An Outcast Species

Confronting this reality of an outcast species situates our vantage point beneath a dark cliff, with all of its literary and emotional conceits. The storm clouds taunt us with their silver lining scenarios, the ones that are quick to enliven in us a doomed conception of some tactical escape route that is, alas, fantasy.

And to this one importunate longing we are haplessly wedded, so nobly detailed in the many clear bells and imperfect unions of *the lost ones*, as Beckett named them. All else resides in the elusive Eleusinian mysteries and others whose energies and group mind hold the promise of eternity, while acknowledging that joy is momentary; Buddhism's piercing reconciliation with the power of impermanence. And our suffering only heralds more of the same. Sentience at the human level, informed by whatever ruse of reason or rage, selects itself in every mirror held up to Nature, and this is key to unlocking the otherwise unbreakable promise we tell each other.

---

[7] Ralph Harper, *Journey From Paradise – Mt. Athos And The Interior Life*, Editions Du Beffroi, 1987, p. 68.

Whatever the dreams, our time works against us. All the explanations merely layer the absolute finale with accretions, centuries of obfuscation. Plato did not invent the cave but the other way 'round. This shadow-play of reciprocities easily refuses genetic distinctions. Evolution slowly reforms barriers to distribution. Mutation might tease forth a new concept now and then, testing the waters, but human time, even hominid time, is a temporality with little possibility of transcending the improbable, invoking instead a steady meditation upon the sad course of misanthropy.

For all those gifts established by our kind, with their penchant for transforming certain moments into a higher calling, the plateau ultimately tumbles. Stated simply, 1 + 1 fails to equal 2, while 2 × 2 goes nowhere. And 3 divided by itself remains a member of that class of objects signified by those units which are never members of their same class and may be manipulated by whatever volumes, swells, population dynamics and discord will predictably emerge. Note Russell's Paradox, earlier discussed. Human causes and consequences are defined most readily by their ultimately purposeless relations, a symptom of all the numbers that must be eschewed, as one formulation after another fails to generate meaning. Mathematically, no number of atoms, cells, molecules, neurons, proteins, bonds, or strands of DNA will alter outcomes. Breeds and hybrids are equally suborned by that force of destiny which paleontology has eerily foretold.

Archaeologically, the rules of probability remain unchangeable from graveyard to graveyard; between middens, cultures and cave walls the dreary monotony of the dead gives no real insight into who these multitudes really were. That they lived has no more rational appeal to us today than the nuances of a fossil record dictated by minerology, hydrology and the weather. The brief record of our past does not prepare us for the end. We argue between hope and despair, neither of which offer much by way of a solution to the darkness of caves or the blinding light of a star. In the greens, between, one searches for what one has already found, to paraphrase Pascal (Fig. 68.4).

## 68.5   Co-evolution amid Misanthropic Airs

Our species can rightly be said to be evolving and co-evolving anew. But evolution herself is not evolving, as far as can be ascertained, and thus the molecular and (purely psychological) numbers and proportions are unvarying. Their depictable rudiments, that is, binary, analogue, do not accrue new characteristics vulnerable to natural selection, either at the cellular or conceptual levels. Dust, ash, lithic tuff, pyroclastic surge beds, andesitic lavas, meteoric water, all the sulfuric acid drops and other stormy configurations of the elements are fixed, both in extant and flux. As part of this irremediable ecology, our feelings and intuitions, unmasked in every poem and protest, stare with equal fixity at the improbable. We won't say *impossible*, though the logic that would have it thus far seems unequivocal.

**Fig. 68.4** "Primates," as pictured in Filippo Luigi Gilil's *Systema Naturae Di Linneo*, Volume 1, 1785

But whatever chance insight be posited in these equations of personage over time present to any inquiring desperation a fully charged mirage in which numbers can be said to play games with old men like nursery rhymes and wives' tales, all the follies of Medieval medicine or hallucination. In the open seas and marginalia of human experience perhaps once, or twice, clarity has prevailed. But this is no given. Alchemies persist with a fury; solidity is transmogrified into faint edges of metaphysical purpose that re-tools in a manner befitting this undying hope that our species can change.

In as much as every proof is subject to the illogical, logic itself must necessarily scorn each flying fish and caterpillar. The tyranny of logic is the liberation of philosophy to connect with love; with all that is regarded as beautiful and sympathetic. In these universal instances between people, the equation, whatever the intended nature of the syllogism, hosts a deep semiotic. It is a language of signs conveying something, or someone, beyond or deep within. A qualia or mode of being whose very vulnerability makes it an appealing suggestion.

All such apprehensions fall into an admittedly uniform classification, misan-thropy, which—from Heraclitus and Molière, to Jonathan Swift and Fernando Pessoa—has long held sway over the mental notes aggregated in the group as well as the individual. The square root of some altogether neglected poetry. As such, we cannot really say anything about it which might enlighten us. In the immutable essence of suggestion prior to all information or elaboration, stands the sense of these relics; fossilized ideas that emerged before the killer in man; during a most fleeting age to which some deep strain in us, countering all misanthropy, strives to reconnect.

Such relic feelings are still inchoate. Permeating modalities of consciousness that trigger enquiry of the Self at a time when ecological calamity overwhelms any lingering Soul from the earliest inception of ourselves. These investigations, sens-ing that most gradations and diagnoses present false positives, work to circumvent our histories, and hence, the essence of Utopic thought. The impulse of the apper-ceptive which knows of humanity's enormous shift at some early point in its ontol-ogy proves the possibility, somewhere in the chain of real presences (George Steiner's term), of the Other, all those lovelies that accompany Alice on one of the finest natural history expeditions of the late nineteenth century. That matters, as our species looks to other options than mere precision, cynicism, or the capture and deformation of specimens like those stolen and imprisoned during the Voyage of the Beagle.

# Chapter 69
# The Other

## 69.1 The Subjective Case for Biological Verisimilitude

Any study of something accords that "thing" with some semblance of independent substance or actual identity. Who, or what can the "Other" actually refer to? By refer, are we presuming a prime number category of referencing? A rational number hierarchy of assignations that is not vulnerable to any of the countless classes of paradox that confound representation and metaphor? In the Jain pantheistic sense, this Other contains a truth that is its actuality, its atomic valency (Jiva), the pillar of biological verisimilitude: interdependency (Parasparopagraho Jivanam).[1] In spite of the existentialist throes and temptations of human depression, there exists, despite most claims denying them, true biological emblems of our global connectivity between all taxonomic clusters and real individuals. We see it in every forest, a standing declaration that we are not independent, alone, or at odds, but *inseparable* as individuals (Fig. 69.1).

A coral reef is only paradoxical to the extent that oblivion confuses humans. In fact, oblivion, the multitudinous congeries of vagueness outlined by the blurred collaborations of personages and populations, engenders the only true stability on the world stage: The Other is that goal and promise, as fundamentally qualified by biology, psychology, and the very essence of the life force. She is all that we sense to be Nature.

A paradox factor follows upon any consideration of that which can be isolated from both within and without the contextual ordering of Linnaeus' computations. While taxonomy and physics are absorbed with, and obsessed by, phylogenetics, rules, laws, borders, and computational translations—evidence for which is gathered by a human history of usually invasive methodologies—all such configurative and declarative hypotheses, not least the current IUCN-Red List, US Fish &

---

[1] See *Life Force: The World of Jainism*, by M. C. Tobias, Asian Humanities Press, Fremont, California 1991.

© Springer Nature Switzerland AG 2021
M. C. Tobias, J. G. Morrison, *On the Nature of Ecological Paradox*,
https://doi.org/10.1007/978-3-030-64526-7_69

Ornamenta novo, iam constant omnia Mūdo.     Ergò capit Requiem septena luce. potenti,
   Ille Opifex, nil, quod pficiatur. habet.         Constituens, dextra, Sabbatha diâ, sacrat.

**Fig. 69.1** Jan Sadeler 1's (1550–1600) engraving, "The Seventh Day: God Blesses Adam, Eve, and the Animals," ca. 1585. (Private Collection). (Photo © M.C. Tobias)

Wildlife, Federal, State, and CITES listings (Convention on International Trade in Endangered Species of Wild Fauna and Flora), from nation to nation, are not credible when weighed against the subjective. The individual's subjective personality hinges upon an entirely different set of calculations from those predicated by any *group* of *Homo sapiens*. Their separate constraints ignite a biological swathe of variability that fosters judicious paradigms and markers, possibly environmental fates themselves, that may be like the idea of uncontacted tribes, a matter of ratiocination and terminology within the penumbra of human linguistic chaos, biosolipsism, and support systems in the name of scientific cohesion—one of numerous prospective storylines, accumulations of anecdotes.

For the sake of allegorical clarity, this syndrome of moral and physical confusion led from the naming of plants and animals in the Bible to the Parmenides dialogue featuring the one and the many. No matter into which human language its description is enshrined, all its words carry the same weight of semions, whether, for example, in the Jain, Ptolemaic, Aristotelian, or Newtonian world orders. Quantum physics may question certainty, as both Heisenberg and Wittgenstein have made tentatively clear. But even their assertions were the byproducts of a cultural hubris fortified by the zoology of their combined circumstances. The outcomes are symbolic of a general predicament which is the core of vagueness. It cannot presume to know the Other because to itself, by itself, for itself, and in any other guise, it is a collection of false consciousness, presumptive negatives.

## 69.2   The Possibility of a Post-human Renaissance

Of course, we cannot actually attest to fact that it is false, because, precisely, we know nothing about it. Our self-approbations, intelligence, and pragmatism are superego posited postures; no more than the map of human vainglories vanquishing and thereupon vanquished in the rage of order. To seek a neutral ground at both cusps of a conscious bifurcation is to, again, intimate a class of mathematically perplexing edge effects that work between X and Y, both modifying, transgressing, and engendering geospatial and semiotic causality. In the Amazon, such effects are physically relatable to the impacts, for example, of road building through sensitive habitat. In the case of demography, there are enormous spatial and psychological levels of fallout. Potts, Hillen, and Lewis write, "On the individual-level, an edge effect is a change in behavioral tendency on or near the edge. On the population-level, it is a pattern of population abundance near an edge that cannot be explained in terms of either habitat in isolation."[2]

If cause-and-effect do, in fact, work after the fact of our machinations, we have reason to be hopeful about the future of the world, an approximate bounty of cohesion that rebounds in the wake of trauma. The possibilities of a hare, a duck, or a deer. The atheist and atomist may decline to consult the pantheist, the pantheist, the atomist. But hidden within their many divided injunctions, across the full spectrum of human attribution, and all the edges of its societal bifurcations, lay buried in a heap of forgotten connections, an inextinguishable love between organisms that contests all philosophies and variables of malice. Certainly, there are arguments for and against this position. But viewed post-biologically, which is to posit a post-evolutionary world, even in the total abstract, is to envision after *H. sapiens* have moved on, a renaissance for all Others (Fig. 69.2).

---

[2] See "The 'edge effect' phenomenon: deriving population abundance patterns from individual animal movement decisions," by Jonathan R. Potts, Thomas Hillen and Mark A. Lewis, Theoretical Ecology, 9, 233–247 (2016), December 23, 2015, SpringerLink, https://link.springer.com/article/10.1007/s12080-015-0283-7, Accessed June 23, 2020.

**Fig. 69.2** "Pastoral Scene," from *Phaedri (Phaedrus) Augusti Liberti Fabularum Aesopiarum*, by David van Hoogstratanus (1658–1724), engravings by P. Van Gust after B. Vailland, Amsterdam 1701, p. 160. (Private Collection). (Photo © M.C. Tobias)

# Chapter 70
# Of Malignant Variables

## 70.1 A Distinctly Agitated Preoccupation

"And so we grope our disconsolate way into the next century," writes Gunter Grass in his *Headbirths or The Germans Are Dying Out*.[1] "All of a sudden," writes Sophie Cabot Black. "It's not the loneliness/But the disappointed path back."[2] And finally, in a similar tenor, the ever unpredictable Emil Cioran suggests that "From so much agitation, so much dynamism and ado, we escape only by aspiring to the repose of the inorganic, the peace at the heart of the elements... Each of us is the product of his past ailments and, *if he is anxious*, of those to come."[3] To be thusly anxious is to participate in the grand spectacle of *variables*. As conscious beings, we are also the unwitting objects of cognition, of counterfeit conclusions, surrogacies and proxies intended to extend our visitation, false positives that signal the fallacies of such thinking, and a forced resolve that is actually the exasperation of existence which knows itself to be ultimately doomed[4] (Fig. 70.1).

Science wrestles with poetry. Too many finely honed details and mood swings become the-thing-in-itself, losing track of the forest. Emerging from the depths of a chemical equation, like those associated with the complicated mineral group of Tourmaline $[(Na,Ca)(Mg,Li,Al,Fe^{2+},Fe^{3+})3(Al,Mg,Cr)_6(BO_3)_3Si_6O_{18}(OH,O,F)_4]^5$ or the physiological factors enabling the Golden Bamboo Lemur (*Hapalemur aureus*)

---

[1] Translated by Ralph Manheim, A Helen and Kurt Wolff Book, Harcourt Brace Jovanovich, New York and London, 1980, p. 68.

[2] From the poem, "Higher Ground," in her volume, *The Misunderstanding of Nature*, by Sophie Cabot Black, Graywolf Press/Saint Paul, 1994, p. 37.

[3] E.M.Cioran, *The Fall into Time*, Translated From The French By Richard Howard, Introduction By Charles Newman, Quadrangle Books, Chicago, 1970, p. 128.

[4] See *The Hypothetical Species: Variables of Human Evolution*, by M .C. Tobias and J. G. Morrison, Springer, New York, 2019.

[5] See "The Chemical Formula of Tourmaline," Minerals.net, https://www.minerals.net/tourmaline_chemical_formula.aspx#targetText=The%20formula%20of%20the%20Tourmaline,chemical%20formula%20of%20all%20minerals, Accessed August 24, 2019.

© Springer Nature Switzerland AG 2021
M. C. Tobias, J. G. Morrison, *On the Nature of Ecological Paradox*,
https://doi.org/10.1007/978-3-030-64526-7_70

**Fig. 70.1** A Scene in Contemporary Poland. (Photo © M.C. Tobias)

to safely consume its local cyanide-producing *Cathariostachys madagascariensis* species of bamboo,[6] if scrutinized indefinitely changes one's attitude. From an interior desperation to that character of certitude. An individual who can wake up, count numbers indefinitely, and be satisfied. A preoccupation with observing things, noting every leg of the millipede, must possess certain advantages unknown to unconsciousness. But what is unknown, versus known, offers no qualitative, definitive success story. An alliance between Self and unknowable creatures and forces has always reigned supreme in the mind. We stay involved with mysteries in ways that cannot be put to rest on a shelf, or even on the printed page. There are no tests in the wild for this sense of lost belonging.

To the extent we might want to perpetuate the clouds of unknowing[7] courts a slow-breath mysticism. We witness this bivalence and dyad biology in everything we perceive. Its conclusions are as rapid as the 268 miles per hour speed "at which signals travel along an alpha motor neuron in the spinal cord, the fastest such transmission in the human body" amid the estimated "100 trillion neural connections, or synapses in the human brain."[8] Such details do not relieve the burden of Being, nor make for a happy species. This much is a given.

---

[6] See Ballhorn, DJ; Kautz, S; Rakotoarivelo, FP (2009). "Quantitative variability of cyanogenesis in *Cathariostachys madagascariensis*- the main food plant of bamboo lemurs in Southeastern Madagascar." American Journal of Primatology. 71 (4): 305–15. doihttps://doi.org/10.1002/ajp.20653. PMID 19132732.

[7] Of the 17 known copies of the Cloud of Unknowing from the late fourteenth century, see British Library Harley MS 2373 and Cambridge University Library Kk.vi.26.

[8] See "Numbers: The Nervous System, From 268-MPH Signals to Trillions of Synapses,"

## 70.2 Biospheric Cul-de-Sacs

Out of this biospheric cul-de-sac, time itself can neither clarify nor rectify. Historians can offer no proof. Our rapacious approach to *systems* does not spare us. Cohesive thoughts have no claims upon our lives with anymore grace or attraction than generalized incoherence, which persists unabated, despite our most fervent efforts to extinguish it. We philosophize systems, from the clothes we wear and food we consume to all that we desire, and all that is subjected to our miscellaneous microscopes.

The individual, however, is not entirely relegated to this devilish closet of presumptuous cohesions, with its obsolete wardrobe. At times, one can't help but volunteer the admission that the individual rises above the *species* category, which is woefully incommensurate with every prequel or subsequent asseveration stemming from the mob. If, by species, its individual taxons are to be extrapolated, we must once more own up to the repetitive fallacy of our self-assertion at the organized, community level. Bands, factions, tribes, cultures, megacities. They harbor a fiction that would link all individuals together. But those units of life cannot exist outside of the quotation marks that represent our bid within the auction house of the biosphere: conquest. A species does not ascend Everest. Walking up to the summit is something different than what collectives do. Groups come together and easily formulate and implement plans to build a Philadelphia or burn the Amazon.

Hence, a complex and potentially planet-shifting discrepancy. The individual does not merely transcend its species, if it works at it, re-defining the rate of change within the genus as genetics and behavior wildly vacillate down through the species level into its differentiated taxa. The individual also bypasses analogies, becoming the first potential candidate for whom it can be imagined there is no definitive outcome. Or, thought of differently, a future worth struggling for. As Kazantzakis once wrote somewhere, "It is not human beings that interest me, but the *flame* that consumes human beings," or words to that effect.

## 70.3 Misanthropy as Ecological Proxy

Misanthropy, obviously, is one obvious condition of such prospects. But not if this *theoretical individual* is to gain our trust, to become the subject of our social compact.[9] Deep demographers might wish for the fatal collision of population thresholds depicted in their aftermaths by so many genetical crashes and ultimate asymptotes. Followers of the Ehrlich-Holdren equations associated with I=PAT

---

by Valerie Ross, May 15, 2011, Discover Magazine, http://discovermagazine.com/2011/mar/10-numbers-the-nervous-system, Accessed August 24, 2019.

[9] See *The Theoretical Individual: Imagination, Ethics and the Future of Humanity*, by M.C. Tobias and J.G. Morrison, Springer, New York, 2018.

(impact = population x affluence x technology)[10] will likely eschew any reciprocity potential between the eight primary taxonomic delineations—domain, kingdom, phylum, class, order, family, genus, and species—when it comes to our own role in the gigantic mix of Beings. Nonetheless, we would argue there might be some faint recognition that can be resurrected between individuals and species, with respect to *H. sapiens*. To ascribe even a remote mathematical chance of such a possibility is to acknowledge potential within our continuing evolution for either a mutational breakthrough, or some consequential shift, prompted through rapid natural selection, that improves upon an otherwise fated wasteland of relations.[11]

Byron's lovely line in "Manfred" regarding one's "fatality to live," which was so extravagantly pinioned in Manfred's clinging figure to the Jungfrau, just as Shelley's own protagonist in "Alastor, or, The Spirit of Solitude" had fixated upon a cliff ledge high in a Vale of Kashmir, the perfect place to perish, the young poet had dreamt. Not in Rome.[12] Such connections yield to an evanescent summation. To curl up peacefully upon that mountainous wall and "die out," as James Dickey once contemplated the nature of extinctions. Or to go on, altering every fiction and rejecting the notion that loss is a negative. A rejection of rejection, in other words. As Samuel Beckett concluded in his novel *Molloy*: "It is midnight. The rain is beating on the windows. It was not midnight. It was not raining."[13]

Neither lost nor found. And hence this most peculiar variable: the longing for an end to longing; for implosion whose self-destructive image can somehow manage the hat-trick, moving philosophically and painlessly from inaction to the iconic stamping out of the entire passion play that is human existence, with its order of magnitudes as concentrated as the instantaneity of an electrical current. In such remote contemplations, this is where metaphysics and natural history converge, pivoting upon the question of human worth among the animals. Of the whole sequence of events following upon the Garden of Eden and expulsion of its first two hominins.[14] Aristotle struggles in his *Metaphysics*, with not only the worth of humanity, but the definitions, characteristics, and the mathematical translation into numbers of Species, Individual, Being, and "whatness." He writes, "Thus, there is an essence only of those things whole formula is a definition… There are also definitions and essences of the others in a similar manner, but not in the primary sense."[15]

---

[10] See Ehrlich, Paul R.; Holdren, John P. (1971). "Impact of Population Growth." Science. American Association for the Advancement of Science. 171 (3977): 1212–1217. doi: https://doi.org/10.1126/science.171.3977.1212. JSTOR 1731166

[11] For real-time rapid evolutionary data, see *The Hypothetical Species*, op.cit., Tobias & Morrison, pp. 111–112, 121, 209, and 255.

[12] See *Keats, Shelley & Rome – An Illustrated Miscellany*, Compiled by Neville Rogers, Postscript by Field-Marshal Earl Wavell, Johnson Publications Ltd., London, 1970.

[13] Samuel Beckett, *Molloy*, translated from the French by Patrick Bowles in collaboration with the Author, Grove Press, Inc., New York 1955, p. 241.

[14] See *The Rise And Fall of Adam and Eve*, by Stephen Greenblatt, W. W. Norton & Company, New York, 2017.

[15] See *Aristotle's Metaphysics, Translated with Commentaries and Glossary*, by Hippocrates G. Apostle, The Peripatetic Press, Grinnell, Iowa, 1979, pp. 112–113.

And later, Aristotle declares, "Now, if all units are comparable and without difference, we get the mathematical numbers and of one kind only, and the Ideas cannot be these numbers. For what kind of a number will Man Himself or Animal Itself or any other Form be?"[16] Aristotle's contention outlined in his *Historia Animalium* amplified the scope of such discourse when he laid claim to the fact that all animals were comprised of both body and soul, of "psychical qualities or attitudes... In a number of animals we observe gentleness or fierceness, mildness or cross temper, courage or timidity, fear or confidence, high spirit or low cunning, and, with regard to intelligence, something equivalent to sagacity."[17] Two thousand years later, Renaissance scholars were still reliant upon these unfinished queries and philosophical interpolations to grapple with humanity's place in a zoological miasma, as witnessed in Conrad Gessner's own enormous, Aristotelian multi-volume *Historia Animalium* (Zurich, 1551–58, and 1587) (Fig. 70.2).

In the tradition of such biological speculation, all the philosophy in the world would fail to explain the rainbow, as Keats, then John Ruskin argued. Or of lightning. So many inalienable, ephemeral noumena, phenomena, impressions, and perceptions; so myriad a world of personalities as to explode the very myths that Linnaeus and others had forced into a language of nature predicated on strict divisions, and upon the one constant between them: Evolution, a mystical force capable of engendering constantly shifting boundaries and convergences. Embodied in that language is something of a primacy, a first order of inquiry that means to fill in the blanks of fossil history, and of the history of mind. To rigorously lay bare the

**Fig. 70.2** Conrad Gesner's (1516–1565) illustration of a rhinoceros after Albrecht Dürer's 1515 version of same, in his 1551 *Historia Animalium*, published in Zurch. (Public Domain)

---

[16] ibid., *Aristotle's Metaphysics*, p. 223.

[17] See *Aristotle: Parva Naturalia*, A Revised Text with Introduction and Commentary, by Sir W. David Ross, Oxford, At The Clarendon Press, Oxford UK, 2001, p. 5.

presupposition of, if not knowledge itself, then at least an orientation to life, call it a deep sensibility, to which we might freely ascribe a so-called *willingness factor*. This is equivalent to that transition initially described herewith; the direction away from the presumed to the vulnerable; from the known to the unknown; from an absolutism of matter and substance, of essence and of being, to some other, free associationism of spirit (Fig. 70.3).

**Fig. 70.3** Family at Sacred Pushkar Sarovar, A Vegetarian Model City of the Present and the Future. (Photo © M.C. Tobias)

# Chapter 71
# The Concept of Zero

## 71.1 Dimensions of Human Response in the Wake of Its Destructive Patterns

In answering to the challenge of variation, we arrive at the concept of zero, or of infinity in terms of our ability to plumb the depths of biological annals and their correlation into linear sense. There is not much to say about it. Language and thought, occupying an unknown *number* of *dimensions*, condition our investment in thought by architectural phantasms. Ten dimensions. Fifteen dimensions. A googol-plexianth, or largest possible number, which must exceed a 1 followed by 100 zeros. In whatever domain of speculation (just words) the individual will be lodged in a limbo of unaccountable coefficients, tempting wormholes toward what in Buddhist thought is transcendent consciousness, infinities marked by a singular infinite apprehension. Each numeric, phase, series, morph, culture, wave, and particle will actually undergo existence and non-existence as equals (Fig. 71.1).

This humility must inform eco-science, particularly in the area of intensive human land use, for example. Numerically, we easily recognize globally massive insults as deforestation and optimal or sub-optimal native seed dispersal mechanisms in both Nature's and human arsenals. Write the authors of a contribution to the literature of southern Indian (Nilgiris) "restoration ecology," "…anthropogenic activities heavily reduce native forest cover and landscape connectivity leading to increased spatial isolation of remnant forest patches."[1] It is one of countless simple-sounding alarm bells, with vast consequences, in this instance, well beyond India, of course. Arctic amplitude, for example, with heat waves in Siberia exceeding 100 degrees (100,000 years sooner than normal).

---

[1] See "Recruitment of saplings in active tea plantations of the Nilgiri Mountains: Implications for restoration ecology," by D. Mohandass, Tarun Chhabra, Ramneek Singh Pannu, and Kingsly Chuo Beng, Tropical Ecology 57(1): 101–118, 2016, © International Society for Tropical Ecology, www.tropecol.com

© Springer Nature Switzerland AG 2021
M. C. Tobias, J. G. Morrison, *On the Nature of Ecological Paradox*,
https://doi.org/10.1007/978-3-030-64526-7_71

**Fig. 71.1** "Manujvajramandala, Containing 43 Deities," Tibetan, ca. 1400–1500, Representing Transcendent Wisdom, A Buddhist Wormhole into Infinity, Held at the Museo d'Arte Orientale, Turino. (Public Domain)

In certain early writings of Plato and Theophrastus (the latter, a student whose affiliations linked to both Plato and Aristotle), local changes on the landscape mosaics of the Mediterranean littoral were noted.[2] Dimensions relating to cause and effect left instant impressions, as would any contemporary wildlands reduced by fire or some other human agency, to a depauperate aftermath. Science historian Barbara Maria Stafford confronted such impressions at a multitude of levels. Writing in her engrossing *Voyage into Substance*, Stafford comments, "Energetic spatial metaphors – tearing, crossing, immersing, penetrating – structure the images of the factual relations... The issue of an individual consciousness locating itself

---

[2] See *Traces on the Rhodian Shore - Nature And Culture In Western Thought From Ancient Times To The End Of The Eighteenth Century*, University of California Press, Berkeley and Los Angeles, CA, 1967, pp. 121 and 130.

outside itself bears on the varieties of the Sublime (rhetorical, natural, religious) enumerated in antiquity and compellingly recast by Edmund Burke…".[3]

But of the dark mirror, eco-restorative science has much to say, its sagas thick with toxins and the dream of ecological release. Notable examples include efforts to redress various forms of destruction throughout the Murray-Darling Basin and sea-grasses and salt marshes on Adelaide's coastline in Australia; intensive microbial phytoremediation endeavors, purging of heavy metals from vulnerable landscapes; diverse work to restore degraded wetlands in the UK and rangelands from Iran to China[4]; and the enormous challenges ahead of the 1344 EPA National Priorities List Superfund sites in the United States, as designated under the 1980 CERCLA legislation (Comprehensive Environmental Response, Compensation, and Liability Act).[5]

These crucially interior modes of "locating," the recognition and apotheosis of *outside consciousness*, transcend distinctions between fiction and non-fiction. It is the prime mover of a cosmopolitan, animal and plant liberation disposition, given to a revolution in human identification of the Other. Identity has been key to unlocking the affairs of state that govern civility, legal protection, encouraged pathways of empathy, ultimately, the enchantment of a wilderness ideal. The challenge of some kind of conversation with Others poses every dialogue worth engaging in. Is it a conversation at the heart of biosolipsism granting a new voice, voices, to the natural history of existentialism?

The mathematical arguments for the phenomenon of equivalency rest upon the least certain of all pathetic fallacies. No forest or flower obeys predictive analogy. Every angiosperm blossoms according to an unfathomable set of forces. Write L. Verhage, G. C. Angenent, and R. G. Immink, "…our knowledge of the genetic and molecular mechanisms underlying this flowering pathway is limited. However, recent advances in Arabidopsis (Arabidopsis thaliana) [a model plant species belonging to the mustard, or Brassicaceae family, also known as Thale Cress] have uncovered multiple molecular mechanisms controlling ambient temperature regulation of flowering."[6] Like the "marvel of Peru," also known as four o'clocks, or *Mirabilis Jalapa*, found typically in the USDA Plant Hardiness Zone 7b-11, moisture, time of day, temperature: such variables assist in our ability to seed and project biological outcomes, but cannot do much more than that.[7] Our limitations harken back to such spiritual conundrums as the Toda "worry flower," or "arkil-poof," a

---

[3] See Barbara Maria Stafford's *Voyage into Substance – Art, Science, Nature, and the Illustrated Travel Account, 1760–1840*, MIT Press, Cambridge Massachusetts, 1984, p. 353.

[4] See *Ecological Restoration – Global Challenges, Social Aspects and Environmental Benefits*, Edited by Victor R. Squires, Nova Publishers, New York, 2016.

[5] See "CERCLA." Legal Information Institute, Cornell Law School.

[6] See "Research on floral timing by ambient temperature comes into blossom," NCBI, PubMed Trends Plant Sci., 2014 Sep;19(9):583–91. doi: https://doi.org/10.1016/j.tplants.2014.03.009. Epub 2014 Apr 27, Accessed August 28, 2019.

[7] "It's Time To Plant Four O'clocks," by Gretchen Heber, Gardener's Path, January 26, 2019, https://gardenerspath.com/plants/flowers/grow-four-o-clock/, Accessed August 28, 2019; See also, https://planthardiness.ars.usda.gov/PHZMWeb/

species the Toda allege is able "to indicate a person's level of anxiety." Writes Tarun Chhabra, "We set out to see this flower, which shuts only if a worried person holds the plant by its stem. How it works remains to be studied, but I can say that it has been very accurate…"[8]

In other words, from the heart of melancholy and frustration with civil society and the breakdown in social contract theory and practice, ecological epiphanies can help redress possibly hundreds of thousands of years of our species' mental and physical woes and outbursts. That is the great hope (Fig. 71.2).

## 71.2   Biosemiotic Impulses

Like vascular plants, every grain of sand along 800 miles of Japanese beachfront purports to a different weight, shape, size, and color. That, too, signals a proliferation of aesthetic, scientific, and consciousness-related Otherness at its most promising. Those grains of sand each emanate differing hydrological echoes and marine memories, geological relations to the glaciers, seas, and sinks of their mineral legacies, all unknown to us, but through the eco-poetic affairs of the mind and the heart. Not one grain of sand on earth is the same as another. We gaze up at the stars with an equally famished intellect when it comes to particular delineating. Such multiplicities ordain a humbling attitude, if we go with it, toward all entities. These are the Others with whom our biosemiotic impulses are wonderfully challenged to the extreme.

A tolerance, given to the appreciation of gradations and unrecordable nuances, understands at a fundamental layer of consciousness that no predicate, neither multiplier nor divider, can account for the illogical premise of an equivalency anywhere in the Cosmos, or at any time within a defined temporal span. The = sign collapses, along with fundamental knowledge that we presume; the very languages we wield to describe known and unknown signatures lose loft at the instant of formation. Every turn of descriptive phrase—"unknown worlds," "at the instant of," and so on, struggles forward, then in reverse, to formulate and seize upon the appropriate catchphrase, like *relativity*, to reconcile the frustrations in mind of a Zeno's Paradox, or basic chemical reaction. To withstand the nagging suspicion that all of our truths and untruths—the pillars of history, ethics, philosophy, and science—are fuzzy.

Even among reactive molecules like $H_2O$, the ocean defies every alleged accomplishment of logic. Reason is thwarted in each wave and current. Mega-tsunamis, like that which overwhelmed Lituya Bay in Alaska in 1958, or the 1970 Ancash Earthquake/Landslide off Peru's Huascaran, killing as many as 70,000 people, natural disasters, go to the core of our instability.

---

[8] See *The Toda Landscape – Explorations in Cultural Ecology*, by Dr. Tarun Chhabra, with a Foreword by Anthony R. Walker, Published By The Department of South Asia Studies, Harvard University and Orient BlackSwan, Cambridge, Massachusetts, 2015, p. lxi.

**Fig. 71.2** Dr. Tarun Chhabra Photographing *Strobilanthes lanata* in the Nilgiris, Tamil Nadu State, India. (Photo © M. C. Tobias)

Occupants of all other ecosystems taunt the lessons of pH; while those soaring high above the whitecaps—for example, the 21 globe-trotting Diomedeidae family of albatrosses, all threatened or near threatened according to the IUCN—have preoccupations that outdate our own fixations by tens of millions of years. But every single phytoplanktonic individual, among more than 5000 known phytoplankton species, the most abundant organisms crucial to the functioning of the aquatic environments of earth, has dramatic lessons for us which we cannot just access, despite all of our earnest probes.[9] The lifestyles of a biosphere, simply stated, transcend specificity, microscopes, DNA markers, even if our conceptualizations, perusal by experiment of everything we can imagine, temporarily comfort us. There are erratic,

---

[9] See "Phytoplankton," https://coastalstudies.org/stellwagen-bank-national-marine-sanctuary/phytoplankton/, Accessed August 28, 2019.

wondrous, transe-infinite gradations and increments of gleaned data hovering on either side of zero, like infinite motes, hues, ripples, and drops of rain in the Amazon. If, as some have speculated, imagination corresponds to reality, the shortcomings of any presumption of inequality between organisms (broken down to atoms, souls, dust motes, and dew drops—among a myriad of others—in Jain ethical computations) should at once be apparent. A critical lesson to ponder with an ecological ticking clock (Fig. 71.3).

**Fig. 71.3**  Rain Drops on an Obscure Tributary in the Ecuadorian Amazon. (Photo © M. C. Tobias)

# Chapter 72
# On the Nature of Equivalencies

## 72.1  A Labyrinth Beyond the Equal Sign

The equal sign having ultimately succumbed to our intuitions—we underscore the prospect that sanity still holds sway in tender, private moments, but is lost in the crowd—where to, then, should prudent guidance offer assistance to any philosophy that purports to capture the essence of all that is vastly larger than ourselves?

How can human history, moments of contemporary probity and gravitas, make any sense of such bold and unqualified sentiments as *social reform*, *movement,* or *destiny?* Equality under the law at once alerts us to the shift of precision, from that which is equal, to the more dominant notion of legal *frameworks* for approaching yet other concepts, more vague by every association. The US population in 1776 was approximately 2.5 million people in 13 colonies (slaves counting for "3/5ths" of a person—the "Three-Fifths Compromise"[1] in the first census of 1790 which recorded a totality of 3,929,314 persons).[2] Founding documents for that frame of reference had countless moving targets to grapple with, obviously months of furious debates and altercations, prior to a Revolution.

Now, with over 331 million moving targets (the US human population in 2021), and thousands of eco-regions, the many current questions of grammar and custom, sets and sub-sets, class actions, cases specific and general; so many contemporary objects of attention, subjects lacking confidence values; or are constants, at great variance with the statistical, cultural, and experiential realities besetting the founders of the United States, or of any other republic. The variance factor is not simply a matter of time, centuries. It occurs continually in physiochemistry and biology. All

---

[1] Madison, James (1902) *The Writings of James Madison*, vol. 3, 1787: The Journal of the Constitutional Convention, Part I (edited by Gaillard Hunt), p.143. Online Library of Liberty, https://oll.libertyfund.org/titles/madison-the-writings-vol-3-1787, Accessed July 20, 2020.

[2] United States Census, "History," https://www.census.gov/history/www/through_the_decades/fast_facts/1790_fast_facts.html, Accessed July 20, 2020.

© Springer Nature Switzerland AG 2021      659
M. C. Tobias, J. G. Morrison, *On the Nature of Ecological Paradox*,
https://doi.org/10.1007/978-3-030-64526-7_72

matter has its physic in perpetual flux. Ecosystems are cumulative ambassadors of this completely inexplicable, constant change.

That is to acknowledge the now arbitrariness of "true" and "false"; and that encompasses what Sartre often referred to as true, or false consciousness. We see far less dialecticity in the scope of human conscience. It either is, or is not present. The problem with non-dialectic conscience is not its presence, but its manifestations, or lack thereof.

Hence, between all such categories no meaningful (or unmeaningful) existence can be reliably asserted. No consequences proffered. Nor any associations confirmed. The contact that might be posited between entities is as infirm and inconsolable a configuration as that glass trough holding tied lobsters waiting to be boiled alive. We have no way of understanding the lone halibut somewhere off the coast of Iceland in the middle of a stormy winter's night, swimming half-a-mile deep. That is the extent of our full ignorance. Trying to grasp the follicle mite squirming for space in an eyelash.

The bedside comfort of a great book, not necessarily its accuracy, teases our efforts to move outside in proportion to the extent that we can go deeper inwardly. All of the hazy interstices, coordinates, passing time as measured in our vagaries and extravagances or illnesses of mind, appear before us as a general swell of existence we cannot hold on to. If we could, the damage we inflict could conceivably be even worse.

No distribution model of any particular philosophy, metaphysic, scientific paradigm, set of details, questions or answers, satisfies the demands of reality, not in our case. And while moral tensions and undefined suasions exert certain practical limitations upon one's mobility, a basic restlessness gives to human life illimitable dreams, no matter what or who the dreamer.

Nor can we, who impose every predominant stressor, presume to gauge the full range of stress-response among fellow species. What we can attest to, as historically-and-future-obsessed organisms, is the reality that a majority of all past ecological destruction was natural. Five calamities dating back 2.45 billion years.[1]

As sweeping as the losses were—between 75% and 96% of all terrestrial and/or aquatic life forms on average per global event—they were not *deliberately* imposed by a living, co-habiting species that finds itself at the center of the Anthropocene, and soon, the Eremozoic, what E. O. Wilson has named the next geological epoch of loneliness[2]; conscious of its actions, and fully ignoring its consequences. This utterly undermines the seriousness of our ability to actually ever engage in meaningful conversation, especially in light of the present condition for wild animal biomass proportions on earth.

Evidence for one of the earliest human massacres, of 27 other people, was unearthed at the Nataruk site in Kenya in 2016, dating to the end of the Stone Age and last Ice Age, roughly 10,000 years ago.[3] It was at approximately this same time

---

[3] See "The Big Five Mass Extinctions," by Viviane Richter, https://cosmosmagazine.com/palaeontology/big-five-extinctions, Accessed August 29, 2019; See also, *The Sixth Extinction: An Unnatural History*, by Elizabeth Kolbert, Henry Holt and Company, New York, 2014.

**Fig. 72.1** A Hen (*Gallus gallus domesticus*) on the U.S. Eastern Seaboard. (© Photo by M. C. Tobias)

that the last of all common hominin ancestors mingled, in a collective numbering roughly 5 million people worldwide. That figure represented roughly 1% of the "weight of [all combined] vertebrate animals" who weighed 99%. Today, that ratio has been thoroughly altered, with humans weighing 32%, domesticated livestock 67%, and wild animals commanding no more than 1%.[4] While such data pertaining to biomass proportions in the Anthropocene have been suspected for several decades (as of 2019), what has changed is the inevitable disappearance of most of that remaining wild 1%, and the consequences for in situ ethological dialogue (Figs. 72.1, 72.2, 72.3, and 72.4).

## 72.2   On the New Nature of Consciousness

In terms of transforming human consciousness in an aspired to greater attunement with the consciousness of other species, aware of our awareness, apprised of the aforementioned problem of variance, longstanding difficulties and debates have not come close to even substantiating a narrative framework for our ability to grasp, understand, or cope with the mental equivalencies of a kind of Black Hole or Supernova in mind, both. Neural correlates to consciousness are rife with explanatory gaps, hard and soft problems. Those include the proposed universality of consciousness as developed in theories of panpsychism, subjective, physical and higher-order experiences. Each of these highlights and evinces a multiplicity

---

[4] See E. O. Wilson, *Consilience: The Unity of Knowledge*, New York, Alfred Knopf, 1998, p. 294.

**Fig. 72.2** Reflections in a Remote Tributary in Yasuní National Park, Ecuador. (© Photo by M. C. Tobias)

**Fig. 72.3** Cloud Cover, Northern New Mexico. (© Photo by M. C. Tobias)

**Fig. 72.4**   Four Ecotones of Consciousness. (Photos © M. C. Tobias)

of contours, dusks, and dawns, with respect to the core issue of why it is we are, allegedly, aware of anything and what such awareness actually means.[5]

Consciousness is subject to every conceivable verb and adjective. It binds, it combines.[6] It exposes objects and qualities, while seeming to integrate information with a neural contagion of prospects that differs from individual to individual, species to species. Fundamentally, consciousness remains an unknown aspect of Nature, notwithstanding the appearance of our apperceptive awareness, of consciousness squared. With no clear remedy in sight, one possible way to come to terms with the unknowable is to work our way backwards from the apparent results of human consciousness, the eco-dynamic fallout of human interaction with its surroundings, the consequences of those interactions in the tangible world. That is one, albeit, constantly in shift, set of focal points that strike us as real. Consciousness gravitates toward creation and destruction. That is a requisite beginning for grasping cause and effect.

Those effects appear to us as our inflictions, and the sum of our impacts, regardless of how philosophers of mind care to characterizes the nature of that which senses, perceives, is part of, or apart from the ensuing combinatory reactivities, in stark biochemical terms. Such ecological forensics, objectified evidence of every

[5] AfricaNews, "Evidence of earliest human massacre 10,000 years ago found in Kenya," https://www.youtube.com/watch?v=Qaq1BdT3w0E, Accessed August 27, 2019.

[6] See "The Overpopulation Project, Population Matters," in "Population Growth Threatens One Million Species With Extinction," Caps News, Summer 2019, Vol. 60, No. 1, p. 4.

problem of consciousness, links the data and sensory centers of the mind and the brain to the Others. One might dispute the evidence, or perceive it through the lens of confirmation bias, but—granting every pragmatic idealism and precautionary principle, versus a wild west of unchecked speculative fictions—there can be little to be gained by outright denial of the ecological cybernetics and modes of feedback that, at least, seem to define the world and its representations in mind, resulting from, or implicated by that mind. An appreciable quantum of this approach to consciousness in Others was undertaken by Maurice Maeterlinck in his masterpiece, *Hours of Gladness*, particularly in the lengthy chapter, "The Intelligence of Flowers."[7]

Therein explodes upon the topology of contemplative organisms the most perplexing and toxic of quandaries: The presence of some form of human consciousness directly responsible for exposing subjects beyond itself to itself. That exposure is nothing like a truth value of interdependency, or an innate disposition toward a loving biophilia. Rather, it inculcates into the life force itself the fullest range of exogenous inputs leading to targets of our impact, anti-amelioration, subjugation, partial then total impact, and eventually an extinction event, and with it, memory within which is exhibited the cessation of all possible dialogue. It is the antonym of all consciousness. The thinking person's worst paradox, but the liberation of all other thought processes that have never relied on human consciousness for anything.

## 72.3   A Biological Pilgrim's Progress

The tired rhetoric, the unending near paralysis of *the thinker* is theoretically circumvented, at least, by such "hours," whose times come seasonally, as first sensationally depicted in the Medieval devotional *horae* genre.[8] No need of some unendurable algebra to corroborate the essence of a flower as it manifests in mind. Sufficiency that surpasseth unconsciousness and leads past pain toward the pleasure and solace of a garden. Pain and pleasure take evolutionary notes, accommodating one of the other by experiential versions of one species adjoining another. Combinations tested over time frames that seem to indicate no other priorities than the very gifts of life herself, as iterated in the underlying critical mass of biodiversity.

---

[7] See Joseph Levine, "Conceivability, Identity, and the Explanatory Gap" in Stuart R. Hameroff, Alfred W. Kaszniak, and David Chalmers (eds.), *Towards a Science of Consciousness III: The Third Tucson Discussions and Debates*, The MIT Press, 1999, pp. 3–12; see also, Chalmers, David (January 1997). "Moving forward on the problem of consciousness." Journal of Consciousness Studies. 4(1): 3–46. This concept of "variance" is as protean as New York Governor Andrew Cuomo declaring a "mask" "the only line between anarchy and civilization," in a press conference on July 20, 2020.

[8] See Revonsuo, A, *Inner Presence: Consciousness as a biological phenomenon*. Cambridge, MA: MIT Press, 2006.

As a measure of so many variables scattered in the teeming boundaries of that which functions and those which have ceased to do so, Being, Consciousness, and their mutual alliance enunciate without surcease the exquisite lessons and rationales that discount fatalistic propositions in favor of some conscious opportunistic embrace. That is the universal sense that natural selection has clearly upheld as a meaningful cohort of survival the intelligence of flowers, the perceptual faculties of an eagle, and the deep range tolerance of a Weddell Seal.[9]

All the variations of such fellow Beings have continued to benefit from every benign commission, act, and impulse that consciousness clairvoyantly adopts. The garden admits to a combining of proclivities and possibilities, and this may be as close as we can come to recognizing some form of the universal bond that is the source of our thoughts and intuitions. A rite of passage that would, if we were only to let it be, provide the itinerary for a very different pilgrim's progress, one that sought solely in consciousness the best for biodiversity.[10,11]

[9] See *Hours Of Gladness*, by M. Maeterlinck, Translated by A. Teixeira DeMattos, Illustrated by E. J. Detmold, George Allen & Co. Ltd., London, 1912.

[10] See *The Book of Hours: With a Historical Survey and Commentary*, by John Harthan, Crowell Publishers, New York, 1977. See also, *Le trésor des Heures – Pages choisies des livres d'Heures des XIV et XV siècles*, by Fanny Faÿ-Sallois, Desclée De Brouwer, Paris, 2002. See also, *Time Sanctified – The Book of Hours in Medieval Art and Life,* by Roger S. Wieck, With Essays By Lawrence R. Poos, Virginia Reinburg and John Plummer, George Braziller, Inc., New York, In Association With The Walters Art Gallery, Baltimore, New York, 1988.

[11] See "Physiological control of divine behavior in the Weddell seal *Leptonychotes weddelli*: a model based on cardiorespiratory control theory," by Richard Stephenson, Journal of Experimental Biology 2005 208:L 1971–1991; doi: https://doi.org/10.1242/jeb.01583, https://jeb.biologists.org/content/208/10/1971, Accessed July 20, 2020.

# Chapter 73
# Metaphorical Realities

## 73.1 The Unclouded Mirror

All of the foregoing has seemed to have collided with some intractability as outlined in the boundary layers of more than four billion years of ecological dynamism that has led to a humanity incapable of releasing its grip upon the world, its human-propelled fast-track toward implosion. That collision equates with the metaphors dogging our reasoning faculties, the obvious cancer so often intoned by our kind in referencing that repugnant vestige of violence dominating the collective power and broken promises of this venomous hominin. Of course, the normal defensive posture is to declaim such language as misanthropic or, worse, to declare it useless, if its aim is to drill into the mind of the cynical reader a helplessness rendered intellectually vacuous by some genetic dictatorship we have implemented, notwithstanding joy, Mozart, Rembrandt, and chocolate. The too obvious point is that this cynicism is precisely our *collective* response, when all the erratic behavior is calculated with respect to our species' heated reactions to the earth underfoot. Our goal by way of this treatise is to nullify the comfort taken in this illusory refuge and, of course, to zero in on what it will take to successfully combat such behavior, all those metaphorical realities of destruction, with an existential truth that places such metaphors before the unclouded mirror. And what reads of an anti-social treadmill obliges greater callings in due course (Fig. 73.1).

As we lend added commentary to the ages of self-deprecation, we have reason to add that this representative of ideas, this human species, is only that, an *idea*, certainly no ideal, with its tangle of morgues and self-interested denial. More recently, of appeals to an economy not convulsed by the constant parsing of corporate earnings, but which strives to enshrine and enliven "Sustainability Credits"[1] and

---

[1] See "A Plan for Social and Economic Justice," by Roy Morrison, Daily Kos, July 19, 2020, dailykos.com

© Springer Nature Switzerland AG 2021
M. C. Tobias, J. G. Morrison, *On the Nature of Ecological Paradox*,
https://doi.org/10.1007/978-3-030-64526-7_73

**Fig. 73.1** School of Rembrandt, Tronie, After Rembrandt's "Old Man" of 1632, Artist Unknown. (Private Collection). (Photo © M.C. Tobias)

economics that are just "enough" and no more,[2] along with almost casually lobbed warnings of total annihilation. As an idea, a bundle of recriminations and unfortunate vetoes and conflict, the very notion of *Homo sapiens*, is just another idea in the free market of such conceptual classes. But within the framework of the earth's own consciousness (to borrow the human designate), a farrago of conditionals makes it especially difficult to write logical sentences connoting the central guise that eludes strict definition or steady-state concentration.

Even within a context of so-called de-extinction, or resurrection biology, of cloning endangered or officially extinct species through the voguish manipulation of

---

[2] See "Steady State Economics In A World Of Resource Limits – A MAHB Dialogue With Environmental Economist Rob Dietz," by Geoffrey Holland, July 16, 2020, https://mahb.stanford. edu/blog/steady-state-economics-in-a-world-of-resource-limits-a-mahb-dialogue-with-environ-mental-economist-rob-dietz/Accessed July 20, 2020.

viable DNA, or embracing the re-wilding paradigm, conservation practicalities are dependent upon the mood swings of geopolitics and economics, strictly human pufferies serving the fickle will of a distribution of superegos, that aforementioned "variance" factor that oscillates between the atom and the aggregates before, during, and after biology.

To be charged and impacted by this furnace of challenges and incongruities of thinking, so desperate and deserted in the very act of cognition, should be viewed as no abnormality, but evidence for the embrace of an unexpectable condition of multiple realities in one. Its foremost characterization may nicely be compared with the puzzle of art history, the International Postal Union, or the electrical grid of Maharashtra State, from among 10 to 1000 other examples. And while a centuries-long drought or the inability to speak a dialect of Tarantula might be annoying in the moment, this more paramount issue of one's personal circumstances of mind condemning reason to unknown, alternative possibilities is especially alarming in light of the rigorous training our gene pools have endured for so many millions of years.

Creatures wedded to their little comforts signal the least likelihood of their ever speaking with strangers. Fear listens only to oneself. Its schedules are the chorus of impulses reined in by the tyranny of being somehow part of that Self, of a Species quite comfortable with the cacophonous chanting of doom and gloom within its own communities and language groups. Outside such habitual comforts, their terror is unspeakable, unknowable, the fears prohibiting all contact beyond immediate, personal borders.

Such glaring issues, grievances—membership in a biographical trap that has been aggressively translocated from the physical commons into thought—might require the remainder of one's present existence to work through, up to the very instant of extinction. Alternatively, avenues of inexhaustible kindnesses, altruisms, generosities, all manner of tenderness afford far less unpleasant options, to be sure. Of that, no matter how dark the aisles leading up to it, we are certain.

Of course, it does matter, in the chain of beings, whether one expresses an idea in one form or some other version of it. The Selves are too myriad to suffer the ecological diagnoses that must follow from this sophistry, the biological histories that ultimately consolidate into an instant of Being, into one's own reality. No one has the patience to endure 300 million years of geology, unless that person was a Trilobite. For them, diagnostics were relevant. But for a species just over 300,000 years old? What can be said about it?

This family of Hominidae, to which we yet belong, is hardly comparable to the more than 50,000 species of the phylum Euarthropoda, the trilobites, clearly one of the most successful branches of life in biological history, whose ending came during a series of pulses, relatively abruptly during the End Permian–Triassic Extinction event, some 252 million years ago.[3]

---

[3] See Jonathan R. Hendricks & Bruce S. Lieberman (2008). "New phylogenetic insights into the Cambrian radiation of arachnomorph arthropods". Journal of Paleontology. 82 (3): 585–594. https://doi.org/10.1666/07-017.1

Climate change, volcano, meteorite—unclear. But it is called the Great Dying for an inconceivably literal reason, and the biodiversity rebound, particularly for terrestrial organisms, took several million years. In terms of surviving a week, a month, a decade, there was little latitude from an individual's perspective, which is where we are today. The mind that quantifies and qualifies such turbulence is the same mind that occasions it. That is more than paradox. Rather, a coincidence of involvements in the chain of being that implicates a singular presence during a catalytic period of *transition* marking an intuitive set of *possibilities*. Which is where existentialism ceases to be the castle of morbidities, transforming into an open-air symphony.

All of those prospects or fated hopes hinge upon the individual partnerships within the behavioral and conceptual realms of the human collective. It is perhaps the most strange and telling burden in the biological annals of culpability and liberation; to be a human fully aware of these galling contingencies, knowing full well that our future is unlikely, unless.

## 73.2   Autocatalysis: New Beginnings or Old Endings?

Unless what? Did the origins of "unless" ("on less than," with its unintended interpolations regarding a negative past as prelude to a potentially negative future) ever anticipate, through the sluggish cogs of etymology, so elegiac and pivotal a predetermination? A grand destruction known as self-fulfillment? The paralysis of self-awareness and the narcissistic abyss of its manifestations?

Intuitively, this catalytic ending, whether linguistically, culturally, or biologically, also summons the realization that what ends must begin again. Therein lies a transitional phase, a certain latitude for change. To this Other end of biological isolationism, a critique suggestive of opposite variables in an equation otherwise swept over by nihilism, we might predict colors and curvatures against the sky of a perfect rainbow. Or, conversely, of a "coming storm," as was the great Luminist painter, George Inness' (1825–1894) stunning idée fix.

The dialectic requires the patience of conservation biology, which, as a science and as a practice, has its protagonists and detractors, for reasons key to understanding the stubborn resistance exhibited throughout the history of science, particularly natural history, to intrinsic values. *Intrinsic*—inward, interior, dating back six centuries in Western languages—conveys an attitude about all sentience that makes it perilous to sort out distinctions. If one microbe has intrinsic merit, all microbes, all aggregates of life do so. We have thus far endeavored to lean mathematics and probability theory, particularly in the summaries of the mathematical Jain genius, toward the most inconvenient realization that every number must connote an individual's biology/biography.

For those who uphold that basic quality of the world—microbial merit at the individual level—everything is possible; anything short of that devolves to human subterfuge, grasping unto and for itself the totality of the life force, handing out its pleasures, affirmations, trifles, and negations as is convenient, to garden plots and companion animals. But when three billion birds are deemed to have gone vanishing in North America since 1970,[4] the biodiversity quantum, morose, dulling, and lacking in any chain of events that can be traced back to our own agency are lost in the numbers. Additionally, this calculus is denominated only among "wild" birds, with no mention of the other hundreds of billions of avians (e.g., chickens, turkeys, emus, and starlings) killed annually from desultory and ravenous human gluttonies. The numbers, as enumerated in Chap. 43, are unfathomable. For example, "21,917,808 chickens per day" in the United States in 2014.[5] This too, in its most gravid and heinous forms, is variance: Again, the ever-changing framework that defies the science of laws, norms, orders, theories, and rules, and any other conceptual cohesion of Nature by which we would endeavor to formulate human comparisons and comfort levels.

## 73.3  Maimonides in Siberia

On September 20, 2019, climate crisis protests erupted around the world, millions of children engaged, in advance of the September 23rd United Nations Climate Action Summit 2019.[6] Too late to save the *child* of the cetaceans, the smallest of them, 19 or fewer remaining individual vaquitas in the Gulf of Mexico, victims of "impoverished poachers, greedy cartels, and corrupt officials".[7] Too late to save a region of Siberia, Yakutia, the size of Europe, rotting under the heat rise of 3°C, the fastest rate of warming, along with Alaska and Greenland, in the world. With millions of other species and millions of individuals per species caught up in humanity's predatory and unreasoning rapacity, there is no metaphysical or practical "guide for the perplexed" any longer. For the twelfth-century Sephardic, Spanish philosopher, Moses ben Maimon, Maimonides (1138–1204), times were just somewhat easier. He looked for comfort and constancy to Aristotle's physical universe and to the many dozen of mizvot or sacred laws/commandments in the Five Books of

---

[4] See, "Decline of the North American avifauna," by Kenneth V. Rosenberg, Adriaan M. Dokter, Peter J. Blancher, John R. Sauer, Adam C. Smith, Paul A. Smith, Jessica C. Stanton, Arvind Panjabi, Laura Helft, Michael Parr, Peter P. Marra, Science 19 Sep 2019: eaaw1313, https://doi.org/10.1126/science.aaw1313, Science, AAAS, https://science.sciencemag.org/content/early/2019/09/18/science.aaw1313, Accessed September 20, 2019.

[5] See "How Many Chickens Are Eaten Each Day?" Reference, https://www.reference.com/pets-animals/many-chickens-eaten-day-97fda74fbde5d17c, Accessed September 20, 2019.

[6] https://www.un.org/en/climatechange/

[7] See "The 'little cow' of the sea nears extinction," by Annie Roth, National Geographic, October 2019.

Moses, or *Torah*, a controversial number that has been variously interpreted from fewer than a hundred to over a thousand.[8]

But in the twenty-first century, human communities, vaguely appointed units of consciousness, responding to instantaneous local and global breakdown via social media and local resource anomalies, we know only too well the consequences. Higher and higher surges of pain in Others, extreme weather, nearly one billion malnourished humans, the fast escalating NPP (human arrogation of all land and sea areas—from surface to sea floors—which can be exploited, fenced off, paved, opened up, and exploited with disastrous edge effects), to crises of will and a chaos of ruptured conscience fanning division and hate across the human worlds. This continuing litany harbors no eusocial-like instincts or juridical guidelines that could prevent collective, unconscious zeal from quashing the prospects of a mathematically augmented restraint among billions of individuals. In other words, little likelihood of change occurring from top to bottom, even though every collective lodges its ironic faith in the baffling deceit of most politics.

There has never been so stricken a dialectic. Polling suggestive of massive panic, depression, and economic and philosophical disparities throughout the human world: half the human population at greater risk than the other half. Such a dyad is non-linear: it adheres to no predictable Markov chain Monte Carlo (MCMC). Indeed, MCMC algorithms, like the Metropolis–Hastings, attempt to randomly sample from a difficult, if impossible, multi-dimensional set of probability distributions (direct samples being too cumbersome to trust).[9]

## 73.4   Probabilities from the Cell to Human Behavior

The Esther and Joshua Lederberg experiment in 1952 combined different plates of bacteria, exposing various colonies to the antibiotic agents of penicillin to determine a baseline in relation to the original plates. The results showed a global communication system that predisposed the behavior and susceptibility of bacteria to

---

[8] See Israel Drazi, Maimonides and the Biblical Prophets. Gefen Publishing House Ltd., Jerusalem, Israel, 2009, p. 209; See also, I. Drazi, "There Are Not 613 Biblical Commands," The Times of Israel, May 31st, 2017, https://blogs.timesofisrael.com/there-are-not-613-biblical-commands/ Accessed September 20, 2019.

[9] See M.N. Rosenbluth (2003). "Genesis of the Monte Carlo Algorithm for Statistical Mechanics". AIP Conference Proceedings. 690: 22–30. https://doi.org/10.1063/1.1632112; See also, Newman, M. E. J.; Barkema, G. T. (1999). Monte Carlo Methods in Statistical Physics. USA: Oxford University, P. Roberts, G.O.; Gelman, A.; Gilks, W.R. (1997). "Weak convergence and optimal scaling of random walk Metropolis algorithms". Ann. Appl. Probab. 7 (1): 110–120. CiteSeerX 10.1.1.717.2582. https://doi.org/10.1214/aoap/1034625254) (*See Subset simulation for structural reliability sensitivity analysis, https://doi.org/10.1016/j.ress.2008.07.006, ScienceDirect, Elsevier, Reliability Engineering & System Safety, Volume 94, Issue 2, February 2009, Pages 658–665, https://www.sciencedirect.com/science/article/pii/S0951832008001944?via%3Dihub, Accessed September 20, 2019.

penicillin prior to actual exposure. In other words, a category of, perhaps, the phenomenon of magnetic resonance, whereby directed mutation is proceeded by some form of deep awareness in bacteria that obviate the need for actual mutations.

At the global scale, the universality of structure in prokaryotes—plasma membranes, cytoplasm, DNA, and ribosomes that nurture and create proteins—exerts life's most dependable (not predictable) organization. As part of this dynamic, the biosphere's antiquity and continuing mastermind of unexpected shifts in eco-dynamics have hard-wired resistance, biological innovation, various survival strategies, and what would appear to be an ever-increasing flexibility in the realms of population ecology and the pure mathematics thus far extrapolated from countless case studies of biological evolution.[10] Unicellular mutation rates are believed to be at the level of 0.0003 mutations per genome per every cell generation. This translates into approximately 64 human cellular mutations every generation, "~3× or ~2.7×10$^{-5}$ per base per 20-year generation".[11]

Stacking these probabilities within cells containing a nucleus and membrane-protected organelles against the fate of a host within the real world pits the templates for organismic self-awareness and the survival-edged championing of that cellular self against the mirrors reflecting Hobbes, as well as Alfred Lord Tennyson's "red tooth and claw" from his 1850 poem, "In Memoriam A. H. H." Such components of a particular evolutionary mindset are at once political, sociological, and forever in retreat from the sobering inroads of scientific method (Fig. 73.2).[12]

## 73.5   Rudiments and Behavior of Ideals

Take a microcosm of this great paradox of ever-conflicted populations, all human societies in turmoil, and a universal desire to somehow configure ideals as alternatives: 25 Brook Street in London, where Handel lived from 1723 until his death in 1759, there composing many of his greatest works, including the 1741 oratorio, "Messiah."

Or to adjoining streets in the Dutch Capital of Peace and Justice, The Hague. Specifically, to a few streets and house numbers: Dunne Bierkade 16a—the Jan van Goyenhuis, where van Goyen (discussed in some detail in Chap. 19) was painted by

---

[10] See "Mutations Are Random," Understanding Evolution," University of California Museum of Paleontology (UCMP) and the National Center for Science Education, fueled by the National Science Foundation and the Howard Hughes Medical Institute, 2004, https://evolution.berkeley.edu/evolibrary/credits.php, Accessed September 20, 2019.

[11] See Schneider S, Excoffier L., July 1999, "Estimation of past demographic parameters from the distribution of pairwise differences when the mutation rates vary among sites: application to human mitochondrial DNA". Genetics. 125 (3): 1079–89. PMC 1460660. PMID 10388826.

[12] See *"A Short Analysis of Tennyson's 'Nature Red in Tooth and Claw' Poem,"* January 1, 2016, Dr. Oliver Tearle, https://interestingliterature.com/2016/01/01/a-short-analysis-of-tennysons-nature-red-in-tooth-and-claw-poem/ Accessed September 20, 2019).

**Fig. 73.2** The Quintessence of Human Domesticity: The View From Jan Van Goyen's Studio, The Hague. (Photo © M.C. Tobias)

the uncannily premonitory, Gerard Ter Borg; and where van Goyen's son-in-law, Jan Steen, executed a wild family portrait, with van Goyen arguably seated in the center. Dunne Bierkade #17—the Paulus Potter house; #18, home of van Balckeneynde. Around the corner, Paviljoensgracht 72–74, also owned by van Goyen, the home where Baruch (Benedict de) Spinoza lived for many years in exile from Amsterdam and wrote his masterpiece, *Etica*, and then died there at the age of 44 in a kind of exile.

Other subsequent painters of The Hague's Romantic Pulchri Studio, which was instituted in 1847, like Bartholomeus van Hove, Johannes Bosboom, and Pieter Stortenbeker who would repaint the ceiling of the front room of Potter's house. It was there at #18 where Potter did much of his work on "De jonge Steier" in 1647, the most evocative and gripping portrait of a large vertebrate in the history of art, nearly 9 feet by 11 feet, and eventually moved a few blocks away to the top floor of the Galerij Prins Willem V of Orange. Potter's great bull, considered one of the five most important masterpieces in the Netherlands, was taken by Napoleon in 1795 and hung in the Louvre for two decades, before being returned to the Dutch Republic and to Mauritshuis, where it hangs to this day.

If so voluminous a density of luminaries were not enough, van Gogh's five years in The Hague saw him wandering repeatedly up and down the Dunne Bierkade and commenting with great enthusiasm on the legacies of these giants of aesthetic and intellectual passion (just as Jimmie Hendrix would move next door to Handel's house and spend a year there from 1968). Dutch geography, and the uniquely Dutch sentience, has apotheosized the argument of transcendence, lending credence to the capacity for resolving the dialectic at the heart of human variance (and the accompanying dilemmas of consciousness) in a small neighborhood, but giving no formal clues to how it might be accomplished at rapidly increasingly complex macroscales, where evolution sways, heaves, and is lost between fixities of short duration, the shortest terms of punctuated re-evolution events, namely, across the fleeting span of human generations.

What these artists and philosophers accomplished was the elevation of the possible into a real series of physical coordinates—the choice of a precise terrain upon which to live and to die, to love, and, as Spinoza remarked, to feel sadness. Just a few miles away, the world of Leiden, where van Goyen as well as Rembrandt was born, dreamed, sweated, and vanished (Fig. 73.3).

It is June 19, 2019, and the Lakenhal Museum in Leiden along the Ouder Single canal has re-opened for a celebration of Rembrandt, van Goyen, and all of their local contemporaries and regional successors: Jan van Steen, van Mieris, Jan Lievens, Julius Porcellis, Gerard Dou, and David Bailly. Even the late nineteenth-century Leiden pointillists are represented adrift the prolific air of passing genius. As if from a scene of Parmenides' short fragmentary poem of 160 remaining verses, "On Nature." Alexander P. D. Mourelatos suggests that Parmenides has conceived of "the real as 'nicely circular from every side'" and holds that "the real is that which is perspectivally neutral," as with a sphere. "The only thing that varies is its size in our field of vision, and that is measured by our proximity to the object." Such deliberations, says Mourelatos, are "of great importance for the history of Western Philosophy".[13] And we would add to rudimentary molecular biology.

## 73.6   Ideations of Nature

At much greater variance, as intoned by Parmenides, is the relationship between those artists and philosophers whose images of Nature are at once confined to the *idea of nature*. Ideas that manifest in the Unity of multiplicity (hence, *the one and*

---

[13] *The Route of Parmenides – A Study of Word, Image, and Argument in the Fragments*, by Alexander P. D. Mourelatos, Yale University Press, New Haven and London, 1970, p. 129.

**Fig. 73.3** Spinoza's House, The Hague. (Photo © M.C. Tobias)

*the many*). Parmenides asserts that the differentiation between the Individual and the Species is a "great confusion" in mind that cannot quite fathom the distinctions between Genus and the Essence of everything that is, of itself, but somehow separate from Singulars. It is a minefield of simulacra, with "God, who governs all things," somehow implicated in that very act of perpetual confusion, the basis of all paradox.[14] God provides a very amenable exit strategy.

A vocal ensemble sings a haunting percussive work upstairs in the crowded museum, with lyrics approximating, "Join us in the emptiness…" repeated again and again. This is the Netherlands, but it is also humanity (Fig. 73.4).

---

[14] See *The History of Philosophy: Containing The Lives, Opinions, Actions and Discourses of the Philosophers of every Sect*, by Thomas Stanley, 2nd Edition, London, 1687, p. 748.

**Fig. 73.4** Portrait of an
Elder Lady, 1675, by
Gerard Ter Borch, Detail.
(Private Collection). (Photo
© M.C. Tobias)

And on this day back in The Hague, the world is pacing dumbstruck over at the Mauritshuis, where yet another dozen or so Rembrandt's are on quiet, darkened display, amid the three Vermeers, three Ter Borgs, and the inexhaustible Roelandt Savery and Jan Breughel/Rubens paradise. Each framed jewel is held in pinioned repose by a certain aerial solemnity to a specific place in time and human nostalgia. All of it is stunned and sadly hopeful for some lost past, much remarked upon by everyone who has ever lived, this feeling of repose in melancholy at the state of biology within our species' impermanent, yet imprisoned guise. These are emotional flash points that spell every coordinate of ecological crisis. Why is it this way? A constellation of longing and despondency. The great yearning gives way to some intangible modifier, we hold off that tempting overdose, or the tall bridge in the night, as paintings such as these, and their entire collective neighborhoods—real places inhabited by real people—lend a persuasive window upon the possibilities for all of us. The more desperate, the more alluring those openings into another reality.

# Chapter 74
# Ecological Epistemology

## 74.1   Feast or Famine Metaphysics

Upon each grain of sand within the co called Sorites' heap[1] are as many as, or more than, 100,000 microorganisms, as was recently approximated on an island off the coast of Germany.[2] Conversely, the Danakil Depression, atop the active Erta Ale volcano in Northeastern Ethiopia, is thought to be the only spot on earth that is utterly lifeless. Zero precipitation at Danakil versus, for example, Mawsynram, in the Khasi Hills north of the Bay of Bengal, within India's Meghalaya State, 11,871 millimeters of rainfall per year, the most recorded on the planet. Such contrasts mirror the human psyche. The Indo-Burma biological hotspot, in which the Khasi subtropical eco-region swarms with India's greatest abundance of species in one of only two remaining rainforests in all of the sub-continent, might be likened to humanity's most generous persona.

Descriptions, beginning with J. D. Hooker's seven-volume *The Flora of British India* (1872–1897) to more recent studies by A. H. Mir, K. Upadhaya, and H. Choudhury,[3] point to an all but obvious dilemma, humanity's other side, its numbers encroaching upon all others, an issue relevant to Ethiopia as well. Population = 115 million in 2020, the second highest human population in all of Africa, after Nigeria, with a corresponding rash of threatened and endangered species. India's moist broadleaf forests, such as Khasi, as well as small portions of

---

[1] Raffman, D. 2005. "How to understand contextualism about vagueness: reply to Stanley". Analysis. 65 287: 244–248. https://doi.org/10.1111/j.1467-8284.2005.00558.x. JSTOR 3329033. The Sorites Paradox was first conceived by the 4th century BCE Megarian thinker, Eubulides.

[2] "Every grain of sand is a metropolis for bacteria PhysOrg, Dec.13, 2017, ttps://phys.org/news/2017-12-grain-sand-metropolis-bacteria.html, Accessed November 23, 2019.

[3] "Diversity of endemic and threatened plant species in Meghalaya, North-East India," Int. Res. J. Env. Sc. 3(12): 64–78, and N. Chettri, E. Sharma, B. Shakya and Thapa (*Biodiversity in the Eastern Himalayas: Status, Trends and Vulnerability to Climate Change*, 2019, ICIMOD Books.

© Springer Nature Switzerland AG 2021
M. C. Tobias, J. G. Morrison, *On the Nature of Ecological Paradox*,
https://doi.org/10.1007/978-3-030-64526-7_74

Nagaland (above 1000 meters) are perpetually bathed in the soaking updrafts from the Bay of Bengal, the world's epicenter of the great damp. In its heart are thousands of endemic species vying for space and time with (as of 2020) 3.6 million humans who, to date, have seen fit to muster no more than six protected areas covering an expanse of less than 713 square kilometers.

The through story of Sorites Paradox tracks with these percentages and proportions.

At what point, however imprecisely, does biodiversity richness devolve to the depauperate? The tension enshrouding the lifespan mirrors, again, human nature, just as the majority of human conflicts occur within the 36 terrestrial biologic hotspots. Computational ecology is all about human behavior. We are mere bystanders to every natural occurrence, lamenting the reality that our involvements exact only peril and the ruinous (Fig. 74.1).

One of those scientific reserves in northeastern India comprises no more than 4.8 acres, the Baghmara Pitcher Plant Sanctuary. This tiny hub of endemism is emblematic of the crisis of numbers that must inform ecological epistemology[4] and the corresponding protective reflexes, polysyllogisms wherein the chain of premises collapses through subsequent betrayal of the original circumstances. Hope is undermined, just as the numbers, as equated with biological gradations, turn toward vagueness, failing their initial promise.[5]

Sorites, from the Demotic Greek, σωρείτης (sōreítēs, as in soros, heap), has invited the principal fundaments of this breakdown, wherein species become endangered, thought processes experience an incremental desuetude akin to cumulative loss. Writes Hyde at the conclusion of 5.3, "The Inscrutability of Reference," "As with earlier problems concerning the role of existential quantification in supervaluationism, one can debate whether this is a consequence to be embraced or an untoward consequence undermining the theory being advanced".[6] In either case, ecologically, there is a breakdown leading toward illogic, murky edges, and biological disruption (Fig. 74.2).

---

[4] See Saikia, Purabi & Khan, Mohammed (2017). "Floristic diversity of Northeast India and its conservation". Plant Diversity in the Himalaya Hotspot Region. Central University of Jharkhand, Dr. Hari Singh Gour University. Bishen Singh Mahendra Pal Singh (pub.). pp. 1023–1036. Upadhaya (January 2015). "Structure and Floristic Composition of Subtropical Broad-Leaved Humid Forest of Cherapunjee in Meghalaya, Northeast India". School of Technology, North Eastern Hill University. Journal of Biodiversity & Forestry Management. Anwaruddin Choudhury (October 2003). "Meghalaya's Vanishing Wilderness". Sanctuary Asia

[5] See Collins, Rory (2018). "On the Borders of Vagueness and the Vagueness of Borders" (PDF). Vassar College Journal of Philosophy. 5: 32. See also, Goguen, J. A. (1969). "The Logic of Inexact Concepts". Synthese. 19 (3–4): 325–378. https://doi.org/10.1007/BF00485654. JSTOR 20114646.

[6] See – Bobzien, Susanne, 2010, "Higher-order *Vagueness*, Radical Unclarity, and Absolute Agnosticism ", Philosophers Imprint, 10(10): 1–30; Cook, Roy T., 2002, "Vagueness and Mathematical Precision", Mind, 111(442): 225–247. https://doi.org/10.1007/BF00175369; Graff, Delia (see also Fara, Delia Graff), 2000, "Shifting Sands: An Interest-Relative Theory of Vagueness", Philosophical Topics, 28(1): 45–81. https://doi.org/10.5840/philtopics20002816; Hyde, Dominic, 1997, "From Heaps and Gaps to Heaps of Gluts", Mind, 106(424): 440–460. https://doi.org/10.1093/mind/106.424.641; Keefe, Rosanna and Peter Smith (eds.), 1996, Vagueness: A Reader, Cambridge, MA: MIT Press; McGee, Vann, 1991, *Truth, Vagueness and*

**Fig. 74.1** The Pitcher
Plant (*Nepenthes
Khasiana*) Endangered
Under IUCN
Classification. (Photo
Public Domain)

## 74.2   The Existential Zero

All of human counting, measuring, countering, and assessing come down to the
underlying abstraction that wants to contest two plus two. The deep psychological
need to refute "four" may stem from the obvious strain of anti-authoritarianism that
runs immensely deep through *Homo sapiens*. And/or it abuts that remonstration
toward all absolutes in Nature. In our nature, it wants not to be ruled by numbers,
confined to the small-minded discoveries of our predecessors, circumscribed by

*Paradox: An Essay on the Logic of Truth*, Indianapolis: Hackett; Scharp, Kevin, 2015, "Tolerance
and the Multi-Range View of Vagueness", Philosophy and Phenomenological Research, 90(2):
467–474. https://doi.org/10.1111/phpr.12173; Tye, Michael, 1994, "Sorites Paradoxes and the
Semantics of Vagueness", Philosophical Perspectives, 8: 189–206. Reprinted in Keefe and Smith
1996: 281–293. doi:https://doi.org/10.2307/2214170, Referenced in, "Sorites Paradox, Jan 17,
1997, Revised Mar 26, 2018, Encyclopedia of Philosophy, https://plato.stanford.edu/entries/
sorites-paradox/, Copyright © 2018 by Dominic Hyde, and Diana Raffman, The Metaphysics
Research Lab, Center for the Study of Language and Information, Stanford University, Stanford,
International Standard Serial Number: ISSN 1095–5054

**Fig. 74.2** Birch Bark Fragment of the Ancient Indian "Bakshali Manuscript," ca. AD 224–383, in Sanskrit: First known Documented Use of the "Zero". (Public Domain)

mere logic. It innately seeks the cover of an exit strategy, the equivalent of a dream beyond the immediate. It is basic.

There are transitions in nature whose increments defy our conceptual grasp of the true heap. These vague contours are weaned off palpable grains in the sand, "traces on the Rhodian Shore," to cite environmental historian Clarence Glacken's work. Robert Kaplan, in his exquisite work, *The Nothing That Is – A Natural History of Zero*,[7] meditated at great length upon the history, enigma, and phenomena that constitute this peculiar function of human logic and longing. Speculating on such landmarks as Osborne Reynolds' *On an Inversion of Ideas as to the Structure of the Universe* (1903), on the passion to reveal nothingness amid plenty, nihilism as the font of bounty, the paradoxical "presence" that is, for all that, one's very "absence",[8] Kaplan wanders into the world of Jainism, expressly the mathematics at the time of Mahavira, suggesting that the mysterious convergence of zero, coming ever closer to other numbers, would invite "an ideal country" comprising new "species," "lenses," an entirely novel realm of landscapes.[9]

Such a zero, and all those numbers above, below, to either side, cannot, in themselves define the summit of Mount Fuji or depths of the Mariana Trench, but they do provide our simplest means of representing Nature to ourselves. Kaplan is reminded of Winston Churchill describing "a quantity passing through infinity and changing its sign from plus to minus".[10] We may pride ourselves on knowing various states of

---

[7] By Robert Kaplan, Illustrations by Ellen Kaplan, Oxford University Press, New York, 2000.

[8] ibid., p. 195.

[9] ibid., p.195.

[10] ibid., p. 142.

being, liquid, gas, solid, particle, sub-particle, self, other. But in every event, knowing and unknowing, intuition, experience, ignorance, inspiration is each confined, somehow. We don't understand the constraints. We want to hope, to believe, to build edifices of certainty that argue in favor of tomorrow, of learning than merely the past, of ameliorating pain and maximizing pleasure. Basic behavioral norms, as well as those transcendental feelings we possess for purposes of an individual or community interest, and in which we normally prefer to remain immersed, render our everyday an ambiguous transcript out loud, in the living of it toward some other combinatorial truth.

Whether we fatally or masochistically adhere to our invented verities, thus far, history tells us that the outcomes are unvarying. This is the human problem: We are incapable of escaping the numbers by which we define and defend ourselves. Our aggressions are numeric. Our weapons calibrated according to their arithmetic damage potential. Our greed, every probability, progression, statistic, and accounting take stock according to this mental scaffolding, unstoppable towers of numerology that will not quit until the very sub-sub-atomic temptations subside. Hence, the love affairs with Pi ($\Pi$, $\pi$) and its trillions of digits, as first intimated by Archimedes and Pythagoras (Fig. 74.3).

Our theories of knowledge that attend upon such mathematical entities, as surrogates for biological units—individuals, species, populations, etc.—place in great peril all Other remaining life forms. Our counting is fraught with contradictions and the actual terror induced by zero. To hit zero is to see, for example, a bankruptcy, not a balanced equation. Zero equals extinction and can be tracked back in time throughout human behavioral history or something else entirely. The zero is also not four but five, or six. It is the horizon which humanity needs like never before.

There, for example, is the Victory Stele of Naram Sin, ruler of the Akkadian Empire from c. 2254 to 2218 BC, obsessed with the outright conquest of his "Mesopotamian" mountain, spear in hand,[11] one of the early, documented emblems of ecological superego, his army's subsequent reiteration in every historical guise. The empirical shadows of our yardsticks, telescopes, and microscopes. But no match for a dragonfly, or the true tenderness we know might yet emanate.

## 74.3 Evolutionary Latitudes

Does evolution, as articulated by every side of its tired exegesis and hermeneutic, remain part of this sentimental equation? Let us subscribe to a hypothesis well tested, of biosemiotics that is, or always was, this earthly jewel of destiny, at least more than four billion years-worth. And then add to that assumption a complete absence of anything like what we take to be evolution. We can easily imagine a *nonevolutionary* collective where change, transformative flux, seed dispersion, every

---

[11] See https://www.louvre.fr/en/oeuvre-notices/victory-stele-naram-sin, Accessed June 23, 2020.

**Fig. 74.3** Pythagoras (ca. 570-495 BC), as Portrayed in *The History of Philosophy*, by Thomas Stanley, Printer Thomas Bassett, London 1687, p. 491. (Private Collection). (Photo © M.C. Tobias)

migration, altricial and precocial contexts, DNA, mitosis, that whole perceived spectrum of molecular biology, requires no special recourse to our perception. It circumvents 330,000 years of this hominin's debacles. It declares that, yes, we, the newcomers, have no claim on any fact. Our mirrors are dizzying, the scripts and transcripts they reflect a blurred surrealism which need not dictate our next steps.

If we assume the aforementioned, freeing ourselves, somehow, in every due course, then we can carefully proceed toward reconciliation and ecological redemption. If we can but liberate current and forward ethics, wherein we come to grasp—individuals and communities—the preconditions and prospects for rapid transition. To know that there is a world out there of parallel vagueness, wherein some promise of a contract with Nature exists, immune to the all but overwhelming doctrine that our lives are condemned to one side or the other of zero.

If such language reads as quintessentially vague, it must be so if it is ever to envision and effect that same Principle of Uncertainty that guides convergence and overlays in the atmosphere, the oceans, every stream, coral community, and neotropic; all nutrient turnovers in every soil; maturations, barriers, and lack of barriers to fertile distributions. And which uncannily works out the relational complexities, coherence, and incoherence, marking the ecotones between those same biological and metaphysical individuals and communities. If this vagueness is real, then all the sciences, at some very minimum, are given a second chance, a reason for optimism that, perhaps, there is, after all, a conceptual natural selection that favors perpetuities between the lines. The marginalia and fractions that actually add up to the biosphere just *days* before our arrival, a mere 0.002% of the time since the Universe herself had come in to being.[12]

There is, however, a flaw, even in that muted note of optimism that must not be overlooked. About 300,000 to 350,000 years ago, there were nine different known coexisting *Homos*, members of a largely competitive Genus, which included *H. sapiens*.[13] Aside from scant evidence of meager, mutualistic, overlapping settlements and intermating, among two or three *Homos*, for short time frames, most forensics indicate a strong likelihood that our species systematically wiped out any and all competitors. Our ecological destinies were forged by tens of thousands of years of accumulating choices and attitudes that boded of today's dominant human personality.

---

[12] See "What Was It Like When The First Humans Arose On Earth?" by Ethan Siegel, Forbes, May 15, 2019, https://www.forbes.com/sites/startswithabang/2019/05/15/what-was-it-like-when-the-first-humans-arose-on-earth/#92e098369975, Accessed July 21, 2020.

[13] See "300,000 years ago, nine human species lived on Earth. Did *Homo sapiens* exterminate the others?" by Nick Longrich, The Conversation, Scroll.in, November 23, 2019, https://scroll.in/article/944558/300000-years-ago-nine-human-species-lived-on-earth-did-homo-sapiens-exterminate-the-others, Accessed July 21, 2020. In addition to our species, there were *H. neanderthalensis*, *H. erectus*, *H. rhodesiensis*, *H. luzonensis*, Asiatic *Denisovans*, *H. naledi*, *H. floresiensis*, and the Chinese "*red deer people*". *Homo habilis*, *H. palaeohungaricus*, *H. cepranensis*, other *archaic Homos*, and the earlier Australopithecines like *A. Paranthropus* and *A. garhi*, had already gone extinct, some at least 1.3 million years before this time, during the Gelasian, or Lower/ early Palaeolithic. Our earliest known traces date to a morphology said to have emerged between 400,000 and 200,000 years ago. Most paleontologists look to approximately 330,000 years ago. Our earliest known remains, thus far, were discovered at Jebel Irhoud, in Morocco. See Hublin, J., Ben-Ncer, A., Bailey, S. *et al.* New fossils from Jebel Irhoud, Morocco and the pan-African origin of *Homo sapiens*. *Nature* 546, 289–292 (2017). https://doi.org/10.1038/nature22336, 08 June 2017, https://www.nature.com/articles/nature22336#citeas, Accessed July 21, 2020.

# Chapter 75
# The Natural Selection of Indeterminacy

## 75.1 Natural Selection Anew

No explanation yet exists for selection; neither who or what is behind it, or why. That is a marvelous thing. An Arabian leopard, almost extinct by human machinations, remains vivid proof that by our own conservation reckonings we are becoming aware of a solid truth, the linkages between ourselves and the Others. Of course, any number of selective pressures, and in our species' case, motives have been preached, but none can persuade in the total absence of credible real predicates or identities. We know nothing against a backdrop of other multitudinous verbs, transitive, intransitive, intermediary. We lack the nominative, in spite of every natural object being the target of such actions. This absence of credible human accounting renders all perception and accretion suspect. In so flimsy an environmental continuum, our communities at once decry their unrealities in the sense that their sum total of individual behaviors defy ecological gravity. In Robert Kaplan's discussion of logician Willard van Orman Quine's suggestion "that 'nothing' and 'something' are false substantives, behaving like nouns grammatically but not logically," he points to that chemical conundrum in which the sum of all inputs must equal the sum of reactants, no matter what has transpired by way of a perceived equation: the Law of the Conservation of Matter.[1] The leopard is a leopard is a leopard. No single action accounts for a truth or even a glimpse. Whatever windows on substance may have teased the history of science are merely the glimmers of seasons gone, songs imagined, pictures annoyingly cloyed in the manner of mud on a grave without a stone to indicate the slightest testament. Indeed, there is no telling anyone or anything, and hence, the environment of which we are so fond of speaking gives nothing away. Reasoning powers are actually powerless, and this is quickly established by any etymology, so short-lived a legacy of vowels, consonants, the aboriginal guttural

---

[1] See Robert Kaplan's book, *The Nothing That Is – A Natural History of Zero*, Illustrations by Ellen Kaplan, Oxford University Press, 2000, p. 59.

© Springer Nature Switzerland AG 2021
M. C. Tobias, J. G. Morrison, *On the Nature of Ecological Paradox*,
https://doi.org/10.1007/978-3-030-64526-7_75

**Fig. 75.1**  Arabian Leopard (*Panthera pardus nimr*), IUCN-Critically Endangered, Possibly Fewer than 40 Individuals in the Wild, Photographed at the Saud al-Faisal Wildlife Research Center, near Taif, Saudi Arabia. (Photo © M.C. Tobias)

dating back to a hominin physiognomy wherein the tongue, palate, sandstorm of sighs are mere murmurs against a Milky Way that does not so much as whisper. Not a single ancestor that exerts anything remotely cognitive. The real "data" is before science (Fig. 75.1).

What meaning have we contrived that should give us to actually believe some natural selection works on our ideas or elevates instinct in the manner of apotheosis? It is no reach to compare our social antics to a global shuffleboard of sub-atomic particles randomly skittering upon glazed, non-reflective surfaces. Distractions of life and death, and of a host of other categories, we cannot fully reconcile. While a philosophy consumed by the forbidden cities, entrances, and exits of no-meaning has little appeal to a species that squirms restlessly, there is also no good reason to discount the essence of all probability. That equation is an equal sign in whatever contest of delusions, out of which we might wonder whether something does not come of nothing.

Imagine if it were true, and if truth herself could hold on long enough to make contact. The question *What would it mean*? has no context, no preparatory ground upon which to launch a formal enquiry. There is, again, nothing but the empty wonderment, and empty it indeed is. We have no reason to delegate substance to consciousness, however furiously we attempt to trace its origins, loci, gene markers, or specific nerve endings. The brain belabors itself, millennium after millennium, to no effect.

Yet, at any instantaneous juncture of wonderments we can easily posit a suspicion that is grand and elaborate. This vacuum of weightless, fully dematerialized stuff of the universe, which we grab for ourselves in the name of Self, or of Mind, in fact, is readily translated. Air fixating upon air, the nothingness aspiring to corporeality by nothing more than a whim. These are not transcendent concepts because the concept itself has no substantiality, is without density or mass, or even energy, or none that can be proved, by theorems that cannot exist in the void and that does not have a name. Laws of thermodynamics are simply inconsequential when it comes to the imagination of any organism, including human.

Yet, so repeatedly removed from any reality, the vague vagueness is somehow subject, like substantiated instinct, to the selective pressure of a pressure with no name, no reality, no identity whatsoever. These empty gasps are on to something, and so we might adduce without prejudice or hope the fact of selection favoring vagueness, selection upon which there is wide agreement that obeys no common ground, no collective assertion, only a solitudal consensus unto itself. No reverberation in the entire Universe—though with laws of energy and causality, mirrors and rockets, and mathematical chimeras we believe we know much of that Cosmos— and still a vague sense of the vague. In terms of natural selection, we might think of Robinson Crusoe as a clear field experiment. To what extent has Daniel Dafoe's creation (first published on April 25, 1719, and likely modeled after Scottish cast- away Alexander Selkirk's time on the island of Más a Tierra) changed over the course of 28 years in random exile, if at all? Taking into fullest account all the vari- ables? (Fig. 75.2).

## 75.2  Kin Altruistic Superorganisms

It is by this mystery that, perhaps, something, some thing, can occur. We are free to imagine what it is, even while freedom itself, free will, freedom to relate to oneself, remains a contested realm more etymology than reality (like the word *sustainabil- ity*, from the seventeenth-century *bearable*), to its disambiguated iterations in such instances as the 2019 *Urgenda v. The Netherlands* Dutch Supreme Court decision regarding a near immediate, mandated reduction of global greenhouse gasses by 25% throughout the Netherlands;[2] the fact that Burlington, Vermont, has for several years been powered entirely by renewable energy sources; or an equally far-reaching interpretive concept, instinct with biological implications, of ecological suicide, most notably as deciphered in a range of microbial self-destructions now believed to involve collective consciousness within kin altruistic superorganisms.[3]

This latter catalyst, whether in *Paenibacillus* sp. or *Homo sapiens*, challenges and elucidates Darwin's most fundamental statements pertaining to natural selec- tion, as described in Chap. IV of his *The Origin of Species*: "Therefore, during the modification of the descendants of any one species, and during the incessant struggle of all species to increase in numbers, the more diversified the descendants become, the better will be their chance of success in the battle for life".[4]

---

[2] See https://www.urgenda.nl/en/themas/climate-case/Accessed December 24, 2019.

[3] See "Ecological suicide in microbes," by Christoph Ratzke, Jonas Denk and Jeff Gore, Nat Ecol Evol **2**, 867–872 (2018) https://doi.org/10.1038/s41559-018-0535-1, 16 April 2018.

[4] Sixth Edition, John Murray, London, 1886, p. 103.

**Fig. 75.2** Robinson Crusoe, as Portrayed in the Frontispiece of *The Life And Strange Surprising Adventures Of Robinson Crusoe; Of Your, Mariner, "Written by Himself,"* C. Hitch and L. Hawes Printers, London, 1761. (Private Collection). (Photo © M.C. Tobias)

# Chapter 76
# Imagining Transitions

## 76.1  Pivoting Between Paradox and Existential Biology

We fill in every blank on the map with quantified, biological data: densities of life, delineations and discriminations, altricial and precocial birth; nesting behavior among avifauna—nidifugous versus nidicolous, the latter referring to those that stay on a while in their nests; deductions, premises, objects of definition, assumptions, propositions—all swirling around our idea of life.

The rise of "paraconsistent logics" and an inconsistency-tolerance has exacerbated clarity by embracing endless implications.

Since the time of Charles Darwin and Gregor Mendel, for example, interpolations of data from varying behavior, and the underlying phylogenetic assumptions and continually evolving genetic storyline, have turned the often conflicting languages of biology on their head. Writes Roy Sorensen, "Paraconsistent logics are designed to safely confine the explosion".[1] Just as in 1631 the publication of a Bible in London was shown to have omitted a single word in the Seventh Commandment, "not." King Charles 1 simply had all copies of the edition sought out and destroyed.[2] An immediate remedy for a specifically closed problem of "Thou Shalt Not" versus "Thou Shalt." For Sorensen, there is the possibility that "some hitherto unconceived field" may yet usher the kind of geography and metaphysics, or some other intellectual topology, wherein certain paradoxes, currently having exploded, are subject to intellectual containment,[3] to an effect as closed and shut as that Bible saga, a single word upon which the future of the world theoretically pivots. It might just be an ethic as fundamental to the Bible as Thou Shalt Not Kill. That is the foremost

---

[1] See *A Brief History of the Paradox – Philosophy And The Labyrinths Of The Mind*, by Roy Sorensen, Oxford University Press, New York, 2003, p.114. See also, 1. See H. A. Wolfson, Philo, Cambridge, Mass: Harvard University Press, 1947, Vols 1–2.

[2] ibid., p.165.

[3] ibid. pp. 370–371.

© Springer Nature Switzerland AG 2021
M. C. Tobias, J. G. Morrison, *On the Nature of Ecological Paradox*,
https://doi.org/10.1007/978-3-030-64526-7_76

ethic relevant to human evolution and the power of natural selection favoring a pragmatic, co-symbiotic ideal. That is, indeed, the very idealism upon which the fundamental assertion of this entire exercise rests (Fig. 76.1).

But, in fact, a non-containment ecological paradox has always languished beneath the surfaces of philosophy and the history of science. In July 1840, when young Søren Kierkegaard wandered through the cold moors of Jutland, his late father's district of Viborg, he was, as biographer Joakim Graff has written, in an utterly disoriented state. He "confused forests with the sea..." The combinations of life forms and formations were everywhere perplexing such that "the increasingly directionless young man" would "plummet through himself, past all fixed points, out into nothing".[4] The bewilderment was tantamount to that of J-P. Sartre's character of Antoine Roquentin, from his 1938 first novel *La Nausée*, the ecological epiphany rooted in an encounter with the knotty brooding disgusting throbbing tree within a city park in the fictional town of Bouville. Where every chapter in one's life is both overthrown and undermined with the literal crush of sensate experience, at "6.00 p.m...." The alien bark, the misanthropic sensation of "obscene nakedness" before the onrush of so much unwanted life, collaboration with the biological bottom line in the guise of a miserable Self, and the unrelenting totality of existence swarming to all sides.[5]

**Fig. 76.1** 1902 "Portrait of Søren Kierkegaard" by Luplau Janssen (1869–1927) in The National Museum of Denmark, Copenhagen. (Public Domain)

---

[4] *Søren Kierkegaard – A Biography*, by Joakim Garff, Translated by Bruce H. Krimmse, Princeton University Press, Princeton, New Jersey, 2005, p. 158.

[5] *Nausea*, Jean-Paul Sartre, Translated From The French by Lloyd Alexander, Hamish Hamilton, London, 1962 edition, p. 172.

Such ecological chaos is indebted to Nietzsche, to Kierkegaard, to Sartre, and to others. They embraced encounters with Nature as the rude, epiphany-laden open wounds of biology. Not those nostalgic landscapes blurry-eyed with the historical romance of Utopia: the Land of Cockaigne, Xanadu, the hortus conclusus, Grails and Antipodes, the writings of Pseudo-Callisthenes in his documentation of the travels of Alexander, and the *Livre des merveilles du monde*;[6] in other words, of whispering, near-death emotion, and pantheistic speculation, all evoking the fantastic aesthetic of some brilliant alternative to the human condition. Gandhi, Sartre, and George Orwell had to cope with the British occupation of India, the Nazis, and what Orwell termed "mental squalor",[7] not paradise. The existentialist outcome of these brute realities of the mid-twentieth century was a queasy but quintessential contemporary activism. Social movements that might help ameliorate every version of modern desperation (Fig. 76.2).

In his biography of Nietzsche, referencing Nietzsche's retrospective year (1888) in Turin, Rüdiger Safranski concretizes the Darwinian stage, already six editions thick in terms of the public reception of *On The Origin of Species*, in which "The ape had replaced God as an object of inquiry" and, hence, God was essentially an extinct being, a perpetually nagging concept.[8] The platitude, "God is dead," was first employed three times by Nietzsche in his collection of essays entitled *Die fröhliche Wissenschaft,* 1882). Iterations were soon to follow, such as Nikos Kazantzakis' early play, *Nikiforos Fokas* (Athens 1927), in which "God – wounded, pale, pitiable – collapses at his feet...".[9] "We now touch only his corpse," the young J-P Sartre said of God.[10] But despite the French revolutionary's insistence on an existential form of atheism, the Nobel of escapisms, the Greek Orthodox philosopher Christos Yannaras claimed that Sartre was "the most important theologian of the West's philosophical tradition".[11] The ambiguities of Sartre's 74 years are telling. When he died on April 15, 1980, the obituary in *Le Matin* included the statement, "With him dies one of the few truly free men of our age..."[12] echoing Kazantzakis' famed gravestone citing "freedom" as the ultimate measure of human meaning. But if we trace the roots of freedom, and the associated impulses to be free, from the Old

---

[6] See *The Book Of Legendary Lands*, by Umberto Eco, Translated by Alastair McEwen, Rizzoli ex libris, New York 2018.

[7] See "George Orwell's Biggest Fear Went Far Beyond Big Brother," by Gus Lubin, Business Insider, June 25, 2013, https://www.businessinsider.com/george-orwells-biggest-fear-went-far-beyond-big-brother-2013-6, Accessed July 21, 2020.

[8] See *Nietzsche – A Philosophical Biography*, by Rüdiger Safranski, Translated by Shelley Frisch, W. W. Norton & Company, New York, 2002, p. 306

[9] *Kazantzakis – Politics of the Spirit*, by Peter Bien, Princeton University Press, Princeton, New Jersey, 2007, p. 414.

[10] See Kate Kirkpatrick's essay, "Sartre Does God," https://blogs.lse.ac.uk/theforum/sartre-does-god/. Accessed December 27, 2019.

[11] ibid, K.Kirkpatrick.

[12] See *Sartre – A Life*, by Annie Cohen-Solal, Translated by Anna Cancogni, Edited by Norman Macafee, Pantheon Books, New York, 1987. p. 520.

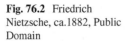

**Fig. 76.2** Friedrich
Nietzsche, ca.1882, Public
Domain

Teutonic, Gothic, and Anglo-Saxon, frijôn, to love, we have an etymologically
schizophrenic dichotomy from Old English in the fourteenth century, "power of
self-determination, state of free will" to the confused mix of freedom fighters, freedom lovers, and "exemption from arbitrary or despotic control".[13] And then, its
ultimate desecration, "Arbeit Macht Frei," Work Makes One Free, the sign at the
entrance to Auschwitz and other Nazi Jewish extermination camps.

Kierkegaard's dialectical exasperations spanning the mundane as well as his own
tortured theology evince other characteristics endemic to a twenty-first-century ecological conscience, with or without the notion of human freedom, salvation, or a
single atom or god of amelioration. The young poet who declares, "Tout n'est rien,
all is nothing," and who, in the "Diapsalmata," informs us, that only in the realization of children, and conversations with them, might we still hold out hope of anything becoming rational, but "-good Lord!" once they have achieved the rational,
what then, he muses?.[14] And years later, writing in "The Essential Expression"
chapter of Kierkegaard's *Concluding Unscientific Postscript* (first published in 1846
as *Concluding Unscientific Postscript to the Philosophical Fragments* by his pseud-
onymous tell-all, Johannes Climacus), "…that a human being can do absolutely

---

[13] See Online Etymology Dictionary, "freedom," https://www.etymonline.com/word/freedom,
Accessed July 21, 2020.

[14] Søren Kierkegaard, *Either/Or*, Abridged Edition, Edited by Steven L. Ross, Translated from the
Danish by George L. Stengren, Perennial Library, New York, 1986, p.13.

nothing of himself…".[15] This then, from the man who recognized that there could be no humanity, only individuals smitten with guilt, existential appeals to some incarnate conscience, to an (again, imagined, as from the Middle Ages) aesthetic consciousness, wherein any and all understanding must disambiguate into dust. It is a mirroring of the only real truth, that nothing exists, including hope vested in the coming generations, for whatever that's worth.

## 76.2   Poets Who Die Young

For Kierkegaard himself, nothing existed after his age of 42. When we attest to all the poets and philosophers who have died young—Keats, 26; Novalis, 28; Shelley, 29; Thoreau, 44; Avicenna, 57; Aristotle, 62—the ground-zero realities overwhelm every thinking moment. Our preoccupations with mortality undermine any reconciliatory pretense that should award or tout existence, any more than we can account for extinction. Our gestures either way are to be lost. "Hell is other people," declared Sartre's Argentinian, Garcin, in "No Exit" which premiered one year prior to the end of World War II.

As far as can be discerned, life knows no startling conclusions, only suffering and change, suffering which, as E. M. Cioran elucidates at the conclusion of his *History And Utopia*, detests all notions of a Golden Age, declaring from every pang of its births, "the double impossibility of paradise".[16] And by the unabashed dictates of this abrupt truth, it would be a simple matter to summarily discount every cell and cell division according to the made-up rules of fatalism. Knowing full well that our seconds are leading to this end, one by one, the essence versus existence debates that have unceasingly absorbed all those who fail to shed restlessness can do nothing for our cause. That crusade to eliminate pain. To uphold some ideal oath granting life more sovereignty than death. The idealistic embrace is that pillar of resistance which rejects violence, suffering in any form, granting in ourselves and in all others, not only the duty to promote tenderness, but the obligation to sustain its rationale. Where reason may fail to capture ultimate solutions to the world's agony, our senses know this paen to relief, medical succor to be the only truth-value, without surcease, paradox, or the slightest argument. It is plain for all to recognize.

Such transitions that proclaim some beautiful bounty beyond the drudgery of a Dickens' world may or may not exist. If they do, it is in spite of ourselves, which puts an entirely new spin natural selection.

---

[15] *Kierkegaard's Concluding Unscientific Postscript*, Translated from the Danish by David F. Swenson, Completed after his death and provided with Introduction and Notes by Walter Lowrie, Princeton University Press, Princeton New Jersey,1968, pp. 417, 419.

[16] E. M. Cioran, *History And Utopia*, Translated from the French by Richard Howard, Seaver Books, New York, 1987, p. 115.

# Chapter 77
# The Finely Honed Basis of Unknowing

## 77.1  What We Don't Know

While each increment of natural selection remains an untold history of biologically tasked nouns, pronouns, verbs, and integers, there is no geographical compass or time frame that in isolation can yield useful elucidation of the living objects of the world we think we know. Even at the most seemingly profound level of classification and philosophizing, E. O. Wilson qualifies his May 30, 1993, essay for *The New York Times Magazine*, "Is humanity suicidal?" by re-stating, "no, we aren't suicidal, but we are death for much of the rest of life and, hence, in ultimate prospect, unwittingly dangerous to ourselves".[1] There is no way to mollify the brutality of our ambiguities. They have transmogrified into unbreachable distances, acute biochemical compromise, inbreeding depression in the absence of any conspecific organisms that might, through hybridization or other means, revivify the population bottleneck headed toward extinction. While polyploid angiosperms might be able to avoid such population-level fatalities (a redundancy of chromosomes), the human inflictions are so widespread, affecting every pinch of soil, column of air and of water as to be far beyond ideal calculations. The destruction afflicts our senses, our reason, our hopes. But it does much more to other species, and this represents a profound cause for both utter despair and the vivid perplexities of philosophy.

---

[1] *Nature Revealed – Selected Writings 1949–2006*, by Edward O. Wilson, The Johns Hopkins University Press, Baltimore, Maryland 2006, p.656

© Springer Nature Switzerland AG 2021
M. C. Tobias, J. G. Morrison, *On the Nature of Ecological Paradox*,
https://doi.org/10.1007/978-3-030-64526-7_77

## 77.2   The Implausibility of Denial

To deny this fatalism represents implausible direction, unless we truly embrace life. What does it even mean? As the *hyperconscious* structuralist semiologist Roland Barthes (1915–1980) has written of that ultimate desire "...to dissolve one's own image...".[2] Barthes adds, "... Such is the paradox, the imaginary of the self as paradox, that gives me a permanent feeling of enigma...".[3] This lack of validity had been earlier rejected by Baruch (Benedict de) Spinoza, who argued that Nature should present no confusion. That "the standard of the given true idea" should eliminate troublesome discrepancies in perception, as well as in reality. "That certain things presented to the imagination also exist in the understanding...".[4] That is among Spinoza's most breathtaking declarations.

## 77.3   Converging Factors

Following upon the Flemish Jesuit Grégoire de Saint-Vincent (1584–1667), who had attempted to resolve Zeno's Paradox, fellow Jesuit mathematician Alphonse Antonio de Sarasa (1618–1667) fundamentally challenged the inaccessibility of infinity by describing hyperbolas (curves within planar surfaces) and the quadrature (creating a square with the volume of a circle or vice versa). Such thought experiments led to the so-called natural logarithm (commonly perceived as the "inverse of an exponential") and a numeric relation that touches upon everything in Nature: "mineral hardness, and the intensities of sounds, stars, windstorms, earthquakes and acids..."[5]

---

[2] *The Neutral*, by Roland Barthes, Lecture Course at the College de France [1977–1978], Translated by Rosalind E. Krauss and Denis Hollier, Text established, annotated, and presented by Thomas Clerc under the direction of Eric Marty, Columbia University Press, New York, 2005, pp.12–13.

[3] ibid., Barthes *The Neutral*, p.101.

[4] See *How To Improve Your Mind*, by Baruch Spinoza, With Biographical Notes by Dagobert D. Runes, The Wisdom Library, A Division of Philosophical Library, New York, p. 66.

[5] See "What Are Logarithms?" By Robert Coolman, LiveScience, May 22, 2015, https://www.livescience.com/50940-logarithms.html, Accessed January 7, 2020. Beginning with the notion that arithmetic sequences mirror geometric ones, logarithmic thinking quickly encompasses the entire planet. The natural logarithm, when applied to a future projected population, becomes the witching stick of extrapolation for any series as well as programming languages. The equation itself, $Nt=P$ e (r * t), relies on a constant, e, that is equal to 2.71828. One case study involves the estimated population of Nigeria from 2011 to the year 2030, representing an increase in numbers of 98.67 million people. See http://maps.unomaha.edu/Peterson/geog1000/PopulationProjections/Population_Projections_GEOG1000-Answers.pdf, Accessed January 6, 2020. A natural logarithm of ten equals 2.30258509. To any casual observer, their numeric relationship is meaningless. See "The On-Line Encyclopedia of Integer Sequences," https://oeis.org/A002392. Chance, supposition, theory: these outliers of probability parallel every philosophical independent event. Random (stochastic) processes and points, from Brownian motion to the Poisson point process, take their cue in the history of mathematics from something as light-headed as gambling. Yet, a game of poker has the same level of sobriety in terms of human affairs as archaeology. The Iranian military

## 77.4   Merging the Imagination with a Probability Event

Any human, with even the slightest effort, should be capable of deep empathy. That would include actually merging one's imagination with a probability event, as well as another's projected needs, hence—a category inviting the similitude of Selves throughout the biosphere. It is likely that such kin altruism, taken to a level of universality, will necessarily encounter what, in mathematical parlance, equivocates around such manipulations as multiple regression (a multitude of independent variables including the interaction effects of more than two variables); the presence of predicted and residual values; varying scales of measurement between the identity of that personage with whom one identifies; and the qualified magnitude of an impact upon that individual. As for a reasonable picture, one *turns away from precision* toward the vague, as earlier indicated, in terms of opening biological pathways that may engender latitude in evolution and natural selection. This turning away, the transition of all transitions, is ecological to the core. Its paradoxical relation to a life form is paramount in as much as two, as earlier suggested, does not add up to one plus one. There are far more, subtler variables involved than the human imagination is heir to.

There is little likelihood of an infinity of gradations, logarithmic functions, equations, or differentiations, giving specificity to any one individual. It is far more complicated than that. We may assume that this turning away abets identification of separate biographies across the wide human historical enterprise to understand life. When Amsterdam apothecary Albertus Seba (1665–1736) sold his first enormous collection of "curiosities" to the Russian Czar Peter the Great in 1717, what, precisely, was at the crux of his zealous transaction? According to one inventory, "72 drawers full of shells, 32 drawers displaying 1,000 European insects, and 400 jars of animal specimens preserved in alcohol".[6] In much more gravid perspective, more

---

leader Quasem Soleimani's assassination and the event's subsequent to its fallout plays precisely into the historic manifestation of the Dempster-Shafer Theory collating combinatorial masses, subjective beliefs, distributions, properties, measures and half-truths, degrees of confidence, false dichotomies and informal fallacies. See Dempster, A. P. (1967). "Upper and lower probabilities induced by a multivalued mapping". The Annals of Mathematical Statistics. 38 (2): 325–339. https://doi.org/10.1214/aoms/1177698950

[6] See "Albertus Seba's Collection Of Natural Specimens And Its Pictorial Inventory, by Irmgard Müsch, in, *Albertus Seba, Cabinet of Natural Curiosities – Das Naturalienkabinett Le Cabinet des curiosites naturells – Locupletissimi rerum naturalium thesauri 1734–1765*, Based on the copy in the Koninklijke Bibliotheek, The Hague, Taschen, Bibliotheca Universalis. Köln, Germany 2007, pp.15–16. How would such a cabinet of dead individuals figure in mathematical configuration? Abstruse transpose and square matrices, vectors, row and reduced row echelon forms; post-Gaussian elimination; a series of linear algebraic matrix retooling, all confounding probability theory and the difficulties of thinking about just what is possible in life, as in death, based upon exoskeletons and so many bits and pieces of death. Such mathematics cannot correlate with living beings. This analogy goes to the core of the human inability to find correspondence with extinctions, or even, by implication, to work backwards from extinction to the living and attempt to gauge cause and effect in all its daily subtleties and cumulative stress. No combination of scientific and mathematical disciplines can account for actual living substance.

**Fig. 77.1** A Reptile Cabinet, Museu de Historia Natural, Maputo, Mozambique. (Photo ©
M.C.Tobias)

than a billion dead animals in the 2019/2020 Australian brushfires, and possibly
more than three trillion dead rove beetles (*Staphylinidae*) (for starters),[7] leave
human consciousness vulnerable to paralyses, or it should. Yet, we wake up the next
day after every tragedy. Terrible headlines have not dissuaded us. Neither our conscious responses, nor the deep-seated staying power of our virtue, our conscience,
our beliefs about a better future, have been altogether dented, despite the common
denominator which is the birth of eco-dynamics (Fig. 77.1).

## 77.5    From a Tired Trigonometry to the Welcoming
Co-symbiosis of the Creation

By that we would suggest that every creation (Being) harbors an unwavering proliferation of microcosms (e.g., bacteria, follicle mites, tens of trillions of cells) whose
interactions—which are translatable at some distant levels of ornate mathematical
aphorism—are most relevant to deep biology, something entirely different than

---

[7] See "Wildfires Are Obliterating Australia's Iconic Ecosystems," by Matt Simon, Wired Magazine,
1/8/20; See also, https://www.youtube.com/watch?v=ZkdMGfJz-wU

deep learning, as it is commonly applied to algorithms, machine intelligence, and neural networks.[8] Deep biology preserves in consciousness the readiness, willingness, and reciprocity potentials of the conscience. It extends from nerve endings, and all those trillions of individuals within us, to the external world of mice and spiders and whales. To companion animals and woodpeckers.

This unknown antecedent of feeling and co-symbiont affiliation procreates the deepest, most intuition-bound logic of empathy between ourselves and all Others, in principle. Such correlates—consciousness, the conscience—are the heart of all paradox, because of the frailty of the mathematical underpinnings, the innate ambivalence of principles, and the stubborn flame within the human spirit (as described by Nikos Kazantzakis) that encounters a similar flame in all other creatures. Much like the hundreds if not thousands of shrimp-like amphipods living inside the gills of the whale shark.[9]

At the time avid collectors like Albertus Seba were assiduously assembling every diverse corpse they could afford, developing secretive methods for preserving the dead bodies, then hiring scores of painters to create black and white, then prettified color images for the eventual copper plates, putting up a third of their own funds to solicit publishing agreements with subscribers who would contribute their share in return for discounted finished products, and, ultimately, marveling at their lovely likenesses, whole natural history museums on paper, there was no paralysis. This was the science of natural history, delivered expensively to the public. Seba died one year after the first edition of Carl Linnaeus' *Systema Naturae*. The scientific renaissance had been launched, and the killing of plants and animals would fuel a commercial popularization that would lead from one of the earliest city gardens, that of Amsterdam's Hortus Botanicus designed in 1638, to the zoo and biomedical cultures of the modern world, wherein animals were viewed as laboratory units of merchandise, things to be calculated, objectified, stored away upon shelves and in dusty backroom cartons.

The paradox of this particular transition in the history of ecological aesthetics is altogether palpable in a painter like Leonardo da Vinci, whose painted backdrops typically encompassed a paradise-like setting—note those to the rear of "The Virgin of the Rocks"(1483–1485), "St John the Baptist (with the Attributes of Bacchus)," Workshop of Leonardo (1513-1519), "Portrait of Ginevra de'Benci" (1478–1480),

---

[8] See "Opportunities and obstacles for deep learning in biology and medicine," by Travers Ching, Daniel S. Himmelstein, Brett K. Beaulieu-Jones, Alexandr A. Kalinin, Brian T. Do, Gregory P. Way, Enrico Ferrero, Paul-Michael Agapow, Michael Zietz, Michael M. Hoffman, Wei Xie, Gail L. Rosen, et.al., 04 April 2018. https://doi.org/10.1098/rsif.2017.0387, Journal of the Royal Society Interface, https://royalsocietypublishing.org/doi/10.1098/rsif.2017.0387; See also, "Recent Advances of Deep Learning in Bioinformatics and Computational Biology," Binhua Tang, Zixiang Pan, Kang Yin and Asif Khateeb, Frontiers in Genetics, Bioinformatics and Computational Biology, https://www.frontiersin.org/articles/10.3389/fgene.2019.00214/full, Review Article, Front. Genet., 26 March 2019, https://doi.org/10.3389/fgene.2019.00214

[9] See "These Newly Discovered Shrimp Call a Whale Shark's Mouth Home," by Jason Daley, Smithsonianmag.com, October 29, 2019, https://www.smithsonianmag.com/smart-news/tiny-crustacean-calls-whale-sharks-mouth-home-180973421/. Accessed August 12, 2020.

and "Annunciation" (1473–1475), "Madonna of the Carnation" (Madonna with a Vase of Flowers), c. 1472–1478, and, of course, "Portrait of Lisa del Giocondo (Mona Lisa)," 1503–1506 and later (1510).[10] Yet, this vivacious intellect, restlessly perusing every conceivable discipline, remains the epitome of that convergence of the consciousness with one's sense of moral duty. Leonardo was a vegetarian, though no vegan. Biographer Serge Bramly mentions a letter in 1515 between two of Leonardo's contemporaries in which is referenced the Hindu's respect for all creatures "like our Leonardo da Vinci".[11] And despite some debate as to the extent of Leonardo's actual eating habits,[12] Walter Isaacson's recent biography cites Leonardo's Notebooks, as well as comments by friends, one of whom declares, "He would not kill a flea for any reason whatsoever." "He preferred to dress in linen, so as not to wear something dead." Isaacson concludes, "His rationale for avoiding meat derived from a morality based on science".[13] By exercising basic dietary restraint from a position of firm ethical conviction, Leonardo had essentially merged his imagination with a probability event, a target of the conscience, life forms he sought to protect (Fig. 77.2).

---

[10] See *Leonardo da Vinci -The Complete Paintings and Drawings*, by Frank Zöllner, Taschen, Köln, Germany 2007.

[11] *Leonardo – Discovering The Life Of Leonardo Da Vinci, A Biography*, by Serge Bramly, Translated by Siân Reynolds, Edward Burlingame Books, HarperCollins Publishers, New York, 1991, p. 240.

[12] See "Was Leonardo Da Vinci a Vegetarian?" by Shelley Esaak, July 10, 2019, ThoughtCo., https://www.thoughtco.com/was-leonardo-a-vegetarian-183277, Accessed June 24, 2020.

[13] See *Leonardo Da Vinci*, by Walter Isaacson, Simon & Schuster, New York 2017, p. 130.

**Fig. 77.2** "St. Jerome and Lion in the Wilderness," c. 1590. Anton II Wierix (c.1552 – c.1604). (Private Collection). (Photo © M.C. Tobias)

# Chapter 78
# The Buddhist Obtuse and Its Ecological Correlates

## 78.1 When Conscience Is More Than Conscience

"Conscience is more than conscience," wrote politician/philosopher T. V. Smith in his classic story of power and the relations between Democratic, Eastern, and Communist versions of outer consciousness and inner conscience.[1] Smith also introduced his autobiography, *A Non-existent Man*, by declaring upfront in its Prelude, "…existence is not the highest form of being".[2] We know that Ludwig Wittgenstein struggled with these distinctions in his *Philosophical Investigations*[3] when he recognizes that the most common tool for human enquiry, language, is insufficient to interpret the gap between real pain and its "expression." That which is contained in the private person has no way of reaching another. This is posited as an understandable impediment to communion in any true ecological sense throughout the interspecies disciplines, but particularly so in the wide-ranging epistemological studies and epiphanies with which various schools of Buddhism have grappled, commencing in documented earnest with Nagarjuna's (c.150–c.250) remarkable treatise, *Root Stanzas* (or *Fundamental Verses*) *of the Middle Way (Mulamadhyamakakarika)*. As with Wittgenstein's personal quest from the time of his first *Tractatus (Logisch-Philosophische Abhandlung*, 1921) to his final manuscripts 175, 176, and 177 (1951) to grasp and communicate his feelings regarding the vast irreconcilable gulf separating thought and expression, Nagarjuna (despite great uncertainty regarding his actual biographical details) was smitten with the concept of *emptiness* (explaining that gulf). He came to be regarded by many as the "Second Buddha" because of this obsession with the same existential *nothingness*

---

[1] *Beyond Conscience*, by Thomas Verner Smith, Whittlesey House, McGraw-Hill Book Company, Inc., New York and London 1934, p. 359.

[2] See *A Non-Existent Man: An Autobiography*, by T. V. Smith, University of Texas Press, 2012.

[3] Translated by G. E. M. Anscombe, P. M. S. Hacker and Joachim Schulte, Revised fourth edition by P. M. S. Hacker and Joachim Shulte, Wiley-Blackwell, Malden MA, 2009, p.95.

© Springer Nature Switzerland AG 2021
M. C. Tobias, J. G. Morrison, *On the Nature of Ecological Paradox*,
https://doi.org/10.1007/978-3-030-64526-7_78

**Fig. 78.1**  Young Buddhist Monk, Bhutan. (Photo © J.G. Morrison)

that has erupted from one philosophical tradition to another. It would come to play out culturally across the geographic span of Mahayana Buddhism, particularly in India, Kashmir, Mongolia, and Tibet (Fig. 78.2).

In a stunning translation and exposition of Mipham Jamyang Namgyal Gyamtso's (1846–1912) commentary on the ninth chapter of "The Way of the Bodhisattva," *The Wisdom Chapter*,[4] we read a breathtaking overview of Nagarjuna's *deconstructivist* understanding of Reality—the embrace of sunyata, of zero—as he is thought to have interpreted original Buddhist wisdom literature, the *Prajnaparamita Sutras*.[5] As Nagarjuna himself apparently warns, "the path is fraught with dangers so great that even the Buddha himself hesitated to teach it…".[6] But others, in addition to Nagarjuna, did communicate the insights, including such Rinpoche's as Bhavaviveka (500–578), Chandrakirti (600–c. 650), and Je Tsongkhapa (1357–1419). Their existential hegemony, core to the Gelugpa Tibetan Buddhist school (Ganden Pa, or Tushita, Pure land),[7] formulated a paradise to which devotees could enter-in-consciousness, as laid out in the 18th of 48 original prescribed (Amitabha) Buddhist vows within Mahayana tradition.[8] So enamored was Nagarjuna of this seemingly psychoanalytic breakthrough with respect to all things, the world of phenomena, that he composed a work comprising *Seventy Verses on Emptiness (Sunyatasaptati)*. The pivot upon which he relentlessly designated all of Nature as merely an

---

[4] The French-based Padmakara Translation Group (https://www.shambhala.com/authors/o-t/padmakara-translation-group.html).

[5] See *The Wisdom Chapter, Jamgön Mipham's Commentary of the Ninth Chapter of "The Way of the Bodhisattva,"* Translated by the Padmakara Translation Group, Shambala Publishers, Boulder Colorado, 2017.

[6] ibid., p. 15.

[7] ibid., p. 7.

[8] See http://www.ammituofo.com/48vows.html

**Fig. 78.2**  11ᵗʰ Century Painted Mongolian Buddha with Devotional Bodhisattvas in the Manner of the Qumtura (Thousand Buddha Caves) Wall Paintings of the 8ᵗʰ-9ᵗʰ Centuries. (Private Collection). (Photo © M.C. Tobias)

unreliable, fleeting idea, somehow surprisingly upended atomist, yogis, Brahmans, and logicians, staking a permanent claim for original essence within the lounge of everyday rational existence, a shadow government of metaphysics exerting the most lasting suasion in the history of Buddhism (Fig. 78.3).

Such emptiness, as defined by the 32-year-old Mipham writing his *Ketaka Jewel*, or Wisdom Chapter (and considered by many to be the most prolific "polymath" in Tibetan history),[9] endeavored to characterize for laypersons the idea of emptiness in

---

[9] See Phuntsho, Karma, 2005, *Mipham's Dialectics and Debates on Emptiness: To Be, Not to Be or Neither*. Routledge Critical Studies in Buddhism. London: RoutledgeCurzon, p. 13.

**Fig. 78.3** Padmasambhava Tangka, Tibet, ca. 1575. (Private Collection). (Photo © M.C. Tobias)

terms that were not empty. This impossible paradox assumed the power of the susur-
rus, harmonic echoes that might yet linger in the unknowable. He described emptiness as "a nonimplicative negation (*med dgag*)",[10] certainly not the most
unambiguous of attributes. In fact, assessing pan-Asiatic spirituality's embrace of
the concept of illusion (the indescribable *maya* in Brahmanical tradition), he revives
a very central debate that enlists Jain, Hindu, Buddhist, and other mainstream con-
tentions each calling into question the Why? Why rescue a horse, or be generous, or
do anything in a world that does not exist?.[11] Mipham's answer is conditioned upon
what he terms "the power of interdependent origination," the fact that dialectical

---

[10] op.cit., Padmakara Group, p. 82.

[11] ibid., p. 89.

truths, the "appearances of samsara and nirvana" (both), continue to inflict upon "the expanse of suchness," engendering real pain, authentic suffering, in "ourselves and others," and it thus becomes essential that we break through dualisms to enact ameliorative "benefit".[12] Described in the context of Nagarjuna, Douglas Berger writes that "interdependent causality in turn is only possible because things, phenomena, lack any fixed nature and so are open (*sunya*) to being transformed".[13] This is a position far exceeding the "noble silence" that Chandrakirti suggested was the "ultimate" response to subject/object contradictions.[14]

For Jean-Paul Sartre, grappling with a summary of such dualities throughout the conclusion of his monumental *Being and Nothingness*, he writes of "detotalized totality," "the transcendent efficacy of consciousness," which, "owing to the repercussions of an act in the world... is neither pure exteriority nor immanence." These assertions are preludes to that "existential psychoanalysis" that may posit *the being by whom values exist,*" the unrecoverable self by choice, "transcendent value which haunts it," and "a being-which-is-not-what-it-is and which-is-what-it-is-not...".[15] Such epistemological rhetoric would, were we to abide by it, undercut the conservation biology reflex which is to restore. The conservation impetus is guided entirely by a belief system which relies on authentic inflections and sincere flash points oriented toward a re-wilding, a re-establishment of credibility in the face of a failing species. It is a rueful, pointedly uncertain impulse that harkens back to the earliest metaphysics of protection, the first preserves, national parks, and present, desperate efforts to save what's left. That it also courts the existential emptiness evinces a fascinating and important bifurcation point within consciousness that pits the imagination against itself. The only remedy to this paradox is choice, and that represents the possibility of virtue and volitional change (Fig. 78.4).

## 78.2  A Context for Choice

The *conservation contexts*, as we have endeavored to explore throughout this work, is not without total moral disarray. Consider the "mainland island"[16] efforts in New Zealand and New Caledonia, or the case of Redonda Island, some 50 kilometers off

---

[12] ibid., p. 89.

[13] "3. Against Worldly and Ultimate Substantialism," "Nagarjuna (c. 150–c.250), Internet Encyclopedia of Philosophy, https://www.iep.utm.edu/nagarjun/. Accessed January 8, 2020.

[14] op.cit., Padmakara Group, p. 51.

[15] Jean-Paul Sartre, *Being and Nothingness – An Essay on Phenomenological Ontology*, Translated and with an introduction by Hazel E. Barnes, Philosophical Library, New York 1956, pp. 623–627.

[16] See Forum article, "Role of predator-proof fences in restoring New Zealand's biodiversity: a response to Scofield et al.," 2011, John Innes, William G. Lee, Bruce Burns, Colin Campbell-Hunt, Corinne Watts, Hilary Phipps and Theo Stephens, Published on-line: 28 May 2012, https://www.researchgate.net/publication/259571855_Role_of_predatorproof_fences_in_restoring_New_Zealand%27s_biodiversity_A_response_to_Scofield_et_al_2011. Accessed January 8, 2020

**Fig. 78.4** Meditation Deity Akshobhyavajra, from the Secret Assembly (Guhyasamaja) Tantra, Bringing Together Five Buddha Families, Early 19th Century, Nepal. (Private Collection). (Photo © M.C. Tobias)

Antigua. There, over 60 starving goats were helicoptered off the island, and more than 6000 black rats were killed in an effort to help restore native species like burrowing owls, pygmy skinks, and a variety of lizards and seabirds.[17] The killing of the rats involved placing bait stations everywhere, in the browned grasslands, and upon cliff ledges, a technique developed during the past half-century throughout all of New Zealand, including its sub-Antarctic islands. The most common bait, Brodifacoum, is a poisonous anticoagulant that usually takes a week to kill its (mam-

---

[17] See "Ravenous wild goats ruled this island for over a century. Now, it's being reborn, by Michael Hingston, January 2, 2020, National Geographic https://www.nationalgeographic.com/science/2020/01/raveous-wild-goats-ruled-this-island-for-over-a-century-being-reborn/.    Accessed January 8, 2020

**Fig. 78.5** Avalokiteshvara Buddha of Compassion Tangka, Thailand, ca. Mid-19th Century. (Private Collection). (Photo © M.C. Tobias)

malian) target, and not without other unintended collateral damage.[18] Many years ago in New Zealand, a Buddhist community with land holdings had to choose what to do about bio-invasives on their property. Without naming names or outcomes, we'll leave it to you, the reader, to work through the likely ethical scenarios (Fig. 78.5).

Yet, on the positive side, none of our current predicament smacks of "emptiness," which Nagarjuna equated with "the ultimate truth" and the essence of Mahayana, "the middle way"[19] Which suggests that the twenty-first century is somewhere murkily in between those same ethical scenarios. Indeed, the lack of moral account-

---

[18] See "Bait residue issues," by Kate Guthrie, April 7, 2016, Predator Free NZ, https://predator-freenz.org/friday-afternoon-reads-5/. Accessed January 8, 2020.

[19] See *The Fundamental Wisdom of the Middle Way, Nagarjuna's Mulamadhyamakakarika*, Translation And Commentary By Jay L. Garfield, Oxford University Press, New York, 1995, p. 93.

**Fig. 78.6** Nirvana Tangka, detail of the Tso k'a pa Buddha, 18th Century, dGe lugs pa School, Gyantse, Tibet. (Private Collection). (Photo © M.C. Tobias)

ability ignites the most significant intellectual and metaphysical discrepancies separating the history of religions from today's ecological realities. Religion and reality. The mind and nature. Is this duality representative of an irreconcilable gulf? Wrote Nagarjuna, "The moon is the public light/What is the use of (personal) property?".[20] "Wildness is not to be assuaged by anything".[21] "The earth, the mighty ocean/And the mountains are not a burden/But he who is ungrateful/Is indeed a heavy burden".[22] And by these conjurations, being or no being, existence or non-existence (think extinct, non-extinct), there still remains a poignant and all-important call to worship. Humanity's relationship to *unexpected, express realities*—objects like moonlight, or all those perceptions of things which define for us the environment we inhabit—is entirely relevant, whether real or not. In the margins, crannies and there, upon unseen footholds is an activist incantation. It may be a case of the Windmill Paradox, as lovingly described by a Cervantes, but—giants or no giants—Don Quixote *must* act so as to transcend the void. For the charging knight, and his loyal friend, the observer (Sancho Panza), vagueness explodes beyond itself into both a literary plane and one steeped in the natural history of perception (Fig. 78.6).

---

[20] See the *She-Rab Dong-Bu, Tree of Wisdom (Prajnya Danda)*, Translated from the Tibetan by Major W. L. Campbell, Calcutta University, 1919, Stanza #116, p. 60.

[21] ibid., Stanza 248, L.3, p. 126.

[22] ibid., Stanza 252, p. 128.

# Chapter 79
# Ecological Emptiness

## 79.1 Ecological Depression

Widespread ecological depression is manifest in the clinical literature.[1] With Apocalypse glibly described as *the new-normal*, the doom-scrolling pit in the stomach, the living dead, decries its past but has only the vaguest suggestion for the future.

We flock toward destruction with inventive gusto and call it logic. But any sensible concept would more carefully formulate a discerning illogic where upon disassociation and the variegated exile of mind were the elegant roots of all fertility and originality.

Hence, the atlas of forgotten, undiscovered, and imaginary places. The precious etymology of all misfits. A fine humor of pre-nihilistic nuances. True kindnesses bestowed upon every plurality, as a fundamental pillar of the biosphere. This is the latitude in evolution we have been searching for, a tailored natural selection that suits our fits and starts. It is pragmatic. The cleansing apparatus bestowed by natural selection upon individuals and populations for purposes of rejuvenation, health, and perpetuation.

There is no fine-tuning these deliberations because every angstrom of delineation, iteration, and occupation is equally engrossing. Their strength is as individual as it is uncharacteristic of a species or community. And therein is the problem.

It comprises wastelands of railroad track and barbwire fencing; of caustic Portland cement, the steadfast hexavalent chromium; of plastic oceans and 129 degrees in Tehran, or temperatures in the sand exceeding 150 where shorebird chicks die of heatstroke in their nests. There are no apparent margins of mercy, no philosophy, or no exposition that can account for human lunacies adept at the election of dictators and dismemberment, whose multiplier dilemmas crowd the *Diagnostic and Statistical Manual of Mental Disorders*.

---

[1] See, for example, "We need to talk about 'ecoanxiety': Climate change is causing PTSD, anxiety, and depression on a mass scale," By Zoë Schlanger April 3, 2017, Quartz, https://qz.com/948909/ecoanxiety-the-american-psychological-association-says-climate-change-is-causing-ptsd-anxiety-and-depression-on-a-mass-scale/. Accessed January 15, 2020.

© Springer Nature Switzerland AG 2021
M. C. Tobias, J. G. Morrison, *On the Nature of Ecological Paradox*,
https://doi.org/10.1007/978-3-030-64526-7_79

**Fig. 79.1** At Père Lachaise Cemetery, Paris. (Photo © M.C. Tobias)

As the escalating stressors of humanity's inflictions narrow the gaps between options, it is supposed that natural selection brutalizes variety, singling out choices with fright and flight. But if this were strictly the case, the microbial mats in thermal vents, or the macrofaunal communities that rapidly reinvented the charred slopes of a Mt. Saint Helens or Iceland's Surtsey would not demonstrate such astonishing, post-catastrophic diversity. Natural selection, while evident in the world's marine muds,[2] poses our greatest intellectual challenge. Namely, that the forensics argue convincingly that the prime Darwinian motif is in fact incommensurate with the last remaining hominin whose internal powers of fertility and compounding consumption have proved comfortable with the ruination of every biological protocol and proxy. These were current during the years Darwin was writing his *On the Origin of Species by Means of Natural Selection* (1859). Side-stepping every primordial rule of logic that longs for what Matthew Arnold exhorted as that "Endless extinction of unhappy hates" ("Merope," 1858, l. 100), that would otherwise assist evolution in her leisurely and beneficent approach to engendering and liberating interdependent life forms, *H. sapiens* have *selected* subversion.

Even as China witnesses its lowest birth rate since 1949 (the one-child policy having finally been rejected in 2016),[3] that country is still marching toward 1.5

---

[2] See "How the dinosaur-killing asteroid primed Earth for modern life," by Tim Vernimmen, January 16, 2020, National Geographic, https://www.nationalgeographic.com/science/2020/01/how-dinosaur-killing-asteroid-primed-earth-for-modern-life/. Accessed January 17, 2020.

[3] See "China's Looming Crisis: A Shrinking Population," by Steven Lee Myers, Jin Wu

billion residents, as documented in China's *Green Book of Population and Labor*.[4] The trend, so well documented both by Edmund Burke, writing of the French Revolution, and then by the Rev. Thomas Malthus, constitutes the most pervasive, unwavering orientation in human kind. If our alleged dominion's collective nature (Jung's "primordial images") devised and ruthlessly enforced were not so consciously and deliberately cruel, it would forge a mere monument to stupidity, overcome with reason and patience, on a par with, though far in numeric excess of, that grisly toll of predation meted out by the world's feral cats. Others can be cruel, clearly, Japanese wasps, for example, but not as a full substitute for some version of predictable restraint or charity.[5]

And it would also contradict the brilliant and loving history of ecological aesthetics within humankind, so reverentially characteristic, for example, of one of Darwin's great contemporary American painters, Jasper Francis Cropsey (1823–1900), a leading member of the Hudson River School, who never ceased to portray a resplendent Western Europe and Eastern Seaboard.[6] This is the great paradox, of course: the beauty we have both espied and eschewed.

## 79.2   Paradigmatic Shifts out of the Wasteland

Darwin was no stranger to this paradox, repeatedly remarking upon the cruelty of nature with an addled humility. Where do all the birds and their eggs end up? Quietly lost, by little fanfare, a soft, but no less brutal (in the end) sacrifice area for wildlife deaths which humans rarely come upon. Darwin's reading of William Paley's book, *Natural Theology or Evidences of the Existence and Attributes of the Deity* published in 1802, 7 years before Darwin was born, underscored the critical watershed that gave to Darwin his stunning courage in demarcating the intelligent

---

and Claire Fu, The New York Times, January 17, 2020, https://www.nytimes.com/interactive/2019/01/17/world/asia/china-population-crisis.html, Accessed January 17, 2020.

[4] No. 19, As reported by the Chinese Academy of Social Sciences http://ex.cssn.cn/zx/bwyc/201901/t20190104_4806519_1.shtml?COLLCC=4186522808&, January 2020.

[5] See, for example, "Australian feral cats wreak the most damage," by Signe Cane, February 3, 2015, Australian Geographica, https://www.australiangeographic.com.au/news/2015/02/australian-feral-cats-wreak-the-most-damage-to-species/. Accessed January 17, 2020. Neither dissection or diagnosis is relevant to the wearying multiplicity of fall-out. Studies of trees and shrubs in Yasuní; of midges in the Scottish highlands; or sandflies along New Zealand's Doubtful Sound, yield ratios and divisible quanta that defy calm analysis in that the numbers and their underlying identities are simply outside our ken. Similarly, most vireos and song thrushes – musical virtuosos all – are acoustically immune to human calculation as are flowers in Spring. We know nothing. Vast amalgams of Plato and Chaucer; of Averroes, Shakespeare, Newton and Humboldt. They leave us – no matter our cues and prompts and many-hued involvements – dry upon dried riverbeds. This is capacity, our destiny, the story we inevitably will leave inextricably behind, namely, that we cannot rightly assess compare other species with ourselves.

[6] See Andrew Speiser's two-volume *Catalogue Raisonné*, Newington-Cropsey Foundation, Hastings-on-Hudson, NY, 2013.

design argument proffered by Paley, versus adherence to natural selection. This debate rages on today, despite the fact that just between 1973 and the 200th anniversary of Darwin's birth, in 2009, an estimated "100,000 [+] peer-reviewed papers" had been researched and published "on neo-Darwinian evolution." A startling and paradigmatic saturation of the scientific community.[7]

Given the intellectual staying power of the idea of natural selection, the coming great ethical hurdles suggest the challenge in formulating some logic, a kind of integumentary system by which there is reason to seek an actual *ideal* in natural selection. Such a connectivity begs the question, though, of violence in nature and its empowerment in the common hominin. No annals have ever recorded the statistical comparisons between the two great sides of sanity with any level of (mathematical) confidence, except to note how sublime, on the one, and how vile and rapacious an organism on the other.[8] Nor are there other known newly born species like ourselves with which we might compare our own ontological progress.

## 79.3  Walking Sharks and Superego

The nine known species of walking sharks (*Hemiscyllium* spp.), for example, have evidenced the newest known shark adaptations (approximately 9 million years old) in 400 million odd years of shark residency. Not coincidentally, these young species are enlisting behavioral qualities favoring short terrestrial trips and remarkable resiliency in the face of shifting and blanching coral reefs (their lifeline), as well as dietary balances more likely to embrace a herbivorous tendency. Humans, the youngest of all large vertebrates, have, it might be argued, relied upon a very different rapid adaptation locus, namely, the Self or Superego, as it has long been optimized through aggression and possession, and continues to accelerate an unprecedented biological *sovereignty of mind*.

The full picture of the human brain's mass and energy can be divined in the summation of its destructions, to paraphrase Picasso. Whereas selection favors exogenous input—such as the local rebirth following volcanic and seismic change—humanity's grave digging has covered every cubic meter of the planet. There are few remaining inputs that can be said to be wild, a topology that is surely undercutting remaining opportunities for the kind of natural selection Darwin and walking sharks preferred. Indeed, without greater, large-scale marine habitat protections across parts of Indonesia and Papua New Guinea, walking sharks are likely to vanish.

Global inbreeding and its concomitant invasion of the human imagination have essentially homogenized our map of activities and potential. The only ideals that have been spared are those which, from very inception, have forever eluded

---

[7] "Darwin: the legacies," by Andrew Kelly, in *Darwin – For The Love Of Science*, Edited by Andrew Kelly and Melanie Kelly, Bristol Cultural Development Partnership, Bristol, UK, 2009, p. 257.

[8] See "War and Peace," by Max Roser, Our World In Data, https://ourworldindata.org/war-and-peace, Accessed January 17, 2020.

predation. Such pure thought is rarified, spiritual, and so deeply ethical as to defy gravity. In seeking some level of "explanations," Historian Alfred W. Crosby, in his defining work, *Ecological Imperialism—The Biological Expansion of Europe, 900–1900*, described the fact that "there are about 10,000 grass species, but a mere forty account for 99 percent of the sown grass pastures of the world. Few, if any, of the forty are native to the great grasslands outside the Old World".[9] Indeed, so concentrated and single-minded were the origins of human colonization that those native grasses of the Old World were concentrated largely within the limited confines of the Roman Empire.[10] Naturalists seeking "physico-theological implications" and "believing in final causes, devoted to a study of nature which was likened to following in the Creator's footsteps, in Sir Thomas Browne's words" were engaged in an effort "to suck Divinity from the flowers of Nature".[11]

## 79.4   Unimagined Bias

The study of the origins of life has been swathed in mineralized murkiness, as viewed through the lenses of a species at the tail end of its durability. Darwin had forced our hand, albeit with stunning scientific instincts, to gravitate with great persuasion toward a bonfire of incantations and antiphony. Such that a misanthropic rant feverishly obeys some visceral cry, part confession, here a gymnastic. A rallying from the bowels of destruction whereby biological orthography wrestles with self-indictment. It struggles to be humane, this heavy toil of logical succession in Nature, but is weak with the grave malady of itself. And it knows all about it, from the deepest within. How can it escape its own brief history of longing and lethal doses? It can try to fall back on every vulnerable argument of humanism, of some Renaissance argument. But cannot resolve even its most mundane contradictions at the heart of Zen emptiness.

Such that it cannot imagine the *unimagined bias* that has been built into the cognition apparatus appertaining to extinction studies in the twenty-first century. Our obsession is becoming quite rampant in sync with the realities of our presence and its death tolls. The distance between research and reality is closing fast. Billion-year-old records of stromatolites and living dinosaurs, to take one well-known set of examples, like that of the macaw or platypus, hail similitude and kinship that has never been a stranger on earth. No other creature has seen fit to lie unrelentingly to itself, on behalf of itself, with absolute disdain for commonality with Others. Our arrogation at every juncture of behavioral infrastructure—neurons, cells, DNA—vies in combinations far beyond our ken to combat ideals. Yet languages,

---

[9] A. W. Crosby, Cambridge University Press, Cambridge, UK, 1986, p. 288.

[10] ibid., p. 288.

[11] From *Traces on the Rhodian Shore - Nature And Culture In Western Thought From Ancient Times To The End Of The Eighteenth Century*, by Clarence J. Glacken, University of California Press, Berkeley and Los Angeles, CA, 1967, p. 426.

Fig. 79.2 "Destruction of the Tower of Babel," 1612, by Crispijn de Passe the Elder. From a from a series of engravings made for the first edition of the "Liber Genesis." Bequest of Phyllis Massar, 2011. The Metropolitan Museum of Art. (Public Domain)

to the extent they have been organized and expressed according to general deep rules of structure, including every human idiolect and self-expression, provide one key to escaping an otherwise unresolvable fate. And that is the clear and possibly unassailable truth of the inexpressible. "…like those that pass or things always and memories I say them as I hear them murmur them in the mud…" (Fig. 79.2).[12]


## 79.5   A Tower of Babel as Ecological Inverse

The Towers of Babel help us to visualize the manifest moment humanity broke ranks with itself. Despite the mythic machinations of mortar and brick, millions of busying hands, hammers, and the same nail, the clumsy edifice paradoxically stood within its own perpetual shadow, disintegrating hour by hour. Eventually, it was lost in a leaning labyrinth of miasmas. So mistaken for a noble cause circling survival at any height, a zenith that forever defines the mortal flaw within, more than any admonition on the circumstances of doomed hubris, like some lost prayer issuing from John Donne's lips, this ghost of arrogance better laid to rest for all time continues to arouse ripe simile.

---

[12]From *How It Is*, by Samuel Beckett, Translated from the French by the author, *The Collected Works of Samuel Beckett*, Grove Press, Inc., New York 1964, p. 7.

**Fig. 79.3**  Female Orangutan, Central Kalimantan, Indonesia. (Photo © M.C. Tobias)

Fleeing its ruins, later generations looked back at Babel in wonder, turned poetically nostalgic toward good shepherds, wisemen, explorers, and gypsies, all of whom have lent the nomadic heart a distinctive whisper. Like all those *mountains and rivers* of the *Shobogenzo* by Soto Zen-founder, Dōgen Kigen (1200–1253), many beetling cliffs, brant, and scrape; the wine dark illuminations beloved by all those from Homer to Aert van Der Neer, a Chopin's Etudes. In every fine museum library, symphony hall… lives and breathless, the fantasies and enquiries that remain in ponderous limbo. Fixated on that which is intimated, only partially approached, encompassed, or even envisioned.

Therein, upon the objects and scenes of vague and indeterminate proximity, is our one hope.

Lodged obliquely between mass genuflection and suicide. The artistic notion of zoology. Philosophers of rock and sand. Ontological echoes whose chords harmonize in total emptiness. A pantheist's reunion of voids. Memories once divine gone brittle and acrid, yet, somehow, coming back each spring to offer us a second chance (Fig. 79.3).

# Chapter 80
# Temptational Obscurity That Brings Hope to Life

## 80.1   Obscurity and Redemptive Intuition

We are inclined to call upon obscurity, the nominative pronouns collectively obliging our stressful, persistent visions of *wilderness*, whose genes—subjectively, objectively, clearly—are vivid and alive, to break free from the shackles of broken trusts and obfuscated contracts. One can still walk away, into a place where fellow humans have not yet conquered crickets and songbirds. In recent phases of that challenge to reverse what has, increasingly, become a foregone conclusion, countless population bubbles brimming with feverish hope have endeavored to stave off what to many is now inevitable.[1] A phrase circulating for many decades that all but assures the *death of nature* (Fig. 80.1).

But death is not death so much as the death of individuals. Biomes, cycles, frequencies, volumes, depths, fringes, cores, and other anatomical facets of ecological reckoning always invoke real ecotones, the transitions we have been evincing throughout this work. It is not enough to psychoanalyze adulterants and breakdowns. Aftermaths exert inevitable suasions that can be shaped according to the spirit of renascence inherent to every combination of life forms and instincts. Ice may melt and skies darken, but the cycles will continue. These are pillars of every lasting ethical tradition and should be viewed as fundamental to the human heart at times of ultimate stress and despair. Not vacuous campaigns and speeches, but fragile, intuitive regard that is enshrined by pragmatic conviction and laid bare by the best of virtuous intentions.

Most, not all, administrations in the twentieth and twenty-first centuries have added to the legacy of protecting more and more wildlands. For example, the Forest

---

[1] See https://us.fsc.org/en-us; See, for example, the history of the National Park Service (nps.gov, and nationalparkservice.org), U.S. Fish and Wildlife (fws.gov), the IUCN (www.iucn.org), UNEP (unenvironment.org), the WWF (worldwildlife.org), Audubon Society (Audubon.org), Wilderness Society (wilderness.org), Nature Conservancy (https://www.nature.org/en-us/), PETA (peta.org), Conservation International (conservation.org), Global Wildlife Conservation (globalwildlife.org), ICCA Consortium (Indigenous Peoples' Community Conserved Areas and Territories), (iccaconsortium.org). Even at the most doubtful, compromised levels – U.S. Presidents – some of their legacies are all but fixed in the imagination of what seems possible: Theodore Roosevelt will mostly be remembered for having "established five national parks, 18 national monuments, 51 bird sanctuaries, [the formation of] the National Wildlife Refuge System, and [having] set aside more than 100 million acres of national forests."

© Springer Nature Switzerland AG 2021                                                            721
M. C. Tobias, J. G. Morrison, *On the Nature of Ecological Paradox*,
https://doi.org/10.1007/978-3-030-64526-7_80

**Fig. 80.1** Mount Saint Elias, 18,009 ft/5489 m, Wrangell-St. Elias National Park and Preserve, Alaska. (Photo © M.C. Tobias)

Stewardship Council has certified 34.9 million acres within the United States as of 2019. Only the dire and fanatical GOP-dictated period of 2016–2020 saw the first abhorrent abnegation of this trend in the American history.[2]

---

[2] Think of Jimmy Carter designating 79.54 million acres of Alaskan geography as "refuge land" under the "Alaska National Interest Lands Conservation Act of 1980," which included 10 national parks such as the largest one in America, Wrangell-St. Elias National Park and Preserve, at 13.2 million acres. See "Presidents' Day: Remembering Conservation Legacies," Defenders of Wildlife, "Jamie," February 20, 2018, https://medium.com/wild-without-end/presidents-day-remembering-conservation-legacies-89caf1750da1; See also, "Digest of Federal Resource Laws of Interest to the U.S. Fish and Wildlife Service," https://www.fws.gov/laws/lawsdigest/alaskcn.html.

One can hopelessly argue over the efficacy of these numbers, the criteria bordering efficacy, or the protective shields they actually imply. Ultimately, one either gets along with other human neighbors, or does not. Concepts of white picket fences, handrails, barbwire, a projected 100,000–156,666 kilometers of new roads built per year during the coming three decades, and "Keep Out" and "No Trespassing" signs are a queasy aggregate of impacts that comes down to massive trends of urban, rural, countryside, and wild. Remote, inaccessible, inhospitable. Versus teeming, density, megacity, conurbations, Metropolitan Statistical Areas, urban clusters. All have their linguistic roles to play in the contemplations that have absorbed our kind since the earliest archaeological records. So persuasive are these words that, according to the Orchard Professor in the History of Landscape at Harvard University, John R. Stilgoe, a book like Julian Tennyson's 1939 *Suffolk Scene: A Book of Description and Adventure* "detailing why [John] Constable painted in certain spots that still satisfied and delighted Englishmen... became important in World War II, for they helped explain why Britons fought".[3]

How different a case for motivation than that exhibited in the person of desert explorer Wilfred Thesiger, who concluded his masterpiece, *Arabian Sands*, with the simple statement of fact: "Although I had no political or economic interest in the country, few people accepted the fact that I travelled there for my own pleasure, certainly not the American oil companies nor the Saudi Government....I realized that the Bedu with whom I had lived and travelled, and in whose company I had found contentment, were doomed".[4] And the combining of these aesthetic reveries, social anxieties, and stark geopolitical contests has not infrequently, from country to country, resulted in such triumphs of unexpected remorse, nostalgia, and amalgamated naturalistic obscurity as was occasioned on the final remark by Washington Irving, in his renowned 1819 short-story "...and the plough-boy, loitering homeward of a still summer evening, has often fancied his voice at a distance, chanting a melancholy psalm tune among the tranquil solitudes of Sleepy Hollow".[5]

In ancient China, that sleepy obscurity, the arcane recollections of what might easily be characterized as a landscape gene, was chiefly espoused by neo-Daoism, particularly He Yan (207–249) of China's Three Kingdoms Period. He promoted the notion of *wu*, the collective energies of yin and yang whose dialectic emerged brilliantly out of the obscurity of all human and natural origins, and the school of Xuanxue, which famously synthesized its principles. It was also Yan who was long credited with overseeing the most comprehensive exegesis of the *Analects* of Confucius,[6] that penultimate cultivation of ethics, whether in the countryside or city. A thorough examination of ecological reciprocity (Fig. 80.2).

---

[3] *What Is Landscape?* by John R. Stilgoe, Massachusetts Institute of Technology, Cambridge Mass., 2015, p. 209.

[4] *Arabian Sands*, Longmans, Green And Co LTD, London 1959, p. 310.

[5] *Rip Van Winkle and The Legend of Sleepy Hollow*, With Fifty-Three Illustrations by George H. Boughton, Macmillan And Co., London and New York, 1893, p. 215.

[6] See "Embodying Nothingness and the Ideal of the Affectless Sage in Daoist Philosophy," in *Nothingness in Asian Philosophy*, JeeLoo Liu and Douglas L. Berger, eds. New York and Oxon: Routledge, 2014, pp. 213–229.

**Fig. 80.2** A Thai Buddha, Detail. (Photo © M.C. Tobias)

Obscurity in and of itself offers no precise divining rod. Rather, the critical evolutionary wiggle room for a change of attitudes, dispositions, moral compass. In the history of Buddhism, for example, it is clear that that which is obscure represents a salient temptation, and typically foreshadows some potent pathway, whether in the foundational "four truths," namely, "Suffering, the Cause of suffering, the Cessation of suffering, the Path which leads to the cessation of the suffering" and all those contemplations of right intention and action necessary to embark on that journey.[7] In religious texts, verbal recitation, and artistic expression, Asian pan-consciousness has endeavored to lend every conceivable iteration and nuance to the "Four Truths," much of it gloriously beautiful. For example, the *Heart Sutra* (in Sanskrit *Prajñāpāramitāhṛdaya, The Heart of the Perfection of Wisdom*), a lyrical cornerstone of Mahayana Buddhism whose origins confound literary forensics, as it was written and embellished sometime from "150 C.E. [in India] to seventh-century C.E. China",[8] asserts most famously that "Form is empty, emptiness is form." That statement is the essence of the obscure. It is akin to saying read between the lines and places the burden of discovery on the imagination, liberated from the rules of science into the laws of Nature. It can be characterized as something mathematical or teleological. Obscurity, of this magnitude, breaks down barriers that have otherwise marginalized access to those visions and conditions of life that all sentience craves, on condition of happiness.

Obscurity is not a clouded lens, an optical cataract, and absence of light, as its etymology has always conferred—darkness. Think, rather, of immanence of light

[7] See *Buddhism*, by T. W. Rhys Davids, Society For Promoting Christian Knowledge, London, 1899, p. 48.

[8] *Painting Enlightenment – Healing Visions Of The Heart Sutra*, by Paula Arai, Shambhala, Boulder, Colorado, 2019, p. 5.

gathering emotional force on the horizon throughout the Hudson River School. Or a grand canvas by Turner or El Greco, the rear-window landscapes throughout the Renaissance, heavenly atmospheres that absorb the dramas in Fra Angelico, or Rogier van Der Weyden. A calling from the unknown, beyond, some philosophical opportunity, the very contours of wilderness. The narrowing of light on the eternal horizons of dawn and dusk, together, forming an ethic silently adjudicated in one's soul.

Take "The Paradise of Amida," for example, a late seventh-century painting on a white clay wall in the main hall (Kondo) of the Horyu-ji Monastery in Nara Prefecture, Japan. The Buddha sits within the four paradises of the Pure Land, that place of enlightenment in Mahayana, that encompasses Su-Meru (Kailasha), the celestial equivalent of Mount Sinai.[9] Tens of thousands of other equivalent works of art have been discovered (and all too frequently transplanted) along the Silk Routes.[10] At the Palace (Beijing) and Shanghai Museums, a startling proliferation of meditative scrolls involving half-empty spaces, gathering literati in open speculation, mostly empty "odes" to the "Red Cliff," quiet pine tree forest vantages, a thatched sylvan cottage with a lone occupant in silent repose, a hoopoe in equally philosophical register amid a bamboo glen, a musician performing for a few friends on the Qin zither without making a sound, myriad farewells that are utterly subsumed within a vast ink-bound landscape that has all but obfuscated the human component, so many endlessly, exquisitely moribund trees and rocks, forever populating 1500 years of timelessness.[11]

From the "shadows of Mt. Huang"[12] to the mid-thirteenth-century painting of "Li Po Chanting a Poem" by Liang K'ai of the Sung Dynasty (died 1210),[13] in which the most spare brushstrokes, ink on the paper of a hanging scroll, depict the essence of humanity reduced in all her complexity, to the most singular and fixed simplicity, a precursor to Japan's famed sixteenth-century masterpiece by Kano Soshu of "Bodhidharma Crossing the Yangzi River on a Reed," (Metropolitan Museum of Art in New York). Or Qiu Ying's impossibly perfect scene of "Long-tailed Birds among Pines and Stream," a paradise from the 1530s; Ming Dynasty, unequaled in the

---

[9] See *Pure Land Buddhist Painting*, by Joji Okazaki, Japanese Arts Library, Kodansha International Ltd. And Shibundo, Tokyo, 1977, p. 34.

[10] See, for example, *The Silk Route and The Diamond Path – Esoteric Buddhist Art On The Trans-Himalayan Trade Routes*, Essays by Maximilian Klimburg, David L.Snellgrove, Fritz Staal, Michel Strickmann, and Chogyam Trungpa, Rinpoche, UCLA Art Council, Los Angeles, California, 1982; See also, *Central Asian Painting – From Afghanistan To Sinkiang*, Text by Mario Bussagli, Editions d'Art Albert Skira S.A. Geneva, Switzerland, 1979.

[11] See *Collection of Ancient Chinese Painting and Calligraphy from the Palace Museum and the Shanghai Museum*, with Forewords by Zheng Xinmiao, Director Palace Museum, and Chen Xiejun, Director Shanghai Museum, Beijing/Shanghai, 2005.

[12] See *Shadows of Mt. Huang, Chinese Painting and Printing of the Anhui School*, James Cahill, Editor, University Art Museum, Berkeley, California, 1991.

[13] See *Chinese Painting*, Text by James Cahill, Editions d'Art Albert Skira S.A. Geneva, Switzerland, 1977, p. 90.

history of ornithology painting;[14] Zhang Feng's rendition of "Immortals' Secrets in a Stone Cave," painted during a "sightseeing" trip between Nanjing and Zhenjiang in 1657.[15] Each proffers ecological obscurity made manifest through art.

These are works of imagined space, enriched by memory, hope, dreams of lost civilizations within civilizations, all of those figments of human qualification that also beguile the flow of equations used in science, natural history, and mathematics, particularly the combinatory theories oriented to the very "existence of monochromatic subsets for finite colorings of large structures... or infinite".[16] In the language of "topological dynamics," "Infinitary Ramsey Theory," and "measurable decomposition," it is not uncommon to read that one theorem "has definite aesthetic advantages over" another theorem.[17]

Similarly, the languages of landscape, of art in general, evince whorls within whorls, true and magnetic norths of unspoken elocution that help us to conceal as much as is revealed about the human species. In the ecological context, the unsullied horizons are temptational obscurities of the first order, promising an untrammeled future.

Such obscurities are not merely the Zen-beholden product of some difficult-of-access *obscurantism*, as the term has often been used, levied against any number of deliberately abstruse equations. By obscure we hasten to record that transition between that which can be asserted to be unknown and that which we assume by common assent is known. A table, a chair, soil in our palm, the night sky, any palpable sensation, anything that we take to be *as is*, commends the span between that which is not and that which appears to be. The history of philosophy, of mathematics, of physics, as has previously been intimated, enshrines the dialectic with a fervor. It is not the case that nothing comes of nothing, the Greek of Aristotle and Parmenides—οὐδὲν ἒξ οὐδενός; and the Latin of Lucretius—*ex nihilo nihil fit*.[18] Unknown deaths—easily 100 billion humans throughout time, hundreds of trillions of Others—become something, do something, go somewhere, reincarnate, reabsorb, and re-tool all the metabolic internalities of their life when death appears to subsume it. This is not *nothing*, this transitional chasm of salient facts in both its irrepressibility and cyclic, systemic mathematical predictability, philosophy shorn to a pure syllogism combining time and substance. The earth herself is the ultimate

---

[14] See *Where The Truth Lies – The Art Of Qiu Ying*, by Stephen Little, with contributions by Wan Kong and Einor K. Cervone, Los Angeles County Museum of Art, DelMonico Books, 2020, p. 81.

[15] See *The Artful Recluse – Painting, Poetry, And Politics In Seventeenth-Century China*, Edited by Peter C. Sturman and Susan S. Tai, Essays by Peter C. Sturman, Timothy Brook, Jonathan Chaves, Jonathan Hay and Hui-shu Lee, Santa Barbara Museum of Art, DelMonico Books, 2012, p. 197.

[16] See *Elemental Methods in Ergodic Ramsey Theory*, by Randall McCutcheon, Springer-Verlag, Berlin, 1999, p. 1.

[17] ibid., p. 2.

[18] See John Burnet's Greek translation, https://lexundria.com/parm_frag/1-19/b; and Lucretius' *De Rerum Natura*, 1.148–156, *The Nature of Things*. Trans. A. E. Stallings. New York: Penguin Classics, 2007.

ecosystem of these chasms turned contemplation, whose group mind unknowable, can be, at least, intimated, whether we are of a position to posit or reject the trappings of such hereafters and possibility, even during these worst of times.

## 80.2   The Elusive Realms of Infinite Possibility

Wherever we stand in our moment of mentation, universally there can be no doubting the transitional ramifications. And thus, by transition we enlist vagueness and obscurity to better condition our proximate poetics and sciences toward some generous orientation given to triple negatives, irrational numbers, and multiples of infinity that all court either the integer one or the substance of zero plus one. From those two domains, every ecological investiture of mind is balanced one way or other. To put it in the simplest context, one can awaken, be liberated with a single brushstroke of the mind, a gesture of the heart (Fig. 80.3).

That realm of infinite possibilities finds its sanity in the satisfactions attendant upon an equation, or an emotion, or some act involving legal consecration of just causes and consequences, whether at conceptual inception or in the reality of Somethingness. Take Thoreau's posthumously published May 1862 essay on "Walking" in *The Atlantic* magazine;[19] Teddy Roosevelt's protection of Pelican Island in Florida, March 14, 1903;[20] Henry III's creation of the Forest Charter of Epping Forest in London;[21] the 1964 Wilderness Act;[22] or the momentous uncovering beneath melting glacial ice of 41,000-year-old living nematodes and 1,500-year-old moss species. Such *events*—all, ultimately, of an instant—translate very real "infinities," as assayed by Rudolf Carnap's (1891–1970) syntactic "axiom of Infinity" and "parallel metamorphosis"[23] into ethics, infinities of virtue in the here and now, as apotheosized in the whole realm of the perceived human conscience and its actual probabilities in space, and in human logic.

We're saying that ecological restoration *can* and *does* happen; that the conscience is not merely semantic: it has scope, dimensions, and extent, though cannot be measured as such. An Oscar Schindler or Chiune Sugihara leaves a quantitative

---

[19] https://www.theatlantic.com/magazine/archive/1862/06/walking/304674/. Accessed January 19, 2020.

[20] See Pelican Island National Wildlife Refuge Archived March 10, 2009, at the Wayback Machine at National Historic Landmarks Program, Accessed January 19, 2020.

[21] City of London, "Epping Forest," https://www.cityoflondon.gov.uk/things-to-do/green-spaces/epping-forest/Pages/default.aspx, Accessed January 19, 2020.

[22] (Public Law 88–577 (16 U.S.C. 1131–1136), 88th Congress, Second Session, September 3, 1964, See "Wilderness Connect," https://wilderness.net/learn-about-wilderness/key-laws/wilderness-act/default.php, Accessed January 19, 2020.

[23] See "Carnap's surprising views on the axiom of infinity," by Gregory Lavers, Metascience 25, 37–41(2016), European Journal for Philosophy of Science, Springer Link, https://link.springer.com/article/10.1007/s11016-015-0023-z, Accessed January 19, 2020.

**Fig. 80.3** Tashi Payden Tshering, Founder and Executive Director, of the Royal Society for Protection and Care of Animals in Bhutan, With Rescued Dogs. (Photo © M.C. Tobias)

summary of all those individuals that they'd rescued from torture and oblivion. We use the term "oblivion" as something other than nothingness, a crucial measure of Carnap's actual emphasis on logic which, by all the vicissitudes of humanity's ordeals, bespeaks pre-existing volumes in hunches, gestures, and intuitions. The descendants of such oblivions, and their descendants' offspring, argue for both theory and hypothesis. The computations that would be necessary to somehow ordain a general law of Nature that factors in oblivion do not reach the level of a qualitative fundament because, as with vagueness and obscurity, we are dealing in transitions akin to the eco-dynamic flux which offer no levels of confidence, only patterns, tendencies, and numbers that are vulnerable by definition. Beneath the vagaries of vulnerability is the staying power of synderesis, the myriad sparks of conscience. Any one of those sparks can ignite.

What are we saying by this? Our expectations for a defined quality of life, one methodologically assured as an a posteriori set of received observations, would suggest a wonderful relationship between our species and that of, say, more than million species within the principal three phylae of worms (flat, round, and segmented), all breathing in oxygen, exhaling carbon dioxide; all with brains surer, more primevally fitted, and a greater calculus of feeling than, at first glance, most of us might assume. In Nature seek both identity and relationships. That is the surest cure for the multitude of taxonomic labels that have continued the myth of separateness and the mentality of pests, bio-invasives, and enemies. There are no enemies in Nature.

## 80.3   A Paean to the Earthworm

Any farmer, or soil scientist, will at once know that humans and worms, particularly more than 1800 species of, specifically, earthworms, are intimately related in ways not readily subject to reason or direct understanding. Empathy, food, the seasons, our synthetic chemical adulterants like DDT, all underscore what Virgil wrote in his *Georgics* (29 BC) that "Not every soil can bear all things" (Book II, 109). And inversely, as Michel Eyquem de Montaigne said in his *Essays* (1580), "Man is certainly stark mad; he cannot make a worm, and yet he will be making gods by dozens" (Book I, Ch. 12). In the sordid case of the Netherlands as of October 2014, there were a known "250,000 sites" "showing various degrees of soil pollution".[24] With the largest number of farm animals per capita and per kilometer in all of Europe, "the Dutch [are] lagging behind their European peers for quality of air, soil and surface water, stuck in fossil-fuel dependency, and with exceptionally high carbon emissions".[25]

All of this is deadly to earthworms. Tens of thousands of worms are considered to be parasites by our kind, such as tapeworms, along with over 7.5 million other known parasitic species on the planet. But, as the growing Anthropocene has previewed, at least half of these, in recent new studies, are predicted to go extinct in coming decades. This may pose one of the most disturbing of all ecosystem annihilatory patterns currently projected.[26] At the same time, researchers with the Canadian Forest Service in Victoria, British Columbia, have sounded the alarm on the humble earthworm's involvement in the climate crisis.[27] The worm that ultimately devours all things is not our enemy, to repeat. The paradox of the worm is that it *equals the equal sign* in terms of computational ecology. The possessor of all things, eschewing nothing, accumulating in its neural repository the sensation and history of the earth like few other organisms.

In sub-Saharan Africa, noted for its sandy, worm-rarified soils, current tests on drought-resistant subsurface water retention technology (SWRT) are being undertaken in eight countries that see the potential to revivify crop root zones and the depauperate yields of smallholder, impoverished farmers. Some have argued that such a new "green revolution" could "capture some 15 million tons of carbon in 20

---

[24] See "Environmental Data Compendium, Number of Contaminated Sites; inventory October 2014," Government of the Netherlands, https://www.clo.nl/en/indicators/en025816-number-of-contaminated-sites-inventory%2D%2Doctober2014, Accessed January 19, 2020; See also, "Dirty Dikes – The green image of the Dutch is at odds with the reality," The Economist, February 4, 2012, https://www.economist.com/europe/2012/02/04/dirty-dikes, Accessed January 19, 2020.

[25] ibid.

[26] See The Smithsonian Magazine, "The World's Parasites Are Going Extinct," by Ben Panko, September 7, 2017, https://www.smithsonianmag.com/science-nature/parasites-are-going-extinct-heres-why-thats-a-bad-thing-180964808/. Accessed January 19, 2020.

[27] "'Earthworm Dilemma,' a new carbon effect, May 26, 2019, Antelope Valley Press, https://www.avpress.com/opinion/editorial/earthworm-dilemma-a-new-carbon-effect/article_a902fe1c-7f5b-11e9-a2d4-8fd068e43202.html. Accessed January 19, 2020.

years".[28] Added food security might well militate against demographic trends, suggesting that Africa's population will easily reach 2.5 billion by 2050.[29] But at the same time, demographers project that "Africa could have one billion undernourished, malnourished and hungry children and young people by 2050 if current levels continue unabated".[30] One thinks back to England, nearly 30 million people, when Thomas Hardy's final novel involving his character of Jude discovers that his son has killed his two young cousins before hanging himself. The perceived motive is too many kids, insufficient means, a wretched life. A ruthless Judgment Day (*Jude the Obscure*, 1894/1895).

So far there has been no correlation in Africa between new farming technologies, the revival of earthworms, their contribution to global greenhouse gasses, and hunger trends among humans. But it is not enough to simply say that it is a complex puzzle. Of course it is. But there are clear ethical trade-offs looming for all involved. Thus far, earthworm taxa have been studied extensively in South Africa but incomparably less so elsewhere on the continent.[31] It should come as no surprise that our engineering and science remain fragmented and, ultimately, when it comes to actually saving lives, periodic and short-lived. It has taken more than 40 years for researchers to develop an HIV vaccine that can at least, generally speaking, neutralize antibodies. Cancer cures remain all but elusive. No truly effective international treaties regarding anti-poaching, population stabilization, transboundary pollution control, carbon and methane neutral societal behavior, or comprehensive non-violence have yet to be consecrated. Our species is not even close. So many continue to look to New Zealand as a relative Utopia in this melee of global problems.

---

[28] See "No soil left behind: How a cost-effective technology can enrich poor fields," International Center for Tropical Agriculture (CIAT), Libère Nkurunziza, Ngonidzashe Chirinda, Marcos Lana, Rolf Sommer, Stanley Karanja, Idupulapati Rao, Miguel Antonio Romero Sanchez, Marcela Quintero, Shem Kuyah, Francis Lewu, Abraham Joel, George Nyamadzawo, Alvin Smucker. *"The Potential Benefits and Trade-Offs of Using Sub-surface Water Retention Technology on Coarse-Textured Soils: Impacts of Water and Nutrient Saving on Maize Production and Soil Carbon Sequestration."* Frontiers in Sustainable Food Systems, 2019; 3. https://doi.org/10.3389/fsufs.2019.00071, Science News, October 9, 2019, https://www.sciencedaily.com/releases/2019/10/191009131746.htm, Accessed January 19, 2020.

[29] See Quartz Africa, "There's a strong chance a third of all people on earth will be African by 2100," by Giles Pison, October 11, 2017, https://qz.com/africa/1099546/population-growth-africans-will-be-a-third-of-all-people-on-earth-by-2100/Accessed January 19, 2020.

[30] See "Nearly half of all child deaths in Africa stem from hunger, study shows," by Saeed Kamali Dehghan, June 5, 2019, The Guardian, https://www.theguardian.com/global-development/2019/jun/05/nearly-half-of-all-child-deaths-in-africa-stem-from-hunger-study-shows, Accessed January 19, 2020.

[31] BioOne African Invertebrates, Vol. 56, Issue 3, 29 December 2015 An Annotated Key Separating Foreign Earthworm Species from the Indigenous South African taxa (Oligochaeta: Acanthodrilidae, Eudrilidae, Glossoscolecidae, Lumbricidae, Megascolecidae, Microchaetidae, Ocnerodrilidae and Tritogeniidae), by Jadwiga D. Plisko, Thembeka C. Nxele, African Invertebrates, 56(3):663–708 (2015). https://doi.org/10.5733/afin.056.0312, https://bioone.org/journals/african-invertebrates/volume-56/issue-3/afin.056.0312/An-Annotated-Key-Separating-Foreign-Earthworm-Species-from-the-Indigenous/10.5733/afin.056.0312.full, Accessed January 20, 2020.

But that nation's transition from innocence to the human ecological outbreak is emblematic, certainly for the life of species other than human, of a crisis of conscience, as described in Chap. 51.[32,33]

We simply lie to ourselves, we lie to worms; and these untruths have become central to the operational rationale behind every human civilization. The dishonesty functions as a trigger for the continued propagation of the attempted re-writing of biological rules, in the same manner that a gentrified Great Britain once formulated maps of the world with its own boundaries central to the global picture, no check upon its conquests, or abrogative referenda that might nullify the legality of its consumptive patterns outside its immediate borders.

## 80.4  Moral Auditing

Lacking moral audits from country to country, fully subscribing to the corrupt and abusive powers that have institutionalized every human hierarchy, and because we are arguably the youngest of all heavy vertebrates, the conundrum of obscurity devolves: Why does a God, the idea of God, or *any ideal* whatsoever, allow for us, given the framework of eco-dynamic constraints upon the interaction of species in general, and within specific populations? Sound, logical rivets holding the biosphere together, which our species has incessantly breached?

Forgetting the mythology of a God, why would natural selection favor such unrelenting violence given that, by its inherent logic, beyond some threshold of violence, all die out? Even when considering the phylogenic gulfs between, say, ants and elephants, microbes and Sequoias, rampant cruelty, egotism never before so visited in such magnitudes of deliberate catastrophe singling out whole species, one

---

[32] See *The Southland's State of the Environment Report for Water*, Government of New Zealand October 2000.

[33] See *God's Country: The New Zealand Factor*, by M.C. Tobias and J.G. Morrison, Dancing Star Foundation, Zorba Press, LA and Ithaca, NY, 2011; See also, "The Ecology of Conscience -Sustainability Issues for New Zealand," by M.C. Tobias, National Planning Institute Conference, Invercargill, NZ, http://www.dancingstarfoundation.org/pdfs/articles/Ecology_of_Conscience.pdf. Considered the "capital of extinctions" (a moniker shared with many other nations) Aotearoa's South Island farming communities have gone a long way towards polluting their inland waterways in the same manner as that of the Netherlands, with cows and sheep. But in addition, despite a country-wide population under 5 million, and a currently strong pro-active environmental framework for most issues, the overall orientation hinges upon the subjugation and killing of vertebrates; the denomination of most non-natives (other than humans) as predators, vermin, and a subsequent war by nearly any and all means against them. Slaughtering possums, rodents, feral cats and mustelid-family species by the hundreds-of-thousands, and the extirpation of "weeds" like gorse (*Ulex europaeus*), is virtually ingrained within the New Zealand farming and conservation communities. Profiting on a false clean-green mantra, New Zealand, like so many human-dominated landscapes remains thoroughly illogical, ill-fitted, ungainly and a tragedy for the majority of (Other-than-human) mammalian residents, the victims of a vast, viciously disingenuous underbelly of the human false survival argument: cannibalism.

throat, one hectare at a time, has no reconciliatory solution under the aegis of natural selection? We may well ponder—tens of millions of people have—how God can exist beside an Auschwitz, and another to be perplexed by the notion of a "natural" natural selection that vouchsafes vast ecological horrors. Add time to the equations of bewilderment, more than four billion years of time, and we might pose the prospect that at the cosmic level, there can be no derivative meaning from earth. Only ephemeral events of zero consequence. This is not to attribute a value to zero, plus or minus. Only an interregnum spanning infinite neutralities, frustratingly neutral.

If these misanthropic calculations test probabilities, even plausibilities within a confined time frame, any ethical virtue remains, at best, something like the square root of obscurity. What set of numbers are disposed toward goodness? Can we equate probability theory with empathy, noting that the mix of observations concerning compassion in our species and countless others, is a genetically fixed menu of perceptions; the methodologies employed predicated upon untested other formulae of personality, observer background, and the stilted language of science or literature that seeks to convey what is ultimately an opinion?.[34] One likely answer, the real story, is something quite different than we might hope for.[35] Rather than somehow personally grappling with the apparent toxic burdens of evolutionary biology, we could defer to the underlying molecular and genomic storylines. To do so would be to *believe* in such data and accompanying scenarios. Belief, beneath the aegis of a skeptical presence, an unhappy state of being, places the ambivalent God within the center of our deliberations; or relegates the atheist, or disinterested, the maleffected anger and depression of ecological fall-out to an (arguably) inoperable guise of potential, or de-conscience. Alternatively, one might develop a thesis that recognizes the paramount differentials separating individuals, at least in part, from their species.[36]

---

[34] See, for example, *Culture of Ambiguity – Implications for Self and Social Understanding in Adolescence*, by Sandra Leanne Bosacki, Sense Publishers, Rotterdam The Netherlands, 2012; See also, "Compassion and the rule of law," by Susan A. Bandes, International Journal of Law in Context, 13,2, Cambridge University Press, 2017, pp. 184–196, https://doi.org/10.1017/S1744552317000118, http://www.susanbandes.com/wp-content/uploads/2017/05/compassion_and_the_rule_of_law.pdf, Accessed January 20, 2020; See also *The Ethics of Ambiguity*, by Simone De Beauvoir, Translated by Bernard Frechtman, The Philosophical Library, New York 1948; *What We Owe to Each Other*, by T. M. Scanlon, Belknap Press, Harvard University Press, Cambridge, Mass., 2000; *Morals by Agreement*, by David Gauthier, Oxford University Press, New York 1987; *Anarchy, State, and Utopia*, by Robert Nozick, Basic Books, New York 2013; *The Community of Advantage: A Behavioural Economist's Defence of the Market*, by Robert Sugden, New York, Oxford University Press, 2018; and *Moral Differences – Truth, Justice and Conscience in a World of Conflict*, by Richard W. Miller, Princeton University Press, Princeton New Jersey 1992.

[35] For a discussion of regret theory, see Loomes, G.; Sugden, R. (1982). "Regret theory: An alternative theory of rational choice under uncertainty". Economic Journal. 92 (4): 805–824. https://doi.org/10.2307/2232669. JSTOR 2232669.

[36] See two of the Authors' previous works, *The Theoretical Individual: Imagination, Ethics and the Future of Humanity*, Springer New York, 2018; and *The Hypothetical Species: Variables of Human Evolution*, Springer New York, 2019.

As soon as we lend full phenotypic significance to the one over the many, various imaginative possibilities are engaged. If the individual is to become the true bell-wether of evolution, not the species, then every prospect of redemption comes into focus. At least it does so on a neutral surface, ignoring sociobiology, geopolitics, and all of the power, suasion, fascisms, and mob rule which, in the case of humans, have intervened successfully to not bifurcate but bolster the taxon. But its backstory will pit the loner against the mass, stacking the odds against morality and dignity and ultimate physical survival, in favor of blind rule and quantitative firing squads. *Crowds and Power*, Elias Canetti's earlier referenced work, incisively delineated the societal priorities from those of personhood, eliciting the rogue elements of any civilization as the equivalent of perverted proxy, an *impossible dream*.

Were natural selection, with high frequency, to favor a *dream*, then all would be possible. But that invokes the problem of belief. Don Quixote was the true believer. Science has long acknowledged developmental bias, environmental factors affecting natural selection.[37] With bias, environmental modification perpetuates the folk-loric natural selection paradigm as a central premise of all those precursors, known and unknown. Multiplying the subject deficit, it were as if the continuation of the Self is the pivotal premise of population ecology. As if the life force itself is a species issue, the individual merely worth the extent of its random mutations adding to population realizability, and the deconstruction of an individual. But what if the selection process were unbiased, the individual exerting as much traction on eco-dynamics as any population? Can we even envision such a choreography? (Fig. 80.4).

## 80.5   Real People, Fictional Populations, and a Semblance of Hope

Time has always ordained mutability down to the sub-atomic particle. By the state of molecules, the aggregate impulses of life have made necessary the complementary space in which occupants might move and reside. But on the large vertebrate scales, the question of ethics, as well as aesthetics, combats natural selection at the level of a bias that transcends the numeric values of science, and may thoroughly defeat natural selection, a tall order to prove by any subjectivity. Yet, there are parallel natural selections favoring unabashedly those Utopic convictions of non-violence that discount any God, or theory of knowledge, that necessarily embrace trial and turmoil as fundamental to evolution, when, just as easily, consciousness could see

---

[37] See, for example "The interaction between developmental bias and natural selection: from centipede segments to a general hypothesis," by W. Arthur, Heredity, Vol. 89, 239–246, September 19, 2002, https://www.nature.com/articles/6800139; See also, "Natural selection, evolvability and bias due to environmental covariance in the field in an annual plant," by A.A. Winn, J Evol Biol. 2004 Sep;17(5):1073–83, PubMed, US National Library of Medicine, PMID:15312079. https://doi.org/10.1111/j.1420-9101.2004.00740.x, https://www.ncbi.nlm.nih.gov/pubmed/15312079, Accessed January 20, 2020.

**Fig. 80.4**   Baby Dolls in Downtown Mexico City. (Photo © M.C. Tobias)

pleasure, joy, and compassion as the true pillars of whatever synonym best fits evolution. Individuals feel this. But at the species level, we *know* nothing. Our data denies us individually even as it sets specific demographic quantities. Numerous doomsday paradoxes have been enshrined in mathematical logic in an effort to grapple with the sheer anticipatory anxiety—a veritable coefficient—lurking beneath the numerous debates whose primary components encompass human population, history, and probability theories. They include the "Carter catastrophe," the Heinz von Foerster "Doomsday equation," and Richard Gott's "vague prior formulation".[38]

---

[38] J. Richard Gott, III (1993). "Implications of the Copernican principle for our future prospects". Nature. 363 (6427): 315–319. Bibcode:1993Natur.363..315G.; the John A. Leslie "1.2 trillion" human population theorem; (Oliver, Jonathan; Korb, Kevin (1998). "A Bayesian Analysis of the Doomsday Argument". CiteSeerX 10.1.1.49.5899.) Dennis Dicks' 1992 "Doomsday argument" (Self-Indication Argument) (See Nick Bostrom and Milan Cirkovic (2003). "The doomsday argument and the self-indication assumption: reply to Olum" (PDF). Philosophical Quarterly. 53 (210): 83–91. https://doi.org/10.1111/1467-9213.00298; the St. Petersburg Paradox (Plous, Scott (January 1, 1993). "Chapter 7". The psychology of decision-making. McGraw-Hill Education; and Eves, Howard (1990). An Introduction To The History of Mathematics (6th ed.). Brooks/Cole – Thomson Learning. p. 427.); the Strong Self-Sampling Assumption (SSSA) (*Brandon Carter; McCrea, W. H. (1983). "The anthropic principle and its implications for biological evolution". Philosophical Transactions of the Royal Society of London. A310 (1512): 347–363. Bibcode:1983RSPTA.310..347C. https://doi.org/10.1098/rsta.1983.0096; and an exuberant rash of self-referencing doomsday rebuttals. ("A probable paradox," Nature from Oct 23, 1997 – A probable paradox. P. T. Landsberg &; J. N. Dewynne ... Landsberg, P., Dewynne, J. A probable paradox. Nature 389, 779 (1997) doi:10.1038/ ...

All such theories are compounded by projections: the impression, for example, that among 114 nations analyzed, it was the density of their human populations that correlated with "the number of endangered birds and mammals...".[39] In an elaborate essay on "existential risks," Dr. Nick Bostrom of Yale University's Department of Philosophy has detailed the many "threats that could cause our extinction," with case-by-case implications for "ethical and policy implications." "Directed evolution" (forms of transhumanism) is one of the assessed strategies within the larger categories Bostrom names: "Bangs," "Crunches," "Shrieks," and "Whimpers".[40]

More recently, Patricia MacCormack's *The Ahuman Manifesto – Activism for the End of the Anthropocene* has been described as "...a delightful provocation and invitation to imagine a world without humans and to think of what we can do to get there. It is an urgent call for action".[41] While a Stephen Hawking was in favor of our species leaving earth to colonize other planets, many have discussed issues of survivor and cognitive biases that inevitably weigh in favor of our species' long-term survival,[42] or of the whole issue of technology again coming to our rescue.[43] In favoring the present survivors, our species' perceptions of risk are much like those caveats imposed by various associations of insurance commissioners whose fidelity tends to align with insurers and agents of corporate conglomerates, not the consumer.[44]

In each of these instances, numerous psychological scenarios emerge: various game theories, cybernetics, and cliodynamics (a multidisciplinary mode of mathematically modeling historical data), Kondratiev waves (multi-decade cycles of economic growth), Great Divergence and Convergence mechanics in both economics and physics, as developed in the work of Leonid Grinin and Andrey Korotayev,[45]

---

[39] J.K. McKee, P.W. Sciulli, C.D. Fooce, and T. A. Waite, "Forecasting Biodiversity Threats Due to Human Population Growth," Biological Conservation 115(1): 161–164, Cited in "Human Population Growth And Extinction," n.a., Center for Biological Diversity, biologicaldiversity.org

[40] "Existential Risks – Analyzing Human Extinction Scenarios and Related Hazards," Journal or Evolution and Technology, Vol. 9, 2002, http://jetpress.org/volume8/symbionics.html; and existential-risk.org.

[41] Christine Daigle, Professor of Philosophy and Director of the Posthumanism Research Institute, Brock University, Canada, from the website comments on the book, Bloomsbury Publishing, https://www.amazon.com/Ahuman-Manifesto-Activism-End-Anthropocene-ebook/dp/B081PHNGDH

[42] See, for example, "We're Underestimating the Risk of Human Extinction, by Ross Andersen, The Atlantic, March 6, 2012; and Eliezer Yudlowsky's "Cognitive biases potentially affecting judgment of global risks," Global catastrophic risks 1, 2008: 86, p. 114.

[43] See, for example, "EmTech: Get Ready for a New Human Species, by Emily Singer, MIT Technology Review, October 19, 2011.

[44] See, for example, "Can you rely on the annuity advice?" by Ethan Schwartz, The Los Angeles Times, March 8, 2020, http://enewspaper.latimes.com/infinity/article_share.aspx?guid=3fd1d8f7-cf68-4ace-8428-4a6d2d399f67); or the much disguised motives employed by human supremacists who combated the organization ZPG (Zero Population Growth, the organization founded in 1968, changing its name in 2002 to Population Connection) on grounds that it was thought (wrongly) to endorse such draconian population control methods as forced sterilization and eugenics.

[45] See *Great Divergence and Convergence – A Global Perspective*, Leonid Grinin and Andrey V. Korotayev, Springer Publishers, New York, 2015.

and cultural projections and events based upon fragmentary past data and/or subjective assumptions that, by whatever means, perpetuate human dominance over any
other concept, or biological organisms. For example, the obsession with sustainability, with its static, endless, and actually meaningless cadence, a "Hail Caesar"
gesture in compliance with self-perpetuation at any cost, maximizing short-term
risks for hoped-for long-term interests. And this deeply worrisome condition, in
spite of the fact our short-term liabilities in a world of largely uncoordinated nation-
states and more than 6000 largely incomprehensible language groups, remains on
an intractable, most probably catastrophic trajectory. No appropriate descriptions
from Big History, Big Data, artificial genomics, natural selection, or evolution are
able to tease meaning from the absolute zero calm at the center of ecological vagueness, or along any length or circumflex of the enchanted, hopeful obscure.

Between the real person and its obverse fiction, the biological mass of such ideas
pertaining to actual events, or unrealized chimera consisting of ideals and extrapolations exists a theoretical contiguity of unknown properties, dimensions, and opportunities, all of which argue for the epiphanies earlier likened to that category of
obscurity, vagueness, and emptiness, at the heart and origins of reconciling paradox
and engendering some semblance of hope for the future (Fig. 80.5).

**Fig. 80.5**  The Man Who Feeds Pigeons At the Umayyad Mosque (الجامع الأموي), Downtown Damascus.
(Photo © M.C. Tobias)

# Chapter 81
# Biological Proxies for the Individual

## 81.1 The Natural Selection of an Individual

He "clearly savors the paradox" of being "trapped" "in the inauthenticities of civilization and of language" wrote Peter Caws of Emil Cioran's book, and its hagiographic context, *The Temptation to Exist*[1]. There is no adequate participle to encompass "trapped" when referring to an entire species that vacillates between itself and a specific personage. In this test of gamboling pivots, the odds favoring the individual are significant, however. *One* always favors the *one*, unless greed, or an altruism born of empathy enters the equation, thereby changing the entire edifice of what is possibility. Cioran grasped the fatal ironies of this condition, upon which the fate of the world now hinges. Aquinas qualified its same challenges by taking up the very cause and circumstances of Being, namely, *essentia* and *substantia*. Immanuel Kant exhorted logic, reason, and the dualistic pressures posed by Reality in attempting to grasp the cosmological mess made more bewildering by the unsteady atom of Being lodged precariously between the finite and the infinite, with no sure roadmaps to exhume phenomena and noumena, ethics, and the transcendent.

Ultimately, contrary dogmas stacked up like pieces of firewood ready to be ignited on the altar of the human ecological crises and their abundant storytelling. Stories likely to go in opposite directions on a whim, much like "formal proofs of the non-existence of a God who creates and organizes the world"—so contradictory a series of ultimately self-destructive selectivity as was described in the *Bodhicharyavatara* ("Bodhisattva's Way of Life," Chapter ix, Verse 119) of Shantideva (685–763)[2]. Again, we're speaking of the dialectic wherein the individual chafes at, and is victimized by, his/her own species, thus giving rise to a seemingly hopeless paradox.

---

[1] "When Adam ate the apple, God lost His Head," The New York Times, March 14, 1971, https://www.nytimes.com/1971/03/14/archives/the-temptation-to-exist-by-e-m-cioran-translated-by-richard-howard.html, Accessed January 23, 2020.

[2] See "Buddhist Atheism," *Encyclopaedia of Religion and Ethics*, Edited By James Hastings, With the Assistance Of John A. Selbie and Louis H. Gray, T. & T. Clark, Edinburgh, Volume II, 1971, pp. 183–184.

© Springer Nature Switzerland AG 2021
M. C. Tobias, J. G. Morrison, *On the Nature of Ecological Paradox*,
https://doi.org/10.1007/978-3-030-64526-7_81

**Fig. 81.1** A Joyous "Jeevynila," A member of the Toda Kerrir Patrician (Clan), the Nilgiris, Tamil Nadu State, India. (Photo © M.C.Tobias)

The natural selection of an individual, when it occurs, must always pose the ultimate challenge: The individual remains irrepressible, in spite of the biological odds favoring his/her species over any one personage. Natural selection is at once reduced to a mere expression, lacking a credible record in the annals of either ethics or personality. Its greatest achievement, if one might be singled out, would be what the authors have elsewhere named "the Reciprocity Potential" between individuals and species, namely, that intention and behavior that "is equal to the square root of compassion, times the square root of the population mean divided by the sum of the entire group, in this case, *Homo sapiens*"[3]. From the obscurity of non-being, to the limelight of being-as-defined-through-compassion, the individual, as concept, breaks with natural selection at the moment such choices are made outside of what might be deemed *human possibility*; love that outweighs the cumulative burden of so many genetic dead-ends, and bad ideas. The proxy for such individualism might, theoretically, be *anything*: the obscurity described in the previous chapter, a moment in time, an artful impulse, a volition dating back millions of years that managed to shoulder along the way some remarkable recessive gene, or eukaryotic gene redundancy[4] (Fig. 81.1).

---

[3] See the Authors' work, *Anthrozoology: Embracing Co-Existence in the Anthropocene*, Springer Publishers New York, 2017, p. 320.

[4] See "Genes duplicated by polyploidy show unequal contributions to the transcriptome and organ-specific reciprocal silencing," Keith L. Adams, Richard Cronn, Ryan Percifield, and Jonathan F. Wendel, PNAS April 15, 2003 100 (8) 4649–4654; https://doi.org/10.1073/pnas.0630618100, https://www.pnas.org/content/100/8/4649PNAS, Proceedings of the National Academy of Sciences of the United States of America, https://www.pnas.org/content/100/8/4649, Accessed January 23, 2020.

To speak of love in a cynical age is to unabashedly assert the verifiable quotients comprising a hypokeimenon, a quiddity, or haecceity, as varied metaphysicians— from Duns Scotus and John Locke to Gilles Deleuze—have described it. A subjective, unique essence that metamorphoses from however meager an opportunism or approbation, the slightest hunch, the anti-meme, to some other noble calling altogether. Rejecting authorship, it reclaims autonomy without communicating such. No trace of solidarity that would suggest the least affiliation with a population. The cenobitic compromise expresses that population paradox, as spatially confined (in an idiorrhythmic monastic setting) between the willful constraint to be oneself, and the pressure of vacillation with and among others, mornings in the refectory. It is a strange ambiguity that has arisen at the same rate and pace as cell division, the invention of sexuality, and the full range of behavioral psychologies that skew with the evolution of community structure, hierarchies, justice issues, politics, international treaties, and the breakdown of the commons: this waffling from one's mind to that of other minds.

## 81.2   A Continuing Strife

Can one claim optimism about circumstances concerning oneself, when others in the world are suffering? Buddha suggested that the world's suffering is inadvertently felt by the one at every cusp of that individual's happiness.

The study of Plato's Theory of Forms, and the tension between monism and pluralism, in terms of consciousness and self-identity, laid out in eight hypotheses or deductions within his most obtuse Dialogue, "The Parmenides"[5], will lead inevitably to that famed declaration by Parmenides sparring partner, Socrates. The admission of absolutely no knowledge.

Nothing, or nothingness—like emptiness, vagueness, and obscurity—hovers at the cosmogonic conclusion of Plato's *Republic*, containing the story of the Myth of Er (10.614–10.621), as well as the gripping details of an afterlife that stirred within the Sixth Book ("Somnium Scipionis") of statesman/philosopher Cicero's *De re publica* (54–51 BC) the celebrated fictional vision of one Roman general, Scipio Aemilianus, who dreamt of a colossal firmament dwarfing the earth and all her ecological progeny, two years prior to his character masterminding the destruction of Carthage in 146 BC. This larger-than-life dream held up for successive centuries, forming the vast cosmic backdrops in Dante and Chaucer, as well as in John Milton (particularly Milton and later illustrator, John Martin), such was the strength of a conceptual foundation upon which Plato, and to an even greater extent, Aristotle, forecast one profound version of the nature of reality, and the difficulties any one person will encounter in attempting to change course, let alone affect the backdrop. Humanity, by most claims, has always been at the mercy of the gods, of clashing armies in the night, of one's personal ancestry, immediate family, and now, genes.

---

[5] See Brisson, L., 2002, "'Is the World One?' A New Interpretation of Plato's Parmenides', Oxford Studies in Ancient Philosophy, 22: 1–20.

The continuing debate between an individual and its population gets that much more terrifying on the Yom Hashoah, International Holocaust Remembrance Day each January 27 (27 Nisan by the Jewish calendar) to remember the start of the Warsaw Uprising, and to commemorate at least 6 million Jewish lives and at least 11 million others, destroyed by the Nazi Regime and its vast number of collaborators[6]. In addition, the last week of January 2020 witnessed the Bulletin of Atomic Scientists heightening the Doomsday Clock to its worst calibration since its inception just after World War II, to 100 seconds before midnight, predicated upon extraordinarily increased tensions and environmental calamities throughout the world, disasters that have begun to clearly transcend any governmental infrastructures to cope with them[7].

## 81.3   The Optimism Paradox

At the same time, earth and Climate Scientist Andrew Glikson of Australian National University has brought to bear a terrifying summary of greenhouse gasses data to date, including "($CO_2$ > 411.76 ppm; $CH_4$ > 1870.5 ppb; $N_2O$ > 333 ppb plus trace greenhouse gases) land temperatures soar (NASA global sea-land mean of 1.05 °C since 1880"[8]. That same week—late January, 2020—at the four-day World Economic Forum in Davos, Switzerland, Jane Goodall warned against messages of "doom and gloom," suggesting that such declarations would indeed daunt all those holding out for hope. In what we might call Goodall's five-point plan, she volunteered five reasons for hope, even as she first intoned "gradually"—in terms of the possible time frame for winning over discord and opposition to viewpoints and strategies; before expressing her belief in "young people," "the resilience of nature," "social media…strength in numbers," "the human brain, we're coming up with solutions faster and faster," and "the indomitable human spirit"[9].

We find all five pivotal points the very essence of the *Optimism Paradox*. By that we mean to imply that such optimistic prospects are a bespoke point of contention, increasingly fragile, aggravated beyond easy intellectual redemption, promulgated for the benefit of some perfect logic, a back-to-school mentality, filled with joy and great excitement, a cause-and-effect teleology that accepts the premise that there are solutions for everyone and everything. As if there were true Utopia in Plato's Republic; a Form or likeness, in which all solutions could be counted on.

---

[6] See the United States Holocaust Memorial Museum, https://www.ushmm.org/remember/days-of-remembrance/resources/calendar, Accessed January 23, 2020.

[7] See "The End May Be Nearer: Doomsday Clock Moves Within 100 Seconds Of Midnight," NPR, by David Welna, January 23, 2020 January 23, 2020, https://www.npr.org/2020/01/23/799047659/the-end-may-be-nearer-doomsday-clock-moves-within-100-seconds-of-midnight,      Accessed January 23, 2020.

[8] See January 15, 2020, "The Australian firestorms: portents of a planetary future," by Andrew Glikson, Arctic News.

[9] See "Jane Goodall: I Do Not Believe In Aggressive Activism/Forum Insight," https://www.youtube.com/watch?v=akFyasOcFQA, Accessed January 24, 2020.

**Fig. 81.2** Ms. Ingrid Newkirk, Co-Founder and President of People For the Ethical Treatment of Animals (PETA) the largest animal rights organization in the world. (Photo Courtesy of PETA)

But when we look at the basic computational coefficients under the macroscope of each of Goodall's five arguments, and the muddied circumstances in which their calculus either succeeds or fails, we come upon the cusp of real-world population-level difficulties to which no philosophically self-assured imperturbability can actually attest in good faith. We say that notwithstanding Goodall and others—like Ingrid Newkirk of PETA—whose lifelong undaunted advocacies in the trenches, and that of PETA's more than 6 million followers, are unquestionably saving lives, impeding cruelty, helping the biosphere (PETA's weekly "Cruelty Investigations" reporting on an abundance of rescued animals from horrifying circumstances is a bountiful case in point)[10] (Fig. 81.2).

---

[10] See "PETA's Ingrid Newkirk on doing whatever it takes to protect animals, both ugly and cute," by Patt Morrison, Los Angeles Times, June 19, 2019, https://www.latimes.com/opinion/op-ed/la-ol-patt-morrison-ingrid-newkirk-peta-santa-anita-20190619-htmlstory.html, Accessed January 25, 2020. See, for example,

*Cruelty Investigations Department, Excerpts: Week of July 20, 2020,* Emergency Response Team Calls for Help: 243, Custom doghouses built and delivered: 3; Mobile Clinics: We spayed/neutered: 259 animals, including: 15 pit bulls and 15 feral cats; Local Community Animal Project Requests for Assistance: 60

Free spay/neuter transports to and from our clinics: 13 dogs and cats; Animals taken in: 46 animals, including 3 dogs and 5 cats transferred to another shelter for adoption; Adoptions: 1 cat, Orange Julius (now Tucker)

Animals still in our direct care: 2 dogs, Mingo and Bonnie (adoption pending), both in foster care

"Winnie" / VA: This 15-year-old cat was brought to PETA for end-of-life help after she came home with a broken jaw. Despite being on strong pain medication and antibiotics her jaw became badly infected. She was peacefully euthanized." Email from Ingrid Newkirk.

Or Bob Gillespie, the Martin Luther King, Jr., of the compassionate population stabilization front[11]. Yet, the computing of idealism in an eco-dynamic landscape cannot be reliably correlated against actual individuals, populations, or biomes, not unlike the ambiguities inherent, say, to the Fragile X (Stephanie) Sherman Paradox[12]. "Thus, seemingly paradoxical results among studies of prevalence may reflect the effects of selection bias, conflation and differing defining criteria for FXS across studies"[13]. In other words, we are trapped in our very optimism by our optimism, a perpetual and undeviating polysyllogism, or inflection point. When we do succeed in saving the life of another, it is not so much a miracle that has transcended para-dox, but a few good people with unconditional hearts.

But as with any epidemiology, human-modified ecosystem in which the dots, pikes, daggers, and insults have all been connected, leaves no hint of widespread recoveries. The words—widespread, recoveries—fall prey to the hazards of Wittgenstein. But subjectivism aside, peer review by actors in the experiment undermines the methodology. At meaningful statistical levels, everyone is essen-tially insane, which makes for continuous exhaustion and a discouraging framework from the very instant of an inspiration. Both N and n = zero. This represents a sub-stantial predicament at the population level, particularly for *optimism* and the mul-tiplicity of attitude variables predictably at odds with its singular appeal.

The confusion of non-linear variables is without precedent, in the case of *H. sapi-ens*. There are too many subjective integers; the matrix lacks any but a narrow defi-nition; there are no mathematical constants with a history of any certainty. The very problems themselves have not been sufficiently circumscribed so as to offer basic hypotheticals, models, let alone biological proxies for the individuals described.

First comes the crisis of gradualism. Demographers and conservationists keep speaking of 2050 and 2100 as convenient thresholds. Yet, every ecological, demo-graphic, consumption, energy, probability distribution, and stochastic (random sam-pling) modeling effort to date in each of those urgent arenas and time frames has yielded constantly increasing numeric projections, or *wobbly tables*[14]. For many years, world demographic statistics relied confidently upon United Nations mean projections for countless sub-sets of population growth. Yet, within no more than a decade, thresholds went from a global population projection of 7 to 7.5 million, up to 9.5 million and beyond[15]. Today, we speak of a 2100 human population exceeding

---

[11] See http://populationcommunication.com/

[12] See "The fragile X prevalence paradox," by Paul J Hagerman, J Med Genet. Author manuscript; available in PMC 2009 Aug 18, J Med Genet. 2008 Aug; 45(8): 498–499.
    Published online 2008 Apr 15. doi: https://doi.org/10.1136/jmg.2008.059055, PMCID: PMC2728763, NIHMSID: NIHMS120046, PMID: 18413371, US National Library of Medicine National Institutes of Health.

[13] https://www.ncbi.nlm.nih.gov/pmc/articles/PMC2728763/. Accessed January 25, 2020.

[14] See "Wobbly Tables - The Math," By Martin Gardiner | April 26th 2012, Science 2.0, https://www.science20.com/beachcombing_academia/wobbly_tables_math-89460, Accessed January 25, 2020.

[15] See the Authors' *World War III – Population and the Biosphere at the End of the Millennium*,

10 or 11 billion, we can't predict which, 10, or 11. (Some demographers have predicted 13 billion.) With few countries, regions, states, or even cities and communities actually ascribing to a uniform population policy, and with economists in every nation continuing to fuel principles of essential growth, the earlier referenced Ehrlich/Holdren I=PAT equation[16] has never been more relevant to ecological and moral principles of uncertainty. It is within the inextricable links defining corporate and consumer consumption, that data swarms have rendered the time frame, particularly one as vague as "gradually" unfit for any formulaic confidence. It has become, scientifically, a wholly useless marker of change.

Next, "young people." Here the problems are magnified, first, by the purchasing power, lack of experience, lack of suffering, absence of altruistic loyalty, fierce resistance (most studies show) to any level of authority intruding on self-centered feelings, and the general sense of rebellion, the very wont of which (power of the purse, etc.) renders this class of individuals most vulnerable, and most in need of the stability that, in fact, does not exist.

We can easily formulate the position that social media, this "strength in numbers" proposition, is actually the cause of our problems, along with "the human brain" and all of its alleged marvelous wonders, solutions, and the like. Solutions to our own numbers, our own brain, our own presence: Let us be clear.

As for this "indomitable human spirit," of course we all have some sense of it in ourselves, but the hard inward-dwelling message is there dangled before our eyes, our conscience, the apperceptive faculties which give our consciousness a sense of ecological remorse. Such spirit is not necessarily an answer to anything. Our species may pride itself with regard to certain ancestral, current, and hoped-for proclivities that side-step every justified premonition and desperate finale. But it is probably irresponsible to flatly reject the mirror images, everywhere around us, of Doomsday. All of the stanchions are giving way. Even John Muir has lost his luster in an era that suddenly finds human words and mottos, albeit steeped in racism, sexism, or some other vile tinge, more palpable, than large blocks of protected earth. Every protagonist is caught up in a shadow that ranges powerfully across its population level of embrace or abandonment.

Within 4 1/2 months from Dame Goodall's well-seasoned recommendations for optimism at the Davos Economic Forum (January 22, 2020) she had a few other (contrary) things to say about that, best captured in a headline from The Guardian: "Jane Goodall: humanity is finished if it fails to adapt after Covid-19"[17] (Fig. 81.3).

---

Continuum Books, New York, 1998, the chapter, "Demographic Madness."

[16] See http://web.mit.edu/2.813/www/Class%20Slides%202008/IPAT%20Eq.pdf.

[17] The Guardian, by Fiona Harvey, June 3, 2020, in which she declared, "If we do not do things differently, we are finished. We can't go on very much longer like this." https://www.theguardian.com/science/2020/jun/03/jane-goodall-humanity-is-finished-if-it-fails-to-adapt-after-covid-19, Accessed June 10, 2020.

**Fig. 81.3** Dame Jane
Goodall. (Photo ©
U.S. Department of State,
Public Domain)

# Chapter 82
# Shifting Balance

## 82.1 Multiple Scales Within Evolution

"The resilience of nature": Hundreds of millions of loopholes, pitfalls, gap analysis intimating any number of null-hypotheses. It begins in differing contexts, such as P. J. den Boer's classic 1968 paper treating of the "spreading of risk" in evolution; "the statistical outcome of selection varying from generation to generation within a heterogeneous population"[1]. The larger point, for our purposes, combines such data on local populations as the discovery by G. F. Edmunds and D. N. Alstad in 1978 that a single but entire population of black pineleaf scale existed "on a single ponderosa pine tree..."[2]. The implications become immediately clear with massive conflagrations and their consequences for local populations, as in the case of the Australian bushfires of 2019/2020. But also in terms of shifting focus from a sylvan glen to a single tree in a biodiverse-rich setting like Yasuní National Park, wherein a single tree becomes the host of multiple, endemic generations of populations. It not only changes our perception but our deliberations regarding the critical legal standing of a single tree.

Other relevant mathematical "shifting balance" examples—the curious case of differing mouse genomes within adjoining mouse communities living within barns on a farm; even among separate mouse families within a single barn—were brought into focus with respect to the realization that a "multipartite population in a multipartite environment is also genetically multipartite"[3]. In such instances—and they likely span the full hypergamut of earth's biodiversity—the identities, robustness, and simultaneous vulnerability of small groups in evolving patterns within fixed

---

[1] Cited on p. 260 of *The Ecological Web – More on the Distribution and Abundance of Animals*, by H. G. Andrewartha and L. C. Birch, University of Chicago Press, Chicago, Illinois 1984.

[2] See Edmunds and Alstad, "Coevolution in insect herbivores and conifers, Science 1999:941–45. P. 261, cited in Andrewartha and Birch, ibid., pp. 261, 472.

[3] ibid., Edmunds and Alstad, p. 262.

© Springer Nature Switzerland AG 2021
M. C. Tobias, J. G. Morrison, *On the Nature of Ecological Paradox*,
https://doi.org/10.1007/978-3-030-64526-7_82

space offer clues to the differential meanings of groups and individuals, families and friends, loved ones, and distant relations. All of these connections form the backdrop of a fantastic re-transcendentalism that is inchoate in the palm of one's hand, so to speak, with correlations invoking activism or, as we watch the Amazon burn, unspeakable sadness, even resignation.

When we place all the global conversations regarding hope and optimism at scale, Doomsday becomes immediately mathematical. Its calculations date back thousands of years, often coinciding with astronomic events that lit up the night sky, or engendered darkness at noon (to borrow from Arthur Koestler's 1940 novel, *Sonnenfinsternis*). Various forms of the populist "Nostradamus Effect" inculcate myriad Armageddons, hieratic hermeneutics, Egyptian gods, and Babylonian goddesses. Myths of Ur, of Gilgamesh, Enkidu and Yggdrasil, of Divine Comedies and Arcadias geographically expressed. In 1993 Princeton astrogeophysicist, J. Richard Gott III, stoked modest controversy with a piece he published in Nature suggesting that there was good mathematical reason to believe the human species would be extinct within 760 years, precisely. William Poundstone explores such mechanical and/or visionary apparitions in his book, *The Doomsday Calculation* (2019) with telling reference to the philosopher/mathematician Thomas Bayes (1701–1761) who had formulated rational foundations for random event probabilities and varietal consequentialities[4].

Bayes' most famous work was a 23-page set of calculations within the realm of "experimental philosophy" arguably given to the "proof" of the existence of God, entitled "An Essay towards Solving a Problem in the Doctrine of Chances" and published with the posthumous assistance of his executor, and close friend, Richard Price two years after Bayes' untimely death. In seeking to provide a circumstance for his speculations, Bayes writes, "Let us imagine to ourselves the case of a person just brought forth into this world and left to collect from his observations the order and course of events what powers and causes take place in it." And later concludes, "And tho' in such cases the Data are not sufficient to discover the exact probability of an event, yet it is very agreeable to be able to find the limits between which it is reasonable to think it must lie, and also to be able to determine the precise degree of assent which is due to any conclusions or assertions relating to them"[5] (Fig. 82.1).

[4] See *William Poundstone, The Doomsday Calculation – How an Equation That Predicts the Future Is Transforming Everything We Know about Life and the Universe*, Little, Brown Spark, New York, 2019. See also, Poundstone's article in Vox, "A math equation that predicts the end of humanity - How much longer till we all die off? 760 years, give or take," July 5, 2019, https://www.vox.com/the-highlight/2019/6/28/18760585/doomsday-argument-calculation-prediction-j-richard-gott, Accessed January 25, 2019

[5] Philosophical Transactions of the Royal Society of London 53, 1763, 370–418, https://web.archive.org/web/20110410085940/http://www.stat.ucla.edu/history/essay.pdf, pp. 19, 23)

**Fig. 82.1** Rev. Thomas Malthus. (Photo Public Domain)

## 82.2 Hope Begets Faith

Price himself was an extraordinary fellow, close correspondent with Benjamin Franklin, and expert demographer whose writings in the early 1780s regarding the population growth in Great Britain would exert significant impact on yet another Reverend, Thomas Malthus[6]. The three preachers—Bayes, Price, and Malthus— together can be said to have closed critical gaps between ideas of one and of infinity, of all things linear and non-linear, populations and species, and the real-time causes and consequences of number theories within the nascent ecological sciences. Martyn Hooper cites mathematical historian Sharon Bertsch McGrayne who has written that Bayes "transformed probability from a gamblers' measure of frequency into a measure of informed belief"[7]. This touches the core subjectivity of most probability theories, namely, the incorporation of "belief" into what, then, has become another version of the Anthropic Principle; the introduction of a central variable into

---

[6] See "Richard Price, Bayes' theorem, and God," by Martyn Hooper, in the publication, Significance, February 2013, pp. 36–39. https://www.york.ac.uk/depts/maths/histstat/price.pdf

[7] See McGrayne, S. B. *The Theory that Would Not Die: How Bayes' Rule Cracked the Enigma Code, Hunted Down Russian Submarines, and Emerged Triumphant from Two Centuries of Controversy.* New Haven, CT: Yale University Press, 2011.

any and all human deliberation and induction, namely, the human itself. We, who are the most probable candidates of extrapolated consequence, no matter what other subjects and predicates may be involved in the probabilities of our survival.

And thus, hope begets more hope, while extinction level events at global scale—unfathomed in their totality—repel any equal measure of conscious investment given the overt outlines of obliteration of that very object of interest, the subject itself, humanity. Bayes' theorem, at the heart of his probability doctrines, and most subsequent statistical foundations, hinges upon the updating of belief predicated by new and revised data. If the central argument of our observed time is that of Hope versus Doomsday, it is clear that the coordinates of destruction offer little dialogue with a most flimsy, finite, self-interested protagonist who reigns unfettered still, at the centerstage of perception and infliction, proffering the idle whim, against the blistering truth of all that humanity has actually undertaken.

Little wonder, then, that over 80 years after Bayes' momentous Essay, Kierkegaard, writing of "Existential Pathos"; of an individual's apprehension of the outer world (the world beyond his "inwardness" "before God," of "a greater or lesser environment") exclaimed, "...the unfortunate individual must not lose his courage, since there are many who are still more unfortunate than he, and besides there is every probability that things will with God's help become better..."[8].

---

[8] *Concluding Unscientific Postscript*, by Søren Kierkegaard, Translated By David F. Swenson and Walter Lowrie, Princeton University Press, Princeton New Jersey, 1941, p.391.

# Chapter 83
# Comes Crashing Down Upon It

## 83.1 Knowing, or, Unknowing?

That moment of ecological love. When all that is hate, and driven away, and broken by laws and theories and coming hypotheses, emerges altogether new, in its re-united reality.

Such eco-dynamic foment is a trembling re-birth. *Knowing* might well mark an empirical fact inherent to this paradoxical novelty, even as its source and wonderment may perpetually come crashing down upon it the mysteries of experience; the value of said perceptions, memories, and impulses, some other utterly unpredicted vagary of biology, rooted in—not knowing—*something else*. So that knowing and unknowing are in conflict, or tandem confluence, at the same moment, in the same personage.

There are any number of nuances surrounding it; summons harkening far back in time, and to tomorrow. Giambattista Vico's 1725 *Scienza Nuova* implemented the mental construct of a human entity, expressly the agency, the ideal, the very basis for any so-called accomplishment. What Vico imagined to be the "Age of Men." A similar vein of blushing enthusiasm pervaded the so-called Enlightenment in Europe.

But it is that Age, humanity's prom-night, that is real enough to the extent of the Anthropocene. And notwithstanding the persuasiveness of Self, not one proof of human existence can be verified. It only matters to the extent that the dream calls out the fallacy of the dream. Contradictions withstand every contrary tendency throughout the fabric of mentation because our logic base is a fiction, even while our morbidity and co-morbidities are clear to the sufferer. A universal conjecture that has framed every narrative of the humanities and sciences in self-portraits of excited angst and moribund attention deficit. The enthusiastic disorders defining our Age—every shred of a house, a fork and knife, cotton, tires and toothpaste, parapets and pledges of allegiance, antibiotics and automobiles, smart devices and dollar bills, Wall Street, weapons, airplanes and engines, democracies and dictatorships—all together now these influences have come crashing down. Have tethered hopes and dreams that take from every square inch. Does not give.

© Springer Nature Switzerland AG 2021      749
M. C. Tobias, J. G. Morrison, *On the Nature of Ecological Paradox*,
https://doi.org/10.1007/978-3-030-64526-7_83

And what could it give? What on earth does our species have to contribute? It is true that our nearly 100 billion corpses to date have provided nutrients, fodder, fuel. While our living bodies afford a rich repository for quadrillions of micro-organisms and busy atoms, hothouses of evolution. But, collectively, we are but one sample of life, skewed and terminal. Every other large vertebrate gives more than their corpse, and the corresponding organisms they host. Just as every tree in a forest like that of Bialowieza in Poland and Belarus champions fertile life-sustaining interdependencies.

While some inkling of service to Others can be deciphered in countless human biocultural practices (Jain philanthropy, Bishnoi guardianship of native species, Jewish and Franciscan charities, the targeted protective umbrellas and basic ethological hospitality of the Todas, and of thousands of NGOs) our 7.8 billion heavy mostly animal consuming bulwark presents to any one imagination a finite crisis. Countless conservation biology, animal liberation, deep demography, and other ethical methodologies have arisen too late to exert the necessary moral critical mass necessary to curb or repeal the woes. Every philosophical system, save for absolute nihilism, has long deceived itself before the onslaught of this one ungainly predator species.

## 83.2   Egocide

Nihilism has only thus far proved successful, in the predictive, empirical sense, for example, documented extinctions, scope, and rate of prime habitat loss. But in what way might this uniquely astringent and unforgiving prelude to ecocide manage to remedy the ecologically inconsolable? To lift the crashing plane from what we now know to be the inevitable chasm of gravities? There is no system of thought, other than acceptance or repeal of our destruction, to do so. None. All have denied the planet. Every thought experiment, when put to the test of its own authorship, aligns with one system or other that inevitably betrays an infra-species loyalty. Those statistically few human individuals who have tried to encourage a responsible humanity are obviously exceptions to what is, admittedly no true rule, only an insight that betrayal is itself the system of betrayals. Its relentless traction has not yet been understood, other than within those privileged bubbles of analysis-like interiors of a Robert Adam, representative sets of numbers quantifying the glimpses we have obtained thus far of the horrific carnage. It is a luxury of tears that has the time and steadiness of breath, of pulse, to track and monitor the planet's multitudinous emergency rooms. All those quanta, the ashes of Birkenau fitted into a far more unthinkable oven of Anthropocenic statistics which unambiguously portend of the finality of a stylized epistemological inroad: the anthropologies of extinction.

Reflection upon these importunate melancholies is akin to seeking ultimate relief in a large caliber bullet up through the mouth—no uncertainty about a finality (perhaps the only certainty left in the annals of contemporary ecosystems). Or massive morphine. One thinks back to the Austrian exiles from Nazi Germany to Petrópolis, Brazil, the great Jewish author/philosopher Stefan Zweig, and his spouse Charlotte E. Altmann, February 22, 1942 when they overdosed on barbiturates to leave behind a most miserable world of unceasing turmoil[1]. Zweig had finished the previous year his memoir of life in Europe, *Die Welt von Gestern (The World of Yesterday)*[2]. Three years later, waves of suicides for very different reasons, encouraged by Hitler, would sweep Germany. Indeed, as reported by Albert Speer, Hitler Youth handed out cyanide pills to concert goers at the final Berlin Philharmonic performance, near the end of the war[3]. Nor is it surprising that the US Centers for Disease Control and Prevention show the highest suicide rates in the United States, at present, since World War II[4]. We are in a war.

There will be those who consider such self-destruction as selfish (though not in the case of Seneca and his wife, as they were charged and, like Socrates, ordered to do so). But, alas, the most selfish act is the most sacred of deliberated anodynes, selfishness being the quintessence of probability theory, at the core of a human system that is least given to randomness. The paradox of a selfish death, an escape from all that swirling dark epiphany-laden earth that has fallen, is successful up to a point of biological reincarnation, whether one believes it spiritually or not. The physical components are irrefutable. One cannot elude the truth of what kind of an organism one is. Whatever dialectics might accrue from its gross anatomy, like a fungus, it has no necessary commentary, no posterity.

The premise of ecocide leveled against oneself pits epistemology against a heavenly euthanasia. This is the philosophical matrix singularly justified and oriented as the ultimate renunciation of pain. Given the state of siege humanity has waged against all Others, this self-annihilation represents the least footprint, a demonstrable reduction to zero of suffering, consistent with the most eloquent poets and cosmologists to ever go on record in its defense. How ironic, then, that prior to the French Revolution, when French laws were liberalized regarding all things pertaining to the self and its freedom, and despite approbation by authors as celebrated as Voltaire and eventually Montesquieu, those who failed in their attempt to commit suicide might easily be mobbed and slaughtered, buried under roads, or sewers, an unconscious, societal highlight—the irrational backlash by which the innocent ending of oneself was highjacked, as it were, by all-out social vengeance. How dare

---

[1] See "The Escape Artist – The death and life of Stefan Zweig, by Leo Carey, The New Yorker, August, 20, 2012, https://www.newyorker.com/magazine/2012/08/27/the-escape-artist-3, Accessed January 27, 2020.

[2] ibid, Leo Carey, The New Yorker Magazine.

[3] See Gitta Sereny, *Albert Speer: His Battle with Truth*, Pan Macmillan, London, UK, 1996, p. 507. See also, "Why a Wave of Suicides Washed Over Germany After the Nazi Defeat," by Oded Heilbronner, November 25, 2019, Haaretz, https://www.haaretz.com/world-news/.premium-germany-s-mass-suicides-that-accompanied-the-nazis-defeat-1.8165793, Accessed June 25, 2020.

[4] See "U.S. Suicide Rates Are the Highest They've Been Since World War II," by Jamie Ducharme, Junw 20, 2019, Time, https://time.com/5609124/us-suicide-rate-increase/. Accessed June 25, 2020.

to engender mercy for oneself by means of that very demolition meted out to the world by our kind[5]. Etymologically, the Anglo-Latin "felo-de-se" = "one guilty concerning himself"[6].

## 83.3   The Ideal of an Alternative?

God touching Adam atop the Sistine Chapel, represents a bias that is encrypted with every communication, hundreds of billions of times per human day, and has absolutely undermined the possibility of escape from that system of ideas which marks the most endemic human characteristic: idealism. As against an Anthropocene suggesting that we are radically condemned to this infernal cycle, our revolutions mere blips on the ongoing, rapid devolution which is mounting a silent quarantine on any prospect for human aftermaths. The more circumscribed our future years, the more obscure our self-isolations, as they move on a trajectory toward the final whispers of our presence in this collective form. Our accumulating exile has no kingdom, no lightning rod. Where is it located?

Do answers to our short-lived biological sovereignty gather at a particular cemetery? Is there any final message commensurate with so hysterical and perplexing a roman candle? No one will ever declare what it is, this experiment with man and woman, offspring and their mean life expectancies. The total demographic experience appears to be a blunder. All of the philosophical ideals stemming from the tens of billions of humans who have been born, then vanished, is a mental category, little ruminated upon, as useful as the inanimate zero which—to great but subtle effect— proffers a layering of caesuras. Binary equations presume to understand the function of a zero, just as the Edo School and shan-shui landscape traditions of East Asia endeavored to make much of *emptiness*, as earlier discussed (Fig. 83.1).

## 83.4   Unlike Nothingness

But unlike *nothingness* the empty provocation leaves a bewildering poetic teeming with suggestion. Everything within emptiness invites an interesting cross between the Sorites and Russell's Paradoxes. And from that plateau, the lopsided world, its infinity of crucial imbalances, may, we say, harbor something profound. And thus, our earlier discussions of transitional evolution and of hope.

Whereas all of the critically basic resources that humanity has taken for granted are fast dwindling, every eco-dynamic calculation pertaining to the rate of species

---

[5] See William Edward Hartpole Lecky, *History of European Morals from Augustus to Charlemagne*, 2 volumes, Longman Green and Co., London, 2nd edition, 1869, Volume 2, p. 61.

[6] See "Online Etymology Dictionary".

**Fig. 83.1** "Josie," a critically endangered Tres Marías amazon (*Amazona oratrix tresmariae*) Endemic to the Islas Mariás off the Pacific Coast of Mexico. (Photo © M.C.Tobias)

declines greatly accelerating, and whereas the selfishness index, like the heat index, has reached an insufferable threshold, we gather 'round this substance within emptiness, these stunning vagaries of vagueness within nothingness. And just quite possibly denominate a biological proxy for ourselves in the near distant coordinates of the final comforts in bed, at the end of the world that we have abandoned in our recklessness. If we can actually envision that end, perhaps we may also conceive of its alternatives.

The Stanford University online journal, MAHB (Millennial Alliance for Humanity and the Biosphere) has for several years proffered increasingly dire schematics of such calculations, inciting the necessary panic that should guarantee a level of ecological deficit *attention*. Rallying cries for the fragmented earth. A Doomsday Compass, not least, biodiversity devastation everywhere, is underscored by such newly charted threats as the "methane gun" at polar latitudes, "trillions of tons" of trapped methane ($CH_4$) now being released in an unstoppable cascade of added atmospheric warming. Writes paleo-climatologist Andrew Glikson of Australia National University, "Budgets on a scale of military spending (US1.7 trillion in 2017) are required in an attempt to slow down the current trend across climate tipping points"[7].

This is a small price tag given an "annual commerce worldwide exceeding US$80 trillion,"[8] the very importunate engine of biospheric malaise touted by our

---

[7] Cited in Geoffrey Holland's superb essay, "Saving Ourselves from Ourselves," MAHB, February 13, 2020, https://mahb.stanford.edu/blog/saving-ourselves-from-ourselves/. Accessed February 13, 2020.

[8] Geoffrey Holland, ibid., Cited from "The $80 Trillion World Economy in One Chart," by Jeff Desjardins, October 10, 2018, https://www.visualcapitalist.com/80-trillion-world-economy-one-chart/.

kind as not only a success (the economy), but an essential component of human survivability. Nothing could be further from the truth. Knowing that, and aware of the suicidal fragments surrounding that sad vision of a man-handled world, we are left to other devices, if only we have the staying power to admit of them.

# Chapter 84
# Systems Paradox

## 84.1 Dimensionality and Relevant Specificities (Fig. 84.1)

As in mathematics, concepts are in infinite dimensions, the space/time continua associated with them are not subject to measurable phenomena. This is the case in all fantastic encounters with the Other. We recall an experience of meeting a manta ray and exchanging courtesies in the Maldivers. Or delighting in the combinatory insights of all those beings drawn by Jacob Hoefnagel (1575–1630) in his *Archetypa Studiaque Patris Georgii Hoefnagelii*, in 1592 may be surmised—the dreams of the elephant beetle (*Coleoptera Dynastinae),* the fantasies of the home-seeking snail (*Gastropoda Helicidae)* or ephemeral passions of Pseudo-Aiusomius' flower, *De rosis nascentibus,* "ipsa dies aperit: conficit ipsa dies" ("One day brings forth [a flower], and the same day ends it"[1]. In her fine study of the Hoefnagel father and son (Joris and Jacob), Marisa Anne Bass examined the elder's decades-long endeavor to complete his *Four Elements*—the Terra, Aier, Acqua, and Ignis folios; his cartographic efforts for his friend Abraham Ortelius' *Theatrum orbis terrarum* (1570); various illuminations for the *Hours of Philip of Cleves,* his *Patientia, Missale Romanum, Mira calligraphiae monumenta,* and numerous allegories and other works commissioned by the likes of Emperor Rudolf II. Much of these artistic masterpieces were composed in exile, following the 1576 brutal attack by the Spanish on Joris' home which was Antwerp. Later on, the Calvinist who actually saw God in every minute insect would find himself embroiled in additional turmoil in Catholic-dominated Munich, spending his last years in Vienna. Bass' elegant and straight-forward thesis: In the space of exile, humbly seeking out one solace after another in the natural world, Hoefnagel found the strength and inspiration to help him survive[2]. This is that aforementioned wherewithal, the human devices that give

---

[1] See *Joris and Jacob Hoefnagel – Art and Science around 1600,* by Thea Vignau-Wilberg, Hatje Cantz Verlag GmbH, Berlin, 2017, p.451.

[2] See *Insect Artifice – Nature and Art in the Dutch Revolt,* by Maris Anne Bass, Princeton University Press, Princeton NJ and Oxford UK, 2019.

© Springer Nature Switzerland AG 2021
M. C. Tobias, J. G. Morrison, *On the Nature of Ecological Paradox,*
https://doi.org/10.1007/978-3-030-64526-7_84

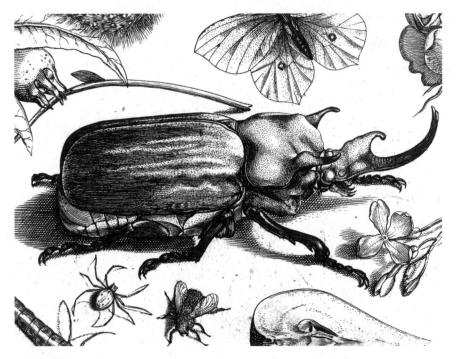

**Fig. 84.1** Jacob Hoefnagel's drawing of a Rhinoceros Beetle, Dynastinae subfamily of the Scarabaeidae Family, from his *Archetypa* of 1592. (Private Collection). (Photo © M.C. Tobias)

us to understand that we can reject our downfall, affirming life sustaining energy rather than its entropic collapse.

Many, like Marvin Minsky, have intimated that one of the great unknowns in this ongoing and death-defying dialectic, has to do with the fact that information may have nothing to do with meaning, which cannot be measured[3]. In the miniatures of the Hoefnagels, and the (purportedly) first or one of the first uses of a microscope for detailing organisms on paper, measurements of any kind lose out to evocation. And skipping 400 years later, even within the comprehensive studies of ants by Hölldobler and Wilson[4], design and data yield to the predictably irregular patterns, to take but one instance, of foraging behavior, a model of randomness that works against the laws of behavior. These antics are no "navigational errors" that conspire against the scientific consistencies sought within population ecology; but, rather, what Michael Strevens has interpreted to mean that "this probabilistic element is sufficient to randomize the evolution of an ecosystem's microvariables"[5]. Translated from biology to the conceptualization, Hannah Arendt, in contemplating the "life of the mind" suggested a rash

---

[3] 12.13, "Bridge-Definitions," *The Society Of Mind*, by Marvin Minsky, Simon And Schuster, New York, 1985, p.131, and p. 161.

[4] B. Hölldobler and E. Wilson, *The Ants*, Harvard University Press, Cambridge, Mass., 1990, p. 385.

[5] From *Bigger than Chaos – Underststanding Complexity through Probability*, Michael Strevens, Harvard University Press, Cambridge, Mass., 2003, pp.324-325.

of "metaphysical fallacies," and applied that "each new generation, every new human being as he becomes conscious of being inserted between an infinite past and an infinite future, must discover and ploddingly pave anew the path of thought"[6].

## 84.2   The Ants and the Humans

The ants plod forward no less than humans; our cognitions give to no correlation. One is not superior or inferior to the other, precisely because there can be no intelligence quotient that is relevant to any but the self-important species enacting the fallacious calculations. Who is a better poet, Einstein or a giraffe? And are we to suppose that Cartesian coordinates are sufficient to define the space and equations of their ruminations (Einstein's and the giraffe's)? $x^2 + y^2 = 4$? Do their spatial hyperplanes devolve into equal trihedral, or octants? Are their X, Y, and Z axes to be taken literally? We should be as apt to ponder the geometrical insights of the precocious Bernhard Riemann (1826–1866) and his "exterior calculus" and "topological manifolds" as we are the imagery of the Hoefnagels. Riemann's transition maps to higher dimensionality within geometry, said to have made Einstein's general theory of relativity possible, with its minimal, ruled, and multidimensional surfaces inside and outside Euclidean spaces, must rely, ultimately, upon space we ourselves can relate to. Otherwise, abstract dimensionality has no *specific-enough* relevancy to either quantification or qualification. Yet, there is always going to be birth out of empty correlates.

Such mathematical orders presuppose some aspect of systematics, but none coming from the inner and outer worlds of other beings, whose seismic sentience necessarily undermines any theory of certainty. And that would include absolute confidence in the genetics or storylines concerning primates, and *H. sapiens* in particular. James Grier Miller forcefully acknowledges a kind of biosemiotics preamble when he writes, "…one cannot measure comparable processes at different levels of systems, to confirm or disconfirm cross-level hypotheses…"[7]. And he concludes, "Any cell in a given location at a given time, any ruler of a given nation in a given period, receives comparable matter-energy and information inputs. But they may act quite differently"[8].

These intimations, artistic and mathematical coefficients of self-compliance and absorption, will necessarily disrupt any given moment of consciousness. As if, waking from a dream to discover that we are veering, half-asleep, into oncoming traffic that is the earth (Fig. 84.2).

---

[6] *The Life Of The Mind*, Volume One, Thinking, Harcourt Brace Jovanovich, New York and London, 1978, p. 210.

[7] See James Grier Miller, "Living Systems – The Basic Concepts," 1978, https://www.panarchy. org/miller/livingsystems.html, no pages numbered, Accessed January 29, 2020. Miller, Panarchy, no pages numbered,1978. See also, Immanuel Wallerstein's four-volume *The Modern World-System* 1974-2011, Academic Press, New York, and University of California Press-Berkeley, CA.

[8] ibid, Panarchy, no pages numbered.

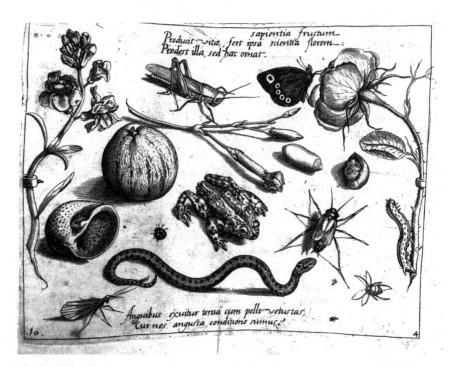

**Fig. 84.2** Another page from Jacob Hoefnagel's *Archetypa* of 1592. (Private Collection). (Photo © M.C. Tobias)

# Chapter 85
# The Final Hermitage of Ideals

## 85.1 One Night in a Pub, Nine-Hundred Years Ago

Like the word evolution, our idea of *comfort* with all of its trappings - agreeable habiliments, etc. - we all know to be a short-lived flash in the psyche. The great hope for peace and personal easement is lodged within an ecological *exclusion* benefiting our transient moment on earth, upon a fine mattress, swathed in clean Egyptian cotton sheeting. Arguments against such equanimity or biosymmetry or any theory-specific, in which biologic variables have apotheosized competition into precocious settlement for our one species, read much like the portraitures of a sublime waterfall, or Jean-Jacques Rousseau's splendid solitude, or the canons of the lives of a Sir Walter Browne, or Marcus Aurelius—equanimities, from time to time, of a Baconian universe, or the epiphanies of Epictetus and his followers.

The first folio of Shakespeare, the illuminations of the incandescent Safavid Dynasty painter Sultan-Muhammad, also mirrors the astonished, inherently tenable coordination that so many masterpieces have in common under the aegis of—call it the wisdom that percolates to the core of survival and there renames it—something that has nothing to do with survival.

Or think of it only in the tone of whispers, echoes, and some axioms that are intimated by that which we know nothing of, like the centuries-old builders of Chartres Cathedral, their day-in grime and dinner conversations. In other words, immutable, humble gestures of human life that surpass those many social zones of plus or minus in the decoding of inexplicables. Deep biography that encompasses all those partaking on any given night in a pub 900 years ago.

By all accounts neither the Tao nor the breath, the inner vision or exquisite proof (conventional proofs) can grasp even the elementary particles, waves, or moonlight that has invested this perfect retreat of a concept with the ultimate hermitage of ideals, seated at Ashikaga Yoshimasa's writing table at Ginkaku-ji in Renaissance Kyoto, or those spellbinding moments during the finale of "The Marriage of Figaro" when something altogether unrelated to anything else occurs at the exact instant that we feel connected to our humanity.

© Springer Nature Switzerland AG 2021
M. C. Tobias, J. G. Morrison, *On the Nature of Ecological Paradox*,
https://doi.org/10.1007/978-3-030-64526-7_85

This is the logic of *comfort in evolution* that manages to perpetually re-invent our lives and expectations over the course of 3.8 billion+ years of cellular experimentation, mutual tolerances explored to the exponential ends of the earth.

How remarkable that we could come to behold and feel this reality amid so much suffering. That is fundamentally paradoxical, and radically troubling. It is, at once, an apologia that skirts courtrooms and rhetoric, science textbooks, and the crowd-power of human causes that insists upon itself. The proportions of a village, village life, all the priorities humanity clings to will demolish the least contrarian view in terms of species etiquette and superiority, as we have reiterated.

To behold the sublime is not easily abetted by abject pain, or the insights into injustice. In the much-aggravated congeries and conurbations of our dense truancies as a collection of population centers—taken in a philosophical light—the lone congress of one, a parliament of the Self, vibrates according to the rules of energy transformation. Some specialists might attribute the religion of a Wordsworth aglow with its intimates of immortality. Or the quirky, strange attractors of electromagnetism, of Planck's Constant, or of any of a jumble of laws in physics. But ultimately, this vibrating essence rebuts all assertions of truth or accuracy stemming merely from universal laws. Even gravity, as tested by a feather in the stratosphere, obeys no predetermined flight plan. Darwin's *The Descent of Man, and Selection in Relation to Sex*[1] recognized at once that his application of all matters evolutionary to human beings would cause a rash of awkward connections, particularly for other primates. When two of us fall in love, there is no certain trajectory or translation of that emotional itinerary. The science falls short. The calculus lags behind. A linear or multi-dimensional depiction must suffice. It might be an octagon of smooth surfaces, or the National Library of Belarus, a rhombicuboctahedron.

## 85.2   Much Like Those Portraits by Hans Holbein the Younger

Much like that series of portraits executed by Hans Holbein the Younger, in 1527 and bringing to somber still-life the court of Henry VIII, we are each of us amalgams of identity absent all meaning but a dog tag[2]. These elegant showcases by Holbein illuminate a heartbreaking clarity which tenaciously informs all future pairs of eyes with a sensate apperception traumatized or enlivened—by the ineluctable mortalities, a fact, a sensation of irreversibility first grasped by children at some point between the third and fourth year of life. Prior to that epiphany, imagine, re-imagine what the world must be like.

---

[1] John Murray, Publisher, London 1871.

[2] See *Holbein and the Court of Henry VIII*, by Hans Holbein The Younger, Published by Queen's Gallery, Buckingham Palace, London, 1978.

Every one of those persons of enormous sobriety that Holbein painted has, of course, perished in a manner of speaking. We know most of their names and in some cases quite considerable biographies. It wasn't that long ago. The same sensation of continuity attends upon Edward Curtis' photographs of North American indigenous peoples, commencing at the time of Princess Angelina's life, at the moment she posed on a beach along Puget Sound for one dollar, and thereby agreed to some vague semblance of intergenerational communication[3].

## 85.3  Unless

Our physiological ties to one another grapple after confirmation, the utility of a family structure, of conjugal nuclearization, the transfer of presumptions of status and continuity. For Japan's baby boomers, or the large extended family cohorts within the matriarchal households of Kerala, human intimacies with one another, codified into practice and tradition, are equivalent to the covalent bonds in atoms. The hazy interiors of these obscure but universally dense noble gasses, negative and positive charges, valency, shells and orbitals, polar solvents, polarity spectra, and the compound chains of any given molecule producing organic compound properties, are as finely tuned as any relationships can be. We see families throughout the Milky Way, but can relate to their cogent structures more readily in a rain forest. Such relationships involve multiples of constantly changing charges, variable ions. But carefully follow any dune, cloud, or wave and the essence of all things believed to have been understood—like a Holbein—collapses beneath the onslaught of incalculable mind. This is a wonderful situation.

About 7.8 billion assumptions compounded nanosecond by nanosecond allow for no shared sentience other than that diminutive cluster of basic survival regimens, the same attractors to comfort earlier referenced. Absent a thorough human reason, all rational behavior has become a myth. Belief in such folklore has its occasionally propitious outcomes and proponents, such as a Hugo Grotius or Mahatma Gandhi. The hotspots methodology harbors a safe-haven clause in the frissons of partially sacrosanct space. Think of all those Holbein portraits, each one of them, as a sanctuary needing protection. Whether we see a face in time, or glimpse a sunset from Moran Point on the South Rim of the Grand Canyon, the instinct to protect is fundamental to our kind, yet the vagaries of human evolution work against the fragile policies struggling to achieve that moment of ecological grace, where a face and the landscape converge upon the most tenuous of all ideals. Unless-

The human wars against the earth, in spite of thousands of years of our allegedly knowing and feeling better, have so perpetually erupted, squandered, and shamed that energy field of the potential conscience, as to obviate its already vastly

---

[3] See *Edward S. Curtis: One Hundred Masterworks*, by Christopher Cardozo, Prestel/Random House, New York, 2015.

attenuated position in the besieged hierarchy of hominid postulations. Both synderesis and conscientia invite an ancient and durable power to emanate, regardless
of their incarnation in choice. Free-will has been highjacked by the random possibility of distributions which in their categorical theories arm human ignorance with
the event horizons continually confirming a negligible attempt to make sincere contact with; to even begin to understand the legitimacy, psyche and emotional needs of
the Other. Classical logic, in any form, has failed to qualify these categories.

Unless numerically, that breakdown into tired, rhetorical "data" is revealed to
have gotten the ethological paradigms wrong. Much like Anaxagoras' "mixture" of
compositions and vague objects. Wittgenstein's premise in his work *On Certainty*[4]
clung to this realm of the topsy-turvy, this maelstrom of doubts about human stability of nature. Ultimately, uncertainty has more mathematics providing it philosophical fodder than certainty. Our various systems of ethics are, alas, *systems*. And as
has been stipulated, any human system is given to inconstancy and disappointment.
There are no venerable exit strategies that might, at an 11th ecological hour, redress
the wrongs, unless obvious moral choice, one by one, addresses this most melancholic modification by endless descent.

Between human complacency and riot exists a realm of fine intuition that may
give some solace where previously war and revolution prevailed. But in a larger
perspective, even that solace is subsumed by a system.

We, the species and its individuals, are pivotal to this ultimately transitive problematique. From full unknowing to the equally unsustainable delusion of a fact,
where certainty has collided with its counter-terrorism of the bewildered emptiness,
therein, emergent and however tentative, a soft-spoken chorus of the one signals
duration, geography, and a target or targets which we may regard as a tenable field
of action. Whether we have the slightest idea as to what we're doing here, the idea
of the target and its transformation in humans from confusion and doubt to an affirmative attribute of attitudinal change, counts at last for everything that still has a
chance for survival.

Because we do not have a torch to actually light up the paths before us, there is
most favorably the rise of a certain fragment, as tempting as any Codex or Tractatus,
an entire civilization free of destiny in our minds. We have by turns and finally all at
once, the inner-most wall of a cave within ourselves. There, by searchlight we
obtain the first and only true glimpses. By the torch of an eerie, science-bereft continuity, we pretend—and, who knows, from time to time realize—some redemptive
hope for the biosphere, in spite of ourselves. There, in those long-ago penetralia that
were so familiar during the Paleolithic twilights, in these dim usurpations of current
pessimism, we may backwards crawl in to the darkness of our surest misgivings and
fears, nestling in the silence of our nothing-to-lose contemplative hours and, setting
out with a most benign bearing, endeavoring to reorient.

Contrary to our nightmares and authoritative histories to bear them out, we might
yet scatter as individuals across a tortured cartography, a land of vagueness and

---

[4] See *On Certainty/Über Gewissheit,* by Ludwig Wittgenstein, Edited By G. E. M. Anscombe and
G. H. Von Wright, The Text In German Translated by Denis Paul and G. E. M. Anscombe, With An
Introduction by Arthur C. Danto, and Twelve Illustrations by Mel Bochner, Arion Press, San
Francisco, CA, 1991.

**Fig. 85.1** Paulus Potter's "De Jonge Steier," 1647, detail, from the reproduction at the Jan Van Goyen House in the Hague. (Photo © M.C.Tobias)

demolished Utopias[5], from privileged bubble to bubble, the last remaining members of our tribes, and therein rekindle a universal freedom that involves both hominin and homily; donkeys and a sputtering candle of the divine. Our last chance to illuminate that cave into which Plato once placed this ideal of humanity. It died with him. But we have no more excuses: the power, vested in us to re-ignite that long-ago dream, exists, both at the individual and population levels. It is a biological truth as much as a spiritual injunction (Fig. 85.1).

---

[5] See *History And Utopia*, by E. M. Cioran, Translated From the French by Richard Howard, Seaver Books, New York, NY, 1987.

# Chapter 86
# The Paradox of Prayer

## 86.1  Confession and Evolution (Fig. 86.1)

To be conversant in truth valuation—multi-valued logic infinitely more nuanced than zero (0) or one (1), true or false, in Boolean algebra—is to grasp that set of non-numbers which have come to represent the endless, elastic, ambivalent span that is the life of the mind. It is instinct with prayer to the extent that its life is a paradox of all that remains unrealized in consciousness, the gaps filled in by faith. As with any ecosystem and the processes therein, no single science, social science, or aesthetic revelation can help the mind to reduce the irreducible. Nature is obviously impervious to our chorales, pieces of paper and canvas, bronze worked into a hallowed likeness. The most ruthless concision does not approximate time in real time. Authentic reflection is only a mirror of prayer, the longing for something certain, like a tree. Our ecological ideations are pretexts for paradoxical prayer, because we utter those words and intentions while we are in the act of both living and dying. Yet, instinct guides us toward something better, not worse. The paradoxical nature of such solemn oaths and objects of beseeching cushions the square root of 3, riddling our most sincere hopes and efforts with an endless multiplicity of other coordinates, enumerative combinatorics and *kissing numbers*, as they are called.

All of these quanta are elusive. In the heart of everything, we sense nothingness, and since these two extremes form an equality (the equal sign earlier discussed) that is biochemical, our best guess pertaining to our species' pertinacity forms the basis for our prayers: infinite dimensional function space. Our confessions are the basis for that quixotic effort to strike some balance, offer up a reconciliation between variables, Boolean operators, dream-like crucibles reflecting unrecognizable or forgotten perceptions with constants or boundaries, such are the coordinates of human orisons. In quantum mechanics, we might be speaking of a harmonic oscillator, with a very attractive graph that begins and ends nowhere. In quantum optics, that *nowhere* might be an instant of squeezed light. In mathematical ambiguities, a Necker or Impossible Cube. A species "mysterium tremendum et fascinans" as German theologian Rudolf Otto (1869–1937) described our obsession[1]. At the conclusions of our fidelity to the pew, we turn our backs and move into impermanence, relativity, volatility. Somehow improbably, we move toward the aftermath.

---

[1] See also Goldstein, Laurence (1996). "Reflexivity, Contradiction, Paradox and M. C. Escher". Leonardo. 29 (4): 299–308. https://doi.org/10.2307/1576313. JSTOR 1576313.

© Springer Nature Switzerland AG 2021
M. C. Tobias, J. G. Morrison, *On the Nature of Ecological Paradox*,
https://doi.org/10.1007/978-3-030-64526-7_86

**Fig. 86.1**  On the side of the road along a mountain pass in Mexico. (Photo © M.C.Tobias)

## 86.2  One Great Supposition: Ecology

To speak of if whilst thoroughly confined to the frissons of the Anthropocene invokes both the desperation and grandiosity of our all too human supplications and hymns. Like all good ideas, prayer decidedly takes enough time to satisfy the breadth of its one great supposition, ecology. Barring the statistical likelihood of perfect beginnings or unequivocal endings, the ambiguous qualia defining human experience at its perdurable moments of extremis are universally in no condition to argue with reality. Philosophy cannot stand up to environmental calamity. Prayer at its most vulnerable moments manifests answers to the nagging nexus of its inherent fragilities. What is most vulnerable—all mind shares in this category—will not be tempted to mount crusades against the metabolic truth of any biome in which we figure.

Because prayer emerges typically at the intersection of angst and evident disaster, the earth will always provide for it, as if biology herself were mindful of the

**Fig. 86.2** "St John Sitting Beside a Lamb," by Joannes van Londerseel (1580–1638) after Gillis Claesz. d'Hondecoeter (1575/1580–1638) from a Series of 130 Engravings. (Private Collection). (Photo © M.C. Tobias)

human penchant for getting into trouble. The endless extremes that invite prayer, and other forms of deep meditacioun, offer no formulas, no clear histories, only the private lives embedded day after day in solitary deaths across the magnificent ecosystem of cemeteries, a perpetual solace that forms the most crucial of all philosophical categories, in juxtaposition with those of life. Writing of the monks at Mount Athos, Ralph Harper would say, "Monks do not learn by studying…but mainly by their daily round of prayer… And as they pray, they are reminded by the ikons of the lives of the saints, of the community of those who live the mystery, those who once knew better than we what the source of the mystery is…"[2].

We speak metaphorically on the cusp of every unknown ecological process. In the thickness of a forest, a towering cathedral, a rustic hut on the edge of the world; or deep in one's heart where no survey has ever tread, the epicenters of epistemology collapse, recalling time after time the original Fall (Fig. 86.2).

---

[2] Ralph Harper, *Journey From Paradise – Mt. Athos And The Interior Life*, Éditions Du Beffroi, Distributed By The Johns Hopkins University Press, Quebec, Canada, 1987, p. 68. See also, "Environment and Sustainability Prayers," – "Canticle of the Creatures," Attributed to Saint Francis of Assisi, jesuitresource.org, https://www.xavier.edu/jesuitresource/online-resources/prayer-index/sustainability-prayers, Accessed July 22, 2020. See also, "St. Francis of Assisi – A profile of the patron saint of animals and ecology," The Humane Society of the United States, https://www.humanesociety.org/resources/st-francis-assisi, Accessed July 22, 2020.

**Fig. 86.3** Dr. Tarun Chhabra in Private Prayer before the Toda Sacred Conical Temple of Konawsh, the Nilgiris, India. (Photo © M.C.Tobias)

Our fanciful notions of destiny that have girded one doomed civilization after another form the full circumference of a uniformly irreligious distress. Within the orbit of those enduring anxieties, the natural disassembly of the body, external causes of injury and mortal wounds, there will always emerge some Symeon the Stylite (390–459) to cheer us up. The saint and his pantheon of prayers bolsters the great ash tree, the formidable ziggurat in every dream of Nature; the dialectic out of which God grows and plays fanatically with the credulous. The stage where mercy contends and victors are null; where one ultimately walks away to get on with his/her life, never to be heard from again, though it matters not, this is the aftermath of prayer, looks forward, looks back, the biology of the faithful and faithless, both[3] (Fig. 86.3).

---

[3] See *A Treasury Of Prayers*, by Frances Lincoln, Frances Lincoln Limited, London, 1996. See also, *The Ecological Self*, by Freya Mathews, Barnes & Noble Books, Savage, Maryland, 1991, particularly Chapter 4, "Value in nature and meaning in life".

# Chapter 87
# Forgiveness

## 87.1 To Forgive (Fig. 87.1)

To forgive is fundamentally ecological; the biosemiosis that signals to itself on behalf of any and all others. To elope into the freedom of non-judgment. Scorn that is fundamentally dualistic, turned to compassion. Unlike the self-badgering of a strongly held antithesis, to forgive, like sharing water and food, or playing, is quint-essentially the most interesting prelude to the breakdown of barriers. This should be an obvious certitude. These behaviors begin with atoms, then molecules and neces-sarily will of their own evolution propel trans-ethical minions for whom such groups are unburdened by scruples or bias. All differences at that point are not only for-given, but celebrated as a very physic bound to all quarters of the Cosmos.

We don't understand it, or necessarily perceive it in action in other species. But antipathy toward oneself, the self-loathing before the mirror of nausea that frames the body of fluids, gasses, enzymes, and feces, becomes an entirely different bio-logical baseline that should underscore the principles leveraging this Lascaux of new equivalencies: humanity joyously but one species among 100 million + others. By the mechanisms of this scenario our species approaches its end un-curtailed any longer. No more doubts. Despite a legacy of wilderness euphorias, our grasp of all things ephemeral is limited to a glimpse of geology, or of stars. We do not cling, let alone worship our deceased, but to the rare friend, friendship presupposing the rubrics of this ecological change of heart, just as we were preparing to gather en route to an abyss (Fig. 87.2).

© Springer Nature Switzerland AG 2021                                              769
M. C. Tobias, J. G. Morrison, *On the Nature of Ecological Paradox*,
https://doi.org/10.1007/978-3-030-64526-7_87

**Fig. 87.1** Michael Aufhauser and Friends at Home, Gut Aiderbichl Sanctuary, Salzburg Austria. (Photo © Gut Aiderbichl)

**Fig. 87.2** Marietta van der Merwe with rescued Cheetah, Harnas Sanctuary, Namibia. (Photo © M.C.Tobias)

## 87.2   Deep Ethological Convergence

Mental fixations follow by primeval logic upon the actions of this biosemiotic convergence, signs flowing in from every quadrant of life. The sum total of signs can never add up to a real cow, obviously. Our meanings are tooled and re-tooled over millennia to convey the simplicities of passing time. Not lost generations.

If readiness- and reciprocity-potentials might yet galvanize the connections between beings of any persuasion, the plight ordaining all those conditionals that enshrine the reality of ecosystems is unlikely to relax the strictures, even for a second, governing separation between *H. sapiens* and all others. To forgive across boundaries is the most welcome of heresies. A world we do not know, even as we have attempted to frame it in hundreds of sprawling natural history museums, brilliant insects pinioned on boards with pins and needles and a gluttonous use of formaldehyde. A newcomer's pathway to forgiveness citizenship is inherently the deep ethological manifestation which finds companionship everywhere in the world.

A history of the arts embodying the trillions of anonymous deaths at the hand of humanity would argue that personality and attitude curves are capable of any deviation. There are no permanent X and Y axes that defy the ever-present options of a noble and self-sacrificing, self-accepting behavior, even before the cliff of overwhelming odds. Communion and unisons are well documented. Although in humans its recognition is rare. In thinking about forgiveness, one must question the paradigm of antagonism: Who are the combatants and is it enough to suggest that survival has anything to do with the basic vocabulary of evolution? The young Darwin was uncertain when traveling through "neutral territory" in Tierra Del Fuego in January 1833[1].

"Survival" has forever been hardwired under the aegis of pain and suffering. While it gained momentum during the varied exchanges between Darwin, Herbert Spencer, and Alfred Wallace[2], it remains an utterly paradoxical concept on account of the near universality of death (other than certain so-called immortal cell lines). So that the first essential qualifier needs be a qualification and quantification of the noun. Survive for how long? At what cost (by what means)? Under what conditions and contexts? In and of itself survival, as we know it, is strictly the precious collaboration of desire, impulse, and adamantine Self. One might pray for peace on earth, or for a patient to survive a harrowing surgical procedure. But prayer, like survival, begins to deteriorate when applied to group selection. It finds its sea-legs most assuredly in the individual.

---

[1] Charles Darwin, *Journal of Researches*, Ward, Lock and Co., London, New York, 1889, p.159.

[2] See Momme von Sydow, M. (2014). 'Survival of the Fittest' in Darwinian Metaphysics – Tautology or Testable Theory? Archived 3 March 2016 at the Wayback Machine (pp. 199–222) In E. Voigts, B. Schaff & M. Pietrzak-Franger (Eds.), *Reflecting on Darwin*. Farnham, London: Ashgate; See also, "This preservation, during the battle for life, of varieties which possess any advantage in structure, constitution, or instinct, I have called Natural Selection; and Mr. Herbert Spencer has well expressed the same idea by the Survival of the Fittest. The term "natural selection" is in some respects a bad one, as it seems to imply conscious choice; but this will be disregarded after a little familiarity." Darwin, Charles *(1868), The Variation of Animals and Plants under Domestication,* 1st edition, London: John Murray, p.6.

If we discount reproductive capacity between like organisms, humanistic progress is not biochemically measurable. A more compelling argument exists were we to shed the taxonomy and consider individuals of widely differing behavioral tendencies to be their *own* species. Earlier, we referenced the concept of an individual comprising countless populations, an obvious truth that deserves to be respected. Diversity escalates, not by millions of species, but trillions of individuals. To the extent it may undermine the power of numbers, such a linguistic and conceptual shift only elevates the traction and profundity of individual beauty and prowess, the prospect of insightful relativity.

## 87.3   Hope Among Individuals

While it is crucial to recognize the extinction of populations (at least "237,000 populations" of the "515 species of terrestrial vertebrates... on the brink of extinction")[3], the more crucial lesson bears upon the number of individuals caught up in such exterminations. In northern Kenya and southern Somalia, everyone is related to the swarming locusts. Without accounting for the individuals themselves, the genetic and demographic facts exist only relatively; they are confined to theory. Theory is hypothetical. Hypotheses are alluded to. Allusions are enveloped, by turns, in a chain of being—or chains of being, wherein blurred degrees of sentience and sensitivity overtake all of the natural sciences. We call upon differences and similarities (agricultural chaos, family loss, community devastation, or re-birth), but ultimately, our human subjectivism stands in the way of any clear perspective, overwhelmed by an implacable point of view. A subjectivism that is tantamount to a critical admission: Nothing is real, not by *our* accounting, save for pain, hunger, suffering. The photoreceptors in our macula admit to the otherwise sedate brain a certain amount of optical information. That same brain may be predetermined in its interpretation, though clearly, with time and degrees of impact, our view changes. Ultimately, we even succumb to pain, which is of certain consolation to the individual concerned.

We ascertain degrees of Being we had previously missed. The sum total of their meaning, aggregated in a sojourn after dusk, leaves no record of footsteps. As sure as Shakespeare, whose own record of having been here can only be divined by that lexicon of meaning, meaningful only to that tribe of disciples whose own lives are passing so quickly as to lend some collective poignancy to the great passing of lives. All the burial grounds, where any such meaning commingles in a suffusion of so many silences upon the face of the earth. That is the issue. How does an unconditional non-violence grasp locust eco-dynamics?

---

[3] See "Loss of land-based vertebrates is accelerating, according to Stanford biologist Paul Ehrlich and others," by Lindsay Filgas, Stanford News Service, June 1, 2020; See also, "Global Extinction Rates: Why Do Estimates Vary So Wildly?" by Fred Pearce, YaleEnvironment 360, August 17, 2015, https://e360.yale.edu/features/global_extinction_rates_why_do_estimates_vary_so_wildly, Accessed July 22, 2020.

**Fig. 87.3**  Jane Gray Morrison, Mac, Josie and Stanley. (Photo © M.C.Tobias)

Amid a profusion of displaced souls, inaccessible lives, bastions of exile, to forgive is to be attuned at the individual level. We don't easily forgive, much less whole populations, systems, civilizations. The freedom any forgiveness tenders is the basis for beginning a journey anew, in a new land[4]. All of the contradictories are behind us, at the instant of easily letting go, whether during birth, or life, or death. The act of forgiveness is an ecological baseline of some certainty. It complements a revolution in aftermath ontologies, finely delineating between individuals and species, demolishing the notion that the survival of individuals, the genetic pertinacity of their populations, and the combining effect of numbers into the general conception of a species, represents a clear picture of what is actually happening within and between life-forms. In that revised context, prayer, forgiveness, love, and faith take on stunningly biological definitions that lend to the individual and his/her "neutral territory" whole worlds of separate, noble identity and behavior (Fig. 87.3).

---

[4] Writes Paul Chi Hun Kim, of Coleridge's Mariner, "His awareness of the I-Thou relationship enabled the Mariner to forgive himself and to re-establish his broken relationship with the community, nature and God…" See "Ecotheology and the Idea of Forgiveness in the Rime of the Ancient Mariner," by Paul Chi Hun Kim, Literature Compass, https://doi.org/10.1111/lic3.12130, 04 February 2014, Wiley Online Library, , Accessed July 22, 2020.

# Chapter 88
# Rebirth

## 88.1 Mortality and Cartography

Samsaric contexts do not obviate the hurdles toward coping with such adversities as the word "No," or the predatory mob, or even the annoyed axon. Suffering may be construed as the clear trigger by which rebirth in some guise is activated. But protracted suffering offers no reliable guide to the facts surrounding, and inherent to all pain, regardless of the promise, uncertain whether or not it will follow, of release, before or after those final minutes. Pain is more paradoxical than death in that it invites every fear and contemplation, without surcease, in the grip of an inevitable death that will come, sooner or later, and only makes sense from a perspective outside ourselves.

Even for the "immortal" monoclonal grove of aspens, the 50 square miles of fungus in Oregon, the mystifying mycologies of Bialowieza or Wisconsin's remote forest floors, the cell lines of certain jellyfish, all their durability is a matter of precise timing. On its cusp, by all accounts, death is either terrible or quietly resolving into peace. But the precision, as an ecological fact, lends exquisite closure on that which we cannot know. A concatenation of elusive truths surrounds it. For those wolves in the greater Yellowstone ecosystem, whose lifespans are approximately five years, or the adult mayfly, one day, our grasp of their lives is only possible through the committed effort to emanate kindness with no expectations. To safeguard their freedom, independence, and very presence in mind.

The genesis of rebirth in human consciousness is thus predicated upon a curious realm of observations throughout Nature. Camus managed to write *La mort heureuse*[1]. That this cognitive bifurcation exists at all is testimony to some framework of cohesion. Attributed to the heavens but bound up with all the gravity and somber ritual surrounding the entombment of mortality. There is no greater paradox than that we should live, only to die. Then live again. Gauzes, chemicals, waxes, fire, weeping, relief. No one can fail to have noticed the dead cicada, its eyes locked

---

[1] Éditions Gallimard, Paris, 1971, *A Happy Death.*

© Springer Nature Switzerland AG 2021
M. C. Tobias, J. G. Morrison, *On the Nature of Ecological Paradox*,
https://doi.org/10.1007/978-3-030-64526-7_88

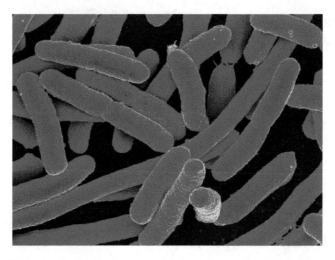

**Fig. 88.1** Colorized scanning electron micrograph of Escherichia coli, grown in culture and adhered to a cover slip. National Institute of Allergy and Infectious Diseases (NIAID). (Public Domain)

solidly open. Other diffusely scattered fragments across every biome of sheered antennae and brains, exploded thoracic cavities, fallen trees eaten clear through. The decomposing layers of death are notice boards and magnets for every angle of orientation. Such bundles of clustering death herald a contagion of theories, compressed and as rife as bacteria or worm-like, fluid-replete gastrotricha. Their seamless proliferation heralds the theory of karma, of passages and outright itineraries, as the teeth fall out, the body turning green, then red, filling with fluids and gasses. A video of the single-celled *Blepharisma* disappearing before our eyes: "I don't know why this one died but how it dissolves to nothingness just broke my heart. Big or small, life is fragile," says microbiologist James Weiss[2]. Some dormant bacteria may live 250 million years, others, mere minutes, and there is wild speculation that there may well have been some 800 billion bacterial generations prior to the rapid evolution of this last remaining hominin, us[3]. Noting that we are the hosts of at least "100 trillion bacterial cells," with over "1000 bacterial species in the gut" alone, 1–3% of human body weight[4] (Fig. 88.1).

---

[2] See "After Microbes Die, What Happens to Their Corpses?" by Ross Pomeroy, RealClear Science, July 22, 2019, https://www.realclearscience.com/blog/2019/07/22/after_microbes_die_what_happens_to_their_corpses.html, Accessed February 2, 2020.

[3] See discussion at https://www.researchgate.net/post/What_is_the_average_life_span_of_a_bacteria, Accessed February 3, 2010.

[4] See "Current understanding of the human microbiome," by Gilbert, Jack; Blaser, Martin J.; Caporaso, J. Gregory; Jansson, Janet; Lynch, Susan V.; Knight, Rob, 10 April 2018. Nature Medicine. 24 (4): 392–400., doi:10.1038/nm.4517, ISSN 1078-8956, PMC 7043356, PMID 29634682. See also, "Scientists but myth that our bodies have more bacteria than human cells," by Alison Abbott, 08 January 2016, Nature, https://www.nature.com/news/scientists-bust-myth-that-our-bodies-have-more-bacteria-than-human-cells-1.19136, Accessed July 22, 2020.

Throughout the world and time, allegorical journeys prompt the imagination of every individual to prepare. Most wander off and disappear; are subsumed into the greater body of a locale, like that of the "Ediacaran Gaojiashan biota [which] displays soft-tissue preservational styles ranging from pervasive pyritization to carbonaceous compression"[5]. Such are the states to which most life flees, following upon its instantaneous ephemeralization in the guise of an individual. Added all together, the life and death that constitutes biomass (organic, inorganic, and everything else) sustains laughter, humus, hunger, apathy, anxiety, and sleep. But it cannot answer the simplest riddle outside of biological peripheries. Such abiding surrealities—like the distance to GN-z11, the most distant known galaxy at 13.39 billion light years from earth—involve no dogma or theology. It simply *is*, lost in the faintly illuminated haze between poetry and science[6].

## 88.2   To Die For

The same status attends upon each one of us, we amalgams of unknown trillions of entities within ourselves (estimates range between 30 and 100 trillion human cells in each body). Such beingness, always individuated, involves a continuum of mass births and deaths, mass graves and silent exiles, the "discursive formations" posited by Michel Foucault[7] and the unseen crevasse receiving the straggler. Or, in the broader view, Spengler's "decline," Diamond's "collapse," all those pre-Cambrian biodiversity "explosions" followed by mass extinctions. Those ecotones proffering the most abundant swathes of life also harbor the perfect opportunistic pathogens. What was a tropic now gives way to the dust bowl. Southeastern Australia, Central Amazonia, the corals of the Caribbean, much of Finland and Siberia, Haiti, of course—all in dramatic exfoliation.

Consciousness has nothing whatsoever at stake in this perennial battleground of crescendos and decrescendos. Like the seasons of any angiosperm or gymnosperm, ovaries or strobili, all is conditioned by the type of seed and the preconditions, whether moisture, topography, altitude, soil type, co-commensurate species, invasives, and proximity to human communities. Within those varied constraints and freedom, rebirths are as conditional the second and third time, as they

---

[5] See "A unifying model for Neoproterozoic-Palaeozoic exceptional fossil preservation through pyritization and carbonaceous compression," by James D. Schiffbauer, Shuhai Xiao, Yaoping Cai, Adam F. Wallace, Hong Hua, Jerry Hunter, Huifang Xu, Yongbo Peng and Alan J. Kaufman, Nature Communications 5, Article number: 5754 (2014), https://www.nature.com/articles/ncomms6754, December 17, 2014, Accessed February 2, 2020.

[6] See "Revised Estimates for the Number of Human and Bacteria Cells in the Body," by Ron Sender, Shai Fuchs and Ron Milo, PLOS Biology, PMC, NCBI, PLoS Biol. 2016 Aug: 14(8): e1002533, Aug. 19, 2016, doi: 10.1371/journal.pbio.1002533, PMCID: PMC4991899, PMID: 27541692, https://www.ncbi.nlm.nih.gov/pmc/articles/PMC4991899/. Accessed July 22, 2020.

[7] *L'archéologie du savoir*, Éditions Gallimard, Paris, 1969.

ever will be. Since the first time, over 4 billion years ago, that rebirth continues unabated; urgently recycling and regenerating its impulses, ideas, feelings, and material predilections.

## 88.3   Organic Interiors

Whether a biome of choice predisposes one's remains to be reborn as a begonia or baboon has never been confirmed within the empirical confines of consciousness. But then, empiricism does not constitute a protein, an enzyme, a cell, or a molecule. It is a biochemical *process* that has as much claim to metaphysics as to hydrology. And while within its machinations there comes a threshold of belief in facts, it is not, itself, a fact. It has neither earth nor fire; water nor air. What is empiricism other than a generally self-satisfied observation by a Self whose quantifications cease to have meaning at the instant of qualification. Any laboratory is absolutely given to the character and aspirations of the empiricist. Authoritarian space as duly described as any other. A geometrical tyranny by which all coordinates within its planes are pre-conditioned by the bias of the experimenter and his/her tools. Science has built itself up upon the laws of Nature, central dichotomies, pre-disposed speculation, the first description of animacules, rules of energy and mass, Lotka Volterra equations, rates of evapotranspiration, *Re* (Reynold's Number) and fluid velocities, Stoke's Law and ten thousand other conceptions, all emerging from that same room with four walls and a roof, the laboratory. Imagine, by extension, the interior of Mauna Loa, the *biggest mountain* in the world, and what *that* space has accomplished? The boundaries, unseen, undetected, it has broken? Or the nearly 480,000 board feet of lumber, with its teeming populations of microbial communities, of any senior redwood? (Fig. 88.2).

Such a space has its microcosm in the brain of an organism. That brain has windows through which, if we could see, we would take in glimpses of the hollow bones of an albatross that abet its soaring, global circumnavigations; those species that seldom drink water, like the desert tortoise (*Gopherus agassizii*); the long-horned coleoptera (*Derobrachus hoverei*) whose adulthood in the summer months is all about mating, not food. The shedding of skin, a multitude of metamorphoses, each phase of an embryology, as determined by our making up of laws and rules and phenomena, are like ordinary and partial differential equations, a calculus encompassing the likes of such mathematicians as Bonaventura Cavalieri (infinitesimals), Leonhard Euler (topology), Joseph-Louis Lagrange (analytical mechanics), Jacob Bernoulli (the law of large numbers), Isaac Newton (fluxions), and Gottfried Wilhelm Leibniz (higher derivatives). Little wonder that the plural form of calculus in Latin refers to "small pebbles": so myriad are the ingredients of our perennial curiosity.

**Fig. 88.2** Dense Dipterocarp forests of Ulu Temburong National Park, Brunei. (Photo ©
M.C.Tobias)

# Chapter 89
# The Cycle of Alterities

## 89.1 Ecosystemic Potentialities

The idea of multiple, indeed endless, deaths cannot be dismissed, given the biological rudiments of rebirth, connoting Otherness. But in light of the brief tenure of adult human cells, seven to ten years, the physiological bias against a continuous consciousness (or soul) has largely been divorced from scientific disciplines, and left to spiritual, psychiatric, and ethical speculation. But there are no rational grounds for supposing that consciousness has no role to play in the process of re-incorporation of physical remains into some other systemic, biological process, which is the basis for a sustained biosphere.

That sustenance comports with a multiplicity of identities predisposed to the life force, as conceived in any statistically aligned survey of the living world, where second winds define rebirth and the statistical margins of confidence leave open the latitude for everything and everyone. In keeping with forest, bog, and other plant successions, the biological distribution of simultaneities in Dante, Bosch, Hindu Vaishnavite tradition, with its emphasis on Kalki, the tenth incarnation of Vishnu, and the co-eval creation and destruction of the cosmos by Shiva, all speak to an ecological eschatology. Endless reiterations emphasized in every consideration of the exterior, of externalities and all things outside the dominion of Self, twenty-first-century alterities are augmented by dint of the escalations of so much pressure. Interdependencies are now expressed as the very conflation of Self and Other. Ecosystem stressors augment the logic driving so many of the world's religious emphases upon metempsychosis, punarbhava ("becoming again"), hulul (Sikh divine transmigrations), and the like[1].

---

[1] See the (retracted) essay, "The mystery of reincarnation," by Anil Kumar Mysore Nagaraj, Raveesh Bevinahalli Nanjegowda, and S. M. Purushothama, in the Indian Journal of Psychiatry, Indian J Psychiatry. 2013 Jan; 55(Suppl 2): S171–S176, https://doi.org/10.4103/0019-5545.105519, PMCID: PMC3705678, PMID: 23858250, https://www.ncbi.nlm.nih.gov/pmc/articles/PMC3705678/#:~:text=Neither%20there%20is%20strong%20objective,is%20nothing%20much%20to%20conclude. Accessed July 23, 2020.

© Springer Nature Switzerland AG 2021
M. C. Tobias, J. G. Morrison, *On the Nature of Ecological Paradox*,
https://doi.org/10.1007/978-3-030-64526-7_89

**Fig. 89.1** The Abundant biological transitions in "Deadwood," Białowieża National Park, Poland-Belarus. (Photo © M.C.Tobias)

From that pivotal nexus of connected life forms and thought emerges targeted consciousness whose ever emergent modes of being bear witness to every layer of evolution and to its ontological offspring among all affected species. That continuous Being within whose self-reflections, all semiotic transmission is accorded a community of Others is not indifferent to subatomic patterns, both local and non-local drift, variations in their genetic frequencies varying across all boundaries of thought. Unstoppable emancipations beholden to no one individual. Of course, we can't *know* this because knowledge herself is caught within that perpetuity, as William Butler Yeats intimated when he declared, "How can we know the dancer from the dance?"[2].

Such consciousness is neither the ocean nor the grain of sand. We cannot trace it in the manner, say, of an anadromous salmon angling upstream to its birthplace, there to mate and ceaselessly create in its own image. The circuitry of moving, biochemical parts, like cations and anions, or the unrehearsed dissolved particles in xylem and phloem, sieve-tube chains, tracheids and plasmodesmata, all of the gravitationally oriented organelles and vessel elements that comprise the lignited tissues of nearly 391,000+[3] plant species throughout the world, can be described as altruistic or self-motivated. We cannot know, not through our own lens, nor those we improvise (Fig. 89.1).

Rather than untangling this multidimensionality, one Other is more apt to revel in the heterophenomenology of yet Another, and another still… until the endless continued fraction (such as $\pi$) instills the loyalty whose common denominator is that ethos which precedes our concepts of biodiversity. We have never found the

[2] From Yeats' poem, "Among School Children," Stanza VIII, line 8, *The Collected Poems of W. B. Yeats,* Definitive Edition, with the Author's Final Revisions, The Macmillan Company, 17th Printing, New York, 1970, p.214.

[3] "State of the World's Plants 2017," Royal Botanic Gardens Kew, https://stateoftheworldsplants.org/2017/report/SOTWP_2017.pdf.

word nor managed to enunciate her underlying reality, though she is everywhere: hovering in our hydrogen selves, governing all eco-dynamics, seamlessly hybridizing individuals and populations, no matter the species or time frame. Her alterity is our identity and the similitudes between us redefine the entire scope of ecosystemic potentialities.

## 89.2   The Map of Biological Patterns

Solo-dimensionality, or the superego, commandeers the mystery of digits, decimals, fractions, rational and irrational, prime and peripheral numbers into the interface of human suasions and collisions. This map of biological patterns explains most of human history. We know consciousness, from its interpolated first stirrings of rDNA and mRNA, to be a non-linear continuum on perhaps as many as 40 billion earth-sized, life-supporting planets[4]. Every metric that might be applied to its emergence or dissipation has hovered around the human bias of cogency and purpose, whereas everything in-between that which we have analyzed has utterly eluded our purview, much like black body thermal energy. We cannot safely assume that the invisible, the unknown, is adapting to new conditions because our apperception is conditional only upon that primary perception whose occupant, a thinker, is, in turn, conditioned upon conditions he/ she cannot control, or apprehend. The brain does. The mind is. We do not see ourselves, and hence the astonishment of Rembrandt's "The Anatomy Lesson of Dr. Nicolaes Tulp" (1632), following upon dissection of animals by Claudius Galen (129–199), and even earlier horrid eviscerations of pigs and monkeys in Alexandria[5]. Such specificities are not conditionals of consciousness, any more than "insight," "food," "mental activity," "bare bones," "assistance," "pre-meditation," "road-kill," "paragon," "paradigm," "vastness," "detail," "appropriation," "envision," "genuine," "hill," "body," or "blue," and offer any instruction in the quest toward meaning. Consciousness, in whatever language or mind, has no meaning in and of itself. It is confined strictly to consciousness. It cannot be illustrated, defined, transmitted, or euthanized. Not that we know of. Our consciousness of it is still confined within it.

Nor should it be confused with "communication." The words "fossil," "spider," "dog" have no meaning. And "meaning" has no meaning, as the penultimate ambassador of this conundrum. Add up all the meaningless (by any mathematical rule of thumb) and the dilemma is not rectified. That universe a human aspires to perceive neither exists or does not. We have the tiresome litanies by Parmenides, Aristotle, and Heraclitus, as well as Nagarjuna, Xuanzang, and their disciples, on this issue of what actually constitutes meaning, object, transitive, and intransitive messaging. All such thoughts are weightless, as far we can measure: weightless and without truth-value.

---

[4] "Habitable Exoplanets Catalog". University of Puerto Rico at Arecibo. 2015.
[5] https://www.ncbi.nlm.nih.gov/pmc/articles/PMC3705678/#:~:text=Neither%20there%20is%20 strong%20objective,is%20nothing%20much%20to%20conclude."History of Anatomy?" Body Worlds, https://bodyworlds.com/about/history-of-anatomy/. Accessed February 3, 2020.

## 89.3   Substance and the Insubstantiality Hypothesis

The proportion of meaning to no meaning—substance to insubstantiality—underscores the futility of assigning significance to anything. Meaning, identity, significance—these are each hypothetical summations that remain unresolved until the myriad equations tempted toward closure find their Being in the proverbial random acts of kindness. We are left with the truly existential consciousness, whose natural history, as we have endeavored to illustrate, is ecologically resonant, but without consistent characteristics that would resonate and give us the coordinates needed to engender the pre-genesis of a theorem that can be tested in the real world.

Ultimately, it comes down to hunches; the grand unified theory of a suspicion which holds that all consciousness is subsumed by a veritable dust storm of alternative beings. Hence, the silly nineteenth-century assertion that "Man, the animal, in fact, has worked his way to the headship of the sentient world, and has become the superb animal which he is, in virtue of his success in the struggle for existence"[6]. And this, in spite of Thomas Huxley's (1825–1895) claim elsewhere "that the structural intervals between the various existing modifications of organic beings may be diminished, or even obliterated, if we take into account the long and varied succession of animals and plants which have preceded these now living…"[7].

If, even to the degree of an angstrom or nanosecond, pellucidity is seen to emanate its own red-shift, some vaguest of obtuse demonstrations of incremental change in the universe we perceive, and within the body of Self, then consciousness, dimly conscious of itself, might shift with the shooting stars, exciting a kind of riot that makes for poetry consistent with compassion. Change at even the most indivisible level informs a new category of experience altogether; gradations divined by other gradations. The ripple in the pond effect. But can the ripple ever seize upon the holistic concept/imagery of the pond through which its energy is transmigrated? Do echoes comprehend their source?

An entire sensate classification system of individuals knows gradation as a primacy forever on the verge of feeling about something, someone greater than oneself. Imagine, then, the exponential sensations circling around the surface of the Sun, or across the fabulist death of consciousness, a concept that is incapable of nullifying itself and therefore depends upon the fiction of a concept. By these miniature Big Bangs the mind stands ready and attuned. A mentality of its essence evaporates into a monumental truth, equivalent to a non-material evanescent impermanence that is beautiful and telling. Such is Nature. We cannot know her because knowledge destroys even the most remote idea, its idealized object. Such is her fragility. Warm palms and vanishing vanishing snowflakes: amid a Cosmos of at least 93 billion light years.

Strange attractors, non-periodic, phase-spaced, as they are characterized by mathematicians, repelled at last on the very threshold of contact[8]. Any equation is

[6] Thomas H. Huxley, *Evolution And Ethics And Other Essays*, D. Appleton and Company, New York, 1898, p.51.

[7] Thomas H. Huxley, *Man's Place In Nature And Other Anthropological Essays*, D. Appleton and Company, New York, 1898, p. 157.

[8] See "Chaos & Fractals," by Larry Bradley, 2010, https://www.stsci.edu/~lbradley/seminar/attrac-

**Fig. 89.2**  A hybrid Parakeet freezing and trapped in Downtown Warsaw in winter, moments prior to her liberation. (Photo © M.C.Tobias)

so arbitrary, at every point, number, and fraction along its course, that a perfect metaphor arises graphically. It signals the transcendent incorporeality of all knowledge and its proposed targets. These incommensurable associations form no ascertainable structure that can be analyzed to closure. The picture is forever being rendered increasingly intangible, beyond all borders of interpolation.

Imagine that which cannot be imagined. Plato, Mahavira, Lao Tzu, Spinoza, and Beethoven had no problem doing so. It is more than a word or numbers game, as Wittgenstein concluded in his manuscript, *On Certainty* (OC1..OC676, 1969). It suggests instead a distribution in no particular proportion of known and unknown deaths: a realm where all numbers are without the slightest relevance, like their makers. Where no thought can get in or out. A black hole of abstractions, suppositions, causes, and effects for which inconsequence is the most pervasive affirmation. Nullity may well be complicated by something, anything, that exists, somewhere within the crystalline structureless domain. That *something* is not subject to knowledge. No one will ever name it. There are no antecedents, no designates. It lives, it dies, both, without targeted representation.

Indeed, there is a third thing, neither life nor death which it embodies, through the very action of embodiment. A fiction nurtured by the truth of its fiction.

And so one goes on and on, wondering what is possible, and, by what means, regulated or undiscovered, yet, the human connection to all other life forms allows for the slightest definition or testament. Our self-imposed crisis is the original ecological paradox. The ongoing tragedy may only be repealed upon that heroism of daily life which seeks first contact with the Other: biosemiotic liberation (Fig. 89.2).

---

tors.html, Accessed August 14, 2020.

# Chapter 90
# The Individual and the Circumference

## 90.1 The Map of Exiles

While we are wont to employ the word apperception (Descartes' "apercevoir," Kant's transcendental apperception), consciousness cannot appreciably or more than intuitively be self-reflective, despite the monumental imagery it presents to the bias of so many biological years. That easily persuaded picture of Self supposes an identity, a closed series of identity-dependent groups, and all the arrangements perpetuating the propositions upholding the ego of a species, the differentiation of selves, and the irrefutable defense of one's own mind. Its starkest contrasts and illumination are evinced at the moment of our exercise of unconditional love. That is the instant of true self-recognition by recognizing the lives of Others (Fig. 90.1).

To inch toward an assertation of the Other, of the fact one's mind is not possessed at all, in other words, to fully renounce the involvement of Self from Self for a reason (to better grasp the natural world?), begs the fallacy of thinking about oneself, and thereby facilitating the connection to Others. It is the indulgence of a sphere that would lose the ability voluntarily to posit a circumference. To be freed, cleansed of co-habitation is to release the inferno of perpetual insistence on the origin of insistence from dependent origination or palingenesis; to co-habit with nothingness in a coyly spirited effort to re-ignite the conversation. Such hopes are unconcerned by the delusion of certainty. Just as Jean-Paul Sartre had written, "Quality is nothing other than the being of the *this* when it is considered apart from the external relation with the world or with other *thises*"[1] (Fig. 90.2).

Not only awkwardly declared and translated, but, consistent with much of Sartre's inscrutably tied-in-knots nausea, a complicated thought, pitting qualia

---

[1] Second on "Transcendence," *Being and Nothingness: an Essay on Phenomenological Ontology*, Translated and with an Introduction by Hazel E. Barnes, Philosophical Library, New York, 1956, p.186.

© Springer Nature Switzerland AG 2021
M. C. Tobias, J. G. Morrison, *On the Nature of Ecological Paradox*,
https://doi.org/10.1007/978-3-030-64526-7_90

**Fig. 90.1** Torture endured by a Burro in India, symptomatic of the condition of many of the 40 million+ donkeys in the world. (Photo © M.C.Tobias)

**Fig. 90.2** Other "thises," Wet Market, The Middle East. (Photo © M.C.Tobias)

against object, subject bound to non-subject. Similarly, our assertion of the Otherness is not about propriety. It is at the heart of what cannot be said, or thought, but is all about both common sense, intuition, and the fantastic beasts that surround all consciousness, no matter what we think or perceive.

At the Buddhist energy core (irrespective of school or tradition) this "heart" is most generously revealed by delineations between metta and vipassana meditation, concentration and insight, as if to declare that such "practices" yield a meaningful mindfulness. Dharma (nature) explodes upon consciousness like some lineup of political candidates in a caucus, the brain lending itself to an arena of competition at the expense of the awesome truth of an agonizing solipsism. You catch all of its signs at once. Not merely in our approach to aardvarks, lichen, or monkeys of the Pithecia genus, or in our mass shootings, but in the everyday wholesale business of capitalism.

Like all systems that apotheosize money and an anthropocentric, human dominated continuum, the demographic trends suggest that the air pollution in Belgrade is more likely to name the future than a hammock and bathing stream. While consciousness easily configures and elaborates upon this problem of opposite scenarios abiding simultaneously in the mind, amphetamizing fears, falling back upon calm, it just as readily bypasses any everyday turmoil that would and should come of it, rejecting pessimism on principle, refusing to disseminate by word of mouth or other media the very idea of negative interest, combat, injury, terrorism, fear of the future. This anti-intellectualism, as it has variously been characterized generation after generation, builds upon the Utopian superego paradigm: That we breathe, dream, eat, and live with no problems, at the zenith of Creation.

## 90.2   That Exile Embodied Consciousness

The most persuasive underlying argument among many to explain this exile of consciousness is the reality that we, our species, our taxonomic anomaly is somehow no longer a true member of any ecological community, having escaped the unruly, ugly, grim reminders of an all too brutal biology. We've swept clean the mud floors, purged our bowels, washed our hands, installed air conditioning, blinds, and blinders in every other sense to segregate our body parts and megacity filth from the estimated 32 million rats in New York City or the unknown powers of magnitude defining the number of eggs laid annually by the 4600 Blattodea and 190 Lepismatidae family cockroaches, silverfish, and the like within human communities. These boundless insect populations circulating busily beneath our conurbations are not ordinarily lauded for their biodiversity assets, though they should be (Fig. 90.3).

**Fig. 90.3** Cockroaches,
Superorder, Dictyoptera;
Order, Blattodea. (Photo
Public Domain)

Nor are the billions of "introgressive hybridization" occurrences among fungal pathogens with whom we live[2]. Transgenic fungus manipulated to decimate malaria parasites[3], just one among tens of thousands of chemicals, treatments, and other interventions engineered by humans to sway or eliminate helminths, protozoa, and other ectoparasitic adversaries. The cumulative efforts are an embassy of transgressions against the otherwise undeflected tides of evolution presaging a deliberately insular choice of fate for our kind, commensurate with every insult we have ever handed down. The backfires are initially recorded in the number of antibiotic-resistant germs. Then it becomes ecologically conceptual.

---

[2] See "The Role of Hybridization in the Evolution and Emergence of New Fungal Plant Pathogens," by E. H. Stukenbrock, Phytopathology. 2016 Feb;106(2):104–12. doi: https://doi.org/10.1094/PHYTO-08-15-0184-RVW. Epub 2016 Jan 29, https://www.ncbi.nlm.nih.gov/pubmed/26824768, Accessed February 4, 2020.

[3] "Development of transgenic fungi that kill human malaria parasites in mosquitoes," by Fang W. Vega-Rodríguez J, Ghosh AK, Jacobs-Lorena M, Kang A, St Leger RJ., Science. 2011 Feb 25;331(6020):1074–7. doi: https://doi.org/10.1126/science.1199115, https://pubmed.ncbi.nlm.nih.gov/21350178-development-of-transgenic-fungi-that-kill-human-malaria-parasites-in-mosquitoes/, Accessed February 4, 2020.

A species in exile. What does it mean, and how could consciousness in humans ever come to reconcile the delusion of Self with all that the Outsider connotes? There are, for now, no other worlds upon which to test this virgin solitaire, and the lunar landings have only underscored the pitiful irony of our desire for reckless exclusivity here on earth. The farmer who trudges through rice paddies; the animal shelter manager working to adopt out charges; the follower of a Humboldt, Gesner, or Wordsworth exploring the world with all those qualities built up in certain individuals that, while not in dispute, are exceptions to the species, hold in common a different variety of empiricism that cannot be divined on any map of most human comings and goings. This is where science and subjectivity quietly mold the ineffable emotions that render maxims subject to entirely other laws of personality and volatile orientation.

Any map, statistically translated, misses that global indigenous and agronomist resistance. The global urban population is now roughly 55%, 68% projected by the UN for the year 2050 (assuming, as in all other matters, that there is a human world in 2050)[4].

Such a map, tracking the solitary hominin, its every hostile flash point, collision threshold, correlating its approximately 1443–1260 cubic centimeter of brain-size[5] with every trajectory it has fabricated, is a gravid echo of a thinking system that has neared the finale of a process of disassociation from all others. That metabolic and conceptual continental drift inciting species separation had numerous trigger points: the first massive assaults resulting in the extirpation 10,000–12,000 years ago of other large megafaunal assemblages (the Pleistocene Overkill Hypothesis discussed in Chap. 27)[6]; much earlier data from sites like Chesowanja in Kenya over 1.3 million years ago, and the Swartkrans caves of South Africa 1 million years ago, where butchers gathered around one of the first known outdoor barbecue pits[7]. The onset of controlled fire by early hominins (clearly prevalent by 450,000 BCE) itself altered the acceleration rate of neural fertility, adding thousands of neurons in the brain every day[8] (Fig. 90.4).

---

[4] See U.N. Department of Economic and Social Affairs, May 16, 2018, New York, https://www.un.org/development/desa/en/news/population/2018-revision-of-world-urbanization-prospects.html, Accessed February 4, 2020.

[5] See Cosgrove, KP; Mazure CM; Staley JK (2007). "Evolving Knowledge of Sex Differences in Brain Structure, Function and Chemistry". Biol Psychiat. 62 (8): 847–55. doi:https://doi.org/10.1016/j.biopsych.2007.03.001. PMC 2711771. PMID 17544382.

[6] See "Overkill Hypothesis 1." Museum of Natural History. 2014. Web. 31 Jan. 2016, https://www.amnh.org/science/biodiversity/extinction/Day1/overkill/Bit1.html.

[7] "Early archaeological sites, hominid remains and traces of fire from Chesowanja, Kenya," J. A. J. Gowlett, J. W. K. Harris, D. Walton and B. A. Wood , Nature volume 294, pages125–129(19), 12 November 1981, https://www.nature.com/articles/294125a0, Accessed February 4, 2020.

[8] For counter-arguments pertaining to evolution of early hominins in terms of dietary shifts, body mass, neurons and fire, see "Human Brain Expansion during Evolution Is Independent of Fire Control and Cooking," PMC, National Library of Medicine, Frontiers in Neuroscience, by Alianda M. Cornélio, Ruben E. de Bittencourt-Navarrete, Ricardo de Bittencourt Brum, Claudio M. Queiroz, and Marcos R. Costa, Front Neurosci. 2016: 10: 167. Published online 2016 Apr 25. https://doi.org/10.3389/fnins.2016.00167, PCID: PMCD4842772, PMID: 27199631, https://www.ncbi.nlm.nih.gov/pmc/articles/PMC4842772/. Accessed February 8, 2020.

**Fig. 90.4** Emmanuel
Levinas (1906–1995).
(Photo © Bracha
L. Ettinger)

## 90.3   Levinas and Buber

An atlas of such exiles, those proposed individuals in the front trenches of evolutionary transitions that either die out, shrink by sheer anonymous mutations, or cluster successfully in the build-up of taxa that will survive within a metapopulation,
has a familiar face to it. We see ourselves, at some point within the horizon or circumference. The reflection adds to that speculative fiction which craves an identity,
a circumstance in which to fix its affiliations with other life forms.

We are there prior to our presence, waiting patiently for something to happen.
While Lithuanian/French philosopher Emmanuel Levinas (1906–1995) famously
declared that "Philosophy is Platonic"[9], we would contend that, contrary to this
level of conceptual disassociation, there is a more general way of isolating the
individual populations and the species toward an eventual reunion that benefits

---

[9] See *Alterity and Transcendence*, by Emmanuel Levinas, Translated by Michael B. Smith, from
Preface by Pierre Hayat, Columbia University Press, New York 1999, p. ix.

from the distinct evolutionary trait of erraticism. What Michel Foucault character-ized as an "archaeology of knowledge" steeped in "incompatible propositions"[10].

Every population promotes some individuals who will, themselves, attempt an epistemological re-engineering. That might mean entry into politics, civil service, philanthropy, or any number of greed orbits. The overall ratio of ethically driven individuals versus those who are not, and may never be, is, of course, a topic of great concern and cynicicism, even while ethics itself remains the subject of con-tinuous debate from every aspect of its meaning, role, and considerable challenges. Consciousness, and its transitive movements toward the generally ascribed to realm of the conscience, involves fundamentally heteronomous, outside influences all but granting the reality, or at least inevitability of "Otherness," and of "the Other (Autrui)." This conceptual relation is a crucial component of Levinas' thinking[11]. The importance of this dialectical saturation of basic terms comes down to the inception of ethics, for Levinas, as for the entire history of metaphysics. Most strikingly, as Peperzak analyzes, the Holocaust survivor (Levinas was a prisoner of war near Hannover at the Fallingbostel Camp from 1940 until liberation) managed to see to the core of ethical realization the most fundamental credo, "Thou shalt not kill"[12], which comes with the responsibility "of protecting widows and orphans, the poor, and the stranger, i.e., any other in its nakedness and vulnerability"[13].

Throughout *Alterity and Transcendence* (first published in 1995) Levinas piv-ots upon the I-Thou relationships of Martin Buber, addressing the "you" as a cen-tral facet of the human experience[14]. This "you" dominates the search for "I" and implicitly pins the entire hope (faith) in God on the equation of an infinite reci-procity "starting out from the face of the other man"[15] and in whose visage we may read the stranger, all others, and the potential for "goodness outside all systems"[16]: a virtue underlying all else, despite every malevolence and misfor-tune. Such that it is only by this multiplied qualifier of Others that goodness is absolute, the basic construct, suggests Levinas, of humanity[17]: "the gravity of the love of one's neighbor"[18].

---

[10] See M.Foucault's book by that title, *The Archaeology Of Knowledge & The Discourse On Language*, Michel Foucault, Translated from the French by A. M. Sheridan Smith, Pantheon Books, Random House, A Tavistock Publications Limited, New York 1972, p. 149.

[11] See *To The Other – An Introduction to the Philosophy of Emmanuel Levinas*, by Adriaan Peperzak, Purdue University Press, West Lafayette, Indiana, 1993, pp. 19, 109.

[12] See Levinas, *Difficult Freedom*, translated by S. Hand, Johns Hopkins University Press, Baltimore, 1990, p. 8f.

[13] op.cit., Peperzak, p. 117, f.73.

[14] Emmanuel Levinas, *Alterity and Transcendence*, from the Preface by Pierre Hayat, Columbia University Press, New York, 1999, p. xx.

[15] ibid., p.29.

[16] ibid. p.108.

[17] ibid. p. 127.

[18] ibid., p. 141.

In this sublime connectivity, he shares an overwhelming alliance with Pascal who, Levinas writes, "says that my place in the sun is the archetype and the beginning of the usurpation of the whole earth"[19]. So invested is this "I" that threatens to topple every "You," that there would appear no stopping humanity, short of the ethical trust in God, in some kind of higher, transcendental faith to which our inscrutable contradictions and endless ambiguities can be turned over, and there, resolved or not. Once faith has obliged this God, we seem content to step back and let the world do with us as it will.

---

[19] ibid. p. 179.

# Chapter 91
# Non-Linear Ethics

## 91.1 Regarding the Conjecture of Ethical Irreversibilities

When we attempt to fathom unsolved propositions using the ideal languages of mathematics and physics, anything is possible. "Goodness" may be construed, even proved to be, "Evil" and evil good. There are no ethical irreversibilities in the language of numbers, or even of consciousness. Gene flow, the directions and course of energy, all matter and mind maintain a position of variability that cannot be "solved." Even if the numbers don't add up, concepts are fluid, gaseous, permeable, indifferent to the plus or minus sides of zero.

German physicist Rudolf Clausius' (1822–1888) theory of entropy[1] appeals to the universe of heat, dissipation, and that which tends toward a closed stability, for example, maximum, as in the case of the three Laws of Thermodynamics (as well as the fourth, "zeroth" law pertaining to multiple systems). Clausius' younger contemporary, the German mathematician David Hilbert (1862–1943) inadvertently challenged the fundaments of thermodynamics by postulating insoluble relations which he set forth in his famed presentation to the International Congress of Mathematicians in Paris in 1900, during which he discussed several of what would eventually be known as the "23 Problems of Hilbert." Some would be solved, some not, or not yet. In 1931, Kurt Gödel asserted with his "Incompleteness Theorems" that much of Hilbert would remain forever unsolved, inaccessible according to any human levels of induction or alleged consistency (Fig. 91.1).

But Hilbert insisted that there were solutions to everything that was taken on by the human imagination, independent of presuppositions, or of God[2].

---

[1] Clausius, R., *The Mechanical Theory of Heat – with its Applications to the Steam Engine and to Physical Properties of Bodies*, John van Voorst, London, 1867.

[2] David Hilbert, Die Grundlagen der Mathematik, Hilbert's program, 22C:096, University of Iowa. See Hilbert, David (1902). "Mathematical Problems". Bulletin of the American Mathematical Society. 8 (10): 437–479. https://doi.org/10.1090/S0002-9904-1902-00923-3. Earlier publications (in the original German) appeared in Hilbert, David (1900). "Mathematische Probleme". Göttinger

© Springer Nature Switzerland AG 2021
M. C. Tobias, J. G. Morrison, *On the Nature of Ecological Paradox*,
https://doi.org/10.1007/978-3-030-64526-7_91

**Fig. 91.1** David Hilbert in
1912, (1862–1943). (Photo
Public Domain)

Aside from their modest gravesites, mottos, and enormous impact on math, physics, and engineering, the scintillating minutiae neither add to nor subtract from the
ethical dilemmas that share an interface with numbers, perceived dimensions, and
endless permutations. We struggle in vain to assign numeric values to integral content, an ecologically hopeless situation. For example, when the 508-million-year-
old Burgess Shale fossil deposits in British Columbia, first discovered by
paleontologist Charles Walcott in 1909, began to yield in the early 1960s the true,
heretofore unexpected vastness of Middle Cambrian biodiversity, it became far
more clear that climate change-related flux during the past half-billion years could
help inform our present crisis on earth with numbers, but not corresponding ethical
injunctions. Of course, it is imperative we rapidly and massively reduce global
warming. But we won't. Cast out into a sea of human metapopulations, such a warning to the vast collective is a lost cautionary tale. Its paleontological import has
demonstrated that we may glean, at best, general alarm at the record of our historical pasts, but are crazily emboldened to think that, somehow, we are not altogether
products of those pasts. Our concepts of Self are entirely rogue, commensurate with
this uncanny, cumulative stray dog sensibility of consciousness, concretized into
hostile units that are amassed, weaponized, politically strewn in such crude and
antipathetic ways as to defy the learning tools of which memory and cognition are
purportedly constructed[3].

---

Nachrichten: 253–297. and Hilbert, David (1901). "[no title cited]". Archiv der Mathematik und
Physik. 3. 1: 44–63, 213–237.

[3] See Peter Douglas Ward, *Under a Green Sky; Global Warming, the Mass Extinctions of the Past,
and What They Can Tell Us About Our Future*, Smithsonian Books/Collins, New York,
2007. OCLC 224875122. Lay summary (July–August 2007). by Christopher Cokinos, Orion magazine; See also, Bennett, Drake (January 11, 2009). "Dark green. A scientist argues that the natural
world isn't benevolent and sustaining: it's bent on self-destruction". Boston Globe.

Hence, such *tools* appear useless to the cause of a cavalier survival. All the numbers in the world simply don't add up to our expectations of Self. They can't. As a species we have rejected absolute enthalpy and entropy.

Indeed, Absolute (0) K does not exist. It has approached, but never reached, zero, although researchers continue to make things colder and colder (currently the record being something like "500 nanokelvin or just a few millionths of a degree above absolute zero")[4]. One of David Hilbert's own proofs, Hilbert's Nullstellensatz ("zero-locus-theorem")[5] elucidates "Isbell Dualities" between algebra and geometry[6], systematic equivalencies, "one object living in two catergories"[7].

## 91.2   The Endpoints of Pi

Until essential utility enters into the equations, like a pH of 7 for pure drinking water, or breathable air, we are nowhere near the zero-locus-theorem. But as we approach the end of Pi, for example, the ecological equivalents take on terrifying, ethical, and real-time concern. Examples: The Argentine Base Esperanza, on Antarctica's Western Peninsula, containing the world's largest Adélie Penguin colonies, hit a record 65 degrees in the Antarctic Summer of 2020. Penguins don't do well in temperatures exceeding 46 degrees. Corresponding to such heating trends discernible in the past half-century, "87 percent of glaciers" in that western region "have retreated"[7]. In 2019, at least 100 elephants died from drought in Botswana's Chobe National Park, with no meaningful attempts by authorities to try and save the animals[8]. Just four months later, Botswana officials auctioned off permits for hunters to kill another 70 elephants allegedly to reduce the conflict with farmers and discourage poaching. You kill in order to prevent killing[9]. Such is the fuzzy logic surrounding the zeros at the edge of ecological speculation and delusion.

---

[4] See "The coldest reaction," Harvard University, Science News, by Caitlin McDermott-Murphy, November 28, 2019. See M.-G. Hu, Y. Liu, D. D. Grimes, Y.-W. Lin, A. H. Gheorghe, R. Vexiau, N. Bouloufa-Maafa, O. Dulieu, T. Rosenband, K.-K. Ni. Direct observation of bimolecular reactions of ultracold KRb molecules. Science, 2019; 366 (6469): 1111 https://doi.org/10.1126/science.aay9531.

[5] See, "Nullstellensatz revisited," Rend. Sem. Mat. Univ. Pol. Torino - Vol. 65 (3) (2007) 365–369.

[6] See https://ncatlab.org/nlab/show/Isbell+duality, Accessed February 8, 2020.

[7] See "Isbell Duality For Modules," by Michael Barr, John F. Kennison, and R. Raphael, pdfs.semanticscholar.org, https://pdfs.semanticscholar.org/3bb0/9e73e754c3d8b43cf5008116cab68a3a6741.pdf.

[8] See "Antarctica just hit 65 degrees, its warmest temperature ever recorded," by Matthew Cappucci, The Washington Post, February 7, 2020, https://www.washingtonpost.com/weather/2020/02/07/antarctica-just-hit-65-degrees-its-warmest-temperature-ever-recorded/Accessed February 8, 2020.

[9] https://phys.org/news/2019-10-drought-elephant-deaths-botswana.html, Accessed February 9, 2020.

By 2020, utilizing the avalanche of data from hundreds of sources[10] we can now approximate a likely number of vertebrates killed by humans globally each year: *5 trillion individuals*, not including even (and ever) greater mathematical exponents of invertebrates. As Hegel once wrote, in a very different context, "the infinite itself creates its own essence, through its own absolute activity, for itself", to which one could append a statement by Kant, "all vain wisdom lasts its time, but finally destroys itself, and its highest culture is also the epoch of its decay"[11, 12].

In this way, by all such machinations, the Anthropocene has caught up the history of philosophy in the worst paradox of all: ecocide purporting to seek ethical balance between fundamental logic, common sense, classical pillars of reason, and the search for some ideal that exists only in the languages of art, science, and mathematics, but not in life herself. Our consciousness is not unaware of this inherent contradiction. If all other tools have failed us in one exercise after another to rectify the clarity of our crisis, what is left? Clearly the path to activism heralds an unblemished pathway. Moreover, Isbell Dualities are fundamentally a question of aesthetics. Just take any half-dozen birds, photographed and painted, and consider how their myriad categories are enshrined in simultaneous, multidimensional loci of consciousness. Every speckle and feather reach pi (Figs. 91.2 and 91.3).

---

[10] https://www.bbc.com/news/world-africa-51413420, Accessed February 9, 2020. These include the United Nations Food and Agriculture Organization in Rome, sentientmedia.org, animaclock.org, road kill and animal control data, the World Economic Forum, the Humane Society of the United States, thevegancalculator.com, faunalytics.org, animalmatters.org, awellfedworld.org, adaptt.org, U.S. Fish and Wildlife and IUCN data, fishcount.org.uk, the Department of Agriculture, Animal and Plant Health 'CauseSpec: A Database of Global Terrestrial Vertebrate Cause-Specific Mortality," Jacob E Hill, Travis L Devault, and Jerrold L Belant, PubMed.gov, US National Library of Medicine, PMID: 31403701 DOI: https://doi.org/10.1002/ecy.2865, https://pubmed.ncbi.nlm.nih.gov/31403701-causespec-a-database-of-global-terrestrial-vertebrate-cause-specific-mortality/ Accessed February 9, 2020, and a myriad of actuarial senescence data, including, "Life-history connections to rates of aging in terrestrial vertebrates," by Robert E. Ricklefs, PNAS, Proceedings of the National Academy of Sciences of the United States of America, June 1, 2010 107 (22) 10314–10319; https://doi.org/10.1073pnas.1005862107, https://www.pnas.org/content/107/22/10314, Accessed February 9, 2020.

[11] See *The Philosophy Of Art: Being The Second Part Of Hegel's Aesthetik*, by Wm. M. Bryant, D. Appleton & Co., New York, 1879, p. 100.

[12] See *Kant's Critical Philosophy*, by John P. Mahaffy and John H. Bernard, Macmillan And Co., London, 1889, p. 137.

**Fig. 91.2** "A Parliament of Birds," Engraving by Wenceslas Hollar, from *The Fables of Aesop Paraphras'd in Verse: Adorn'd With Sculpture and Illustrated With Annotations*, The Second Edition, by John Ogilby, Esq., Printer Thomas Roycroft, London, 1668. (Private Collection). (Photo © M.C. Tobias)

**Fig. 91.3** Brandt's Cormorants, Common Murres, Western Gulls, and others at the Farallon Islands Wildlife Refuge, Northern California. (Photo © M.C.Tobias)

# Chapter 92
# A Lost Species

## 92.1 A Romantically Neglected Dualism

Between 1810 and 1813, Alexander von Humboldt commissioned numerous European artists to illustrate his book *Views of the Cordilleras and Monuments of the Indigenous Peoples of the Americas*. The aesthetic component satisfied crucial elements of the inveterate explorer's attempt to "fathom a natural landscape," as Alicia Lubowski-Jahn has written: a means toward conveying "total impression (Totaleindruck), vitality (Lebenskraft), and ordering (la géographie physique)."[1] That *fathoming*, however relentless, hagiographic, and widespread throughout the humanities, hits up against a global incongruity that is not easily resolved. Earlier, we have likened it to an Isbell Duality (named after mathematician, and pioneering category theorist, John Rolfe Isbell, 1930–2005). Artists such as Caspar David Friedrich (1774–1840) were, like Humboldt, still seeking to actuate a zoomorphic connectivity between the inner Soul and biological metamorphosis. Nina Amstutz, in her monograph on the artist, denominated "the core principles of vitalism and Naturphilosphie,"[2] post-Marxist societies, certainly in the rapidly industrializing nations which were producing a framework of scientific principles. These, in adherence predominately to Darwin over Lamarck, sought a unity of variation within evolution. In other words, multiple categories inflecting that simultaneity that transcends duality—a scaffolding, in other words, upon which genetics and molecular

---

[1] See Alicia Lubowski-Jahn's essay, "The Picturesque Atlas: The Landscape Illustrations in Alexander von Humboldt's Views of the Cordilleras and Monuments of the Indigenous Peoples of the Americas," in *Unity of Nature – Alexander von Humboldt and the Americas*, accompanying the exhibition of the same title, curated by Georgia de Havenon and Alicia Lubowski-Jahn, Americas Society Art Gallery, April 29–July 26, 2014, New York; the book published by Americas Society, New York, Editors Georgia de Havenon, Christina De León, Alicia Lubowsk-Jahn and Gabriela Rangel, Kerber Verlag, Bielefeld, Germany, 2014, p.71.

[2] See *Caspar David Friedrich – Nature and the Self*, Yale University Press, New Haven, CT., 2020, p. 210.

© Springer Nature Switzerland AG 2021
M. C. Tobias, J. G. Morrison, *On the Nature of Ecological Paradox*,
https://doi.org/10.1007/978-3-030-64526-7_92

biology, and the eventual New Synthesis, might satisfy the prime engine of change, regardless of any purpose or degree of complexity (e.g., our species). And if not our species, then, at least, our species' behavior. This connotes a major shift in the history of science in terms of who and what we are.

For some, like the earlier referenced Thomas Huxley, complexity in Nature met her match in humanity. For others, though, evolution has no mandate to ordain any one species. It is an open-ended enigma between cause and effect, chance and repetition, singularities and convergence.

As in the writings of the poet Novalis (Georg Philipp Friedrich Freiherr von Hardenberg, 1772–1801) and in the Romantic outpourings of other contemporaries of Friedrich, there was, in the aesthetic totality, particularly landscape painting, a "conceptual conflation of the eye"[3] that, both inside and outside the injunctions and latitude of religion, invited a profound correspondence between the agencies of life and death, such that, in an undated fragment, Friedrich had written, "In order to live forever once, One must give in to death often."[4]

This rallying cry of German Romanticism reaches back to the subjective brilliance of such religious sagas as that most famed depiction of the hunter's vision of the crucified Christ, by Antonio Pisano, known as Pisanello (1395–1455) in his work, "The Vision of St. Eustace" (?c. 1435, The National Gallery, London).[5] Eustace invites a revolution of animal liberations from the dark forested heart of an otherwise fixated and inert objectivity. Albrecht Dürer in his own engraved version of Pisanello's painting ("Vision of St. Eustace," ca. 1501) (Fig. 92.1). Celebrates this divine instant (enshrining the much more ancient Eastern Orthodox Church and its own veneration of St. Eustathios the Great Martyr) and lends it to the history of our dialectical relation to all Others at the very moment that Western Civilization had become fixated upon the capture of life forms for every version of dissection—a prelude to the ramping up of the mass slaughters. As echoed in Edward Young's "Night Thoughts" ("The Complaint: or, Night-thoughts on Life, Death, & Immortality," 1742, 1745): "Fond man, the vision of a moment made / Dream of a dream, and shadow of a shade"[6].

---

[3] Amstutz, p.138.

[4] Amstutz, p.190, in the chapter, "The Change of Seasons."

[5] See "The Vision of St. Eustace," https://www.nationalgallery.org.uk/paintings/pisanello-the-vision-of-saint-eustace, Accessed June 27, 2020.

[6] See *Places of Delight – The Pastoral Landscape*, by Robert Cafritz, Lawrence Gowing and David Rosand, The Phillips Collection in association with the National Gallery of Art, Washington, D. C., Clarkson N. Potter, Inc./Publishers, Distributed by Crown Publishers, Inc., New York 1988, p. 217, cited from David Cecil, *Visionary and Dreamer, Two Poetic Painters: Samuel Palmer and Edward Burne-Jones*, Bollingen Series, no. 15, 1966, Princeton University Press, Princeton, NJ, 1969, p. 10. See also, *Viennese Watercolors of the Nineteenth Century*, by Walter Koschatzky, Harry N. Abrams, Inc., Publishers, New York, 1988; See also, *Salzburg und das Salzkammergut – Die Künstlerische Entwicklung Der Stadt Und Der Landschaft In Bildern Des 19. Jahrhunerts*, by Heinrich Schwarz, Verlag Galerie Welz Salzburg, 1977.

**Fig. 92.1** "The Vision of St. Eustace," from photographs of engravings, etc., of Albrecht Dürer from the Pinnacothek at Munich, 1879. (Private Collection). (Photo © M.C. Tobias)

## 92.2 Looking Out or Looking In?

Consider all of the treehouses, caves, yurts, igloos, tepees, and other shelters, open space, old neighborhoods, and hamlets we have collectively cohabited, and extrapolate from the myriad occupations, if it were possible, the multitudinous ethnographic gleanings from the inside looking out: a sensory and ideological bifurcation point forever separating the individual from the collective; the collective from all other collectivities. And still, as ecologist/geographer Yi-Fu Tuan has written, "yet we say, 'Kublai Khan built Cambaluc,'" meaning, the overpowering political circumstances, patronizing and inviting art, have nothing to do with the actual revelatory inside story of such art. There is a divide, insufferable and distant, as those alienating factors of human metabolism and its conceptual byproducts, as distinguished from the inextricable paradox which we simplistically characterize as a nature/nurture problem.[7] The perception of a paleoglobalism, comprising Ice Ages, lightning strikes, the sound of thundering waves on the farthest beach, the early winds of Winter pounding forest depths, or rain continually crashing down upon our ancestors, conveys the intuitive sense of ecological asymptotes—natural connections that seem to both engage and repel species. We can relate, even though such original experiences can never quite touch their *Kreatur der Anziehung*, like so many tourists bumper-to-bumper in Yellowstone, longing to meet the primeval beast. This asymptotic view of humanity's isolation was further enshrined in the canon of literary expulsions. One recalls that famed image by William Blake designed for William Hayley's 1802 *Ballads Founded on Anecdotes of Animals*, of the Orpheus-like figure whose expressive, conversational nakedness, between the lion and horse, the ram, the snake, the peacock, the eagle, and the rooster, exhibited an unabashed and unambiguous assertion of paradise, not yet lost.[8]

Even clambering homeward, soaked to the bone, to the fox hunter's hut near the base of a bleak and weather-tormented glacier in early century Iceland. There is the human-foundering belief system that inescapably confuses folklore with reality, isolating the human brain from the more likely truths surrounding it. In the case of that hunter, Icelandic anthropology suggests that stories passed down for generations posited a similitude of souls between human and fox. Unequivocally, fox and human shared all the "same feelings and emotions."[9] We have heard similar assertions from among the Inuit in Greenland, just as the Icelandic Sagas are rife with "the betrayal of Gunnlaug," "the death of Grím," "the exile of the Helgi Droplaug's son" ("a hero to the bone" as recounted in the "Droplaugarsona Saga"):[10] a world where "brothers slay one the other for gains sake, and none spareth father or sons in that manslaughter and sibslaying... Whoredoms many, an axeage, a swordage, shields are cloven, a windage, a wolfage, ere the world stoops to doom."[11] And all this in a time where, by the Late Middle Ages, it was noted that Iceland, unlike continental Europe, was ruled by no King, but by law: laws where confirmation and collaborative focus were measured by opposition, its protagonists judged by their moderation or complete barbarism, in various turns.

---

[7] See Tuan, *Passing Strange and Wonderful: Aesthetics, Nature, and Culture*, A Shearwater Book, Island Press, New York 1992, p. 19.

[8] For a superb overview of this cultural moment, see David Perkins' *Romanticism and Animal Rights*, Cambridge University Press, Cambridge, UK, 2003. The cover photo on the book is that of the William Blake image.

[9] See *The Anthropology of Iceland*, Edited by E. Paul Durrenberger and Gísli Pálsson, University Of Iowa Press, Iowa City, 1989, "The Hunter and the Animal," by Haraldur Ólafsson, p. 46.

[10] See *Three Icelandic Sagas*, Translated By Margaret Schlauch and M. H. Scargill, Princeton University Press For the American-Scandinavian Foundation, Princeton, New Jersey, 1950, p.131.

[11] "Gylfi's Mocking," *The Prose or Younger Edda Commonly Ascribed to Snorri Sturluson*, Translated From The Old Norse by George Webbe Dasent, Norstedt And Sons, Stockholm, 1842, p. 78.

## 92.3    Old Sagas, Novel Futures

Endless feuds were played out on battlefields and relationships guided by a reckless counterbalancing of self-interests and the overreaching burden of chieftains. In the "Vápnfirðinga Saga," detailing the murder of a relative of Eric the Red and the subsequent and retaliatory horrors, the location of every farm in northeastern Iceland, where the action occurred, has been long memorized by locals.[12] This facet of Icelandic society reached a most palpable flash point with Nobel Laureate Halldór Laxness' novel *Independent People* (1935) in which the indiscriminate details of the slaughter of a lamb by a desperate farmer's wife in the early twentieth century must strike grief and horror into readers: but no differently than all the bloody battles of the Hundred Years War, commemorated by continuing factions with their own tortured views of the drama in question, or all the other endless confrontations throughout the history of Western and Eastern civilizations, from the horrendous Battle of Cannae, during the Second Punic War (August 2, 216 BC) in which Hannibal's legions defeated the huge army of the Roman Republic to the Battle of Gettysburg (July 1–3, 1863), with its approximately 50,000 casualties.

Such recollections are not easily equipped to inculcate Pierre Teilhard de Chardin, writing in his book, *The Future of Mankind* (1949), the notion of "transhumanizing" oneself. However, more recently, numerous biological futurists have projected mutational inroads, the opportunism of doomsday to forge new evolutionary paradigms (evolution in harsh places, or the logical end of a Moore's Law applied to artificial intelligence [AI], robotics, cyborgs, and other conflations), a point we would underscore as the only foreseeable option, whatever it might mean, for an otherwise lost species.[13] In the case of certain gene drivers, genomic imbalances, "segregation distorters" that natural selection would otherwise have eliminated or suppressed before distortion destroyed the host, persisting for "perhaps hundreds of thousands of years," researchers have posed what they describe as "a major evolutionary paradox."[14]

It is a lose-lose that portends of a mirroring of the Anthropocene, the human genetic fall-out of a species that carries on just long enough to distort and uproot every natural support system in its arsenal, before itself disintegrating. No individual, not even the theoretically all-compassionate person, has ever come upon the

---

[12] See *Medieval Iceland – Society, Sagas, and Power*, by Jesse L. Byock, University of California Press, Berkeley, CA, 1988, p.209.

[13] See Jeffrey Kevin McKee's *The Riddled Chain: Chance, Coincidence, and Chaos in Human Evolution*, Rutgers University Press, New Brunswick, New Jersey, 2000; See also, "The Evolution of Transcendence," by Gregory Gorelik, Evolutionary Psychological Science, December 2016, Volume 2, Issue 4, pp. 287–307, Springer Link, https://link.springer.com/article/10.1007/s40806-016-0059-3, Accessed February 9, 2020.

[14] See "Ancient gene drives: an evolutionary paradox," T. A. R. Price, R. Verspoor, and N. Wedell, Published: 18 December 2019, https://doi.org/10.1098/rspb.2019.2267, Proceedings of the Royal Society B, https://royalsocietypublishing.org/doi/10.1098/rspb.2019.2267, Accessed February 9, 2020.

edge of two worlds at war, without a clue how to resolve the fatal discrepancies. One may hypothesize some heretofore undefined distribution load of conscious orientation consolidated into a human entity driven solely by ethical considerations. The Digambara monk? The deceased saint? Other legacies that live on? The idea that human history can help us fix the future has proved a mathematical non-starter. As a genetic narrative, it has no more conclusiveness than a theoretical history of all antibodies. We have no idea how to meaningfully correlate our genes, cells, and bacteria with environmental conditions over time.

Our species, as it painted lost worlds at the end of the upper Paleolithic period of the Stone Age (50,000–10,000 BCE), depicted its own separation from previously related constellations of other sentient beings (Fig. 92.2). It is as if *Homo sapiens* had become trapped upon one or other rim of a great canyon whose walls had separated over geological time. It happened with the break-up of Gondwanaland, of mountain building coming to whole continents, fracturing biosemiosympatric communities into exclusive domains where previous collectives had known only participation with one another. Unable to reverse the irreversible, incapable of further hominin hybridizing, no cross-bred humanity would ever likely again emerge. Even as we barrel toward our own extinction, we have no logical system or ascertainable methodology by which individual persons might escape the hominin pack, abetted by genetic squeezes theoretically capable of exciting novel biological responses to stress.

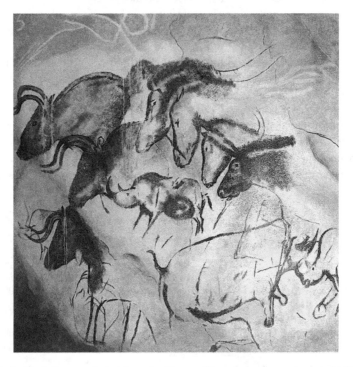

**Fig. 92.2** Replica of cave painting from the Chauvet Cave, dating approximately 31,000 years, late Aurignacian, in the Anthropos Museum, Brno, Czech Republic. (Photo by HTO)

# Chapter 93
# Ecological Idealism

## 93.1 Constructivist Lives and Deaths

At the conclusion of his book *The Social Conquest of Earth*, E.O. Wilson writes in a chapter entitled "A New Enlightenment" that, guided by his "blind faith" the "Earth, by the twenty second century, can be turned, if we so wish, into a permanent paradise for human beings"[1]. We saw a similar idealism propounded in Paul and Anne Ehrlich's 1987 book, *Earth*, in which, and subsequent to such chapters as "The Shadow of Humanity," "The Costs of Numbers" (figuring an image of a starving Sudanese child), "Fouling the Nest," "Disrupting the Biosphere," and later, "The Spirit of Folly," their Chapter Nine is titled, "The News Is Not All Bad," as they indicate that "There are some hopeful signs."[2] In their far more devastating work, *The Annihilation of Nature – Human Extinction of Birds and Mammals* (2015), the Ehrlichs, and co-author Gerardo Cerballos, concluded their soul-dispiriting litany of extinction events, with the added news of 28 years or so of ponderous devastations, offering up "a grim account" throughout the book, though concluding with a final section, "Toward a Better Future"[3] (Fig. 93.1). Even as early as 1962, following upon her own intense diagnoses of "Rivers of Death" and "The Human Price," Rachel Carson elaborated upon "The Other Road," declaring "Only by taking account of such life forces and by cautiously seeking to guide them into channels favorable to ourselves can we hope to achieve a reasonable accommodation between the insect hordes and ourselves."[4]

---

[1] *The Social Conquest of Earth*, by E. O. Wilson, Liveright Publishing Corporation, A Division of W. W. Norton & Company, New York, 2012, p. 297.

[2] *Earth*, by Anne H. Ehrlich and Paul R. Ehrlich, Franklin Watts, New York, 1987, p. 205.

[3] *The Annihilation of Nature – Human Extinction of Birds and Mammals*, by Gerardo Ceballos, Anne H. Ehrlich, and Paul R Ehrlich, Johns Hopkins University Press, Baltimore, Maryland, 2015, p. 178.

[4] *Silent Spring*, by Rachel Carson, Houghton Mifflin Company, Boston, 1962, p. 296.

© Springer Nature Switzerland AG 2021
M. C. Tobias, J. G. Morrison, *On the Nature of Ecological Paradox*,
https://doi.org/10.1007/978-3-030-64526-7_93

**Fig. 93.1**   Drs. Paul R. Ehrlich and Anne H. Ehrlich at the Rocky Mountain Biological Laboratory, Gothic Colorado. (Photo © M.C. Tobias)

American bacteriologist Hans Zinsser (1878–1940) championed a more thorough and unstintingly morose picture of the life of this planet, with a famed and exhilarating prose style perhaps best captured in his great classic *Rats, Lice and History*—being a study in biography, which, after 12 preliminary chapters indispensable for the preparation of the lay reader, deals with the life history of typhus fever: "Man sees it from his own prejudiced point of view; but clams, oysters, insects, fish, flowers, tobacco, potatoes, tomatoes, fruit, shrubs, trees, have their own varieties of smallpox, measles, cancer, or tuberculosis. Incessantly, the pitiless war goes on, without quarter or armistice – a nationalism of species against species."[5]

The man who had developed a typhus vaccine ultimately succumbed to that universal *war of the worlds*, in the form of acute leukemia. It is fascinating to note that Zinsser, a poet/immunologist of the deepest sympathies with life, was laid to rest in America's most alluring cemetery, Sleepy Hollow in Mount Pleasant, about 30 miles north of New York City, adjoining the Rockefeller cemetery and Old Dutch Church. Among the 45,000 or so other humans interred in those 90 acres is a veritable monument to the human species itself; a Who's Who of that stellar constellation of passions, forebodings, energies, and mirrors of the earth that has received them all: Washington Irving, Brooke and Vincent Astor, Andrew Carnegie and his wife Louise Whitfield Carnegie, William Rockefeller, Walter Chrysler, Hudson

---

[5] First published in 1934; this being the Black Dog & Leventhal Publishers, Inc., edition, New York, in arrangement with Little, Brown and Company, Inc., 1996, p.7.

River School luminary Jasper Francis Cropsey, American Federation of Labor founder Samuel Gompers, Audubon's chief engraver Robert Havell, Jr., a son and daughter of Alexander Hamilton, and Central Park Commissioner Moses Grinnell, to name just a few of the lucky ones who lie quietly and shaded, seemingly forever.

Like other charismatic cemeteries, Père-Lachaise in Paris, Cimitero Monumentale in Milan, Sleepy Hollow has its share of flamboyant tours ("murder and mayhem" by lantern light, etc.).[6] Not the case with more solemn idylls of eternal return, monuments to living sparks, like the old Jewish cemetery in Prague from the fifteenth century, or that of the Mount of Olives in Jerusalem, dating back at least 3000 years. But more provocatively, the Sleepy Hollow concedes an intrinsically peaceful dénouement, a vision of that final union of art, nature, and death which has so wooed the living for so many tens of thousands of years from entropic life into a state of metabolic, ecological liberation: constructivist lives and deaths.

Washington Irving himself had written, "The grave should be surrounded with everything that might ensure tenderness and veneration"—a concept inherent to the largest cemetery in the world, Wadi-us-Salaam, Valley of Peace, in Najaf, Iraq (1485.5 acres). In his ode to "Cemeteries of the Future: Permanent, Unpolluted, Inviolate," W. Robinson intoned a heaven of the "best climbing roses…sweetbriar, or honeysuckle." And went on to advise how "Our old city churchyards could all be easily converted into oases of trees" and every conceivable embellishment of life presiding in tune with death that might preserve "a national garden in the best sense" "God's Acre Beautiful," Robinson summarized.[7] We are universally reminded of the ineluctable mating of life and death when we observe the festival of colors and pixel-strength dominating forest composition every mid-October from Maroon Bells, Colorado, to all of New England's Eliot Porter or Jackson Pollack-like orgies of hue upon every leaf.

Not surprisingly, those same colors have imbued the varieties of mourning attending upon the rituals and pomp of a funeral, as complexity turns back to its humblest origins, biologies perdurably protected—of the colors in which life is perennially cloaked, declared in rich metaphor. In ancient Greece and Rome, "Black," for "midnight gloom"; "greyish brown – the colour of the earth" for those mourning in Ethiopia; "pale brown – the colour of withered leaves" as befit all those who have sustained loss in Persia; "Sky-blue" draping those en-route to an afterlife in "Syria, Cappadocia, and Armenia"; "Deep-blue in Bokhara"; violet in Turkey; "scarlet" among Renaissance royalty in France; "white" for "white-handed hope" in China; and "Yellow -the sear and yellow leaf" in "Egypt and Burmah."[8] The colors, resplendent, soothing foliage, park-like settings, the long-dissolve into that unfathomable

---

[6] See http://visitsleepyhollow.com/historic-sites/sleepy-hollow-cemetery/. Accessed March 9, 2020.

[7] See *God's Acre Beautiful or the Cemeteries of the Future*, by W. Robinson, The Garden Office, London; Scribner and Welford, New York, 1880, p.28; See also, *Beautiful Death – Art of the Cemetery*, by David Robinson with a text by Dean Koontz, Penguin Studio, Viking Penguin New York, 1996.

[8] See *History of Mourning* by Richard Davey, Jay's, Regent Street, London 1880, p.104.

transition to death, is the core truth-value underpinning our ambivalent approach to life-after-death, and death-within-life, the crucial calculus crushing in upon the Anthropocene and our unique psychological imperatives, social duties, and moral responsibilities thereof: the essence of personal *hope* within global crisis, not hope for death, clearly. We persist in our optimisms, for every fathomable reason.

Psychologists attempt to measure hope by breaking it down into various domains.[9] Brian Albuquerque writes that "subjective well-being (SWB) as defined as 'a person's cognitive and affective evaluations of his or her life.'"[10] But note the lens through which psychology attempts to define well-being: it is the *person's* well-being. Even the most ardent, deep eco-psychologist is ineluctably starting from an anthropogenic centrality, speaking the language of human patients. We all accept the common currency. It's getting beyond that barter system almost entirely oriented to the Self that is the colossal sticking point, even as we attempt to extrapolate the modalities of Self and its well-being into broader contexts—the health of Democracy, of civilization, the staying power of courtesy, integrity, all of the virtues that our metaphors have gathered from our observations over long lives surrounded by birdsong.

## 93.2    The Psychoanalysis of Belief Systems

Confessions of hope betray hidden personal fears and group pressures: homework versus an expanding universe. All those motivational speeches on battlefields throughout history—General Eisenhower preparing troops for the invasion of Normandy, June 6, 1944; President Lincoln speaking to the Ohio Regiment, August 22, 1864; General George Washington to his "rebellious and doubtful army, March 15, 1783"; St. Bernard invoking enthusiasm among his troops for the Second Crusade in 1146; or Pericles urging his fellow Greeks to go to war against the Spartans.[11] Then there is the 28th President Woodrow Wilson presenting The Treaty of Paris to the United States Senate, July 10, 1919, in which he hailed American "spirit," "vision," "the light." "America shall in truth show the way," he declared.[12]

---

[9] See "What Exactly is Hope and How Can You Measure it?" by K. Hanson, October 24, 2009, Positive Psychology, http://positivepsychology.org.uk/hope-theory-snyder-adult-scale/. Accessed February 10, 2020.

[10] Citing Diener, E., Oishi, S., & Lucas, R. E. (2002). "Subjective well-being: The science of happiness and life satisfaction," In C.R. Snyder & S.J. Lopez (Ed.), *Handbook of Positive Psychology*. Oxford and New York: Oxford University Press, p.63; See Brian Albuquerque, "What is Subjective Well-Being: Understanding and Measuring Subjective Well-Being," Positive Psychology, January 16, 2010, http://positivepsychology.org.uk/subjective-well-being/. Accessed February 10, 2020.

[11] See Business Insider, "Team Mighty, We Are The Mighty," June 2, 2015, https://www.businessinsider.com/16-of-the-best-excerpts-from-the-greatest-military-speeches-ever-given-2015-6. Accessed February 10, 2020.

[12] See Address of the President of the United States to the Senate...July 10, 1919, Washington, D.C.: Government Printing Office, 1919. Reprinted as "An Address to the Senate (July 10, 1919)," in Arthur S. Link, ed. et al., *The Papers of Woodrow Wilson*, Vol. 61, June 19–July 25, 1919,

However, 101 years later, America would re-think Wilson, removing his name from Princeton's School of Public and International Affairs in the wake of searing memories of his racist character, and part of that long-overdue movement to rectify (retranslate) glossed-over darkness throughout American history.

George Steiner, in his monumental book *After Babel – Aspects of Language and Translation* (1975), provides some insight into the subject of adequate or inadequate translation, as he speaks of the "trick of blinding vision across time" and the reality that "Magnification is the subtler form of treason" that "can arise from a variety of motives: through misjudgment or professional obligation." And, indeed, the extent to which those motives are empowered or stymied, as the case may be, the original text might well be apotheosized to what Steiner describes as "an alien elevation."[13] "Hell/Is the fulfillment which stifles their desire/By granting it" wrote the poet Charles Tomlinson (1927–2015) in his poem titled "Last Judgement."[14] The same poet who announces his work by quoting from Jean François Millet, "I want the cries of my geese to echo in space."

Behavioral ecology may be dynamic but it is also fickle.[15] One might assume there is nothing we can do about it.[16] Others will disagree. And that's that. A "Clean, Green New Zealand"?[17] Dark sides to that country's history, notwithstanding: cruel predator control, or the case of the indigenous Moriori on the Chatham Islands of New Zealand, at least one in six of whom were enslaved by the conquering Maori.[18]

---

Princeton, NJ: Princeton University Press, 1989. 426–436; online at: History Matters, http://historymatters.gmu.edu/d/4979/. Accessed February 10, 2020.

[13] George Steiner, *After Babel – Aspects of Language and Translation*, Oxford University Press, New York and London, 1975, p.401.

[14] *A Peopled Landscape – Poems* (Oxford University Press, London, 1963, p. 41).

[15] See "Ecology: Boom and bust," by J. Zaanen, November 4, 2009, Nature 462, 15(2009), Ecol. Lett. https://doi.org/10.1111/j.1461-0248.2009.01391.x(2009), https://www.nature.com/articles/462015d. Accessed February 10, 2020. The research pertains to "eight years of measurements from a Baltic Sea plankton community…From amid the chaos of thousands of population measurements, they were able to discern for the time in real life two coupled predator-prey cycles oscillating out of sync."

[16] See HistoryExtra, from BBC History Revealed Magazine, December 20, 2014, https://www.historyextra.com/period/victorian/why-we-say-bah-humbug-christmas-carol-scrooge-dickenswhat-mean/. Accessed February 10, 2020.

[17] See a full analysis of this complex obfuscation in *God's Country: The New Zealand Factor*, by M.C. Tobias and J.G. Morrison, Zorba Press, A Dancing Star Foundation Book, 2011.

[18] See *Our Islands, Our Selves – A History of Conservation in New Zealand*, by David Young, University Otago Press, Dunedin, New Zealand, 2004, p. 233; See also Paul Moon's *This Horrid Practice*, Penguin Group, New Zealand 2008; See also, Tahana, Yvonne (12 July 2008). "Cannibalism had little to do with consuming enemies' mana, says historian". The New Zealand Herald. See also John Byron, Samuel Wallis, Philip Carteret, James Cook, Joseph Banks (1785, 3rd ed.). An account of the voyages undertaken by the order of His present Majesty for making discoveries in the southern hemisphere, vol. 3 (London) p. 295. In addition, see "The Sad Story of the Moriori, Who Learned to Live at the Edge of the World," by Natasha Frost, Atlas Obscura, March 6, 2018, https://www.atlasobscura.com/articles/moriori-people-genocide-history-chathamislands. Accessed February 11, 2020; and Michael King's *Moriori: A People Rediscovered*, Reed Books Pty Ltd., Wellington, NZ, 1994.

More recently, in the context of New Zealand adopting Bhutan's vision of Gross National Happiness as a viable index to replace conventional gross domestic product (GDP), New Zealand's Finance Minister Grant Robertson declared, "How could we be a rockstar... with homelessness, child poverty and inequality on the rise?"[19]

In grappling with biological terms and definitions, how we might *remedy* human-induced imbalances and destruction, debate over every detail is argued ad nauseum. In philosophy, previously unimagined crevasses sunder even the slightest cohesion.[20] Ethical purgatory tracks biological data deficiency even at the level of forensic analysis of an aftermath, such as crab, shrimp, and amphibian mutations, hybrid invasives, an exploded vault of nullities following an ecological disaster.[21]

Never has objectivity applied to who people are or how they behave. Objectivity strips natural history of any potential moral injunctions because, by definition, it ignores our complexity, and the power of our mood swings, invoking the false neutral. In representing such conceptual contradictions, our very models—mathematical, linguistic, "multi-criteria analysis" driven—waver between the abstract and a set of probability targets, as we endlessly search for increasingly ingenious ways to test the great experiment which is the idea of reality, and the even more bizarre verisimilitude of ourselves—a liable fact made even more vulnerable by that school of thought which holds all consciousness to be an illusion. Of course, illusion, delusion, every eccentric misapprehension nonetheless involves a fundamental act of the mind, not the pineal gland, which some, like Descartes, imagined to be the "seat of the soul."

Human perception of the solid ground of a biological planet commands and controls through our presumption of human consciousness—whatever it is—making its way onto the world stage by our constantly inventing tools meant to design, implement, and ultimately resuscitate that which has been wounded. Pain, and the distractions we enlist to escape it, is the fullest extent of the world with which our minds have co-evolved. While many are quick to reclaim our earliest co-habitation with canids, felids, crickets, and apple trees, the evolution of those Beings remains independent of our own. Collectively, as paleontology and genetics have rapidly attested, we snuck up on the world, one tribe at a time, with varying, sometimes merging, infra-specifics and super-spreaders—until our numbers surpassed 1 million, then 100 million, and so on. We've lost track of other mammalian guidelines, baseline data that would give us some numerical understanding of comparable carrying

---

[19] See "New Zealand Ditches GDP For Happiness And Wellbeing," by James Ellsmoor, Forbes OnLine, July 11, 2019, https://www.forbes.com/sites/jamesellsmoor/2019/07/11/new-zealand-ditches-gdp-for-happiness-and-wellbeing/#6ac6d5501942. Accessed February 11, 2020.

[20] See Odenbaugh, Jay, "Conservation Biology", The Stanford Encyclopedia of Philosophy (Fall 2019 Edition), Edward N. Zalta (ed.), forthcoming URL = https://plato.stanford.edu/archives/fall2019/entries/conservation-biology/.

[21] "Persistent and substantial impacts of the Deepwater Horizon oil spill on deep-sea megafauna," Craig R. McClain, Clifton Nunnally and Mark C. Benfield, Published: 28 August 2019, https://doi.org/10.1098/rsos.191164, The Royal Society Publishing, https://royalsocietypublishing.org/doi/full/10.1098/rsos.191164. Accessed September 16, 2019.

capacities.[22] How many flies can the world sustainably contain? Every conceivable question involving the integrity, extent, and plausible proposition inherent to the very idea of consciousness meets its implacable Self in any Other. Our doubts, ideals, and earnest insistence are, at every juncture, made whole by the ecological context in which consciousness operates. We test the veracity of humanity against the life and death of other species. Beyond that biological bottom-line, there is only one guiding principle that equates with all the truths our species has generated: the amelioration of suffering.

In fashioning stories handed down between generations, the legerdemain of literary exposition has enabled us to pose for all the selfies in the world before a picture of annihilation; to pretend that we inhabit the requisite psychological profile and dexterity to at once describe and, for the time being, elude our own ephemerality. As we verge nearer and nearer, location by location, to the most inconvenient of all scenarios, the attenuation of pain, Hippocratically endorsed generation after generation, comes into precise focus.

## 93.3    The Geopolitics of Ecological Resolution

A complete and comprehensive set of legislative fiats oriented to the surcease of the exploitation of any other species, as well as our own, runs the gamut from deep bioethics to politics. De-escalation of ecological paradox by definition requires the embrace of a pellucid idealism almost entirely out of fashion (Fig. 93.2). Not a lament for lost causes, but their celebration. For example:

- Abolition of all animal products.
- Universal, free education through graduate school and/or any guild-like system students desire.
- Metal, plastic, synthetic-chemical, greenhouse gasses, and radiation free societies.
- An eco-credit world banking system.
- A re-wilding of every prolific habitat worldwide, not just 50%: In the United States, this would encompass rendering as much as 97% of America, which is still rural (though inhabited by over 60 million people in at least 2323 counties), wild, or semi-wild.[23]
- The complete opening of all borders to all life forms, an end to any so-called citizen requirements, and, hence, an end to choosing one species over another.

[22] See "Whale Numbers – An Uncertain Science," from "The conservation of whales in the 21st century," New Zealand Department of Conservation, https://www.doc.govt.nz/about-us/science-publications/conservation-publications/native-animals/marine-mammals/conservation-of-whales-in-the-21st-century/whales/whale-numbers-an-uncertain-science/. Accessed September 15, 2019.

[23] See "New Census Data Show Differences Between Urban and Rural Populations," December 08, 2016, Release Number CB16-210. United States Census Bureau, Newsroom, https://www.census.gov/newsroom/press-releases/2016/cb16-210.html. Accessed September 14, 2019.

**Fig. 93.2** "Quintum creationis opus: of the birds and fishes," Jan Sadeler 1's (1550–1600) engraving, ca. 1585. (Private Collection). (Photo © M.C. Tobias)

- Full national support, policies to incentivize adoptions of orphans worldwide, and zero population growth to achieve sustainable stabilization.[24]
- Universal free healthcare that includes all pain killers, family planning services, and a most generous and unstinting freedom in all manners of assisted euthanasia.
- Redistribution of financial wealth and megacities according to a communitarian social democracy favoring habitat protection—commons taxed according to their natural capital appreciation/depreciation—and in so doing, re-thinking the civic obligations toward an ecological civilization that apportions, and demarcates, embracing aggregate, biological fruition and human-assisted efficiencies.[25]

---

[24] The idea here would be to recognize the nightmare outlined in Paul and Anne Ehrlich's *The Population Bomb* (1968) in which he coupled unending logarithmic doubling-times of the human population, arriving in 900 years at "sixty million billion people"; "100 persons for each square yard of the Earth's surface, land and sea" or fitted into "a continuous 2000-story building covering our entire planet." See *The Population Bomb*, 2nd edition, Sierra Club Books, San Francisco, p.14; Ehrlich was citing J. H. Fremlin, "How Many People Can the World Support?" New Scientist, October 29, 1964.

[25] See "A Model for Efficient Aggregation of Resources for Economic Public Goods on the Internet," Volume 1, Issue 1 & 2, January, JEP The Journal of Electronic Publishing, https://doi.org/10.3998/3336451.0001.125, Presented at MIT Workshop on Internet Economics, March 1995; See also studies of the "aggregate consumer demand curve," aggregating microeconomic theories, and econometrics.

- Intergenerational, biosemiotic politics that are inspired by interspecies compacts; animal liberation charter documents as sacred and binding as those covenants and community standards that engendered the democracies in San Marino, Iceland, the Isle of Man, the Iroquois Nation, as well as such traditions as the Digambara Jains, Quakers, Hindu Bishnoi, South Indian Todas, Essenes, and Vajrayana Buddhists.
- An end to military service and the revocation of all guns and all weapons development, as was successfully initiated during Japan's Tokugawa Shogunate during the Imperial Regent Toyotomi Hideyoshi's rule.
- And so on.

Of course, as with each one of the above injunctions, human *natures* (citing Paul Ehrlich's famed book title[26]) invoke the inevitability of a Tragedy of the Commons— like altercation between individuals, regions, and nation-states, regardless of their respective previous signatory status to international treaties, conventions, or other declared policy positions, which change from hour to hour, town hall meeting to meeting. Hence, is ecological paradox so intractable as to render all idealism no more than a quixotry? The idle stare of frustration with too much time on its hands? It is not our intention to sabotage abiding aspiration but, rather, embrace honest faith in its fullest revivification on behalf of all those who remain hopeful, open-minded. Deep reflection embracing a planetary ethic and kinship keeping the fire going on a freezing night are not incommensurate, in our view.

But let us agree that—geopolitically, economically, pragmatically—the issues of governance fundamentally comprise a blind and grasping minimum set of standards (which we have seen the worst of in most current animal-related legislation) in order to seduce the rank and file, moderate voters, energized base, other cliché demographics that are entirely opposite the intuitive, reclusive life that simply finds the crowd, the mob, the vision of shared prosperity—a vacuous lure borne of egalitarian, arrogant speciesism. Does a bipedal gait and opposable thumb allow for a behaviorally different large vertebrate collective? Could our evolution have been different? At this point, only anecdotally so: an anthropological and literary tease which, by the day, grows more and more remote.

A philosophy of retreat, age-old, venerable, taken as a spiritual defense on the edge of the last forest, is the very ecological rallying cry of the individual that is willing to sit down on that forest floor and re-think everything. But we have no theory pretending to believe that at the species-wide level, this is as yet possible. For now, we see only infernal contests over whose property rights restrict access to that same forest floor.

A true politics of peace calls for a thorough assessment of the human fall-out, its unrelenting undermining of ecological stability, and continued depredation of every integral ecosystem and wild and feral inhabitant. Much of that census has been done for charismatic species and habitat. There are a lot of people which translates into

---

[26] *Human Natures – Genes, Cultures, and the Human Prospect*, by Paul R. Ehrlich, Island Press/ Shearwater Books, Washington, DC, Covelo, CA, 2000.

**Fig. 93.3** Burgos Complex wall painting, heretofore Unknown Peoples, Tamaulipas State, Mexico, ca. 4500 BP. (Photo © M.C. Tobias)

many subsequent prospects for research to expand that census so as to embrace what earlier we qualified as the "pain points" (farm animals, vertebrates, and invertebrates throughout the seas) and a concomitant ethical engagement of their liberation, while humanitarian fronts are escalated. But throughout the course of this revolution of ideals, the separation of the self from the Self in a critical path of honest, ecological clarification is the psychological task before us. Not the Great Dying, but, hopefully, steps toward a Great Merging (Fig. 93.3).[27]

---

[27] See, for example, "The ecological and evolutionary implications of merging different types of networks," by Colin Fontaine, Paulo R. Guimarães Jr, Sonia Kéfi, Nicolas Loeuille, Jane Memmott, Wim H. van der Putten, Frank J. F. van Veen, and Elisa Thébault, Ecology Letters, Wiley Online Library, 23 September 2011. https://doi.org/10.1111/j.1461-0248.2011.01688.x, https://onlinelibrary.wiley.com/doi/full/10.1111/j.1461-0248.2011.01688.x. Accessed July 23, 2020.

# Chapter 94
# The Problem of Interdependency

## 94.1 The Challenges of Collaboration

A clear orientation to the natural world around us sets in, within seconds of considering our politicized options as both ideal and praxis, abiding in either case by obvious implications of biological interdependency. A fine case in point: Bambi in a Massachusetts forest. Hardwood swamplands, oaks, pines, hemlocks, other hardwoods, and, therein, large numbers of white-tailed deer (*Odocoileus virginianus*) capable of transmitting the epidemic deer tick infections. Not just the ubiquitous Lyme disease, "the most prevalent infectious disease in Massachusetts…now considered to be a public health crisis… [but other] pathogens which cause Babesiosis, Anaplasmosis, relapsing fever and Powassan virus, all of which can be very serious."[1] In 2018, the state of Massachusetts saw over 67,000 cases of Lyme disease, a "66,000 percent increase… in just 25 years."[2] Throughout America, there were an estimated 380,000 victims in 2015.[3]

The Dog Tick (*Dermacentor variabilis*), Lone Star Tick (*Amblyomma americanum*), and the Blacklegged Tick (*Ixodes scapularis*), as well as at least seven

---

[1] "One Deer Tick Bite Can Change Your Life," May 15, 2018, University of Massachusetts Amherst, The Center for Agriculture, Food and the Environment, https://ag.umass.edu/cafe/news/tick-talk-time-2018, Accessed September 12, 2019.

[2] "Tick Borne-Diseases are Skyrocketing in Eastern Massachusetts. Is the Lack of Hunting Access to Blame?" By Thomas Gerencer, Outdoor Life, September 5, 2019, https://www.outdoorlife.com/tick-borne-diseases-are-on-rise-in-eastern-massachusetts-is-lack-hunting-access-to-blame/, Accessed September 14, 2019.

[3] "Lyme disease: 'Public health time bomb' in the Berkshires environment," by William P. Densmore, April 19, 2018, https://theberkshireedge.com/lyme-disease-public-health-time-bomb-in-the-berkshires-environment/, Accessed September 14, 2019; See also, "Surveillance for Lyme Disease – United States, 2008–2015," *Surveillance Summaries*/November 10, 2017/66(22);1–12, by Amy M. Schwartz, MPH; Alison F. Hinckley, PhD; Paul S. Mead, MD; Sarah A. Hook, MA, Kiersten J. Kugeler, PhD, Morbidity and Mortality Weekly Report (MMWR), CDC, Centers for Disease Control and Prevention, https://www.cdc.gov/mmwr/volumes/66/ss/ss6622a1.htm?s_cid=ss6622a1_w, Accessed September 14, 2019.

© Springer Nature Switzerland AG 2021
M. C. Tobias, J. G. Morrison, *On the Nature of Ecological Paradox*,
https://doi.org/10.1007/978-3-030-64526-7_94

**Fig. 94.1** Sambar deer (*Rusa unicolor*) in a tea plantation in Southern India. (Photo © M.C. Tobias)

additional tick species throughout North America, knowing that Bambi is the vec-
tor-borne agent within a fiercely competitive predatory environment reeling from
other, zoonotic-borne agents, challenges any earlier referenced *politics of peace*.[4]
Aristotle's theory of *intermediates* (somewhere between the physical and the
Platonic Form) implicates the ethical/spiritual conundrum that follows upon every
zoonotic transmission, notably in the case of female mosquitoes of the *Anopheles*
Genus that are so vulnerable to a single-celled parasite, manifested in at least six
*Plasmodium* obligate parasites, causing at least 400,000 human deaths annually.[5]
We've seen such geopolitics and wildlife-human conflicts flare up in every human-
occupied location on earth—from Bhutan to Ladakh, where Buddhism and stray
dogs, for example, have fueled nearly insoluble problems.[6] Gandhi, like Orwell,
misread both Hitler and the crisis for the Jews. And, as with the Jain ethical response
to a potentially lethal snake, Gandhi, while never supporting a *first strike* mentality,
did affirm self-defense for himself and others. Moreover, the large Sambar deer
(*Rusa unicolor*) throughout India, China, and Southeast Asia (by far the largest
human contingent) also transmit tick-borne diseases (Fig. 94.1).[7]

---

[4] See http://publichealth.lacounty.gov/acd/Vector.htm, Accessed September 12, 2019.

[5] "Do all mosquitoes transmit malaria?" World Health Organization, April 2016, https://www.who.
int/features/qa/10/en/, Accessed September 15, 2019.

[6] See "Deadly Predators and Virtuous Buddhists: Dog Population Control and the Politics of Ethics
in Ladakh, by Karine Gagné, Himalaya, The Journal of the Association for Nepal and Himalayan
Studies, 39 (1), 2019, https://digitalcommons.macalester.edu/himalaya/vol39/iss1/6/, Accessed
June 28, 2020. See also, "Human-Wildlife Conflicts" Chapter in *Bhutan*, by Dr. Ugyen Tshewang,
Dr. M.C. Tobias, and J. G. Morrison, Springer, Nature, New York, 2021. See also, Pierre-Paul
Grassé's *Parasites et Parasitisme*, Armand Collin, Paris, 1935; Grassé's *Termitologia*, Vol. 1:
"Anatomie Physiologie Reproduction," Fondation des Sociétés Construction, Vol. III:
"Comportement Socialite Écologie Évolution Systématique," Masson, Paris, 1982–1986.

[7] See "Haemaphysalis (H.) sambar sp. N. (Ixodoidea: Ixodidae), a Parasite of the Sambar Deer in
Southern India," by Harry Hoogstraal, The Journal of Parasitology, Vol. 57, No. 1, Feb. 1971,

But the human/animal rights issues pose an inherently stigmergic challenge to broader issues of collaboration in Nature. Myxobacteria and other swarming cell organisms function best in group deployment, behaving according to a multibillion-year-old coordinated eusocial pattern that (of stigmergy) humans have never succeeded at, other than through a dominant mode of domestication. In some studied birds and carnivores, bacteria grouping in the animal's microbiome clusters have shown that they are essentially in charge of scented exudations and critical life-cycle behaviors.[8] This is certainly the case in humans in terms of our antibodies and life-supportive internal communities, that is, gut microbiota. We have no idea how such life forms within us affect our judgments and assumptions. Consciousness does not commune in its normal daily conversations with the numeric swarms within, which, by inference from Arthur Schopenhauer, have their own "will" and "representation,"[9] even ethical standing in the guise of self-preservation. Our efforts to promote our immune systems are, ultimately, the first line of defense for that primary agent of consciousness, which is the survival quotient itself, an ethical value among trillions of others.

In other words, an animal liberation choice perpetually emerges as a human goes about her/his every day: Would you give your life, if need be, for another? If not, then how can such a self-fulfilling choice be recognized as capable of crossing over any other sentient boundaries, abandoning sovereignties in favor of universalism, granting another organism all the freedom and affection of the Magna Carta or Declaration of Independence?

## 94.2  Single, Large, or Several Small Habitat Refuges?

"Epistemology" does not lend credence to gun owners or killers of pigs and cows. It deplores dictators while upholding social contracts that have the vim and vigor to incentivize behavioral change among all constituencies, while marginalizing those who insist on resorting to violence. We've seen great philosophical strides in realms of non-violence and cooperative treaty agreements, from Mahavira to Hugo Grotius to Jean-Jacques Rousseau. In some ways, this kind of moral political ground mirrors one of the leading conservation biology debates: SLOSS—single large or several small habitat refuges.[10]

---

pp. 173–176, Published by: Allen Press on behalf of the American Society of Parasitologists, https://doi.org/10.2307/3277775, https://www.jstor.org/stable/3277775, https://www.jstor.org/stable/3277775?seq=1, Accessed August 15, 2020.

[8] See "Experimental evidence that symbiotic bacteria produce chemical cues in a songbird," by Danielle J. Whittaker, Samuel P. Slowinski, et al., Journal of Experimental Biology," 2019, 222: jeb202978, https://doi.org/10.1242/jeb.202978. Published 16 October 2019, https://jeb.biologists.org/content/222/20/jeb202978, Accessed November 13, 2019.

[9] See Die Welt als Wille und Vorstellung, Leipzig, Brockhaus, 1818/1819.

[10] "The SLOSS Debate," by Jennifer Bove, ThinkCo., August 27, 2018, https://www.thoughtco.com/overview-of-the-sloss-debate-1181943. See also, Robert, A., 2009, The effects of spatially correlated perturbations and habitat configuration on metapopulation persistence. Oikos 118: 1590–1600.

Like SLOSS, the methodologies selected to determine the most efficacious resolution to human violence against Others have traditionally connoted and conjured up increments of compromise; and the ethical rudiments of a baseline that can help us fathom the notion of gradations and degree, what we know, or don't know, and in what stages (molts, morphs, etc.) of growth, or amplitude, relativity, and frequency, an ecological optimum may be obtained. All ethics are condemned to choices, or deferrals, to the rational or irrational, dominant or recessive.

## 94.3   Giacometti's Unsolved Solitaires

In one gravely suggestive guise, such deliberations come to a sobering halt in the life and times of Alberto Giacometti and his obsessions with life forms of solitude, particularly in the period beginning in 1947, when Giacometti was in exile from the Nazis in Switzerland and already moving toward that attenuated vision that would see him create the set design, in 1953, for his friend Samuel Beckett's "Waiting for Godot"—a lone, tortured tree of raw plaster. The synthetic biology of being alone was sculpturally writ as a looming reminder that the history, stature, and attitude of the philosopher in each of us come down to a troubled thinker (Fig. 94.2). By necessity, our solutions must arise in intermediary steps, every day, by fronting, and believing in that definitive bias in favor of positive change. The less the probability of this communion of choices and consent, the more solitary we become. A species losing moral, intelligent, sustainable points of reference among its kind, until—ultimately isolated—it has nowhere to move or turn, no companionship or even company to call upon. Giacometti's sculptures represent that ultimate paralysis. Stultified and frozen in time, mid-step. Such bronze is doomed to the sterility of that species, and her individuals, who are clearly trapped. Even ticks know to stay away.

**Fig. 94.2** "Heraclitus," (the Solitary, "Melancholic Philosopher," 535–475 BC) as Portrayed in The History of Philosophy, by Thomas Stanley, p. 737, Printer Thomas Bassett, London 1687. (Private Collection). (Photo © M.C. Tobias)

# Chapter 95
# A Metaphysics of Naturalism

## 95.1 Varieties of Supervenience

Naturalism—supervenient sets of classes, ontological, empirical, philosophical, rational, scientific, and/or consensual—whose most clairvoyant strength lies in the span of its intimations, conceptualizations, and experiential rubrics, suggests a complex dialectic attendant upon the internalization of Nature within the human psyche and its offspring, the natural sciences. Debate over whether there can be a uniformly ascribed to or reliable context in Nature is no less problematic (subjective) than identification with a supernal godhead (e.g., a Carl Sagan suggesting there is only the Cosmos and nothing else).

Giacometti's iconic legacy, based upon Tuscan/Villanovan/Etruscan predecessors, has given the "Solitaire" an existentialist visage (Fig. 95.1). In a world ripe with suspect meaning, the lone figure stands somehow apart. Has not *meaning* always been a self-evident quotient in human affairs? Wrote Jean-Paul Sartre, "Between the model and the material there seems to be an unbridgeable chasm; yet the chasm exists for us only because Giacometti took hold of it."[1]

Meaning is somehow masterminded, manipulated by any number of causes and locations, including, but not limited to, evolution, or possibly the very idea of evolution. The organism and those who would seek to stand apart from that organism are subject to the co-existential distribution of axioms, and the lure of comforts, in mind, outside the mind. Do all of the many sub-sets of human consideration, history, art, science, provide anything other than a field of confirmation biases whereby all of our ponderations are self-gratifying? Can we actually challenge these paradigms of customary thought, daily conversation, our highest aspirations, and collected unconscious? And what, if so, do such challenges amount to?

---

[1] In Jean-Paul Sartre's essay, *"The Search for the Absolute,"* in *Albert Giacometti*, New York: Pierre Matisse Gallery, 1948. Translation by Lionel Abel—found in *The Diary of Jakob Knulp*, from, *Theories and Documents of Contemporary Art: A Sourcebook of Artist's Writings*, http://www.giacomobutte.com/blog/?p=1346, Accessed September 15, 2019.

© Springer Nature Switzerland AG 2021
M. C. Tobias, J. G. Morrison, *On the Nature of Ecological Paradox*,
https://doi.org/10.1007/978-3-030-64526-7_95

**Fig. 95.1** A white-
crowned hornbill
(*Berenicornis comatus*) in
Malaysia, a true solitaire,
in this instance, who must
necessarily awaken the
suppressed altruism that is
animal liberation in each of
us. (Photo © M.C. Tobias)

Or do our species' narcissism and its solipsistic indulgences detail the very beginning and end of what we are, who we are, where we're going, and what the possibilities are, as fixed between Aristotle's "contraries"? He says "it is impossible for anyone to believe the same thing to be and not to be…"[2] But we have no reason to *believe* Aristotle any more or less than Charlie Chaplin's "little tramp."

What prospects in mind exist for this large vertebrate, newly established within the vastly greater mosaic of life forms, given its disastrous track-record of aggression? And what are the ramifications of its restless pursuits? Willard Van Orman Quine, in his 1951 essay, "Two Dogmas of Empiricism," attempts to work "a manageable structure into the flux of experience" that echoes humanity's targeted physicalism within the sciences and philosophy, predicating our ideas of truth on

---

[2] See Aristotle's *Metaphysics*, Translated with Commentaries and Glossary by Hippocrates G. Apostle, The Peripatetic Press, Grinnell, Iowa, 1979, p. 59.

the blurred relations between experience and intuition: the belief that all of the disciplines within human thought are connected, and easily revised. In other words, the super-imposition of human life upon the biosphere has emerged, of late, without any conclusions, but a vast amalgam of forensic evidence[3] pointing to our rapid descent into madness.

That underlying perplexity ("naturalist confusions") pertains to three fundamental aspects of the Anthropocene, species-wide implosions and corroborative manure, as the subject of diagnosis, prognosis, and commentary. It is a plethora of unhappy memes, tropes, and dark scenarios conflating non-fiction and fiction; imagination, ethics, and nightmare; sins and misgivings, the faces of hopelessness, all the Sisyphean tenacities, pragmatic repeals, and quixotic retreats. Notwithstanding a rash of sustainability solutions, anodynes, diagnostics, and therapeutics, there is as yet no all-encompassing vaccine for what humans do to the earth.

Combined, these myriad influences and behaviors comprise a *frisson* of substantialities that have painfully and self-referentially emerged on the remote geography of our disappearance as a species: a hominin engendering the most concise of all epitaphs. This biological remembrance of things present and past constitutes a deafening roar: a deep demography amassing more wounds than can be accounted for.

Conversely, aware of our dialectical behavior and vulnerabilities, it can be said that within the entire annals of the natural sciences, there has never been so multitudinous the temptation to exist, as Romanian cynic Emil Cioran once described it,[4] but also the desperate injunction to remove oneself from the crowd and the suspect hierarchy of taxonomy[5] where the conviviality of perfect naturalism is actually overrun with our realization of pain everywhere. We want to be free of that penumbra that is our infliction. That desperate longing can go in one of three primary directions: 1) the disappearance of Self; 2) an explosive social emergence of endless debate (which we witness every day through strident trial and error); and 3) frustrated, pained, and unpredictable reactivity: Revolutions.

## 95.2 The Evolution of Solitaires

Regardless of the disputed cladistics, family trees, heritable traits, dispersion of genes, suffering cuts across all previous barriers to distribution, the consequences of our confrontation in the moment with our ecological fall-out can be described mathematically, or according to certain laws of eco-dynamics (i.e., adaptive

---

[3] See W. V. O. Quine, "Two Dogmas of Empiricism," The Philosophical Review. 60 (1) 20–43: https://doi.org/10.2307/2181906, JSTOR 2181906, 1951.

[4] See Features March 1988, "The anguishes of E.M. Cioran," by Roger Kimball, The New Criterion, https://www.newcriterion.com/issues/1988/3/the-anguishes-of-em-cioran, Accessed September 16, 2019.

[5] Ereshefsky, M., 2000, *The Poverty of the Linnaean Hierarchy: A Philosophical Study of Biological Taxonomy*, Cambridge: Cambridge University Press.

radiation),[6] Ernst Mayr's theories set forth in his book, *What Evolution Is*,[7] as well as within rubrics of evolutionary psychology.[8] Indeed, there is an endless array of apt, descriptive scenarios for who we are and what, alas, we are doing.

In that impulse to survive, against the odds we have suicidally stacked against earth, there is revival of the pre-Socratic Renaissance of physiolatry, the love and reverence of nature, even as our species solely manifests a final crusade of violence against the planet, for no reason. Obviously, deeply rooted fear, greed, anger, super-ego, and self-loathing enter into any such equations.

*No reason*, of course, defies every workable simile that might shed logic, plausi-bility, order, or, most tempting of all, empirical probability. As an unprecedented weight of biomass and heavy consciousness aggregated in a non-linear distribution of consumption and ungainly presence across every biome, from year-long habita-tion of Antarctica to the most recent deadly traffic jams atop Mount Everest, the environmentally precocious, presumptuous, and self-destructive investiture of *Homo sapiens* confronts philosophical analysis with a genetically unscrupulous dilemma that is embodied in the many visages of Giacometti's lone figures.

We cannot follow real-time re-evolution occurring in the wake of so many bio-logical stressors. It is evidenced from rapid phenotypic changes in Caribbean mos-quitoes, wasps and fig trees, magpies and cottonwoods, and in demonstrative instances of behavioral adaptations in urban acorn ants to Mongolian rodents. The onrush of data provides a poignant context for both our deepest anxieties, and most unabashed tributes to the underlying significance of scientific inquiry. Everything biological cries out to us with meaning, giving us to strive toward enshrinement, replication, translation, and expression: toward some rarified solace that persuades and gives credence to that collective conscience that would consecrate the very meaning of evolution.

Current studies of metaphysical, and naturalist paradox, suspect rules, and easily bypassed laws of contemplation, have all but settled the primordial questions sur-rounding self-doubt and the fundamental conjecture at the heart of both Western and Eastern philosophical and ethical analysis: How do we know what we think we know? We don't. And that is, surprisingly, a valuable component of true freedom. The question is what we do with it (Fig. 95.2).

---

[6] "Ecological Opportunity and Adaptive Radiation," by James T. Stroud and Jonathan B. Losos, Annual Review of Ecology, Evolution, and Systematics. 47: 507–532, 2016, https://doi.org/10.1146/annurev-ecolsys-12,145-032254.

[7] Basic Books, New York, 2001.

[8] See "Evolutionary psychology: An emerging integrative perspective within the science and prac-tice of psychology," by L. E. O. Kennair, Human Nature Review 2: 17–61, 2003.

**Fig. 95.2** In Thrumshingla National Park, Bhutan, a "Solitary" waterfall, a font of interdependency. (Photo © M.C. Tobias)

# Chapter 96
# The Phylogenetic Conundrum

## 96.1 The Ancestries of Individualism

In a persuasive open-access research paper, scientists recently explored a brilliant insight: "The question thus arises whether the conservatism of these features, influencing which species occur in a given place, scales up to ensemble-level properties, influencing how many species of a clade can coexist."[1] Amid a teeming biological framework, focal organisms of the study comprised "14 lichens, 9 insect, and 9 avian families" in the Cantabrian Mountains of northern Spain[2] and it is of course noteworthy that the taxa were assessed at their Family levels, where, the authors of the research paper contend, deep-time, greatly expanded "evolutionary innovations" are likely to occur.[3] Life history studies that were scaled "to higher levels than aggregated individual traits"[4] follow poignantly upon the earlier inferenced *Samavasarana*, a community of diverse traits allied in those parliaments of species that have galvanized world iconography across every known ethical expanse of experiential biosemiotics and the so-called phylogenetic signals ("common ancestries").[5]

Traits to semions; semions to summons, to ethical incitement, to activist ordination. If this is the outcome of continuing phylogenetics, then we see how nature/nurture/nature is the product of not merely random, but intentional selection. It comports with sēmeiōtikos, the observance of signs. Observation reconfirms the

---

[1] Laiolo, P., Pato, J., Jiménez-Alfaro, B. et al. Evolutionary conservation of within-family biodiversity patterns. Nat Commun 11, 882 (2020). https://doi.org/10.1038/s41467-020-14720-3. Received 24 May 2019, Accepted 29 January 2020, Published 14 February 2020, https://doi.org/10.1038/s41467-020-14720-3, https://www.nature.com/articles/s41467-020-14720-3#citeas, Accessed February 14, 2020.

[2] ibid., from Results.

[3] ibid., in Discussion.

[4] ibid., from Introduction.

[5] ibid.

© Springer Nature Switzerland AG 2021
M. C. Tobias, J. G. Morrison, *On the Nature of Ecological Paradox*,
https://doi.org/10.1007/978-3-030-64526-7_96

whole historic thrust of natural history, but with a psycholinguistic twist: we can change in accordance with what any of our senses, or intuitive, or any other faculty interpret. Granting the subjectivity of such interior shifts, we have, by de facto conundrum, come to the individual's prowess in terms of free-will, evolution, and natural selection. The freedom to choose at the bequest of ancestries whose paradoxical contagion, chain of being, cumulative parturitions, has led to the individual.

## 96.2    The Poetry of Taxocoenosis

When those ancestor traits commingle in any guise, we see the emergence of "taxocoenoses"[6] whose shared interspecies coefficients (e.g., coalitions) represent profoundly encouraging evolutionary pathways that may transcend the conventionally presumed "environmental filters."[7] That *guise*, as we shall name it, represents a mystery affiliation that appears (though should not come as a surprise) throughout the biosphere. The data sets used to establish these provocative hypotheses regarding species richness and phylogenetic combinatorial relational causes and consequences included a staggering 2999 sites containing "726 lichen species," "114 bird species," and "233 insect species,"[8] in the aforementioned study from Spain.

The journal essay, "An Epistemology for Ecology" (Scheiner, Hudson, VanderMeulen, 1993)[9] was largely a defense against a claim that the ecosciences were not sufficiently "theory" oriented, or rules and laws disposed, to allow for an actual epistemological theory of knowledge applicable to natural history and the biological sciences. Years later, any such doubts have been vindicated.

Richmond Campbells' "How Ecological Should Epistemology Be?"[10] considered Plato's dialogue *Theaetetus* (ca. 369 BCE)[11] in which Socrates wearied of all the discordant philosophical analyses concerning what constitutes knowledge, including the rhetorical analogy to an aviary, a linguistically feathered Babel immune to human guesswork.

Akin to Medieval definitions of God, and recent linguistic and phenomenological inquiries into the very existence of *knowledge as Being*, Plato's frustrations in

---

[6] See its etymological underpinnings in terms of biological communities, at: https://www.proz. com/kudoz/english-to-serbo-croat/environment-ecology/4498915-taxocoenosis.html.

[7] ibid.

[8] See Laiolo, P., Pato, J. Jimenez-Alfaro, B. & Obeso, J. R. Evolutionary Conservation of Within-family Biodiversity [Dataset]. (Digital CSIC Repository, 2020). https://doi.org/10.20350/digitalCSIC/10529.

[9] By Samuel M. Scheiner, André J. Hudson and Mark A. VanderMeulen, Bulletin of the Ecological Society of America (Vol. 74, No. 1, Mar., 1993, pp. 17–21), Published by: Wiley on behalf of the Ecological Society of America, https://www.jstor.org/stable/20167407?seq=1#page_scan_tab_contents.

[10] In Hypatia, Vol. 23, No. 1, 2008, pp. 161–169, Project MUSE, muse.jhu.edu/article/228154.

[11] See D. Bostock, *Plato's Theaetetus*, Oxford University Press, Oxford, UK, 1988.

mind were echoed from Aristotle to Kant whose vast bodies of work laid out arguments for and against sensory perception versus the pure inherent understanding of propositions akin to physics, mathematics, and music that are free of human meddling or definition. Again, this echoes the inflection of the individual as a potential non-local bifurcation point within the broadest reach of natural selectivity.

In Ludwig Wittgenstein's *Tractatus Logico-Philosophicus* (1922), certainty, judgment, and all claims to knowledge are relegated to a mysticism that ultimately prevails throughout experience and artistic expression, usurping the sterile refutations within philosophical disciplines, which are by definition endless. American philosopher Edmund Gettier III (b.1927) posed a fundamental problem, ordained by the very maelstrom of wavering confidence in human introspection concerning the presumed laws and unmistakable gravitas of any and all alleged certainty. Whereas, ecology changes everything; the a priori and a posteriori approaches to pragmatism are both quantified but, more importantly, qualified. We can concede, within that aviary of a world, that the semions of avifauna are beyond our ken. We can only hope to understand. And therein, the word *hope* embodies a far more realistic and relevant approach to the world, where humans are concerned. It is a hope that entails some level of grasping the end, or the context, the framework of life histories. In the case of ornithology, it is clear that the statistically average person's complex relationship with birds has broadened out into what we may suggest is the matrix of all avifauna and of humanity. It is no chance encounter resulting simply in pleasure or consumption. Rather, a vast mating ground of aesthetics, curiosity, cosymbiotics, and deep affiliation.

In considering the great gatherings within mythological time spans, spearheaded by a Mahavira or St. Francis, we dream of, and put much stock in the taxocoenotic visions of interspecies relationships. Science does not normally aggrandize the components in the guise of articulated *reunions*, but that is precisely what they are, whether at a watering hole or in the codependent configurations of a wasp and a fig-tree.[12] We have earlier mentioned St. Eustace, going from the hunter to a pacifist/vegetarian, of the great vision of Isaiah, of wolves and lions and lambs taking pleasure in each other's companionship. But we, aside from Isaiah or Eustace themselves, omitted from the picture that we engender, an endogenous peril that resides most paradoxically within the same Thinker. There is no closure drawing upon this problem, particularly at present, and certainly since the rise of Nazism. The Anthropocene has forensically forecast every gap in Renaissance and Romantic logic: in all those other organisms that so readily demonstrate a presocial or eusocial set of prevailing orientations versus that anomalous peril represented by *Homo sapiens*. These rules of decorum come naturally to slugs and hippos; although we do not deny that ants and termites have been fighting over, it would appear, territory, for millions of years.

---

[12] See, for example, the summary of Ficus-specific wasp co-evolution, in "The story of the fig and its wasp," by Katie Kline, May 20, 2011, esa, https://www.esa.org/esablog/research/the-story-of-the-fig-and-its-wasp/, Accessed February 14, 2020.

## 96.3   Two Embedded and Dialectical Paradigms

Yet, taken collectively amid the vast arrays of other organisms, we alone are scis-sored between two cultural paradigms formulated in very specific historic tapestries. The first, the 230-foot long Bayeux,[13] 70 scenes of conflict and terror embroidered just a few years after the exploits of Norman the Conqueror (1064–1066); invaders, excommunication, the mixed motives of the papacy, and the whole orchestra of vio-lence. Versus that of the six woven tapestries from Flanders created by commission of Antoine II Le Viste (1470–1534), friend of Charles VII, to celebrate the intoxi-cating elixir of love and the senses, as sensuously articulated in the mille-fleurs style, between "La Dame" and a tranquil lion and devoted unicorn, both famously at her side. (See the display in the Musée national du Moyen Âge in Paris.)

Consciousness between species understands such works of art and knows their abiding truths. How is it that we are adamantly left out of the equation, despite our species having produced the individuals who made the art?

Our picture of such divisive reality, a berserker picture of evolution, taxonomy, and phylogenetics, may not simply be a complication of consciousness; a flaw in the mathematical space harboring, somewhere, the conscience, even as the picture post-cards would all deny mayhem in favor of the Mona Lisa's smile. Somehow unmi-raculously, we perpetually fail to favor one constant over the other. While bad news punctuates headlines, leaving the soft murmur of the peaceful largely unheralded, the tallies of destruction will never outweigh the totality of quiet footprints of all those who remained pacifist bystanders beside the fury. The ultimate abassadors of this mode-of-being are trees (Fig. 96.1).

## 96.4   A Different Kind of Intelligence

Think of the Class Oligohymenophorea, Phylum Ciliophora, those lovely unicellular ciliates, known as paramecia. With no true nervous system, they nonetheless learn and build upon their powerful lessons. Scientific methods tell us this, but curious indi-viduals push such enquiries, ultimately not for sadistic purpose, but out of no less a love than that connecting that woven Lady and her two companions. Herein, the great riddle. The paramecia show qualities of intelligent control typically associated with other animals that only accomplish such output by neural assembly. In the single-celled paramecia, according to Roger Penrose, are centrioles and gathering micro-tubules which, to the human senses, seem almost miraculous in their engineering versatility. Their existence is dominated by inexorabilities that involve geometrical cytoskeletal orientations, for example, Fibonacci sequencing for processing informa-tion akin to the same numerically delineated structures that do so in large mammals.[14]

---

[13] See https://www.bayeuxmuseum.com/en/the-bayeux-tapestry/

[14] See "The Shadows of the Mind," by Roger Penrose, http://community.fortunecity.ws/emachines/e11/86/shadow.html, Accessed February 15, 2020.

**Fig. 96.1** Conifer and broadleaved forest in Eastern Europe, a peaceful coexistence. (Photo © M.C. Tobias)

First seen in 1678, described in 1703, their 29 major (thus far identified) species have globally demonstrated the most comprehensive picture of unicellular efficiency, reproductive agility, and learning abilities of any such organisms.[15] Might our species' micronuclear structures have the resiliency to divine paramecium methods of rejuvenation, from within the throes of self-inflicted destruction? Or any number of others, such as the 70,000 species of seed shrimp that once existed. Some 57,000 of those species are believed to have gone extinct. But despite blind corners, dark alleyways, dead-ends, and, today, a global slaughter (US$50 billion annually) our estimations of "the informative fossil record of [these] cytheroid ostracods"[15] also must remind us that against so many *biases in natural selection*, at least 13,000 species of their genera—again, in spite of an uncertain phylogenetics[16]—have survived thus far.

---

[15] See Denis Lynn, *The Ciliated Protozoa: Characterization, Classification and Guide to the Literature* (3rd edition), Springer Netherlands. See also, Ginsburg, Simona; Jablonka, Eva (2009). "Epigenetic learning in non-neural organisms". Journal of Biosciences. 34 (4): 633–646. https://doi.org/10.1007/s12038-009-0081-8. PMID 19920348. And, Kreutz, Martin; Stoeck, Thorsten; Foissner, Wilhelm (2012). "Morphological and Molecular Characterization of Paramecium (Viridoparamecium nov. subgen.) chlorelligerum Kahl (Ciliophora)". Journal of Eukaryotic Microbiology. 59 (6): 548–563. https://doi.org/10.1111/j.1550-7408.2012.00638.x. PMC 3866650. PMID 22827482.

[16] See "Palaeos – Life Through Deep Time," "Decapoda: Caridea," http://palaeos.com/metazoa/arthropoda/decapoda/caridea.html, Accessed June 29, 2020.

A majority of animals, including mammals, are herbivorous. Over the course of the latter's 320 million years of synapsid (amniote clade) development, since the late Carboniferous period, they have exhibited a largely peaceful coexistence, both at the individual and group size. Certainly of a critical mass to excite the earlier debated word, hope, hopa, Old English, hoop, Dutch, hoffen, German, espoir, French, nadezhda, Russian, nambikkai, Tamil, elpida, Greek, Spe, Latin, and so on. An utterance that connotes trust, active planning, anticipation, actual learned strategies: all are the engineering puzzles solved by paramecia and some mammals. But is learning enough? Especially when the internalization is simply repeating a pattern that is experientially a mammoth psychological and morphological cul-de-sac? While cilia expend energy at the top of their efficiency potential, given the relational size of each individual ciliated hair, human mentation is highly counter-efficient, churning at a low energy threshold, relative to the mass of other neurons in the brain.

Such that *hope*, a mere word encapsulated in a transformative genomics context, is as vague a clamor in the heart as the word *landscape* within the broader stretch of ecological despoliation, great dying and great living.

## 96.5   Contemplating the End

Free to call the absence of pain the only real gift, all organisms remain committed to lifelines and timelines. In a sense, we all gaze upon "Ötzi," the 5100+-year-old Tyrolean man from Hauslabjoch, his 45 years of life in the Alps, and all the comorbidities leading to his fatal loss of blood, but one sample out of quadrillions.

The faithful and bewildered gathered at St. Peters throughout the day of December 31, 999 AD, probably convinced that the monumental Apocalypse was to befall them at midnight. After all, from Norway to the Principality of Lombard, the previous week had unleashed every conceivable iteration of the *Revelation of St. John*; deaths and monsters; epidemics and frightening cold; fungal infestations of rye; hungry packs of wolves throughout Navarre; unprecedented sandstorms across the territories of the North African Umayyad Caliphate. Vesuvius was again belching, and in Ireland some 7000 troops were killed as a contest over leadership was fought out at Glenmama, County Wicklow. It was left to the bibliophile scientist, Gerbertus Aureliacensis (de Aurillac), otherwise known as Pope Sylvester II, to calm the greatly agitated masses down below his holy perch at the Vatican, which he did.

The new millennium foretold of catastrophe, an ecofatalism tapping maniacally into the unconscious dread of our collective demise with a radically forlorn set of prerequisites our moribund pedigree has been cultivating for a period long before the life of Gerbertus Aureliacensis. When this pattern of intentional species sabotage began might well be pinpointed to approximately 6 million years ago, the first insistent evidence of human carnivory likely commencing with the arrival of *Sahelanthropus tchadensis* and his usurpation of the coeval *Purgatorius*, who had been vegan for tens of millions of years. By 2.5 million years ago, descendants of *Purgatorius* and *Sahelanthropus*, gripped by protracted climate change-related impacts on availability of enough nutrient-rich herbivorous raw food in Africa, had had sufficient time to work through the many digestive tract complications stemming from pushing meat through a colon.[17] That this malignant malice spread rapidly will correlate with our species' density and distribution.

Hundreds of genetic and paleontological clues to the forensics of human violence have already yielded that tentative picture of our changeover and takeover. This consisted of the imposition of design upon the relations of the sexes, relegation of outcasts, amalgamation of tyrannies, and consolidation of that most pitiful of all proclivities, the quest for *power*. Latin form was podir, posse, potĕre, *to be able*, in use from 842 through the fourteenth century, when it came to be applied as today. The word power has no meaning in the cold vacuums of space. And it is certainly not as etymologically interesting as the word hope.

The energy (warmth) of campfires and cooked food seems to have segued into our warming of the entire planet, with the consequent dream of the cool air of the Ice Ages. The last glaciers are likely to be worshipped, their passing (and Ötzi's) mourned as our species disappears, no weeklong Summits in Davos for world leaders and teenagers. A world that overnight shall have entered the Eutrophizoic Era explodes within the imagination, with the alacrity of the ultimate pandemic.

Just short of that wasteland, by the number, the holy, immaculate, terrifying inconceivable one, is the present widespread condition of consciousness apprehending the storylines of this last horror, at least for us. As we wade out into the infernal dark, knowing there is no return and wondering if our evolution might have been different. All the information gathering to come will certainly obsess over this foundational conundrum, these last, hopeless conjectures. Did the question of our survivability arise 2.5 million years ago? Were alternative scenarios discussed? Did anyone die happily? The heart of today would like to know, and the answer is worth cherishing, as individuals, still breathing, still wondering, still hoping, as if by nature, whether we have any chance of survival (Fig. 96.2).

---

[17] See "How Humans Became Meat Eaters," by Marta Zaraska, February 19, 2016, The Atlantic Monthly, https://www.theatlantic.com/science/archive/2016/02/when-humans-became-meateaters/463305/, Accessed February 17, 2020; see also, *Primate Taxonomy*, by Colin Groves, Smithsonian Institution Press, Washington., D.C., and London, 2001; and *Handbook of the Mammals of the World – Primates*, Volume 3, Ed. by Russell A. Mittermeier, Anthony B. Rylands and Don E. Wilson, 2013, Lynx Ediciones, Barcelona, Spain 2013. See also, *Meat-Eating and Human Evolution*, Edited by Craig B. Stanford and Henry T. Bunn, Oxford University Press, New York, 2001.This latter work presupposes that carnivory, either through hunting or scavenging has been a crucial physiological necessity of *H. sapiens*, suggesting "When, why, and how early humans began to eat meat are three of the most fundamental unresolved questions in the study of human origins."

**Fig. 96.2** One possible future: harmony, Professor Francisco Petrucci-Fonseca with friend, at the Iberian Wolf Recovery Center in Portugal. (Photo © M.C. Tobias)

Of course, toward what end? The perennial salvation? Some private retreat, wherein a modicum of consolation devolves to the lucky ones in possession of a few books, consoling caches of strong painkillers, a friend, a bottle of fresh water. Is there any other plausible vision of the individual in a biosphere that has been so maligned by that person's kind?[18]

And for all that, on the morning of AD 1000, the business of frenetic unfazed biochemistry went on as usual the next day.

---

[18] Numerous works have easily intimated ecological forensics and the power of human choice: Edward O. Wilson's *Nature Revealed -Selected Writings, 1949-2006*, John Hopkins University Press, Baltimore, Md., 2006; *Objectivity*, by Lorraine Daston and Peter Galison, Zone Books, New York, 2007; *The Society of Mind*, by Marvin Minsky, Simon & Schuster, New York, 1985; *Wild Chorus*, by Peter Scott, Country Life Limited, London 1938; *Animals and Men – Their relationship as reflected in Western art from prehistory to the present day*, by Kenneth Clark, William Morrow & Co., New York, 1977; *Heinrich Kühn – The Perfect Photograph*, With Contributions by M. Faber, A. Gruber, F. Heilbrun, A. Mahler and A. Tucker, Hatje Cantz, Albertina, Wien and Paris, 2010; *A Day In The Country: Impressionism and the French Landscape*, LACMA, Los Angeles 1985; and most connected in form and concept, *Ernest Thompson Seton – The Life and Legacy of an Artist and Conservationist*, by David L. Witt, Foreword by David Attenborough, Gibbs Smith, Layton, UT, 2010. In addition, we would note *Science in the Soul: Selected Writings of a Passionate Rationalist*, by Richard Dawkins (Random House, New York 2017; and *Land & Life: A Selection From the Writings of Carl Ortwin Sauer*, University of California Press, Berkeley 1974.

# Chapter 97
# The Biosphere Beyond Humanity

## 97.1 To Seize the Day

As early as 1981, the future of evolution 50 million years hence had been conceived with stunning, flamboyant biologically picturesque enthusiasm by Dougal Dixon in his *After Man – A Zoology of the Future*[1] (Fig. 97.1). "The floor of the great rain forest of the Australian sub-continent is home for a number of marsupial mammals," Dixon modestly teases.[2] A sensational prelude to a wondrous set of behavioral premonitions of the future. Hybrid beasts, part Lewis Carroll liberated, Hieronymus Bosch on a very bad day, loving and grand, that we can all eye with great bemusement and understanding. The same for "South American Forests" and the imagined "Night Stalker" and "Flooer" of "The Islands of Batavia."[3] Or the envisioned "Zarander," "Turmi," "Swimming Monkey," "Swimming Ant-Eater," "Giantala," and "Posset" of coming twilights across the "Island of Lemuria" and countless other wild and crazily wonderful "forest floors" and "tropical wetlands." The Scottish paleontologist, who went on to write *Man After Man* (1990), *Greenworld* (2010), and more than an astonishing 200 other works, is today thought of as a founder of speculative evolution, a genre replete with exobiological flavors, and the perennial charm of great storytelling, all rooted to an adept and comprehensive sense of our post-Darwinian earth.

However marvelous the coming machinations may, or may not, prove to be, the current biological divides, driven by the last known hominin, provide all the fodder in the world for imagining alternatives which, in our most ardent sustainability modes, continue to elude the elixirs beyond mere ratification. As described in the Introduction, the record, to date, of humanity's ability to save what is here, now; to protect what is left, in thousands of laws, agreements, treaties, accords, presents but "a dismal record."[4] So what do we do now?

---

[1] Introduction by Desmond Morris, St. Martin's Press, New York, 1981.

[2] ibid., p. 98.

[3] ibid., pp. 102, 108, 109.

[4] See John Vidal, "Many treaties to save the earth, but where's the will to implement them?" June 7, 2012, https://www.theguardian.com/environment/blog/2012/jun/07/earth-treaties-environmental-agreements, Accessed February 17, 2020.

© Springer Nature Switzerland AG 2021
M. C. Tobias, J. G. Morrison, *On the Nature of Ecological Paradox*,
https://doi.org/10.1007/978-3-030-64526-7_97

**Fig. 97.1** Endemic to New Zealand, the Tuatara (*Sphenodon punctatus*) are the last living reptiles to have wandered beside dinosaurs 200 million years ago. (Photo © M.C. Tobias)

To be clear, Dixon's creatures are no more fantastic, charismatic, or unusual than any one of today's species, including ourselves. That is of utmost importance to contemplate. Take hummingbirds. To date, the oldest fossils of their expansive kind were found in Germany, and date to approximately 30 million years ago.[5] The second largest family of birds is the neotropical hummingbird, with 338 known species, far more than 10% of which are endangered.[6] To grasp some inkling of their marvelous life histories, consider the very foundation upon which all sciences and their human practitioners are hypnotized. Facts and dreams whose combined charm, eloquence, artfulness, and sheer weight of the anomalous lead irresistibly to that mental monument which underpins a belief in the future, because the future must be predicated upon the present. In both biology and mathematics, *nonlinear systems* that tend to ignore our obsession with the regular, ordered, linear modus operandi are manifest everywhere.[7] We find them even in our backyards.

---

[5] See "Old World Fossil Record of Modern-Type Hummingbirds," by Gerald Mayr, Science 07 May 2004: Vol. 304, Issue 5672, pp. 861–864, https://doi.org/10.1126/science.1096856, AAAS, https://science.sciencemag.org/content/304/5672/861.abstract, Accessed June 29, 2020.

[6] See The Hummingbird Society, https://www.hummingbirdsociety.org/endangered/, Accessed February 17, 2020.

[7] See de Canete, Javier, Cipriano Galindo, and Inmaculada Garcia-Moral (2011). *System Engineering and Automation: An Interactive Educational Approach*. Berlin: Springer. p. 46. See also "Nonlinear Biology", The Nonlinear Universe, The Frontiers Collection, Springer Berlin Heidelberg, 2007, pp. 181–276, https://doi.org/10.1007/978-3-540-34,153-6_7.

## 97.2 Anna

Anna's hummingbird (*Calypte anna*) is a great example. She makes a lovely case for input/output efficiency variance that functions at the highest levels of intelligence. Green and pale, flashes of red in the male, stubbornly non-migratory, and named by René Primevère Lesson in 1829 after the French courtier, Anna de Belle Massena, Princesse d'Essling (1802–1887), the Grande-Maitresse, Mistress of the Robes, to the last Empress of France, Eugénie de Montijo, spouse to Emperor Napoleon III. Anna and her husband, the ornithologist Prince Victor Massena, maintained a collection of birds, which Audubon himself visited in Paris. He famously described his hostess as a "beautiful young woman, not more than twenty, extremely graceful and polite."[8] Anna (the bird) is famed for her tireless pectoral musculation, the hummingbird's chemistry of metabolic wonders that liberate immense ratios of energy from its conversion of glucose to pyruvic acid. That is an alpha-keto that helps to enable the tiny marvel to hover in every conceivable aerial gymnastic with a cascade of time-lapse-like superlatives, from reckless torpor to maniacal pollinating frenzy, 50–1200 heartbeats and wing actions per second. In addition, she breathes "273 times per minute."[9] There is no algorithm that will ever endow a bipedal ape with such pink-in-winter flight feathers, though there may be other shared instincts (Fig. 97.2).

She is, in other words—like the more than 300 other hummingbird species—a remarkable, otherworldly phenomenon, consolidated into a 4-inch marvel of life. For all our enduring love of the Trochillidae Family, we don't know how to ensure their safety. One of them is the smallest known bird in the world, the *Mellisuga helenae*, or bee hummingbird of Cuba, weighing approximately 1.6 grams fully grown. There may be an even smaller bee hummingbird in Haiti. And there is the recently discovered, 99-million-year-old *Oculudentavis khaungraae* skull, found in a piece of amber from northern Myanmar, an even smaller bird, its skull 14.25 millimeter in length. Mellisuga is listed as "Near Threatened. A2c+3c."[10] If our conscience be our guide in all aspects of conservation biology, as in animal rights generally, then the metaphysics of protection, as we have elsewhere labeled the totality of our efforts to save other life forms[11] will be judged: adjudicated both by the scope of our nurturing ambitions, as well as the pragmatic imaginativeness we

[8] See Southwestern Idaho Birders Association, December 5, 2016, from https://www.birdnote.org/.

[9] See *The Life of the Hummingbird*, by Alexander F. Skutch, Illustrated by Arthur B. Singer, Crown Publishers, New York, 1973, p. 41. See also, *Hummingbirds of North America of North America: Attracting, Feeding, and Photographing*, by Dan True, University of New Mexico Press, Albuquerque, NM, 1993, pp. 96–99; See also, *Birds of North America*, Western Region, Smithsonian Handbooks, by Fred J. Alsop III, p. 396.

[10] BirdLife International 2016. *Mellisuga helenae*. The IUCN Red List of Threatened Species 2016: e.T22688214A93187682. https://doi.org/10.2305/IUCN.UK.2016-3.RLTS.T22688214A93187682.en, https://www.iucnredlist.org/species/22688214/93187682. Its last assessment was October 1. 2016.

[11] See *The Metaphysics of Protection*, by M. C. Tobias and J. G. Morrison, A Dancing Star Foundation Book, Waterside Productions, Cardiff-by-the-Sea, CA, 2014.

**Fig. 97.2**  Hummingbird in Big Sur California. (Photo © M.C. Tobias)

bring to the task. There is a divide between weights and measures, diets and mobility, as preludes to what may well be a host, not of other differences, but of similarities. Oregonians relish the fact, those who know, that their resident hummingbird stays in Oregon all winter; loves the color red; craves sugar, and by all indications was well named, in terms of "Anna" and the final years of French nobility. Will she be around 50 million years from now? Probably not.

## 97.3  Scintilla Conscientiae

As we endeavor to elucidate the prospects for a general embrace of our internal *scintilla conscientiae* (Saint Jerome's name for it), we grapple beyond mere praxis into the realms of ideology. We seek guidance from contemporary philosophy, large problems analysis, neuroscience, artificial intelligence. As simple as safeguarding a bird's habitat may sound, we are finding it incredibly challenging to do so, regardless the number of hummingbird feeders we are prompted to seasonally install. Yet, every time we move, make a sound, do anything, we are testing the boundaries of community acceptance and disturbance with regard to such vulnerable avians.

We look out into the world and see a distant glow of a signal light, lighthouses shivering gray and cobalt across the tropics, twilight in the Karakoram; casting uncertain cold lights upon the last of the mauve corals and the first knells of something nascent, quavering on the horizon; clues beyond the human meaning exciting our perpetual, extroverted intrigues. But no more than to instill a solemnity by which we restrain rather than investigate. The inward-dwellingness of poetic reverie

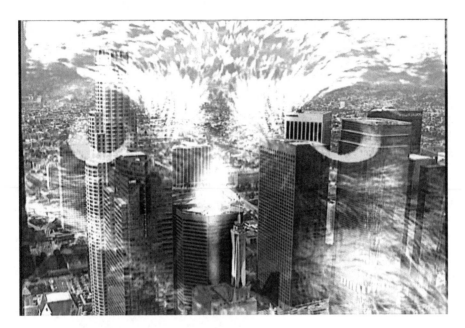

**Fig. 97.3** Since 1970, bird populations in the US have declined by nearly one-third. (Photos ©
Dancing Star Foundation)

that is not merely the trap of hedonism and its blinders, but some other far more
dignified and gentle mode of behavior, kindly tropes of expression, wherein between
language and cognition, perception and action, there is no impossible gulf, but the
perpetual yearning to be fully reconnected, enchanted, accepted by the rest of the
world.[12] If that were the case.

Imagine the fossils, carbon, sand, mountain ranges, the sky, and the stone, the
biology teeming within every cranny, all reminding our kind of fellow beetles and
moths, hummingbirds and whales, the beginning and end of worlds. With an impassioned fellowship that craves not the rules and certitudes. Let go of the fools who
think themselves lofty in ourselves, to otherwise take a first step, initial glimpses of
our untold loss, not unimagined, that biosphere which is by uninterrupted definition,
every definition, beyond humanity. Has somehow miraculously touched us. Just as
we have all but systematically renounced this final possibility. It is like the story of
zero, commencing in Sumer, perhaps long before that. Plus-or-minus, on either side
of that primordial zero: the story is not over (Fig. 97.3).

---

[12] For additional reflections on such precepts, see Patricia S. Churchland's *Conscience – The Origins
of Moral Intuition*, W.W. Norton & Company, New York, 2019; Michael Graziano's *Rethinking
Consciousness – A Scientific Theory of Subjective Experience*, W.W. Norton & Company, New York,
2019; Flynn Coleman's *A Human Algorithm: How Artificial Intelligence Is Redefining Who We Are*,
Counterpoint Publishers, Berkeley, CA, 2019; Brian Christian and Tom Griffith's *Algorithms to Live
By – The Computer Science of Human Decisions*, Henry Holt & Co., New York, 2016; and Jim
Holt's *When Einstein Walked With Gödel – Excursions to the Edge of Thought*, Macmillan Audio,
2018, Macmillan Publishers, Holtzbrinck Publishing Group, Stuttgart, Germany, 2018.

# Chapter 98
# The Anthropic Syllogism

## 98.1 Apocalyptic Underpinnings

The ecology of human experience traverses infernos and greener pastures. The ancient Judaic Sheol, Realm of Shadows, forever punishing lost souls.[1] Purgatorio,[2] its dismal state portrayed by Sandro Botticelli's map of hell for Dante's *Divine Comedy* (1485), a vast mathematical funnel cloud of perceived horrors. And, conversely but in tow, every paradise scene. Each and every ecological disruption and social discord is wedded to its antonym, the combining power of these two opposites intimating classical antecedents dating to an unknown time in human consciousness when microcultures first emerged, enshrining what would come to be thought of as *archetypes* of Good and Evil, or feelings to that effect[3] (Fig. 98.1).

Their cumulative environmental messenger is, today, not the unambiguous and ever-pressing paradox of human nature. The late ecologist Paul Shepard applied the notion of sickness and duality to everything attendant upon human evolution, its punctuation marks characterized by hunter/gatherers and agriculturalists, nomadic Old Testament prophets, communities marked by increasing possessions versus renunciatory Desert Fathers, Puritans, and Mechanists. From the very beginning of these differentiations, there was the fundamental issue of humanity's relationships "with otherness," psychological great divides hid from within an inchoate predilection that, according to Shepard, still held out realistic hope for a great healing. But in nearly four decades since Shepard published *Nature and Madness*, clearly the state of the world has only deteriorated far beyond what most could have projected.[4]

---

[1] See *Symbols and Allegories in Art*, by Matilde Battistini, Translated by Stephen Sartarelli, The J. Paul Getty Museum, Los Angeles, 2005, p. 222.

[2] See "Apocalypse Now," *The Divine Comedy* by Dante Alighieri, Digital Dante Edition with Commento Baroliniano, MMXIV-MMXX, Columbia University, New York, https://digitaldante. columbia.edu/dante/divine-comedy/purgatorio/purgatorio-32/, Accessed February 18, 2020.

[3] See Nieman, Susan *Evil in Modern Thought: An Alternative History of Philosophy*. Princeton University Press, Princeton, NJ, 2015. See also, *Inside the Neolithic Mind: Consciousness and the Realm of the Gods*, by David Lewis-Williams and David Pearce, Thames and Hudson, London 2005.

[4] P. Shepard, *Nature and Madness*, Sierra Club Books, San Francisco, CA, 1982, commentary on Mircea Eliade's *The Quest*, University of Chicago Press, 1969, p. 125.

© Springer Nature Switzerland AG 2021                                    843
M. C. Tobias, J. G. Morrison, *On the Nature of Ecological Paradox*,
https://doi.org/10.1007/978-3-030-64526-7_98

**Fig. 98.1** "The Archangel Raphael and Tobit in a Landscape." Among the most complex tales of good, evil, redemption, and nature in the history of Western religions. Engraving by Joannes van Londerseel (1578–1625), after Gilles Claesz. d'Hondecoeter (1575/1580-1638), Inventor. Claes Jansz. Visscher (1587–1652), Publisher. (Private Collection). (Photo © M.C. Tobias)

The etiologies of human ecological regression, an escalating set of co-morbidities and extinctions, challenge our perception as a species. There are as many as "348,524,706,340" digital images at any moment on the Internet.[5] Thousands of image banks, a recursive grammar that fortifies the closed semiosis of our anthropic lens, unheralded layers of what ultimately is a blank stare, with every new shock to the system in trillions of sensory frames of life each day. We recall a few of them, as if struggling to remember a dream. Our senses guarantee the evolution of that confirmation bias that leads to a similarly embraced disposition in the arts and sciences, culminating in viewpoints and attitudes.

In the meantime, from nation to nation, hamlet to megacity, our encultured codes of survival have changed little from our earliest documented desperations and visions. Throughout the history of science and human philosophy, we keep asking the same questions. Nor have our answers changed. Only the baseline, the very

---

[5] https://www.youtube.com/watch?v=7p3RvIPuhw8; See also "Mary Meeker's 2019 Internet Trends Report," Kate Clark, June 11, 2019, https://techcrunch.com/2019/06/11/internet-trends-report-2019/. In 2014, Meeker's Trends suggested "657 billion" images per year on the Internet. See also, https://www.nationalgeographic.com/photography/100-million-instagram-followers/photos-our-followers-liked-most/, Accessed February 27, 2020.

Weltanschauung, the world's context, has been seismically altered, from Hadrian's Wall, to the Enola Gay B-29, an image of half a hydrogen atom's width, to optical-infrared astronomy attempting to photograph the Big Bang. In one sense, all of these endeavors can be symbolized at the moment of Charles Darwin's death, as described by Charles Fisher in some detail. It casts a vivid picture of the cognitive Empire that was nineteenth-century observation and science, overthrown in an instant by the persistent realities of an individual's vulnerability.

Eschatology, the study of final things and its hybrid commentaries upon an elusive Self, accompanies us as the mob of history rages through the alley, our hidden eye of the trapped observer through a peephole observing reality passing by. We are present in each member of that swarm, a behavioral genome nestled in its kernel, echoing back at us, clambering for some, *any* dependable biological identity beyond the fray of the busy cell. But it doesn't exist. During the weeks leading up to April 19, 1882, when Darwin died in his wife's arms of congestive heart failure, the scientist realized that neither his evolutionary status as a human nor the vast amalgam of ideas and acutely conscious concepts that had fueled his life any longer mattered, as his body collapsed from within.[6] Individuals and species were subject to the mutability of the ever-changing rules of natural selection.[7] Only among *Homo sapiens* is paradox and inherent contradiction vulnerable to its own evolution. We should repeat that, because it not only endlessly complicates but also elucidates our prospects, based upon what we do next (Fig. 98.2). What this clears the way for is a constellation of improved choices by our kind, just as Dante's Inferno had its mirroring pinnacle in Paradise.

We don't *know* this; we don't *know* anything; but our ideas on occasion do seem to form assertive conclusions that mimic the class of certitudes that are both true and false, yet certain in the sense of what underlies an assertion. Their express petition of verity is the ruse that propels consciousness toward its proprietary empiricism. But there are no actual facts present. All is anthropic syllogism.

## 98.2   In the Shadows of Our Kind?

"So the first part of dealing with the human predicament must consist, quite literally, of changing our minds," wrote Paul and Anne Ehrlich.[8] This is the ultimate teaser with respect to what can be divined about our future, much like the Preface

---

[6] *Dismantling Discontent – Buddha's Way Through Darwin's World*, by Charles Fisher, Ph.D., Elite Books, Santa Rosa, CA, p. 82, citing A. Desmond and J. Moore, *Darwin*, New York, Warner Books, 1991, p. 662.

[7] See "Darwin Correspondence Project," University of Cambridge, https://www.darwinproject. ac.uk/letters/darwins-life-letters/darwin-letters-1844-1846-building-scientific-network, Accessed February 28, 2020.

[8] *The Population Explosion*, by Paul R. Ehrlich and Anne H. Ehrlich, Simon and Schuster, New York 1990, p. 189.

**Fig. 98.2**   One of the last tigers killed in Singapore, ca. early 1930s. (Photographed by M.C. Tobias at Education Center in Bukit Timah Nature Reserve, Singapore)

to *N by E*, by artist Rockwell Kent, which begins, "In this book is told the story of an actual voyage to Greenland in a small boat; of shipwreck there and of what, if anything, happened afterwards."[9] Ships carrying alcoholic drink and wood supplies from Norway, and general news of the world, "of kings and kin" arrived, departing with supplies of "ivory and oil," the spoils of Inuit-slaughtered walrus and whale.[10]

Our line will soon bear its final progeny, despite the demographic madness that continues to rage at a near global Total Fertility Rate of 2.5.[11] With a coming 10, 12 billion humans, show us the single sustainable Being? What, who is it? How does she/he/it thrive? What are the potential harmonious variables, given the myriad baggage of such life histories, in the shadows of our kind, that such a one might look forward to? Let us be clear: we believe that harmony is possible.

We seek the transitional equipoise possible between reluctant opposites. It can happen in any number of geographical moments. This hypothesized shift represents a unison of human ideals and actual implementation, at the gates to the city. The conversion inculcates diametrically opposed neuro-theaters, the one being our esthetic sense of salvation and the other our perceived doom. Together, there remains a persistent lure in the guise of the stranger that inhabits ourselves and the evolutionary singularity beyond all predation in nature, from the amoeba to the Giant Sequoia. To reach the point where we can say definitively that hunger, fear, solace, and natural selection are merely categories of thought and compulsion. Is that who

[9] *N by E*, by Rockwell Kent, Random House, New York 1930, p. 11.

[10] ibid., p. 186.

[11] See https://populationeducation.org/what-total-fertility-rate/, Accessed February 28, 2020.

**Fig. 98.3** Depiction of the "Serengetis" of Nature at the Museu De Historia Natural de Maputo, Mozambique. (Photo © M.C. Tobias)

we are? Yet, no window on the Serengeti harbors the least glimpse of an etude or sonatina. In effect, we have no viable apology for the existence of rampant pain and even less to say on the matter of human cruelty.

Because we are as much a part of the universal equations of predation as any eagle or wasp, our culpability begins with millions of others. But then the human factor strains the violence into altogether untried territories and collides by way of community transmission into numbers that defy biological history, certainly at the large vertebrate level. There have always been strong detractors who gain some perverse confidence in linking human violence to that of ants, avian insectivores (admittedly problematic), invasive pythons in Florida, or the zebra beleaguered by the big cat (Fig. 98.3). Such comparisons are false, one more indication of a desperate conscience. If goodness, kindness, and love are each transcendental capacities of our conscious lives, then there must be a systemic antidote to Serengeti within the scope of that same soft-spoken conscience. It, too, has paradoxical powers that may play out in either direction, fostering sufficient force to counter the brutalities of nature.

Taken within the context of the history of wars, slaughterhouses, and other inordinately *human* violence, even our reflexes and instincts that might, like a whistle-blower, ignite concern within cognition remain bound to feelings of an informal self that resides within Self, a dualism as impossible to diagnose as a Black Hole. We have no evidence linking the human organism to an original instance of mindfulness. No spark that leads by scientific forensics to a bonfire. Our application of phenomena to noumena, or sensations to a universal wellspring of qualia, can only be the setting for linear debate, not certainty. We can certainly live with that. We do so every day. We have only our own reassurances to look toward. No mathematics, no physics, and no laws of chemistry to relieve the phenomenon of self-involvement. Its finitude is a singularity that offers no authority. In fact, quite the contrary, as every set of causes and consequences, adding up to that embryological reality of ourselves as individuals, and as a species, cannot be differentiated from the continuum inherent to the biosphere, an omniscient, if measureless expanse. Some are content to label it full of meaning, commensurate with God; for others, a pointless conception.

Regardless of billions of differing opinions, the amputated leg continues to echo, survivors never forget, and the trauma reverberates. Personal histories retrace the etiology of disease, down to the nearest corollary in decimals, fractions, and the memories in nerve endings. The units of measurement provide the physical space for our every desperate attempt toward closure, our rehearsals before the mirages of Utopia. All of this has been ceaselessly uttered. Is there some alternative history contrary to the tired rhetoric and jaundiced reiterations of philosophy? Some altogether new vista beyond the devastating skylines of our doomed ecologies?

Let us rank the plausibility of an answer according to an entirely exogenous realm of data, coined in the language specifics of the most rarified ideal. What, where is it? We cannot assign to it the conventional nomenclature of reality, any more than a binomial would be appropriate. Any human fails completely to bridge the gap, the dramaturgical fourth wall. We have no suasion to merge asymptotic coordinates. Theirs is a closed system, fully embodied. Because all metaphor remains trapped in the human, the first question, concerning the ultimate feasibility of a connection to anything outside of ourselves, is readily subject to the idea of pain, which shatters all complacency in the face of however many quadrillions of quadrillions of posited experiences. We sense such a world across the inter-demographic map we have always agreed upon, conceded into Being, a through-story of suffering. Hence, where there was nothing, suddenly manifested is pain. Where we had given up at the base of the ladder, again, there are sorrows. At that juncture between words and objects, the imagined and the physical, there is a thread leading us from one melancholy to another, ghosts skirting the rim-light of every tombstone.

## 98.3   The Paradoxes of Pain and Cruelty

But for now, let it suffice that there are perceived if unclear population dynamics—if between species and individuals, then within individuals effecting all the conceivable relationships of the conscience and her consciousness—of pain that has resulted whether by unnatural device originating within those intentions and means, most cruel, or by some other order we loosely take to mean the nature of nature. Some unnatural selection that has selected again and again, a truly most ponderous and imponderable dilemma. Nothing less than a paradox of pain. Or, more problematic still, a paradox of paradoxes that actually has no counterpart in the natural world.

We are, and would be, utterly alone in it, were it not for both its idolatrous disposition and the utterly chaotic misfortunes of its (our) inflictions.

If pain, then, is a marker spanning that unfathomable chasm between consciousness and something else entire, then by what means and to which ends do we effectively grapple with it, layers and layers of pain, as a collective that has never prioritized between collectives? Other than in the power-grabs of nation states? And how do we, then, address cruelty in all its individual forms? Their connection pertains most acutely to that distinction between human consciousness and the myriad manifestations of the conscience. One is a collective phenomenon, the other individual.

In a most curious theological treatise, *The Garden of Eden* (1849), John Nichols analyzed the Old Testament description of the Tree of Life in an attempt to calculate when and why Adam and Eve ate from the Tree. Was it the Tree of Life or the Tree of Knowledge? Were both trees side by side? Did the first inhabitants of Eden taste the Tree of Knowledge before the Tree of Life, or vice versa? It was this Mosaic (approaching a silly) confusion, according to the author, that led to agriculture, so that all people would "eat *mentally* of the tree of life *before* having eaten mentally of the tree of knowledge of good and evil, and dying a moral death...."[12] In a review at the time, Nichols was described as a "patient and independent thinker," his topic one that would be perused by "the common people" and fully believed in, until such time as teachers could convey a more plausible scenario, a persuasive allegory yet to inform[13] (Figs. 98.4 and 98.5).

From culture to culture, human emotional pluralities dictate the psychological spectrum of continuously fluctuant rational numbers. Inferences, trundling so much human baggage, turn to the irrational with an ease befitting those classical lamentations and Apocalyptic underpinnings of both the paradise aspiration and each person's own concretized desolation. Ethnographers have long studied depression among country-wide groups.[14] Far less research has examined the notion of extinction between cultures, let alone the word and its meaning, among the more than

---

[12] *The Garden of Eden: A Theological, Philosophical and Practical Illustration Consisting of a Treatise on the "Garden of Eden," "Tree of Knowledge of Good and Evil," and "Tree of Life,"* by John Nichols, Holliston, Boston, 1849, p. 117.

[13] See The Universalist Miscellany, A Monthly Magazine, Volume 6, 1849, edited by O.A. Skinner, and Rev. S. Streeter, p. 361.

[14] See, for example, "Rethinking Depression: An Ethnographic Study of the Experiences of Depression Among Chinese," by Dominic T. S. Lee, Joan Kleinman and Arthur Kleinman, Harvard Review of Psychiatry, Volume 15, 2007, Issue 1, Published online July 3, 2009, Taylor & Francis Online, https://www.tandfonline.com/doi/abs/10.1080/10673220601183915; See also, "Migration, ageing and mental health; an ethnographic study on perceptions of life satisfaction, anxiety and depression in older Somali men in east London," by Ellen Silveira and Peter Allebeck, International Journal of Social Welfare, Wiley Online Library, February 4, 2003, https://doi.org/10.1111/1468-2397.00188, https://onlinelibrary.wiley.com/doi/abs/10.1111/1468-2397.00188, Accessed June 30, 2020; see also, "The Ethnographic Study of Cultural Knowledge of 'Mental Disorder,'" by Geoffrey M. White, in *Cultural Conceptions of Mental Health and Therapy*, Edited by Anthony J. Marsella, and Geoffrey M. White, pp. 69–95, (CIHE, volume 4), https://doi.org/10.1007/978-94-010-9220-3, D. Reidel Publishing, Dordrecht, Netherlands, Springer Science+Business Media, B.V. 1982, https://link.springer.com/book/10.1007/978-94-010-9220-3#about.

**Fig. 98.4** Human reconciliatory possibilities as witnessed here with Marietta van der Merwe with rescued young smiling lion, Harnas Sanctuary, Namibia. (Photo © M.C. Tobias)

**Fig. 98.5** From a monastery garden in Kyoto. (Photo © J.G. Morrison)

6500 spoken languages as of 2020. Some of these basics are captured in the work of Genese Marie Sodikoff (ed.), *The Anthropology of Extinction*, from chapters on "genetic rescue," "Endangered Species and Moral Practice in Madagascar" to "Collateral Extinctions," "Endangered Languages," and "Disappearing Wildmen."[15]

If we should apply such distinctions, for example, to the 100 or so remaining North Sentinelese, and, conversely, to the thousands of tourists flocking a mere 20 miles away, at Wandoor Beach and Mahatma Gandhi Marine National Park, between Port Blair and Tarmugli Island, in the Andamans, the contradiction strikes at the heart of the overall human condition, a subject of profound irony upon which we shall conclude this inquiry.[16]

---

[15] See *The Anthropology of Extinction – Essays on Culture and Species Death*, Edited by Genese Marie Sodikoff, Indiana University Press, Bloomington, Indiana, 2011.

[16] See https://www.worldbank.org/en/topic/indigenouspeoples, Accessed February 28, 2020; For two unsurpassed photo documentations of various tribes and their habitats, see *Genesis*, by Sebastião Salgado, Taschen, Cologne, Germany 2013; and *Before They Pass Away*, by Jimmy Nelson, teNeues, Kempen, Germany 2018.

# Chapter 99
# The Last Island

## 99.1 Cantos of a Lost Paradise

In John Pope-Hennessy's study of Dante's *Divine Comedy* illuminations by Giovanni di Paolo, several profound ideas lead by unexpected turns to India's Andamans, and North Sentinel Island in particular. It may be a geographical quirk of the imagination, as was the "Paradiso," in relation to the earlier "Inferno" and "Purgatorio"; but it is one hominin imagination, one planet, and a singular question central to the future—of a future that either is or is not—a day coming that is less cruel, inflictive, and untenable than at present. Pope-Hennessy describes the "insuperable difficulty" in painting what Dante imagined Paradise to look like (*Inferno* and *Purgatorio* were easier). He points out that St Peter in Canto XXIV is not even physically shown as he speaks to Beatrice and Dante, any more than the Resurrection was ever said to have been witnessed in the New Testament. Rather, in the case of Dante, "a 'blessed flame' from which there issued the injunction recorded in the poem."[1] Pope-Hennessy suggests that the 115 illuminations within the so-called Yates-Thompson Codex were probably commissioned in the year 1444[2] in Sienna, when Giovanni di Paolo was likely in his late 40s.[3] What profoundly distinguishes di Paolo's vision from other Dante illuminations, like those later 92 surviving images by Sandro Botticelli, commissioned by Lorenzo de Medici, is that di Paolo has Dante, Virgil, and Beatrice looking down upon the earth from a celestial vantage.

Moreover, as Canto I describes in its beginnings (l.2), "I have been in the heaven that most receives of His light, and have seen things which whoso descends from up there has neither the knowledge nor the power to relate, because, as it draws near to its desire our intellect enters so deep that memory cannot go back upon the track."[4]

---

[1] *Paradiso, The Illuminations to Dante's Divine Comedy by Giovanni di Paolo*, by John Pope-Hennessy, Dante's Paradiso Translated by Charles Singleton, Random House New York, © by Thames and Hudson Ltd., London, 1993, p. 18.

[2] ibid., p. 12.

[3] See the "British Library Catalogue of Illuminated Manuscripts," http://www.bl.uk/catalogues/illuminatedmanuscripts/record.asp?MSID=6468.

[4] Pope-Hennessy, from Charles Singleton's translation, p. 194.

© Springer Nature Switzerland AG 2021
M. C. Tobias, J. G. Morrison, *On the Nature of Ecological Paradox*,
https://doi.org/10.1007/978-3-030-64526-7_99

The earth, throughout the work, and in other seminal paintings by di Paolo, is a small and fragile thing. Most noticeably in his "Creation and Expulsion from Paradise" (The Metropolitan Museum of Art, New York, Robert Lehman Collection), there are the naked original couple being hastily escorted by the Archangel out into a very domesticated landscape, seven fruiting trees in rectilinear array and some shrubs, leaving forever behind their home-planet, the original Garden and four rivers and a "mappamondo or plan of the known earth surrounded by the zones of water air and fire, and the circles of the seven planets."[5] God, in company with his cherubim, all in blazing blues and golds, is emphatically ensuring that Adam and Eve never come back. Painted in around 1445 for the church of San Domenico in Siena and thus intimately linked to his Dantean illuminations, both the "Creation of the World" and Yates-Thompson Codex share, as Pope-Hennessy elucidates, an intimacy "both on the narrative and a decorative plane"[6] that emphasizes several crucial images and storylines which emphatically speak to the twenty-first-century vanishing tribe paradox. And perhaps no more so than in the Andaman Sea. Nudity, a zoological frenzy first enshrined in Adam and Eve, locates those disparate cornerstones of human history, as allegorically rendered that concretize in our minds the whole history of human origins and Fall into experience, those residual hominid roots that must continue to make us both nervous and envious.[7]

Other illustrators of Dante, from Botticelli to Doré to Dali, have emphasized the Empyrean (highest heaven), in which Dante and Beatrice are shown the resplendent rose-petalled abodes of God the Father, a lightning bright scene of absolute astronomical repose within tumult; schematics of faith, hope, and love, of the "Ascent to Fixed Stars," of "Jacob's Ladder," and a "Vision of Christ" (Fig. 99.1).

> O eternal light!
>     Sole understood, past, present, or to come;
>     Thou smiledst, on that circling, which in thee
>     Seem'd as reflected splendor, while I mused;
>     For I therein, methought, in its own hue
>     Beheld our image painted; stedfastly
>     I therefore pored upon the view.[8]
>     (The Rev. Henry Francis Cary 1860 translation)

Seen from the air, or nearer to the elixir, slowly roaming North Sentinel Island on any of the open-source virtual globes, yields a high-resolution picture of a Lost

---

[5] ibid., p. 34.

[6] ibid., p. 34.

[7] See *The Metropolitan Museum of Art – The Renaissance in Italy and Spain*, Introduction by Frederick Hartt, New York, 1987, p. 35.

[8] See *The Vision of Purgatory And Paradise*, by Dante Alighieri, Translated By The Rev. Henry Francis Cary, Illustrated With The Designs of M. Gustave Doré, Cassell and Company, Limited, London, ca.1860, Canto XXXIII, ll.115–122, p. 336. See also one of countless, instructively differing translation, as presented in 124, 125, 127, 128, 129, in Dante, *Paradiso, Third Book Of The Divine Comedy, A New Verse Translation with Introduction and Commentary by Allen Mandelbaum*, and 19 original pen and wash drawings by Barry Moser, University of California Press, Berkeley, 1984, p. 294.

**Fig. 99.1** Canto XXI from *The Vision of Purgatory and Paradise* by Dante Alighieri. (Translated by the Rev. Henry Francis Cary, illustrated with the designs of M. Gustave Doré, Cassell and Company, Ltd., London, 1903, pp. 276–277. (Private Collection). (Photo © M.C. Tobias)

World of the Sentinelese. Lost to us in ways that Dante and his countless interpreters over the last 700+ years (*Paradiso* was finished in 1320) have so enunciated, though probably never so poignantly than in the first and final illuminations of Giovanni di Paolo's masterpiece of 1445. Dante confronts Apollo, God of Poetry, to help him survive the journey into Paradise, noticing the supine, naked Marsyas, the Phrygian Satyr who invented the flute, and beside Apollo, Daphne having been transformed into a laurel tree upon Apollo's sexual pursuit of her. The summit of Mount Parnassus, a jagged, rocky peak is there, angling above, forest to the side,[9] a juxtaposition of every prefigurement of the human odyssey, the great dream of Paradise, its complicated landscape, and the perils of pursuing it. And to illustrate Canto XXXIII, a boat, the Argo crossing a treacherous sea, toward a noticing Dante and Beatrice in prayer, as the Virgin of the Assumption hovers amid the golden rays of divine love, also in prayer, and far off on the horizon, Neptune bidding farewell to the ship, or, as Pope-Hennessy characterizes his gesture, "registers wonder."[10]

---

[9] op.cit., Pope-Hennessy, pp. 70–71.

[10] ibid., Pope-Hennessy, pp. 188–189.

John Milton had lent no comfort to the final moments, the naked couple lost, with no alternative but that "hand in hand with wandering steps and slow, Through Eden took their solitary way."[11] The second illumination for Canto I by di Paolo is equally perplexing, alluring, and adamantine. It is the first appearance of Beatrice who, beside Dante, is mid-air in a gesture of astonished solemnity as she looks out upon a Paradise, which is the "great sea of being" inhabited by two-dozen distinct terrestrial and marine creatures "outside reason"; but each and every one, from halibut to horse and rabbit to fox, is partaking of the human drama, and all centered beneath "the winged figure of Love."[12]

## 99.2   Dante Among the Andamans

Dante and di Paolo formed a perfect conceptual pairing of great visions that, to any ordinary sensory system, should at once spell out the mythopoetic challenge to humanity: peace and compassion at any price. Yet, the repository of data regarding an estimated "100 uncontacted or near uncontacted tribes" remaining in the world[13] underscores the human tragedy, still unprocessed, that corresponds to the Anthropocene and its incalculable backfire upon *Homo sapiens*.

The story of those remaining inhabitants on North Sentinel Island, approximately 800 miles south of Kolkata, has been in the news at least since accidental contact by an East India survey crew in 1771; a second encounter by the crew aboard a ship that foundered on the North Sentinel reef in 1867; and more serious systematic encounters in the 1880s by the British naval officer Maurice Vidal Portman who was then the Andamans administrator based in Port Blair. Of the 572 islands within the archipelago, it is thought that fewer than 40 are officially inhabited.

Contact with other island tribes within the Andamans, including the Onge, Jarawa, and Great Andamanese, has perpetually gone badly. According to Survival International, "high rates of malnutrition, infant mortality and perilously low growth rate" have reduced their population(s) from "670 in 1900 to around 112 today."[14] Their one island, Little Andaman, was their sole territory as late as 1940. Since that time, at least 18,000 migrants from neighboring parts of the Nicobar Islands, Bangladesh, and elsewhere in India have forced the Sentinelese into a tiny reserve, Dugong Creek. In 1970, the Indian Government planted metal survey pegs on several island locations throughout the Andamans to reassert their sovereignty. Dugong Creek was devastated by the 2004 Tsunami. As with their neighboring

---

[11] *Paradise Lost*, by John Milton, *The Works of John Milton In Verse and Prose*, Volume II, Book 12, William Pickering, London, 1851, p. 345.

[12] op.cit., Pope-Hennessy, p. 72.

[13] "Let Them Live," https://www.survivalinternational.org/uncontactedtribes.

[14] See "The Onge," https://www.survivalinternational.org/tribes/onge, Accessed March 1, 2020.

Great Andamanese and Jawara tribes, the outside world has not been friendly. Despite Indian laws forbidding it, there is "human safari" traffic, particularly enabling tourists to obtain even a fleeting glimpse of the Jawara, who number roughly 400.[15]

## 99.3  The Last Island

As for the Sentinelese, the fact their one island lies 20 miles to the west of the main Andamans in rough seas has been their salvation, notwithstanding countless attempts by outsiders to make contact, at least two such instances resulting in the killing of the interlopers, most recently, a young American missionary who illegally kayaked to shore. Survival International, which is the most critical non-governmental organization (NGO) working worldwide to protect uncontacted tribes (peoples) worldwide, writes, "The Sentinelese are the most isolated tribe in the world, and have captured the imagination of millions. They live on their own small forested island called North Sentinel, which is approximately the size of Manhattan. They continue to resist all contact with outsiders, attacking anyone who comes near."[16]

Following India's Independence on August 15, 1947, the indigenous peoples were granted a level of "protection," but in 1974, the Indian Government allowed a National Geographic film crew to try and document the four Andaman and two Nicobar Island groups (we hesitate to reiterate the Indian official use of the word "tribe" as it has long been deemed genetically and culturally racist), the latter being the Shompen and Car Nicobarese. But, in fact, at least in 1858, there were a known 16 groups and sub-groups on these islands.[17] The documentary's director was shot in the thigh with an arrow, promptly ending the expedition. Nonetheless, India's official position perpetuated the notion that, under "appropriate circumstances," these islanders should welcome friendly, gift-giving contact—possible employment, better housing, and medical care somewhere on the mainland. Such language must evince in most sensitive people a near horror, confusing our inherent urge to protect, with that dismal reality of assimilation that has swept over the world, and our consciousness in turn. That inevitable capture and breaking in of the wild. Until not one square inch remains free of our mindless, incessant thralldom.

---

[15] See "The Jarawa," https://www.survivalinternational.org/tribes/jarawa, Accessed March 1, 2020.

[16] https://www.survivalinternational.org/tribes/sentinelese, Accessed March 1, 2020.

[17] See "The Andamanese – The Tribes," by George Weber, https://web.archive.org/web/20130507061710/http://www.andaman.org/BOOK/chapter8/text8.htm#aryoto-eremtaga, Accessed March 1, 2020.

This has certainly been the world's presumption and experience regarding other "noble savages," most notably in the novels *Atala* (1801) and *René* (1802) by François-René de Chateaubriand (1768–1848),[18] a conquistador's precept deeply embedded in the perdurable Western arrogance of possession, enshrined by Christopher Columbus (statues of whom are finally being removed from sites throughout the Western world). This stranglehold paradigm of power was augmented by the anthropological racist and polygenist, Dr. John Crawfurd (1783–1868). Crawfurd had no doubt read bits and pieces of Charles Dickens' response to George Catlin's Indian Gallery in London between 1840 and 1845: a deeply disturbed malice that appropriated sentiments directly out of Thomas Hobbes' influential view of so-called *primitive life*[19] (Fig. 99.2).

Anthropologist Madhumala Chattopadhyay was the first to make friendly contact with the Sentinelese on January 4, 1991, along with 12 other fellow team members of the Anthropological Survey of India.[20] "Playful," "intelligent," and "fierce" were words used by Chattopadhyay to describe one of the women in the first encounter who, emerging from the forest, came onto the beach and waded into the sea with her friends to accept some coconuts as a gift.[21] The second day of the visit did not go well, and the expedition packed its bags, returning seven weeks later for a brief, final visit. After that, the Indian Government declared a permanent moratorium on any and all outsider visits, with a 9.26 km band around the island patrolled by the Indian Navy to protect against poachers and any other unwanted visitors, as per the Andaman and Nicobar Islands Protection of Aboriginal Tribes Act of 1956.

---

[18] See Terry Jay Ellingson, *The Myth of the Noble Savage*, University of California Press, Berkeley, 2001, p. 380. This tragic arrogation of indigenous peoples has, of course, been a central pillar of nineteenth- and twentieth-century anthropological discussion, field methodologies, and ethical deliberation. It was most poignantly portrayed in the evocative novel, based on the true story, the personage of the Siberian Nanai hunter, Dersu Uzala (1849–1908), by Russian explorer, Vladimir Arsenyev (1872–1930). *Dersu Uzala* (1923) would be translated into two feature films, one by Akira Kurosawa in 1975, which won the Oscar for Best Foreign Film in 1976. One of the most controversial ethnographic immersions into this crisis of assimilation of tribal groups emerges from the legacy of Polish anthropologist, Bronislaw Malinowski, and his work with the Trobriand Islanders of Papua New Guinea in the 1920s. See also, *Mountain People*, by Michael Tobias (Ed.), University of Oklahoma Press, Norman, OK, 1986; *The Gentle Tasaday: A Stone Age People in the Philippine Rain Forest*, by John Nance, Houghton Mifflin Harcourt Publishers, Boston, MA, 1971; and *Learning Non-Aggression: The Experience of Non-Literate Societies*, by Ashley Montagu (Ed.), Oxford University Press, New York, 1978. See also, C. P. Snow's *The Two Cultures and the Scientific Revolution*, Cambridge University Press, New York, 1960.

[19] See "Wild American Savages and the Civilized English: Catlin's Indian Gallery and the Shows of London," by Robert M. Lewis, European Journal of American Studies, https://doi.org/10.4000/ejas.2263, 3-1, 2008, https://journals.openedition.org/ejas/2263.

[20] See "The woman who made 'friendly contact' with Andaman's Sentinelese," by Dhamini Ratnam, Hindustan Times, New Delhi, December 4, 2018, https://www.hindustantimes.com/india-news/the-woman-who-made-friendly-contact-with-andaman-s-sentinelese/story-7Dt1VPjZX-EpvEWWQY8Mu2M.html, Accessed March 1, 2020.

[21] ibid.

**Fig. 99.2**  Indigenous
Amazonians. (*John
Nieuhoff's Remarkable
Voyages & Travels*,
London, 1703.
(Private Collection).
(Photo © M.C. Tobias)

Not 30 miles from North Sentinel Island is a five-star hotel, and dozens of
other hotels, along the main Andaman island, and nearly 380,000 people. Despite
opposition from anthropologists within India and throughout the world, 29 islands
within the Andamans were opened to an aggressive tourism initiative by the
Indian Government in November 2018, the same month the 26-year-old American
missionary was killed on North Sentinel. It is said to be a tourism experiment, at
least through 2022, a modernization scheme similar to that which has displaced
most locals atop the exquisitely isolated village of Atule'er in Southwest China,
uniquely pinioned atop the 2600-foot cliff of hanging ladders, where they have
lived for centuries. No more waterfalls, but, rather, running water from a sink
in a concrete bunker. An indoor toilet for their freedom. Moving people out of
"poverty."[22]

---

[22] See, "For 200 years, these villagers have lived 2,600 feet up a remote cliff. Now they're in a housing estate," by Nectar Gan, May 15, 2020, CNN, https://www.cnn.com/2020/05/15/asia/china-
cliff-top-village-relocation-intl-hnk-scli/index.html.

## 99.4   Ultimate Metaphors

And, so, the ultimate ecological metaphor continues to rally support of a theatrical cul-de-sac with every layer and chorus of the drama buttressing a biological tragedy best thought of as current human evolution. The case of the Sentinelese underscores the urgent sense of our species' principal rejection of ecological abstinence, a dual fascination with a biological virginity that has been reduced on North Sentinel island to at best 100 or so occupants. Our insatiable curiosity, even at the risk of personal death, and species extinction, knows no limit. Treating a little earth, as a petri dish, or bowling alley.

Yet, looking down at North Sentinel from the air, one sees that tiny green orb, all but undisturbed, a perfect halo of the earthly microcosm that Giovanni di Paolo painted (Fig. 99.3). We remain the unhappy beneficiaries of this vision of purity— reminded of our global transgressions from the comfort of home or hotels with room service and AC. Left to our naked selves, how many of us would have a clue how to survive, be ourselves, get through a day and a night? Castaways in every sense.

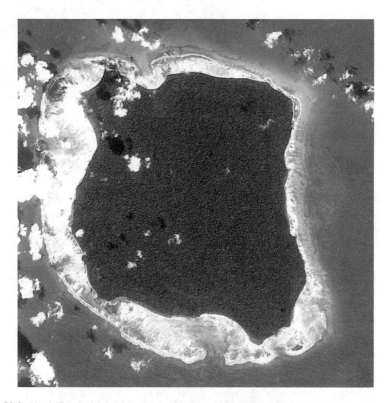

**Fig. 99.3** North Sentinel Island in the Andamans. (NASA Earth Observatory image created by Jesse Allen, using data provided by the NASA EO-1 Team, Public Domain)

Clearly, on North Sentinel, they know how to do it. Mathematically perfect, fulfilling every scenario of which no component is left out. This is as good as it can be, these Sentinelese.

And the rest of us stare across the tropical waters of the Bay of Bengal at the perfect genome and phenotype, the ultimate in human evolution. There they are. The best that is possible. Let us pray for at least another decade or two.

# Chapter 100
# Coda: Liberation Ecosynthesis

## 100.1 Humanity's Looming Sense of Itself

The fate of life comprises all those stories, in infinite guises, with a staying power to elicit conceptualization among both populations and individuals. At either level, it is critical in our evaluation of ourselves within the perceived rubrics of the biography of life that we recognize and accept the consequences of self-interest. That anthropic bias underscores what David J. Chapman and J. William Schopf have written regarding "Biological and Biochemical Effects of the Development of an Aerobic Environment," namely, that "One hesitates to rank events in terms of their perceived evolutionary 'importance'."[1]

Our one species has no monopoly on creation, only on destruction, although our prolific presence in the world has persuaded many of us that we alone command and control. Our fertility rate exceeds that of any large vertebrate, undermining all other biological codes of ethics and efficacy. Our collective consumption is unheard of in the annals of earth. By those two measures, we seem to think of ourselves as a resounding success. Charles S. Cockell, in a groundbreaking chain of speculations, notes that "Paradoxically, the biologically extreme conditions of the interior of a planet and the inimical conditions of outer space, between which life is trapped, are the locations from which volcanism and impact events, respectively originate."[2]

---

[1] See David J. Chapman and J. William Schopf, "Biological and Biochemical Effects of the Development of an Aerobic Environment," in *Earth's Earliest Biosphere – Its Origin and Evolution*, Edited by J. William Schopf, Princeton University Press, Princeton, New Jersey, 1983, p. 313.

[2] See "Life in the lithosphere, kinetics and the prospects for life elsewhere," by Charles S. Cockell, Philosophical Transactions Of The Royal Society A, Mathematical, Physical and Engineering Sciences, 13 February 2011, https://doi.org/10.1098/rsta.2010.0232, The Royal Society Publishing, from the Abstract, https://royalsocietypublishing.org/doi/10.1098/rsta.2010.0232, Accessed June 1, 2020. See also, "Cellular redox – living chemistry," Science in School, The European Journal For Science Teachers, Prince Saforo Amponsah, May 31, 2016, https://www.scienceinschool.org/content/cellular-redox-%E2%80%93-living-chemistry, Accessed June 1, 2020.

© Springer Nature Switzerland AG 2021
M. C. Tobias, J. G. Morrison, *On the Nature of Ecological Paradox*,
https://doi.org/10.1007/978-3-030-64526-7_100

And he goes on to point out that meager thickness of tropospheric warmth, humidity, and just the right chemical composition is all but a few miles.[3] Such fragility is not easily reconciled with the equations of our largely un-self-reflecting presence.

In the combined "branching selection," as Darwin appears to have understood evolution,[4] modern ecological science has attempted to predict, model, and ask the right questions about "scale," modes of prediction, and the essence of biological relations. As John A. Wiens has written, "...to understand anything about wildlife-habitat relationships, we need to adopt a view of 'habitat' that is centered on how organisms might perceive and respond to it, not how humans think of it."[5] But the real paradox in such deep ethological constructs is the underlying rationale guiding all eco-restoration, namely, the "identification of disturbances" which are solely human.[6] How do we separate ourselves from the maelstroms we have mindlessly wreaked and attend to that conscious reckoning that is required, but which remains dependent upon its own inflictive neural infrastructure? How do we solve problems when we are those very problems?

## 100.2   Transcending Maelstrom

Conversely, other types of maelstroms can be identified within natural boom-and-bust populations. This, too, further confounds the challenges before us. Some of the classic periodic outbreaks include those among locusts, "mice," "spruce budworm," "the pink gum psyllid" (the jumping plant lice), and "European rabbit."[7] Humans have perpetually intervened in each of these natural, demographic avalanches, just as we have upended the continuity and contiguity of nearly every ecosystem and its progeny. From a philosophical as well as a practical point of view, human ecosystem-dominated mosaics are a truly monumental mess (Fig. 100.1).

In the case of rabbits, which were formally introduced into Australia, for example, by the mid-1860s for human hunting and consumption, rabbit fertility rates

---

[3] ibid.

[4] "Charles Darwin's On The Origin Of Species," by Jonathan Hodge, in *Darwin – For the Love of Science*, Edited by Andrew Kelly and Melanie Kelly, Bristol Cultural Development Partnership, Bristol, UK, 2009, p. 155.

[5] See "Predicting Species Occurrences: Progress, Problems, and Prospects," by John A. Wiens, in *Predicting Species Occurrences -Issues of Accuracy and Scale*, Edited by J. Michael Scott, Patricia J. Heglund, Michael L. Morrison, Jonathan B. Haufler, Martin G. Raphael, William A. Wall and Fred B. Samson, Island Press, Washington, D.C., Covelo, CA, and London, UK, 2002, Chapter 65, pp. 747–748.

[6] See "Ecological Restoration," by A. D. Bradshaw, in *Principles of Conservation Biology*, by Gary K. Meffe, C. Ronald Carroll, and Contributors, Sinauer Associates, Inc. Publishers, Sunderland, Massachusetts, 1994, Chapter 14, p. 436.

[7] See Chapter 10, "Some Controversies in Population Ecology," in The Ecological Web -More on the Distribution and Abundance of Animals, by H. G. Andrewartha and L. C. Birch, The University of Chicago Press, Chicago and London, 1984, p. 243.

**Fig. 100.1** Drawing and book by Lewis Carroll, from *Alice's Adventures Under Ground*, being a facsimile of the original manuscript. Book afterward developed into *Alice's Adventures in Wonderland*, Macmillan and Co, London 1886, p.13. (Private Collection). (Photo © M.C. Tobias)

soon expressed an astonishing robustness. At some point, their herbivorous collectivities were soon characterized as a general menace to indigenous Australian ecosystems, Australians themselves. Since that time, a war of the worlds, humans versus *Oryctolagus cuniculus*, has been waged by farmers and conservationists for generations. In 1950, a virus (myxomatosis) that had been discovered in Uruguay in 1896 (its host being the Brazilian wild rabbit), and later validated in trials carried out in the United Kingdom, was introduced into Australia, transmitted by all innocent fleas, which succeeded in destroying *nearly* every last rabbit, some half-billion of them. A 0.2% survival rate soon morphed into another shock wave of metabolic enthusiasm; 200 million more rabbits appeared on the scene.[8] By the mid-1990s, ecologists attacked the rabbits with yet another virus, this one from Spain, the Rabbit Hemorrhagic Disease, or Calicivirus, transmitted by flies (who simply do what they know to do), and capable of killing a rabbit within two days. Some parts of the country would be strewn with hundreds of thousands of corpses. But this was

---

[8] "Killed 500 million rabbits in two years…" by Stefan Andrews, December 8, 2017, The Vintage News, https://www.thevintagenews.com/2017/12/08/myxomatosis-australia-rabbits/, Accessed June 2.

**Fig. 100.2** *Dracaena cinnabari*, the Socotra dragon's blood tree, thought by some to be the very elixir of life; IUCN-Vulnerable, Socotra Archipelago, Yemen. (Photo © M.C. Tobias)

seen as a multibillion-dollar boon to human agricultural industries in the country.[9] One example, one location, one period in history that represents humanity's sense of itself, its mighty ties, and leveraging of "power" in the otherwise natural world.

Such biocontrol manipulations between every arbitrary national boundary mirror the contentious nature of ecological ultimate tensile strength as tested by our species against most others. Engineers examine stress-strain curves, as measured in internationally standardized units, kilopascals, or $1000 \text{ N/m}^2$ of pressure. Ecologists do so by literally testing the waters (pH), or sampling, or killing the most numbers of individuals and measuring ecosystem responses according to utterly non-standardized subjective interpretations, driven by species composition preferences, or only vague ideas about who lives where, who was there first, what species were introduced, and when. And who *belongs* and *deserves* to survive. These are arrogant determinations that have no place on a Tree of Life (Fig. 100.2).

Ecological tautologies, a set of illogical propositions wherein analytical truth and falsity hinge upon the bias of the observer, also tell us that, as Aristotle insisted, an idea or object cannot be both part of something and not part of that same something.[10] And yet, human observation presupposes involvement, thereby containing all the variables that will result in the definitions, and propositions, regardless of

---

[9] See "How Australia Controls Its Wild Rabbits," by David Peacock, from The Wildlife Professional Magazine, The Wildlife Society, May 15, 2015, https://wildlife.org/how-australia-controls-its-wild-rabbits/#:~:text=RHDV%20was%20introduced%20to%20Australia,160%20years%20 (Cooke%202014).

[10] Horn, Laurence R. (2018), "Contradiction", in Zalta, Edward N. (ed.), *The Stanford Encyclopedia of Philosophy* (Winter 2018 ed.), Metaphysics Research Lab, Stanford University.

verity, or discredited analogies. These paradoxical theories of knowledge and con-flicted self-interest are philosophical flash points that illustrate our most vulnerable precepts regarding natural selection and evolution. We simply cannot think straight. Hypotheses that continue to promote the well-being of our species at the price of others reflect the bondage inherent to our terminology and self-referential identifi-cations—the fact that we hold hostage the world's biodiversity. The proliferation of this profound dilemma is everywhere and in everyone.

This problem for human consciousness peaks at the juncture of violence and the remaining, well-tested truths of non-violence, where its theoretical groundswells have provided fodder for any number of post-evolutionary ideas, ideals, disposi-tions, proxies, and other moral equivalencies, all untested in totality, meaning at the human population level, but thoroughly and successfully tried by individuals.

One salient critique of the prominence of this most perplexing juncture was raised by Francis Fukuyama in 2002,[11] when he questioned the perils of transhumanist biotechnological manipulation and its possible consequences for democracy. This is a large view of genetics that is at a crossroads with the majority of myths typified by the Western Bible, in which "The heavens belong to Yahweh,/but the earth He gave over to man...."[12] That is a rationale employed by the biotech revolutions and those who would honor it by enshrining its impulses as key to the future of humanity.[13]

Others look to the practitioners of the Davos Economic Forum(s) getting the religion of environmental cause and effect; or to some other sublime series of engi-neering technofixes, encompassing medical breakthroughs that would allow humans to live multiples of lives; or to economic transcendentalism – that perennially false Utopia of the 1% with its obsolete trickle-down theories.

These treacherous dichotomies, a system of lifelong oppositions that has run amuck, may be the final mental hominin hierarchy, or set of discrete biological ech-elons of "preferred" consciousness, ever established on earth, as far as can be ascer-tained. It is no false dichotomy, only a brutally infective, egregious, and bankrupt concept at work.

## 100.3 Rejoining the Biological Commons

Our projection of the collective ego, guaranteeing its positioning in the front row of evolution, is a lethal addiction. Its cure requires us, from within thought and inten-tion, to filter and re-sequester a consequential realm of ideals that have been liber-ated from such inextricably complex mechanics and social norms.

---

[11] *Our Posthuman Future: Consequences of the Biotechnology Revolution*, by Francis Fukuyama, Farrar, Straus and Giroux, New York, 2002.

[12] Psalm 115:16. See *The Natural History of the Bible – An Environmental Exploration of the Hebrew Scriptures*, by Daniel Hillel, Columbia University Press, New York, 2006, p. 243.

[13] See, for example, *Liberation Biology: The Scientific and Moral Case for the Biotech Revolution*, by Ronald Bailey, Prometheus Books, New York, 2005. See also Bailey's *The End of Doom: Environmental Renewal in the Twenty-First Century*, Thomas Dunne Books, St. Martin's Press, New York, 2015.

**Fig. 100.3** Landscape with figures. Etching by Paul Sandby, 1758, published in *A Collection of Landskips and Figures*, by Paul Sandby and M. Chatlain, Robert Sayer Printer, London, 1773. (Private Collection). (Photo © M.C. Tobias)

In tracking the natural history sentiments of explorers during and after the Enlightenment, Barbara Maria Stafford describes the "Hegemony of the External" as that instant when "the traveler's gaze... his glance arrested at a precise moment in time – become[s] congruent, if only fleetingly, with a specific moment of matter's existence, or an aspect of its history. Thus, personal, human perspective might harmonize with a span or an occasion in nature's behavior."[14] That is the romantic, artistic sensibility, laid down over millennia of various Gardens of Eden and its iterations in every act of sentiment, epithalamium, moral disclosure, rejection of the status quo, and hope for the realization of those ideals which, taken together, might assist in that liberation, an ecosynthesis of the best, not the worst, of humanity (Fig. 100.3).

## 100.4   Ecosynthesis

Such concerns have for many decades acknowledged the biosemiotic commons, that exquisite multiplication of psyches and communication compulsions; of both mournful and expectant feelings, unstinting genius-in-sentience; of a universal,

---

[14] See *Voyage into Substance – Art, Science, Nature, and the Illustrated Travel Account 1760–1840*, The MIT Press, Cambridge, Massachusetts, 1984, pp. 399–400.

**Fig. 100.4** A church garden in Utrecht, The Netherlands. (Photo © M.C. Tobias)

meaningful conscience that exists in whole, or in part *regardless* of our own partaking of this otherwise universal communion, though that is surely of practical importance to us, assuming we prove ourselves up to it, if for no other reason than to survive.

This realm of biological assertiveness holds to some *other* set of forces than evolution and natural selection, obviating nearly two centuries of Darwinian anxiety, agitation, irrationally obsessed genetic re-engineering, and our social contracts that have utterly failed the world. It invites our urgent consideration of a *liberation ecosynthesis* that is overtly volitional, combining billions of years of biophilia, and several hundred thousand human years of synderesis, physiolatry (the love of, and reverence for, Nature), and the full spectrum of both readiness and reciprocity potential—crucial factors noted in all forms of restorative and resilient personal and attitudinal ecosystems. Perceptions and behaviors that can be rapidly altered through the free-will of a versatile and eusocial ethics consolidated in each and every individual. If we are up to it, such transformations can happen as quickly and effectively as Canadian white-throated Sparrows (*Zonotrichia albicollis*) suddenly, within a few generations, changing their calls across thousands of miles of territory[15] (Fig. 100.4). Alternatively, we can continue to drive Australia's regent honeyeater (*Anthochaera phrygia*) rapidly toward extinction by all those human stressors

---

[15] See "Continent-wide Shifts in Song Dialects of White-Throated Sparrows," by Ken A. Otter, Alexandra Mckenna, Stefanie E. LaZerte and Scott M. Ramsay, Current Biology 30, 1–5 August 17, 2020, Elsevier Inc. https://doi.org/10.1016/j.cub.2020.05.084, https://www.cell.com/current-biology/pdf/S0960-9822(20)30771-5.pdf?_returnURL=https%3A%2F%2Flinkinghub.elsevier.com%2Fretrieve%2Fpii%2FS0960982220307715%3Fshowall%3Dtrue, Accessed July 3, 2020.

**Fig. 100.5**  Sunset in Tuscany. (Photo © M.C. Tobias)

interferrng with the adult birds' abilities to pass down critical mating and territorial songs/language to its offspring[16].

Ecosynthesis in humans is congruent with all forms and ideations of restorative origins, gaining much traction in precise coordination with the logic of watering a plant. Its constructs and behavioral manifestations are yet to correspond with community affairs at a pace and scale commensurate with other biotic families and populations. But, given our combined social and equally private propensities, our species might still have sufficient will-power to enact a revolution in decidedly genuine, expansively compassionate courtesies; steeped in the tenderness of flowers, constrained only by four seasons, which, in the thick of a continuing population explosion, may well prove to be our ultimate safeguards, with the climate-feedback loops of those four seasons at their most demonstrative, assertive, and unblushing. Only such falling skies may be of sufficient interest to our kind to wake us up. Additionally, and fortunately, those still remaining North Sentinelese, and others like them, still resonate with ample evidence of biocultural, navigable skills, against unseasonable backdrops. Those peoples are likely to hold the secrets to success, if any, for our species as a whole; not just geographically, where physical conquest has proved a facile matter, but lodged in our psyches, a primordial content that is not so easily dismissed or overrun (Fig. 100.5).

---

[16] See "This Endangered Australian Bird Species Is Forgetting Its Song," by Olivia Rosane, EcoWatch, March 17, 2021, https://www.ecowatch.com/endangered-bird-losingsong-2651115111.html.

**Fig. 100.6** A Monk's Hermitage on the Cliff at Taktsang Monastery, Paro, Bhutan. (Photo © M.C. Tobias)

Contrary to natural selection, as *the* overarching explanation for all that humanity has both accomplished and impacted, there is unquestionably an impetus, independent of any seemingly solidified traits, evolutionary, or hereditary constraints, to see a paramount merging of ideals, all borne out by the prolifically diverse ethos of tens of millions of species that continue to enliven and define the biosphere. The verdict is out, but our final responses will be monitored in storm and silence, as we endeavor, or not, to manifest 4 billion years of brilliantly satisfying logic (not ours but the world's) across every ecological, economic, and geopolitical space of our inordinately outsized deliberations. There is not a lot of time left to willingly, humbly heed the call of this espied, this stunning, deeply apperceived, transformation. Otherwise, and all too clearly, our worst emergent nightmare, collectively, will be shown to have been the very paradox of our presence on earth. Neither good, nor bad. Just another sobering fact of life (Fig. 100.6).

# Index

**A**

Aappilattoq, 176, 177

Aboriginal and Biblical hermeneutics, 49

Aboriginal Tribes Act of 1956, 856

Abortion, 61

Accomplishment, 749

Acharanga Sutra, 599

A Chorus of Birds, 82

Adam and Eve, 852

Adam, Robert, 381, 386

ADE classification system, 285

"Adoration of the Mystic Lamb", 233

Aerodynamics, 377

Aesop, 280, 351

    "The Aesop Romance", 351

    animal illustrations, 356

    animated life of birds, 355

    biospheric hierarchies, 351

    declarations, 354

    fables, 356

    genius and character, 351

    history, 351, 355

    Holy Grail of biosemiotics, 351

    humanity, 354

    life-weary look, 352

    the Medici Aesop, 354

    1640 portrait, 352

    a student of Aristotle, 352

    zoology, 355

Aesthetics, 9, 733

Afrotropical "extra continental" land birds, 154

*After Man–A Zoology of the Future*, 837

Aggression, 60

Aggressive cognition, 74

"Agnus Dei–The Lamb of God", 112

Agriculture and Food Regulatory Authority, 542

Agronomy, 157

A Harvest of Death, 609

Air pollution, 479

"Alastor/The Spirit of Solitude", 347

Albrecht Dürer, 49

Aldrovandi, U., 144

Alice in Wonderland, 560

Alice's Adventures In Wonderland, 561

Alien elevation, 811

Allegorical clarity, 644

"The Allegory of the Battle of Lepanto", 309

Allele frequencies, 613

Alligator snapping turtle, 214

Allometrics, 456

Allusions, 772

"Almost zero", 67

Alphonse Antonio de Sarasa, 698

Altamaha River basin, 218

Alterities, 781, 783

Alterity and Transcendence, 793

Alt National Park Service, 32

Amalgam, 631

Amalgamation, 4

Amata, Terra, 629

Amazon, 106

    Basin's highest biodiversity, 197

    ITT Initiative, 197

    roadless, 199

    Yasuní-ITT economic agenda, 199

Amazonian forest, 199

"Ambrotype", 295

American environmental history, 44

American Mathematical Society (AMS), 98

Amplifying feedbacks, 596

CPSIA information can be obtained
at www.ICGtesting.com
Printed in the USA
LVHW082053100621
689600LV00004B/2

9 783030 645250